T0093512

Quantum Phases of Matter

This modern text describes the remarkable developments in quantum condensed matter physics following the experimental discoveries of quantum Hall effects and high-temperature superconductivity in the 1980s. After a review of the phases of matter amenable to an independent-particle description, entangled phases of matter are described in an accessible and unified manner. The concepts of fractionalization and emergent gauge fields are introduced using the simplest resonating-valence-bond insulator with an energy gap, the Z_2 spin liquid. Concepts in band topology and the parton method are then combined to obtain a large variety of experimentally relevant gapped states. Correlated metallic states are described, beginning with a discussion of the Kondo effect on magnetic impurities in metals. Metals without quasiparticle excitations are introduced using the Sachdev–Ye–Kitaev model, followed by a discussion of critical Fermi surfaces and strange metals. Numerous end-of-chapter problems expand readers' comprehension and reinforce key concepts.

Subir Sachdev is the Herchel Smith Professor of Physics at Harvard University. He has also held professional positions at Bell Labs and Yale University. He has been elected to national academies of science in India and the United States and is a recipient of several prestigious awards, including the Dirac Medal from the International Centre for Theoretical Physics, and the Lars Onsager Prize from the American Physical Society.

Quantum Phases of Matter

SUBIR SACHDEV
Harvard University, Massachusetts

Shaftesbury Road, Cambridge CB2 8EA, United Kingdom

One Liberty Plaza, 20th Floor, New York, NY 10006, USA

477 Williamstown Road, Port Melbourne, VIC 3207, Australia

314–321, 3rd Floor, Plot 3, Splendor Forum, Jasola District Centre, New Delhi – 110025, India

103 Penang Road, #05–06/07, Visioncrest Commercial, Singapore 238467

Cambridge University Press is part of Cambridge University Press & Assessment,
a department of the University of Cambridge.

We share the University's mission to contribute to society through the pursuit of
education, learning and research at the highest international levels of excellence.

www.cambridge.org
Information on this title: www.cambridge.org/9781009212694
DOI: 10.1017/9781009212717

First published 2023

Printed in the United Kingdom by TJ Books Limited, Padstow Cornwall

A catalogue record for this publication is available from the British Library.

Library of Congress Cataloging-in-Publication Data
Names: Sachdev, Subir, 1961- author.
Title: Quantum phases of matter / Subir Sachdev, Harvard University, Massachusetts.
Description: Cambridge, United Kingdom ; New York, NY : Cambridge
University Press, 2023.
Identifiers: LCCN 2022043286 | ISBN 9781009212694 (hardback) |
ISBN 9781009212717 (ebook)
Subjects: LCSH: Condensed matter–Textbooks.
Classification: LCC QC173.454 .S224 2023 | DDC 530.4/1–dc23/eng20221121
LC record available at https://lccn.loc.gov/2022043286

ISBN 978-1-009-21269-4 Hardback

Additional resources for this publication at www.cambridge.org/9781009212694.

To

Usha

for making it possible

Contents

Preface

This book aims to give an overview of the remarkable developments in the theory of quantum condensed matter physics that were initiated by three experimental discoveries: the integer quantum Hall effect in 1980 [144], the fractional quantum Hall effect in 1982 [282], and high-temperature superconductivity in the cuprates in 1986 [24]. I was fortunate to be present as a young postdoc at the famous "Woodstock" meeting of the American Physical Society in March 1987 in New York, where the excitement was palpable. The dominant view appeared to be that the problem of high-temperature superconductivity would be quickly solved, in the same manner that Laughlin's wavefunction had already clarified some aspects of the fractional quantum Hall effect in 1983 [151]. Instead, these experimental discoveries launched a paradigm shift that would take several decades to evolve, and which continues to this day. My aim here is to give a pedagogical overview of the many theoretical developments that followed. Of course, it is impossible to present a comprehensive treatment in a book of this length, and I have chosen topics that I view to be most relevant to the many experimental developments that have also followed.

The basic principle underlying early developments in statistical mechanics and solid-state physics was that of "mean-field" theory, in which the motion of each particle could be described in an average potential created by the other particles. In the theory of metals, these ideas found a refined form in Landau's Fermi liquid theory, which is presented in Part I in Chapter 2. New ideas based on emergent collective degrees of freedom, and theories of "order parameters" were developed in the theory of phases and phase transitions, and these found a pinnacle in 1972 in the Wilson–Fisher theory [310] of the critical point of the classical three-dimensional Ising model. The implications of the Wilson–Fisher theory for quantum systems, especially at nonzero temperature, have been described in my earlier book, *Quantum Phase Transitions* [234], which I will refer to hereafter simply as the QPT book. Some of the material covered in the QPT book is needed background for the present book, and so is included, and expanded on, here in Part I in Chapters 3, 10, 11, and 12. Part I also describes other important developments in traditional solid-state physics needed for the modern theory; these topics can also be found in many other textbooks, but my presentation attempts to relate them to observations in modern quantum materials.

In the context of quantum condensed matter physics, ideas moving beyond the independent-particle paradigm require theories of many-particle quantum entanglement. In its interpretation as the critical theory of the quantum Ising model in two spatial dimensions, the Wilson–Fisher theory provides an example of a system without any particle-like excitations, and with long-range many-particle quantum entanglement. Theories of many-particle quantum entanglement directed at the fractional quantum Hall effect built upon the physical picture provided by the Laughlin wavefunction. However, there are two distinct ingredients in the theory of the fractional quantum Hall state: the first is the *fractionalization* of the electron and the associated appearance of *emergent gauge fields*, and the second can now be referred to as *band topology*. The distinction between these ingredients was not clear in these earlier studies, and their effects were often conflated in a manner that was confusing to many (including the author). In the present book, I will make the distinction quite explicit, and cover the central ideas of fractionalization without band topology in Part II, while band topology is covered in Part III. Armed with these theories, we are able to apply both sets of ideas to a wide variety of quantum phases of matter in Part IV, including the fractional quantum Hall states.

Part II presents the theory of fractionalization "on its own" in two spatial dimensions. The theoretical developments start from Pauling's resonating-valence-bond (RVB) wavefunction, proposed in 1949 in his paper entitled "A resonating-valence-bond theory of metals and intermetallic compounds" [201]. However, Pauling's approach to simple metals was superseded by mean-field "density functional" theories building upon electrons in Bloch waves. In 1973, Anderson pointed out [8] that the quantum correlations in Mott insulators could be described by the RVB wavefunction, and applied similar ideas to cuprate high-temperature superconductivity in 1987 [9]. Soon after, Kivelson et al. [139] pointed out in 1987 that the RVB wavefunction also features fractionalization of the electron. The stability of the RVB state, and of its fractionalized excitations, to quantum corrections beyond an ansatz for a wavefunction, was first established in the theory of the odd $\mathbb{Z}_2$ spin liquid [119, 219, 303], which is presented in Chapters 15 and 16.

Part III presents a relatively brief exposition of what has now become a vast subject: the different varieties of "topological" band structures that can be realized by non-interacting fermions. The different realizations of band topology are distinguished by the presence of protected gapless excitations on the boundaries of the sample. It is now standard terminology to refer to materials with non-trivial band topology as topological insulators, metals, or superconductors; I refer the reader to the books by Bernevig and Hughes [28] and Vanderbilt [284] for a more complete treatment of the theory of *band topology* of non-interacting fermions. Unfortunately, this now widely accepted terminology of "topological insulators" etc. is apt to be confused with the characterization of quantum states with fractionalized excitations as possessing *topological order*, and I have therefore de-emphasized the latter terminology.

Part IV combines fractionalization and band topology, beginning with an overview in Chapter 21 on "parton theories." The electron (or a spin or a boson) is presumed to fractionalize into "partons," and the partons can then move in a possibly topological band structure. Combining the rich sets of possibilities of parton fractionalization and band topology, we obtain a powerful toolbox to describe a wide variety of quantum phases.

Almost all of the discussion of novel quantum phases with fractionalization in Parts II and IV is in the context of insulators. But in experimental realizations, novel phases are far more common in metals, which feature a Fermi surface of gapless excitations. We turn to a discussion of correlated metals in Part V. I will begin with a review in Chapter 29 of the Kondo impurity model, in which a localized spin degree of freedom is coupled to a Fermi surface. I then turn in Chapter 30 to the Kondo lattice model, in which a lattice of spins is coupled to a Fermi surface of conduction electrons. The most common fate of the Kondo lattice model is the "heavy Fermi liquid," in which the spins "become part of" the Fermi surface, and we obtain a large Fermi surface of electron-like quasiparticles, albeit with a large effective mass. However, another possible fate of the Kondo lattice model is the "fractionalized Fermi liquid" which features coexistence of a (small) Fermi surface of electron-like quasiparticles with fractionalized parton excitations; I describe this state and its applications to intermetallic compounds in Chapter 31. A more subtle point is that the fractionalized Fermi liquid state can also appear in single-band models, such as those are appropriate for the cuprate superconductors; this is discussed in Section 31.4, along with applications to the "pseudogap metal" phase of the cuprates

Part V also discusses metals that do not have *any* quasiparticle excitations, neither electron-like nor in any parton form. Much insight on such states of matter has been gained from the Sachdev–Ye–Kitaev (SYK) model, which I describe in Chapter 32. Remarkably, the SYK model also provides a description of the universal low-energy physics of quantum gravity for a wide class of black holes. I do not describe these holographic connections here: I refer the reader to my book, *Holographic Quantum Matter*, written with Sean Hartnoll and Andrew Lucas [103], and the review article with Debanjan Chowdhury, Antoine Georges, and Olivier Parcollet [46]. Insights from the SYK model are employed in the discussion of quantum-critical metals without quasiparticles in Chapter 34. This last chapter also explores connections of such theories to the ubiquitous "strange metal" or "Planckian metal" phase of numerous modern quantum materials.

A significant unifying theme in this book is the crucial role played by the spin Berry phase in (18.23) and (A.38) (sometimes, also referred to as the Wess–Zumino–Witten term in $0+1$ spacetime dimensions). In recent developments, the consequences of such Berry phases are phrased in terms of "anomalies" associated with symmetry-protected topological states in higher dimensions, but these connections will not be explored in this book. The spin Berry phase is absent in the relativistic field theories that were the focus of the QPT book; such field theores are continuum representations of basic models of statistical mechanics, such as the XY model in (14.1). With the Berry phases present, gapless and fractionalized phases or critical points become much more likely, as will become clear from many examples in this book. In an imaginary time representation of quantum spin models with spin S on each site, the spin Berry phase is shown in Appendix C to lead to a staggered background gauge charge for collinear antiferromagnets. It leads to the Berry phase factor shown in (14.36) for the XY model, and shown in (15.49) for the relativistic $O(4)$ model describing non-collinear antiferromagnets. In the $\mathbb{Z}_2$ gauge theory of spin liquids, this Berry phase is realized by the Gauss law constraint $G_i = (-1)^{2S}$ in (16.13), and was first obtained in a dual form in Ref. [119]. As the analysis of $\mathbb{Z}_2$ gauge theory in Chapter 16 makes clear, this Berry phase can prevent confinement even at very strong coupling on certain lattices, expanding the stability of fractionalized phases. In $U(1)$ gauge theories, the Berry phases suppress monopoles, leading to large confinement length scales or deconfined critical points, as will be described in Chapters 26 and 28. The Chern–Simon term of the chiral spin liquid discussed in Chapter 22 can also be viewed as a consequence of this spin Berry phase. In random spin systems, it is the Berry phase which distinguishes quantum rotor models from quantum spin models, and leads to the appearance of the gapless spin liquid with SYK correlations, as described in Chapter 33. The spin Berry phase is also an ingredient in the Kondo model, and important in the derivation of the renormalization group equations in Section 29.3. Finally, I note the connection of the spin Berry phase to the Luttinger relation of Kondo lattice models described in Section 31.3.1.

This book was initially developed from lectures notes for my courses at Harvard in 2016, 2018, 2020, and 2021. It took nearly its final form in a course I taught in the fall of 2021, jointly at the Institute for Advanced Study (IAS) in Princeton and the Tata Institute of Fundamental Research (TIFR) campuses in Bengaluru, Hyderabad, and Mumbai. This novel arrangement was an unexpected benefit of pandemic technology: I lectured using a blackboard either at the IAS or the International Centre for Theoretical Sciences, TIFR campus in Bengaluru. Video recordings of all lectures are freely available online on the IAS channel on YouTube at tinyurl.com/rwey5fyt, or following links at my website, sachdev.physics.harvard.edu. Readers are encouraged to use the videos for more informal and intuitive discussions of the topics covered in this book.

I have designed the book to be used in graduate courses on quantum condensed matter theory. Prior graduate courses in quantum mechanics and statistical mechanics are prerequisites. The book does not cover the technology of second quantization and finite temperature Green's functions; this is covered in numerous excellent textbooks, and on some occasions I have lectured on this material before turning to the topics covered here. Most of Parts I, II, and III can be covered in a one-semester graduate course, followed by a selection of material from the remaining chapters. Please see my IAS/TIFR lectures for the set of choices I made most recently.

Some parts of this book have been adapted from previous publications. Apart from the chapters from the QPT book noted earlier, portions of Chapters 15 and 16 are drawn from the review in Ref. [227], Chapter 32 and parts of Chapter 34 from the review in Ref. [46]. I thank my co-authors of Ref. [46], Debanjan Chowdhury, Antoine Georges, and Olivier Parcollet for permission to include the material here.

I owe a significant debt of gratitude to all the students and participants in my lectures for their interest, engagement, and stimulating and clarifying discussions. I sincerely thank the teaching assistants of my Harvard courses, Debanjan Chowdhury, Wenbo Fu, Haoyu Guo, Aavishkar Patel, and Rhine Samajdar for many helpful contributions and corrections; Appendix B was written by Rhine Samajdar. I thank Subhro Bhattacharjee, Kedar Damle, Rajesh Gopakumar, Kabir Ramola, Sandip Trivedi, and Spenta Wadia at TIFR, and Nathan Seiberg at IAS, for their valuable help and encouragement in the intercontinental course, which was made possible by support from IAS, Harvard, and ICTS. The structure of the book owes much to the feedback I received during this course. I had useful discussions on the content of the book with Maissam Barkeshli, Michael Levin, Leo Radzihovsky, Nathan Seiberg, T. Senthil, and Ruben Verresen. I thank Darshan Joshi, Henry Shackleton, Yanting Teng, Maria Tikhanovskaya, Ruben Verresen, and especially Leo Radzihovsky for their careful reading of many chapters.

I am grateful to the US National Science Foundation for extended support of my research over many years, most recently under Grant No. DMR-2002850. I also thank the US Department of Energy for support of my research on "Quantum simulation of correlated quantum matter" under grant No. DE-SC0019030.

I thank the staff at Cambridge University Press, Sarah Armstrong, Simon Capelin (who also helped with my QPT book), Jane Chan, Nicholas Gibbons, Subathra Manogaran, and Zoë Lewin.

Finally, I thank my late mother Usha Sachdev, my late father Dharmendra Kumar Sachdev, my wife Usha Pasi, and my daughters Monisha Sachdev and Menaka Sachdev for their constant support.

Subir Sachdev
Cambridge MA, Princeton NJ, Bengaluru

1 Survey of Experiments

A survey of key experiments exhibiting the quantum phases of matter studied in this book. Experiments on metals, band insulators, magnetic order and spin liquids in Mott insulators, ultracold atoms, the heavy fermion compounds, and the cuprate superconductors are described.

The theory of electronic quantum matter began in 1928, soon after the formulation of quantum mechanics, when Sommerfeld [265] proposed a theory of metals, using independent electrons obeying Fermi–Dirac statistics. Many experimental and theoretical developments followed, leading to a very successful theory of metals, semiconductors, superconductors, and insulators. This theory accounted for the Coulomb interactions between the electrons, but nevertheless the independent-electron paradigm survived in the form of a theory of nearly-independent electronic quasiparticles with the same spin and charge as an electron.

Beginning in the early 1980s, observations on new "quantum materials" could not be easily fit into such an independent-electron paradigm, and this stimulated the development of new theories of quantum matter in which the many-particle wavefunction is fundamentally different from a product of single-particle states, that is, the particles are "entangled." Many of these theoretical developments are described in this book, beginning in Part II. In this chapter, I present a selective survey of some experimental observations that motivated these studies. There is no attempt at completeness here, this is just a selection of experiments that highlight the main phenomena.

1.1 Metals and Band Insulators

In the independent-electron theory, the electrons occupy states in Bloch bands specified by a crystal momentum $\boldsymbol{k}$ and a band index n (which we will often drop because of our focus on single-band models). The values of $\boldsymbol{k}$ extend over the first Brillouin zone of the lattice. By considering a finite lattice, most conveniently with periodic boundary conditions, we can discretize the values of $\boldsymbol{k}$ and count the number of $\boldsymbol{k}$ states: the total number of $\boldsymbol{k}$ states in the first Brillouin zone equals the total number of unit cells in the finite lattice. Consequently, the density of electrons in a fully filled band is one per unit cell and per spin. A crystal with an odd number of electrons per unit cell, and

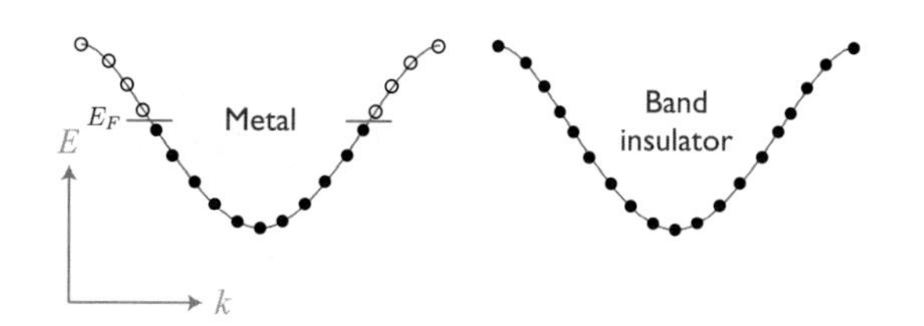

Figure 1.1 Electrons occupying states in a Bloch band. Each filled circle represents two electrons, after accounting for the spin degeneracy. A metal has a partially filled band, and this allows for excitations with vanishing energy near the Fermi surface separating the occupied and empty states. The Fermi energy is $E_{\rm F}$.

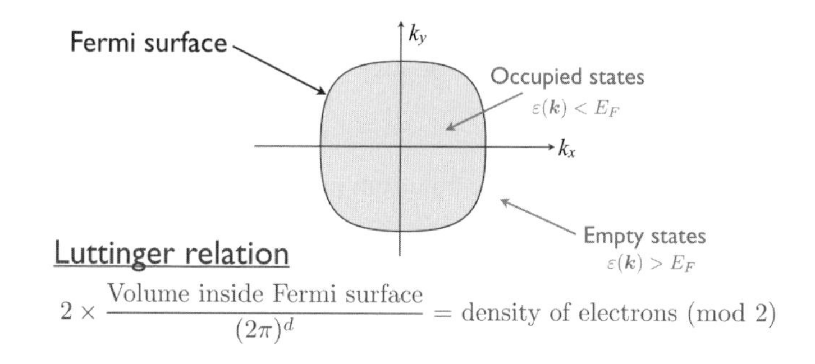

Figure 1.2 The Fermi surface and the Luttinger relation of Fermi liquid theory. The modulo 2 (mod 2) accounts for the electrons in fully filled bands, and assumes that electron spins are unpolarized.

with spin unpolarized, cannot fully fill a Bloch band, and so must be a metal in the independent-electron theory – this is illustrated in Fig. 1.1.

When there are an even number of electrons per unit cell, then a band insulator is a possible outcome, as also shown in Fig. 1.1. All bands in a band insulator are either completely empty or fully occupied. There is an energy gap towards creating excitations, which require moving an electron from a filled band to an empty band. However, when there is overlap in the energy eigenvalues of different bands with different $\boldsymbol{k}$ values, we can have multiple bands partially filled, and so a metal is another possible outcome with an even number of electrons per unit cell.

Going beyond the independent-electron theory, interactions in a metal can be accounted for in the context of Fermi liquid theory, as discussed in Chapter 2. The independent-electron theory of a metal implies the existence of a Fermi surface in $\boldsymbol{k}$ space, which separates the occupied and empty states of a partially filled band, as shown in Fig 1.2. The concept of a Fermi surface withstands the presence of interactions, and its position remains precisely defined in a Fermi liquid. A crucial result of Fermi liquid theory is that the volume enclosed by the Fermi surface does not change as the interactions are turned on – we prove this Luttinger relation in Section 30.2. Indeed, the concept of a Fermi surface is remarkably robust, and even holds in metallic states without quasiparticle excitations, as discussed in Chapter 34.

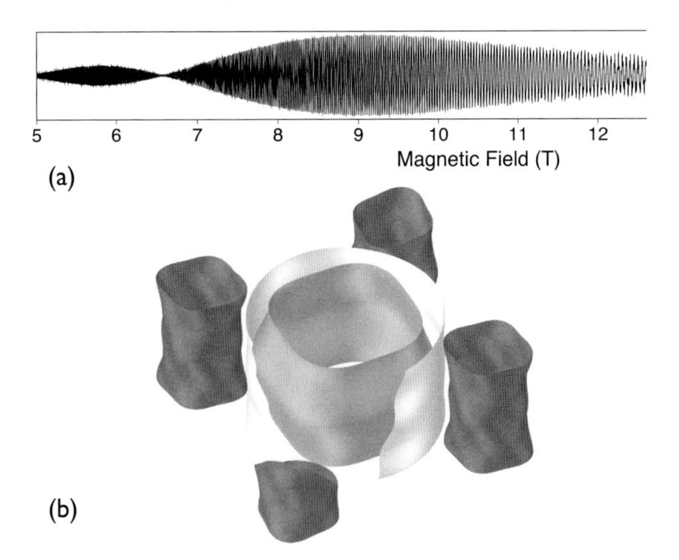

Figure 1.3 (a) Oscillations in the magnetization (dHvA oscillations) of Sr_2RuO_4 in a magnetic field. Such data are used to construct the multiple sheets of the Fermi surface shown in (b) is a quasi-two-dimensional material and the ripples in the vertical direction have been exaggerated by a factor of 15. From Ref. [26]. Reprinted with permission from APS.

The presence of a Fermi surface in a metal and the associated low-energy excitations across the Fermi surface dominate the observable properties of a metal. Among the consequences of a Fermi surface are

- The entropy S and the specific heat C of a metal vanish as temperature $T \to 0$ as

$$S = C = N\gamma T\,, \tag{1.1}$$

  where N is the total number of electrons. The Sommerfeld coefficient γ is proportional to the density of quasiparticle states at the Fermi level.
- The response to an applied magnetic field with a Zeeman coupling to the electrons is given by the Pauli spin susceptibility $\chi(T \to 0) = \chi_0$, where χ_0 is also proportional to the quasiparticle density of states at the Fermi level.
- In the presence of impurities, the resistivity ρ of a metal has the T dependence,

$$\rho(T) = \rho_0 + AT^2\,, \tag{1.2}$$

  where the A coefficient is controlled by the quasiparticle interactions.

The presence of a Fermi surface also induces oscillations in various observables in the presence of an applied magnetic field, as shown in Fig. 1.3. Such oscillations can be used to deduce remarkably precise information on the shape of the Fermi surface.

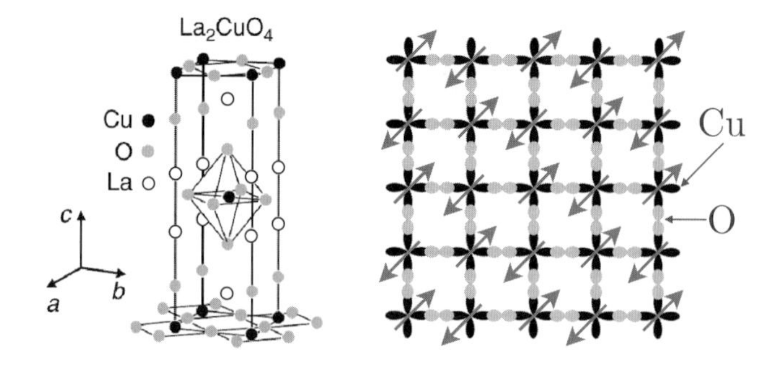

Figure 1.4 The Mott insulator La_2CuO_4. The active band resides on CuO_2 planes. At low T, there is broken lattice and spin rotation symmetry, with the electrons forming the Néel state, with a non-zero staggered magnetic moment centered on each Cu site.

1.2 Mott Insulators

The precise definition of a Mott insulator is a bit ambiguous in the literature. But, generally speaking, a Mott insulator is a crystal which is an insulator even though band theory requires the crystal to be a metal. Specifically, band theory requires a spin-unpolarized system with an odd number of electrons per unit cell to be a metal, as we noted above. Nevertheless, electron–electron interactions can drive such a system to be an insulator. The ambiguity arises when electron–electron interactions also induce some form of lattice symmetry breaking at low temperatures, and the size of the unit cell is at least doubled, so that there are now an even number of electrons in the new unit cell. Then, strictly speaking, band theory would allow the formation of a band insulator. Nevertheless, we will follow common practice, and continue to label such an insulator a Mott insulator. The influence of electron–electron interactions is paramount in such an insulator, as the energy gap to charged excitations is usually much larger than the energy gap that is deduced from a band-theory analysis.

The best-studied Mott insulator is the compound La_2CuO_4, shown in Fig. 1.4. The great interest is largely due to the appearance of high-temperature superconductivity when this compound is doped, as discussed below. We can describe the low-energy properties of this material by a single-band model with the orbitals centered on the Cu atoms on the vertices of a square lattice. In the insulator, this band has one electron per unit cell of the square lattice, and so band theory predicts this material must be a metal. However, electron–electron interactions, dominated by an on-site repulsion U between two opposite-spin electrons, make this system an insulator. The single-band model with interaction U is the Hubbard model, which is discussed further in Chapter 9.

At low T, the unit cell in La_2CuO_4 is doubled by the onset of Néel order shown in Fig. 1.4. Now there are two electrons per unit cell, and so the Néel state are allowed to be an insulator by band theory. In other words, we can adiabatically connect the Néel state to an insulator in the small U limit, provided we maintain the broken symmetry of the Néel state. However, the energy scales of excitations in the large U Néel state are

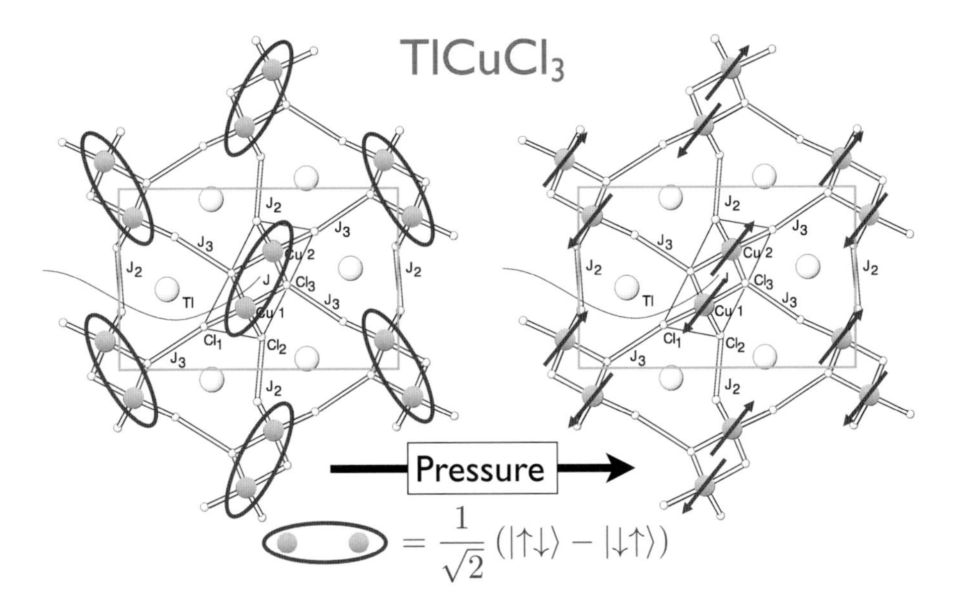

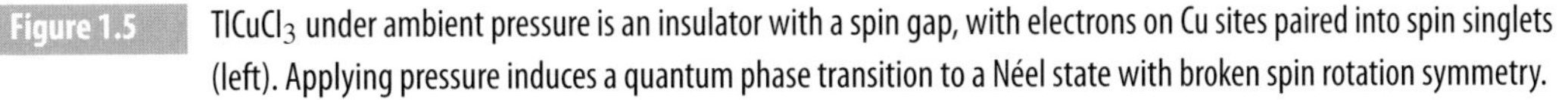
Figure 1.5 $TlCuCl_3$ under ambient pressure is an insulator with a spin gap, with electrons on Cu sites paired into spin singlets (left). Applying pressure induces a quantum phase transition to a Néel state with broken spin rotation symmetry.

very different from the small U band insulator with Néel order, which is why we prefer to label the large U state a Mott insulator. Specifically, the charge gap in the large U case is of order U, and is independent of the strength of the Néel order; in contrast, the charge gap in the band insulator is determined by the strength of the magnetic Néel order.

Another interesting insulator is $TlCuCl_3$, shown in Fig. 1.5. This insulator has one unpaired electron on each Cu atom, and an even number of Cu atoms per unit cell. Consequently, it can be adiabatically connected to a band insulator. However, it is better to think of it as a Mott insulator, in which the electrons on the Cu sites remain immobile, and pair up into spin singlets between neighboring Cu atoms, as shown in the left panel of Fig. 1.5. Note that the pairing pattern preserves all the symmetries of the crystal structure. Upon applying pressure, a quantum phase transition is observed at a critical pressure, above which the ground state becomes a Néel state, similar to that found in La_2CuO_4. The spins are now polarized in a staggered pattern, as shown in the right panel of Fig. 1.5.

The theory for the quantum phase transition in $TlCuCl_3$ is reviewed in Section 11.2. It is a relativistic field theory for the Néel order parameter in $3+1$ dimensions (the configuration of Cu atoms is three-dimensional in $TlCuCl_3$, unlike the two-dimensionality of La_2CuO_4). Section 11.2 also describes the evolution of the excitation spectrum across the quantum phase transition, and such a spectrum has been measured by neutron scattering, as shown in Fig. 1.6.

Another example of a pressure-induced transition from a gapped quantum paramagnet to a Néel state appears in the compound $SrCu_2(BO_3)_2$ [323], and is shown in

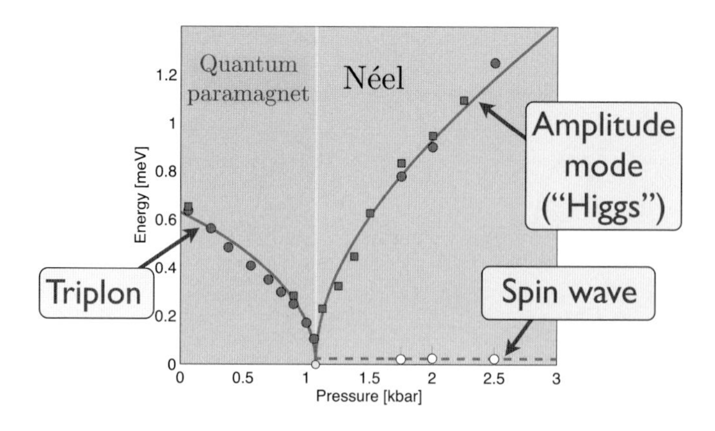

Figure 1.6 Neutron scattering observations [225] across the pressure-induced quantum phase transition in $TlCuCl_3$. The triplon particle is a spin-triplet excitation on a pair of Cu sites, which hops on the pairs shown on the left in Fig. 1.5; a field-theoretic description of this excitation is in Section 11.2.1. The spin-wave and amplitude-mode excitations of the Néel state are described in Section 11.2.3. Reprinted with permission from APS.

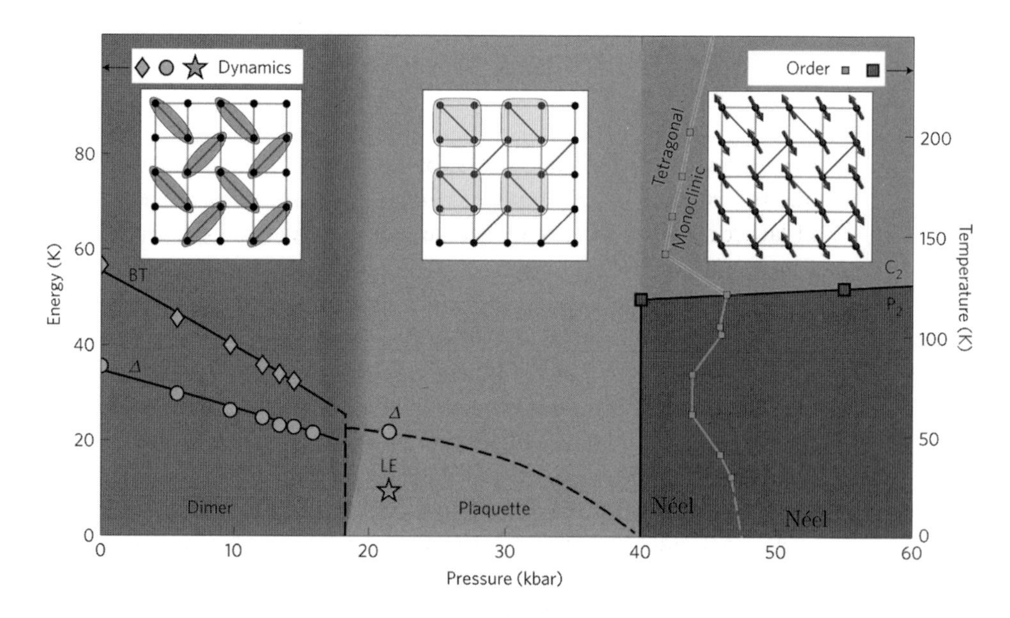

Figure 1.7 Phase diagram of $SrCu_2(BO_3)_2$ under pressure [323]. The dimer phase breaks no symmetries of the lattice, while the Néel phase breaks spin rotation symmetry. The intermediate plaquette phase is a VBS, and it breaks lattice symmetries while preserving spin rotation symmetry. Reprinted with permission from Springer Nature.

Fig. 1.7. At ambient pressure, this compound is a gapped quantum paramagnet in which the spins on the Cu sites form singlet pairs in a manner that does not break any lattice symmetries, just as in $TlCuCl_3$. At large pressure there is a collinear Néel state, which breaks spin rotation symmetry, also as in $TlCuCl_3$. The new phenomenon here is

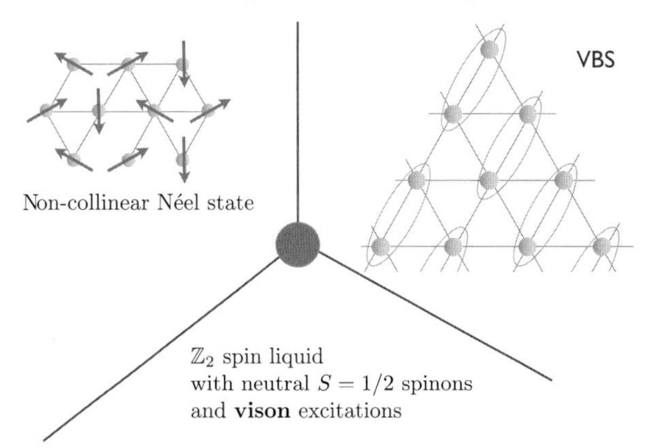

Figure 1.8 Schematic of possible phases of systems with $S = 1/2$ spins on the sites of a triangular lattice, and coupled with antiferromagnetic exchange interactions. The transition between the $\mathbb{Z}_2$ spin liquid and the non-collinear Néel state is discussed in Section 15.4.1, near Fig. 15.5. The transition between the $\mathbb{Z}_2$ spin liquid and VBS is discussed in Sections 16.5.2 and 26.2.3, and Chapter 28. The transition from the VBS to the non-collinear Néel state is discussed in Refs. [120, 268, 270].

the intermediate plaquette valence-bond solid (VBS) phase, in which four spins in a plaquette form a spin singlet, and the choice of the four-spin plaquettes requires a breaking of lattice symmetry. I discuss such collinear Néel–VBS quantum phase transitions in Chapter 28.

We turn now to insulators in which the immobile electrons reside on triangular or kagome lattices. Some possible fates of the electronic spins are sketched in Fig. 1.8. If the exchange interactions between the spins are antiferromagnetic, then such lattices are "frustrated," because it is not possible to find classical spin configurations in which the energy of each bond is minimized. The classical antiferromagnet has non-collinear Néel order, as shown in Fig. 1.8. Such Néel states can also be found for quantum $S = 1/2$ spins, and have been observed in $Ba_3CoSb_2O_9$ [166], as shown in Fig. 1.9.

Another state in Fig. 1.8 is the VBS. The spins pair into spin-singlet bonds as in the gapped state of $TlCuCl_3$. But this pairing breaks the symmetry of the triangular lattice, and there are six equivalent patterns of the columnar pattern of singlet bonds; this is similar to the plaquette VBS state of $SrCu_2(BO_3)_2$. This symmetry breaking has important consequences for the quantum phase transitions out of the VBS phase, as investigated in Sections 16.5.2 and 26.2.3, and Chapter 28. An example of a VBS state in κ-(BEDT-TTF)$_2$Cu$_2$(CN)$_3$ [178] is shown in Fig. 1.10.

The remaining phase in Fig. 1.8 is the $\mathbb{Z}_2$ spin liquid. This does not break any symmetries, either lattice or spin rotation. So this state is a true Mott insulator, with no caveats. It realizes a resonating valence-bond spin liquid, and will be the focus of much discussion starting in Chapter 13. There is "long-range quantum entanglement" in a spin liquid, and so it cannot be studied by the traditional perturbative approaches of

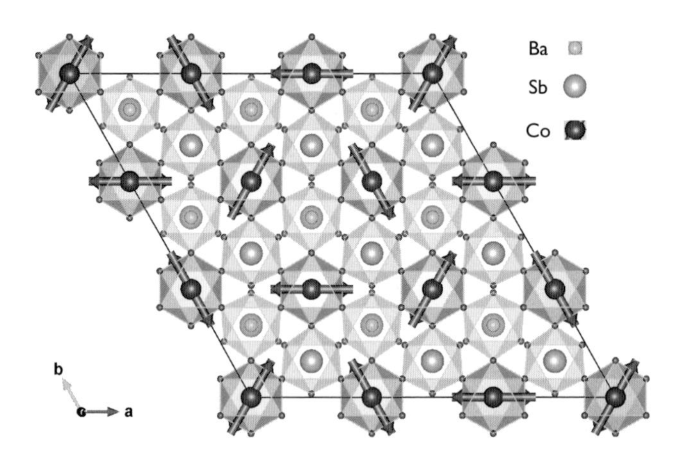

Figure 1.9 Non-collinear Néel order of $Ba_3CoSb_2O_9$ [166]. The arrows indicate the direction of the magnetic moment on the Co sites.

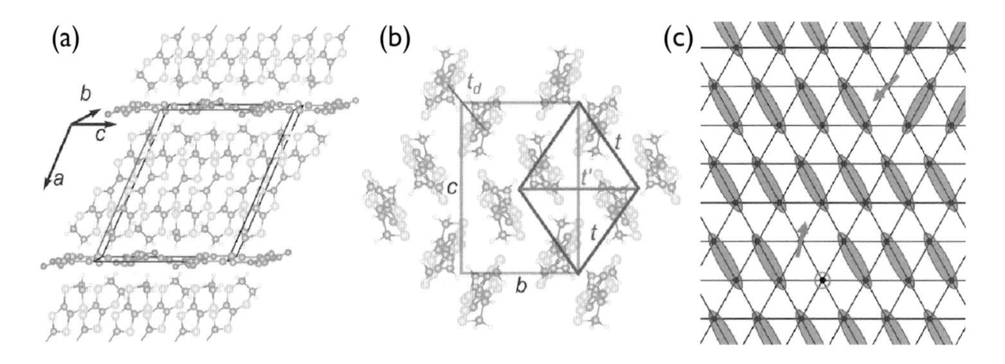

Figure 1.10 A valence bond solid state in the organic compound κ-(BEDT-TTF)$_2$Cu$_2$(CN)$_3$ shown in (a) [178]. There is a single unpaired spin on each organic molecule, and the molecules reside on a distorted triangular lattice distorted triangular lattice shown in (b). These spins pair up to form the valence bond solid shown in (c). Reprinted with permission from AAAS.

quantum many-body theory to be reviewed in Part I. As an experimental example, note the observations [77] on $Cu_3Zn(OH)_6FBr$ in Fig. 1.11, in which the spins reside on the Cu atoms on a kagome lattice.

1.3 Ultracold Atoms

While the focus of this book is on electronic quantum matter, we will also consider studies of ultracold atoms. Remarkable advances in cooling and trapping atoms have opened in a new frontier in the experimental study of quantum many-body systems. These offer a new set of experimental tools to study novel quantum phases, often with single-site resolution, and the ability to control the couplings in the Hamiltonian.

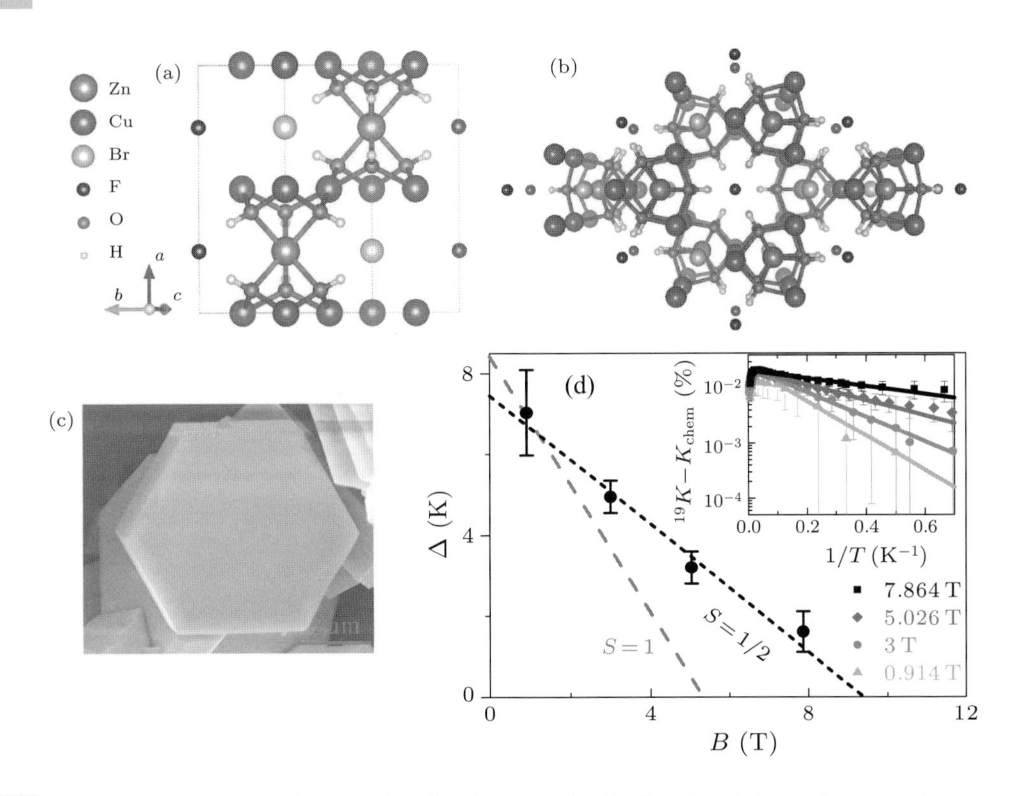

Figure 1.11 (a) and (b) Crystal structures of the gapped spin liquid candidate $Cu_3Zn(OH)_6FBr$ [77]. An image of a crystal is shown in (c). (d) The spin gap Δ is measured in an applied magnetic field, and its slope indicates the presence of fractionalized spin $S = 1/2$ excitations. Reprinted with permission from *Chinese Physics Letters*.

Connecting such studies to those in electronic quantum matter is a promising avenue for future research.

The first studies of ultracold atoms focused on bosonic atoms, and cooling them to a superfluid state: the theory of the dilute Bose gas is presented Chapter 3. Subsequently, the atoms were placed in a periodic potential, and the superfluid–insulator quantum phase transition was observed [95] as shown in Fig. 1.12. This is a Mott insulator, because of repulsive interactions between the bosons, and an integer density of bosons per unit cell. The transition of bosons between a Mott insulator and a superfluid are discussed in Chapter 8.

More recent experiments have also examined Mott insulators of spin-1/2 fermionic atoms, which form a Néel state on the square lattice, similar to La_2CuO_4 in Fig. 1.4. By varying the density of atoms, these studies have now explored the phase diagram of the doped Mott insulator. This is similar to the doped cuprate compounds, which are discussed in Section 1.5. However the accessible temperatures in the ultracold-atom systems are rather high on the relevant microscopic energy scales, in comparison to the cuprates. Superconductivity has not yet been observed, but new information has been revealed on the microscopic correlations at higher temperatures, as shown

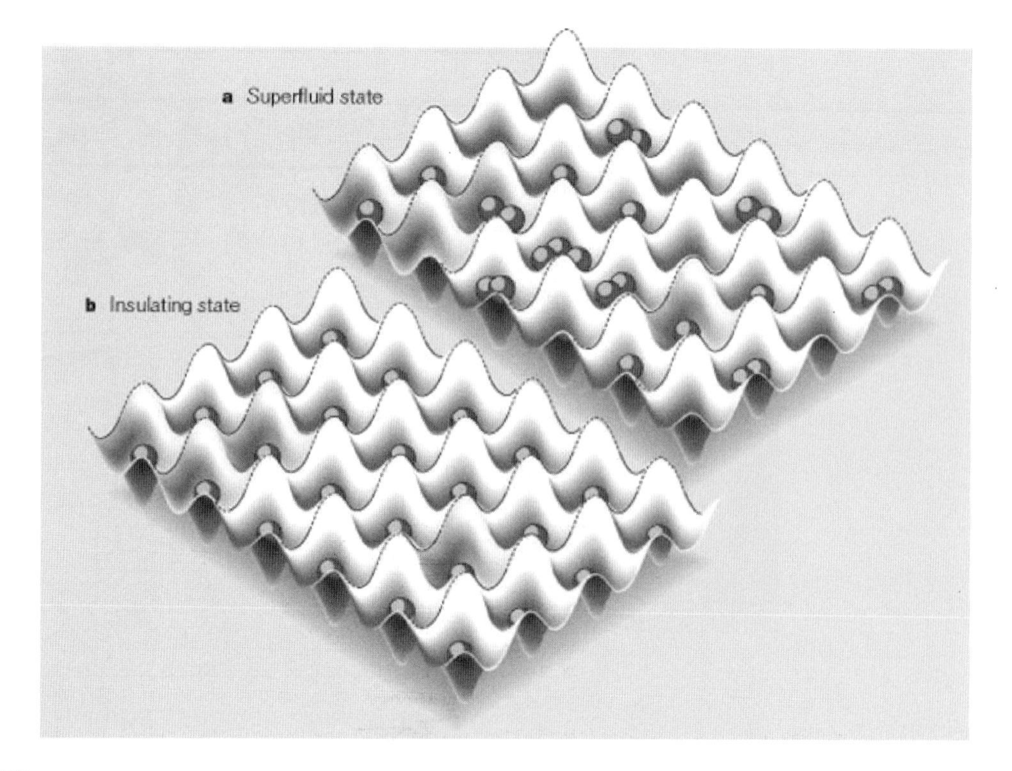

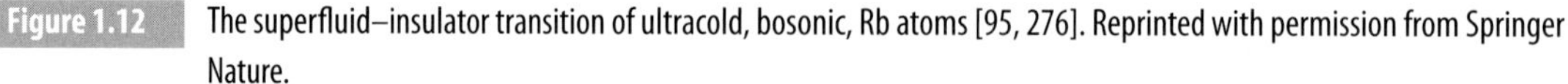
Figure 1.12 The superfluid–insulator transition of ultracold, bosonic, Rb atoms [95, 276]. Reprinted with permission from Springer Nature.

in Fig. 1.13. We will begin our study of the fermionic doped Mott insulators in Chapter 9, with a focus on the low-temperature superfluidity. This superfluidity is believed to be in the same class as the Bardeen–Cooper–Schrieffer (BCS) theory of the condensation of fermion pairs, which is described in Chapter 4. At higher temperatures, the fermionic doped Mott insulators exhibit the "pseudogap metal" and the "strange metal," as shown in Fig. 1.13: these are phases of matter which acquire non-trivial many-body correlations, similar to those of spin liquids. A theory of the pseudogap metal is described in Section 31.4, and that of the strange metal in Chapter 34.

A different class of ultracold-atom experiments trap atoms in laser tweezers, as shown in Fig. 1.14. A separate laser excites the atoms into Rydberg states, which have a much larger size, and have large repulsive interactions with neighboring atoms excited to Rydberg states. This off-site interaction induces quantum correlations between the different Rydberg atoms, and this leads to rich possibilities for novel correlated phases. Figure 1.15 shows theoretical predictions [244] for the phases of Rydberg atoms on a square lattice. The phases are distinguished by the wavevectors of modulations in the density of Rydberg excited states, which are absent only in the "disordered" regime. Unlike the optical lattice experiments in Fig. 1.12, there is no superfluid phase here because the number of atoms in the Rydberg states are not conserved. Experiments on

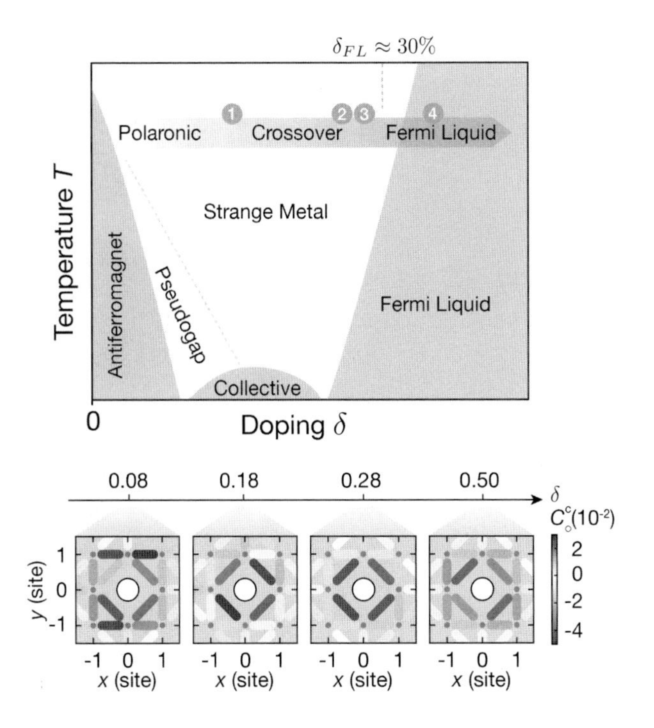

Figure 1.13 Phase diagram of the doped Mott insulator realized by ultracold fermionic atoms on a square lattice [145]. The experiments have explored spin correlations at high temperatures as a function of the doping δ. The bottom panel sketches the evolution in the spin correlations around a mobile hole with increasing δ. Reprinted with permission from AAAS.

Figure 1.14 Atoms trapped by laser tweezers. The atoms can be excited to a Rydberg state by separate lasers. Figure by Jacob P. Covey.

square-lattice arrays [65] are in close correspondence with the theoretical predictions, and have also explored the quantum phase transitions between the states.

A more recent experiment [254] using Rydberg atoms studied Rydberg atoms arranged on the links of a kagome lattice, as shown in Fig. 1.16. This experiment

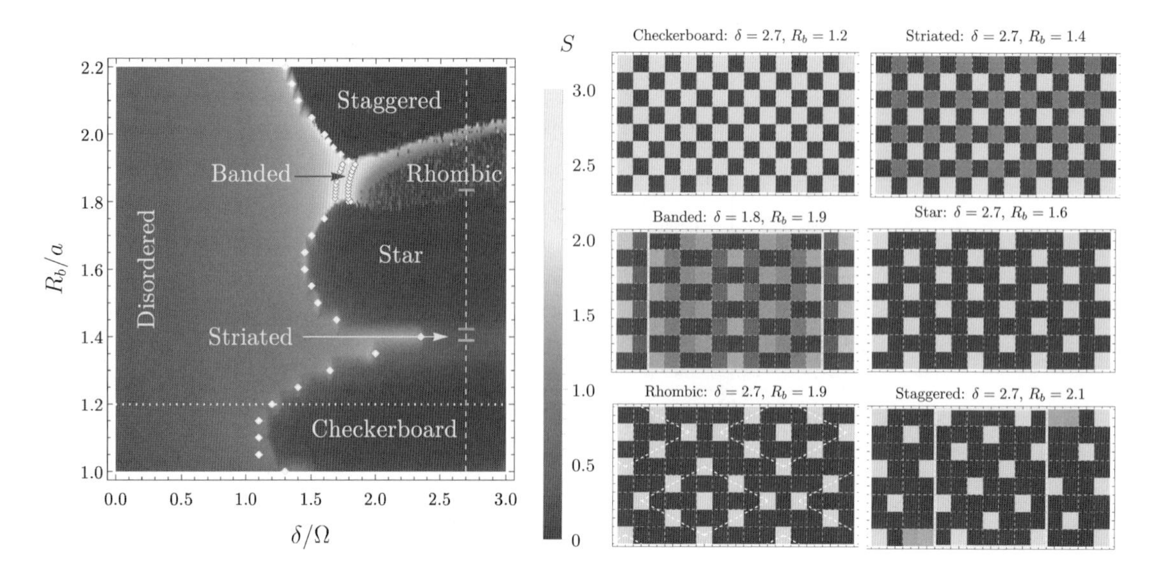

Figure 1.15 Density-matrix renormalization group results [244] for the phases of a square-lattice array of Rb atoms in laser tweezers shown in Fig. 1.14. The lattice spacing is a, the Rydberg blockade radius is R_b, the laser detuning is δ, and the Rabi frequency for the ground state to Rydberg state transition is Ω. Reprinted with permission from APS.

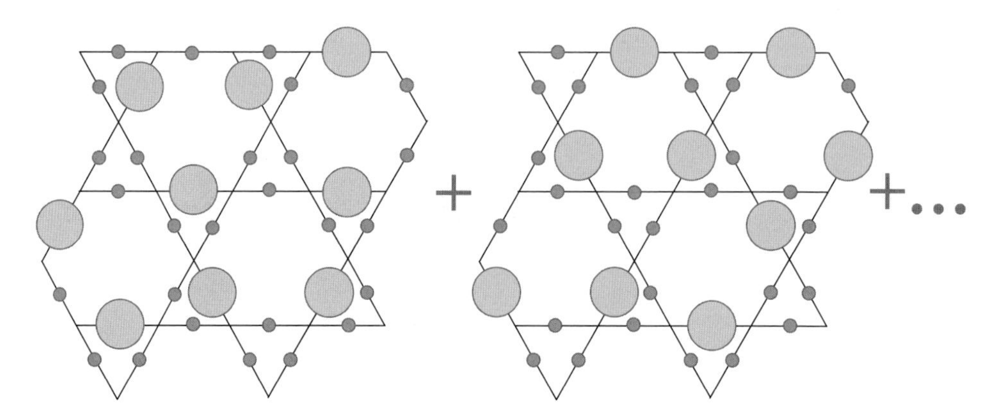

Figure 1.16 A quantum liquid of excited Rydberg atoms (indicated by the larger circles), along with atoms in the ground state (indicated by the smaller circles). The state is a coherent superposition of many such configurations, only two of which are shown above.

shows evidence for the quantum correlations of a "spin liquid" state that is discussed in Section 16.6. Similar to the spin liquid noted in Fig. 1.11, this is a state which is the coherent superposition of many configurations of atoms excited into the larger Rydberg state. The excitations "resonate" with each other, similar to the resonating-valence-bond state to be discussed starting in Chapter 13.

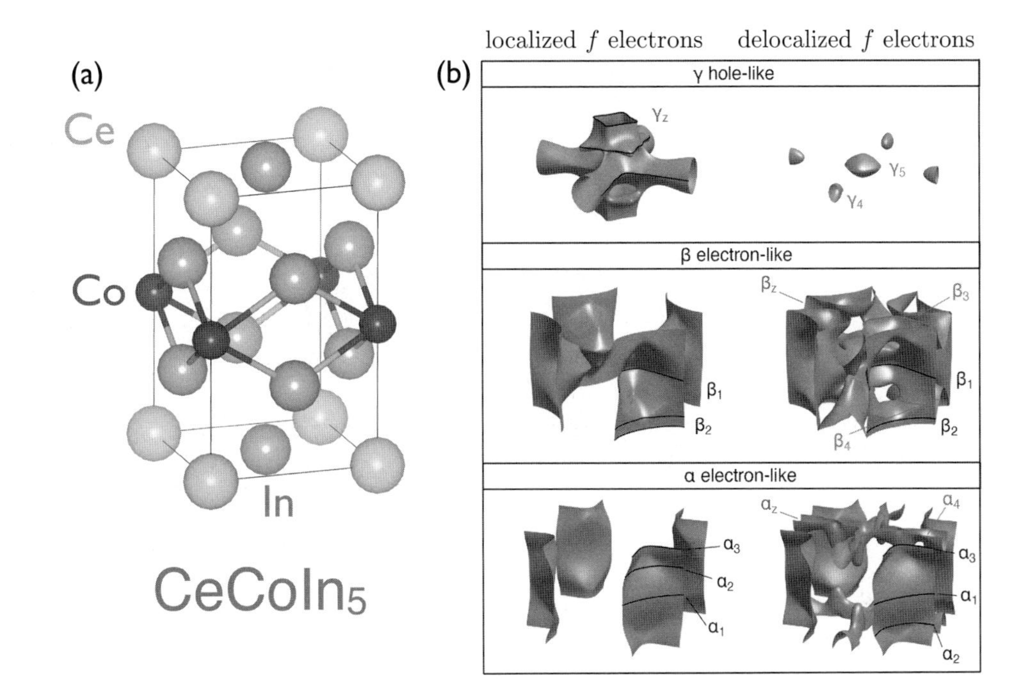

Figure 1.17 (a) Crystal structure of $CeCoIn_5$. (b) Fermi surfaces of $CeCoIn_5$ as measure by quantum oscillations [169]. In pure $CeCoIn_5$, the Fermi surfaces are computed by a theory which assumes the f electron on Ce is localized and not part of the Fermi surface. After doping with 0.33% Sn, the Fermi surface includes the f electron, which has now become mobile. Reprinted with permission from AAAS.

1.4 The Heavy-Fermion Intermetallic Compounds

Finally we turn to metallic electronic states with strong correlations, distinct from the metallic Fermi liquid states discussed in Section 1.1. The heavy-fermion compounds are described by Kondo lattice models, which are discussed in Chapter 30.

An example of such a compound is $CeCoIn_5$, shown in Fig. 1.17. The rare-earth element Ce has an electron in the f band which is nearly localized on the atomic site. However, the spin of this f electron is an active degree of freedom, which interacts via an exchange coupling with the mobile electrons from the other bands: this leads to a description as a Kondo lattice model. At generic electron densities, the ground state of the Kondo lattice model is a metal. A key question is whether this metal is in the class of Fermi liquid states discussed in Fig. 1.2. Such a state must obey the Luttinger relation, which dictates that the f states must also be included in the count of states contributing to the volume of the Fermi surface. And indeed many intermetallic compounds realize such a "heavy Fermi liquid" state, in which the main consequence

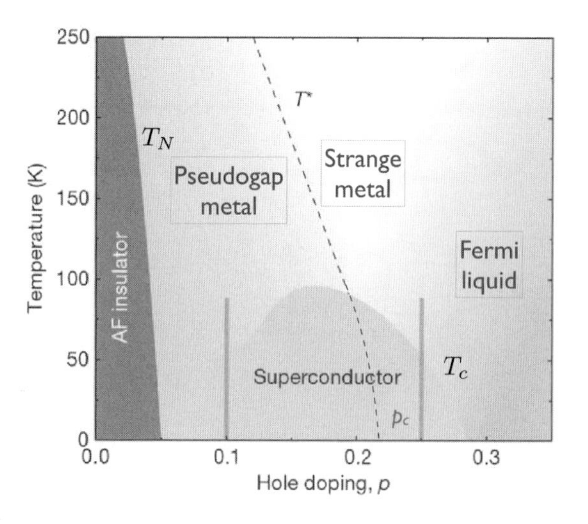

Figure 1.18 Schematic phase diagram of the hole-doped cuprate superconductors. The $p = 0$ line is a Mott insulator with antiferromagnetic (AF) order, as in Fig. 1.4. Additional phases with charge-density wave order at low T are not shown.

of the nearly localized nature of the f band is that the quasiparticle excitations on the Fermi surface have a large effective mass. Such a heavy Fermi liquid state is also realized in $CeCoIn_5$ but only after doping with a small concentration of Sn – the Fermi surface of this state is shown on the right of Fig. 1.17.

But without Sn doping, $CeCoIn_5$ realizes a novel metallic state that we call the "fractionalized Fermi liquid," which is discussed in Chapter 31. In this state, the f electrons are not included in the computation of the size of the Fermi surface, and the conventional Luttinger relation is violated. Instead, the f electrons form a spin liquid, similar to that in the Mott insulator in Fig. 1.11. The absence of the f states in the Fermi volume computation implies the Fermi surface is "small" in the fractionalized Fermi liquid, in contrast to the "large" Fermi surface of the heavy Fermi liquid. The localized f-electron column in Fig. 1.17b shows the "small" Fermi surface of the fractionalized Fermi liquid state of $CeCoIn_5$.

1.5 The Cuprates

It was the discovery of the celebrated copper oxide-based high-temperature superconductors in 1987 that launched the modern era in the study of the quantum phases of matter, which is the focus of this book. The superconductivity is obtained by doping a Mott insulator, similar to La_2CuO_4 in Fig. 1.4, by changing the composition of elements away from the CuO_2 plane. The low-energy physics is described by an electronic Hubbard model, similar on the square lattice to that discussed for fermionic ultracold atoms in Fig. 1.13. A schematic phase diagram as a function of electron density $= 1 - p$ and temperature is shown in Fig. 1.18. Note the similarity to the ultracold-atom phase

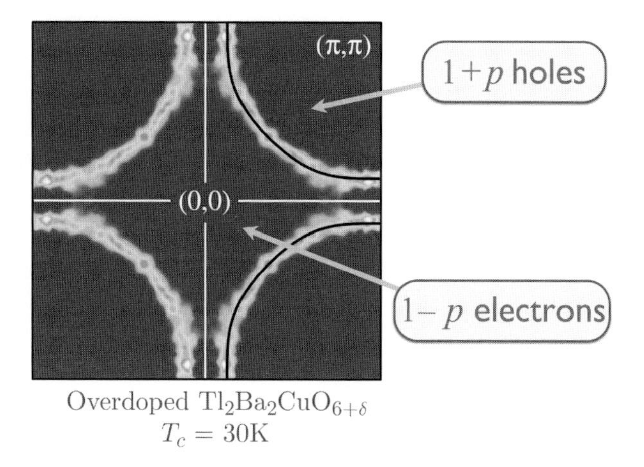

Figure 1.19 Photoemission observations [205] of the Fermi surface of in $Tl_2Ba_2CuO_{6+\delta}$ in the large-p Fermi liquid regime of Fig. 1.18. The Fermi surface obeys the conventional Luttinger relation of Fig. 1.2. Reprinted with permission from APS.

diagram in Fig. 1.13. The nature of the low-temperature phases of the antiferromagnet and superconductor is discussed in Chapter 9.

Much of Part V is directed towards a theory of the metallic phases that appear at higher temperatures, shown in Fig. 1.18 (and also in Fig. 1.13). The large-p metallic state (the "overdoped" regime) is a conventional Fermi liquid, and this is confirmed by photoemission experiments, which show a Fermi surface obeying the conventional Luttinger relation; see Fig. 1.19.

In contrast, the "pseudogap metal" at small p does not fall into the traditional Fermi liquid paradigms. A great deal of experimental effort has been devoted to carefully characterizing the remarkable properties of this pseudogap metal, and associated ordered phases at low T. I do not survey this work here, and only note the analog of the photoemission observations in Fig. 1.19, but now carried out at low T. Figure 1.20 shows that the "large Fermi surface" of Fig. 1.19 is not present at small p. Instead there is a gap in the electronic spectrum near the "antinodes" (this is the region of the Brillouin zone near $\boldsymbol{k} = (\pi, 0), (0, \pi)$). The Fermi surface appears to only be present in "arc"-like form near the "nodal region" (this is the region of the Brillouin zone near $\boldsymbol{k} = (\pm\pi/4, \pm\pi/4)$). There are strong theoretical constraints that the Fermi surface must be a closed curve, and so a natural hypothesis is that the Fermi arcs are only the front portion of small hole-pocket Fermi surfaces in the nodal region, and the photoemission intensity on the "back side" of the pockets is suppressed by a small quasiparticle residue (see Ref. [73] for transport evidence for such pocket Fermi surfaces). An attractive interpretation of such Luttinger-relation-violating small Fermi surfaces is that the pseudogap metal is a fractionalized Fermi liquid, similar to that discussed in Section 1.4 for the intermetallic compounds. However, in the cuprates there is no analog of the f band in which some of the electrons can localize and so reduce the volume enclosed by the Fermi surface. There is only one single band that crosses the Fermi level, as is amply clear from the Fermi surface observed in Fig. 1.19,

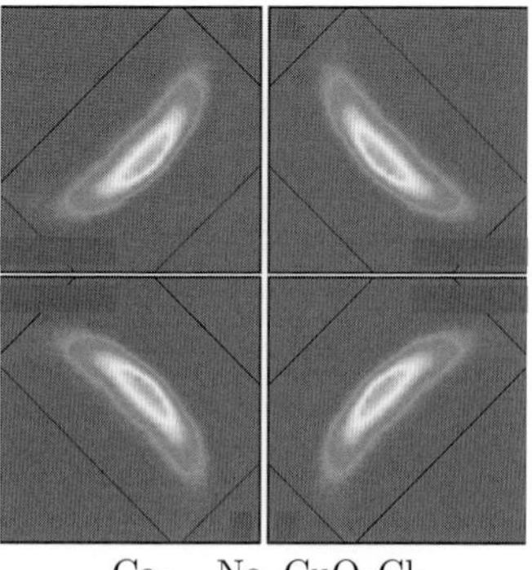

Figure 1.20 Photoemission spectrum of $Ca_{2-x}Na_xCuO_2Cl_2$ [264] in the underdoped pseudogap metal region of Fig. 1.18. Reprinted with permission from AAAS.

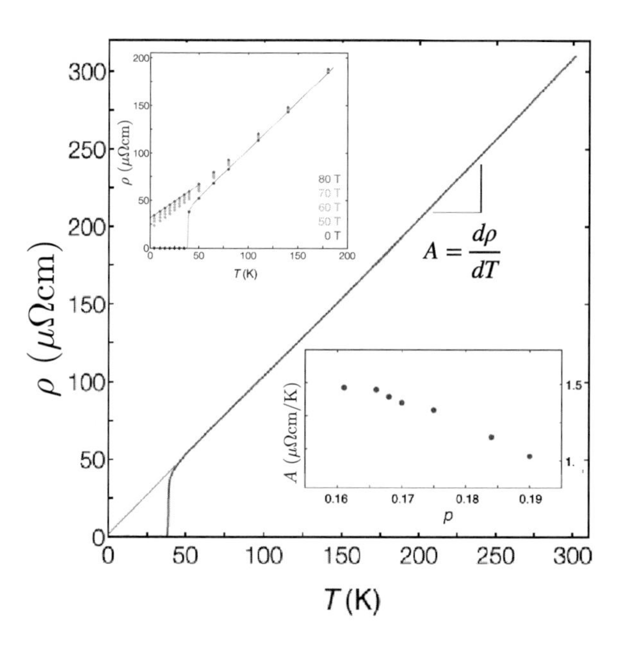

Figure 1.21 Resistivity of $La_{2-x}Sr_xCuO_4$ [94] in the strange-metal regime of Fig. 1.18. Reprinted with permission from AAAS.

and partially localizing some of the electrons in this band seems difficult to achieve. In Section 31.4 I will present a theory for the formation of a fractionalized Fermi liquid state in a single-band model: this theory is able to successfully model the photoemission spectra as a function of momentum and energy [173].

The last remaining metal in Fig. 1.18 is the "strange metal." This also does not fall into the Fermi liquid paradigm, and its well-known characteristic is the linear-in-temperature resistivity shown in Fig. 1.21. Photoemission experiments and other observations show that there are no well-defined fermionic quasiparticles in the strange metal, in contrast to both the pseudogap metal and the Fermi liquid. However, as discussed in Chapter 34, the breakdown of quasiparticles does not exclude the presence of a sharp Fermi surface in momentum space. In Chapter 34 I also discuss theories of strange metals without quasiparticles, along with an analysis of their transport properties.

PART I

BACKGROUND

2 Fermi Liquid Theory

The basic principles of Fermi liquid theory are reviewed, including the definition of the Fermi surface for interacting electrons, and the divergence in the quasiparticle lifetime at low temperature.

The conventional theory of metals starts from a theory of the free-electron gas, and then perturbatively accounts for the Coulomb interactions between the electrons. Already at leading order, we find a rather strong effect of the Coulomb interactions: a logarithmic divergence in the effective mass of the single-particle excitations near the Fermi surface. Further examination of the perturbation theory shows that this divergence is due to a failure to account for the screening of the long-range Coulomb interactions. Formally, screening can be accounted for by a simple modification of the perturbative series: introduce a dielectric constant in the interaction propagator, and sum only graphs that are irreducible with respect to the interaction line. Once screening is accounted for by this method, the effective mass of the single-particle excitations becomes finite.

In this initial chapter we ask: Is it possible to give a description of the interacting-electron gas that is valid to all orders in the Coulomb interactions? By "all orders in perturbation theory" we are assuming the validity of the perturbation theory, and cannot rule out non-perturbative effects, which could lead to the appearance of new phases of matter. Indeed the study of such new phases of matter is the focus of a major part of this book. But in this chapter, we present an all-orders description of the electron gas. This starts by formalizing the definition of a "quasiparticle" excitation, as a central ingredient in the theory of many-particle quantum systems.

2.1 Free-Electron Gas

Let us start by recalling the basic properties of the free-electron gas. We work in a second quantized formalism with electron annihilation operators $c_{\boldsymbol{p}\alpha}$, where $\boldsymbol{p}$ is momentum and α, β is the electron spin. The electron operator obeys the anti-commutation relation

$$[c_{\boldsymbol{k}\alpha}, c^{\dagger}_{\boldsymbol{k}'\beta}]_{+} = \delta_{\boldsymbol{k},\boldsymbol{k}'}\delta_{\alpha\beta}. \tag{2.1}$$

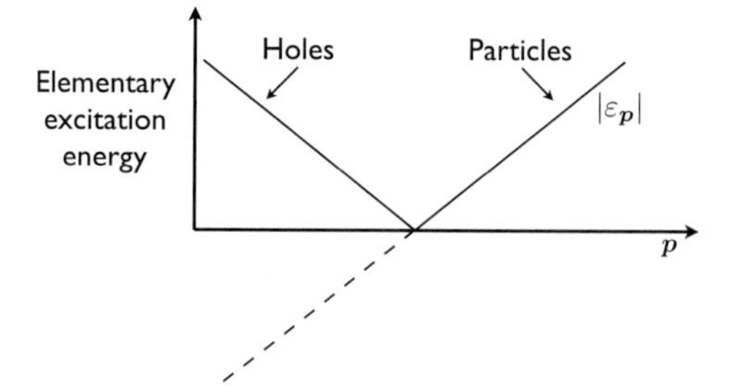

Figure 2.1 Fermionic excitation spectrum of a Fermi liquid as a function of momentum $\boldsymbol{p}$ along a fixed direction from the origin.

We assume the dispersion of a single electron is ε_p. The chemical potential μ is assumed to be included in ε_p; so for the jellium model $\varepsilon_p = \hbar^2 p^2/(2m) - \mu$. Then the Hamiltonian is

$$H = \sum_{p,\alpha} \varepsilon_p c^\dagger_{p\alpha} c_{p\alpha} \,. \tag{2.2}$$

The $T = 0$ ground state of this Hamiltonian is

$$|G\rangle = \prod_{\varepsilon_p<0,\alpha} c^\dagger_{p\alpha}|0\rangle. \tag{2.3}$$

The equation $\varepsilon_p = 0$ defines the *Fermi surface* in momentum space, separating the occupied and unoccupied states.

The elementary excitations of this state are of two types. Outside the Fermi surface we have particle-like excitations

$$\text{Particles:} \qquad c^\dagger_{p,\alpha}|G\rangle, \quad \boldsymbol{p} \text{ outside Fermi surface,} \tag{2.4}$$

while inside the Fermi surface we have hole-like excitations

$$\text{Holes:} \qquad c_{p,\alpha}|G\rangle, \quad \boldsymbol{p} \text{ inside Fermi surface.} \tag{2.5}$$

The energy of these excitations must be positive (by definition), and is easily seen to equal $|\varepsilon_p|$, as illustrated in Fig. 2.1.

From these elementary excitations, we can now build an exponentially large number of multi-particle excitations. In the independent-electron theory, their energies are simply the sum of the energies of the elementary excitations $\sum_{p,\alpha} |\varepsilon_p|$.

2.2 Interacting-Electron Gas

Our basic assumption is one of adiabatic continuity from the free-electron gas. We imagine we can tune the strength of the Coulomb interactions, and slowly turn them on from the independent-electron theory. Alternatively, we can assert that there is no quantum phase transition as the strength of the interactions is increased: note this is an assumption, and we will meet situations where this is not the case. In this adiabatic

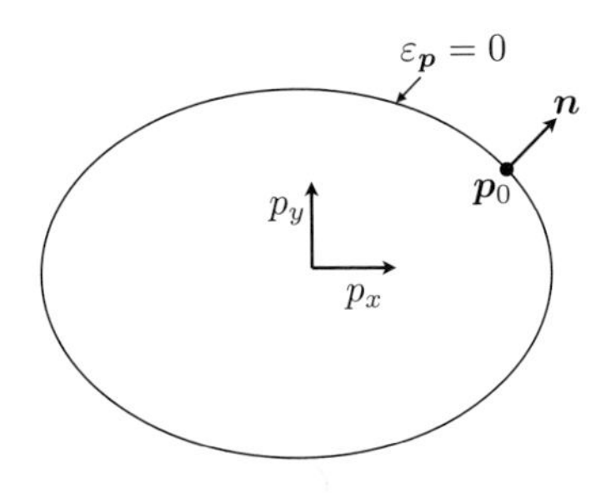

Figure 2.2 A point $\boldsymbol{p}_0$ on the Fermi surface, and its unit normal $\boldsymbol{n}$.

process, we assume that there is a correspondence between the ground states and the elementary excitations of the free- and interacting-electron gas. So the state $|G\rangle$ in (2.3) evolves smoothly to the unknown ground state of the interacting-electron gas. And, importantly, there is also a correspondence in the excitations. In the "jellium" model, with continuous translational symmetry and a uniform background neutralizing charge, this correspondence is simply one-to-one: a particle excitation with energy ε_p evolves into a "quasiparticle excitation" with a modified value of ε_p; and, similarly, for a "quasihole" with modified energy $-\varepsilon_p$. An important assumption is that ε_p remains a smooth function through the Fermi surface, and the energies of both particles and holes is given by $|\varepsilon_p|$.

In the presence of a lattice, the process of adiabatic evolution is more subtle, because we cannot assume that ε_p is only a function of $|\boldsymbol{p}|$. Consequently the *shape* of the Fermi surface can change in the adiabatic evolution, and a particle with momentum $\boldsymbol{p}$ can be inside the Fermi surface for the free-electron gas, and outside the Fermi surface for the interacting-electron gas. The crucial Luttinger theorem states that even though the shape of the Fermi surface can evolve, the volume enclosed by the Fermi surface is an adiabatic invariant; I defer discussion of this theorem to Section 30.2 in Part V. In the presence of a lattice, our basic assumption is that there is a smooth function ε_p so that the Fermi surface is defined by $\varepsilon_p = 0$, and the excitation energies of the quasiparticles and quasiholes is $|\varepsilon_p|$. Near the Fermi surface, we assume a linear dependence in momentum orthogonal to it: at a point $\boldsymbol{p}_0$ on the Fermi surface, let the normal to the Fermi surface be the direction $\boldsymbol{n}$ (the value of p_F can depend upon $\boldsymbol{p}_0$, see Fig. 2.2), and so we can write for $\boldsymbol{p}$ close to $\boldsymbol{p}_0$

$$\varepsilon_p = v_F(\boldsymbol{p} - \boldsymbol{p}_0) \cdot \boldsymbol{n}, \quad v_F = |\nabla_p \varepsilon_p| \equiv p_F/m^*, \tag{2.6}$$

where $p_F = |\boldsymbol{p}_0|$. This equation defines the Fermi momentum p_F, the Fermi velocity v_F, and the effective mass m^*, all of which can depend upon the direction $\hat{\boldsymbol{p}}_0$ in the presence of a lattice. Note that $\nabla_p \varepsilon_p = |\nabla_p \varepsilon_p| \boldsymbol{n}$ is a vector normal to the Fermi surface.

A further *assumption* in the theory of the interacting-electron gas is that we can build up the exponentially large number of other excitations also by composing the elementary excitations. (In a finite system of size N, the number of elementary excitations is of order N, while the number of composite excitations is exponentially large in N.) As we are interested in the thermodynamic limit, we can characterize these excitations by

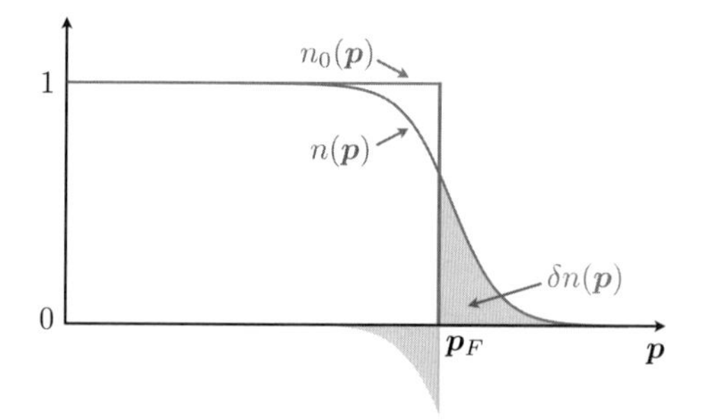

Figure 2.3 Plot of the quasiparticle distribution functions $n(\boldsymbol{p})$ and $\delta n(\boldsymbol{p})$ of an excited state of the Fermi liquid. Note that $\delta n(\boldsymbol{p})$ has a discontinuity of unity at the Fermi surface.

the densities of quasiholes and quasiparticles. In practice, it is quite tedious to keep track of two separate densities, along with a non-analytic dependence of their excitation energy $|\varepsilon_p|$ on $\boldsymbol{p}$. Both these problems can be overcome by a clever mathematical trick; we emphasize that there is no physics assumption involved in this trick – it is merely a bookkeeping device. We *postulate* that the interacting ground state has the same form as the free-electron ground state in (2.3). So the ground state has a density of quasiparticles $n_0(\boldsymbol{p})$ given by

$$
\begin{aligned}
n_0(\boldsymbol{p}) &= 1, \quad \boldsymbol{p} \text{ inside the Fermi surface,} \\
n_0(\boldsymbol{p}) &= 0, \quad \boldsymbol{p} \text{ outside the Fermi surface,}
\end{aligned}
\tag{2.7}
$$

as shown in Fig. 2.3. Then, an excited state is characterized by the density of quasiparticles $n(\boldsymbol{p})$, but the excitation energy will depend only upon

$$\delta n(\boldsymbol{p}) = n(\boldsymbol{p}) - n_0(\boldsymbol{p}), \tag{2.8}$$

where $\delta n(\boldsymbol{p})$ has a discontinuity of unity at the Fermi surface. So for $\boldsymbol{p}$ outside the Fermi surface $\delta n(\boldsymbol{p})$ measures the density of quasiparticle excitations, while for $\boldsymbol{p}$ inside the Fermi surface $-\delta n(\boldsymbol{p})$ measures the density of quasihole excitations. (All of these densities can also depend upon the spin of the quasiparticles or quasiholes, a complication we shall ignore in the following discussion.) So the actual density of excitations with energy $|\varepsilon_p|$ is $|\delta n(\boldsymbol{p})|$. For the total excitation energy, which depends on their product, we can drop the absolute value, as in the first term on the right hand side of (2.9): this is one of the advantages of this mathematical trick.

We assume we are at temperature $T \ll E_F$, where E_F is the Fermi energy so that the density of quasiparticles and quasiholes is small. Our first thought is that, because of the low density, we can ignore the interactions between the quasiparticles and quasiholes, and compute the total energy of the multi-particle/hole excitations simply by adding their individual energies. An important observation by Landau was that this is not correct. If we wish to work consistently to order $(T/E_F)^2$ in the total energy, one (and only one) additional term is necessary; ignoring spin dependence, we have the Landau energy functional

$$E[\delta n(\boldsymbol{p})] = \sum_{p} \varepsilon_p \delta n(\boldsymbol{p}) + \frac{1}{2V} \sum_{p,k} F_{\hat{p},\hat{k}}\, \delta n(\boldsymbol{p}) \delta n(\boldsymbol{k}), \tag{2.9}$$

where V is the volume of the system. At a temperature $T \ll E_F$, $\delta n(\boldsymbol{p})$ is of order unity only in a window of momenta with $v_F|p - p_F| \sim T$, where $|\varepsilon_p| \sim T$. Then, as we perform the radial integral in the first term in (2.3), we pick up a factor T from ε_p, and a second factor of T from the limits on the integral; so the first term is of order T^2. Landau's point is that the second term in (2.9) is also of order T^2; there now are two integrals over radial momenta, and their product yields a factor of T^2. This term describes the interaction between the quasiparticles and quasiholes, and is characterized by the unknown Landau interaction function $F_{\hat{p},\hat{k}}$. To order T^2, we can consistently assume that all the quasiparticles and quasiholes are practically on the Fermi surface in the interaction term, and so $F_{\hat{p},\hat{k}}$ depends only upon the directions of $\boldsymbol{p}$ and $\boldsymbol{k}$.

Although the quasiparticles and quasiholes are assumed to interact in Landau's functional, the interaction is conservative, which means it does not scatter quasiparticles between momenta or change the quasiparticle distribution function. The main effect of the interaction term is that the change in the energy of the system upon adding a quasiparticle or quasihole depends upon the density of excitations already present. Scattering processes of quasiparticles are considered later in Section 2.6: these lead to a finite quasiparticle lifetime, but the correponding corrections to the energy functional are higher order in T.

Landau's central point is that the values of m^* and $F_{\hat{p},\hat{k}}$ are sufficient to provide a description of the low-temperature properties of the interacting-electron gas to order $(T/E_F)^2$, and all orders in the strength of the underlying Coulomb interactions.

2.3 Specific Heat

As a first application of Landau's Fermi liquid theory, let us compute the specific heat. Assuming a thermal distribution of excitations, we have, using the Fermi function $f(\varepsilon) = 1/(e^{\varepsilon/T} + 1)$:

$$\delta n(\boldsymbol{p}) = f(\varepsilon_p) - n_0(\boldsymbol{p}). \tag{2.10}$$

Now using (2.7), the identity $f(-\varepsilon) = 1 - f(\varepsilon)$, and approximating $\varepsilon_p = v_F(\boldsymbol{p} - \boldsymbol{p}_0) \cdot \boldsymbol{n}$ at low T close to the Fermi surface, we can easily show that $\delta n(\boldsymbol{p})$ is an odd function of $(\boldsymbol{p} - \boldsymbol{p}_0) \cdot \boldsymbol{n}$ at each point on the Fermi surface $\boldsymbol{p}_0$. An immediate consequence is that the second term in (2.9) vanishes to order T^2, because it has no further dependence upon momenta normal to the Fermi surface. More physically stated, equal numbers of quasiparticles and quasiholes are excited in thermal equilibrium and, as the Landau interaction depends upon the total density of excitations, it does not contribute to the total energy.

Consequently, the Landau interactions make no difference to the free energy, and so the specific heat of the interacting-electron gas is the same as that of the free-electron gas, after replacing ε_p by the true quasiparticle dispersion. So we have the specific heat

$$C_V = \gamma T, \tag{2.11}$$

where the "γ-coefficient" is given by

$$\gamma = \frac{\pi^2 k_B^2 T}{3} g(0), \tag{2.12}$$

with $g(0)$ being the density of quasiparticle states at the Fermi level. For the spinful jellium model in dimension $d = 3$

$$g(0) = \frac{m^* p_F}{2\pi^2 \hbar^3}. \tag{2.13}$$

2.4 Compressibility

The compressibility measures the change in electron density in response to a change in the chemical potential. As a change in the chemical potential has a different effect on quasiparticles and quasiholes, the Landau interaction parameters now do have an important effect.

With the change $\mu \to \mu + \delta\mu$, we can argue from the Landau theory that the distribution of quasiparticles is

$$\delta n(\boldsymbol{p}) = f(\varepsilon^*_{\boldsymbol{p}}) - n_0(\boldsymbol{p}), \tag{2.14}$$

where

$$\begin{aligned} \varepsilon^*_{\boldsymbol{p}} &= \frac{\delta E}{\delta n(\boldsymbol{p})} \\ &= \varepsilon_p - \delta\mu + \frac{1}{V}\sum_{\boldsymbol{k}} F_{\hat{p},\hat{k}}\, \delta n(\boldsymbol{k}). \end{aligned} \tag{2.15}$$

We have now accounted for the fact that the energy of each quasiparticle depends upon the density of other quasiparticles via the Landau interaction. We did not have to account for such a dependence in the computation of the specific heat because the sum over $\boldsymbol{k}$ vanishes in that case. Note that $\delta n(\boldsymbol{p})$ appears on both the left- and right-hand sides of (2.14), and so has to be determined self-consistently. This is rather similar to Hartree–Fock theory, where expectation values of fermion bilinears appeared both in the Hartree–Fock Hamiltonian and in the self-consistency condition. The difference, of course, is that the present considerations are exact at low T.

As $T \to 0$, we expand the equations (2.14) and (2.15) to linear order in $\delta\mu$. Using the low T identity

$$-\frac{\partial f}{\partial \varepsilon} = \delta(\varepsilon) \tag{2.16}$$

we can conclude that the distribution function is of the form

$$\delta n(\boldsymbol{p}) = A\,\delta\mu\,\delta(\varepsilon_{\boldsymbol{p}}), \tag{2.17}$$

where A is a $\boldsymbol{p}$-independent constant. Inserting (2.17) into (2.14) and (2.15), we obtain

$$A = 1 - \frac{A}{V}\sum_{\boldsymbol{k}} F_{\hat{p},\hat{k}}\,\delta(\varepsilon_{\boldsymbol{k}}). \tag{2.18}$$

Combining all the results, we obtain the compressibility

$$\begin{aligned} \frac{dn}{d\mu} &= \frac{1}{\delta\mu}\frac{1}{V}\sum_{\boldsymbol{p}}\delta n(\boldsymbol{p}) \\ &= \frac{1}{1+F_0}\frac{1}{V}\sum_{\boldsymbol{p}}\delta(\varepsilon_{\boldsymbol{p}}) \\ &= \frac{g(0)}{1+F_0}, \end{aligned} \tag{2.19}$$

where

$$F_0 = \frac{1}{V}\sum_{\boldsymbol{k}} F_{\hat{p},\hat{k}}\,\delta(\varepsilon_{\boldsymbol{k}}) \tag{2.20}$$

is the average of the Landau interaction parameter around the Fermi surface. For the case with full rotational symmetry (as in jellium), we can decompose the Landau interactions into angular-momentum components F_ℓ, and F_0 is then the s-wave component.

So there is a renormalization by a factor of $1/(1+F_0)$ of the compressibility from the Landau interactions.

2.5 Dynamic Response Functions

We can extend the ideas of Fermi liquid theory to include responses to time-dependent perturbations. We place the Fermi liquid in an external potential $V(\boldsymbol{r},t)$, and examine the response of the quasiparticle distribution function. Then (2.15) is modified to

$$\varepsilon_{\boldsymbol{p}}^* = \varepsilon_{\boldsymbol{p}} + \frac{1}{V}\sum_{\boldsymbol{k}} F_{\hat{p},\hat{k}}\,\delta n(\boldsymbol{k},\boldsymbol{r},t) + V(\boldsymbol{r},t), \tag{2.21}$$

where we have also allowed the quasiparticle distribution function to become time-dependent. Provided the external potential is slowly varying in space, we can use semiclassical equations of motion to describe the time and space evolution of the quasiparticles with average momentum $\boldsymbol{p}$ and position $\boldsymbol{r}$:

$$\begin{aligned} \frac{d\langle\boldsymbol{r}\rangle}{dt} &= \frac{\partial\varepsilon_{\boldsymbol{p}}^*}{\partial\boldsymbol{p}} \\ \frac{d\langle\boldsymbol{p}\rangle}{dt} &= -\frac{\partial\varepsilon_{\boldsymbol{p}}^*}{\partial\boldsymbol{r}}. \end{aligned} \tag{2.22}$$

From this we can write down the Boltzmann equation for the evolution of the quasiparticle distribution function:

$$\frac{\partial \delta n}{\partial t} + \frac{d\langle \boldsymbol{r} \rangle}{dt} \frac{\partial \delta n}{\partial \boldsymbol{r}} + \frac{d\langle \boldsymbol{p} \rangle}{dt} \frac{\partial \delta n}{\partial \boldsymbol{p}} = I_{col} . \tag{2.23}$$

The right-hand side of (2.23) is the "collision" term, which scatters quasiparticles around the Fermi surface among themselves. The key assumption of Fermi liquid theory is that this scattering rate is small, as we will see in Section 2.6. Neglecting I_{col}, we can solve (2.23) to obtain various collective properties of Fermi liquids, including the existence of "zero sound."

2.6 Green's Functions and Quasiparticle Lifetime

For further discussion of the properties of the Fermi liquid, and the nature of its corrections when we consider higher temperatures, it is useful to employ the language of Green's functions. We use the standard many-body Green's function defined in Ref. [39]. The most convenient definition starts from the Green's functions defined in imaginary time τ (ignoring the electron spin α)

$$G(\boldsymbol{p}, \tau) = -\left\langle T_\tau c_{\boldsymbol{p}}(\tau) c_{\boldsymbol{p}}^\dagger(0) \right\rangle , \tag{2.24}$$

where T_τ is the time-ordering symbol. We can then Fourier transform this to the Matsubara frequencies $\omega_n = (2n+1)\pi T/\hbar$, n integer, to obtain $G(\boldsymbol{p}, i\omega_n)$. More generally, we can consider the Green's function in the complex z plane, $G(\boldsymbol{p}, z)$, obtained by analytic continuation of $G(\boldsymbol{p}, i\omega_n)$. This Green's function obeys the spectral representation

$$G(\boldsymbol{p}, z) = \int_{-\infty}^{\infty} d\Omega \frac{\rho(\boldsymbol{p}, \Omega)}{z - \Omega}, \tag{2.25}$$

where $\rho(\boldsymbol{p}, \Omega) = -(1/\pi)\text{Im}\left[G(\boldsymbol{p}, \Omega + i0^+)\right] > 0$ is the spectral density. We will also refer to the retarded Green's function $G^R(\boldsymbol{p}, \omega) = G(\boldsymbol{p}, \omega + i0^+)$, and more generally $G^R(\boldsymbol{p}, z) = G(\boldsymbol{p}, z)$ for z in the upper half-plane. Closely associated is the electron self-energy $\Sigma(\boldsymbol{p}, z)$, which is related to the Green's function by Dyson's equation

$$G(\boldsymbol{p}, z) = \frac{1}{z - \varepsilon_{\boldsymbol{p}}^0 - \Sigma(\boldsymbol{p}, z)}, \tag{2.26}$$

where by $\varepsilon_{\boldsymbol{p}}^0$ we now denote the bare electron dispersion before the effects of electron–electron interactions are accounted for.

The postulates of Fermi liquid theory described above have strong implications for the structure of the Green's function in the complex frequency plane. Specifically, the existence of long-lived quasiparticles near the Fermi surface implies that the Green's function has a pole very close to the real frequency axis, at a frequency obeying $\text{Re}(z) = \varepsilon_{\boldsymbol{p}}$ for $\boldsymbol{p}$ close to the Fermi surface. The existence of such a pole implies a free-particle behavior of the Green's function at long times, representing the propagation of

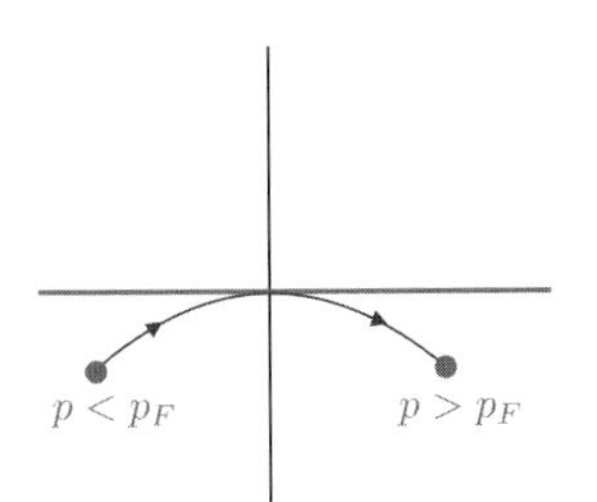

Figure 2.4 The poles of the Green's function $G^R(\boldsymbol{p}, z)$ in the complex z plane. The poles are in the second Riemann sheet, and the horizontal line represents the branch cut implied by (2.25).

the quasiparticle. In this section, I wish to go beyond Fermi liquid theory and include a finite quasiparticle lifetime by taking the pole just off the real axis.

Actually, there is an important subtlety in the statement "there is a pole in the Green's function," which we need to keep in mind. The spectral definition (2.25) implies that $G(\boldsymbol{p}, z)$ is an analytic function for all z, with a branch cut on the real frequency axis, for an interacting system with a reasonably smooth spectral density $\rho(\boldsymbol{p}, \Omega)$. The pole is actually in a different Riemann sheet from the definition (2.25), and is reached by analytically continuing across the branch cut. So the retarded Green's function $G^R(\boldsymbol{p}, z)$ is analytic for all z in the upper half-plane, and the pole is obtained when we analytically continue $G^R(\boldsymbol{p}, z)$ to the lower half-plane (where it is *not* equal to the $G(\boldsymbol{p}, z)$ defined by (2.25)). For $\boldsymbol{p}$ close to the Fermi surface in a Fermi liquid, this pole is at a frequency $z = \varepsilon_p - i\gamma_p$, where $\gamma_p > 0$ is related to the quasiparticle lifetime $\tau_p = 1/(2\gamma_p)$ because it leads to exponential decay for the Green's function in real time (the factor of two arises because we measure the *probability* of observing a quasiparticle a time τ_p after creating it). Note that the pole is in the lower half-plane of the analytically continued $G^R(\boldsymbol{p}, z)$ for both signs of ε_p, that is, for both quasiparticles and quasiholes; see Fig. 2.4.

Ultimately, this complexity can be succinctly captured by initially restricting attention to the G Green's function on the imaginary frequency axis. Then, the existence of the quasiparticle implies that the Green's function defined by (2.25) obeys

$$G(\boldsymbol{p}, i\omega) = \frac{Z_{\hat{p}}}{i\omega - \varepsilon_p + i\gamma_p \,\mathrm{sgn}(\omega)} + G_{inc}(\boldsymbol{p}, i\omega_n), \tag{2.27}$$

where

$$\varepsilon_p = \varepsilon_p^0 + \mathrm{Re}\left[\Sigma(\boldsymbol{p}, 0)\right] \tag{2.28}$$

is the renormalized quasiparticle dispersion, ε_p^0 is the bare quasiparticle dispersion, and

$$\gamma_p = -\mathrm{Im}\left[\Sigma(\boldsymbol{p}, \varepsilon_p + i0^+)\right] > 0. \tag{2.29}$$

Consistency of the above definitions requires that the inverse lifetime of the quasiparticle is much smaller than its excitation energy:

$$\gamma_p \ll |\varepsilon_p|, \tag{2.30}$$

for $\boldsymbol{p}$ close the Fermi surface. The Fourier transform of G has a slowly decaying contribution, which is just that of a free particle but with renormalized dispersion, and an amplitude suppressed by $Z_{\hat{p}}$. Consequently, $Z_{\hat{p}}$ is the quasiparticle residue, and it equals the square of the overlap between the free and quasiparticle wavefunctions. The G_{inc} term is the "incoherent" contribution, associated with additional excitations created from the particle–hole continuum upon inserting a single particle into the system: this contribution decays rapidly in time, and can be ignored relative the quasiparticle contribution for the low-energy physics.

From (2.27), we can now compute the momentum distribution function $n_e(\boldsymbol{p})$ of the underlying electrons;

$$n_e(\boldsymbol{p}) = \left\langle c^\dagger_{\boldsymbol{p}} c_{\boldsymbol{p}} \right\rangle, \tag{2.31}$$

where we are dropping the spin index. For a free-electron gas

$$n_e(\boldsymbol{p}) = \theta(-\varepsilon^0_{\boldsymbol{p}}), \quad \text{free electrons, } T = 0, \tag{2.32}$$

where $\theta(x)$ is the unit step function. So there is a discontinuity of size unity on the Fermi surface in $n_e(\boldsymbol{p})$. For the interacting-electron gas, it is important to distinguish $n_e(\boldsymbol{p})$ from the distribution function of quasiparticles $n(\boldsymbol{p})$ in (2.8). The *quasiparticle* momentum distribution function continues to have a discontinuity of size *unity* on the Fermi surface $\varepsilon_{\boldsymbol{p}} = 0$. For the electron momentum distribution function at $T = 0$, we need to evaluate

$$n_e(\boldsymbol{p}) = \int_{-\infty}^{\infty} \frac{d\omega}{2\pi} G(\boldsymbol{p}, i\omega) e^{i\omega 0^+}. \tag{2.33}$$

Evaluating the integral in (2.33) using (2.27), we find a discontinuous contribution from the pole near the Fermi surface. There is no reason to expect a discontinuity from G_{inc}, and so we obtain

$$n_e(\boldsymbol{p}) = Z_{\hat{p}}\, \theta(-\varepsilon_{\boldsymbol{p}}) + \cdots, \quad \text{interacting electrons, } T = 0, \tag{2.34}$$

where $\ldots$ is the contribution from G_{inc}. A typical plot of $n_e(\boldsymbol{p})$ is shown in Fig. 2.5. Because $n_e(\boldsymbol{p})$ must be positive and bounded by unity, we have a constraint on the quasiparticle residue

$$0 < Z_{\hat{p}} \leq 1. \tag{2.35}$$

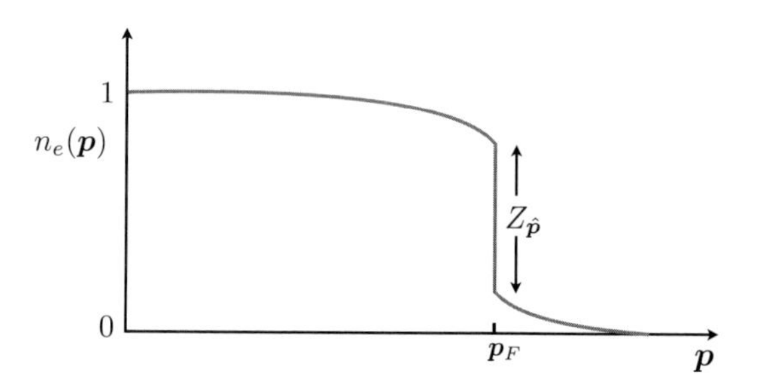

Figure 2.5 The momentum distribution function of bare electrons in a Fermi liquid at $T = 0$. There is a discontinuity of size $Z_{\hat{p}}$ on the Fermi surface.

Note that a small $Z_{\hat{p}}$ is not an indication that the Fermi liquid theory is not robust: it merely indicates a small overlap between the bare electron and the renormalized quasiparticle. Systems with very small $Z_{\hat{p}}$ can be very good Fermi liquids: we will study the heavy-fermion compounds in Chapter 30, which are of this type. Rather, it is a short quasiparticle lifetime, or large γ_p, and the failure of (2.30), which is a diagnostic of the breakdown of Fermi liquid theory. We will turn to "non-Fermi liquids" (which also have $Z_{\hat{p}} = 0$) in Chapters 32 and 34.

For an explicit evaluation of the inverse lifetime γ_p, we have to consider processes beyond those present in Landau Fermi liquid theory. In particular, we have to evaluate the imaginary part of the self energy in (2.29) for $\boldsymbol{p}$ near the Fermi surface. This requires a somewhat tedious evaluation of the relevant Feynman diagrams, and we explicitly compute an example in Section 34.1.2. For now, we are satisfied here by "guessing" the answer by Fermi's golden rule. Assuming only a contact interaction, U, between the quasiparticles, we can write the inverse lifetime as

$$\begin{aligned}\frac{1}{\tau_p} = 2\gamma_p = 2\pi U^2 \frac{1}{V^2} \sum_{\boldsymbol{k},\boldsymbol{q}} & f(\varepsilon_{\boldsymbol{k}})[1 - f(\varepsilon_{\boldsymbol{p}+\boldsymbol{q}})][1 - f(\varepsilon_{\boldsymbol{k}-\boldsymbol{q}})] \\ & \times \delta\left(\varepsilon_{\boldsymbol{p}} + \varepsilon_{\boldsymbol{k}} - \varepsilon_{\boldsymbol{p}+\boldsymbol{q}} - \varepsilon_{\boldsymbol{k}-\boldsymbol{q}}\right).\end{aligned} \tag{2.36}$$

This is obtained by employing Fermi's golden role to the process sketched in Fig. 2.6, and including probabilities that the initial states are occupied, and the final states are empty. The momentum integrals in (2.36) are quite difficult to evaluate in general, but it is not hard to see that the result becomes very small for $\boldsymbol{p}$ near the Fermi surface and small T, because of the constraints imposed by the Fermi functions and the energy-conserving delta function. A simple overestimate can be made by simply ignoring the constraints from momentum conservation, in which case we obtain

$$\begin{aligned}\gamma_\varepsilon \sim U^2[d(0)]^3 p_F^{-d} & \int_{-\infty}^{\infty} d\varepsilon_1 d\varepsilon_2 d\varepsilon_3 f(\varepsilon_1)[1 - f(\varepsilon_2)][1 - f(\varepsilon_3)] \\ & \times \delta(\varepsilon + \varepsilon_1 - \varepsilon_2 - \varepsilon_3) \\ = U^2[d(0)]^3 p_F^{-d} & \times \begin{cases} \pi^2 T^2/4 & \text{for } \varepsilon = 0 \\ \varepsilon^2/2 & \text{for } T = 0. \end{cases}\end{aligned} \tag{2.37}$$

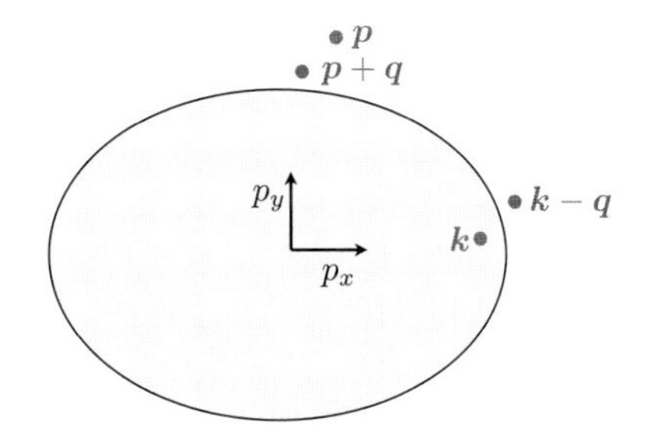

Figure 2.6 Decay of a quasiparticle with momentum $\boldsymbol{p}$ by scattering off a pre-existing quasiparticle with momentum $\boldsymbol{k}$ to produce quasiparticles of momena $\boldsymbol{p} + \boldsymbol{q}$ and $\boldsymbol{k} - \boldsymbol{q}$.

More careful considerations of momentum conservations are needed to obtain the precise coefficients (see Section 34.1.2 for an example), but they show that the power laws above in T and ε are correct. So, at low temperatures, $\gamma_p \sim T^2$ is always much smaller than $|\varepsilon_p| \sim T$, and this justifies Fermi liquid theory.

We can also use these results to give a formal definition of the Fermi surface using Green's functions. Notice that γ_ε in (2.37) vanishes as $\varepsilon \to 0$ at $T = 0$. This follows from the vanishing of the phase space for the decay of an excitation with energy ε as $\varepsilon \to 0$. This is actually a special case of a more general phenomenon following from the stability of the ground state, and does not even require excitations to be close to the Fermi surface. The more general statement is

$$\operatorname{Im}\left[\Sigma(\boldsymbol{p}, \Omega + i0^+)\right] \to 0 \text{ as } \Omega \to 0 \text{ at } T = 0 \tag{2.38}$$

for any $\boldsymbol{p}$, and its validity can be checked by examining the structure of the Feynman graph expansion for Σ. We will see in Chapter 32 that (2.38) applies also to non-Fermi liquids without quasiparticle excitations. We can now define the Fermi surface by the pole in the Green's function which is determined by

$$G^{-1}(\boldsymbol{p}_F, i0^+) = 0 \text{ at } T = 0. \tag{2.39}$$

By (2.38), the left-hand side of (2.39) is real, and so the solution of (2.39) determines a surface of co-dimension 1 in $\boldsymbol{p}$ space, which is the Fermi surface. These definitions are useful in establishing the Luttinger relation constraining the volume enclosed by the Fermi surface, which is discussed in Section 30.2.

Problem

2.1 Consider fermions $c_{k\alpha}$ with a Dirac dispersion in dimensions $d = 1, 2, 3$, described by the Hamiltonian

$$H = v_F \sum_{\boldsymbol{k}} c_{\boldsymbol{k}}^\dagger (\boldsymbol{k} \cdot \boldsymbol{\sigma}) c_{\boldsymbol{k}} + \frac{U}{2V} \sum_{\boldsymbol{k}, \boldsymbol{k}', \boldsymbol{q}} c_{\boldsymbol{k}+\boldsymbol{q},\alpha}^\dagger c_{\boldsymbol{k}'-\boldsymbol{q},\beta}^\dagger c_{\boldsymbol{k}',\beta} c_{\boldsymbol{k},\alpha}, \tag{2.40}$$

where $\boldsymbol{\sigma}$ is a vector of d Pauli matrices. As in (2.36), use Fermi's golden rule to compute the lifetime of a particle with a small momentum $\boldsymbol{k}$ at a low temperature T. Show that

$$\gamma_{\boldsymbol{k}} \sim U^2 \left[\operatorname{Max}(v_F |\boldsymbol{k}|, T)\right]^{2d-1}. \tag{2.41}$$

Unlike the estimate in (2.37) for a Fermi surface, it is important to include the conservation of momentum to obtain this estimate. So Dirac quasiparticles are well defined in $d = 2, 3$, but not in $d = 1$. A full consideration of the situation in $d = 1$ appears in Chapter 12.

3 Dilute Bose Gas

The Bogoliubov theory of the weakly interacting Bose gas is described using both the Hamiltonian and path-integral approaches. The off-diagonal long-range order in the ground state is related to the excitation spectrum of second sound modes.

I described a fundamental quantum liquid state in Chapter 2: that of interacting fermions. This liquid turned out to be smoothly connected to the liquid of free fermions, and we used this to our advantage to develop the basic principles of Fermi liquid theory. This chapter turns to a second fundamental quantum liquid state, that of interacting bosons. In this case, it turns out that the non-interacting Bose liquid is not a particularly good starting point for understanding the ground state of bosons with even a weak interaction. Fundamentally new ideas on broken symmetry are needed to understand the Bose liquid. In this chapter, I present the theory of the weakly interacting Bose gas, which will then form a basis for the more general discussion of broken symmetry in Chapter 5.

3.1 Bogoliubov Theory

We consider bosons $b_{\boldsymbol{k}}$, where $\boldsymbol{k}$ is a wavevector, interacting with a weak, repulsive short-range interaction u_0. The Bose operator obeys the commutation relation

$$[b_{\boldsymbol{k}}, b^{\dagger}_{\boldsymbol{k}'}] = \delta_{\boldsymbol{k},\boldsymbol{k}'} \tag{3.1}$$

and the Hamiltonian is

$$H = \sum_{\boldsymbol{k}} \varepsilon_{\boldsymbol{k}} b^{\dagger}_{\boldsymbol{k}} b_{\boldsymbol{k}} + \frac{u_0}{2V} \sum_{\boldsymbol{k},\boldsymbol{k}',\boldsymbol{q}} b^{\dagger}_{\boldsymbol{k}+\boldsymbol{q}} b^{\dagger}_{\boldsymbol{k}'-\boldsymbol{q}} b_{\boldsymbol{k}'} b_{\boldsymbol{k}}, \tag{3.2}$$

where V is the volume. In continuum free space the boson dispersion is $\varepsilon_{\boldsymbol{k}} = \hbar^2 \boldsymbol{k}^2/(2m)$, where m is the mass of the boson. But our analysis will also apply for other monotonic dispersions with a minimum at $\varepsilon_{\boldsymbol{k}=0} = 0$. We assume there is a high momentum cutoff in the interaction term in (3.2), beyond which the simple contact form of the interaction does not apply.

We will develop a theory for the ground state of H in a perturbation expansion in u_0. At $u_0 = 0$, the lowest energy state has all bosons in the $\boldsymbol{k} = 0$ state, with

$$|G\rangle = \frac{1}{\sqrt{N_0!}}(b_0^\dagger)^{N_0}|0\rangle, \tag{3.3}$$

where $|0\rangle$ is the empty state with no bosons, and N_0 is the number of bosons in the $\boldsymbol{k} = 0$ state. This is the Bose condensate, with all the bosons in the single-particle ground state.

Once we turn on interactions between the bosons, some fraction of the bosons will occupy non-zero momenta even in the ground state. Rather than computing this fraction at a fixed total particle number, it turns out to be far easier to describe the ground state in the grand canonical ensemble at a fixed chemical potential μ. In this case, we need to find the state with smallest value for the "grand energy" $H - \mu N$, where

$$N = \sum_{\boldsymbol{k}} b_{\boldsymbol{k}}^\dagger b_{\boldsymbol{k}}. \tag{3.4}$$

To begin, let us just use the state in (3.3) as a variational trial wavefunction, and evaluate the expectation value of the grand energy

$$\begin{aligned}\langle G|H - \mu N|G\rangle &= -\mu N_0 + \frac{u_0}{2V}N_0(N_0 - 1) \\ &= V\left[-\mu n_0 + \frac{u_0}{2}n_0(n_0 - 1/V)\right],\end{aligned} \tag{3.5}$$

where

$$n_0 = \frac{N_0}{V} \tag{3.6}$$

is the density of particles in the $\boldsymbol{k} = 0$ state. Minimizing (3.5) with respect to n_0 in the thermodynamic limit ($N \to \infty$) we obtain

$$n_0 = \frac{\mu}{u_0} \tag{3.7}$$

and the optimum value of the grand energy is

$$\langle G|H - \mu N|G\rangle = V\left[-\frac{\mu^2}{2u_0}\right]. \tag{3.8}$$

The result in (3.8) is the leading answer in the small u_0 expansion at fixed μ: notice that it diverges as $1/u_0$.

We note an important feature of the computation above: the $1/V$ term in (3.5) dropped out in the thermodynamic limit; this term arose from the non-zero commutator $[b_0, b_0^\dagger] = 1$. The surviving terms would have been obtained if we had just ignored the non-zero commutator, and just replaced b_0 by the number $\sqrt{N_0}$. This is a consequence of having a non-zero *density* of particles at $\boldsymbol{k} = 0$. Going forward, we will directly use the replacement $b_0 \to \sqrt{N_0}$, and this will only discard unimportant terms that vanish in the thermodynamic limit.

We now want to compute the corrections to (3.8) at order $(u_0)^0$. For this, we need to include the contributions of the bosons at $\boldsymbol{k} \neq 0$. We return to the original Hamiltonian in (3.2), replace $b_0 \to \sqrt{N_0}$, and keep all terms which are second order in $b_{\boldsymbol{k}}$, $b_{\boldsymbol{k}}^\dagger$ with $\boldsymbol{k} \neq 0$. This yields

$$
\begin{aligned}
H - \mu N = & V\left[-\mu n_0 + \frac{u_0}{2} n_0^2\right] \\
& + \sum_{\boldsymbol{k} \neq 0} \left(\varepsilon_{\boldsymbol{k}} - \mu + u_0 n_0\right) b_{\boldsymbol{k}}^\dagger b_{\boldsymbol{k}} \\
& + \frac{u_0 n_0}{2} \sum_{\boldsymbol{k} \neq 0} \left(b_{\boldsymbol{k}}^\dagger b_{-\boldsymbol{k}}^\dagger + b_{-\boldsymbol{k}} b_{\boldsymbol{k}} + b_{\boldsymbol{k}}^\dagger b_{\boldsymbol{k}} + b_{-\boldsymbol{k}}^\dagger b_{-\boldsymbol{k}}\right).
\end{aligned}
\tag{3.9}
$$

The first line of (3.9) is the same as (3.5), and we optimized this by choosing n_0 in (3.7). The remaining lines of (3.9) describe the Bogoliubov Hamiltonian for bosons with $\boldsymbol{k} \neq 0$, in which the boson b_0 has been replaced by the number n_0 in (3.7). A notable feature of the Bogoliubov Hamiltonian is that it appears to not conserve the total number of bosons, with the presence of terms that annihilate or create pairs of bosons. Of course, the total number of bosons is actually conserved but, by treating b_0 as a number, we are not keeping precise track of the number of bosons in the condensate: in the thermodynamic limit, we can safely ignore the difference between a condensate with N_0 particles or $N_0 \pm 2$ particles.

We can now proceed to diagonalize (3.9), and so obtain the order $(u_0)^0$ correction to the ground-state energy, and also the spectrum of excitation at non-zero $\boldsymbol{k}$; note that all the terms in the Bogoliubov Hamiltonian are of order $(u_0)^0$ after using (3.7). We can diagonalize (3.9) by introducing a new set of bosons, $\eta_{\boldsymbol{k}}$, which obey the canonical Bose commutation relations

$$
[\eta_{\boldsymbol{k}}, \eta_{\boldsymbol{k}'}^\dagger] = \delta_{\boldsymbol{k},\boldsymbol{k}'}. \tag{3.10}
$$

These are related to the $b_{\boldsymbol{k}}$ by a Bogoliubov transformation

$$
b_{\boldsymbol{k}} = \eta_{\boldsymbol{k}} \cosh(\theta_{\boldsymbol{k}}) - \eta_{-\boldsymbol{k}}^\dagger \sinh(\theta_{\boldsymbol{k}}) \tag{3.11}
$$

with $\theta_{-\boldsymbol{k}} = \theta_{\boldsymbol{k}}$ an arbitrary parameter for now. This transformation has been chosen so that the commutator (3.10) implies the commutator (3.1) and vice versa. Inserting (3.11) into (3.9), we find that we can get all the terms that do not conserve the total number of $\eta_{\boldsymbol{k}}$ bosons to cancel, provided we choose

$$
\tanh(2\theta_{\boldsymbol{k}}) = \frac{u_0 n_0}{\varepsilon_{\boldsymbol{k}} + u_0 n_0}. \tag{3.12}
$$

Then the grand Hamiltonian becomes

$$
\begin{aligned}
H - \mu N = & -\frac{V\mu^2}{2u_0} + \sum_{\boldsymbol{k} \neq 0} \left[E_{\boldsymbol{k}} - \varepsilon_{\boldsymbol{k}} - u_0 n_0\right] \\
& + \sum_{\boldsymbol{k} \neq 0} E_{\boldsymbol{k}} \eta_{\boldsymbol{k}}^\dagger \eta_{\boldsymbol{k}},
\end{aligned}
\tag{3.13}
$$

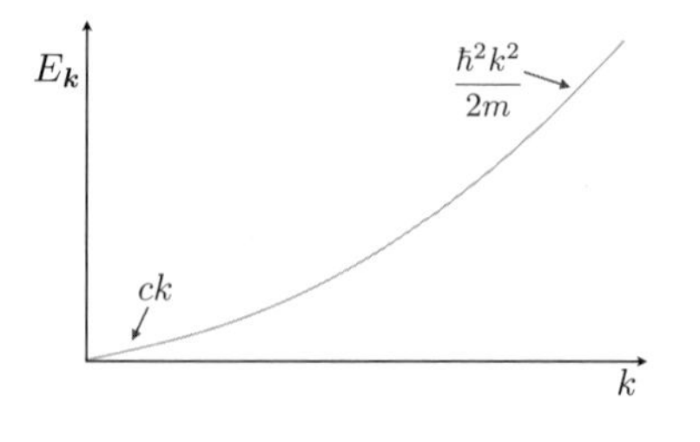

Figure 3.1 Excitation spectrum of Bose gas for $\varepsilon_{\boldsymbol{k}} = \hbar^2 k^2/(2m)$ with $k = |\boldsymbol{k}|$.

where

$$E_{\boldsymbol{k}} = \sqrt{\varepsilon_{\boldsymbol{k}}^2 + 2u_0 n_0 \varepsilon_{\boldsymbol{k}}}\,. \tag{3.14}$$

This is the Hamiltonian for free $\eta_{\boldsymbol{k}}$ bosons with energy $E_{\boldsymbol{k}} > 0$. So the first line of (3.13) is the energy of the ground state, which has zero $\eta_{\boldsymbol{k}\neq 0}$ bosons, that is, the new ground state is defined by

$$\eta_{\boldsymbol{k}\neq 0}|G\rangle = 0. \tag{3.15}$$

See Problem 3.4 for an explicit representation of $|G\rangle$ in terms of the $b_{\boldsymbol{k}}^\dagger$ operators.

The excited states consist of arbitrary numbers of $\eta_{\boldsymbol{k}\neq 0}$ bosons with energy $E_{\boldsymbol{k}}$, which is sketched in Fig. 3.1. At large $|\boldsymbol{k}|$, we have the free dispersion of the underlying bosons with $E_{\boldsymbol{k}} \approx \varepsilon_{\boldsymbol{k}}$. However, at small $|\boldsymbol{k}|$,

$$E_{\boldsymbol{k}} \to c|\boldsymbol{k}|, \quad c = \sqrt{\frac{\hbar^2 u_0 n_0}{m}} \tag{3.16}$$

for $\varepsilon = \hbar^2 \boldsymbol{k}^2/(2m)$. So there is a gapless spectrum of linearly dispersing bosonic particles. It can be shown that this excitation carries longitudinal density fluctuations (we will see this more explicitly in Section 3.3), and so is a "phonon" and c is a *sound* velocity. The path-integral analysis in Section 3.3 also shows that the gaplessness of the spectrum in (3.14) is a direct consequence of a broken symmetry in the ground state of the interacting Bose gas.

The excitation in (3.16) is sometimes called a "second sound" excitation to distinguish it from the usual "first sound" mode found also in a classical gas. The fundamental distinction between them is that a second sound excitation is a coherent excitation in a collisionless regime, while the first sound excitation is a hydrodynamic mode in a collision-dominated regime. Both sound modes exist in the present Bose gas at low T. As in Section 2.6 on the Fermi liquid, we can compute the lifetime of the excitation in Fig. 3.1 using Fermi's golden rule from higher-order terms in H that were omitted in the Bogoliubov Hamiltonian: these will lead to a collision time $\tau_{\boldsymbol{k}}$ which diverges rapidly as $T \to 0$. The second sound mode is present in the collisionless regime $\omega\tau_{\boldsymbol{k}} \gg 1$, while the first sound mode is present in the hydrodynamic collision-dominated regime $\omega\tau_{\boldsymbol{k}} \ll 1$, where we use $|\boldsymbol{k}| = \omega/c$.

3.2 Off-Diagonal Long-Range Order

A peculiar feature of the ground state we have described above is that the $\boldsymbol{k}=0$ wavevector is treated differently from all non-zero $\boldsymbol{k}$ even in the thermodynamic limit. It would be preferable to have a statement of this feature in terms of correlation functions of local operators, as that would help generalize the theory to situations where periodic, or even random, potentials are present and then the plane-wave basis plays no special role. To this end, we introduce the field operator

$$\psi(\boldsymbol{r}) = \frac{1}{\sqrt{V}} \sum_{\boldsymbol{k}} b_{\boldsymbol{k}} e^{i\boldsymbol{k}\cdot\boldsymbol{r}} . \tag{3.17}$$

From (3.1), this obeys the commutation relation

$$[\psi(\boldsymbol{r}), \psi^{\dagger}(\boldsymbol{r})] = \delta(\boldsymbol{r}-\boldsymbol{r}') . \tag{3.18}$$

We now examine the position–space correlator

$$\left\langle \psi^{\dagger}(\boldsymbol{r})\psi(\boldsymbol{r}')\right\rangle = \frac{1}{V}\sum_{\boldsymbol{k}} n_b(\boldsymbol{k}) e^{i\boldsymbol{k}\cdot(\boldsymbol{r}'-\boldsymbol{r})} . \tag{3.19}$$

where $n_b(\boldsymbol{k})$ is the boson momentum distribution function

$$n_b(\boldsymbol{k}) = \left\langle b_{\boldsymbol{k}}^{\dagger} b_{\boldsymbol{k}} \right\rangle , \tag{3.20}$$

the analog of the electron momentum distribution function in (2.31) (assuming translational invariance, $\left\langle b_{\boldsymbol{k}}^{\dagger} b_{\boldsymbol{k}'} \right\rangle$ is non-zero only for $\boldsymbol{k}=\boldsymbol{k}'$). We can evaluate $n_b(\boldsymbol{k})$ by transforming to the $\eta_{\boldsymbol{k}}$ basis using (3.11), and then at $T=0$ we obtain after using (3.15)

$$n_b(\boldsymbol{k}) = \begin{cases} N_0 & \boldsymbol{k}=0 \\ \sinh^2(\theta_{\boldsymbol{k}}) & \boldsymbol{k}\neq 0. \end{cases} \tag{3.21}$$

In the thermodynamic limit ($V\to\infty$, $N_0\to\infty$, $n_0=N_0/V$ fixed), we can write this as

$$n_b(\boldsymbol{k}) = n_0(2\pi)^d\delta(\boldsymbol{k}) + \frac{1}{2}\left(\frac{\varepsilon_{\boldsymbol{k}}+u_0 n_0}{E_{\boldsymbol{k}}} - 1\right) . \tag{3.22}$$

The co-efficient in front of $\delta(\boldsymbol{k})$ can be verified by comparing the integral of (3.22) with the summation of (3.10) over some region of $\boldsymbol{k}$ space including the origin. Finally, we insert (3.22) into (3.19) and obtain

$$\left\langle \psi^{\dagger}(\boldsymbol{r})\psi(\boldsymbol{r}')\right\rangle = n_0 + \int \frac{d^d k}{(2\pi)^d}\frac{1}{2}\left(\frac{\varepsilon_{\boldsymbol{k}}+u_0 n_0}{E_{\boldsymbol{k}}} - 1\right) e^{i\boldsymbol{k}\cdot(\boldsymbol{r}'-\boldsymbol{r})} . \tag{3.23}$$

The important point is that the $\boldsymbol{k}$ integral in (3.23) yields a term which vanishes as $|\boldsymbol{r}-\boldsymbol{r}'|\to\infty$, and so

$$\lim_{|\boldsymbol{r}-\boldsymbol{r}'|\to\infty} \left\langle \psi^{\dagger}(\boldsymbol{r})\psi(\boldsymbol{r}')\right\rangle = n_0 . \tag{3.24}$$

This is the key statement of off-diagonal long-range order (ODLRO): the left-hand side has the interpretation of an off-diagonal element of the one-particle density matrix, and this does not vanish as the $\boldsymbol{r}$ and $\boldsymbol{r}'$ are separated by an infinite distance. It is a fundamental characteristic of the ground state of the dilute Bose gas, and indicates the presence of a broken symmetry. We discuss the broken symmetry features further in Chapter 5, along with the implication for superfluidity and superconductivity. As we raise the temperature, the Bose gas turns into a normal liquid without ODLRO, and the vanishing of ODLRO implies that this must be accompanied by a phase transition.

It is also useful to contrast the one-particle density matrix of the Bose gas in (3.23), with that of the Fermi liquid. In the latter case, the fermion momentum distribution function is given by (2.34) and Fig. 2.5, and it does not have a $\delta(\boldsymbol{k})$ term, unlike (3.22). Instead, it is characterized by a discontinuity at the Fermi surface; after a Fourier transform, we obtain from (2.34), for an isotropic Fermi surface,

$$\begin{aligned}\langle \psi^\dagger(\boldsymbol{r})\psi(\boldsymbol{r}')\rangle &= Z\int_0^{k_{\rm F}} \frac{d^3k}{(2\pi)^d} e^{i\boldsymbol{k}\cdot(\boldsymbol{r}'-\boldsymbol{r})} + \cdots \\ &= \frac{Z}{2\pi^2 r^3}\left(\sin(k_{\rm F}r) - \frac{\cos(k_{\rm F}r)}{k_{\rm F}r}\right) + \cdots . \end{aligned} \tag{3.25}$$

Clearly there is no ODLRO in the Fermi gas. It is replaced by a characteristic term that oscillates with the Fermi wavevector $k_{\rm F}$, and has an envelope which decays as $1/r^3$ in dimension $d = 3$; the ellipses indicate terms that decay faster. There are no such oscillatory terms in the Bose gas.

3.3 Path Integral Theory

The Bogoliubov theory presented above provides a simple intuitive description of the superfluid ground state. However, some of its features appear mysterious at first glance: for example, why is the excitation spectrum gapless, and why does $E_{\boldsymbol{k}}$ disperse linearly at small $\boldsymbol{k}$ in (3.16)? Moreover, the connection of these results to the ODLRO and symmetry breaking is not clear. Many of these issues are clarified in a path-integral theory that I will now describe. This formulation will also be useful for the considerations of symmetry and superfluidity in Chapter 5.

Following the coherent-state path integral described in Appendix A, we can write the partition function of the boson Hamiltonian (3.2) as the path integral over the boson field $\psi(\boldsymbol{r},\tau)$, where τ is imaginary time extending on the thermal circle of circumference $\beta = \hbar/(k_B T)$, and ψ is periodic around the circle. We have, setting $\hbar = k_B = 1$, the path integral for the partition function

$$\mathcal{Z} = {\rm Tr} e^{-\beta H} = \int \mathcal{D}\psi(\boldsymbol{r},\tau) e^{-S}. \tag{3.26}$$

where the action S is

$$S = \int_0^\beta d\tau \int d^d r \left[\psi^* \frac{\partial \psi}{\partial \tau} + \frac{|\nabla \psi|^2}{2m} - \mu |\psi|^2 + \frac{u_0}{2} |\psi|^2 \right]. \tag{3.27}$$

Here, we focus on the case where $\varepsilon_k = k^2/(2m)$.

We will evaluate path integral by finding the saddlepoint of the action, and examining the fluctuations about the saddle point. The saddle point of (3.27) is at

$$\psi = \sqrt{n_0} e^{i\theta_0}, \tag{3.28}$$

where n_0 is as in (3.7), and the angle θ_0 is arbitrary. So by picking a saddle point with an arbitrary value of θ_0, we "break" a global $U(1)$ symmetry of the action under which

$$\psi(\boldsymbol{r}, \tau) \to \psi(\boldsymbol{r}, \tau) e^{i\theta}, \tag{3.29}$$

where θ is independent of space and time.

Let us now examine fluctuations about the saddlepoint (3.28). It is convenient to parameterize these fluctuations by two real fields, $n_1(\boldsymbol{r}, \tau)$ and $\theta(\boldsymbol{r}, \tau)$:

$$\psi(\boldsymbol{r}, \tau) = (n_0 + n_1(\boldsymbol{r}, \tau))^{1/2} e^{i\theta(\boldsymbol{r}, \tau)}. \tag{3.30}$$

It is clear that n_1 represents density fluctuations, and $\theta(\boldsymbol{r}, \tau)$ represents fluctuations in the phase of the condensate shown by (3.28). Inserting (3.30) into (3.27), we obtain

$$S = S_0 + \int_0^\beta d\tau \int d^d r \left[i n_1 \frac{\partial \theta}{\partial \tau} + \frac{u_0}{2} n_1^2 + \frac{n_0 (\nabla \theta)^2}{2m} + \frac{(\nabla n_1)^2}{8m(n_0 + n_1)} + n_1 \frac{(\nabla \theta)^2}{2m} \right]. \tag{3.31}$$

Note that all terms depend only upon the spatial or temporal gradients of θ: this is a consequence of the global $U(1)$ symmetry (3.29), and this feature will help us understand the gapless nature of the excitation spectrum. Let us now transform to momentum space, and retain only terms quadratic in θ and n_1 (this will turn out to be equivalent to the Bogoliubov theory). Then (3.31) yields the following imaginary time action:

$$\mathcal{S}_2 = \int_0^\beta d\tau \sum_k \left[i n_{1k} \frac{\partial \theta_{-k}}{\partial \tau} + \frac{1}{2} \left(u_0 + \frac{k^2}{4mn_0} \right) |n_{1k}|^2 + \frac{n_0 k^2}{2m} |\theta_k|^2 \right]. \tag{3.32}$$

This is precisely the action of a set of harmonic oscillators, one for each $\boldsymbol{k}$, with "coordinate" θ_k, and canonically conjugate "momentum" n_{1k}. So the density fluctuation of a Bose gas is canonically conjugate to the phase of the condensate. The energy spectrum of this set of oscillators is most conveniently obtained by performing the Gaussian functional integral over n_{1k} with action $\mathcal{S}_2$; this yields the action for phase fluctuations

$$\mathcal{S}_\theta = \int_0^\beta d\tau \sum_k \left[\frac{1}{2(u_0 + k^2/(4mn_0))} \left| \frac{\partial \theta_k}{\partial \tau} \right|^2 + \frac{n_0 k^2}{2m} |\theta_k|^2 \right]. \tag{3.33}$$

This is clearly the action of a set of harmonic oscillators with frequency

$$\omega_k = \left[\frac{n_0 k^2}{m} \left(u_0 + \frac{k^2}{4mn_0} \right) \right]^{1/2}. \tag{3.34}$$

Comparing (3.34) with (3.14), we confirm that $\omega_k = E_k$, the energy of the η_k bosons in the Bogoliubov theory: so these bosons are simply the annihilation operators of oscillators representing fluctuations of the phase of the condensate.

Problem

3.1 Consider a weakly interacting Bose gas at a fixed chemical potential μ, and temperature T, described by the Hamiltonian in (3.2). We will compute the total density $\rho = (1/V)\sum_k \langle b_k^\dagger b_k \rangle$ as an expansion in powers of u_0, including terms up to order u_0^0. Following the analysis in Section 3.1, we write

$$b_k = \sqrt{\frac{\mu}{u}}\delta_{k=0} + \tilde{b}_k \tag{3.35}$$

and diagonalize the Hamiltonian by expressing $\tilde{b}_k$ in terms of the Bogoliubov operators η_k. Evaluate the average density at a temperature T, and obtain an expression for the density $\rho(\mu, T)$ in the form

$$\rho(\mu, T) = \frac{\mu}{u_0} + \int \frac{d^3k}{8\pi^3} G(k, T, \mu) + \mathcal{O}(u^0) \tag{3.36}$$

for some function G. Evaluate the momentum integral in the limiting regimes $T \ll \mu$ and $T \gg \mu$.

3.2 Re-examine Problem 3.1 by the path-integral method. Instead of (3.30), we write

$$\psi = \sqrt{\frac{\mu}{u_0}} + \psi_1 \tag{3.37}$$

and expand the Lagrangian to order ψ_1^2. Then, after evaluation of the Gaussian integral over ψ_1, show that

$$\rho(\mu, T) = \frac{\mu}{u_0} + \int \frac{d^3k}{8\pi^3} T \sum_{\omega_n} \frac{\left(i\omega_n + \dfrac{k^2}{2m} + \mu\right)}{\left(\omega_n^2 + \left(\dfrac{k^2}{2m} + \mu\right)^2 - \mu^2\right)} e^{i\omega_n 0^+} + \mathcal{O}(u_0), \tag{3.38}$$

where ω_n is an integer multiple of $2\pi T$. Evaluate the frequency summation, and compare with (3.36).

3.3 Consider a Bose gas described by the Hamiltonian (in first-quantized form)

$$H = -\frac{\hbar^2}{2m}\sum_{j=1}^{N} \nabla_j^2 + \frac{1}{2}\sum_{i \neq j = 1}^{N} u(\mathbf{r}_i - \mathbf{r}_j), \tag{3.39}$$

where u is the interaction between pairs of particles.

(a) Prove that the total momentum

$$\boldsymbol{P} = \frac{\hbar}{i} \sum_{j=1}^{N} \boldsymbol{\nabla}_j \tag{3.40}$$

commutes with H. So every eigenstate of H can also be chosen to be an eigenstate of $\boldsymbol{P}$.

(b) Let ψ be any eigenstate of H and $\boldsymbol{P}$ with eigenvalues E and $\boldsymbol{p}$. Prove that a state boosted by a velocity $\boldsymbol{v}$

$$\exp\left(im\boldsymbol{v} \cdot \sum_{j=1}^{N} \boldsymbol{r}_j/\hbar \right) \psi \tag{3.41}$$

is also an eigenstate of H and $\boldsymbol{P}$, with eigenvalues $E + \boldsymbol{p}\cdot\boldsymbol{v} + Nm\boldsymbol{v}^2/2$ and $\boldsymbol{p} + Nm\boldsymbol{v}$, respectively.

(c) Let the ground state of H be ψ_0 with energy E_0 and momentum $\boldsymbol{P} = 0$. Then a state with a single Bogoliubov excitation has momentum $\boldsymbol{P} = \boldsymbol{p}$ and energy $E_0 + \varepsilon(\boldsymbol{p})$ (where $\varepsilon(\boldsymbol{p})$ is the energy of the Bogoliubov excitation). We now form a state with superflow at velocity $\boldsymbol{v}$ by boosting ψ as in Eq. (3.41). Show that, if

$$\varepsilon(\boldsymbol{p}) + \boldsymbol{p}\cdot\boldsymbol{v} > 0, \tag{3.42}$$

this superflowing state has an energy lower than that of a state with a single Bogoliubov excitation of momentum $\boldsymbol{p}$, also boosted by a velocity $\boldsymbol{v}$. The Landau criterion for the stability of superflow with velocity $\boldsymbol{v}$ is that (3.42) be satisfied for all $\boldsymbol{p}$.

3.4 We described the ground state $|G\rangle$ of a weakly interacting Bose gas by condensing N_0 particles in the zero momentum state, and then requiring

$$\eta_{\boldsymbol{k}}|G\rangle = 0 \quad , \quad \boldsymbol{k} \neq 0, \tag{3.43}$$

where $b_{\boldsymbol{k}}$ and $\eta_{\boldsymbol{k}}$ are related by

$$b_{\boldsymbol{k}} = \eta_{\boldsymbol{k}} \cosh(\theta_{\boldsymbol{k}}) - \eta^{\dagger}_{-\boldsymbol{k}} \sinh(\theta_{\boldsymbol{k}}). \tag{3.44}$$

Show that we can write $|G\rangle$ as

$$|G\rangle = \left(b_0^{\dagger}\right)^{N_0} \exp\left(-\sum_{\boldsymbol{k}\neq 0} f_{\boldsymbol{k}} b^{\dagger}_{\boldsymbol{k}} b^{\dagger}_{-\boldsymbol{k}} \right) |0\rangle, \tag{3.45}$$

where $|0\rangle$ is the empty state with no bosons. Determine $f_{\boldsymbol{k}}$.

3.5 A more general form of the ground state $|G\rangle$ is the state

$$|\phi\rangle = \left(e^{i\phi} b_0^{\dagger}\right)^{N_0} \exp\left(-e^{2i\phi} \sum_{\boldsymbol{k}\neq 0} f_{\boldsymbol{k}} b^{\dagger}_{\boldsymbol{k}} b^{\dagger}_{-\boldsymbol{k}} \right) |0\rangle, \tag{3.46}$$

where the phase ϕ is arbitrary. This arbitrariness corresponds to the phase of the condensate, which we had arbitrarily set to zero earlier. From the states $|\phi\rangle$, we can obtain the state $|N\rangle$ with a definite number of N particles

$$|N\rangle = \int_0^{2\pi} \frac{d\phi}{2\pi} e^{-iN\phi} |\phi\rangle. \tag{3.47}$$

Evaluate the mean and variance of the number of particles in the state $|\phi\rangle$ by computing

$$N = \langle\phi|\hat{N}|\phi\rangle, \quad (\Delta N)^2 = \langle\phi|(\hat{N} - N)^2\phi\rangle, \tag{3.48}$$

where

$$\hat{N} = \sum_{\boldsymbol{k}} b_{\boldsymbol{k}}^\dagger b_{\boldsymbol{k}}. \tag{3.49}$$

It is simplest to do this by transforming to the η_k basis, and using $\eta_k|\phi\rangle = 0$.

3.6 We describe the Anderson–Higgs mechanism in charged superfluid. We consider a superfluid of charged bosons ψ with long-range Coulomb interactions:

$$\begin{aligned} H = & \int d^3x \frac{\nabla\psi^\dagger(\boldsymbol{x})\nabla\psi(\boldsymbol{x})}{2m} \\ & + \frac{1}{2}\int d^3x d^3y \left[\psi^\dagger(\boldsymbol{x})\psi(\boldsymbol{x}) - n_0\right] U(\boldsymbol{x}-\boldsymbol{y}) \left[\psi^\dagger(\boldsymbol{y})\psi(\boldsymbol{y}) - n_0\right]. \end{aligned} \tag{3.50}$$

Here $U(\boldsymbol{x}) = e^2/|\boldsymbol{x}|$ is the Coulomb potential, and n_0 is a background charge that maintains global charge neutrality. You can assume that the Hamiltonian is implicitly normal ordered.

(a) Find the ground state $\psi(\boldsymbol{x}) = \psi_0$ of H.
(b) Derive equations of motion for field $\psi(\boldsymbol{x})$, that is, compute $i\partial_t\psi(x)$.
(c) Let us consider fluctuations on top of the ground state. Linearize the equation of motion $\psi(\boldsymbol{x}) = \psi_0 + \delta\psi(\boldsymbol{x})$ you derived around the ground state. Find the dispersion relation $E_{\boldsymbol{k}}$ of elementary excitations.
(d) What is different from the case with short-range interactions considered in this chapter? What is the physical picture of the low-energy excitations?

4 Bardeen–Cooper–Schrieffer Theory of Superconductivity

The Bardeen–Cooper–Schrieffer theory of superconductivity in a Fermi gas with attractive interactions is described. The spectrum of fermionic excitations is obtained, and their energy gap is related to the critical temperature for superconductivity.

The Bardeen–Cooper–Schrieffer (BCS) theory was first proposed by these authors in 1957, and it resolved the long-standing problem of superconductivity in simple metals. The parent state of the superconductor is a simple metal, well described by the jellium model: it has quasiparticles with dispersion ε_k, which are long-lived near a Fermi surface. The Coulomb interaction between two quasiparticles is screened but remains repulsive. The BCS theory exploits the remarkable fact that the interaction between quasiparticles near the Fermi surface can sometimes become attractive. Studies of the coupling of the electrons to lattice vibrations (phonons) had shown that exchange of phonons can induce a weak attractive interaction between quasiparticles within a window of energies between $\pm\omega_D$ of the Fermi level. Here $\omega_{\rm D}$ is the Debye frequency of the phonons (roughly, the maximum phonon frequency). We have $\omega_{\rm D} \ll E_F$ because the atomic mass is much larger than the electron mass. The BCS theory is quantitatively successful because $\omega_{\rm D} \ll E_{\rm F}$, and so the electrons prefer to form pairs only close to the Fermi surface.

However, we know today that, even without phonons, electrons on lattices can experience a net attractive interaction in certain channels from the Coulomb interactions alone. This interaction is not restricted to the vicinity of the Fermi surface, and so a quantitive theory of superconductivity is much more difficult. Nevertheless, it is generally believed that the BCS pairing theory is at least qualitatively correct in describing the superconducting state of such higher-temperature superconductors, as I will discuss further in Chapter 9.

4.1 The BCS Wavefunction

We start by introducing the BCS wavefunction as a variational state for electrons with an attractive interaction. As the electrons prefer to from pairs, let us assume that a pair of electrons with coordinates $\boldsymbol{r}_1$, $\boldsymbol{r}_2$ and spins σ_1, σ_2 have the wavefunction

$$g(\boldsymbol{r}_1 - \boldsymbol{r}_2)\chi_{12}, \tag{4.1}$$

where $g(\boldsymbol{r})$ is the spatial wavefunction, and the spin wavefunction of a spin-singlet pair is

$$\chi_{ij} = \frac{1}{\sqrt{2}}\left[\delta_{\sigma_i\uparrow}\delta_{\sigma_j\downarrow} - \delta_{\sigma_i\downarrow}\delta_{\sigma_j\uparrow}\right]. \tag{4.2}$$

This separation between the spin and spatial parts of the wavefunction is expected in the absence of spin–orbit interactions. Notice, we assumed that there is no motion of the center of mass of the two electrons, which means the wavefunction is independent of $(\boldsymbol{r}_1+\boldsymbol{r}_2)/2$. For N electrons (N even) we can form $N/2$ pairs, and so the wavefunction, which simply places each pair of electrons into the zero center-of-mass momentum state, is

$$\begin{aligned}\Psi(\boldsymbol{r}_1\sigma_1, \boldsymbol{r}_2\sigma_2, \ldots, \boldsymbol{r}_N\sigma_N) &= \\ C\mathcal{A}\left[g(\boldsymbol{r}_1-\boldsymbol{r}_2)\chi_{12}g(\boldsymbol{r}_3-\boldsymbol{r}_4)\chi_{34}\cdots g(\boldsymbol{r}_{N-1}-\boldsymbol{r}_N)\chi_{N-1,N}\right]&,\end{aligned} \tag{4.3}$$

where $\mathcal{A}$ antisymmetrizes the wavefunction by summing over all permutations, and C is a normalization constant.

This wavefunction is not easy to work with. Fortunately, there is a simple representation using second quantized field operators $\psi_\sigma(\boldsymbol{r})$, which annihilate an electron with spin σ and coordinate $\boldsymbol{r}$:

$$|\Psi\rangle \propto \left[\int d^3r_1 d^3r_2\, \psi_\uparrow^\dagger(\boldsymbol{r}_1)\psi_\downarrow^\dagger(\boldsymbol{r}_2)g(\boldsymbol{r}_1-\boldsymbol{r}_2)\right]^{N/2}|0\rangle. \tag{4.4}$$

It is easy to verify that the electron pairs are in a spin-singlet state when $g(-\boldsymbol{r}) = g(\boldsymbol{r})$. Next, we transform this wavefunction to momentum space

$$|\Psi\rangle \propto \left[\sum_{\boldsymbol{k}} g(\boldsymbol{k})c_{\boldsymbol{k}\uparrow}^\dagger c_{-\boldsymbol{k}\downarrow}^\dagger\right]^{N/2}|0\rangle, \tag{4.5}$$

where the $c_{\boldsymbol{k}\alpha}^\dagger$ are electron creation operators, as in Chapter 2. This is almost the final form of the BCS wavefunction. By comparing to (3.3), it is evident that (4.5) is simply a Bose–Einstein condensate of electron pairs (Cooper pairs) in a zero momentum state. However, this form is still cumbersome to work with, and does not connect in any evident manner to the ordinary metal when there is no pairing.

A more convenient form is obtained by working in the full Fock space with arbitrary numbers of electrons (as was also the case for the Bose gas in Chapter 3). Provided N is large enough, this should make no difference to the thermodynamic properties, for the same reasons that the grand canonical ensemble gives the same results as the canonical ensemble. So we write

$$|\Psi\rangle \propto \mathcal{P}_N \exp\left(\sum_{\boldsymbol{k}} g(\boldsymbol{k})c_{\boldsymbol{k}\uparrow}^\dagger c_{-\boldsymbol{k}\downarrow}^\dagger\right)|0\rangle, \tag{4.6}$$

where $\mathcal{P}_N$ is the projector onto the subspace of N particles. Now we can exploit the fact that operators with different values of $\boldsymbol{k}$ in the sum all commute with each other to write

$$|\Psi\rangle \propto \mathcal{P}_N \prod_{\boldsymbol{k}} \exp\left(g(\boldsymbol{k}) c^{\dagger}_{\boldsymbol{k}\uparrow} c^{\dagger}_{-\boldsymbol{k}\downarrow}\right) |0\rangle . \tag{4.7}$$

Finally, we use the fact that the square of the argument of each exponential vanishes to obtain the BCS wavefunction

$$|\Psi\rangle = \mathcal{P}_N \prod_{\boldsymbol{k}} \left(u_k + v_k c^{\dagger}_{\boldsymbol{k}\uparrow} c^{\dagger}_{-\boldsymbol{k}\downarrow}\right) |0\rangle , \tag{4.8}$$

and we express $g(\boldsymbol{k})$ in terms of two new functions u_k and v_k:

$$g(\boldsymbol{k}) = \frac{v_k}{u_k}. \tag{4.9}$$

In the $N \to \infty$ limit, the projector $\mathcal{P}_N$ can be ignored in most computations because all expectation values of local operators are dominated by terms in a window of order $\sqrt{N}$ about the mean number of particles N. So the BCS state is

$$|\text{BCS}\rangle = \prod_{\boldsymbol{k}} \left(u_k + v_k c^{\dagger}_{\boldsymbol{k}\uparrow} c^{\dagger}_{-\boldsymbol{k}\downarrow}\right) |0\rangle . \tag{4.10}$$

It is easy to show that the wavefunction is normalized when

$$|u_k|^2 + |v_k|^2 = 1 , \quad \forall \boldsymbol{k} . \tag{4.11}$$

A nice property of the wavefunction in (4.10) is that it can also describe the free-electron Fermi surface state with the choice

$$\begin{aligned} u_k &= 0 \text{ and } v_k = 1 \text{ for } \varepsilon_k < 0, \\ u_k &= 1 \text{ and } v_k = 0 \text{ for } \varepsilon_k > 0, \end{aligned} \tag{4.12}$$

where we have absorbed the chemical potential in the definition of ε_k. We will see shortly that the superconducting state is one which in the discontinuity of u_k, v_k at $\varepsilon_k = 0$ is smoothed out on the scale of ω_D.

4.2 Off-Diagonal Long-Range Order

We now show that the BCS state has off-diagonal long-range order (ODLRO) very similar to that of the Bose gas in Section 3.2. The Bose field operator is now replaced by an electron-pair (or Cooper-pair) operator.

What makes the $|\text{BCS}\rangle$ state different from other variational states encountered in quantum mechanics is that it does not conserve the total electron number. Consequently, when we evaluate expectation values of operators that annihilate a pair of electrons in a spin-singlet state, we obtain a non-zero value:

$$\langle \text{BCS}|\, c_{\boldsymbol{k}\uparrow} c_{-\boldsymbol{k}\downarrow} \,|\text{BCS}\rangle \neq 0 . \tag{4.13}$$

Indeed, we can define an infinite family of states, labeled by an angle θ, which differ from the BCS state at $\theta = 0$ only by an overall phase factor after projection to a fixed number of particles:

$$|\mathrm{BCS}_\theta\rangle = \prod_k \left(u_k + v_k e^{i\theta} c^\dagger_{k\uparrow} c^\dagger_{-k\downarrow}\right)|0\rangle\,,$$
$$\mathcal{P}_N|\mathrm{BCS}_\theta\rangle = e^{iN\theta/2}\mathcal{P}_N|\mathrm{BCS}\rangle\,. \tag{4.14}$$

Now the expectation value of the spin-singlet electron-pair operator acquires this phase factor:

$$\langle \mathrm{BCS}_\theta | c_{k\uparrow} c_{-k\downarrow} | \mathrm{BCS}_\theta\rangle \propto e^{i\theta}\,, \tag{4.15}$$

and the energy of the $|\mathrm{BCS}_\theta\rangle$ is independent of the angle θ.

There is an analogy here with the broken spin-rotation symmetry in a ferromagnet, which is worth exploiting. Consider a ferromagnet in which the Hamiltonian is invariant under spin rotations in the XY plane. Then the ferromagnetic state "breaks" this rotation symmetry by picking an arbitrary direction labeled by the angle θ along which the ferromagnetic moment is oriented. Similarly, in the BCS theory, the Hamiltonian is invariant under the "rotation"

$$c_{k\sigma} \to e^{i\phi} c_{k\sigma} \tag{4.16}$$

because it conserves the total number of particles. Under this transformation $\theta \to \theta + 2\phi$, and so by choosing a definite θ, the $|\mathrm{BCS}_\theta\rangle$ breaks this "rotation" symmetry.

The reader might be concerned that the state with a fixed particle number does not break this symmetry because

$$\langle \mathrm{BCS}_\theta | \mathcal{P}_N c_{k\uparrow} c_{-k\downarrow} \mathcal{P}_N | \mathrm{BCS}_\theta\rangle = 0\,. \tag{4.17}$$

However, this also has an analogy in the ferromagnet: we can also choose a state for the ferromagnet in which we average over the different orientations of the ferromagnetic moment, so that the net moment vanishes. Even more explicitly, note that the number projection operator can be expressed as an average over the values of θ

$$\mathcal{P}_N|\mathrm{BCS}\rangle = \int_0^{2\pi} \frac{d\theta}{2\pi} e^{-iN\theta/2}|\mathrm{BCS}_\theta\rangle\,. \tag{4.18}$$

As in the ferromagnet, and in the Bose gas in Section 3.2, the way to avoid sensitivity to these global averages is to characterize the physics in terms of correlation functions of spatially local operators. So let us define the Cooper-pair operator

$$\Psi_\mathrm{C}(\boldsymbol{R}) = \int d^3r\, g(\boldsymbol{r})\, \psi_\uparrow(\boldsymbol{R}+\boldsymbol{r}/2)\psi_\downarrow(\boldsymbol{R}-\boldsymbol{r}/2), \tag{4.19}$$

which annihilates the electron pair in (4.1) centered at the spatial position $\boldsymbol{R}$. Then, in the BCS state, we have a non-zero "condensate" Ψ_0 associated with the non-zero expectation value

$$\langle \mathrm{BCS}_\theta | \Psi_\mathrm{C}(\boldsymbol{R}) | \mathrm{BCS}_\theta\rangle = \Psi_0 \neq 0\,. \tag{4.20}$$

On the other hand, from (4.17), we have

$$\langle \mathrm{BCS}_\theta | \mathcal{P}_N \Psi_\mathrm{C}(\boldsymbol{R}) \mathcal{P}_N | \mathrm{BCS}_\theta \rangle = 0 . \tag{4.21}$$

This strong consequence of the projection operator disappears when we consider two-point correlation functions of $\Psi_\mathrm{C}(\boldsymbol{R})$. A lengthy but straightforward computation shows that

$$\begin{aligned} \lim_{|\boldsymbol{R}|\to\infty} \langle \mathrm{BCS}_\theta | \Psi_\mathrm{C}^\dagger(\boldsymbol{R}) \Psi_\mathrm{C}(\boldsymbol{0}) | \mathrm{BCS}_\theta \rangle &= |\Psi_0|^2 \\ &= \lim_{|\boldsymbol{R}|\to\infty} \langle \mathrm{BCS}_\theta | \mathcal{P}_N \Psi_\mathrm{C}^\dagger(\boldsymbol{R}) \Psi_\mathrm{C}(\boldsymbol{0}) \mathcal{P}_N | \mathrm{BCS}_\theta \rangle . \end{aligned} \tag{4.22}$$

Noting the analogy to (3.24), we may regard this as the formal definition of the ODLRO in the BCS state. Also note that it is only the state *without* number projection that has a nice "clustering" property:

$$\lim_{|\boldsymbol{R}|\to\infty} \langle \Psi_\mathrm{C}^\dagger(\boldsymbol{R}) \Psi_\mathrm{C}(\boldsymbol{0}) \rangle = \langle \Psi_\mathrm{C}^\dagger(\boldsymbol{R}) \rangle \langle \Psi_\mathrm{C}(\boldsymbol{0}) \rangle . \tag{4.23}$$

This is the main reason it is better to work with $|\mathrm{BCS}_\theta\rangle$ rather than $\mathcal{P}_N |\mathrm{BCS}_\theta\rangle$: we can assume that well-separated operators are independent.

4.3 Bogoliubov Theory

At this point, we could proceed as Bardeen, Cooper, and Schrieffer did, and perform a variational computation on a suitable Hamiltonian to find the optimum values of u_k and v_k. However, we follow an alternative, and simpler, route which also gives useful information on the nature of the excitations and the properties at non-zero temperature. In Hartree–Fock theory the variational procedure is equivalent to decoupling the Hamiltonian into all possible combinations of bilinears of the form $\langle c_{k_1\sigma_1}^\dagger c_{k_2\sigma_2} \rangle$. Analogously, given the discussion above, we will now also allow bilinears of the form $\langle c_{k_1\sigma_1} c_{k_2\sigma_2} \rangle$ (and their Hermitian conjugates) and obtain the Bogoliubov Hamiltonian. The justification is ultimately Wick's theorem, which can be shown to hold also for such bilinears. So, in a general notation, we are going to approximate:

$$\begin{aligned} A^\dagger B^\dagger C D \approx & \langle A^\dagger D \rangle B^\dagger C + \langle B^\dagger C \rangle A^\dagger D - \langle A^\dagger D \rangle \langle B^\dagger C \rangle \\ & - \langle A^\dagger C \rangle B^\dagger D - \langle B^\dagger D \rangle A^\dagger C + \langle A^\dagger C \rangle \langle B^\dagger D \rangle \\ & + \langle A^\dagger B^\dagger \rangle C D + \langle C D \rangle A^\dagger B^\dagger - \langle A^\dagger B^\dagger \rangle \langle C D \rangle . \end{aligned} \tag{4.24}$$

The new terms are in the last line, and keeping these terms is equivalent to doing a variational computation with the BCS wavefunction.

We will perform computations with the simple Hamiltonian introduced in BCS theory:

$$H = \sum_{k\sigma} \varepsilon_k c_{k\sigma}^\dagger c_{k\sigma} - \frac{U_0}{2V} \sum_{k,k',q}{}' \sum_{\sigma,\sigma'} c_{k+q,\sigma}^\dagger c_{k'-q,\sigma'}^\dagger c_{k'\sigma'} c_{k\sigma} . \tag{4.25}$$

Here the prime on the momentum summation indicates that all operators are restricted to within ω_D of the Fermi level: $|\varepsilon_k| < \omega_D$, $|\varepsilon_{k'}| < \omega_D$, $|\varepsilon_{k'-q}| < \omega_D$, $|\varepsilon_{k+q}| < \omega_D$. The novel ingredient is the attractive interaction of strength U_0 experienced by electrons within ω_D of the Fermi surface. This is presumed to be a reasonable approximation of the effect of phonons, and will be the driving force for superconductivity.

Applying (4.24) we now drop the Hartree–Fock factorizations in the first two lines, as we assume these have already been absorbed into the definitions of ε_k. Keeping only the third line, we obtain the Bogoliubov Hamiltonian

$$H_B = \sum_{k\sigma} \varepsilon_k c^\dagger_{k\sigma} c_{k\sigma} - \sum_k \left[\Delta c^\dagger_{k\uparrow} c^\dagger_{-k\downarrow} + \Delta^* c_{-k\downarrow} c_{k\uparrow} \right] + \text{constant}, \tag{4.26}$$

where the important parameter Δ will turn out to be the BCS energy gap, one of the key predictions of the BCS theory. The value of Δ is to be determined by solving a self-consistency equation in the ground state (or density matrix, at non-zero temperature) of H_B

$$\Delta = \frac{U_0}{V} \sum_k{}' \left\langle c_{-k\downarrow} c_{k\uparrow} \right\rangle_{H_B}, \tag{4.27}$$

where V is the volume of the system.

To proceed, we need to determine the eigenstates and eigenenergies of H_B. This can be achieved by a Bogoliubov rotation:

$$\begin{pmatrix} c_{k\uparrow} \\ c^\dagger_{-k\downarrow} \end{pmatrix} = \begin{pmatrix} u^*_k & v_k \\ -v^*_k & u_k \end{pmatrix} \begin{pmatrix} \gamma_{k\uparrow} \\ \gamma^\dagger_{-k\downarrow} \end{pmatrix} \tag{4.28}$$

and the inverse rotation

$$\begin{pmatrix} \gamma_{k\uparrow} \\ \gamma^\dagger_{-k\downarrow} \end{pmatrix} = \begin{pmatrix} u_k & -v_k \\ v^*_k & u^*_k \end{pmatrix} \begin{pmatrix} c_{k\uparrow} \\ c^\dagger_{-k\downarrow} \end{pmatrix}. \tag{4.29}$$

Here u_k and v_k are an arbitrary set of complex numbers obeying the normalization (4.11), and not yet related to the parameters in the BCS wavefunction (although they will be soon). The transformations (4.28) and (4.29) ensure that the $\gamma_{k\sigma}$ are canonical fermion operators obeying the same anti-commutation relations as the $c_{k\sigma}$:

$$\begin{aligned} \left[\gamma_{k\sigma}, \gamma_{k'\sigma'}\right]_+ &= 0 \\ \left[\gamma_{k\sigma}, \gamma^\dagger_{k'\sigma'}\right]_+ &= \delta_{\sigma\sigma'} \delta_{k,k'}. \end{aligned} \tag{4.30}$$

Also note that the rotations in (4.28) and (4.29) correspond to unitary matrices, while that in (3.22) does not.

We now insert (4.28) into H_B and demand that the coefficents of terms like $\gamma\gamma$ and $\gamma^\dagger\gamma^\dagger$ vanish. A simple computation shows that this can be satisfied provided we choose

$$\begin{aligned} u_k &= \sin(\varphi_k) \\ v_k &= e^{i\theta} \cos(\varphi_k) \\ \Delta &= |\Delta| e^{i\theta} \end{aligned}$$

$$\sin(2\varphi_k) = \frac{|\Delta|}{\sqrt{\varepsilon_k^2 + |\Delta|^2}}. \tag{4.31}$$

Thus, we have fixed all the u_k and v_k in terms of two real parameters $|\Delta|$ and θ. The value of $|\Delta|$ will be fixed shortly by applying the self-consistency condition (4.27), while the angle θ will remain undetermined. So we are free to choose to θ, corresponding to the broken 'rotation' symmetry we discussed in Section 4.1.

After these transformations, the Bogoliubov Hamiltonian just reduces to a free-fermion form

$$H_{\text{B}} = \sum_{k\sigma} E_k \gamma_{k\sigma}^\dagger \gamma_{k\sigma} + \text{constant}, \tag{4.32}$$

where the fermion dispersion is

$$E_k = \sqrt{\varepsilon_k^2 + |\Delta|^2}. \tag{4.33}$$

Notice the close analogy of these manipulations to those for the Bose gas in Section 3.1. The dispersion obtained there in (3.14) was for bosonic quasiparticles representing phase and density fluctuations; in contrast, (4.33) applies to fermionic quasiparticles. Moreover, the spectrum in (4.33) shows that it requires a minimum energy $|\Delta|$ to create any Bogoliubov quasiparticle excitation, and these minimum energy excitations are on the Fermi surface of the parent metal where $\varepsilon_k = 0$; this is in contrast to the gapless spectrum in (3.14). Here we have fermionic excitations above the ground state in which there are no Bogoliubov quasiparticles:

$$\gamma_{k\sigma} |\text{BCS}\rangle_\theta = 0, \quad \forall \boldsymbol{k}. \tag{4.34}$$

By our notation, we have already indicated that the required ground state is indeed the BCS state in (4.14), as is not difficult to verify from (4.28) and (4.29). So the u_k and v_k above are indeed the required variational parameters in the BCS state.

Note that the BCS state also has a sector with bosonic excitations above the ground state, similar to that of the Bose gas in Chapter 3. These will be discussed in Chapter 6.

4.4 The Energy Gap

Our remaining task is to determine the energy gap $|\Delta|$. From (4.27) we obtain

$$\Delta = \frac{U_0}{V} \sum_k \left[u_k^* v_k \left\langle \gamma_{-k\downarrow} \gamma_{-k\downarrow}^\dagger \right\rangle_{H_{\text{B}}} - v_k u_k^* \left\langle \gamma_{k\downarrow}^\dagger \gamma_{k\downarrow} \right\rangle_{H_{\text{B}}} \right]. \tag{4.35}$$

As H_B is a free-fermion Hamiltonian, these expectation values are easy to evaluate. Also, using the relations in (4.31), we obtain

$$|\Delta| = |\Delta| \frac{U_0}{V} \sum_k{}' \frac{\tanh\left(E_k/(2T)\right)}{2E_k}. \tag{4.36}$$

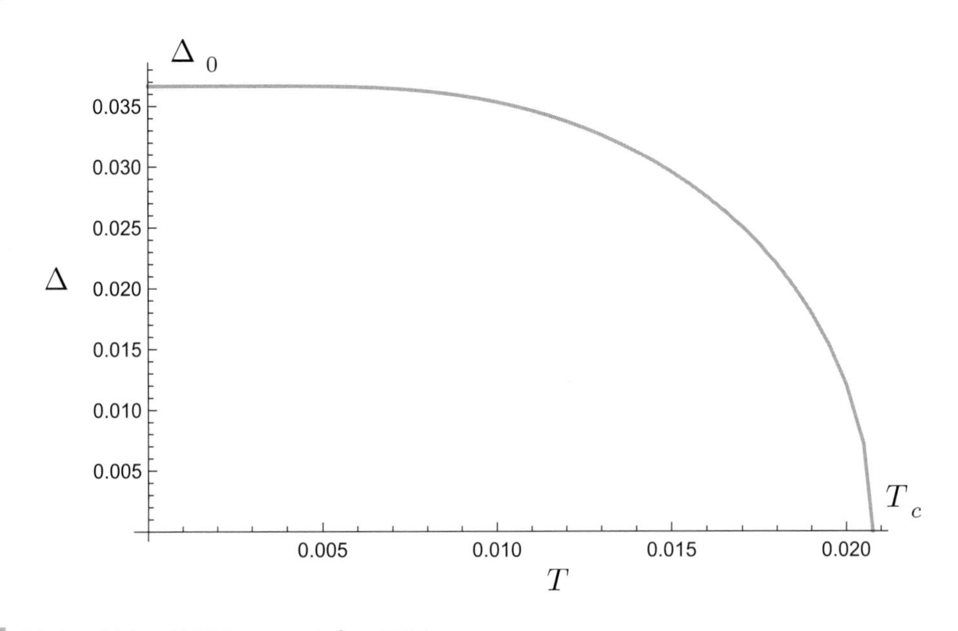

Figure 4.1 Solution of Δ from (4.38) for $\omega_D = 1, \lambda = 0.25$.

Now we insert a factor of $1 = \int d\varepsilon \delta(\varepsilon - \varepsilon_{\boldsymbol{k}})$ on the right-hand side and interchange the order of $\boldsymbol{k}$ summations and ε integration. Then we obtain without further approximation

$$|\Delta| = |\Delta| U_0 \int_{-\omega_D}^{\omega_D} d\varepsilon \, d(\varepsilon) \frac{\tanh\left(\sqrt{\varepsilon^2 + |\Delta|^2}/(2T)\right)}{2\sqrt{\varepsilon^2 + |\Delta|^2}}, \tag{4.37}$$

where $d(\varepsilon) = (1/V) \sum_{\boldsymbol{k}} \delta(\varepsilon - \varepsilon_{\boldsymbol{k}})$ is the density of states. As the integral is focused on the vicinity of the Fermi level, we can approximate $d(\varepsilon) \approx d(0)$, and obtain finally

$$|\Delta| = |\Delta| \lambda \int_{-\omega_D}^{\omega_D} d\varepsilon \frac{\tanh\left(\sqrt{\varepsilon^2 + |\Delta|^2}/(2T)\right)}{2\sqrt{\varepsilon^2 + |\Delta|^2}}, \tag{4.38}$$

with $\lambda > 0$ a dimensionless coupling constant given by

$$\lambda = U_0 d(0), \tag{4.39}$$

where $d(0)$ is the single spin density of states at the Fermi level of the parent metal.

We now need to solve (4.38) for $|\Delta|$ as a function of temperature, λ, and ω_D. Of course, one solution is simply $\Delta = 0$. This is the non-superconducting "normal state." This is the only solution above a critical temperature T_c. However, for $T < T_c$, there is a solution with $\Delta \neq 0$, and this is lower in energy from the normal state, whenever it exists. A plot of the numerical solution for Δ from (4.38) is shown in Fig. 4.1.

We can make analytic progress in two important limits. First consider $T = 0$. Then $\Delta = \Delta_0$ (4.38) reduces to

$$\frac{1}{\lambda} = \int_0^{\omega_D} \frac{d\varepsilon}{\sqrt{\varepsilon^2 + \Delta_0^2}} = \sinh^{-1}\left(\frac{\omega_D}{\Delta_0}\right). \tag{4.40}$$

So as $\lambda \to 0$ we have

$$\Delta_0 = 2\omega_D \exp\left(-\frac{1}{\lambda}\right). \tag{4.41}$$

The energy gap is exponentially small.

On the other hand, at $T = T_c$, $\Delta = 0^+$, and then (4.38) yields

$$\begin{aligned} \frac{1}{\lambda} &= \int_0^{\omega_D} d\varepsilon \frac{\tanh(\varepsilon/(2T_c))}{\varepsilon} \\ &= \ln\left(\frac{\omega_D}{2T_c}\right) - \ln\left(\frac{\pi}{4}\right) + \gamma + \mathcal{O}\left(\frac{T_c}{\omega_D}\right), \end{aligned} \tag{4.42}$$

where γ is now Euler's constant. So at small λ

$$T_c = \frac{2}{\pi} e^{\gamma} \omega_D \exp\left(-\frac{1}{\lambda}\right). \tag{4.43}$$

It is now useful to compute the ratio of Δ_0 and T_c. We can do this by comparing (4.41) and (4.43), but it is instructive to compute the ratio directly in a manner that illuminates the universality of the result. Let us subtract (4.42) from (4.40) and obtain

$$\int_0^{\omega_D} d\varepsilon \left[\frac{1}{\sqrt{\varepsilon^2 + \Delta_0^2}} - \frac{\tanh(\varepsilon/(2T_c))}{\varepsilon}\right] = 0. \tag{4.44}$$

We can now change variables to the dimensionless quantity $y = \varepsilon/\Delta_0$ and write this as

$$\int_0^{\omega_D/\Delta_0} dy \left[\frac{1}{\sqrt{y^2+1}} - \frac{\tanh(y\Delta_0/(2T_c))}{y}\right] = 0. \tag{4.45}$$

Now the key observation is that the integrand is $\sim 1/y^3$ at large y, and so is convergent as $y \to \infty$. The upper limit of the integrand is $\omega_D/\Delta \gg 1$ for small λ, and so we can safely take it to infinity. There is also no singularity in the integral as $y \to 0$. In effect, the integral is dominated by $\varepsilon \sim \Delta_0 \sim T_c \ll \omega_D$, and so it only depends upon universal physics at energies well below those where the details of the phonon band structure become important. So now we have a result that is explicitly an equation only for $z \equiv 2\Delta_0/T_c$:

$$\int_0^{\infty} dy \left[\frac{1}{\sqrt{y^2+1}} - \frac{\tanh(yz/4)}{y}\right] = 0. \tag{4.46}$$

Performing the integral in (4.46), we can solve for z and obtain the famous BCS result

$$\frac{2\Delta_0}{T_c} = 2\pi e^{-\gamma} = 3.52775398\ldots. \tag{4.47}$$

This relationship was verified in early optical experiments, quickly establishing the validity of the BCS theory in the lower T_c superconductors.

Problems

4.1 (a) Compute the expectation values of the number operator

$$\hat{N} = \sum_k \left(c_{k\uparrow}^\dagger c_{k\uparrow} + c_{k\downarrow}^\dagger c_{k\downarrow} \right) \tag{4.48}$$

in the BCS ground state. Do this using the BCS wavefunction in (4.10), or using the Bogoliubov operators in (4.29). Show that

$$\langle \hat{N} \rangle = 2 \sum_k |v_k|^2 . \tag{4.49}$$

(b) Next, compute the variance in the number $\delta N^2 = \langle \hat{N}^2 \rangle - \langle \hat{N} \rangle^2$ by both methods, and show that

$$\delta N^2 = 4 \sum_k |u_k|^2 |v_k|^2 . \tag{4.50}$$

(c) Compute the relative standard deviation $\delta N / \langle \hat{N} \rangle$ for Δ much smaller than the Fermi energy.

4.2 Consider a metal in the presence of a magnetic field field B that couples only via the Zeeman coupling to the electrons. So the complete Hamiltonian is (after absorbing g-factors etc. in the definition of B)

$$H = \sum_k \left(\varepsilon_k \left[c_{k\uparrow}^\dagger c_{k\uparrow}^\dagger + c_{k\downarrow}^\dagger c_{k\downarrow}^\dagger \right] - B \left[c_{k\uparrow}^\dagger c_{k\uparrow}^\dagger - c_{k\downarrow}^\dagger c_{k\downarrow}^\dagger \right] \right) . \tag{4.51}$$

(a) Obtain an exact expression for the average value of the magnetization density

$$M = \frac{1}{V} \sum_k \left(c_{k\uparrow}^\dagger c_{k\uparrow}^\dagger - c_{k\downarrow}^\dagger c_{k\downarrow}^\dagger \right) \tag{4.52}$$

at a temperature T. Express your answer in terms of an integral over the single spin density of states per unit volume, $g(\varepsilon)$.

(b) Take the $B \to 0$ limit, and write $M = \chi_P B$, where χ_P is the Pauli spin susceptibility; what is the value of χ_P?

(c) Now we address the same problem for a BCS superconductor. In the presence of a Zeeman term the Hamiltonian is

$$H = H_B - B \sum_k \left(c_{k\uparrow}^\dagger c_{k\uparrow} - c_{k\downarrow}^\dagger c_{k\downarrow} \right) , \tag{4.53}$$

where H_B is given in (4.26). Express the Hamiltonian in terms of the Bogoliubov operators that diagonalize H_B.

(d) Determine the magnetization density to linear order in B at a non-zero temperature T. Express your final result as an integral over the energy of the single-particle states of the metal and its density of states.

5 Broken Symmetry and Superfluidity

The concept of broken symmetry is introduced, and combined with constraints from gauge invariance to obtain the London equation for superfluids and superconductors. The Meissner effect is described, and the London penetration depth is computed.

The defining property of the Bardeen–Cooper–Schrieffer (BCS) theory is the existence of off-diagonal long-range order (ODLRO), defined by the correlator of the Cooper-pair operator

$$\lim_{|\boldsymbol{r}|\to\infty} \left\langle \Psi_C^\dagger(\boldsymbol{r})\Psi_C(\boldsymbol{0}) \right\rangle = |\Psi_0|^2 \neq 0 \,. \tag{5.1}$$

Note that this correlator can be evaluated either in the fixed N (canonical ensemble) or the fixed θ (grand-canonical ensemble) BCS state, and the same answer is obtained in both cases. We will now investigate some general consequences of the presence of ODLRO. Combined with the constraints of gauge invariance, we will see that the main properties of a superconductor follow from rather general considerations.

Similar ideas apply to the theory of the Bose gas in Chapter 3, simply by replacing the Cooper-pair operator Φ_C with the boson field operator ψ. But I present this discussion in the context of the BCS theory.

The basic idea is that we need to consider the field $\Psi_C(\boldsymbol{r})$ as an "order parameter," which must be included along with other thermodynamic variables (such as pressure, temperature, volume) in a description of macroscopic properties. Linked to this order parameter is a conjugate variable that determines its response to external perturbations: this is the "helicity modulus."

As will become clear below, the helicity modulus of a superconductor (or a superfluid) is the analog of the shear modulus of a crystalline solid. A crystalline solid is characterized by long-range order in the atomic positions. A shear strain is a "twist" in these atomic positions, and the change in energy of the solid from a shear strain is determined by a shear modulus. Similarly, when we twist the order parameter of a superconductor, by making its phase space-dependent, the change is characterized by the helicity modulus.

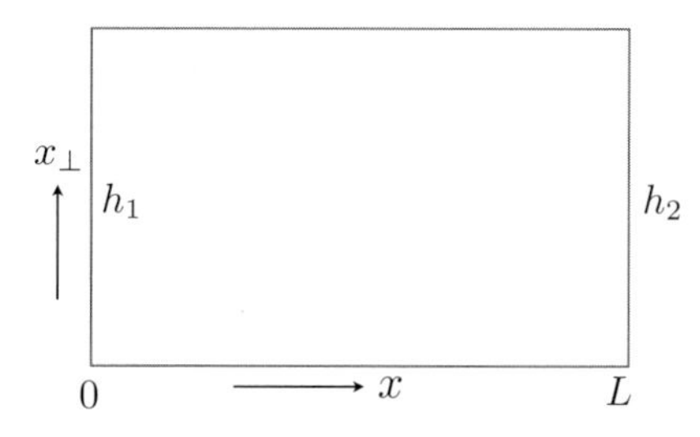

Figure 5.1 System with $d-1$ periodic directions $x_\perp$, and open boundary conditions along the single x direction. External fields h_1 and h_2 are applied on the two boundaries.

5.1 Ising Model and Surface Tension

We begin by describing the simplest case of the Ising model of ferromagnetism. We want to impose an external Zeeman field, which will twist the ferromagnetic order parameter, and ask for the change in the free energy.

Take a $d \geq$ two-dimensional Ising model at non-zero temperature, which has open boundary conditions along one preferred direction $0 < x < L$, and periodic boundary conditions along the remaining $d-1$ dimensions $x_\perp$ of length $L_\perp$ (see Fig. 5.1) We now apply an external field at the $x = 0$ and $x = L$ boundaries so that the Hamiltonian is

$$H = H_{Ising} - \sum_{x_\perp} \{h_1\, \sigma(0, x_\perp) + h_2\, \sigma(L, x_\perp)\}\,, \tag{5.2}$$

where $\sigma(x, x_\perp) = \pm 1$ is the Ising variable. Imagine we can compute the free energy of this Ising model exactly, and then we compute the free-energy difference

$$\Delta F = F(h_1 = -h_2) - F(h_1 = h_2)\,. \tag{5.3}$$

We are interested in the behavior of ΔF as both $L, L_\perp \to \infty$. The total free energy, F, has an extensive term $\sim L L_\perp^{d-1}$. As the external perturbation is only on the boundaries, we don't expect ΔF to be extensive: we ask for the leading sub-extensive term.

The form of ΔF turns out to be dramatically different depending upon whether we are above or below the critical temperature T_c, that is, depending upon whether the ferromagnetic long-range order is absent or present. Above the critical temperature, what happens at the $x = 0$ boundary should be largely independent of what happens at the $x = L$ boundary, as there are no long-range correlations. So we expect

$$\Delta F \sim L_\perp^{d-1} \exp(-L/\xi), \quad T > T_c\,. \tag{5.4}$$

Here, ξ is correlation length, which remains finite as we take the large volume limit. However, below T_c, the imposed boundary conditions will prefer that the $x = 0$ boundary has a ferromagnetic moment in the opposite (same) direction as the $x = L$ boundary for $h_1 = -h_2$ ($h_1 = h_2$). Consequently, somewhere in between there must be a *domain wall* between the oppositely oriented ferromagnets for $h_1 = -h_2$, which is absent for

$h_1 = h_2$. The width of this domain wall is set by the ordered state correlation length ξ. This domain wall will have a non-zero free energy per unit area, and so we expect

$$\Delta F = \Sigma L_{\perp}^{d-1} \quad , \quad T < T_c, \tag{5.5}$$

where the proportionality constant Σ is the *surface tension*, which has dimensions of (energy)×(length)$^{1-d}$. Note that Σ is independent of L, as $L \to \infty$, and also independent of precisely how we choose the boundary conditions. It is an intensive thermodynamic variable (like the pressure or the temperature) that characterizes the new physical properties of the ferromagnetic phase.

5.2 *XY* Model and Helicity Modulus

Let us now go through the same argument for the case of a continuous broken symmetry, rather than a discrete broken symmetry. So we consider a XY ferromagnet, described by an angular variable $\theta(x, x_{\perp})$ that is periodic with period 2π. We apply boundary conditions so that the two-component fields h_1 and h_2 in Fig. 5.1 have an angle Θ between them. This will impose a twist of θ by an angle Θ, with the Hamiltonian

$$H = H_{XY} - h \sum_{x_{\perp}} \{\cos[\theta(0, x_{\perp})] + \cos[\theta(L, x_{\perp}) - \Theta]\} . \tag{5.6}$$

Now the free-energy difference is

$$\Delta F(\Theta) = F(\Theta) - F(\Theta = 0). \tag{5.7}$$

Because the order parameter is continuous, the system will try to rotate the ferromagnetic moment so that it minimizes ΔF. The optimum strategy turns out to be to spread the twist uniformly between $x = 0$ and $x = L$. We expect there to be a local free-energy cost, which depends upon the square of the gradient of the local twist; it can only depend upon the gradient, because a uniform twist costs no energy, and a positive gradient should have the same energy as a negative gradient. So we obtain

$$\Delta F(\Theta) = \frac{1}{2} \rho_s \left(\frac{\Theta}{L} \right)^2 L_{\perp}^{d-1} L, \quad T < T_c, \tag{5.8}$$

where Θ/L is the gradient of the phase. The prefactor ρ_s is the advertized intensive thermodynamic variable: the *helicity modulus*. Depending upon the context, it is also referred to as the superfluid or spin stiffness, and often as the "superfluid density." But it is not a density, and has dimensions of (energy)×(length)$^{2-d}$.

It is useful to express (5.8) in the framework of Landau–Ginzburg–Wilson theory. We imagine that there is a coarse-grained effective Hamiltonian that describes the local thermodynamics of a small region with many microscopic degrees of freedom. We choose to represent all these local degrees of freedom by a single collective variable:

the angular field $\theta(\boldsymbol{r})$, which represents the local orientation of the order parameter. Then the content of (5.8) is the effective Hamiltonian

$$H_{eff} = \frac{\rho_s}{2} \int d^d r \left(\nabla_r \theta(\boldsymbol{r})\right)^2, \tag{5.9}$$

imposing an energy cost proportional to the local gradient squared of the local order parameter.

5.3 Superconductors and Gauge Invariance

We now have all the ingredients to define the helicity modulus of a BCS superconductor. After coarse-graining the electronic degrees of freedom, we treat the phase of the Cooper-pair operator as a classical variable representing the local phase of the Cooper-pair condensate

$$\Psi_C(\boldsymbol{r}) = |\Psi_C(\boldsymbol{r})| \exp\left(i\theta(\boldsymbol{r})\right). \tag{5.10}$$

Then the effective Hamiltonian for $\theta(\boldsymbol{r})$ is (5.9).

We can now use gauge invariance to obtain a very powerful result: the generalization of H_{eff} in the presence of an external vector potential $\boldsymbol{A}(\boldsymbol{r})$. In the presence of an external electromagnetic field, we know that the underlying physics must respect gauge invariance under which the electron field operator $\psi(\boldsymbol{r})$ and the electromagnetic field transform as

$$\begin{aligned} \psi_\sigma(\boldsymbol{r}) &\to \psi_\sigma(\boldsymbol{r}) \exp\left(-i\frac{e}{\hbar c}\chi(\boldsymbol{r})\right), \\ \boldsymbol{A}(\boldsymbol{r}) &\to \boldsymbol{A}(\boldsymbol{r}) + \nabla_r \chi(\boldsymbol{r}). \end{aligned} \tag{5.11}$$

From the definition of the Cooper-pair operator and its phase field, we therefore conclude that

$$\begin{aligned} \Psi_C(\boldsymbol{r}) &\to \Psi_C(\boldsymbol{r}) \exp\left(-i\frac{2e}{\hbar c}\chi(\boldsymbol{r})\right) \\ \theta(\boldsymbol{r}) &\to \theta(\boldsymbol{r}) - \frac{2e}{\hbar c}\chi(\boldsymbol{r}). \end{aligned} \tag{5.12}$$

Now it is clear that there is a unique way to make (5.9) gauge invariant: it must be replaced by

$$H_{eff} = \frac{\rho_s}{2} \int d^d r \left(\nabla_r \theta(\boldsymbol{r}) + \frac{2e}{\hbar c}\boldsymbol{A}(\boldsymbol{r})\right)^2. \tag{5.13}$$

Note we did not need any other new parameters to describe this coupling, only fundamental constants of nature. Gauge invariance precisely quantizes the relative coefficents of the two terms inside the brackets, and this has many dramatic physical consequences.

5.4 The London Equation

We now restrict our considerations to the London gauge $\nabla_r \cdot \boldsymbol{A} = 0$.

From a knowledge of the dependence of the effective action on $\boldsymbol{A}$, we can also compute the physical electronic current

$$\boldsymbol{J}(\boldsymbol{r}) = -c \frac{\delta H_{eff}}{\delta \boldsymbol{A}(\boldsymbol{r})}. \tag{5.14}$$

(This equation is easily seen to be a property of the microscopic Hamiltonian, and so it should also apply to the effective Hamiltonian, provided we preserve gauge invariance.) Taking this derivative from (5.13), we obtain the London equation

$$\boldsymbol{J}(\boldsymbol{r}) = -\frac{\rho_s (2e)^2}{\hbar^2 c} \boldsymbol{A}(\boldsymbol{r}). \tag{5.15}$$

This is the key equation that determines the basic properties of a superconductor.

First, we can take the time derivative of (5.15) and obtain

$$\frac{d\boldsymbol{J}}{dt} = \frac{\rho_s (2e)^2}{\hbar^2} \boldsymbol{E}, \tag{5.16}$$

which implies that the current accelerates in the presence of an electric field, which means there is no resistance. By itself, this is not a remarkable statement, as it also holds for a perfect metal with no impurities. However, (5.16) also holds in the presence of impurities and other perturbations that break translational invariance, as it only relied on two basic features: (i) the presence of long-range order, and (ii) gauge invariance. Therefore, we do indeed obtain the explanation for superconductivity from the London equation.

Next, we can combine (5.15) with Maxwell's equations, and obtain an explanation of the Meissner effect: the expulsion of magnetic flux by a superconductor. We use the Maxwell equation

$$\nabla_r \times \boldsymbol{B} = \frac{4\pi}{c} \boldsymbol{J} \tag{5.17}$$

and the definition $\boldsymbol{B} = \nabla_r \times \boldsymbol{A}$. In the London gauge $\nabla_r \cdot \boldsymbol{A} = 0$ we obtain then that in a superconductor

$$\nabla_r^2 \boldsymbol{B} = \frac{1}{\lambda_L^2} \boldsymbol{B}, \tag{5.18}$$

where λ_L is the London penetration depth

$$\frac{1}{\lambda_L^2} = \frac{4\pi \rho_s (2e)^2}{\hbar^2 c^2}. \tag{5.19}$$

Note that λ_L, which is directly measurable in experiments, depends only upon ρ_s and fundamental constants of nature. So ρ_s is also a measurable variable.

It is easy to see that (5.18) implies the Meissner effect. Consider a superconductor in the half space $x > 0$, with a boundary at $x = 0$ to the free space present for $x < 0$ (see

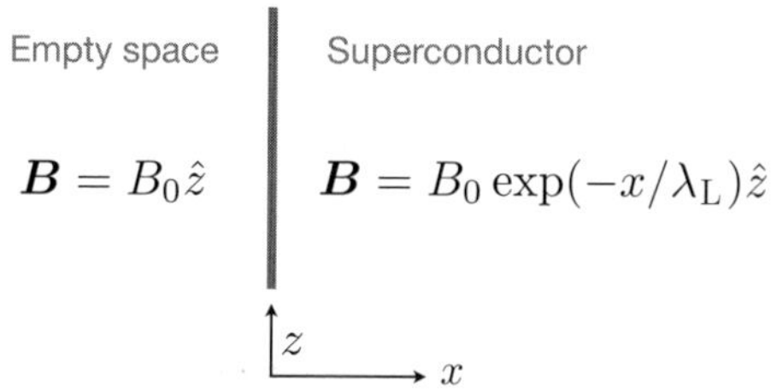

Figure 5.2 Magnetic field penetrates a distance λ_L into the superconductor.

Fig. 5.2). Apply a magnetic field $\boldsymbol{B} = B_0\hat{z}$ in the free-space region. Matching boundary conditions at $x = 0$, we find that the magnetic field inside the superconductor is

$$\boldsymbol{B}(\boldsymbol{r}) = B_0 \exp(-x/\lambda_L)\,\hat{z}, \quad x > 0; \tag{5.20}$$

which means it decays to zero in a distance of the order of the London penetration depth. From (5.17) we see that there is a supercurrent on the surface of the superconductor:

$$\boldsymbol{J}(\boldsymbol{r}) = \frac{B_0 c}{4\pi\lambda_L} \exp(-x/\lambda_L)\,\hat{y}, \quad x > 0. \tag{5.21}$$

It is this supercurrent that "screens" the external magnetic field inside the superconductor. As long as the external magnetic field is present, this supercurrent will flow "forever."

Problems

5.1 Couple the Bose gas theory in (3.27) to a *time-independent* vector potential $\boldsymbol{A}$, via $\boldsymbol{\nabla} \to \boldsymbol{\nabla} - i\boldsymbol{A}$. So the current of the bosons is

$$\boldsymbol{J} = \frac{1}{2mi}\left[\psi^*(\boldsymbol{\nabla} - i\boldsymbol{A})\psi - \psi(\boldsymbol{\nabla} + i\boldsymbol{A})\psi^*\right]. \tag{5.22}$$

Evaluate $\langle \boldsymbol{J} \rangle$ in linear response to $\boldsymbol{A}$, and to leading non-vanishing order in u. Do not assume any particular gauge choice for $\boldsymbol{A}$. You will need to insert (3.35) into (5.22) and expand it as

$$\boldsymbol{J} = -\frac{\mu}{u_0 m}\boldsymbol{A} + \frac{1}{2mi}\sqrt{\frac{\mu}{u_0}}\left(\boldsymbol{\nabla}\tilde{\psi} - \boldsymbol{\nabla}\tilde{\psi}^*\right) + \mathcal{O}(u_0^0). \tag{5.23}$$

Evaluate the second term in (5.23) to linear order in $\boldsymbol{A}$ and so show that

$$\langle J_i \rangle = -\frac{\mu}{u_0 m}\left(\delta_{ij} - \frac{k_i k_j}{k^2}\right) A_j + \mathcal{O}(u_0^0). \tag{5.24}$$

where $\boldsymbol{k}$ is the wavevector Explain why this answer is a sensible generalization of the London equation to an arbitrary gauge.

6 Landau–Ginzburg Theory

The Landau–Ginzburg functional of superconductivity is derived from the Bardeen–Cooper–Schrieffer theory.

The Bardeen–Cooper–Schrieffer (BCS) theory of Chapter 4 works extremely well at zero temperature in the limit of weak attraction between the electrons. It gives a satisfactory description of the ground state, and of the fermionic Bogoliubov quasiparticle excitations. However, it is not too convenient to work with when we are thinking about situations in which the angle θ can be space- and time-dependent, that is, when we have bosonic phase fluctuations present, like those considered in Section 3.3 for the Bose gas. Moreover, as we raise the temperature to approach T_c, we also have to consider fluctuations in the amplitude of the gap $|\Delta|$. Such effects are more conveniently discussed in the Landau–Ginzburg framework, which is expressed directly in terms of the Cooper-pair operator, but allows it to be space- and time-dependent.

Apart from leading to a complete description of the physics near T_c, the Landau–Ginzburg theory was also historically the vehicle for some remarkable theoretical predictions described in Chapter 7: the existence of vortices with quantized electromagnetic flux and a new vortex lattice phase of matter found in type-II superconductors.

6.1 Hubbard–Stratonovich Transformation

In place of the BCS Hamiltonian in (4.25), it is more convenient here to work with a Hubbard model of electrons on a lattice of sites i with a weak on-site attraction $U < 0$:

$$\begin{aligned} H &= H_0 + H_1, \\ H_0 &= \sum_{k\sigma} \varepsilon_k c_{k\sigma}^\dagger c_{k\sigma}, \\ H_1 &= U \sum_i c_{i\uparrow}^\dagger c_{i\downarrow}^\dagger c_{i\downarrow} c_{i\uparrow}. \end{aligned} \tag{6.1}$$

We set up the computation of the partition function in time-ordered perturbation theory using the interaction representation associated with H_0. Then

$$
\begin{aligned}
\mathcal{Z} &= \operatorname{Tr}\exp(-H/T) \\
&= \mathcal{Z}_0 \left\langle T_\tau \exp\left(-\int_0^\beta d\tau \hat{H}_1(\tau)\right)\right\rangle_0 ,
\end{aligned} \tag{6.2}
$$

where the time evolution in the interaction representation is given by

$$
\hat{H}_1(\tau) = e^{H_0\tau} H_1 e^{-H_0\tau}, \tag{6.3}
$$

$\beta = 1/T$, and the expectation values are evaluated in the free-particle ensemble at temperature T with partition function

$$
\mathcal{Z}_0 = \operatorname{Tr}\exp(-H_0/T) . \tag{6.4}
$$

The Hubbard–Stratonovich transformation is the functional generalization of a simple identity of a Gaussian integral

$$
\begin{aligned}
&\int d\psi d\psi^* \exp\left(-|\psi|^2/a - \psi z^* - \psi^* z\right) \\
&\quad = \exp\left(a|z|^2\right) \int d\psi d\psi^* \exp\left(-|\psi|^2/a\right),
\end{aligned} \tag{6.5}
$$

where the integrals are over the complex ψ plane. We apply this identity repeatedly to $\mathcal{Z}/\mathcal{Z}_0$ at each site i and each time τ (which we momentarily discretize). In this manner we obtain

$$
\begin{aligned}
\frac{\mathcal{Z}}{\mathcal{Z}_0} &= \left\langle T_\tau \exp\left(-U \int_0^\beta d\tau \sum_i \hat{c}_{i\uparrow}^\dagger(\tau)\hat{c}_{i\downarrow}^\dagger(\tau)\hat{c}_{i\downarrow}(\tau)\hat{c}_{i\uparrow}(\tau)\right)\right\rangle_0 \\
&= \int \prod_i \mathcal{D}\Psi_i(\tau)\mathcal{D}\Psi_i^*(\tau) \exp\left(-\int_0^\beta d\tau \sum_i \frac{|\Psi_i|^2}{|U|}\right) \\
&\quad \times \left\langle T_\tau \exp\left(-\int_0^\beta d\tau \sum_i \left[\Psi_i(\tau)\hat{c}_{i\uparrow}^\dagger(\tau)\hat{c}_{i\downarrow}^\dagger(\tau) + \text{H.c.}\right]\right)\right\rangle_0 \\
&\equiv \int \prod_i \mathcal{D}\Psi_i(\tau)\mathcal{D}\Psi_i^*(\tau) \exp\left(-\mathcal{S}_{LG}\left[\Psi_i(\tau)\right]\right),
\end{aligned} \tag{6.6}
$$

where the Landau–Ginzburg action is

$$
\begin{aligned}
\mathcal{S}_{LG}\left[\Psi_i(\tau)\right] &= \int_0^\beta d\tau \sum_i \frac{|\Psi_i|^2}{|U|} \\
&\quad - \ln \left\langle T_\tau \exp\left(-\int_0^\beta d\tau \sum_i \left[\Psi_i(\tau)\hat{c}_{i\uparrow}^\dagger(\tau)\hat{c}_{i\downarrow}^\dagger(\tau) + \text{H.c.}\right]\right)\right\rangle_0 .
\end{aligned} \tag{6.7}
$$

So far, everything is exact. We have formally rewritten the partition function of the electronic system in terms of the path integral of the complex field $\Psi_i(\tau)$. This field couples to the local electron Cooper-pair operator, and so we have theory expressed in terms of the Cooper pairs alone. Unfortunately, this theory has a very complicated, highly non-local, action functional in (6.7), and so is usually not a very useful object to work with.

6.2 Expansion near T_c

Gorkov argued that, for temperatures near T_c, we can expand $\mathcal{S}_{LG}[\Psi]$ in powers of Ψ, where the average value of Ψ is expected to be small.

Let us begin by keeping terms only to quadratic order. Then, after Fourier transformations, we can write

$$\mathcal{S}_{LG}\left[\Psi_i(\tau)\right] = \frac{T}{V} \sum_{\boldsymbol{k},\omega_n} |\Psi(\boldsymbol{k},\omega_n)|^2 \left[\frac{1}{|U|} - P(\boldsymbol{k},i\omega_n)\right] + \cdots, \tag{6.8}$$

where $P(\boldsymbol{k},i\omega_n)$ is given by the particle–particle bubble graph shown in Fig. 6.1:

$$P(\boldsymbol{k},i\omega_n) = \frac{T}{V} \sum_{\boldsymbol{p},\varepsilon_n} \frac{1}{(i\varepsilon_n - \varepsilon_{\boldsymbol{p}})(-i\varepsilon_n + i\omega_n - \varepsilon_{\boldsymbol{k}-\boldsymbol{p}})}. \tag{6.9}$$

Performing the sum over frequencies by partial fractions, we obtain

$$P(\boldsymbol{k},i\omega_n) = \frac{1}{V} \sum_{\boldsymbol{p}} \frac{1 - f(\varepsilon_{\boldsymbol{p}}) - f(\varepsilon_{\boldsymbol{k}-\boldsymbol{p}})}{(-i\omega_n + \varepsilon_{\boldsymbol{k}-\boldsymbol{p}} + \varepsilon_{\boldsymbol{p}})}. \tag{6.10}$$

Focusing further on the form at $\boldsymbol{k} = 0$ and $\omega_n = 0$, we can write

$$\begin{aligned} P(0,0) &= \frac{1}{V} \sum_{\boldsymbol{p}} \frac{1 - 2f(\varepsilon_{\boldsymbol{p}})}{2\varepsilon_{\boldsymbol{p}}} \\ &\approx d(0) \int_{-\Lambda}^{\Lambda} d\varepsilon \frac{\tanh(\varepsilon/(2T))}{2\varepsilon}, \end{aligned} \tag{6.11}$$

where $d(0)$ is the single spin density of states at the Fermi level, and Λ is an upper cutoff of the order of the Fermi level. Now we notice from (6.8) that the coefficient of $|\Psi|^2$, at $\boldsymbol{k} = \omega_n = 0$, vanishes at a temperature where $1/|U| = P(0,0)$, and this is *precisely* the same equation for T_c as we obtained in the BCS theory in (4.42). For $T > T_c$, the coefficient is positive, and in this case the action is minimized at $\Psi = 0$, that is, in the normal state. On the other hand, for $T < T_c$, the coefficient will be negative, and then the optimum values of Ψ are non-zero. indicating there is a Cooper-pair condensate, and we obtain a superconductor. To determine the value of $\langle\Psi\rangle$, we need the term in the action which has four powers of Ψ – we will examine this shortly.

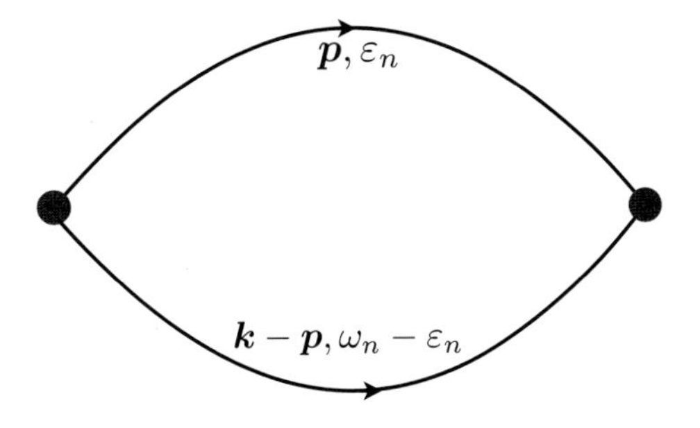

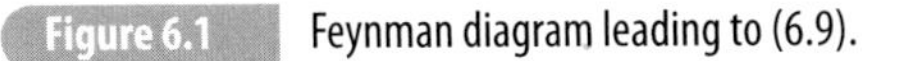
Figure 6.1 Feynman diagram leading to (6.9).

But first, let us examine the consequences of spatial and time-dependent variations in Ψ by expanding P for small ω_n and $\boldsymbol{k}$. Considering first the frequency dependence, we have

$$\begin{aligned}
P(0,i\omega_n) &= \frac{1}{V}\sum_{p}\frac{1-2f(\varepsilon_p)}{-i\omega_n+2\varepsilon_p} \\
&\approx d(0)\int_0^{\Lambda}\frac{d\varepsilon}{\varepsilon}\tanh\left(\frac{\varepsilon}{2T}\right)\left[1-\frac{\omega_n^2}{\omega_n^2+4\varepsilon^2}\right] \\
&\approx d(0)\left[\ln\left(\frac{\Lambda}{T}\right)-\frac{\omega_n^2}{2T}\int_0^{\infty}d\varepsilon\frac{1}{\omega_n^2+4\varepsilon^2}\right] \\
&= d(0)\left[\ln\left(\frac{\Lambda}{T}\right)-\frac{\pi|\omega_n|}{8T}\right],
\end{aligned} \tag{6.12}$$

where we have symmetrized the momentum integral about the Fermi level, and assumed the density of states is energy independent. In a similar manner, we can evaluate the dependence upon small $\boldsymbol{k}$, and obtain

$$P(\boldsymbol{k},i\omega_n)\approx d(0)\left[\ln\left(\frac{\Lambda}{T}\right)-\frac{\pi|\omega_n|}{8T}-\frac{Cv_F^2k^2}{T^2}\right], \tag{6.13}$$

where C is a dimensionless numerical constant of the order of unity, whose value will depend upon the precise shape of the Fermi surface.

Finally, to complete the derivation of the Landau–Ginzburg theory, we need the coefficient of the $|\Psi|^4$ term. This is obtained by evaluating the graph shown in Fig. 6.2. Close to T_c, it turns out only the zero momentum and frequency term is needed, and this evaluates to

$$\begin{aligned}
\frac{T}{2V}\sum_{\boldsymbol{k},\omega_n}\frac{1}{(\omega_n^2+\varepsilon_{\boldsymbol{k}}^2)^2} &\approx \frac{d(0)T}{2}\sum_{\omega_n}\int_{-\infty}^{\infty}d\varepsilon\frac{1}{(\omega_n^2+\varepsilon^2)^2} \\
&= \frac{7\zeta(3)}{16\pi^2}\frac{d(0)}{T^2}.
\end{aligned} \tag{6.14}$$

Note that ω_n is a fermionic frequency, and so there is no $n=0$ term in (6.14).

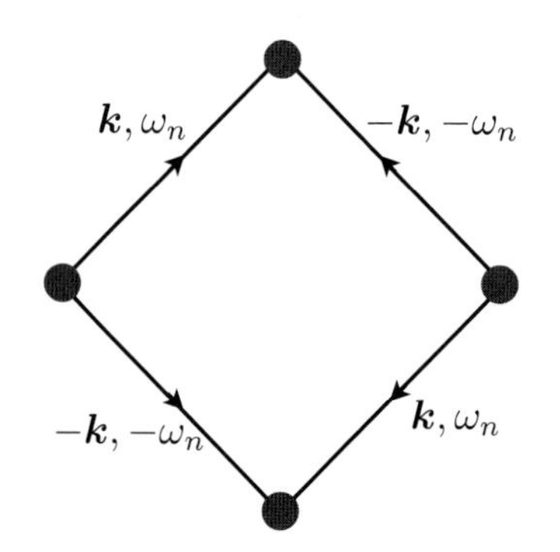

Figure 6.2 Feynman diagram leading to (6.14).

6.3 Effective Classical Theory

We can now collect all the results, and obtain an explicit prediction for the thermodynamics of the superconductor close to T_c.

The equal-time thermal fluctuations of the order of the parameter $\Psi(\boldsymbol{r})$ are expressed in terms of a free energy functional $\mathcal{F}_{LG}[\Psi(\boldsymbol{r})]$ and partition function

$$\mathcal{Z} = \int \mathcal{D}\Psi(\boldsymbol{r})\mathcal{D}\Psi^*(\boldsymbol{r}) \exp\left(-\frac{\mathcal{F}_{LG}[\Psi(\boldsymbol{r})]}{T}\right). \tag{6.15}$$

The Landau–Ginzburg free-energy functional can be written as

$$\mathcal{F}_{LG}[\Psi(\boldsymbol{r})] = \int d^d r \left[D|\nabla_r \Psi(\boldsymbol{r})|^2 + \alpha|\Psi(\boldsymbol{r})|^2 + \frac{\widetilde{\beta}}{2}|\Psi(\boldsymbol{r})|^4 \right], \tag{6.16}$$

From our results in the Gorkov expansion, we can now read off the values of all the parameters in this free energy (we have rescaled $\Psi(\boldsymbol{r})$ by a factor of $1/\sqrt{d(0)}$ for convenience)

$$\begin{aligned} \alpha &= \frac{T - T_c}{T_c}, \\ \widetilde{\beta} &= \frac{7\zeta(3)}{8\pi^2 d(0) T_c^2}, \\ D &= \frac{C v_F^2}{T_c^2}, \end{aligned} \tag{6.17}$$

where T_c has the same value as that computed in the BCS theory in Chapter 4. We have replaced all occurrences of T in the original theory by T_c, except in α. As α vanishes at $T = T_c$, we do have to retain the first-order correction in $T - T_c$ is α. Minimizing the free energy, we see that $\langle\Psi\rangle = 0$ for $T > T_c$, and

$$\langle\Psi\rangle = \sqrt{-\alpha/\widetilde{\beta}} \quad , \quad T < T_c. \tag{6.18}$$

We can now compute the helicity modulus by writing $\Psi(\boldsymbol{r}) = \sqrt{-\alpha/\widetilde{\beta}} e^{i\theta(\boldsymbol{r})}$ and computing the coefficient of $(\nabla_r \theta(\boldsymbol{r}))^2$ to obtain

$$\rho_s = \frac{2D|\alpha|}{T_c \widetilde{\beta}} \quad , \quad T < T_c. \tag{6.19}$$

The results in (6.18) and (6.19) are mean-field predictions. As we move closer to T_c, fluctuation corrections from the functional integral in (6.15) become important: these are treated in the theory of classical critical phenomena, which we do not explore here. We only note that fluctuations about the mean field can be ignored only when the Ginzburg criterion is satisfied:

$$\frac{\widetilde{\beta} T_c}{D^{d/2}} \frac{1}{|\alpha|^{(4-d)/2}} \ll 1. \tag{6.20}$$

So when we approach T_c, this criterion is violated for small enough $|T - T_c|$ as long as $d < 4$. Only for $d > 4$ is the mean field adequate for the critical behavior. We can estimate the prefactor

$$\frac{\widetilde{\beta} T_c}{D^{d/2}} \sim \left(\frac{T_c}{E_F}\right)^{d-1}. \tag{6.21}$$

So if $T_c \ll E_F$, as is the case at weak coupling, the window of strong fluctuations is present only for a narrow window near T_c.

6.4 Classical Dynamics

From the frequency dependence of P, we can also write down a classical theory for the dynamic fluctuations of Ψ, valid near T_c. It is written in terms of a time-dependent Ginzburg–Landau theory, and also known as Model A in the theory of dynamic critical phenomena by Hohenberg and Halperin [111]. This dynamics is described by the Langevin equation

$$\frac{\partial \Psi(\boldsymbol{r},t)}{\partial t} = -\Gamma \frac{\delta \mathcal{F}_{LG}[\Psi(\boldsymbol{r},t)]}{\delta \Psi^*(r,t)} + \zeta(\boldsymbol{r},t). \tag{6.22}$$

This dynamics is dissipative, with damping determined by Γ, and $\zeta(\boldsymbol{r},t)$ is a Gaussian random noise term, which ensures that the relaxation is towards a thermal state at a temperature T. This is achieved by choosing the noise correlator

$$\langle \zeta(\boldsymbol{r},t)\zeta^*(\boldsymbol{r}',t')\rangle = 2\Gamma T \delta^d(\boldsymbol{r}-\boldsymbol{r}')\delta(t-t'); \tag{6.23}$$

(this is a version of the fluctuation–dissipation theorem see Problem 6.1). The damping arises from the $|\omega_n|$ term in P, and is associated with the decay of Cooper pairs into the two-particle continuum of the Fermi liquid. From our result for this coefficient, we obtain

$$\Gamma = \frac{8}{\pi} T_c. \tag{6.24}$$

6.5 Magnetic Field

It is simple to use gauge invariance to add the orbital coupling of the magnetic field to the Landau–Ginzburg functional. Following our earlier argument for the London equation in Section 5.3, we can now write

$$\mathcal{F}_{LG}[\Psi(\boldsymbol{r})] = \int d^d r \left[D \left| \left(\nabla_r + i\frac{2e}{\hbar c}\boldsymbol{A}\right)\Psi(\boldsymbol{r})\right|^2 + \alpha|\Psi(\boldsymbol{r})|^2 + \frac{\widetilde{\beta}}{2}|\Psi(\boldsymbol{r})|^4 \right]. \tag{6.25}$$

Minimization of $\mathcal{F}_{LG}$ in the presence of a non-zero $\boldsymbol{A}$ leads to much rich physics: the appearance of vortices with quantized flux and the theory of type II superconductors, discussed in Chapter 7.

Problems

6.1 Use the Gaussian approximation for $\mathcal{F}_{LG}$ in (6.16) in the Langevin equation (6.22) to compute the equilibrium two-point correlator (the dynamic structure factor) $\langle|\Psi(\boldsymbol{k},\omega)|^2\rangle$. Next, couple Ψ linearly to a space- and time-dependent external field, to obtain the dynamic susceptibility from the linearized version of (6.22). Finally, use the fluctuation–dissipation relation in (11.22) to verify the consistency of (6.23).

7 Vortices in Superfluids and Superconductors

The vortex solutions of neutral and charged superfluids are obtained as saddle points of the Landau–Ginzburg functional.

Our treatment of superfluids and superconductors has focused on spatially uniform saddle points of the action, and a discussion of fluctuations about such saddle points. However these systems also display an important class of spatially non-uniform saddle points: vortices. Usually, these saddle points have a higher energy than the ground state, but are locally stable: they have a topologically non-trivial structure that prevents a runaway decay to the ground state. Despite their higher energy, these saddle points can be crucial at higher temperatures because they are entropically preferred: this is the situation in the Kosterlitz–Thouless transition of two-dimensional superfluids, which is discussed in Section 25.2. In some cases, the spatially non-uniform solution can be the lowest energy state: this happens to type-II superconductors in an applied magnetic field, as discussed in Section 7.4.

7.1 Neutral Superfluids

Let us look for topologically non-trivial saddle points of the Landau–Ginzburg free energy (6.16), which is also the time-independent part of the action in (3.27) for the Bose gas. We will restrict our attention to saddle points whose spatial non-uniformity extends only along two spatial directions $\boldsymbol{r} = (x, y)$. So we consider the free energy

$$\mathcal{F}[\Psi(\boldsymbol{r})] = \int d^2r \left[D|\nabla_r \Psi(\boldsymbol{r})|^2 + \alpha|\Psi(\boldsymbol{r})|^2 + \frac{\widetilde{\beta}}{2}|\Psi(\boldsymbol{r})|^4 \right]. \tag{7.1}$$

The saddle-point condition

$$\frac{\delta \mathcal{F}_{LG}}{\delta \Psi(\boldsymbol{r})} = 0 \tag{7.2}$$

leads to the equation

$$-D\nabla_r^2 \Psi + \alpha \Psi + \widetilde{\beta}|\Psi|^2 \Psi = 0. \tag{7.3}$$

We now examine solutions of (7.3) that have a rotational symmetry about the origin of coordinates, and so we make the ansatz

$$\Psi(\mathbf{r}) = f(r)e^{in\theta}, \tag{7.4}$$

where

$$r = \sqrt{x^2+y^2} \quad , \quad \theta = \tan^{-1}(y/x) \tag{7.5}$$

are polar coordinates, $f(r)$ is an unknown function, and n must be an integer for the solution (7.4) to be single-valued. A non-zero n implies a topologically non-trivial winding of the phase of the superfluid, and this ensures the metastability of such solutions.

Rather than solving (7.3), it is useful to insert the ansatz (7.4) directly into the original free energy (7.1). Then we obtain

$$\mathcal{F} = 2\pi \int_0^\infty r dr \left[D\left(\frac{df}{dr}\right)^2 + Dn^2\frac{f^2}{r^2} + \alpha f^2 + \frac{\widetilde{\beta}}{2} f^4 \right]. \tag{7.6}$$

We are looking for solutions whose deviation from the spatially uniform solution is limited to a region near the origin, and so we have

$$f(r \to \infty) = \sqrt{\frac{-\alpha}{\widetilde{\beta}}}. \tag{7.7}$$

Our task is to minimize the expression in (7.6) for each integer n. For $n = 0$, the spatially uniform solution applies, and (7.7) holds for all r. However, for $n \neq 0$, the global minimum of (7.6) is spatially non-uniform: thus the vortex solution is "protected" by a non-zero phase winding. This is immediately apparent from the fact that for $f =$ constant, the f^2/r^2 term leads to a divergence in the r integral at $r = 0$. The equation satisfied by a spatially non-uniform f can be obtained by taking the variational derivative of (7.6), which yields

$$-\frac{D}{r}\frac{d}{dr}\left(r\frac{df}{dr}\right) + Dn^2\frac{f}{r^2} + \alpha f + \widetilde{\beta} f^3 = 0. \tag{7.8}$$

An exact solution of (7.8) is not possible, but we can deduce the basic features from asymptotic analysis. We focus on the $r \to 0$ region, where the first two terms in (7.8) dominate, and make the ansatz

$$f(r \to 0) = r^p, \tag{7.9}$$

where p is an unknown parameter Inserting (7.9) into (7.8) we obtain from the $r \to 0$ limit

$$p^2 = n^2. \tag{7.10}$$

The solution $p = |n|$ is easily seen to have the lower energy.

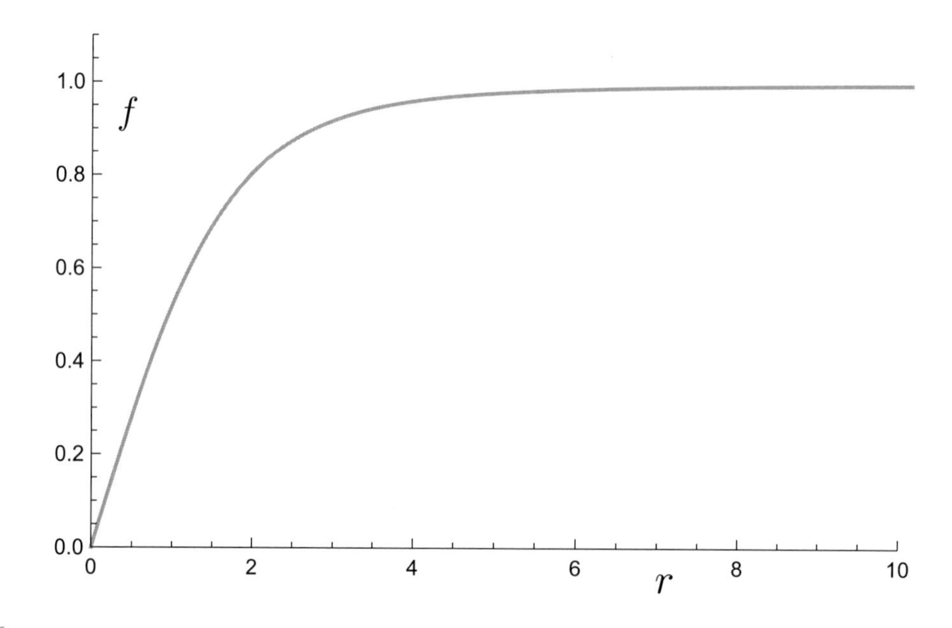

Figure 7.1 The function $f(r)$ for a neutral vortex, obtained by minimization of (7.6) for $n = 1, D = 1, \alpha = -1, \widetilde{\beta} = 1$. The coherence length is $\xi_c = 1$.

The full solution of (7.8) interpolates between the limits in (7.7) and (7.9). This can be easily obtained by a numerical minimization of (7.6), and the result in shown in Fig. 7.1. The linear form in (7.9) holds for $r \lesssim \xi_c$, where ξ_c is the "coherence length"

$$\xi_c = \sqrt{\frac{-\alpha}{D}}. \tag{7.11}$$

The region $r \lesssim \xi_c$ is known as the vortex core, and linear vanishing of f at the origin ensures that there is no divergence of the free energy at $r = 0$, and the core has a finite contribution.

An important feature of an $n \neq 0$ vortex solution is that it carries a persistent supercurrent. This is permitted because a solution with $n \neq 0$ breaks time-reversal and mirror symmetries. We can compute the current by gauging the global $U(1)$ symmetry of (7.1) to (6.25), and then evaluating the functional derivative (5.14) to obtain

$$\begin{aligned} \boldsymbol{J} &= \frac{1}{i}\left(\Psi^* \nabla_r \Psi - \Psi \nabla_r \Psi^*\right) \\ &= 2Dn\frac{f^2}{r}\hat{e}_\theta, \end{aligned} \tag{7.12}$$

where $\hat{e}_\theta$ is a unit vector in the θ direction of polar coordinates. The current pattern is sketched in Fig. 7.2. Note that the current falls off as $1/r$ as $r \to \infty$.

Although the vortex solution removes the logarithmic divergence in the energy of a vortex solution at $r = 0$, there remains a divergence as $r \to \infty$ that we discuss now. From (7.6), we can write

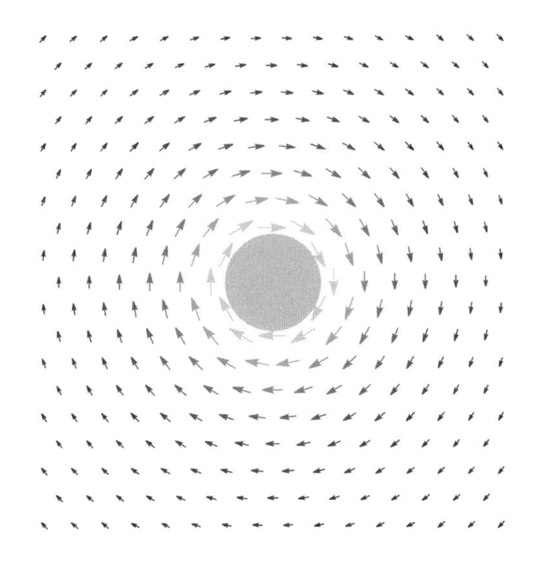

Figure 7.2 Superflow around a vortex core of size $\sim \xi_c$.

$$\mathcal{F}_{vortex} - \mathcal{F}_{no\ vortex} \approx 2\pi D n^2 \int_{\xi_c}^{\infty} r dr \frac{f^2}{r^2} = \frac{2\pi n^2 D|\alpha|}{\widetilde{\beta}} \int_{\xi_c}^{\infty} \frac{dr}{r}. \quad (7.13)$$

For an isolated vortex, this logarithmic divergence is only cut off by the system size L, and so the energy of a single vortex diverges as $\ln(L/\xi)$. But this is a rather mild divergence, and isolated vortices can be detected in moderate-size systems. But in an infinite system, vortices can only occur as vortex/anti-vortex pairs so that there is no net circulation of the current at infinity. The energy of such a pair will then be

$$\mathcal{F}_{vortex/anti\text{-}vortex} \approx \frac{4\pi n^2 D|\alpha|}{\widetilde{\beta}} \ln\left(\frac{R}{\xi_c}\right), \quad (7.14)$$

where R is the separation between the vortices. This implies an attractive $1/R$ force between a vortex and an anti-vortex in two dimensions, a feature that will play an important role in Section 25.2.

7.2 Charged Superfluids

Let us now consider the case of a charged superfluid, where the free energy is given by (6.25) with an additional vector potential $\boldsymbol{A}$, not present in (7.1). The vortex solution has the supercurrent in (7.12), and this current produces a magnetic field; so $\boldsymbol{A}$ is necessarily non-zero in a charged superfluid. As I now describe, the non-zero $\boldsymbol{A}$

has the remarkable consequence of removing the logarithmic divergence in (7.13), and rendering the energy of an isolated vortex finite.

Given the azimuthal symmetry of the vortex, a natural choice for $\boldsymbol{A}$ is

$$\boldsymbol{A}(\boldsymbol{r}) = \frac{\Phi(r)}{2\pi r}\hat{e}_\theta. \tag{7.15}$$

The $\Phi(r)$ is the flux enclosed in a circle of radius r:

$$\Phi(r) = \int_0^r d^2r B_z = \int_0^r d^2r(\partial_x A_y - \partial_y A_x) = \oint \boldsymbol{A}.d\boldsymbol{r}. \tag{7.16}$$

Now we insert the ansatzes (7.15) and (7.4) into (6.25), and obtain in place of (7.6)

$$\mathcal{F} = 2\pi \int_0^\infty rdr \left[D\left(\frac{df}{dr}\right)^2 + D\left(n - \frac{q\Phi}{2\pi\hbar c}\right)^2 \frac{f^2}{r^2} + \alpha f^2 + \frac{\widetilde{\beta}}{2} f^4 \right], \tag{7.17}$$

where $q = 2e$ is the charge of a Cooper pair. The only change is in the coefficient of the f^2/r^2 term. Notice that we can make this term vanish as $r \to \infty$ if and only if

$$\lim_{r\to\infty} \Phi(r) = n\Phi_0, \tag{7.18}$$

where

$$\Phi_0 = \frac{hc}{q} \tag{7.19}$$

is the flux quantum of a superfluid in which the condensate has charge q. The quantized flux value is preferred because, only then, we remove the logarithmic divergence in system size in the energy of the vortex.

The full solution of the vortex configuration specified by $f(r)$ and $\Phi(r)$ is obtained by minimizing the sum of (7.17) and the electromagnetic energy

$$\begin{aligned} \mathcal{F}_{total} &= \mathcal{F} + \frac{1}{8\pi}\int d^2r(\partial_x A_y - \partial_y A_x)^2 \\ &= \mathcal{F} + \frac{2\pi}{8\pi}\int_0^\infty rdr\left(\frac{1}{2\pi r}\frac{d\Phi}{dr}\right)^2. \end{aligned} \tag{7.20}$$

We will not enter into the details of this, but sketch the qualitative form of the solution in Fig. 7.3. The form of $f(r)$ is similar to that in Fig. 7.1, while the magnetic field $B_z(r)$ extends nearly uniformly out to a distance of order the London penetration depth λ_L we encountered in (5.19). In the present situation, the value of ρ_s is given by (6.19), and so

$$\frac{1}{\lambda_L^2} = \frac{8\pi D|\alpha|q^2}{\hbar^2 c^2 \widetilde{\beta}}. \tag{7.21}$$

For $r < \lambda_L$, Φ quickly reaches its asymptotic value in (7.18). Consequently we can expect that the free energy of a vortex is modified from (7.13) to

$$\mathcal{F}_{vortex} - \mathcal{F}_{no\ vortex} \approx \frac{2\pi n^2 D|\alpha|}{\widetilde{\beta}} \ln\left(\frac{\lambda_L}{\xi_c}\right). \tag{7.22}$$

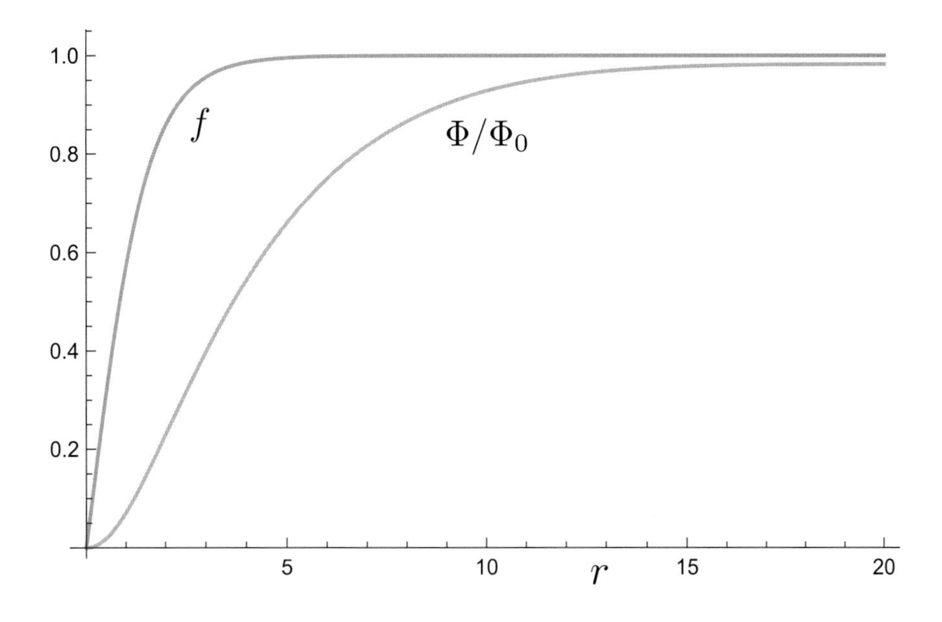

Figure 7.3 Solution of a vortex in a charged superfluid, obtained by minimizing (7.20). Parameters are as for the neutral vortex in Fig. 7.1, along with $q/(\hbar c) = 0.075$. So $\xi_c = 1$ and $\lambda_L = 2.66$.

Morever, the interaction energy of a vortex/anti-vortex pair will be as in (7.14) for $R < \lambda_L$, and have only an exponentially weak R dependence for $R > \lambda_L$.

All of the above analysis assumes that $\lambda_L \gg \xi_c$, which is the case for type-II superconductors.

7.3 Flux Quantization

The remarkable quantization of flux in a vortex observed in (7.18) and (7.19) deserves further comment. It is actually a very robust feature, and applies in far more general situations than that considered so far: in cases where there is no rotation symmetry, and even when translational symmetry is broken by the presence of disorder.

Let us assume that far from the core of the vortex we have

$$\Psi(\boldsymbol{r}) = |\Psi_0| e^{i\theta(\boldsymbol{r})}, \tag{7.23}$$

where $|\Psi_0|$ takes a possibly space-dependent value that minimizes the local free energy. Then there is a term in the free energy of the form

$$\left| \left(\nabla_r - \frac{iq}{\hbar c} \boldsymbol{A} \right) \Psi \right|^2 = |\Psi_0|^2 \left(\nabla_r \theta - \frac{q}{\hbar c} \boldsymbol{A} \right)^2 + \cdots . \tag{7.24}$$

The energy will therefore be minimized if we are able to choose

$$\nabla_r \theta = \frac{q}{\hbar c} \boldsymbol{A} \tag{7.25}$$

as $r \to \infty$. Integrating (7.25) over a large contour C we then have

$$\frac{q\Phi(r \to \infty)}{\hbar c} = \frac{q}{\hbar c}\oint_C \boldsymbol{A}.d\boldsymbol{r} = \oint_C \nabla_{\boldsymbol{r}}\theta.d\boldsymbol{r} = 2\pi n. \tag{7.26}$$

The last inequality follows from the single-valuedness of $e^{i\theta}$, and implies the quantization of the flux in integer multiples of Φ_0.

7.4 Vortex Lattices

Let us now consider a two-dimensional section of a superconductor placed in the presence of a uniform magnetic field B_{ext} oriented in the z direction. We assume that B_{ext} is created by some fixed set of external currents $\boldsymbol{J}_{ext}$, so that

$$\nabla_{\boldsymbol{r}} \times \boldsymbol{B}_{ext} = \frac{4\pi}{c}\boldsymbol{J}_{ext}. \tag{7.27}$$

This field will act on the superconductor, and produce a supercurrent $\boldsymbol{J}$ specified by the analog of (5.14), which will in turn modify the total magnetic field to $B(\boldsymbol{r})$, not equal to B_{ext}. Indeed, the total magnetic field will be given by the Maxwell equation

$$\nabla_{\boldsymbol{r}} \times \boldsymbol{B} = \frac{4\pi}{c}\left(\boldsymbol{J}_{\text{ext}} + \boldsymbol{J}\right). \tag{7.28}$$

Furthermore, the presence of a non-zero $\boldsymbol{B}$ will also produce a spatial variation in $\Psi(r)$ related to that found in the vortex solution.

We would like to reduce the determination of $\boldsymbol{B} = \nabla_{\boldsymbol{r}} \times \boldsymbol{A}$ and $\Psi(r)$ in the presence of B_{ext} to an energy minimization problem. The required free energy turns out to be a simple modification of

$$\mathcal{F}_{total} = \mathcal{F} + \frac{1}{8\pi}\int d^2r(\partial_x A_y - \partial_y A_x - B_{ext})^2, \tag{7.29}$$

where $\mathcal{F}$ is that specified in (6.25). Taking the variational derivative of (7.29) with respect to $\boldsymbol{A}(\boldsymbol{r})$, we obtain (7.28). So the problem is reduced to minimizing (7.29) with respect to $\boldsymbol{A}(\boldsymbol{r})$ and $\Psi(\boldsymbol{r})$.

This minimization problem was first addressed by Abrikosov [2]. For the suitable set of parameters, the minimum energy solution is a *vortex lattice*. This is a triangular lattice of $n = 1$ vortices similar to that found in Section 7.2. Each vortex carries a flux hc/q and so the vortex lattice spacing is given by

$$B_{ext}A_t = \frac{hc}{q}, \tag{7.30}$$

where A_t is the area of the unit cell of the triangular lattice. This prediction by Abrikosov has since been confirmed by the observation of a such a vortex lattice in numerous experiments.

Problems

7.1 Consider a neutral superfluid with a vortex at $\boldsymbol{r} = 0$ and an anti-vortex at $\boldsymbol{r} = (R, 0)$ with $R \gg \xi_c$. Neglect the vortex-core regions of size ξ_c, and assume that outside the core we need only account for variations in the phase of the superfluid order with

$$\Psi(\mathbf{r}) = \sqrt{\frac{-\alpha}{\widetilde{\beta}}}\, e^{i\theta(\boldsymbol{r})} . \tag{7.31}$$

Insert (7.31) into (7.1), obtain the saddle-point equation for $\theta(\boldsymbol{r})$, and solve this equation for the configuration noted; this solution will be a sum of solutions of the form in (7.5). Finally, insert this solution back into (7.1), and so obtain (7.14) to logarithmic accuracy.

8 Boson Hubbard Model

The phase diagram of the boson Hubbard model is obtained in mean-field theory, containing lobes of Mott insulators and a superfluid. The continuum quantum field theories of the quantum phase transitions are obtained.

Our theory of the Bose gas in Chapter 3 was perturbative in the repulsive interaction between the bosons, u_0. This analysis always yields a superfluid ground state for the Bose gas. This chapter begins our discussion of strong interactions, where the ground state need not be a superfluid or, for fermions, a Fermi liquid.

For bosons in the continuum, we know from experiments on helium-4 that it realizes a hexagonal-closed-packed (hcp) crystalline solid under pressure at $T = 0$, and this solid has a zero helicity modulus (neglecting the possibility of a supersolid). However, the $T = 0$ transition from the superfluid to the hcp solid is first order, and no controlled analytic treatment is available in the vicinity of the transition.

In this chapter we therefore consider a simpler situation: bosons moving on a fixed background lattice. This situation is realized in ultracold-atom systems in the presence of an optical lattice created by standing waves of lasers. We show in this chapter that for lattice bosons, for a sufficiently strong, but finite, repulsive interaction, there can be quantum transitions from the superfluid to a gapped non-superfluid phase, which is often referred to in this context as an "insulator." We first restrict our attention to just on-site repulsive interactions, in which case the Hamiltonian realizes a Hubbard model, and we obtain "trivial" insulators at integer densities that do not break any lattice symmetries. The superfluid to trivial-insulator transition can be second order, and Section 8.3 describes an emergent continuum theory for the vicinity of the transition which is distinct from that discussed in Chapter 3 for the dilute Bose gas.

Section 8.4 begins our consideration of insulators at non-integer densities. These cannot be trivial, and here we only discuss the case where there is translational symmetry breaking so that density in the new unit cell is again an integer. More subtle examples of insulators that do not break translational symmetry at non-integer filling will occupy much of Parts II and IV.

8.1 Lattice Hamiltonian

Here, we introduce the boson operator $\hat{b}_i$, which annihilates bosons on the sites, i, of a regular lattice in d dimensions. These Bose operators and their Hermitian conjugate creation operators obey the commutation relation

$$\left[\hat{b}_i, \hat{b}_j^\dagger\right] = \delta_{ij}, \tag{8.1}$$

while two creation or annihilation operators always commute. It is also useful to introduce the boson number operator

$$\hat{n}_{bi} = \hat{b}_i^\dagger \hat{b}_i, \tag{8.2}$$

which counts the number of bosons on each site. We allow an arbitrary number of bosons on each site. Thus the Hilbert space consists of states $|\{m_j\}\rangle$, which are eigenstates of the number operators

$$\hat{n}_{bi}|\{m_j\}\rangle = m_i|\{m_j\}\rangle, \tag{8.3}$$

and every m_j in the set $\{m_j\}$ is allowed to run over all non-negative integers. This includes the "vacuum" state with no bosons at all $|\{m_j = 0\}\rangle$.

The Hamiltonian of the boson Hubbard model is

$$H_B = -w\sum_{\langle ij\rangle} \left(\hat{b}_i^\dagger \hat{b}_j + \hat{b}_j^\dagger \hat{b}_i\right) - \mu \sum_i \hat{n}_{bi} + (U/2)\sum_i \hat{n}_{bi}(\hat{n}_{bi} - 1). \tag{8.4}$$

The first term, proportional to w, allows hopping of bosons from site to site ($\langle ij\rangle$ represents nearest neighbor pairs); if each site represents a superconducting grain, then w is the Josephson tunneling that allows Cooper pairs to move between grains. The second term, μ, represents the chemical potential of the bosons: changing the value of μ changes the total number of bosons. Depending upon the physical conditions, a given system can either be constrained to be at a fixed chemical potential (the grand canonical ensemble) or have a fixed total number of bosons (the canonical ensemble). Theoretically it is much simpler to consider the fixed chemical-potential case, and results at fixed density can always be obtained from them after a Legendre transformation. Finally, the last term, $U > 0$, represents the simplest possible repulsive interaction between the bosons. We have taken only an on-site repulsion. This can be considered to be the charging energy of each superconducting grain. Off-site and longer-range repulsions are undoubtedly important in realistic systems, but these are neglected in this simplest model.

The Hubbard model H_B is invariant under a global $U(1) \equiv O(2)$ phase transformation, as in (3.29), under which

$$\hat{b}_i \to \hat{b}_i e^{i\phi}. \tag{8.5}$$

This symmetry is related to the conservation of the total number of bosons

$$\hat{N}_b = \sum_i \hat{n}_{bi}; \tag{8.6}$$

it is easily verified that $\hat{N}_b$ commutes with $\hat{H}$.

We begin our study of H_B by introducing a simple mean-field theory in Section 8.2. This theory displays superfluid–insulator transitions, and we employ the coherent-state path-integral approach of Appendix A to obtain continuum quantum theories describing fluctuations near the quantum critical points in Section 8.3. Our treatment builds on the work of Fisher *et al.* [83].

8.2 Mean-Field Theory

The strategy, as in any mean-field theory, is to model the properties of H_B by the best possible sum, H_{MF}, of single-site Hamiltonians:

$$H_{MF} = \sum_i \left(-\mu \hat{n}_{bi} + (U/2)\,\hat{n}_{bi}(\hat{n}_{bi} - 1) - \Psi_B^* \hat{b}_i - \Psi_B \hat{b}_i^\dagger \right), \tag{8.7}$$

where the complex number Ψ_B is a variational parameter. We have chosen a mean-field Hamiltonian with the same on-site terms as H_B and have added an additional term with a "field" Ψ_B to represent the influence of the neighboring sites; this field has to be self-consistently determined. Notice that this term breaks the $U(1)$ symmetry and does not conserve the total number of particles. This is to allow for the possibility of broken-symmetric phases, whereas symmetric phases will appear at the special value $\Psi_B = 0$. As we saw in the analysis of H_R, the state that breaks the $U(1)$ symmetry will have a non-zero stiffness to rotations of the order parameter; in the present case, this stiffness is the superfluid density characterizing a superfluid ground state of the bosons.

Another important assumption underlying (8.7) is that the ground state does not spontaneously break a translational symmetry of the lattice, as the mean-field Hamiltonian is the same on every site. Such a symmetry breaking is certainly a reasonable possibility, but we will ignore this complication here for simplicity.

Let us determine the optimum value of the mean-field parameter Ψ_B by a standard procedure. First, we determine the ground-state wavefunction of H_{MF} for an arbitrary Ψ_B; because H_{MF} is a sum of single-site Hamiltonians, this wavefunction will simply be a product of single-site wavefunctions. Next, we evaluate the expectation value of H_B in this wavefunction. By adding and subtracting H_{MF} from H_B, we can write the mean-field value of the ground-state energy of H_B in the form

$$\frac{E_0}{N_s} = \frac{E_{MF}(\Psi_B)}{N_s} - Zw\langle \hat{b}^\dagger \rangle \langle \hat{b} \rangle + \langle \hat{b} \rangle \Psi_B^* + \langle \hat{b}^\dagger \rangle \Psi_B, \tag{8.8}$$

where $E_{MF}(\Psi_B)$ is the ground-state energy of H_{MF}, N_s is the number of sites of the lattice, Z is the number of nearest neighbors around each lattice point (the "coordination number"), and the expectation values $\langle \hat{b} \rangle$ and $\langle \hat{b}^\dagger \rangle$ are evaluated in the ground state of $H_{\rm MF}$. The final step is to minimize (8.8) over variations in Ψ_B. This step has been carried out numerically and the results are shown in Fig. 8.1.

Notice that even on a single site, H_{MF} has an infinite number of states, corresponding to the allowed values $m \geq 0$ of the integer number of bosons on each site. The numerical procedure necessarily truncates these states at some large occupation number, but the

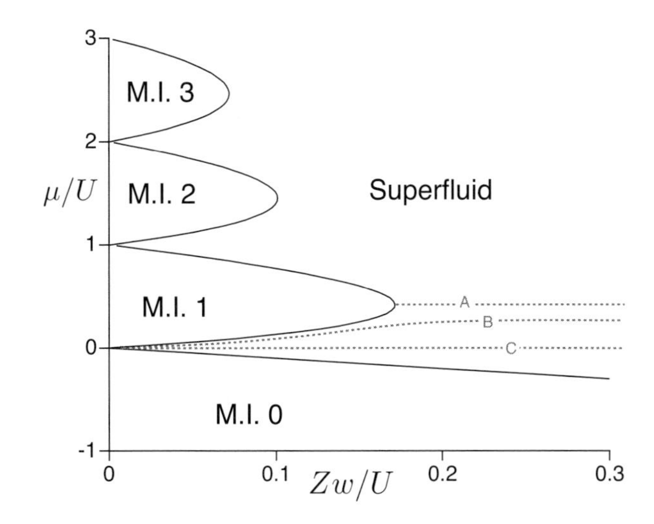

Figure 8.1 Mean-field phase diagram of the ground state of the boson Hubbard model H_B in (8.4). The notation M.I. n refers to a Mott insulator with $n_0(\mu/U) = n$. The dashed lines are sketches of contours of equal density: the line A has density 1, the line C has density 1/2, and the line B has a density between 0 and 1.

errors are not difficult to control. In any case, we will show that all the essential properties of the phase diagram can be obtained analytically. Also, by taking the derivative of (8.8) with respect to Ψ_B, it is easy to show that at the optimum value of Ψ_B is

$$\Psi_B = Zw\langle\hat{b}\rangle; \tag{8.9}$$

this relation, however, does not hold at a general point in parameter space.

First, let us consider the limit $w=0$. In this case, the sites are decoupled, and the mean-field theory is exact. It is also evident that $\Psi_B = 0$, and we simply have to minimize the on-site interaction energy. The on-site Hamiltonian contains only the operator $\hat{n}$, and the solution involves finding the boson occupation number (these occupation numbers are the integer-valued eigenvalues of $\hat{n}$) that minimizes H_B. This is simple to carry out, and we get the ground-state wavefunction

$$|m_i = n_0(\mu/U)\rangle, \tag{8.10}$$

where the integer-valued function $n_0(\mu/U)$ is given by

$$n_0(\mu/U) = \begin{cases} 0, & \text{for } \mu/U < 0, \\ 1, & \text{for } 0 < \mu/U < 1, \\ 2, & \text{for } 1 < \mu/U < 2, \\ \vdots & \vdots \\ n, & \text{for } n-1 < \mu/U < n. \end{cases} \tag{8.11}$$

Thus, each site has exactly the same integer number of bosons, which jumps discontinuously whenever μ/U goes through a positive integer. When μ/U is exactly equal to a positive integer, there are two degenerate states on each site (with boson numbers differing by 1) and so the entire system has a degeneracy of 2^{N_s}. This large degeneracy implies a macroscopic entropy; it will be lifted once we turn on a non-zero w.

We now consider the effects of a small non-zero w. As is shown in Fig. 8.1, the regions with $\Psi_B = 0$ survive in lobes around each $w = 0$ state (8.10) characterized by a given integer value of $n_0(\mu/U)$. Only at the degenerate point with $\mu/U =$ integer does a non-zero w immediately lead to a state with $\Psi_B \neq 0$. We will consider the properties of this $\Psi_B \neq 0$ later, but now we discuss the properties of the lobes with $\Psi_B = 0$ in some more detail. In mean-field theory, these states have wavefunctions still given exactly by (8.10). However, it is possible to go beyond mean-field theory and make an important exact statement about each of the lobes. The expectation value of the number of bosons in each site is given by

$$\langle \hat{b}_i^\dagger \hat{b}_i \rangle = n_0(\mu/U), \tag{8.12}$$

which is the same result one would obtain from the product state (8.10) (which, I emphasize, is not the exact wavefunction for $w \neq 0$). There are two important ingredients behind the result (8.12): the existence of an energy gap and the fact that $\hat{N}_b$ commutes with H_B. First, recall that at $w = 0$, provided μ/U was not exactly equal to a positive integer, there was a unique ground state, and there was a non-zero energy separating this state from all other states (this is the energy gap). As a result, when we turn on a small non-zero w, the ground state will move adiabatically without undergoing any level crossings with any other state. Now the $w = 0$ state is an exact eigenstate of $\hat{N}_b$ with an eigenvalue $N_s n_0(\mu/U)$, and the perturbation arising from a non-zero w commutes with $\hat{N}_b$. Consequently, the ground state will remain an eigenstate of $\hat{N}_b$ with precisely the same eigenvalue, $N_s n_0(\mu/U)$, even for small non-zero w. Assuming translational invariance, we then immediately have the exact result (8.12). Notice that this argument also shows that the energy gap above the ground state will survive everywhere within the lobe. These regions with a quantized value of density and an energy gap to all excitations are known as "Mott insulators." Their ground states are very similar to, but not exactly equal to, the simple state (8.10). They involve in addition terms with bosons undergoing virtual fluctuations between pairs of sites, creating particle–hole pairs. The Mott insulators are also known as "incompressible" because their density does not change under changes of the chemical potential μ or other parameters in H_B:

$$\frac{\partial \langle \hat{N}_b \rangle}{\partial \mu} = 0. \tag{8.13}$$

It is worth re-emphasizing here the remarkable nature of the exact result (8.12). From the perspective of classical critical phenomena, it is most unusual to find the expectation value of any observable to be pinned at a quantized value over a finite region of the phase diagram. The existence of observables such as $\hat{N}_b$ that commute with the Hamiltonian is clearly a crucial ingredient.

The numerical analysis shows that the boundary of the Mott-insulating phases is a second-order quantum phase transition (i.e., a non-zero Ψ_B turns on continuously). With the benefit of this knowledge, we can determine the positions of the phase boundaries. By the usual Landau theory argument, we simply need to expand E_0 in (8.8) in powers of Ψ_B,

$$E_0 = E_{00} + r|\Psi_B|^2 + \mathcal{O}(|\Psi_B|^4), \tag{8.14}$$

and the phase boundary appears when r changes sign. The value of r can be computed from (8.8) and (8.7) by second-order perturbation theory, and we find

$$r = \chi_0(\mu/U)\left[1 - Zw\chi_0(\mu/U)\right], \tag{8.15}$$

where

$$\chi_0(\mu/U) = \frac{n_0(\mu/U)+1}{Un_0(\mu/U)-\mu} + \frac{n_0(\mu/U)}{\mu - U(n_0(\mu/U)-1)}. \tag{8.16}$$

The function $n_0(\mu/U)$ in (8.11) is such that the denominators in (8.16) are positive, except at the points at which the boson occupation number jumps at $w = 0$. The solution of the simple equation $r = 0$ leads to the phase boundaries shown in Fig. 8.1.

Finally, we turn to the phase with $\Psi_B \neq 0$. The mean-field parameter Ψ_B varies continuously as the parameters are varied. As a result, all thermodynamic variables also change, and the density does not take a quantized value; by a suitable choice of parameters, the average density can be varied smoothly across any real positive value. So this is a compressible state in which

$$\frac{\partial \langle \hat{N}_b \rangle}{\partial \mu} \neq 0. \tag{8.17}$$

As we noted earlier, the presence of a $\Psi_B \neq 0$ implies that the $U(1)$ symmetry is broken, and there is a non-zero stiffness (i.e. helicity modulus) to twist in the orientation of the order parameter.

We also note that extensions of the boson Hubbard model with interactions beyond the nearest neighbor can spontaneously break translational symmetry at certain densities. If this symmetry breaking coexists with the superfluid order, one can obtain a "supersolid" phase.

A notable feature of Fig. 8.1 is that states with an energy gap to all excitations, and no broken translational symmetry, only appear at exactly integer-quantized densities. We can ask if this is a generic feature; are there gapped states with no broken symmetry at other densities? In one spatial dimension, the answer is no, and this is one of the consequences of the Lieb–Schultz–Mattis theorem [159]. However, the answer in spatial dimension $d = 2$ (and higher) is yes, and describing examples of such systems is a major focus of Parts II and IV.

8.3 Continuum Quantum Field Theories

Returning to our discussion of the boson Hubbard model, here I describe the low-energy properties of the quantum phase transitions between the Mott insulators and the superfluid found in Section 8.2. We will find that it is crucial to distinguish between the two different cases, each characterized by its own universality class and continuum quantum field theory. The important diagnostic distinguishing the two possibilities is the behavior of the boson density across the transition. In the Mott insulator, this density is of course always pinned at some integer value. As one undergoes the transition to the superfluid, depending upon the precise location of the system in the phase diagram of Fig. 8.1, there are two possible behaviors of the density: (a) the density remains pinned at its quantized value in the superfluid in the vicinity of the quantum critical point, or (b) the transition is accompanied by a change in the density.

We begin by writing the partition function of H_B, $\mathcal{Z}_B = \mathrm{Tr} e^{-H_B/T}$ in the coherent-state path-integral representation derived in Appendix A:

$$
\begin{aligned}
\mathcal{Z}_B &= \int \mathcal{D}b_i(\tau)\mathcal{D}b_i^\dagger(\tau) \exp\left(-\int_0^{1/T} d\tau \mathcal{L}_b\right), \\
\mathcal{L}_b &= \sum_i \left(b_i^\dagger \frac{db_i}{d\tau} - \mu b_i^\dagger b_i + (U/2) b_i^\dagger b_i^\dagger b_i b_i\right) - w \sum_{\langle ij \rangle} (b_i^\dagger b_j + b_j^\dagger b_i).
\end{aligned} \tag{8.18}
$$

Here, we have changed the notation $\psi(\tau) \to b(\tau)$, as is conventional; we are dealing exclusively with path integrals from now on, and so there is no possibility of confusion with the operators $\hat{b}$ in the Hamiltonian language. Also note that the repulsion proportional to U in (8.4) becomes the product of four boson operators above, after normal ordering, and we can then use (A.4).

It is clear that the critical field theory of the superfluid–insulator transition should be expressed in terms of a spacetime-dependent field $\Psi_B(x,\tau)$, which is analogous to the mean-field parameter Ψ_B appearing in Section 8.2. Such a field is most conveniently introduced by the applying the Hubbard–Stratonovich transformation of Section 6.1 on the coherent-state path integral. We decouple the hopping term proportional to w by introducing an auxiliary field $\Psi_{Bi}(\tau)$ and transforming $\mathcal{Z}_B$ to

$$
\begin{aligned}
\mathcal{Z}_B &= \int \mathcal{D}b_i(\tau)\mathcal{D}b_i^\dagger(\tau)\mathcal{D}\Psi_{Bi}(\tau)\mathcal{D}\Psi_{Bi}^\dagger(\tau) \exp\left(-\int_0^{1/T} d\tau \mathcal{L}_b'\right), \\
\mathcal{L}_b' &= \sum_i \left(b_i^\dagger \frac{db_i}{d\tau} - \mu b_i^\dagger b_i + (U/2) b_i^\dagger b_i^\dagger b_i b_i - \Psi_{Bi} b_i^\dagger - \Psi_{Bi}^* b_i\right) \\
&\quad + \sum_{i,j} \Psi_{Bi}^* w_{ij}^{-1} \Psi_{Bj}.
\end{aligned} \tag{8.19}
$$

We have introduced the symmetric matrix w_{ij} whose elements equal w if i and j are nearest neighbors, and vanish otherwise. The equivalence between (8.19) and (8.18) can be easily established, as in Section 6.1, by simply carrying out the Gaussian integral over Ψ_B; this also generates some overall normalization factors, but these have been absorbed into a definition of the measure $\mathcal{D}\Psi_B$. Let us also note a subtlety we have glossed over. Strictly speaking, the transformation between (8.19) and (8.18) requires that all the eigenvalues of w_{ij} be positive, for only then are the Gaussian integrals over Ψ_B well defined. This is not the case for, say, the hypercubic lattice, which has negative eigenvalues for w_{ij}. This can be repaired by adding a positive constant to all the diagonal elements of w_{ij} and subtracting the same constant from the on-site b part of the Hamiltonian. We will not explicitly do this here as our interest is only in the long-wavelength modes of the Ψ_B field, and the corresponding eigenvalues of w_{ij} are positive.

For our future purposes, it is useful to describe an important symmetry property of (8.19). Notice that the functional integrand is invariant under the following time-dependent $U(1)$ gauge transformation:

$$\begin{aligned} b_i &\to b_i e^{i\phi(\tau)}, \\ \Psi_{Bi} &\to \Psi_{Bi} e^{i\phi(\tau)}, \\ \mu &\to \mu + i\frac{\partial \phi}{\partial \tau}. \end{aligned} \tag{8.20}$$

The chemical potential μ becomes time-dependent above, and so this transformation takes one out of the physical parameter regime; nevertheless (8.20) is very useful, as it places important restrictions on subsequent manipulations of $\mathcal{Z}_B$.

The next step is to integrate out the $b_i, b_i^\dagger$ fields from (8.19). This can be done exactly in powers of Ψ_B and Ψ_B^*: The coefficients are simply products of Green's functions of the b_i. The latter can be determined in closed form because the Ψ_B-independent part of $\mathcal{L}_b'$ is simply a sum of single-site Hamiltonians for the b_i: these were exactly diagonalized in (8.10), and all single-site Green's functions can also be easily determined. We re-exponentiate the resulting series in powers of Ψ_B, Ψ_B^* and expand the terms in spatial and temporal gradients of Ψ_B. The expression for $\mathcal{Z}_B$ can now be written as [83]

$$\mathcal{Z}_B = \int \mathcal{D}\Psi_B(x,\tau)\mathcal{D}\Psi_B^*(x,\tau) \exp\left(-\frac{V\mathcal{F}_0}{T} - \int_0^{1/T} d\tau \int d^d x \mathcal{L}_B \right), \tag{8.21}$$

$$\mathcal{L}_B = K_1 \Psi_B^* \frac{\partial \Psi_B}{\partial \tau} + K_2 \left| \frac{\partial \Psi_B}{\partial \tau} \right|^2 + K_3 |\nabla \Psi_B|^2 + \tilde{r}|\Psi_B|^2 + \frac{u}{2}|\Psi_B|^4 + \cdots .$$

Here $V = N_s a^d$ is the total volume of the lattice, and a^d is the volume per site. The quantity $\mathcal{F}_0$ is the free-energy density of a system of decoupled sites; its derivative with respect to the chemical potential gives the density of the Mott-insulating state, and so

$$-\frac{\partial \mathcal{F}_0}{\partial \mu} = \frac{n_0(\mu/U)}{a^d}. \tag{8.22}$$

The other parameters in (8.21) can also be expressed in terms of μ, U, and w but we will not display explicit expressions for all of them. Most important is the parameter $\tilde{r}$, which can be seen to be

$$\tilde{r} a^d = \frac{1}{Zw} - \chi_0(\mu/U), \tag{8.23}$$

where χ_0 was defined in (8.16). Notice that $\tilde{r}$ is proportional to the mean-field r in (8.15); in particular, $\tilde{r}$ vanishes when r vanishes, and the two quantities have the same sign. The mean-field critical point between the Mott insulator and the superfluid appeared at $r = 0$, and it is not surprising that the mean-field critical point of the continuum theory (8.21) is given by the same condition.

Of the other couplings in (8.21), K_1, the coefficient of the first-order time derivative also plays a crucial role. It can be computed explicitly, but it is simpler to note that the value of K_1 can be fixed by demanding that (8.21) be invariant under (8.20) for small ϕ; a simple calculation shows that we must have

$$K_1 = -\frac{\partial \tilde{r}}{\partial \mu}. \tag{8.24}$$

We pause to note the similarity of the argument requiring invariance under (8.20), to that used for establishing the Luttinger relation of Fermi liquid theory in Section 30.2 around (30.28). Both involve constraints that arise when we have a background density of the conserved particle number, and this type of "anomaly" will continue to play an important role in Parts II and IV.

In the present case, the relationship (8.24) has a very interesting consequence. Notice that K_1 vanishes when $\tilde{r}$ is μ-independent; however, this is precisely the condition that the Mott insulator–superfluid phase boundary in Fig. 8.1 has a vertical tangent, that is, at the tips of the Mott-insulating lobes, such as the where the contour A meets the insulator. This is significant because, at the value $K_1 = 0$, (8.21) is a *relativistic* field theory, for a complex scalar field Ψ_B. So the Mott insulator to superfluid transition is in the universality class of a relativistic scalar-field theory for $K_1 = 0$.

In contrast, for $K_1 > 0$, we have a rather different field theory with a first-order time derivative: in this case we can drop the K_2 term as it involves two time derivatives and so is irrelevant with respect to the single time derivative in the K_1 term; then the field theory in (8.21) is identical to the theory (3.27) for the dilute Bose case in Chapter 3. However, in the present situation, the bosons are not necessarily dilute; instead, we have shown that the excess density of bosons over the density of the Mott insulator can be effectively treated as a dilute gas. Similarly, for $K_1 < 0$, we have an essentially identical theory of bosonic "holes" and a depletion of density from the Mott insulator.

To conclude this section, I would like to correlate the above discussion on the distinction between the two universality classes with the behavior of the boson density across the transition. This can be evaluated by taking the derivative of the total free energy with respect to the chemical potential, as is clear from (8.4):

$$\begin{aligned}\langle \hat{b}_i^\dagger \hat{b}_i \rangle &= -a^d \frac{\partial \mathcal{F}_0}{\partial \mu} - a^d \frac{\partial \mathcal{F}_B}{\partial \mu} \\ &= n_0(\mu/U) - a^d \frac{\partial \mathcal{F}_B}{\partial \mu},\end{aligned} \tag{8.25}$$

where $\mathcal{F}_B$ is the free energy resulting from the functional integral over Ψ_B in (8.21).

In mean-field theory, for $\tilde{r} > 0$, we have $\Psi_B = 0$, and therefore $\mathcal{F}_B = 0$, implying

$$\langle \hat{b}_i^\dagger \hat{b}_i \rangle = n_0(\mu/U), \text{for } \tilde{r} > 0. \tag{8.26}$$

This clearly places us in a Mott insulator. As argued in Section 8.2, Eqn. (8.26) is an exact result.

For $\tilde{r} < 0$, we have $\Psi_B = (-\tilde{r}/u)^{1/2}$, as follows from a simple minimization of $\mathcal{L}_B$; computing the resulting free energy we have

$$\begin{aligned}\langle \hat{b}_i^\dagger \hat{b}_i \rangle &= n_0(\mu/U) + a^d \frac{\partial}{\partial \mu}\left(\frac{\tilde{r}^2}{2u}\right) \\ &\approx n_0(\mu/U) + \frac{a^d \tilde{r}}{u}\frac{\partial \tilde{r}}{\partial \mu}.\end{aligned} \tag{8.27}$$

In the second expression, we ignored the derivative of u as it is less singular as $\tilde{r}$ approaches 0; I will comment on the consequences of this shortly. Thus, at the transition point at which $K_1 = 0$, by (8.24) we see that the leading correction to the density of the superfluid phase vanishes, and it remains pinned at the same value as in the Mott insulator. Conversely, for the case $K_1 \neq 0$, the transition is always accompanied by a density change and the excess density can be effectively treated in the dilute Bose gas theory of (3.27).

Let me close by commenting on the consequences of the omitted higher-order terms in (8.27) to the discussion above. Consider the trajectory of points in the superfluid with their density equal to some integer n. The implication of the above discussion is that this trajectory will meet the Mott insulator with $n_0(\mu/U) = n$ at its lobe. The relativistic phase transition then describes the transition out of the Mott insulator into the superfluid along a direction tangential to the trajectory of density n. The approximations made above merely amounted to assuming that this trajectory was a straight line.

For a deeper understanding of the nature of the fixed-density superfluid–insulator transition at the tips of the Mott lobe, where $K_1 = 0$, we need to analyze the relativistic quantum field theory of an $M = 2$ component scalar with $O(M)$ symmetry. The basic features of this relativistic field theory are described in Chapters 10 and 11, and they are applied to physical properties of the superfluid–insulator transition in Section 11.2. We will find a remarkable new feature at the superfluid–insulator quantum critical point in $d = 2$: this is a many-body quantum state *without* quasiparticle excitations. There is much additional discussion on such non-quasiparticle states in Chapters 32 and 34.

8.4 Insulators at Non-Integer Filling

A key feature of the phase diagram in Fig. 8.1 is that all the integers have exactly an integer number of bosons per site. Here, we argue that, once we move away from the purely on-site interactions, insulators are also possible at rational densities, $\langle \hat{n}_b \rangle = p/q$, with p and q rational. The simplest ones are those that break translational symmetry so that the boson density per unit cell is again an integer. Insulators at non-integer rational densities that do not break translational symmetry are far more subtle, and will be discussed extensively in Parts II and IV.

Here we limit ourself to the important case of density per site $\langle \hat{n}_b \rangle = 1/2$. In Fig. 8.1 notice that at $\mu = 0$ there is a superfluid state at any non-zero w at density 1/2 along the contour C. It is not difficult to show that at very small w such a superfluid is unstable to the formation of an insulator in the presence of longer-range interactions. With an eye to future applications, let us consider "hard-core" bosons $\hat{B}_i$ for which the on-site repulsion U is so large that we may ignore all states with more than a single boson on each site

$$\hat{B}_i^\dagger \hat{B}_i = 0, 1 \,. \tag{8.28}$$

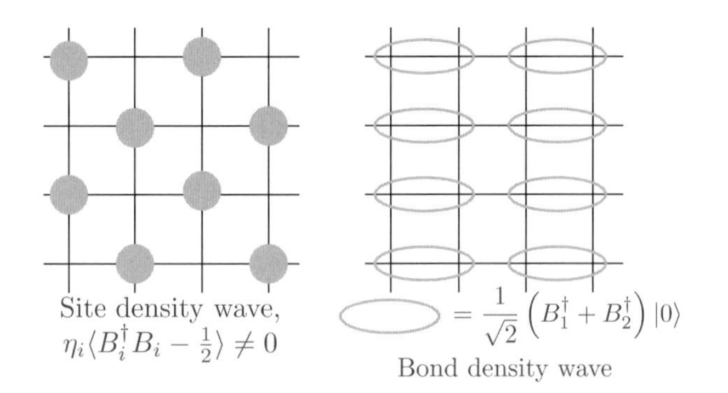

Figure 8.2 Insulators of hard-core bosons at density 1/2. In the site density wave, the site density oscillates in space with $\eta_i = \pm 1$ on the two sublattices. In the bond density wave, the site density is uniform, but translational symmetry is broken by the preferential occupation of certain bonding orbitals.

Let us include a nearest-neighbor repulsion V between such bosons in the Hamiltonian

$$H_{hc} = -w \sum_{\langle ij \rangle} (\hat{B}_i^\dagger \hat{B}_j + \hat{B}_j^\dagger \hat{B}_i) - \mu \sum_i \hat{B}_i^\dagger \hat{B}_i + V \sum_{\langle ij \rangle} \hat{B}_i^\dagger \hat{B}_i \hat{B}_j^\dagger \hat{B}_j . \tag{8.29}$$

Then, for $w \to 0$, $\mu > 0$, and V large, we should maximize the density of bosons without paying the energy cost V of nearest-neighbor repulsion. This is achieved in the site (or "charge") density wave shown in Fig. 8.2. Notice that this state has density of exactly 1/2, and a two-fold degeneracy; there is another state in which the empty and occupied sites are interchanged. Translational symmetry has been broken, the unit-cell size has been doubled, and so the net density per unit cell is 1. Most importantly, this insulator has a gap of order V to all excitations. This means that it is stable to the onset of superfluidity upon turning on a small w.

For more complicated off-site interactions, it is also possible to form the bond density wave state at density 1/2, as shown in Fig. 8.2. This state also doubles the unit cell, and has unit density per unit cell. We meet such states in our study of antiferromagnets in Part IV, where they are usually known as a valence-bond solid (VBS).

Problems

8.1 We investigate the superfluid–insulator transition in the boson Hubbard model with a staggered potential:

$$H_B = \sum_i (-\mu + \varepsilon_0 \eta_i) b_i^\dagger b_i - w \sum_{\langle ij \rangle} \left(b_i^\dagger b_j + b_j^\dagger b_i \right), + \frac{U}{2} \sum_i n_i (n_i - 1), \tag{8.30}$$

where $n_i = b_i^\dagger b_i$ and $\eta_i = +1$ ($\eta_i = -1$) on sublattice A (sublattice B) of the square lattice. Note that $\varepsilon_0 > 0$ breaks the sublattice symmetry, and there are two sites per unit cell. Assume $U > 2\varepsilon_0$. The analysis below will parallel that carried out in this chapter for $\varepsilon_0 = 0$.

(a) First consider the ground state at $w = 0$ as a function of μ. The crucial point is that this ground state is non-degenerate, except at certain isolated values of μ. Find the range of values of μ for which the average density per site is $1/2$, that is, there is one boson per unit cell, and the filling is $\nu = 1$. We will assume μ lies within this range from now on.

(b) Compute the exact on-site boson correlators $\chi_{AA}(\tau) = \left\langle T\left(b_A(\tau)b_A^\dagger(0)\right)\right\rangle$ and $\chi_{BB}(\tau) = \left\langle T\left(b_B(\tau)b_B^\dagger(0)\right)\right\rangle$ at $w = 0$. Compute these using the spectral decomposition, and your knowledge of the exact eigenstates at $w = 0$. After transforming to imaginary frequencies, these susceptibilities should agree at $\omega = 0$ with (8.16) with $\mu \to \mu \pm \varepsilon_0$.

(c) Expand these susceptibilities at small ω, and identify the co-efficients as below

$$
\begin{aligned}
\chi_{AA}(\omega) &= -r_A + iK_{1A}\omega - K_{2A}\omega^2 + \cdots \\
\chi_{BB}(\omega) &= -r_B + iK_{1B}\omega - K_{2B}\omega^2 + \cdots .
\end{aligned} \tag{8.31}
$$

Verify the analog of (8.24):

$$
K_{1A} = -\frac{\partial r_A}{\partial \mu} \quad , \quad K_{1B} = -\frac{\partial r_B}{\partial \mu}. \tag{8.32}
$$

(d) Now consider $w \neq 0$, and proceed as below (8.18). This yields a path integral over a complex field $\psi_i(\tau)$ where i extends over both the A and B sublattices. The quadratic terms in the effective action are

$$
S = \int \frac{d\omega}{2\pi}\left[\sum_{i,j}\psi_i^*(\omega)w_{ij}^{-1}\psi_j(\omega) - \sum_{i\in A}\chi_{AA}(\omega)|\psi_i(\omega)|^2 - \sum_{i\in B}\chi_{BB}(\omega)|\psi_i(\omega)|^2\right]. \tag{8.33}
$$

(e) Now write (8.33) in momentum space; the quadratic form becomes a 2×2 matrix. Argue that when the lower eigenvalue of this matrix at $k = 0$ and $\omega = 0$ crosses zero we reach the superfluid phase. In this manner, determine that the phase boundary between the $\nu = 1$ insulator and the superfluid in the w,μ plane is given by

$$
\frac{1}{Z^2 w^2} = r_A r_B. \tag{8.34}
$$

(f) Near the critical point, focus on the effective action for the lower eigenmode only; for this you will need the form of the lower eigenvalue as a function of k and ω. In this manner, argue that the phase transition at the tip of the $\nu = 1$ Mott lobe is given by the Wilson–Fisher conformal field theory.

8.2 Supersolids. Consider a double-layer boson Hubbard model of bosons $\hat{b}_{1i}$ and $\hat{b}_{2i}$ on two parallel layers 1,2:

$$H_{2B} = -w \sum_{\langle ij \rangle} (\hat{b}_{1i}^\dagger \hat{b}_{1j} + \hat{b}_{1j}^\dagger \hat{b}_{1i} + \hat{b}_{2i}^\dagger \hat{b}_{2j} + \hat{b}_{2j}^\dagger \hat{b}_{2i}) - w \sum_i (\hat{b}_{1i}^\dagger \hat{b}_{2i} + \hat{b}_{2i}^\dagger \hat{b}_{1i})$$
$$+ \sum_i \left(-\mu \left[\hat{n}_{b1i} + \hat{n}_{b2i} \right] + \frac{U}{2} \left[\hat{n}_{b1i}(\hat{n}_{b1i} - 1) + \hat{n}_{b2i}(\hat{n}_{b2i} - 1) \right] \right)$$
$$+ V \sum_i \hat{n}_{b1i} \hat{n}_{b2i} + W \sum_{\langle ij \rangle} \left[\hat{n}_{b1i} \hat{n}_{b1j} + \hat{n}_{b2i} \hat{n}_{b2j} \right]. \tag{8.35}$$

Thus, bosons on the same layer have an on-site repulsion $U > 0$, bosons on opposite layers have a repulsion $V > 0$. Bosons on the same layer also have a nearest-neighbor interaction W, and we will allow W to have either sign. Consider the case where the average boson density per site and per layer is exactly $1/2$, and we take the limit $U \to \infty$: thus, no site can have more than one boson. Use a variational approach to determine the ground state of H_{2B} as a function of V/w and W/w. The proposed mean-field variational wavefunction is

$$|G\rangle = \prod_i \left(\alpha_1 + \alpha_2 \hat{b}_{1i}^\dagger + \alpha_3 \hat{b}_{2i}^\dagger + \alpha_4 \hat{b}_{1i}^\dagger \hat{b}_{2i}^\dagger \right) |0\rangle, \tag{8.36}$$

where $|0\rangle$ is the empty state, and α_1, α_2, α_3, and α_4 are variational parameters. Normalization of the wavefunction implies that

$$|\alpha_1|^2 + |\alpha_2|^2 + |\alpha_3|^2 + |\alpha_4|^2 = 1. \tag{8.37}$$

(a) Show that the average density of 1/2 implies

$$|\alpha_2|^2 + |\alpha_3|^2 + 2|\alpha_4|^2 = 1. \tag{8.38}$$

(b) Compute $\langle G|H_{2B}|G\rangle$ for a lattice with the same-layer coordination number Z. Then minimize this as a function of the $\alpha_{1,2,3,4}$ subject to the constraints (8.37) and (8.38). The results yield a phase diagram, and the phases can be identified as discussed below.
(c) Argue that any phase with $\alpha_1 \neq 0$ must be a superfluid.
(d) Similarly, show that any phase with $\alpha_1 = \alpha_4 = 0$ is an insulator.
(e) The model H_{2B} has a layer interchange symmetry, and our mean field allows this symmetry to be spontaneously broken. Show that this symmetry is broken in phases in which $|\alpha_2| \neq |\alpha_3|$. As such phases break a lattice symmetry, it is natural to refer to them as "solids."
(f) Are there any regimes which are both solids and superfluids? Such a phase would be a supersolid.
(g) Determine the order (first or second) of all quantum phase transitions in the phase diagram.

9 Electron Hubbard Model

The phase diagram of the electron Hubbard model is described. The superexchange interaction of the Mott insulator leads to antiferromagnetic order on the square lattice. The doped Mott insulator is described by a t–J model, and a Bardeen–Cooper–Schrieffer mean-field analysis of the t–J model is shown to lead to d-wave superconductivity. The magnetism of the metallic state is described by a paramagnon theory, and the exchange of paramagnons is also shown to lead to d-wave superconductivity. The metallic antiferromagnet is shown to exhibit Fermi surface reconstruction.

This chapter begins a discussion of the Hubbard model of Chapter 8 applied to spin-1/2 fermions instead of bosons, and the further developments are the focus of much of this book. Both the Fermi statistics and the spin of the lattice particles makes a crucial difference to the physics, and the problem is considerably more complicated than the boson case of Chapter 8. The present chapter contains a description of phases that can be obtained via mean-field descriptions, which are ultimately similar to those employed in Chapter 8: the Fermi statistics and spin already make a significant difference to the phases so obtained. However, the full phase diagrams of Hubbard and related models are far richer than those encountered in the present chapter: Parts II and IV introduce new ideas on fractionalization and emergent gauge fields, which lead to novel phases that are not discussed here.

Our attention will mostly focus on the square lattice, and we will consider the Hamiltonian

$$H_U = -t \sum_{\langle ij \rangle} \left[c_{i\alpha}^\dagger c_{j\alpha} + c_{j\alpha}^\dagger c_{i\alpha} \right] + \sum_i \left[-\mu (n_{i\uparrow} + n_{i\downarrow}) + U n_{i\uparrow} n_{i\downarrow} \right], \tag{9.1}$$

where $c_{i\alpha}$ annihilates a fermion on site i with spin $\alpha, \beta = \uparrow, \downarrow$, and the number operators are

$$n_{i\uparrow} = c_{i\uparrow}^\dagger c_{i\uparrow} \quad , \quad n_{i\downarrow} = c_{i\downarrow}^\dagger c_{i\downarrow}. \tag{9.2}$$

The fermions hop with amplitude t between sites, and have a local repulsion $U > 0$. A more realistic model will also have long-range Coulomb interactions between the fermions, but we assume this is screened, and focus only on the new physics introduced by the on-site interaction U.

A very instructive starting point is the analog of the Mott-insulating phases of bosons in Fig. 8.1, which is shown in Fig. 9.1. As in the boson case, we expect such phases to be present for $U \gg t$, when there an integer number of fermions on each site.

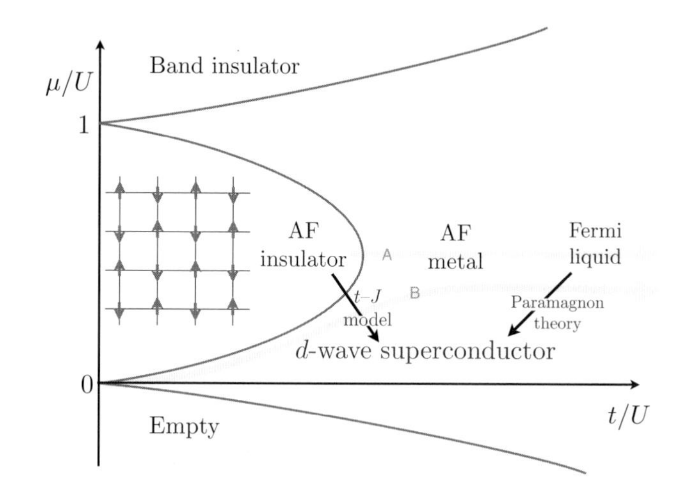

Figure 9.1 Schematic phase diagram of the mean-field states of the electron Hubbard model in (9.1) discussed in this chapter. The evolution of the Fermi surface from the antiferromagnetic (AF) insulator to the Fermi liquid is shown in Fig. 9.7 along contour A with a fixed density of one electron per site. If the electron density is different from unity, then the antiferromagnetic metal is present even for small t/U, and the Fermi surface evolves along contour B as in Fig. 9.8. Most of the metallic region is expected to be unstable to d-wave superconductivity (at least at small U), and we present two complementary approaches to the superconducting instabilities in Sections 9.3 and 9.4.3.

Because of the exclusion principle, only Mott insulators with $n_0 = 0, 1, 2$ are permitted here. It is also easy to see that the Mott insulators with $n_0 = 0, 2$ are trivial: all the states in the Hilbert space of the fermions are either empty or occupied, and the empty/filled state is an exact ground state. Furthermore, the filled state is also a band insulator, as there is only one state at this density

$$\prod_i c^\dagger_{i\uparrow} c^\dagger_{i\downarrow} |0\rangle = \prod_k c^\dagger_{k\uparrow} c^\dagger_{k\downarrow} |0\rangle \,. \tag{9.3}$$

So there is only a single Mott insulator, with exactly one fermion per site. However, its physics is far more subtle than the corresponding boson Mott insulator. In stark contrast to the boson case, this Mott insulator has an exponentially large degeneracy in the limit $U \to \infty$. Each site can be occupied with a spin-up or -down fermion, and so the total number of ground states at $U = \infty$ is 2^N, where N is the number of lattice sites. Section 9.1 shows how this degeneracy is lifted by a superexchange interaction that appears at the first order in the t/U expansion: this leads to the antiferromagnetic insulator state, sketched in Fig. 9.1.

At the right end of Fig. 9.1, where U is small, we obtain the Fermi liquid state of Chapter 2. With increasing U, there is an onset of antiferromagnetic order already in the metal, in the antiferromagnetic metal phase, before there is a transition to the antiferromagnetic insulator at integer filling: this evolution is described using the paramagnon theory in Section 9.4.

Much of the metallic region in Fig. 9.1 is expected to have an instability to electron pairing, and the onset of d-wave superconductivity at low temperatures. This instability is described using the large U approach of the t–J model in Section 9.3, and the small-U approach of the paramagnon theory in Section 9.4.3. Qualitatively, the same d-wave state is obtained from these two approaches.

9.1 The Superexchange Interaction

The exponential degeneracy of the Mott insulator at integer density must be lifted at a large but finite U, and we investigate this here in a perturbation theory in t/U. The expansion in $1/U$ must involve degenerate perturbation theory, and for infinite systems this is best investigated by an effective Hamiltonian method: we explicitly describe this method later in Problem 9.1. The steps in this method are: (i) separate the Hilbert space into the subspace of interest, and "other" states that have a zeroth-order energy well separated from this subspace; and (ii) perform a canonical transformation (often called the Schrieffer–Wolff transformation) to eliminate the matrix elements between the subspace and the other states. For the Hubbard model, this procedure can be implemented order-by-order in t/U, as described in Ref. [168].

We are interested here only in the leading correction to the Hamiltonian at order t^2/U. In this case, we can sidestep the somewhat cumbersome canonical transformation procedure, and obtain the needed answer by a shortcut. It is not difficult to see from the mechanics of the canonical transformation that only pair-wise interactions between nearest-neighbor sites are generated at order t^2/U, and each pair of sites can be treated independently of all other pairs for the purposes of obtaining the effective Hamiltonian. Consequently, we can obtain the needed Hamiltonian by exactly diagonalizing H_U for a pair of sites, and mapping the spectrum onto the states of an effective Hamiltonian.

So we consider the spectrum of H_U for two sites $i = 1, 2$ for a total of two electrons. This model has a total of six states, listed below with their energies E_0 at $t = 0$ (because there is no fluctuation in the total number of electrons, we can safely drop the μ term):

$$
\begin{aligned}
|1\rangle &= c_{1\uparrow}^{\dagger} c_{2\uparrow}^{\dagger} |0\rangle \quad &;& \quad E_0 = 0, \\
|2\rangle &= c_{1\uparrow}^{\dagger} c_{2\downarrow}^{\dagger} |0\rangle \quad &;& \quad E_0 = 0, \\
|3\rangle &= c_{1\downarrow}^{\dagger} c_{2\uparrow}^{\dagger} |0\rangle \quad &;& \quad E_0 = 0, \\
|4\rangle &= c_{1\downarrow}^{\dagger} c_{2\downarrow}^{\dagger} |0\rangle \quad &;& \quad E_0 = 0, \\
|5\rangle &= c_{1\uparrow}^{\dagger} c_{1\downarrow}^{\dagger} |0\rangle \quad &;& \quad E_0 = U, \\
|6\rangle &= c_{2\uparrow}^{\dagger} c_{2\downarrow}^{\dagger} |0\rangle \quad &;& \quad E_0 = U.
\end{aligned} \tag{9.4}
$$

Clearly, the subspace of interest contains the states $|1,2,3,4\rangle$, and we have to eliminate their coupling to the states $|5,6\rangle$. The hopping term, H_t, in H_U has the non-zero matrix elements

$$
\begin{aligned}
\langle 5| H_t |2\rangle &= \langle 6| H_t |2\rangle = -t, \\
\langle 5| H_t |3\rangle &= \langle 6| H_t |3\rangle = t.
\end{aligned} \tag{9.5}
$$

It is now apparent that $|1\rangle$ and $|4\rangle$ are already exact eigenstates of the two-site Hamiltonian with total energy $E = 0$, and we only need to diagonalize the remaining 4×4 Hamiltonian.

We can reduce the work further, by considering eigenstates of parity and total spin. The following state has odd parity under site exchange, total spin $S = 1$, and no non-zero matrix elements associated with H_t

$$\frac{1}{\sqrt{2}}(|2\rangle + |3\rangle) \quad ; \quad E = 0. \tag{9.6}$$

Clearly, this state combines with $|1\rangle$ and $|4\rangle$ to form a spin triplet with energy $E = 0$.
Similarly the state

$$\frac{1}{\sqrt{2}}(|5\rangle - |6\rangle) \quad ; \quad E = U, \tag{9.7}$$

with total spin $S = 0$, odd parity under site exchange, decouples from all other states. This state is not part of the low-energy subspace.

Finally, we have to consider the remaining two states, which have total spin 0 and even parity under site exchange

$$\begin{aligned} |a\rangle &= \frac{1}{\sqrt{2}}(|2\rangle - |3\rangle), \\ |b\rangle &= \frac{1}{\sqrt{2}}(|5\rangle + |6\rangle). \end{aligned} \tag{9.8}$$

The Hamiltonian on these two states reduces to

$$H_{ab} = \begin{pmatrix} 0 & -2t \\ -2t & U \end{pmatrix} \tag{9.9}$$

and so we have the remaining two eigenenergies

$$E = U/2 \pm \sqrt{4t^2 + (U/2)^2}. \tag{9.10}$$

So we have found four low-energy states, a triplet with energy $E = 0$, and a singlet with energy $E = U/2 - \sqrt{4t^2 + (U/2)^2}$. The other two states have energy $\sim U$ in the limit of large U, and we will not consider them further. The four low-energy states reduce to linear combinations of the states $|1,2,3,4\rangle$ in (9.4) as $t/U \to 0$. The total spin of the energy eigenstates shows us that to order t^2/U, the same eigenstates and eigenenergies are obtained from the effective Hamiltonian

$$H_{eff} = -\frac{J}{4} + J\boldsymbol{S}_1 \cdot \boldsymbol{S}_2, \tag{9.11}$$

where

$$\boldsymbol{S}_i = \frac{1}{2} c_{i\alpha}^{\dagger} \boldsymbol{\sigma}_{\alpha\beta} c_{i\beta} \tag{9.12}$$

is the spin operator on site $i = 1, 2$, $\boldsymbol{\sigma}$ are the Pauli matrices, and the exchange constant

$$J = -\left[U/2 - \sqrt{4t^2 + (U/2)^2}\right] \approx \frac{4t^2}{U}. \tag{9.13}$$

So we have reduced the two-site, two-electron Hubbard model to a *spin* model, with $S = 1/2$ spin on each site, coupled to each other with an *antiferromagnetic* exchange interaction $J > 0$. The antiferromagnetism refers to the fact that classically the spins prefer opposite orientations, and quantum mechanically they form a spin singlet in the ground state. It is now a simple step to write down the effective Hamiltonian for the entire lattice Hubbard model, within the Mott insulator with one electron per site:

$$H_J = J \sum_{\langle ij \rangle} \boldsymbol{S}_i \cdot \boldsymbol{S}_j. \tag{9.14}$$

Unravelling the structure of the eigenstates of H_J for spin $S = 1/2$ occupy the next two sections, and a significant portion of Parts II and IV. However, if we consider the model in (9.14) for general S on a bipartite lattice, the ground state for large S is smoothly connected to the classical ground state: this is the Néel state characterized by a non-zero expectation value of the Néel order parameter

$$\mathcal{N} = \eta_i \boldsymbol{S}_i, \tag{9.15}$$

where $\eta_i = \pm 1$ on the two sublattices.

9.2 Insulating Antiferromagnets and Hard-Core Bosons

A first analysis of the phases of the $S = 1/2$ antiferromagnetic Hamiltonian follows from a mapping between the spin-1/2 states, and a hard-core lattice boson B_i. We make a correspondence between the up and down and the boson state as follows (dropping the site index)

$$\begin{aligned} |\downarrow\rangle \quad &\Leftrightarrow \quad |0\rangle\,, \\ |\uparrow\rangle \quad &\Leftrightarrow \quad B^\dagger\,|0\rangle\,. \end{aligned} \tag{9.16}$$

As there are no additional states, the bosons must satisfy the hard-core constraint on every site

$$B_i^\dagger B_i \leq 1\,. \tag{9.17}$$

This mapping between states also implies a mapping between operators:

$$\begin{aligned} S_{i+} = S_{ix} + iS_{iy} \quad &\Leftrightarrow \quad \eta_i B_i^\dagger, \\ S_{i-} = S_{ix} - iS_{iy} \quad &\Leftrightarrow \quad \eta_i B_i, \\ S_{iz} \quad &\Leftrightarrow \quad B_i^\dagger B_i - 1/2. \end{aligned} \tag{9.18}$$

For our convenience below, we have a chosen a phase factor $\eta_i = +1$ on the A square sublattice, and $\eta_i = -1$ on the B square sublattice ($\eta_i = e^{i\boldsymbol{k}\cdot\boldsymbol{r}_i}$ with $\boldsymbol{k} = (\pi,\pi)$). In phases in which the total spin S_z vanishes, the average B boson density is $1/2$.

The hard-core boson representation is clearly tied to states that are eigenstates of S_z, and so does not naturally preserve the full $SU(2)$ rotation symmetry of the H_J. Consequently, in the context of the boson model, it is conventional to consider a more general Hamiltonian with only a $U(1)$ spin rotation symmetry, often referred to as the XXZ Hamiltonian:

$$H_{\rm XXZ} = \sum_{\langle ij \rangle} \left[\frac{J_X}{2} \left(S_{i+} S_{j-} + S_{j+} S_{i-} \right) + J_Z S_{iz} S_{jz} \right]. \tag{9.19}$$

For $J_Z = J_X = J$, (9.19) reduces to (9.14). Inserting the boson representation in (9.18) into (9.19), we obtain the Hamiltoinian for the hard-core bosons

$$H_{XXZ} = \sum_{\langle ij \rangle} \left[-\frac{J_X}{2} \left(B_i^\dagger B_j + B_j^\dagger B_i \right) + J_Z \left(B_i^\dagger B_i B_j^\dagger B_j - \frac{B_i^\dagger B_i}{2} - \frac{B_j^\dagger B_j}{2} \right) \right], \tag{9.20}$$

where we have dropped an additive constant. So we have obtained a model of bosons with a nearest-neighbor hopping matrix element J_X, and a nearest-neighbor repulsion J_Z. The η_i in (9.18) were chosen so that the sign of the hopping prefers that single boson dispersion has a minimum at momentum $\boldsymbol{k} = 0$. This Hamiltonian is of the form considered earlier in (8.29).

9.2.1 Possible ground states

One advantage of the boson representation is that we can use our previous study of boson systems in Section 8.4, and some intuition, to guess possible states of the spin system. This correspondence between boson states and spin states on the square lattice is shown in Fig. 9.2. From our discussion of the Bose gas in Chapter 3, we can expect that the bosons will condense into a superfluid state with $\langle B \rangle \neq 0$, despite the hard-core interactions. Numerical studies have shown that the ground state is indeed a superfluid for $0 \leq J_Z \leq J_X$. In terms of the spin degrees of freedom, this superfluid corresponds to Néel order in the XY plane, as shown in Fig. 9.2. This follows from the operator correspondence in (9.18), where a uniform condensate in B_i translates into a "staggered" expectation value for the spin operator $\sim \eta_i$. The phase θ_0 of the condensate in (3.28) determines the orientation of the spins within the XY plane: a real condensate with $\theta_0 = 0$ leads to spins in the X direction.

The second panel of Fig. 9.2 shows another state that appears for $J_Z \geq J_X$. In this case, there is a strong nearest-neighbor repulsion between the bosons in (9.20), and so the bosons will try to not occupy nearest-neighbor sites. As the average boson density is $1/2$, we have the possibility of a site (or "charge") density wave state in which one sublattice is preferentially occupied. For $J_Z > J_X$, this state is a Mott insulator of bosons, similar to those found in Chapter 8, but with an important difference: translational symmetry is broken by the the checkerboard pattern, so that we obtain the required integer number of bosons in the expanded unit cell, as required for a Mott insulator. The broken symmetry also implies that the ground state is degenerate: there are two equivalent density wave states, depending upon which sublattice is preferentially occupied. In the spin language, this state has Néel order along the Z direction, as shown in Fig. 9.2. At $J_Z = J_X$, the underlying Hamiltonian has a $SU(2)$ rotation symmetry, and so the properties of the Néel state oriented along the Z direction should be the same as those for the Néel state along the X direction, which is not a Mott insulator of bosons; this will become clearer when we consider the excitation spectra in Section 9.2.2.

A third possible state is shown in the bottom panel of Fig. 9.2. This a *bond* density wave of the B bosons, or a valence-bond solid (VBS) of the spins. This state is also a Mott insulator of the B bosons, but now the translational symmetry has been

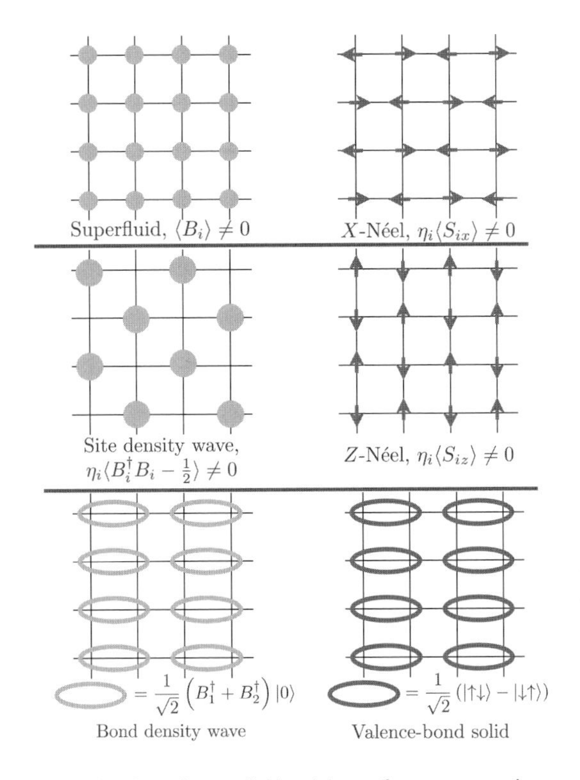

Figure 9.2 Correspondence between states of hard-core bosons (left) and the antiferromagnet on the square lattice. The non-zero expectation values are equated to values independent of i; so the η_i factor leads to the sublattice oscillations in the Néel states. Compare with Fig. 8.2.

broken in a manner that preserves reflection symmetry about the centers of the links of the lattice; the unit cell has doubled so that there are an integer number of bosons per unit cell: in a fully ordered wavefunction, each boson occupies an orbital which is a linear combination of the states on a pair of sites. In the spin language, the spins form singlet valence bonds between nearest-neighbor sites. Unlike the other states in Fig. 9.2, this state preserves full spin rotation symmetry, but does break translational symmetry. The VBS state is not present for the nearest-neighbor Hamiltonian H_{XXZ} we have considered here, but it does appear for models with further neighbor couplings [79, 161, 192, 296]: the VBS state plays an important role in the structure of the gauge theories of fractionalized phases, as we shall see in Sections 16.5.2 and Chapter 26.

An interesting feature of the states I have listed so far is that all of them break either the spin rotation symmetry or a lattice rotation symmetry. We have not found any state that preserves all symmetries of the Hamiltonian H_{XXZ}. In fact, such states are possible, and will be the major focus of Parts II and IV: such states feature fractionalization and emergent gauge fields. We can further ask for states with neither broken symmetry nor fractionalization, that is, trivial states like the Mott insulators at integer filling in Chapter 8. In fact, such trivial states do not exist for H_{XXZ} applied to $S = 1/2$ spins, as proven in Refs. [105, 194].

9.2.2 Excitations of Insulating Magnets

This section will describe the excitation spectra of the states in Fig. 9.2. In the spin language, these excitations are known as "magnons" or "spin waves" for the Néel states, and "triplons" for the VBS state with $SU(2)$ symmetry. All these excitations carry integer spin S_z, or B boson number, even though the underlying spins have $S_z = \pm 1/2$. We will meet excitations with half-integer spin or boson number in Parts II and IV, when we describe states with fractionalization; such excitations are "spinons."

First, let us consider the X-Néel state in Fig. 9.2. This is a superfluid of the B bosons, which has a gapless "phonon" excitation with a linear dispersion at long wavelengths, as in (3.16). For the antiferromagnet, this is a gapless spin wave. The gaplessness is now a reflection of the broken spin rotation symmetry about the z axis in the X-Néel state.

For a systematic derivation of the spin wave spectrum, we need to generalize the boson representations of the spins to general spin S, and then perform a $1/S$ expansion. However, at linear order for $S = 1/2$, we proceed here in the spirit of the dilute gas theory of Chapter 3, in the expectation that the hard-core repulsion is not crucial in the dilute gas limit. The dilute gas limit appears most simply for the Z-ferromagnet, with $J_Z < J_X < 0$, in which case the ground state is simply the B boson vacuum. In the spin language, all spins are polarized with $S_{iz} = -1/2$, as befits a ferromagnet, and it is easy to show that this is an exact eigenstate of H_{XXZ}. To obtain the excitation spectrum, we retain only the quadratic terms in (9.20), and then the Hamiltonian describes bosons

$$H_{XXZ} = \sum_k E_k Bk^\dagger B_k, \tag{9.21}$$

with disperson

$$E_k = 2|J_Z| + |J_X|(\cos(k_x) + \cos(k_y)). \tag{9.22}$$

This dispersion has a minimum at $\boldsymbol{k} = (\pi, \pi)$, but this is an artifact of the η_i factors in (9.18), which are superfluous for a ferromagnet. For $|J_Z| > |J_X|$, these magnon excitations are gapped at all momenta: this reflects the fact that the Z-ferromagnet only breaks the discrete $S_{iz} \to -S_{iz}$ symmetry of H_{XXZ}, and so the gapless excitations of Section 3.3 do not exist here. However, for $J_Z = J_X < 0$, H_{XXZ} has the full $SU(2)$ rotation symmetry of H_J, and then the magnon excitations are gapless: their dispersion $E_{(\pi,\pi)+\boldsymbol{k}} \sim k^2$. So, somewhat surprisingly, we have found *quadratically* dispersing excitations in a ferromagnet with $SU(2)$ spin rotation symmetry, in contrast to the linearly dispersing excitations noted above for the X-antiferromagnet, which is a superfluid of the B bosons. This quadratic dispersion is a consequence of a special feature of the ferromagnet: the ferromagnetic order parameter is also the generator of rotations of $SU(2)$ symmetry [100, 111].

Let us now derive the excitations of the Z-Néel state by the dilute Bose gas approach. By the cases already considered, we expect a gapped magnon spectrum for $J_Z > J_X > 0$, and a linearly dispersing gapless magnon at the $SU(2)$ point $J_Z = J_X > 0$ where it should be the same as the X-Néel state. To obtain a classical ground state that is a boson vacuum, we need to modify the mapping in (9.16) and (9.18). We retain these mappings

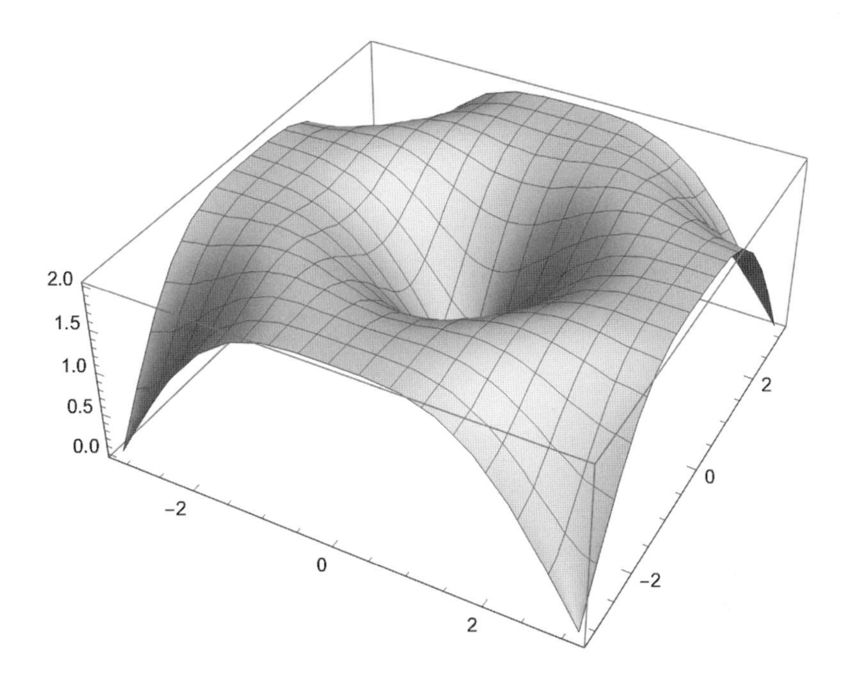

Figure 9.3 Magnon dispersion for the Z-Néel state in (9.25), for $J_Z = J_X = 1$.

for $i \in$ the B sublattice, but for $i \in$ A sublattice, we use instead

$$\begin{aligned} |\downarrow\rangle &\Leftrightarrow A^\dagger |0\rangle, \\ |\uparrow\rangle &\Leftrightarrow |0\rangle, \\ S_{i+} = S_{ix} + iS_{iy} &\Leftrightarrow A_i, \\ S_{i-} = S_{ix} - iS_{iy} &\Leftrightarrow A_i^\dagger, \\ S_{iz} &\Leftrightarrow 1/2 - A_i^\dagger A_i. \end{aligned} \tag{9.23}$$

Then the A and B boson vacuum will have spins $S_{iz} = 1/2$ on the A sublattice, and $S_{iz} = -1/2$ on the B sublattice, exactly as needed for the Z-Néel state. Inserting (9.23) into (9.19), we now obtain

$$H_{XXZ} = \sum_k{}' \left[-J_X \left(\cos(k_x) + \cos(k_y)\right) \left(B_{-k} A_k + A_k^\dagger B_{-k}^\dagger\right) + 2J_Z (A_k^\dagger A_k + B_k^\dagger B_k) \right], \tag{9.24}$$

where the prime indicates the sum is over the reduced Brillouin zone because the unit cell has been doubled. We can diagonalize (9.24) by the Bogoliubov transformation of Section 3.1, and hence obtain the dispersion relation for a pair of degenerate magnons

$$E_k = \left(4J_Z^2 - J_X^2 (\cos(k_x) + \cos(k_y))^2\right)^{1/2}. \tag{9.25}$$

As expected, this has a gap for $J_Z > J_X$, and has a gapless linear dispersion for $J_Z = J_X$. We show a plot of this dispersion in Fig. 9.3 for the gapless case.

Finally, we address the excitation spectrum of the VBS state, which does not break any spin rotation symmetry. The spectrum of this state is always gapped and, for

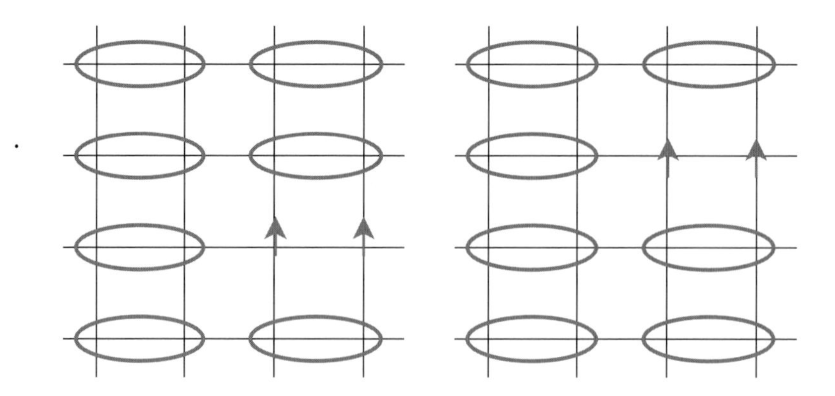

Figure 9.4 A triplon excitation in a VBS state: shown is an $S_z = 1$ excitation on a bond, hopping vertically by a lattice spacing.

$J_Z = J_X$, often referred to as a "triplon": these are the three states of the gapped $S = 1$ excitation plotted in Fig. 9.4. A full derivation of the triplon dispersion spectrum may be found in Ref. [237].

9.3 The t–J Model and d-Wave Pairing

We now move away from the Mott insulator, and consider the case where the density of fermions deviates from exactly one per site. With an eye to the application to the cuprates, we consider doping the Mott insulator with a density of p holes, so the density of fermions is $1 - p$. Then a simple generalization of the insulating effective Hamiltonian H_J in (9.14) is the popular t–J model

$$H_{tJ} = -t\sum_{\langle ij\rangle} \mathcal{P}_d \left[c_{i\alpha}^\dagger c_{j\alpha} + c_{j\alpha}^\dagger c_{i\alpha}\right] \mathcal{P}_d - \mu\sum_i c_{i\alpha}^\dagger c_{i\alpha} + J\sum_{\langle ij\rangle} \boldsymbol{S}_i \cdot \boldsymbol{S}_j, \tag{9.26}$$

where $\mathcal{P}_d$ is a projection operator that annihilates any state in which there are two fermions on any site:

$$\mathcal{P}_d = \prod_i (1 - n_{i\uparrow} n_{i\downarrow}). \tag{9.27}$$

This ensures that

$$\sum_\alpha c_{i\alpha}^\dagger c_{i\alpha} \leq 1, \forall i. \tag{9.28}$$

The t–J model description applies along the arrow sketched in Fig. 9.1: in the large U limit of the original Hubbard model in (9.1) and, for densities $1 - p$ with $p > 0$, the low-energy subspace includes all states that obey (9.28). The effective Hamiltonian in this subspace will contain the hopping term descending directly from that already in (9.28), along with the $\mathcal{P}_d$ operators, which ensures that both the initial and final states are in the low-energy subspace. The order t^2/U computation that led to the exchange term in (9.14) and (9.26) also leads to some additional terms in the effective Hamiltonian, which are not contained in (9.26) (see Ref. [168]), but we will follow common practice and ignore them because they are not believed to be essential to the physics.

A key property of the $p \neq 0$ t–J model is that it is not an insulator, but a conductor. It is possible for electrons to hop from site to site, with amplitude t, while remaining within the low-energy subspace: this is the process where an electron hops from a singly occupied site to one that is empty (a "hole"). The constraint of no double occupancy does not fully restrain motion of the electron charge once $p \neq 0$.

But the construction of a theory of this conductor is difficult because of the no-double-occupancy constraint, that is, the central difficulty in the study of the t–J model in (9.26) is the role of the projection operator $\mathcal{P}_d$. This is clearly crucial in the insulating $p \to 0$ limit, when H_{tJ} reduces to H_J in (9.14): then we obtain a model of $S = 1/2$ spins, not electrons, and we described such models using hard-core bosons in Sections 9.2.1 and 9.2.2; theories of "spin liquid" phases of such spin models are discussed in Parts II and IV. Much theoretical effort has been expended in extending such theories of localized spins to $p \neq 0$, when mobile charge carriers are also present.

In this chapter, we take a "vanilla" point of view [10], and treat the $\mathcal{P}_d$ operator, and the exchange interaction J, in a mean-field manner. This yields a theory of a metallic phase that is a conventional Fermi liquid, with all the properties described in Chapter 2. We also describe the tendency of the quasiparticles to pair as in Chapter 4, and this leads to a theory of d-wave superconductivity. The cuprates do indeed display d-wave pairing, and so this is an important success of the vanilla theory. However, such vanilla approaches do not give a reasonable description of the small, but non-zero, p regime, where we obtain a "pseudogap" metal, and also of quantum phase transitions that extend out of the pseudogap phase. We turn our attention to such matters in Part V, and in Section 31.4.

The main idea of the vanilla approach is that the dominant effect of the $\mathcal{P}_d$ operator is to renormalize the hopping t to a smaller value. This has the effect of increasing the effective mass of the carriers in a p-dependent manner, with the effective mass diverging in the insulating $p \to 0$ limit; this is often called the Brinkman–Rice renormalization [36]. We will see later in Section 31.4 that the mass divergence as $p \to 0$ is incorrect, and so restrict our attention to moderate values of p, where the present approach is reasonable (in the "overdoped" regime of the cuprates). From a variational point of view, the vanilla approach corresponds to the Gutzwiller wavefunction for the metallic state

$$|\text{Vanilla metal}\rangle = \mathcal{P}_d \prod_{\varepsilon_p < 0, \alpha} c_{p\alpha}^{\dagger} |0\rangle \tag{9.29}$$

in which we project out all states with double-occupied sites in the free-fermion wavefunction in (2.3). We can obtain an estimate of the renormalized hopping by computing the dispersion of a quasiparticle added to this wavefunction of the vanilla metal.

Implicitly carrying out this renormalization of t in the metallic state, we now proceed to apply the BCS theory of superconductivity in Chapter 4 to the exchange interaction in H_{tJ} in (9.26).

A transparent view of the origin of pairing is obtained by rewriting the exchange interaction in a different manner. We use the Pauli matrix identity

$$\sum_{\ell=x,y,z} \sigma^{\ell}_{\alpha\beta}\sigma^{\ell}_{\gamma\delta} = -2\varepsilon_{\alpha\gamma}\varepsilon_{\beta\delta} + \delta_{\alpha\beta}\delta_{\gamma\delta}, \tag{9.30}$$

where $\varepsilon_{\alpha\beta}$ is the antisymmetric tensor with $\varepsilon_{\uparrow\downarrow} = 1$. This identity is easily verified by explicit evaluation for the various possibility of the spin indices. Inserting this identity into H_J in (9.14) after using (9.12), we obtain

$$H_J = J\sum_{\langle ij\rangle}\left[-\frac{1}{2}\left(\varepsilon_{\alpha\gamma}c^{\dagger}_{i\alpha}c^{\dagger}_{j\gamma}\right)\left(\varepsilon_{\beta\delta}c_{j\delta}c_{i\beta}\right) + \frac{1}{4}\left(c^{\dagger}_{i\alpha}c_{i\alpha}\right)\left(c^{\dagger}_{j\beta}c_{j\beta}\right)\right]. \tag{9.31}$$

The first term in (9.31) is written as a number operator of a pair of electrons in a spin singlet, that is, a Cooper pair on the link $\langle ij\rangle$. Any such Cooper pair gains energy for $J > 0$, and hence the superconductivity with singlet pairing is connected to antiferromagnetism.

We now return to H_{tJ} in (9.14), and proceed with the Bardeen–Cooper–Schrieffer (BCS) factorization of Chapter 4. We perform the factorization here in real space, rather than momentum space. Then we obtain

$$\begin{aligned} H_{tJ,BCS} = &-t\sum_{\langle ij\rangle}\left[c^{\dagger}_{i\alpha}c_{j\alpha} + c^{\dagger}_{j\alpha}c_{i\alpha}\right] - \mu\sum_{i} c^{\dagger}_{i\alpha}c_{i\alpha} \\ &-\frac{J}{2}\sum_{\langle ij\rangle}\left[\Delta_{ij}\varepsilon_{\alpha\beta}c^{\dagger}_{i\alpha}c^{\dagger}_{j\beta} + \Delta^{*}_{ij}\varepsilon_{\alpha\beta}c_{j\beta}c_{i\alpha} - |\Delta_{ij}|^2\right], \end{aligned} \tag{9.32}$$

where

$$\Delta_{ij} = -\left\langle \varepsilon_{\alpha\beta}c_{i\alpha}c_{j\beta}\right\rangle. \tag{9.33}$$

(For simplicity, we have dropped the pairing contribution from the second term in (9.31) in (9.32).) Now the theory is expressed in terms of pairing amplitudes Δ_{ij}, one for each link in the lattice. We restrict our attention here to pairing amplitudes that are translationally invariant, so that

$$\Delta_{ij} = \Delta_{ji} = \Delta_{i-j}. \tag{9.34}$$

In this case, the momentum-space representation of (9.33) is

$$\Delta_{i-j} = 2\sum_{\boldsymbol{k}}\left\langle c_{-\boldsymbol{k}\downarrow}c_{\boldsymbol{k}\uparrow}\right\rangle e^{i\boldsymbol{k}\cdot(\boldsymbol{r}_i-\boldsymbol{r}_j)}, \tag{9.35}$$

where Δ_{i-j} is the Fourier transform of the pairing amplitude of Chapter 4, which must now be $\boldsymbol{k}$-dependent.

For our model in which the exchange interaction is nearest neighbor, there are only two indpendent pairing amplitudes Δ_x and Δ_y. In terms of these parameters, we define

$$\Delta_{\boldsymbol{k}} \equiv \Delta_x\cos(k_x) + \Delta_y\cos(k_y). \tag{9.36}$$

It is important to note that $\Delta_{\boldsymbol{k}}$ is *not* the Fourier transform of Δ_{i-j} in (9.35). From (9.33), we see that Δ_{ij} is the pairing amplitude between electrons on sites i and j, and this is non-zero for a generic pair of sites under the Hamiltonian in (9.32). However, the Δ_{ij} which appear in the Hamiltonian are only for nearest neighbors, and only these are included in (9.36). So Δ_k is the Fourier transform of $\sum_d \Delta_{i-j}\delta_{r_i-r_j,d}$, where $\boldsymbol{d}$ extends over nearest neighbors.

In terms of Δ_k, we can write (9.32) in momentum space as

$$H_{tJ,BCS} = \sum_k \varepsilon_k c^\dagger_{k\alpha} c_{k\alpha} - J\sum_k \left[\Delta_k c^\dagger_{k\uparrow} c^\dagger_{-k\downarrow} + \Delta^*_k c_{-k\downarrow} c_{k\uparrow} - |\Delta_x|^2 - |\Delta_y|^2\right]. \tag{9.37}$$

Here, the dispersion for nearest-neighbor hopping on the square lattice is

$$\varepsilon_k = -2t(\cos(k_x) + \cos(k_y)) - \mu\,. \tag{9.38}$$

Following the procedure in Chapter 4, we now have to minimize the free energy implied by the Bogoliubov Hamiltonian in (9.37), subject to the self-consistency conditions

$$\begin{aligned}
\Delta_x &= -\left\langle \varepsilon_{\alpha\beta} c_{i\alpha} c_{i+\hat{x},\beta}\right\rangle \\
&= 2\sum_k \left\langle c_{-k\downarrow} c_{k\uparrow}\right\rangle \cos(k_x) \\
\Delta_y &= 2\sum_k \left\langle c_{-k\downarrow} c_{k\uparrow}\right\rangle \cos(k_y).
\end{aligned} \tag{9.39}$$

We now have two variational parameters, Δ_x and Δ_y, representing the pairing amplitude on nearest-neighbor sites oriented along the x and y directions, in contrast to the single pairing Δ in Chapter 4. We can now proceed as in Chapter 4, and the analogs of the self-consistency condition in (4.36) are

$$\begin{aligned}
\Delta_x &= \frac{J}{V}\sum_k \frac{\Delta_k \cos(k_x)}{2E_k} \tanh\left(E_k/(2T)\right) \\
\Delta_y &= \frac{J}{V}\sum_k \frac{\Delta_k \cos(k_y)}{2E_k} \tanh\left(E_k/(2T)\right),
\end{aligned} \tag{9.40}$$

where now

$$E_k = \sqrt{\varepsilon_k^2 + |\Delta_k|^2}. \tag{9.41}$$

We further restrict attention to solutions of (9.37), (9.40), and (9.41) that preserve square-lattice rotational symmetry, in which there is no gauge-invariant local observable that can distinguish between the x and y directions. There turn out to be two such solutions:

$$\begin{aligned}
\Delta_x &= \Delta_y, \quad \text{extended } s \text{ wave} \\
\Delta_x &= -\Delta_y, \quad d \text{ wave}.
\end{aligned} \tag{9.42}$$

The extended s-wave solution is similar to that discussed in Chapter 4, apart from a different spatial dependence in Δ_{i-j}. In Chapter 4, Δ_{i-j} was a monotonically decaying function $\boldsymbol{r}_i - \boldsymbol{r}_j$, largest on-site where $i = j$. In the present case, the largest pairing is at nearest-neighbor sites. Such extended s-wave pairing is found in certain pnictide compounds.

However, the lowest-energy solution in the cuprate case turns out to be the d-wave. In this case, the pairing amplitude changes sign between the x and y directions, and this has been experimentally detected by suitable interference experiments. Moreover, Δ_{i-j}

is identically zero when $i = j$, and this is compatible with the no-double-occupancy constraint associated with the low-energy subspace, which has a unit eigenvalue under $\mathcal{P}_d$.

More generally, we can account for the no-double-occupancy constraint in a manner similar to the theory of the vanilla metal in (9.29). We take the BCS wavefunction in (4.4), and apply the projection operator which removes doubly occupied states

$$|\text{Vanilla BCS}\rangle \propto \mathcal{P}_d \left[\int d^3 r_1 d^3 r_2 \, \psi_\uparrow^\dagger(\boldsymbol{r}_1) \psi_\downarrow^\dagger(\boldsymbol{r}_2) g(\boldsymbol{r}_1 - \boldsymbol{r}_2) \right]^{N/2} |0\rangle \,. \tag{9.43}$$

Here $g(\boldsymbol{k})$ is defined as in (4.9), and the relationship between $u_{\boldsymbol{k}}$ and $v_{\boldsymbol{k}}$, and $\Delta_{\boldsymbol{k}}$ generalizes (4.31). As was the case for the vanilla metal, the projection operator in (9.43) does not significantly modify the properties of the BCS state, apart from a renormalization of the quasiparticle dispersion.

9.4 Paramagnon Theory of Antiferromagnetic Metals

We have so far concentrated on a large-U approach to the electron Hubbard model, focusing on the left end of Fig. 9.1. We now turn to the right end of Fig. 9.1, and describe the physics from a small-U perspective. The main ingredient here will be a bosonic collective mode representing antiferromagnetic spin fluctuations in the metal; this boson is the "paramagnon."

Paramagnons can undergo a condensation transition, and this leads to the appearance of a metallic state with spin density wave (SDW) order. We will focus on the case where the wavevector of the SDW is $\boldsymbol{K} = (\pi, \pi)$ on the square lattice, and so the ordering has the same symmetry as the Néel state of Fig. 9.2 in the antiferromagnetic insulator at $p = 0$: that is, a spontaneous spin polarization that has opposite orientations on the two sublattices. At $p \neq 0$, such a state is a metal, with fermionic quasiparticle excitations, in addition to magnons, descending from paramagnons, which are analogs of the spin-wave excitations of the insulator that were discussed in Section 9.2.2. In Section 9.4.3, we will argue that paramagnon exchange between electrons can lead to a Cooper-pairing instability to d-wave pairing, in a manner analogous to phonon exchange in the BCS theory.

Near the transition from the Fermi liquid to the antiferromagnetic metal, it is possible to derive a systematic approach to the paramagnon modes of a metal. The method is very similar to that followed in Chapter 6 for the Landau–Ginzburg theory of superconductivity. There, we assumed model with $U < 0$ in (6.1), and decoupled it via the Hubbard–Stratonovich transformation in (6.6), using a complex scalar that eventually became the field for the Cooperpair. Here, we are interested in $U > 0$, and so the transformation in (6.6) for decoupling the particle–particle channel does not yield a convergent integral. However, after using the single-site identity

$$U\left(n_{i\uparrow} - \frac{1}{2}\right)\left(n_{i\downarrow} - \frac{1}{2}\right) = -\frac{2U}{3}\boldsymbol{S}_i^2 + \frac{U}{4} \tag{9.44}$$

(which is easily established from the electron commutation relations) it becomes possible to decouple the four-fermion term in a particle–hole channel. So, in decoupling the interaction term in the Hubbard model in (9.1), we replace the Hubbard–Stratonovich transformation in (6.6) by

$$\exp\left(\frac{2U}{3}\sum_i \int d\tau \boldsymbol{S}_i^2\right) = \int \mathcal{D}\boldsymbol{\Phi}_i(\tau) \exp\left(-\sum_i \int d\tau \left[\frac{3}{8U}\boldsymbol{\Phi}_i^2 - \boldsymbol{\Phi}_i \cdot c_{i\alpha}^\dagger \frac{\boldsymbol{\sigma}_{\alpha\beta}}{2} c_{i\beta}\right]\right). \tag{9.45}$$

We now have a new field $\boldsymbol{\Phi}_i(\tau)$, which will play the role of the paramagnon field.

Continuing the strategy of Chapter 6, we can now integrate out the electron field, and obtain an effective theory for the $\boldsymbol{\Phi}$ field. In Chapter 6, this required us to be close to $T = T_c$, so the magnitude of the Cooper-pair ordering was small. In the present situation we can follow the same strategy at $T = 0$ near the transition from a Fermi liquid to an antiferromagnetic metal. We expect the Fermi liquid to be stable for a finite range of $U > 0$ (unless the Fermi surface obeys special nesting conditions), and so we can work near the quantum phase transition for the onset of antiferromagnetism. The path integral of the Hubbard model can be written exactly as (see Appendix B for a discussion of the coherent-state path integral for fermions)

$$\begin{aligned}\mathcal{Z} = \int \mathcal{D}c_{i\alpha}(\tau)\mathcal{D}\boldsymbol{\Phi}_i(\tau) \exp\Bigg(-\int d\tau \Bigg\{ &\sum_{k,\alpha} c_{k\alpha}^\dagger \left[\frac{\partial}{\partial\tau} + \varepsilon_k\right] c_{k\alpha} \\ &+ \sum_i \left[\frac{3}{8U}\boldsymbol{\Phi}_i^2 - \boldsymbol{\Phi}_i \cdot c_{i\alpha}^\dagger \frac{\boldsymbol{\sigma}_{\alpha\beta}}{2} c_{i\beta}\right]\Bigg\}\Bigg).\end{aligned} \tag{9.46}$$

We can now formally integrate out the electrons, and then the analog of the path integral in (6.6) is

$$\frac{\mathcal{Z}}{\mathcal{Z}_0} = \int \prod_i \mathcal{D}\boldsymbol{\Phi}_i(\tau) \exp\left(-\mathcal{S}_{\text{paramagnon}}\left[\boldsymbol{\Phi}_i(\tau)\right]\right), \tag{9.47}$$

where $\mathcal{Z}_0$ is the free-electron partition function. Close to the onset of SDW order (but still on the non-magnetic side), we can expand the action in powers of $\boldsymbol{\Phi}$, and the analog of (6.8) is

$$\mathcal{S}_{\text{paramagnon}}\left[\boldsymbol{\Phi}_i(\tau)\right] = \frac{T}{2}\sum_{q,\omega_n} |\boldsymbol{\Phi}(\boldsymbol{q},\omega_n)|^2 \left[\frac{3}{4U} - \frac{\chi_0(\boldsymbol{q}, i\omega_n)}{2}\right] + \cdots, \tag{9.48}$$

where $\chi_0(\boldsymbol{q}, i\omega_n)$ is the frequency-dependent Lindhard susceptibility, given by the particle–hole bubble graph shown in Fig. 9.5:

$$\chi_0(\boldsymbol{q}, i\omega_n) = -\frac{T}{V}\sum_{p,\varepsilon_n} \frac{1}{(i\varepsilon_n - \varepsilon_k)(i\varepsilon_n + i\omega_n - \varepsilon_{k+q})}. \tag{9.49}$$

Performing the sum over frequencies by partial fractions, we obtain

$$\chi_0(\boldsymbol{q}, i\omega_n) = \frac{1}{V}\sum_k \frac{f(\varepsilon_{k+q}) - f(\varepsilon_k)}{i\omega_n + \varepsilon_k - \varepsilon_{k+q}}. \tag{9.50}$$

From the structure of the $\mathbf{\Phi}$ propagator, it is clear that $\mathbf{\Phi}$ will first condense at the wavevector $\boldsymbol{q}_{max}$ at which $\chi_0(\boldsymbol{q}, i\omega = 0)$ is a maximum, and $\boldsymbol{q}_{max}$ is then the wavevector of the SDW. In the mean-field treatment of (9.48), the appearance of this condensate requires that U is large enough to obey the "Stoner criterion":

$$\frac{3}{4U} - \frac{\chi_0(\boldsymbol{q}_{max}, i\omega = 0)}{2} < 0. \tag{9.51}$$

This wavevector is in turn determined by the dispersion ε_k of the underlying fermions. For simplicitly, we only consider the case of a SDW with wavevector $\boldsymbol{K} = (\pi, \pi)$ below, where the spatial pattern of the $\mathbf{\Phi}$ condensate is the same as the Z-Néel state in Fig. 9.2. The frequency dependence of $\chi_0(\boldsymbol{q}, i\omega)$ also has an important influence on the dynamics of the paramagnon fluctuations. Computation of (9.50) shows that there is a damping term, similar to the $|\omega_n|$ term in (6.12). A key difference from Chapter 6 is that such damping is present also at $T = 0$, and not just near T_c. This damping influences the nature of the transition between the Fermi liquid and the antiferromagnetic metal, as has been discussed in some detail in the QPT book [234].

9.4.1 Paramagnon Hamiltonian

For future applications in Section 31.4, it useful to write down the paramagnon theory in Hamiltonian form, without integrating out the low-energy electron modes (which led to the $|\omega_n|$ term noted above). To make the Φ_i field dynamical, we can integrate some of the *high*-energy electrons far from the Fermi surface. This will induce additional terms in a local potential $V(\mathbf{\Phi}_i)$, which controls fluctuations in the magnitude of $\mathbf{\Phi}_i$. The high-energy electrons will also induce a dynamical term $\omega_n^2 \mathbf{\Phi}_i^2$ in the effective action for $\mathbf{\Phi}_i$ (where ω_n is a Matsubara frequency) so that the on-site $\mathbf{\Phi}_i$ paramagnon Lagrangian on each site is

$$\mathcal{L}_i = \frac{1}{2g}(\partial_\tau \mathbf{\Phi}_i)^2 + V(\mathbf{\Phi}_i). \tag{9.52}$$

Our main assumption is that $V(\mathbf{\Phi}_i)$ has a minimum at a non-zero value of $|\mathbf{\Phi}_i|$ so that the lowest-energy states correspond to the angular momentum $\ell = 0, 1 \ldots$ from rotation motion of the $\mathbf{\Phi}_i$ in a radial state around the $|\mathbf{\Phi}_i|$ minimum. In the following, we will only keep the $\ell = 0, 1$ states.

We can make the analysis more explicit by rescaling $|\mathbf{\Phi}_i|$, and replacing $V(\mathbf{\Phi}_i)$ by a unit-length constraint on each site:

$$\mathbf{\Phi}_i^2 = 1. \tag{9.53}$$

With the constraint (9.53), the dynamical term in (9.52) is simply the kinetic energy proportional to $\boldsymbol{L}_i^2$ of a particle with angular momentum $\boldsymbol{L}_i$ moving on the unit sphere. In this manner, we obtain a Hamiltonian for electrons coupled to a paramagnon *quantum rotor* on each site, illustrated in Fig. 9.6.

$$H_{paramagnon} = \sum_p \varepsilon_p c_{p\sigma}^\dagger c_{p\sigma} + \frac{g}{2}\sum_i \boldsymbol{L}_i^2 + \sum_i \left(\lambda \mathbf{\Phi}_i + \tilde{\lambda} \boldsymbol{L}_i\right) \cdot c_{i\alpha}^\dagger \frac{\boldsymbol{\sigma}_{\alpha\beta}}{2} c_{i\beta}. \tag{9.54}$$

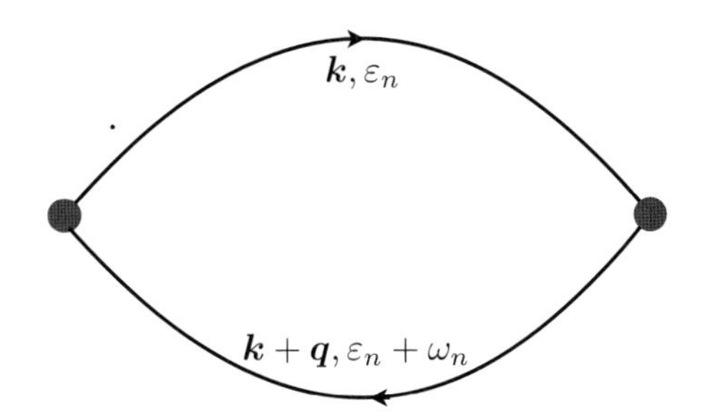

Figure 9.5 Feynman diagram leading to (9.49).

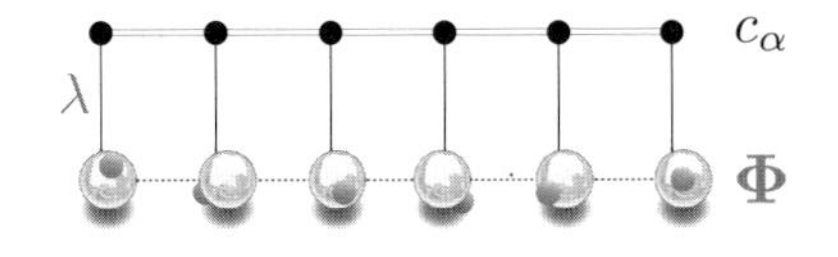

Figure 9.6 Hamiltonian form of the paramagnon theory of magnetism in Fermi liquids: a band of electrons c_α, with each site coupled to a paramagnon quantum rotor. The rotor is a particle of mass $1/g$ constrained to move on the unit sphere with coordinate $\boldsymbol{\Phi}$. The c_α and $\boldsymbol{\Phi}$ reside on a $d >$ one-dimensional lattice, although only one dimension is shown.

The $\boldsymbol{L}_i$ and $\boldsymbol{\Phi}_i$ obey the usual commutation relations of single-particle rotational quantum mechanics

$$[L_a, L_b] = i\varepsilon_{abc} L_c \quad , \quad [L_a, \Phi_b] = i\varepsilon_{abc}\Phi_c \quad , \quad [\Phi_a, \Phi_b] = 0 \,, \tag{9.55}$$

where $a,b,c = x,y,z$, we have dropped the site label i, and ε_{abc} is the unit antisymmetric tensor. Now on each site we have a "particle" of mass $1/g$ moving on a unit sphere with angular momentum $\boldsymbol{L}_i$; this is the paramagnon rotor, which has couplings λ, $\widetilde{\lambda}$ to the low-energy electrons. The coupling $\widetilde{\lambda}$ was also obtained by integrating out the high-energy electrons.

To ensure the consistency of our procedure, let us undo the mappings above, and show how the original Hubbard model can be obtained starting from the rotor–fermion Hamiltonian in (9.54). At $\lambda = 0$, it is a simple matter to diagonalize the rotor spectrum.

On each site, we have states labeled by the usual angular momentum quantum numbers ℓ

$$|\ell, m\rangle_i \quad , \quad \ell = 0,1,2,\ldots; m = -\ell,\ldots,\ell \quad , \quad \text{Energy} = \frac{g}{2}\ell(\ell+1) \,. \tag{9.56}$$

So the ground state has $\ell = 0$ on each site, and there is a three-fold degenerate excited state with $\ell = 1$ and energy g. Turning to non-zero λ, but $|\lambda| \ll g$, we can eliminate the coupling between the electrons and the rotors by a canonical transformation. Note that the λ coupling is only active when the electronic state on a given site i has spin 1/2; so the influence of the λ is to lower the energy of this state with respect to the empty and doubly occupied sites, which have spin 0. To compute this energy shift we need the matrix elements of $\boldsymbol{\Phi}$ between the $\ell = 0$ ground state and the $\ell = 1$ excited states; the non-zero matrix elements are

$$\langle \ell = 0 | \Phi_z | \ell = 1, m = 0 \rangle = \frac{1}{\sqrt{3}}, \quad \langle \ell = 0 | \Phi_x \pm i\Phi_y | \ell = 1, m = \mp 1 \rangle = \sqrt{\frac{2}{3}}. \tag{9.57}$$

We can now use perturbation theory in λ to compute the energy of the spin-1/2 state; in this manner, to second order in λ, we obtain an effective model, which is just the original Hubbard model with

$$U = \frac{\lambda^2}{4g}. \tag{9.58}$$

Note that the coupling $\widetilde{\lambda}$ does not appear at this order because $\boldsymbol{L}_i = 0$ when acting on the rung singlet state.

It is important to note that, so far, all we have done is to cast the paramagnon theory in (9.46) into a Hamiltonian form. No fundamentally new step has yet been taken, but we will do so in our consideration of metallic states with fractionalization in Section 31.4.

9.4.2 Fermi Surface Reconstruction

Let us now move into the antiferromagnetic metal phase, where there is a $\boldsymbol{\Phi}$ condensate at wavevector $\boldsymbol{K} = (\pi, \pi)$:

$$\langle \boldsymbol{\Phi}_i \rangle = \eta_i \mathcal{N} \hat{z}, \tag{9.59}$$

with $\mathcal{N}$ measuring the strength of the Néel ordered moment in (9.15). There is no ferromagnetic moment, and so we have

$$\langle \boldsymbol{L}_i \rangle = 0. \tag{9.60}$$

We wish to describe the excitations of this state. One class of excitations are spin waves, similar to the spin-wave excitations of the insulator that were discussed in Section 9.2.2; these can be obtained by considering transverse fluctuations of $\boldsymbol{\Phi}$ about the condensate in (9.59) using the full action in (9.47). However, there are also low-energy fermionic excitations in the antiferromagnetic metal, which are gapped in the insulator. We can determine the spectrum of the fermions by inserting (9.59) into (9.54) ; using $\eta_i = e^{i\boldsymbol{K}\cdot\boldsymbol{r}_i}$, with $\boldsymbol{K} = (\pi, \pi)$, we can write the fermion Hamiltonian in momentum space

$$H_{AFM} = \sum_{\boldsymbol{k}} \left[\varepsilon_{\boldsymbol{k}} c^\dagger_{\boldsymbol{k}\alpha} c_{\boldsymbol{k}\alpha} - \Delta c^\dagger_{\boldsymbol{k}\alpha} \sigma^z_{\alpha\alpha} c_{\boldsymbol{k}+\boldsymbol{K},\alpha} \right] + \text{constant}. \tag{9.61}$$

This is the analog of (4.26), and the analog of the pairing gap is the energy

$$\Delta = \lambda \mathcal{N}. \tag{9.62}$$

But, in general, the spectrum of H_{AFM} does not have a gap, as we will see below. As in BCS theory, the value of $\mathcal{N}$ has to be determined self-consistently from the mean-field equations.

To obtain the fermionic excitation spectrum, we have to perform the analog of the Bogoliubov rotation in (4.28). Here, this is achieved by writing H_{AFM} in a 2×2 matrix

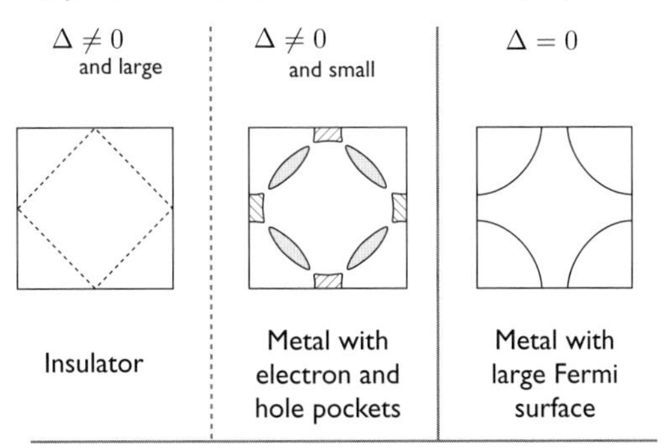

Figure 9.7 Fermi surfaces of the Néel state at $p = 0$. The pockets intersecting the diagonals of the Brillouin zone have both bands in (9.64) empty and so form hole pockets, while the remaining pockets have both bands occupied and form electron pockets. The $\Delta = 0$ Fermi surface is the same as in Fig. 9.12, except it is now centered at $\boldsymbol{k} = 0$. The dashed line in the insulator shows the boundary of the Brillouin zone of the Néel state

form by using the fact that $2\boldsymbol{K}$ is a reciprocal lattice vector, and so $\varepsilon_{\boldsymbol{k}+2\boldsymbol{K}} = \varepsilon_{\boldsymbol{k}}$; correspondingly, the prime over the summation indicates that it only extends over half the Brillouin zone of the underlying lattice, shown in the left panel of Fig. 9.7, which is the Brillouin zone of the lattice with Néel order:

$$H_{AFM} = \sum_{\boldsymbol{k}}{}' (c^{\dagger}_{\boldsymbol{k}\alpha}, c^{\dagger}_{\boldsymbol{k}+\boldsymbol{K},\alpha}) \begin{pmatrix} \varepsilon_{\boldsymbol{k}} & -\Delta\sigma^z_{\alpha\alpha} \\ -\Delta\sigma^z_{\alpha\alpha} & \varepsilon_{\boldsymbol{k}+\boldsymbol{K}} \end{pmatrix} \begin{pmatrix} c_{\boldsymbol{k}\alpha} \\ c_{\boldsymbol{k}+\boldsymbol{K},\alpha} \end{pmatrix}. \tag{9.63}$$

It is now easy to diagonalize the 2×2 matrix in (9.63), and we obtain the analog of (4.33), which is

$$E_{\boldsymbol{k}\pm} = \frac{\varepsilon_{\boldsymbol{k}} + \varepsilon_{\boldsymbol{k}+\boldsymbol{K}}}{2} \pm \left[\left(\frac{\varepsilon_{\boldsymbol{k}} - \varepsilon_{\boldsymbol{k}+\boldsymbol{K}}}{2} \right)^2 + \Delta^2 \right]^{1/2}. \tag{9.64}$$

Unlike (4.33), the spectrum in (9.64) is not gapped, or even positive definite. Rather, it is the spectrum of a metal, in which the negative energy states are filled, and bounded by a Fermi surface. The Fermi surfaces so obtained are shown in Figs. 9.7–9.9 for different values of p.

We observe that the "large" Fermi surface of the paramagnetic metal has been "reconstructed" into small pocket Fermi surfaces in the SDW state. The excitations of the SDW metal are hole-like quasiparticles on the Fermi surfaces surrounding the hole pockets, and electron-like quasiparticles on the Fermi surfaces surrounding the electron pockets. The spin-wave excitations interact rather weakly with the fermionic quasiparticle excitations; this can be see from a somewhat involved computation from the effective action.

Finally, we discuss the fate of the Luttinger relation in this SDW metal. The Luttinger relation connects the volume enclosed by the Fermi surface to the density of electrons per unit-cell modulo-2, as we will establish in Section 30.2. It should be applied in the Brillouin zone of the Néel state, which is half the size of the Brillouin

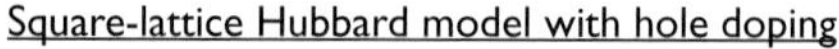

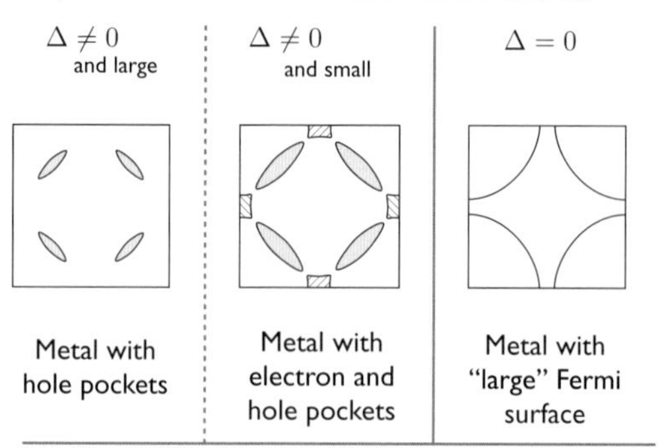

Figure 9.8 Fermi surfaces of the Néel state at $p > 0$. The pockets are as in Fig. 9.7.

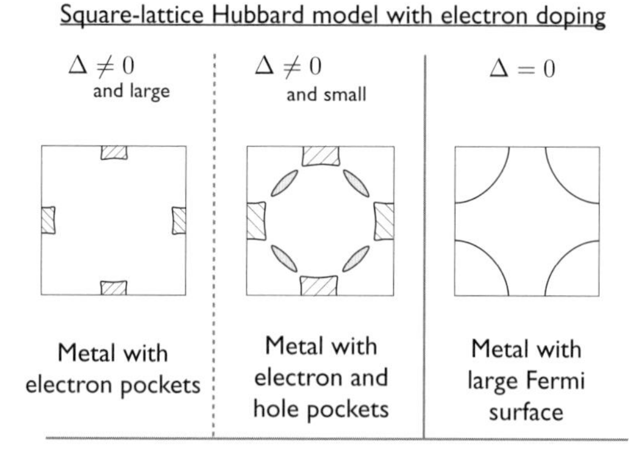

Figure 9.9 Fermi surfaces of the Néel state at $p < 0$, with pockets as in Figs. 9.7.

zone of the underlying square lattice, as shown in Fig. 9.7. In real space, this corresponds to the fact that the unit cell has doubled, and so the density of electrons per unit cell is $2(1-p)$. For spinful electrons, the Luttinger relation measures the electron density modulo 2, and so the density appearing in the Luttinger relation is $-2p$. This has to be equated to twice the volumes enclosed by the electron and hole pockets within the diamond-shaped Brillouin zone in Fig. 9.7. Let $\mathcal{A}_h$ be the area of a single elliptical hole pocket. There are four such pockets in the complete Brillouin zone of the square lattice or two pockets in the Brillouiin zone of the Néel state, as is apparent from Figs. 9.7–9.9. Similarly, let $\mathcal{A}_e$ be the area of a single elliptical electron pocket: there are two such pockets in the complete Brillouin zone of the square lattice or one pocket in the Brillouin zone of the Néel state. These arguments show that the Luttinger relation becomes

$$2 \times \frac{1}{(2\pi)^2/2} \times (-2\mathcal{A}_h + \mathcal{A}_e) = -2p\,. \tag{9.65}$$

On the left-hand side, the first factor is the spin degeneracy, and the second factor is the inverse of the volume of the Brillouin zone of the Néel state. To reiterate, this is the conventional Luttinger relation applied after accounting for the doubling of the unit cell, and it determines a linear constraint on the areas of the electron and hole pockets.

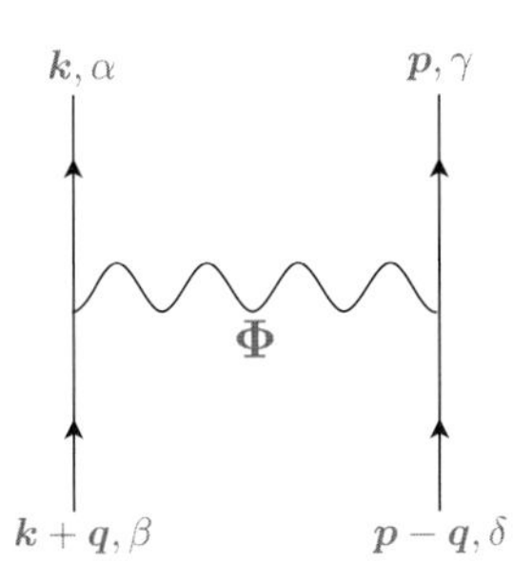

Figure 9.10 Electron interaction from the exchange of paramagnon, **Φ**, in a metal.

9.4.3 Pairing from Paramagnon Exchange

We discussed the appearance of d-wave pairing in the large-U t–J model in Section 9.3. Here, we address the pairing instability of the small-U Fermi liquid using the paramagnon theory [14, 213], as sketched in Fig. 9.1. We assume we are in the Fermi liquid without long-range antiferromagnetic order, but close to the paramagnon condensation transition so that the paramagnon fluctuations are strong. In the spirit of the BCS theory of Chapter 4, we examine the interaction between electrons induced by paramagnon exchange, and study the resulting superconducting state. The choice of d-wave pairing above has been presented as a somewhat mysterious feature that arises from a numerical minimization of (9.37). Clearer insight arises from the present paramagnon-induced pairing approach, which has its origins in studies of the superfluidity of ^{3}He.

The simplest diagram in the theory (9.46) leading to an electron–electron interaction from paramagnon exchange is shown in Fig. 9.10. This leads to an effective interaction

$$H_V = \frac{1}{2V}\sum_{\boldsymbol{q}}\sum_{\boldsymbol{p},\gamma,\delta}\sum_{\boldsymbol{k},\alpha,\beta} V_{\alpha\beta,\gamma\delta}(\boldsymbol{q}) c^\dagger_{\boldsymbol{k},\alpha} c_{\boldsymbol{k}+\boldsymbol{q},\beta} c^\dagger_{\boldsymbol{p},\gamma} c_{\boldsymbol{p}-\boldsymbol{q},\delta}\,. \tag{9.66}$$

We use the renormalized **Φ** propagator from (9.48) to obtain

$$V_{\alpha\beta,\gamma\delta}(\boldsymbol{q}) = J(\boldsymbol{q})\boldsymbol{\sigma}_{\alpha\beta}\cdot\boldsymbol{\sigma}_{\gamma\delta}\,, \tag{9.67}$$

where

$$J(\boldsymbol{q}) = -\frac{1}{4\left[3/(4U) - \chi_0(\boldsymbol{q},0)/2\right]}\,. \tag{9.68}$$

This will lead to a spin-dependent interaction that is peaked at the wavevector $\boldsymbol{q} = \boldsymbol{K}$ determined by the maximum in $\chi_0(\boldsymbol{q},0)$.

We note that a similar interaction appeared in the discussion of the t–J model. A momentum-space presentation of the nearest-neighbor exchange interaction in H_J in (9.14) leads to

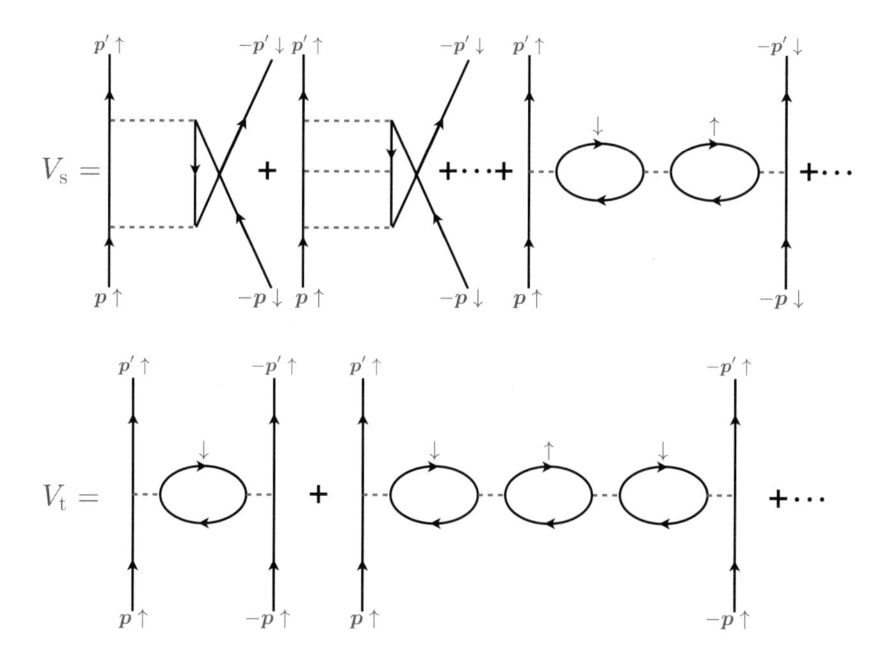

Figure 9.11 Adapted from Ref. [249]: Singlet (V_s) and triplet (V_t) interactions between a pair of electrons with total momentum zero. The dashed line represents the Hubbard interaction U in (9.1).

$$J(\boldsymbol{q}) = \frac{J}{2}\left[\cos(q_x) + \cos(q_y)\right]. \tag{9.69}$$

(Note: we are now including the density–density interaction discarded in (9.32).) Similar to (9.68), the interaction in (9.69) peaks at a negative value at $\boldsymbol{q} = (\pi, \pi)$.

We mention, in passing, another approach to treat the paramagnon-induced interaction. In our starting point in (9.45), we made a choice to decouple the Hubbard interaction in the spin channel only. A more democratic choice is to include all possible channels by summing "ladder" and "bubble" diagrams to all orders in U. Such a strategy has often been used in the context of ^{3}He and the pnictides. While this approach appears more complete, it is difficult to use it as a starting point of a theory of the higher-order effects of the interactions. Summing such diagrams leads to interactions between electrons that can be decomposed into singlet and triplet components, V_s and V_t, and these are shown diagrammatically in Fig. 9.11. Evaluations of these diagrams leads to the interactions [249]

$$\begin{aligned} V_s &= \frac{U^2\chi_0(\boldsymbol{p}'+\boldsymbol{p})}{1-U\chi_0(\boldsymbol{p}'+\boldsymbol{p})} + \frac{U^3\chi_0^2(\boldsymbol{p}'-\boldsymbol{p})}{1-U^2\chi_0^2(\boldsymbol{p}'-\boldsymbol{p})}, \\ V_t &= -\frac{U^2\chi_0(\boldsymbol{p}'-\boldsymbol{p})}{1-U^2\chi_0^2(\boldsymbol{p}'-\boldsymbol{p})}, \end{aligned} \tag{9.70}$$

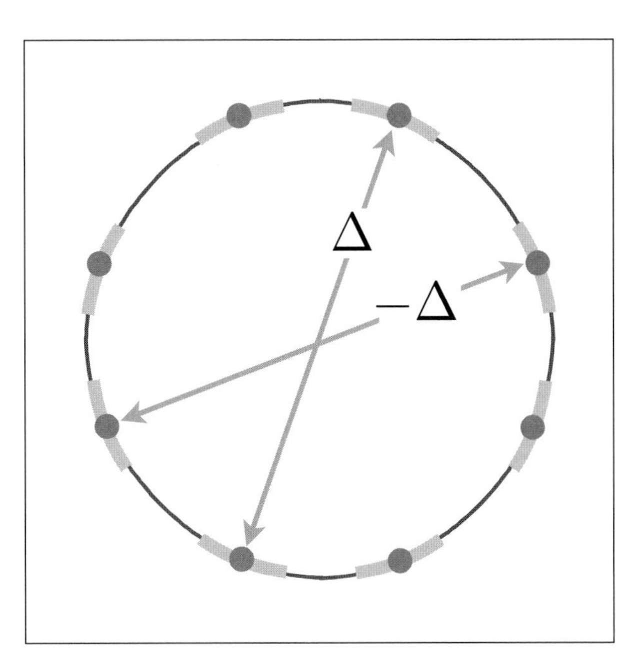

Figure 9.12 Pairing around a single large Fermi surface on the square lattice, as found in the cuprates. The Fermi surface has been centered at $\boldsymbol{K}$.

These interactions are similar to those in (9.67) and (9.68), but notice the different numerical factors in front of $U\chi_0$; this is a consequence of the choices of the decoupling channels.

We now return to the general form of the paramagnon induced interaction in (9.66) and (9.67), and study the pairing instability, as in Chapter 4. Our only assumption is that $J(\boldsymbol{q})$ is peaked at a negative value at $\boldsymbol{q}=\boldsymbol{K}\approx(\pi,\pi)$. We decouple (9.66) and obtain an equation for the pairing amplitude:

$$\Delta_{\boldsymbol{k}}=\frac{1}{2}\varepsilon_{\alpha\beta}\left\langle c^{\dagger}_{\boldsymbol{k}\alpha}c^{\dagger}_{-\boldsymbol{k}\beta}\right\rangle, \tag{9.71}$$

which obeys

$$\Delta_{\boldsymbol{p}}=\frac{3}{2V}\sum_{\boldsymbol{k}}J(\boldsymbol{p}-\boldsymbol{k})\frac{\Delta_{\boldsymbol{k}}}{\sqrt{\varepsilon_{\boldsymbol{k}}^2+\Delta_{\boldsymbol{k}}^2}}\tanh\left(\frac{\sqrt{\varepsilon_{\boldsymbol{k}}^2+\Delta_{\boldsymbol{k}}^2}}{2T}\right). \tag{9.72}$$

This is clearly closely related to (9.40). To optimize the solution, we should have $|J(\boldsymbol{p}-\boldsymbol{k})|$ large when $\boldsymbol{k}$ and $\boldsymbol{p}$ are near the Fermi surface. In the cuprates $J(\boldsymbol{q}=\boldsymbol{K})=-2J$ is maximal, and from (9.72) this requires $\Delta_{\boldsymbol{k}=(\pi,0)}=-\Delta_{\boldsymbol{p}=(0,\pi)}$. This leads to d-wave pairing $\Delta_x=-\Delta_y$ for the cuprate Fermi surface, as illustrated in Fig. 9.12.

A summary for the origin of d-wave pairing is presented in Fig. 9.13. Cooper pairs at points A and B on the Fermi surface are scattered by the exchange interaction J to points C and D, after an exchange of wavevector K. The negative sign of the interaction must be compensated by opposites of the pairing amplitude between A,B and C,D.

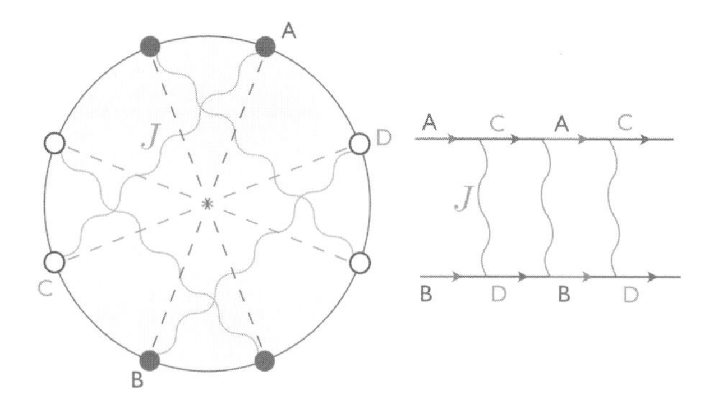

Figure 9.13 Cooper pairing from exchange interactions J at wavevector (π,π). The pairing amplitude is positive (negative) at the filled (open) circles around the Fermi surface.

Problems

9.1 Spin waves in quantum ferromagnets

Consider the quantum Heisenberg ferromagnet with the Hamiltonian

$$H = -J\sum_{\langle i,j\rangle} \vec{S}_i \cdot \vec{S}_j \quad (J>0), \tag{9.73}$$

where the lattice operators $\vec{S}_i$ obey the usual $SU(2)$ commutation algebra

$$\left[S_i^\alpha, S_j^\beta\right] = i\delta_{ij}\varepsilon^{\alpha\beta\gamma}S_j^\gamma, \tag{9.74}$$

with $\alpha,\beta,\gamma = x,y,z$.

Additionally, $\sum_\alpha S_i^\alpha S_i^\alpha = S(S+1)$, that is, the maximum z component of the spin is S. We are often be interested in the case of $S=1/2$.

Our goal is to find the actual ground state and the nature of the excitations above the ground state.

(a) As an ansatz for the ground state, consider the state $|\Omega\rangle = \bigotimes_i |S_i\rangle$, where $|S_i\rangle$ represents a state with maximal spin-z component: $S_i^z|S_i\rangle = S|S_i\rangle$. Compute the energy expectation value of the state $|\Omega\rangle$. Defining *global* spin operators through $S^\mu = \sum_i S_i^\mu$, consider the state $|\zeta\rangle = \exp(i\zeta\cdot\vec{S})|\Omega\rangle$ and verify that the state $|\zeta\rangle$ is degenerate with $|\Omega\rangle$. Hence, the system possesses a global spin-rotation symmetry.

(b) How do we know that $|\Omega\rangle$ is a true ground state? Obtain a lower bound on the ground-state energy from (9.73) and compare it to the energy calculated in part (a).

(c) We now study the excitations above the ground-state. In order to do this, we use the Holstein–Primakoff transformation and find the spin-wave excitations in the semiclassical limit of large S. Considering the Heisenberg uncertainty relation $\Delta S^\alpha \Delta S^\beta \sim |\langle[S^\alpha, S^\beta]\rangle|$, obtain an estimate for the relative uncertainty

$\Delta S^\alpha / S$, where ΔS^α is the root mean square of the quantum uncertainty of the spin component α. Show that quantum fluctuations of the spin become less important as $S \gg 1$.

(d) In the Holstein–Primakoff transformation, the spin operators $S^\pm, \vec{S}$ are specified in terms of bosonic creation and annihilation operators $a^\dagger$, a as

$$S_i^- = a_i^\dagger (2S - a_i^\dagger a_i)^{1/2},\ S_i^+ = (2S - a_i^\dagger a_i)^{1/2} a_i,\ S_i^z = S - a_i^\dagger a_i. \tag{9.75}$$

Using the bosonic commutation relations, confirm that the spin operators satisfy the commutation relations

$$\left[S_i^z, S_j^\pm\right] = \pm\delta_{ij} S_i^\pm, \quad \left[S_i^+, S_j^-\right] = 2\delta_{ij} S_i^z. \tag{9.76}$$

(e) The utility of this representation is clear: when the spin is large, $S \gg 1$, an expansion in powers of $1/S$ gives $S_i^z = S - a_i^\dagger a_i$, $S_i^- \simeq (2S)^{1/2} a_i^\dagger$, $S_i^+ \simeq (2S)^{1/2} a_i$. In this approximation, show that the one-dimensional Heisenberg Hamiltonian takes the form

$$H = -JNS^2 + \sum_k \hbar\omega_k a_k^\dagger a_k + \mathcal{O}(S^0). \tag{9.77}$$

Calculate the dispersion relation of the spin excitations $\hbar\omega_k$ and show that the energy of the elementary excitations vanishes in the limit $k \to 0$. These massless low-energy excitations, known as *magnons*, describe the elementary spin-wave excitations of the ferromagnet.

(f) Repeat the calculation above for two dimensions and compute the dispersion $\hbar\omega_k$.

9.2 Spin waves in quantum antiferromagnets

Now, we consider the quantum Heisenberg antiferromagnet with the Hamiltonian

$$H = J \sum_{\langle i,j \rangle} \vec{S}_i \cdot \vec{S}_j \quad (J > 0). \tag{9.78}$$

We will focus on *bipartite* lattices, that is, lattices for which the sites can be divided into two sublattices, say, A and B, such that the neighbors of one sublattice A belong to the other sublattice B.

(a) The classical ground state can be written as

$$|\mathrm{GS}\rangle_c = \prod_i |S_i^z = \varepsilon_i S\rangle,\ \varepsilon_i = \begin{cases} +1 \text{ on A} \\ -1 \text{ on B} \end{cases}. \tag{9.79}$$

Is this a ground state of the Hamiltonian? For simplicity, it is illustrative to look at the two-site problem $H_2 = J\vec{S}_1 \cdot \vec{S}_2$. What is the true ground state in this case?

(b) To gain some insight into this problem, let us carry out a $1/S$ expansion about $|\mathrm{GS}\rangle_c$ as we did for the ferromagnetic case. Let us define the following mappings on the two sublattices:

$$\text{On A:} \quad S_i^- = a_i^\dagger (2S - a_i^\dagger a_i)^{1/2},\ S_i^+ = (2S - a_i^\dagger a_i)^{1/2} a_i,\ S_i^z = S - a_i^\dagger a_i; \tag{9.80}$$

On B: $S_i^- = (2S - b_i^\dagger b_i)^{1/2} b_i,\; S_i^+ = b_i^\dagger (2S - b_i^\dagger b_i)^{1/2},\; S_i^z = -S + b_i^\dagger b_i.$ (9.81)

In terms of these transformed operators, show that the Hamiltonian is

$$H = J \sum_{\langle i,j \rangle} \left[-S^2 + S\left(a_i^\dagger a_i + b_j^\dagger b_j + a_i b_j + a_i^\dagger b_j \right) \right] + \mathcal{O}(S^0). \tag{9.82}$$

(c) Fourier transforming to momentum space for a d-dimensional hypercubic lattice, diagonalize the quadratic Hamiltonian by a Bogoliubov transformation. The Hamiltonian is then of the form

$$H = -JNdS^2 + \sum_k E_{\vec{k}} \left(\mathcal{A}_k^\dagger \mathcal{A}_k + \mathcal{B}_k^\dagger \mathcal{B}_k \right). \tag{9.83}$$

What is the dispersion $E_{\boldsymbol{k}}$?

(d) For two dimensions, analyze the dispersion in the $k \to 0$ limit and show that, once again, we have gapless magnons associated with the broken rotational symmetry. What is the difference compared to the ferromagnetic case?

(e) What is the zero-temperature value of $\langle S^z \rangle$ in the ground state? Hint: It might be easier to evaluate this in terms of the bogoliubons.
Generically, we find that $\langle S^z \rangle \sim S -$ (corrections). Show, based on dimensional analysis, that the correction term diverges in one dimension – this is a manifestation of the Mermin–Wagner theorem.

9.3 The electron spin density $\boldsymbol{\sigma}(\boldsymbol{r})$ (measured in units of $\hbar/2$) is given by the three operators:

$$\sigma_x(\boldsymbol{r}) = \left[\Psi_\uparrow^\dagger(\boldsymbol{r})\Psi_\downarrow(\boldsymbol{r}) + \Psi_\downarrow^\dagger(\boldsymbol{r})\Psi_\uparrow(\boldsymbol{r}) \right];$$
$$\sigma_y(\boldsymbol{r}) = \frac{1}{i} \left[\Psi_\uparrow^\dagger(\boldsymbol{r})\Psi_\downarrow(\boldsymbol{r}) - \Psi_\downarrow^\dagger(\boldsymbol{r})\Psi_\uparrow(\boldsymbol{r}) \right];$$
$$\sigma_z(\boldsymbol{r}) = \left[\Psi_\uparrow^\dagger(\boldsymbol{r})\Psi_\uparrow(\boldsymbol{r}) - \Psi_\downarrow^\dagger(\boldsymbol{r})\Psi_\downarrow(\boldsymbol{r}) \right].$$

(a) Let

$$\boldsymbol{\sigma}(\boldsymbol{k}) = \int d^3 r \boldsymbol{\sigma}(\boldsymbol{r}) e^{-i\boldsymbol{k}\cdot\boldsymbol{r}}.$$

Express $\sigma_x(\boldsymbol{k})$, $\sigma_y(\boldsymbol{k})$, and $\sigma_z(\boldsymbol{k})$ in terms of the creation and annihilation operators $c_{p\alpha}^\dagger$ and $c_{p\alpha}$ ($\alpha = \uparrow, \downarrow$).

(b) Consider the Hamiltonian

$$H = \sum_{p,\alpha} \varepsilon_p c_{p\alpha}^\dagger c_{p\alpha} - \int \boldsymbol{h}(\boldsymbol{r},t) \cdot \boldsymbol{\sigma}(\boldsymbol{r}) d^3 r,$$

where $\varepsilon_p = p^2/(2m) - \mu$ includes a chemical potential adjusted so that the density of electrons is n, and $\boldsymbol{h}$ is an external, time-varying, spatially dependent magnetic field. We shall consider the situation where h_z is constant in space and time, and *not* necessarily small, while h_x and h_y are small and depend upon $\boldsymbol{r}$

and t. At $t \to -\infty$, the system is in thermal equilibrium in the magnetic field h_z at a temperature T. Suppose that $h_y = 0$ and

$$h_x = \text{Re}\left(\lambda e^{ik\cdot r - i\omega t} e^{\eta t}\right), \tag{9.84}$$

where η is a positive infinitesimal. Write down an expression for the magnetization $\langle \sigma_z \rangle$ at $t = -\infty$, without assuming that h_z is small. Provided λ is small, show that we can write the linear response magnetizations in the x and y directions as

$$\begin{aligned} \langle \sigma_x(r,t) \rangle &= \lambda \quad \text{Re}\left\{\chi^0_{xx}(k,\omega) e^{ik\cdot r - i\omega t}\right\}, \\ \langle \sigma_y(r,t) \rangle &= \lambda \quad \text{Re}\left\{\chi^0_{yx}(k,\omega) e^{ik\cdot r - i\omega t}\right\}, \end{aligned} \tag{9.85}$$

where

$$\chi^0_{xx}(k,\omega) = \frac{1}{V}\sum_p \left\{\frac{f_{p\uparrow} - f_{p+k\downarrow}}{\varepsilon_{p+k\downarrow} - \varepsilon_{p\uparrow} - \omega - i\eta} + \frac{f_{p\downarrow} - f_{p+k\uparrow}}{\varepsilon_{p+k\uparrow} - \varepsilon_{p\downarrow} - \omega - i\eta}\right\}, \tag{9.86}$$

where $\varepsilon_{p\uparrow} = \varepsilon_p - h_z$, $\varepsilon_{p\downarrow} = \varepsilon_p + h_z$, and $f_{p\alpha} \equiv f(\varepsilon_{p\alpha})$ is the Fermi function. Obtain the corresponding expression for $\chi^0_{yx}(k,\omega)$.

(c) Specialize the results of (b) to zero temperature. Assume that h_z is sufficiently strong that the electrons are completely polarized at $t = -\infty$ ($\langle \sigma_z \rangle = n$). Calculate the response functions, χ^0_{xx}, χ^0_{yx} explicitly in the limit $k \to 0$ for arbitrary ω. For what values of ω and k is energy absorbed from the time-varying field h_x?

(d) Consider now a system of *interacting* electrons, which we will treat in the local exchange approximation in a Hartree–Fock theory. Thus, the electrons now see an effective Hamiltonian

$$H^{eff} = \sum_{p,\alpha} \varepsilon_p c^\dagger_{p\alpha} c_{p\alpha} - \int \boldsymbol{h}^{eff}(\boldsymbol{r},t)\cdot \boldsymbol{\sigma}(\boldsymbol{r},t) d^3r,$$

$$\boldsymbol{h}^{eff}(\boldsymbol{r},t) = I\langle \boldsymbol{\sigma}(\boldsymbol{r})\rangle_t + \boldsymbol{h}(\boldsymbol{r},t),$$

where $\boldsymbol{h}$ is the external field and I is the exchange constant. Assume now that $h_z = 0$, but that I is sufficiently large that the system is ferromagnetic and that the magnetization points in the z direction at $t = -\infty$. (Assume that $T = 0$, and that the system is completely polarized, $\langle \sigma_z \rangle = n$ at $t = -\infty$). As in (9.85), we write linear response expressions for $\langle \sigma_x(\boldsymbol{r},t)\rangle$ and $\langle \sigma_y(\boldsymbol{r},t)\rangle$ in terms of response function χ_{xx} and χ_{yx}. Show that χ_{xx} and χ_{yx} can be written in terms of I and the response functions χ^0_{xx}, χ^0_{yx} computed in (b), with the latter evaluated in a field $h_z = In$. Evaluate the resulting response functions χ_{xx}, χ_{yx} when both k and ω are small, and show that there is a pole at a frequency $\omega_k = Dk^2$, for $k \to 0$; calculate the coefficient D. (The pole corresponds to the excitation of "spin waves" in this model).

10 Relativistic Scalar Field: Diagrams

The field theory of a scalar with a quartic self-interaction is used to illustrate perturbation theory with Feynman diagrams.

We encountered a relativistic field theory in Chapter 8 on the boson Hubbard model: at a *fixed* integer density, the transition from the Mott insulator to the superfluid is described by the theory of a complex scalar field in (8.21) at $K_1 = 0$, which is the point at which all displayed terms in the action have an emergent relativistic invariance.

This chapter focuses on the properties of this relativistic field theory, by examining the case of an M-component real scalar field with $O(M)$ symmetry; the superfluid insulator transition corresponds to the case $M = 2$. We develop the diagrammatic perturbation theory of this theory in powers of the quartic interaction, and examine its structure near the critical point in general D spacetime dimensions. The case $D = 3$, with spatial dimension $d = 2$, will be of particular interest to us.

The partition function of the $O(M)$ relativistic scalar is represented by the functional integral

$$\mathcal{Z} = \int \mathcal{D}\phi_\alpha(x) \exp(-\mathcal{S}_\phi), \tag{10.1}$$

$$\mathcal{S}_\phi = \int d^D x \left\{ \frac{1}{2} \left[(\nabla_x \phi_\alpha)^2 + r\phi_\alpha^2(x) \right] + \frac{u}{4!} \left(\phi_\alpha^2(x) \right)^2 \right\}, \tag{10.2}$$

where $\mathcal{S}_\phi$ is an imaginary time action. Here, the symbol $\int \mathcal{D}\phi_\alpha(x)$ represents an infinite dimensional integral over the values of the field $\phi_\alpha(x)$ at every spatial point x. Whenever in doubt, we will interpret this somewhat vague mathematical definition by discretizing x to a set of lattice points of small spacing $\sim 1/\Lambda$. Equivalently, we will Fourier transform $\phi_\alpha(x)$ to $\phi_\alpha(k)$, and impose a cutoff $|k| < \Lambda$ in the set of allowed wavevectors.

We have set the coefficient of the gradient term $\mathcal{K}$ equal to unity in (10.2). This is to avoid clutter of notation, and is easily accomplished by an appropriate rescaling of the field ϕ_α and the spatial coordinates.

An immediate advantage of the representation in (10.2) is that the Landau theory is obtained simply by making the saddle-point approximation to the functional integral. We can also see that, as is described in more detail below, systematic corrections to the Landau theory appear in an expansion in powers of the quartic coupling u. The remainder of this chapter is devoted to explaining how to compute the terms in the u

expansion. Each term has an efficient representation in terms of "Feynman diagrams," from which an analytic expression can also be obtained.

10.1 Gaussian Integrals

We introduce the technology of Feynman diagrams in the simplest possible setting. Let us discretize space, and write the $\phi_\alpha(x_i)$ variables as y_i; we drop the α label to avoid clutter of indices. Then we consider the multidimensional integral

$$\mathcal{Z}(u) = \int \mathcal{D}y \exp\left(-\frac{1}{2}\sum_{ij} y_i A_{ij} y_j - \frac{u}{24}\sum_i y_i^4\right), \tag{10.3}$$

where the off-diagonal terms in the matrix A arise from the spatial gradient terms in $\mathcal{S}_\phi$. In this section, we will consider A to be an arbitrary positive definite, symmetric matrix. The positive definiteness requires that $r > 0$, that is, $K < K_c$. Also, we have defined

$$\int \mathcal{D}y = \prod_i \int_{-\infty}^{\infty} \frac{dy_i}{\sqrt{2\pi}}, \tag{10.4}$$

and are interested in the expansion of $\mathcal{Z}(u)$ in powers of u. Thinking of (10.3) as a statistical mechanics ensemble, we are also interested in the power-series expansion of correlators like

$$C_{ij}(u) \equiv \langle y_i y_j \rangle \equiv \frac{1}{\mathcal{Z}(u)} \int \mathcal{D}y\, y_i y_j \exp\left(-\frac{1}{2}\sum_{k\ell} y_k A_{k\ell} y_\ell - \frac{u}{24}\sum_k y_k^4\right). \tag{10.5}$$

First, we note the exact expressions for these quantities at $u = 0$. The partition function is

$$\mathcal{Z}(0) = (\det A)^{-1/2}. \tag{10.6}$$

This result is most easily obtained by performing an orthogonal rotation of the y_i to a basis which diagonalizes the matrix A_{ij} before performing the integral. Also useful for the u expansion is the identity

$$\int \mathcal{D}y \exp\left(-\frac{1}{2}\sum_{ij} y_i A_{ij} y_j - \sum_i J_i y_i\right) = (\det A)^{-1/2} \exp\left(\frac{1}{2}\sum_{ij} J_i A_{ij}^{-1} J_j\right), \tag{10.7}$$

which is obtained by shifting the y_i variables to complete the square in the argument of the exponential. By taking derivatives of this identity with respect to the J_i, and then setting $J_i = 0$, we can generate expressions of all the correlators at $u = 0$. In particular, the two-point correlator is

$$C_{ij}(0) = A_{ij}^{-1}. \tag{10.8}$$

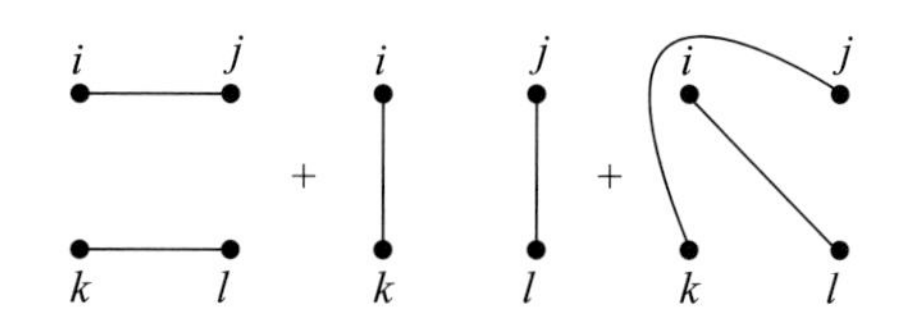

Figure 10.1 Diagrammatic representation of the four-point correlator in (10.10). Each line is factor of the propagator in (10.8).

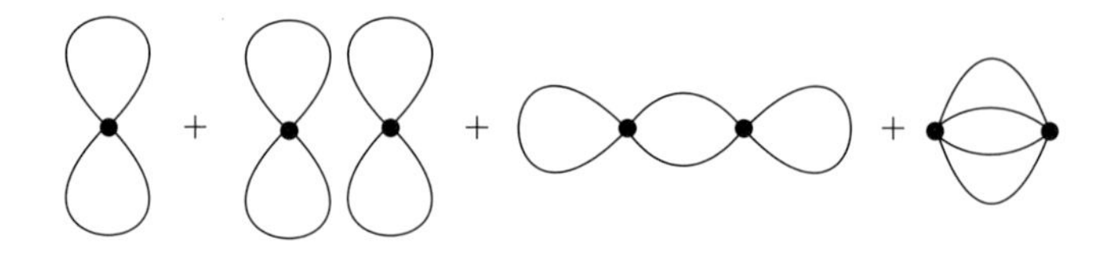

Figure 10.2 Diagrams for the partition function to order u^2.

For the $2n$-point correlator, we have an expression known as Wick's theorem:

$$\langle y_1 y_2 \dots y_{2n} \rangle = \sum_P \langle y_{P1} y_{P2} \rangle \cdots \langle y_{P(2n-1)} y_{P2n} \rangle , \tag{10.9}$$

where the summation over P represents the sum over all possible products of pairs, and we reiterate that both sides of the equation are evaluated at $u = 0$. Thus, for the four-point correlator, we have

$$\langle y_i y_j y_k y_\ell \rangle = \langle y_i y_j \rangle \langle y_k y_\ell \rangle + \langle y_i y_k \rangle \langle y_j y_\ell \rangle + \langle y_i y_\ell \rangle \langle y_j y_k \rangle . \tag{10.10}$$

There is a natural diagrammatic representation of the right-hand side of (10.10): we represent each distinct field y_i by a dot, and then draw a line between dots i and j to represent each factor of $C_{ij}(0)$; see Fig 10.1.

We can now generate the needed expansions of $\mathcal{Z}(u)$ and $C_{ij}(u)$ simply by expanding the integrands in powers of u, and by evaluating the resulting series term by term using Wick's theorem. What follows is simply a set of very useful diagrammatic rules for efficiently obtaining the answer at each order. However, whenever in doubt on the value of a diagram, it is often easiest to go back to this primary definition.

For $\mathcal{Z}(u)$, expanding to order u^2, we obtain the diagrams shown in Fig 10.2, which evaluate to the expression

$$\begin{aligned} \frac{\mathcal{Z}(u)}{\mathcal{Z}(0)} = {} & 1 - \frac{u}{8} \sum_i (A_{ii}^{-1})^2 + \frac{1}{2} \left(\frac{u}{8} \sum_i (A_{ii}^{-1})^2 \right)^2 + \frac{u^2}{16} \sum_{i,j} A_{ii}^{-1} A_{jj}^{-1} \left(A_{ij}^{-1} \right)^2 \\ & + \frac{u^2}{48} \sum_{i,j} \left(A_{ij}^{-1} \right)^4 + \mathcal{O}(u^3). \end{aligned} \tag{10.11}$$

We are usually interested in the free energy, which is obtained by taking the logarithm of the above expression, yielding

$$\ln \frac{\mathcal{Z}(u)}{\mathcal{Z}(0)} = -\frac{u}{8} \sum_i (A_{ii}^{-1})^2 + \frac{u^2}{16} \sum_{i,j} A_{ii}^{-1} A_{jj}^{-1} \left(A_{ij}^{-1} \right)^2 + \frac{u^2}{48} \sum_{i,j} \left(A_{ij}^{-1} \right)^4 + \mathcal{O}(u^3). \tag{10.12}$$

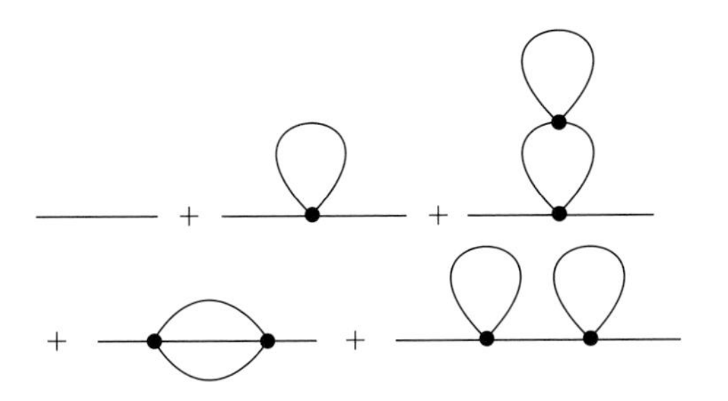

Figure 10.3 Diagrams for the two-point correlation function to order u^2.

Now, notice an important feature of (10.12): the terms here correspond precisely to the subset of the terms in Fig. 10.2 associated with the *connected* diagrams. These are diagrams in which all points are connected to each other by at least one line, and this result is an example of the "linked cluster theorem." We will not prove this very useful result here: at all orders in u, we can obtain the perturbation theory for the free energy by keeping only the connected diagrams in the expansion of the partition function.

Now let us consider the u expansion of the two-point correlator, $C_{ij}(u)$, in (10.5). Here, we have to expand the numerator and denominator in (10.5) in powers of u, evaluate each term using Wick's theorem, and then divide the result series. Fortunately, the linked cluster theorem simplifies things a great deal here too. The result of the division is simply to cancel all the disconnected diagrams. Thus, we need only expand the numerator, and keep only connected diagrams. The diagrams are shown in Fig. 10.3 to order u^2, and they evaluate to

$$C_{ij}(u) = A_{ij}^{-1} - \frac{u}{2}\sum_k A_{ik}^{-1}A_{kk}^{-1}A_{kj}^{-1} + \frac{u^2}{4}\sum_{k,\ell} A_{ik}^{-1}A_{kk}^{-1}A_{k\ell}^{-1}A_{\ell\ell}^{-1}A_{\ell j}^{-1}$$
$$+ \frac{u^2}{4}\sum_{k,\ell} A_{ik}^{-1}\left(A_{k\ell}^{-1}\right)^2 A_{\ell\ell}^{-1}A_{kj}^{-1} + \frac{u^2}{6}\sum_{k,\ell} A_{ik}^{-1}\left(A_{k\ell}^{-1}\right)^3 A_{\ell j}^{-1}. \tag{10.13}$$

We now state the useful *Dyson's theorem*. For this, it is useful to consider the expansion of the inverse of the C_{ij} matrix, and write it as

$$C_{ij}^{-1} = A_{ij} - \Sigma_{ij}, \tag{10.14}$$

where the matrix Σ_{ij} is called the "self-energy," for historical reasons not appropriate here. Using (10.13), some algebra shows that, to order u^2,

$$\Sigma_{ij}(u) = -\delta_{ij}\frac{u}{2}A_{ii}^{-1} + \delta_{ij}\frac{u^2}{4}\sum_k \left(A_{ik}^{-1}\right)^2 A_{kk}^{-1} + \frac{u^2}{6}\left(A_{ij}^{-1}\right)^3, \tag{10.15}$$

and these are shown graphically in Fig. 10.4. Dyson's theorem states that we can obtain the expression for the Σ_{ij} directly from the graphs for C_{ij} in Fig. 10.3 by two modifications: (i) drop the factors of A^{-1} associated with external lines, and (ii) keep only the graphs which are one-particle irreducible (1PI). The latter are graphs that do not break into disconnected pieces when one internal line is cut; the last graph in Fig. 10.3 is one-particle reducible, and so does not appear in (10.15) and Fig. 10.4.

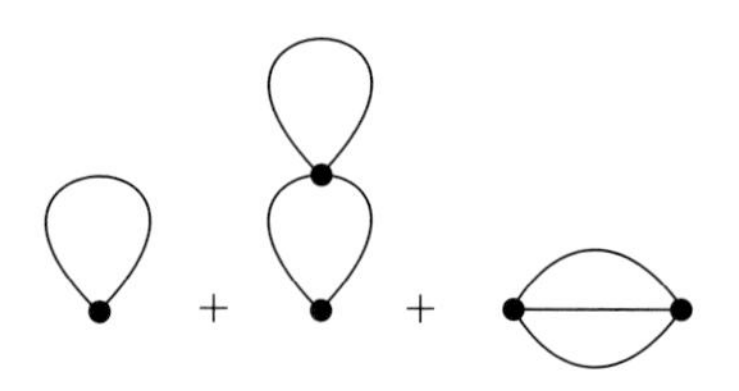

Figure 10.4 Diagrams for the self-energy to order u^2.

10.2 Expansion for Susceptibility

We now apply the results of Section 10.1 to our functional integral representation in (10.2) for the vicinity of the phase transition.

The problem defined by (10.2) has an important simplifying feature not shared by our general analysis of (10.3): translational invariance. This means that correlators only depend upon the differences of spatial coordinates, and that the analog of the matrix A can be diagonalized by a Fourier tranformation. So now we define the correlator

$$C_{\alpha\beta}(x-y) = \langle\phi_\alpha(x)\phi_\beta(y)\rangle - \langle\phi_\alpha(x)\rangle\langle\phi_\beta(y)\rangle, \quad (10.16)$$

where the subtraction allows generalization to the ferromagnetic phase; we will only consider the paramagnetic phase here.

The subtraction in (10.16) is also needed for the fluctuation–dissipation theorem. We discuss the full version of this theorem in Section 11.1.2, but note a simpler version. We consider the susceptibility, $\chi_{\alpha\beta}$, the response of the system to an applied "magnetic" field h_α under which the action changes as

$$\mathcal{S}_\phi \to \mathcal{S}_\phi - \int d^D x h_\alpha(x)\phi_\alpha(x). \quad (10.17)$$

Then

$$\chi_{\alpha\beta}(x-y) = \frac{\delta\langle\phi_\alpha(x)\rangle}{\delta h_\beta(y)} = C_{\alpha\beta}(x-y), \quad (10.18)$$

where the last equality follows from taking the derivative with respect to the field. Below we set $h_\alpha = 0$ after taking the derivative. The Fourier transform of the susceptibility $\chi_{\alpha\beta}$ is

$$\chi_{\alpha\beta}(k) = \int d^D x e^{-ikx}\chi_{\alpha\beta}(x). \quad (10.19)$$

In the paramagnetic phase $\chi_{\alpha\beta}(k) \equiv \delta_{\alpha\beta}\chi(k)$, and the susceptibility $\chi(k)$ will play a central role in our analysis.

We can also Fourier transform the field $\phi_\alpha(x)$ to $\phi_\alpha(k)$, and so obtain the following representation of the action from (10.2)

$$S_\phi = \frac{1}{2}\int \frac{d^D k}{(2\pi)^D} |\phi_\alpha(k)|^2 (k^2 + r)$$
$$+ \frac{u}{4!}\int \frac{d^D k}{(2\pi)^D}\frac{d^D q}{(2\pi)^D}\frac{d^D p}{(2\pi)^D}\phi_\alpha(k)\phi_\alpha(q)\phi_\alpha(p)\phi_\alpha(-k-p-q). \tag{10.20}$$

In this representation it is clear that the quadratic term in the action is diagonal, and so the inversion of the matrix A is immediate. In particular, from (10.8) we have the susceptibility at $u = 0$

$$\chi_0(k) = \frac{1}{k^2 + r}, \tag{10.21}$$

where we have defined $\chi_0(k)$ to be the value of $\chi(k)$ at $u = 0$. Dyson's theorem in (10.14) becomes a simple algebraic relation

$$\chi(k) = \frac{1}{1/\chi_0(k) - \Sigma(k)} = \frac{1}{k^2 + r - \Sigma(k)}. \tag{10.22}$$

We will shortly obtain an explicit expression for $\Sigma(k)$.

Let us now explore some of the consequences of the $u = 0$ result in (10.21), which describes Gaussian fluctuations about mean-field theory in the paramagnetic phase, $r > 0$. The zero momentum susceptibility, which we denote simply as $\chi \equiv \chi(k = 0) = 1/r$, diverges as we approach the phase transition at $K = K_c$ from the high-temperature paramagnetic phase. This divergence is a key feature of the phase transition, and its nature is encoded in the critical exponent γ defined by

$$\chi \sim (K_c - K)^{-\gamma}. \tag{10.23}$$

At this leading order in u we have $\gamma = 1$.

We can also examine the spatial correlations in the $u = 0$ theory. Performing the inverse Fourier transform to $C_{\alpha\beta}(x) = \delta_{\alpha\beta}C(x)$ we find

$$C(x) = \int \frac{d^D k}{(2\pi)^D}\frac{e^{ikx}}{(k^2 + r)} = \frac{(2\pi)^{-D/2}}{(x\xi)^{(D-2)/2}} K_{(D-2)/2}(x/\xi), \tag{10.24}$$

where here K is the modified Bessel function, and we have introduced a characteristic length scale, ξ, defined by

$$\xi = 1/\sqrt{r}. \tag{10.25}$$

This is the correlation length, and is a measure of the distance over which fluctuations of ϕ_α (or the underlying spins σ_i^z) are correlated. This is evident from the limiting forms of (10.24) in various asymptotic regimes:

$$C(x) \sim \begin{cases} \dfrac{1}{x^{D-2}} & , \quad x \ll \xi \\ \dfrac{e^{-x/\xi}}{x^{(D-1)/2}\xi^{(D-3)/2}} & , \quad x \gg \xi \end{cases}. \tag{10.26}$$

As could be expected of a correlation length, the correlations decay exponentially to zero at distances larger than ξ.

An important property of our expression in (10.25) for the correlation length is that it diverges upon the approach to the critical point. This divergence is also associated with a critical exponent, ν, defined by

$$\xi \sim (K_c - K)^{-\nu}, \tag{10.27}$$

and our present theory yields $\nu = 1/2$. In the vicinity of the phase transition, this large value of ξ provides an a posteriori justification of our taking a continuum perspective on the fluctuations.

Let us now move beyond the $u = 0$ theory, and consider the corrections at order u. After mapping to Fourier space, the result in (10.15) for the self-energy yields

$$\Sigma(k) = -u\frac{(M+2)}{6}\int \frac{d^D p}{(2\pi)^D}\frac{1}{p^2+r}. \tag{10.28}$$

Here, and below, there is an implicit upper bound of $k < \Lambda$ needed to obtain finite answers for the wavevector integrals. The M dependence comes from keeping track of the spin index α along each line of the Feynman diagram, and allowing for the different possible contractions of such indices at each u interaction point. We then have from (10.22) our main result for the correction in the susceptibility

$$\frac{1}{\chi(k)} = k^2 + r + u\frac{(M+2)}{6}\int \frac{d^D p}{(2\pi)^D}\frac{1}{p^2+r} + \mathcal{O}(u^2). \tag{10.29}$$

The first consequence of (10.29) is a shift in the position of the critical point. From (10.23), a natural way to define the position of the phase transition is by the zero of $1/\chi$. The order u correction in (10.29) shows that the critical point is no longer at $r = r_c = 0$, but at

$$r_c = -u\frac{(M+2)}{6}\int \frac{d^D p}{(2\pi)^D}\frac{1}{p^2} + \mathcal{O}(u^2). \tag{10.30}$$

Now, let us combine Eqs. (10.29) and (10.30) to determine the behavior of χ as $r \searrow r_c$. We introduce the coupling s defined by

$$s \equiv r - r_c, \tag{10.31}$$

which measures the deviation of the system from the critical point. Rewriting (10.29) in terms of s rather than r (we will always use s in favor or r in all subsequent analysis), we have

$$\frac{1}{\chi} = s + u\left(\frac{M+2}{6}\right)\int^{\Lambda} \frac{d^D p}{(2\pi)^D}\left(\frac{1}{p^2+s} - \frac{1}{p^2}\right). \tag{10.32}$$

We are interested in the vicinity of the critical point, at which $s \to 0$.

A crucial point is that the nature of this limit depends sensitively on whether D is greater than or less than four. For $D > 4$, we can simply expand the integrand in (10.32) in powers of s and obtain

$$\frac{1}{\chi} = s(1 - c_1 u \Lambda^{D-4}), \tag{10.33}$$

where c_1 is a non-universal constant dependent upon the nature of the cutoff. Thus, the effects of interactions appear to be relatively innocuous; the static susceptibility still diverges with the mean-field form $\chi(0) \sim 1/s$ as $s \to 0$, with the critical exponent $\gamma = 1$. This is in fact the generic behavior to all orders in u, and all the mean-field critical exponents apply for $D > 4$.

For $D < 4$, we notice that the integrand in (10.32) is convergent at high momenta, and so it is permissible to send $\Lambda \to \infty$. We then find that the correction to first order in u has a universal form

$$\frac{1}{\chi} = s\left[1 - \left(\frac{M+2}{6}\right)\frac{2\Gamma((4-D)/2)}{(D-2)(4\pi)^{D/2}}\frac{u}{s^{(4-D)/2}}\right]. \tag{10.34}$$

Notice that no matter how small u is, the correction term eventually becomes important for a sufficiently small s, and indeed it diverges as $s \to 0$. So, for sufficiently large ξ, the mean-field behavior cannot be correct, and a resummation of the perturbation expansion in u is necessary.

The situtation becomes worse at higher orders in u. As suggested by (10.34), the perturbation series for $1/(s\chi)$ is actually in powers of $u/s^{(4-D)/2}$, and so each successive term diverges more strongly as $s \to 0$. Thus, the present perturbative analysis is unable to describe the vicinity of the critical point for $D < 4$. This problem is cured by a renormalization group treatment which the reader can find in the QPT book.

Problems

10.1 Use Wick's theorem to obtain (10.29) from the first term in (10.15). Also, obtain the terms to u^2 for the M-component ϕ^4 field theory from the last two terms in (10.15).

11 Relativistic Scalar Field: Correlation Functions

The connection between correlation functions in imaginary and real time is developed using spectral representations and the fluctuation–dissipation theorem. The linear response to external perturbations is described by the Kubo formula. These general methods are illustrated by an application to the ordering quantum phase transition of a scalar field in two spatial dimensions. Quasiparticles are argued to be absent at the quantum critical point.

The relativistic scalar field theory was treated in Chapter 10 in imaginary time in D spacetime dimensions. There was complete isotropy between space and time, and the field theory examined was effectively a classical statistical mechanics problem in D dimensions.

In this chapter we examine some of the consequences of this imaginary time theory for the real-time and frequency correlation functions of a quantum system in $d = D - 1$ spatial dimensions. This quantum system has a continuum Hamiltonian $\mathcal{H}$, and the field $\phi_\alpha(x,\tau)$ corresponds to an operator $\phi_\alpha(x)$, which acts on the Hilbert space of the Hamiltonian; this connection has been explored in much detail in the QPT book [233]. In particular, the two-point correlator of ϕ_α in D classical dimensions maps onto the *time-ordered* correlation function

$$C_{\alpha\beta}(x,\tau_1;y,\tau_2) = \begin{cases} \frac{1}{\mathcal{Z}} \mathrm{Tr}\left(e^{-\mathcal{H}/T} \hat{\phi}_\alpha(x,\tau_1) \hat{\phi}_\beta(y,\tau_2)\right) & \text{for } \tau_1 > \tau_2, \\ \frac{1}{\mathcal{Z}} \mathrm{Tr}\left(e^{-\mathcal{H}/T} \hat{\phi}_\beta(y,\tau_2) \hat{\phi}_\alpha(x,\tau_1)\right) & \text{for } \tau_1 < \tau_2, \end{cases} \tag{11.1}$$

where $\hat{\phi}_\alpha(x,\tau)$ is defined by imaginary-time evolution under the $\mathcal{H}$:

$$\hat{\phi}_\alpha(x,\tau) \equiv e^{\mathcal{H}\tau} \phi_\alpha(x) e^{-\mathcal{H}\tau}. \tag{11.2}$$

This chapter describes the consequences of this mapping for the correlations of the quantum system.

11.1 Spectral Representation

The first step in our analysis is to express C in (11.1) in the so-called spectral representation.

To clean up the notation, we will drop the spin components in (11.1) because they play no essential role, and consider the case $M = 1$. Because the correlator in (11.1) is periodic in time with period $1/T$, it is useful to define its Fourier transform at the "Matsubara" frequency ω_n, which must be an integer multiple of $2\pi T$, $\omega_n = 2\pi nT$, by

$$\begin{aligned}\chi(x,\omega_n) &\equiv \int_0^{1/T} d\tau e^{i\omega_n\tau} C(x,\tau;0,0) \\ &= \frac{1}{\mathcal{Z}}\int_0^{1/T} d\tau e^{i\omega_n\tau} \mathrm{Tr}\left(e^{-\mathcal{H}/T}\hat{\phi}(x,\tau)\hat{\phi}(0,0)\right),\end{aligned} \tag{11.3}$$

where we have used spatial and temporal translation invariance to set the arguments of the second $\hat{\phi}$ at the origin of spacetime.

Now imagine we know all the eigenstates and eigenenergies of the continuum quantum Hamiltonian $\mathcal{H}$. In general, these states occupy a continuum of energies, but by placing the field theory in a d-dimensional cubic box of size L (we will eventually take $L \to \infty$) we can obtain a discrete spectrum in which the exact eigenstates are labeled by the index m. Thus, a complete set of orthonormal eigenstates is $|m\rangle$, and their eigenenergies are m. These eigenstates satisfy the completeness identity:

$$\sum_m |m\rangle\langle m| = \hat{1}, \tag{11.4}$$

where $\hat{1}$ is the identity operator. We now insert this identity before and after the first $\hat{\phi}$ operator to obtain

$$\begin{aligned}\chi(x,\omega_n) &= \sum_{m,m'} \frac{\langle m'|\phi(x)|m\rangle\langle m|\phi(0)|m'\rangle}{\mathcal{Z}} \int_0^{1/T} d\tau e^{(i\omega_n - E_m + E_{m'})\tau - E_{m'}/T} \\ &= \frac{1}{\mathcal{Z}}\sum_{m,m'} \langle m'|\phi(x)|m\rangle\langle m|\phi(0)|m'\rangle \frac{\left(e^{-E_m/T} - e^{-E_{m'}/T}\right)}{(i\omega_n - E_m + E_{m'})};\end{aligned} \tag{11.5}$$

in the last step we used the fact that $e^{i\omega_n/T} = 1$ at all Matsubara frequencies. We can now write this in its final form, known as the spectral representation:

$$\chi(x,\omega_n) = \int_{-\infty}^{\infty} \frac{d\Omega}{\pi}\frac{\rho(x,\Omega)}{\Omega - i\omega_n}, \tag{11.6}$$

where the spectral density $\rho(x,\Omega)$ is given by (see also (2.25) for the Fermi liquid)

$$\begin{aligned}\rho(x,\Omega) \equiv \frac{\pi}{\mathcal{Z}}\sum_{m,m'} &\langle m'|\phi(x)|m\rangle\langle m|\phi(0)|m'\rangle \\ &\times \left(e^{-E_{m'}/T} - e^{-E_m/T}\right)\delta(\Omega - E_m + E_{m'}).\end{aligned} \tag{11.7}$$

This spectral density is the key quantity connecting various correlation functions both in real and imaginary time. Indeed once we know the spectral density, we can easily obtain all needed correlation functions. In particular, from (11.6) we immediately obtain the correlation function at the Matsubara frequencies of imaginary time. The inverse problem is much more difficult: from a knowledge of $\chi(x,\omega_n)$ at all ω_n, it is not easy to find $\rho(x,\Omega)$. Indeed, this problem is ill-posed: very small errors in the values of $\chi(x,\omega_n)$ lead to large errors in $\rho(x,\Omega)$. However, when exact analytic expressions for

$\chi(x, \omega_n)$ are available, it is possible to determine $\rho(x, \Omega)$; we will use this method on a number of occasions.

11.1.1 Structure Factor

Let us now turn to correlation functions in *real* time, t, which are directly observable in the laboratory. We define time evolution of operators in the Heisenberg picture by (compare (11.2))

$$\hat{\phi}_\alpha(x,t) \equiv e^{i\mathcal{H}t}\phi_\alpha(x)e^{-i\mathcal{H}t}. \tag{11.8}$$

Then the real-time analog of (11.1) is the correlation function

$$\widetilde{C}_{\alpha\beta}(x,t;x',t') = \frac{1}{\mathcal{Z}}\mathrm{Tr}\left(e^{-\mathcal{H}/T}\hat{\phi}_\alpha(x,t)\hat{\phi}_\beta(x',t')\right), \tag{11.9}$$

As above, we drop the indices α, β below, and deal only with the case $M = 1$.

The dynamic structure factor $S(k, \omega)$ is defined by a Fourier transform of the real-time correlation (compare (11.3):

$$S(k,\omega) = \int d^d x \int_{-\infty}^{\infty} dt \widetilde{C}(x,t;x',t')e^{-i\boldsymbol{k}\cdot(\boldsymbol{x}-\boldsymbol{x}')+i\omega(t-t')}. \tag{11.10}$$

Notice that the time integration extends over all real values of t, unlike the limited domain between 0 and $1/T$ for imaginary time. Also, there is no time-ordering above, unlike in (11.1).

The dynamic structure factor is the quantity naturally measured in scattering experiments, such as neutron, X-ray, or light scattering of solid-state systems. This becomes clear from the spectral representation. Proceeding as in (11.5) by repeated insertions of the identity (11.4), it is easy to show that

$$S(k,\omega) = \frac{2\pi}{\mathcal{Z}V}\sum_{m,m'} e^{-E_{m'}/T}|\langle m'|\phi(\boldsymbol{k})|m\rangle|^2\delta(\omega - E_m + E_{m'}), \tag{11.11}$$

where V is the volume of the system and $\phi(\boldsymbol{k})$ is the spatial Fourier transform of the operator $\phi(x)$. The expression (11.11) has the structure of a transition rate computed using Fermi's golden rule. The system is initially in the state $|m'\rangle$ with the thermal probability $e^{-E_{m'}/T}/\mathcal{Z}$; an external perturbation (the incoming photon or neutron) couples linearly to the operator $\phi(\boldsymbol{k})$, and (11.11) computes the transition probability per unit time to the final state $|m\rangle$. The result is clearly proportional to the Born scattering cross section of the photon or neutron with momentum transfer $\boldsymbol{k}$ and energy transfer ω. Note that we are only making the Born approximation on the coupling between the probe and the system; in principle, (11.11) treats all interactions within the system exactly.

Comparing the expression (11.11) with the spectral density (11.7), we obtain the exact identity

$$S(k,\omega) = \frac{2}{1-e^{-\omega/T}}\rho(k,\omega), \tag{11.12}$$

where $\rho(k,\omega)$ is the spatial Fourier transform of $\rho(x,\omega)$. This is the first of our needed connections between real and imaginary time correlations, relating the dynamic structure factor to the spectral density, which in turn determines the correlator at the imaginary Matsubara frequencies by (11.6). The identity (11.12) is one statement of the "fluctuation–dissipation" theorem, and the reason for this terminology will become clearer in the following subsection.

11.1.2 Linear Response

Now we consider another experimentally useful quantity: the time-dependent response to an external perturbation. For simplicity, we consider an external time- and space-dependent "field" $h_\alpha(x,t)$ that couples linearly to the field operator $\phi(x)$, and so changes the Hamiltonian by

$$\mathcal{H} \to \mathcal{H} - \int d^d x \phi_\alpha(x) h_\alpha(x,t). \tag{11.13}$$

Because of the presence of $h_\alpha(x,t)$, all observables will now have space and time dependence, and the system will no longer be in thermal equilibrium. We would like to compute the change in the observables from equilibrium to linear order in $h_\alpha(x,t)$. This is given by a very general expression known as the Kubo formula. Without any specific knowledge of $\mathcal{H}$, we can write the shift away from equilibrium for an arbitrary observable $\mathcal{O}(x)$ in the following form:

$$\delta\langle\mathcal{O}(x)\rangle(t) = \int d^d x' \int_{-\infty}^{\infty} dt' \chi_{\mathcal{O}\alpha}(x-x',t-t') h_\alpha(x',t'), \tag{11.14}$$

where the initial δ indicates "change due to an external field," and the expectation value on the left-hand side is evaluated in the density matrix describing the state of the system in the presence of h. The coefficient on the right-hand-side is the dynamic susceptibility χ; it is a characteristic of $\mathcal{H}$ in the absence of h, and so it is invariant under time and space translations. Finally, the expression (11.14) must obey the important constraint of causality:

$$\chi(x,t) = 0 \text{ for } t < 0 \tag{11.15}$$

because the response can only depend upon the values of h at earlier times. This identifies χ as the so-called "retarded" response function.

The Kubo formula is a general result for the susceptibility χ. Its derivation involves a simple exercise in first-order time-dependent perturbation theory. We start from an initial thermal state described by a density matrix $\exp(-\mathcal{H}/T)/\mathcal{Z}$, and compute its evolution under the change (11.13) by integrating the equations of motion to first order in h. Such a computation leads to the main result:

$$\chi_{\mathcal{O}\alpha}(x-x',t-t') = i\theta(t-t')\frac{1}{\mathcal{Z}}\text{Tr}\left(e^{-\mathcal{H}/T}[\hat{\mathcal{O}}(x,t),\hat{\phi}_\alpha(x',t')]\right), \tag{11.16}$$

where $\theta(t)$ is the unit step function, $\mathcal{H}$ is the Hamiltonian in the absence of h, and the time evolution of the operators is specified as in (11.8).

For our subsequent analysis we focus on the observable $\mathcal{O} = \phi$, and drop the α index by considering $M = 1$. Then the susceptibility of interest is

$$\chi(x-x',t-t') = i\theta(t-t')\frac{1}{\mathcal{Z}}\mathrm{Tr}\left(e^{-\mathcal{H}/T}[\hat{\phi}(x,t),\hat{\phi}(x',t')]\right). \tag{11.17}$$

It is useful to consider this susceptibility in momentum and frequency space by defining

$$\chi(k,\omega) = \int d^d x \int_0^\infty dt\, \chi(x,t) e^{-ik\cdot x + i\omega t}. \tag{11.18}$$

Note the limits on the time integration, which are a consequence of (11.16). Because of these limits, if we consider ω as a complex number, the integral in (11.18) is well defined for ω in the upper half-plane. The oscillatory factor $e^{i\omega t}$ becomes a decaying exponential for ω in the upper half-plane, and so the integral (11.18) converges. The function $\chi(k,\omega)$ is therefore an analytic function of ω in the upper half-plane, and we define its value on the real ω axis by analytic continuation from the upper half-plane. Alternatively stated, we map $\omega \to \omega + i\eta$, where η is a small positive number, at intermediate stages of the computation, and take the limit $\eta \to 0$ at the end; this procedure leads to convergent results at all stages.

Let us now obtain a spectral representation of (11.17) and (11.18) as before. We insert (11.4) around the ϕ operators, and perform the Fourier transform to obtain

$$\chi(k,\omega) = \frac{1}{\mathcal{Z}V}\sum_{m,m'} |\langle m'|\phi(k)|m\rangle|^2 \frac{\left(e^{-E_{m'}/T} - e^{-E_m/T}\right)}{\omega + i\eta - E_m + E_{m'}} \tag{11.19}$$

in the limit $\eta \to 0^+$. Now comparing (11.19) with (11.7), we obtain our main result:

$$\chi(k,\omega) = \int_{-\infty}^{\infty} \frac{d\Omega}{\pi} \frac{\rho(k,\Omega)}{\Omega - \omega - i\eta}, \tag{11.20}$$

connecting the retarded response function to the spectral density. The relations (11.6), (11.12), and (11.20) are the key results of this section, connecting the spectral density to the imaginary-time correlations, the real-time dynamic structure factor and retarded susceptibility. Also note that $\chi(k,\omega = 0) \equiv \chi(k)$ is the *static* susceptibility.

A key feature of our results is the close similarity between (11.6) and (11.20). They show that the imaginary-time susceptibility $\chi(k,i\omega_n)$ and the retarded response function $\chi(k,\omega)$ are part of the same analytic function $\chi(k,z)$ defined by

$$\chi(k,z) = \int_{-\infty}^{\infty} \frac{d\Omega}{\pi} \frac{\rho(k,\Omega)}{\Omega - z} \tag{11.21}$$

for a general complex frequency z. For $z = i\omega_n$ on the imaginary axis, $\chi(k,z)$ is the imaginary-time correlation at the Matsubara frequencies. And for $z = \omega + i\eta$, just above the real axis, we obtain the retarded response functions of the Kubo formula. Thus, we can map the imaginary-time correlation to the retarded response function by analytic continuation. Also, our notation for the frequency argument of χ, ω_n vs. ω, will implicitly determine whether we are considering response functions on the imaginary or real axis.

For the case where the Kubo formula like (11.17) involving the commutator of a field with its Hermitian conjugate, the associated spectral density $\rho(x,\Omega)$ in (11.7) is real. Then we can write (11.20) as

$$\rho(k,\omega) = \mathrm{Im}\chi(k,\omega) = \frac{(1-e^{-\omega/T})}{2} S(k,\omega), \tag{11.22}$$

where we used (11.12). The structure factor on the right-hand side is a measure of fluctuations of the field ϕ, while $\mathrm{Im}\chi(k,\omega)$ measures the response of ϕ which is out of phase with the applied field (from (11.14)). As in the damped harmonic oscillator, it is the out-of-phase component that measures the energy absorbed by the system from the external field, thus justifying the name "fluctuation–dissipation" theorem.

11.2 Correlations across the Quantum Critical Point

Let us recall the field theory

$$\mathcal{S}_\phi = \int d^D x \left\{ \frac{1}{2} \left[(\nabla_x \phi_\alpha)^2 + r\phi_\alpha^2(x) \right] + \frac{u}{4!} \left(\phi_\alpha^2(x) \right)^2 \right\}, \tag{11.23}$$

where $D = d+1$ is the number of spacetime dimensions, and $\alpha = 1,\ldots,M$. The superfluid–insulator transition in (8.21) at $K_1 = 0$ corresponds to the case $M = 2$, and the superfluid order parameter is $\Psi_B \sim \phi_1 + i\phi_2$ We are interested in the nature of the spectrum as the field theory is tuned from the superfluid to the insulator with r increasing across the quantum critical point at $r = r_c$. The $M = 3$ case of this theory describes the pressure-induced quantum phase transition in $\mathrm{TlCuCl_3}$ that we discussed in Section 1.2.

11.2.1 Insulator

The insulator is present for large and positive r, and here we just use a perturbation theory in u. We compute the correlation function

$$\chi(k) = \int d^D x \langle \phi_\alpha(x)\phi_\alpha(0)\rangle e^{-ikx}. \tag{11.24}$$

At leading order in u, this is simply $\chi(k) = 1/(k^2 + r)$. Analytically continuing this to the quantum theory in d dimensions, we map $k^2 \to c^2k^2 - \omega^2$, and so obtain the retarded response function

$$\chi(k,\omega) = \frac{1}{c^2k^2 + r - (\omega + i\eta)^2}. \tag{11.25}$$

Taking its imaginary part, we have the spectral density

$$\rho(k,\omega) = \frac{\mathcal{A}}{2\varepsilon_k} \left[\delta(\omega - \varepsilon_k) - \delta(\omega + \varepsilon_k) \right], \tag{11.26}$$

where

$$\varepsilon_k = (c^2k^2 + r)^{1/2} \tag{11.27}$$

is the dispersion, and we have introduced a "quasiparticle residue" $\mathcal{A} = 1$. Thus, the spectrum consists of $M = 2$ "particles," and these correspond to the particle and hole excitations of the Mott insulator.

Now let us move beyond the Gaussian theory, and look at perturbative corrections in u. This is represented by the self-energy diagrams in Fig. 10.4. After analytic continuation, we can write the susceptibility in the form

$$\chi(k,\omega) = \frac{1}{c^2k^2 + r - (\omega + i\eta)^2 - \Sigma(k,\omega)}. \tag{11.28}$$

We continue to identify the position of the pole of $\chi(k,\omega)$ (if present) as a function of ω as a determinant of the spectrum of the quasiparticle, and the residue of the pole as the quasiparticle residue $\mathcal{A}$. The real part of the self-energy $\Sigma(k,\omega)$ will serve to modify the quasiparticle dispersion relation, and the value of $\mathcal{A}$, but will not remove the pole from real ω axis. To understand the possible decay of the quasiparticle, we need to consider the imaginary part of the self-energy.

From rather general arguments (in the next paragraph), it is possible to see that

$$\mathrm{Im}\Sigma(k, \omega = \varepsilon_k) = 0 \tag{11.29}$$

at $T = 0$. Compare this to the result (2.38), which holds only in the limit of vanishing frequency; here it holds for a finite range of momentum and frequency. This can be explicitly verified by a somewhat lengthy evaluation of the diagrams in Fig. 10.4, and an analytic continuation of the result. An immediate consequence is that the dynamic susceptibility has a delta-function contribution, which is given exactly by (11.26). All the higher-order corrections only serve to renormalize r, and reduce the quasiparticle residue $\mathcal{A}$ from unity; the dispersion relation continues to retain the form in (11.27) by relativistic invariance. The stability of the delta function reflects the stability of the single quasiparticle excitations; a quasiparticle with momentum k not too large cannot decay into any other quasiparticle state and still conserve energy and momentum.

However, $\Sigma(k,\omega)$ does have some more interesting consequences at higher ω. We can view ω as the energy inserted by ϕ into the ground state, and so far we have assumed that this energy can only create a quasiparticle with energy ε_k, which has a minimum energy of r. Only for $\omega > pr$, with p an integer, can we expect the creation of p particle states. The global $O(M)$ symmetry actually restricts p to be odd, and so the lowest-energy multi-particle states that will appear in χ are at $\omega = 3r$. Consonant with this, we find that the self-energy acquires a non-zero imaginary part at zero momentum only for $\omega > 3r$, which means there is a threshold for three-particle creation at $\omega = 3r$. The form of $\mathrm{Im}\Sigma(0,\omega)$ at the threshold runs out to

$$\mathrm{Im}\Sigma(0,\omega) \propto \mathrm{sgn}(\omega)\theta(|\omega| - 3r)(|\omega| - 3r)^{(d-1)} \tag{11.30}$$

for ω around $3r$. Taking the imaginary part of (11.28), we obtain the generic form of the spectral density shown in Fig. 11.1.

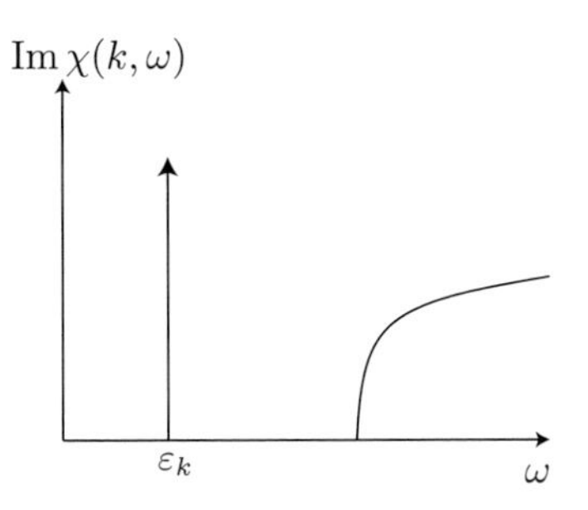

Figure 11.1 The spectral density in the paramagnetic phase at $T = 0$ and a small k. Shown are a quasiparticle delta function at $\omega = \varepsilon_k$ and a three-particle continuum at higher frequencies. There are additional n-particle continua ($n \geq 5$ and odd) at higher energies, which are not shown.

We now present a simple physical argument for the nature of the threshold singularity in Eq. (11.30). Just above threshold, we have a particle with energy $3r + \delta\omega$ that decays into three particles with energies just above r. The particles in the final state will also have a small momentum, and so we can make a non-relativistic approximation for their dispersion: $r + c^2k^2/(2r)$. Because the rest mass contributions, r, add up to the energy of the initial state, we can neglect from now. The decay rate, by Fermi's golden rule is proportional to the density of final states, which yields

$$\begin{aligned}\mathrm{Im}\Sigma(0, 3r + \delta\omega) \propto & \int_0^{\delta\omega} d\Omega_1 d\Omega_2 \int \frac{d^d p}{(2\pi)^d} \frac{d^d q}{(2\pi)^d} \delta\left(\Omega_1 - \frac{c^2 p^2}{2\sqrt{r}}\right) \\ & \times \delta\left(\Omega_2 - \frac{c^2 q^2}{2\sqrt{r}}\right) \delta\left(\delta\omega - \Omega_1 - \Omega_2 - \frac{c^2(p+q)^2}{2\sqrt{r}}\right) \\ & \sim (\delta\omega)^{(d-1)}, \end{aligned} \tag{11.31}$$

in agreement with (11.30). We expect this perturbative estimate of the threshold singularity to be exact in all $d \geq 2$.

11.2.2 Quantum Critical Point

Evaluation of the susceptibility of the classical field theory (11.23) at its critical point $r = r_c$ requires a sophisticated resummation of perturbation theory in u using the renormalization group (RG). To order u^2, at the renormalized critical point, perturbation theory in dimension $D = 4$ shows that

$$\chi(k) = \frac{1}{k^2}\left(1 - C_1 u^2 \ln\left(\frac{\Lambda}{k}\right) + \cdots\right) \tag{11.32}$$

where Λ is a high momentum cutoff, and C_1 is a positive numerical constant. The RG shows that for $D < 4$ we should exponentiate this series to

$$\chi(k) \sim \frac{1}{k^{2-\eta}}, \tag{11.33}$$

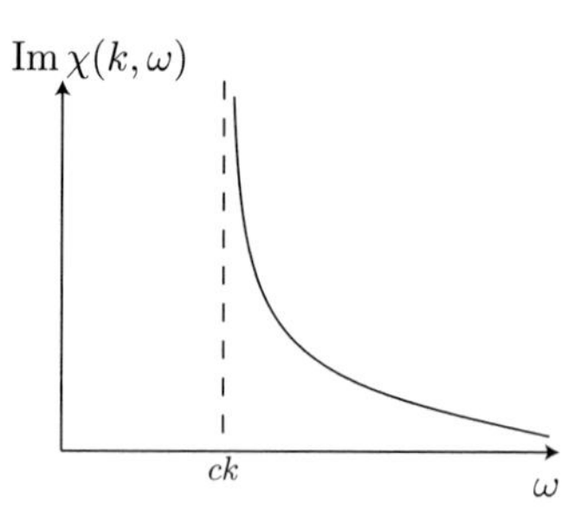

Figure 11.2 The spectral density at the quantum critical point. Note the absence of a quasiparticle pole, like that in Fig. 11.1.

where $\eta > 0$ is the so-called universal anomalous dimension. Precise computations of the value of η are now available for many critical points, including the $M = 2$ case of the field theory (11.23).

For the quantum critical point, we analytically continue the classical critical-point result in (11.33) to obtain the dynamic susceptibility at the quantum critical point at $T = 0$:

$$\chi(k,\omega) \sim \frac{1}{(c^2k^2 - \omega^2)^{1-\eta/2}}. \tag{11.34}$$

The key feature contrasting this result from (11.25) is that this susceptibility does *not* have poles on the real frequency axis. Rather, there are branch cuts going out from $\omega = \pm ck$ to infinity. Taking the imaginary part, we obtain a continuous spectral weight at $|\omega| > ck$:

$$\mathrm{Im}\chi(k,\omega) \sim \frac{\mathrm{sgn}(\omega)\theta(|\omega| - ck)}{(\omega^2 - c^2k^2)^{1-\eta/2}}; \tag{11.35}$$

see Fig 11.2. The absence of a pole indicates that there are no well-defined quasiparticle excitations. Instead, we have a dissipative continuum of critical excitations at all $|\omega| > ck$; any perturbation will not create a particle-like pulse, but decay into a broad continuum. This is a generic property of a strongly coupled quantum critical point. Further discussion of the physics of quantum states of matter without quasiparticle excitations appears in Chapters 32 and 34.

More generally, we can use scaling to describe the evolution of the spectrum as r approaches the critical point at $r = r_c$ from the insulating phase at $T = 0$. Because of the relativistic invariance, the energy gap $\Delta \sim \xi^{-z}$ with $z = 1$, where the correlation length ξ diverges as in the classical model $\xi \sim (r - r_c)^{-\nu}$ (these are definitions of the critical exponents z and ν). In terms of Δ, scaling arguments imply that the susceptibility obeys

$$\chi(k,\omega) = \frac{1}{\Delta^{2-\eta}} \widetilde{F}\left(\frac{ck}{\Delta}, \frac{\omega}{\Delta}\right) \tag{11.36}$$

for some scaling function $\widetilde{F}$. In the insulating phase, the M quasiparticles have dispersion $\varepsilon_k = (c^2k^2 + \Delta^2)^{1/2}$ (the momentum dependence follows from relativistic invariance). Comparing (11.36) with (11.26), we see that the two expressions are compatible if the quasiparticle residue scales as

$$\mathcal{A} \sim \Delta^{\eta}; \tag{11.37}$$

so the quasiparticle residue vanishes as we approach the quantum critical point. Above the quasiparticle pole, the susceptibility of the paramagnetic phase also has p-particle continua having thresholds at $\omega = (c^2k^2 + p^2\Delta^2)^{1/2}$, with $p \geq 3$ and p odd. As $\Delta \to 0$ upon approaching the quantum critical point, these multi-particle continua merge to a common threshold at $\omega = ck$ to yield the quantum critical spectrum in (11.35).

11.2.3 Superfluid State

Now $r < r_c$, and we have to expand about the ordered saddle point with $\phi_\alpha = N_0\delta_{\alpha,1}$, where

$$N_0 = \sqrt{\frac{-6r}{u}}. \tag{11.38}$$

So we write

$$\phi_\alpha(x) = N_0\delta_{\alpha,1} + \tilde{\phi}_\alpha(x) \tag{11.39}$$

and expand the action in powers of $\tilde{\phi}_\alpha$.

The first important consequence of the superfluid order is that the dynamic structure factor

$$S(k,\omega) = N_0^2(2\pi)^{d+1}\delta(\omega)\delta^d(k) + \cdots, \tag{11.40}$$

where the ellipsis represents contributions at non-zero ω. This delta function is related to that in (3.22), and is easily detectable in elastic neutron scattering as a clear signature of the presence of superfluid long-range order.

We now discuss the finite ω contributions to (11.40). We assume the ordered moment is oriented along the $\alpha = 1$ direction. From Gaussian fluctuations about the saddle point of (11.23) we obtain susceptibilities that are diagonal in the spin index, with the longitudinal susceptibility (see Fig. 11.3)

$$\chi_{11}(k,\omega) = \frac{1}{c^2k^2 - (\omega + i\eta)^2 + 2|r|}, \tag{11.41}$$

and the transverse susceptibility

$$\chi_{\alpha\alpha}(k,\omega) = \frac{1}{c^2k^2 - (\omega + i\eta)^2}, \quad \alpha > 1. \tag{11.42}$$

The pole in the transverse susceptibility is the gapless "spin-wave" excitation, corresponding to the phonon excitation of the Bose gas in (3.16). The pole in the longtudinal susceptibility yields a gapped massive particle with mass-squared equal to $2|r|$; this is an analog of the Higgs particle of particle physics. However, this massive particle can decay into multiple spin-wave excitations, and ultimately the Higgs excitation is overdamped [206].

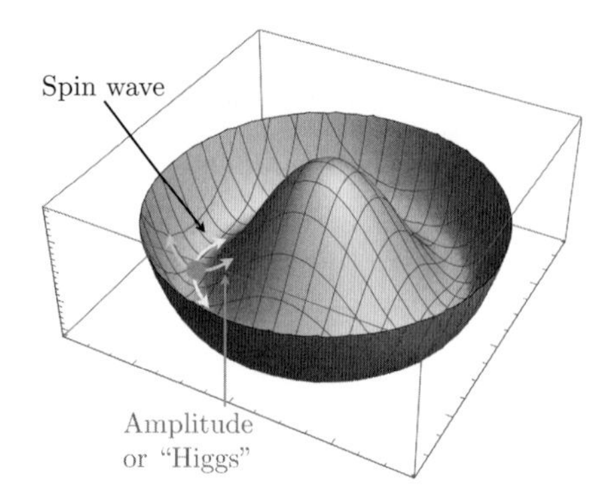

Figure 11.3 Energy as a function of the ϕ_α field for $M = 2$. The filled circle indicates one possible ground state. The transverse susceptibility yields a gapless spin-wave mode, while the longitudinal susceptibility is associated with amplitude or "Higgs" excitations.

Problem

11.1 Obtain a more precise version of (11.31) from the order u^2 expression for Σ which follows from (10.15). Write each propagator using a spectral representation, perform the frequency integrals, and then take the imaginary part.

12 Fermions and Bosons in One Spatial Dimension

The theory of weakly interacting Fermi and Bose gases in one dimension is presented. Gapless phases are described by the Tomonaga–Luttinger liquid theory of a gapless scalar, and its instabilities are obtained via a sine-Gordon field theory. The bosonization relations between relativistic fermions and bosons in one dimension are obtained.

The theory of the weakly interacting Fermi gas in Chapter 2 and the theory of the weakly interacting Bose gas in Chapter 3 are faithful representations of the low-energy quantum physics only in spatial dimensions $d \geq 2$. These theories fail for arbitrary weak interactions in the special case of $d = 1$, as described in this chapter. This one-dimensional failure was long considered of academic interest only, but has since found numerous important experimental applications with the advances in nano-scale experiments. Moreover, as we will see in Parts II and IV, the one-dimensional theory also applies to the edges of numerous interesting two-dimensional systems.

Remarkably, weakly interacting bosons and fermions realize essentially the same phase of quantum matter [165], often called the Luttinger liquid (or the Tomonaga–Luttinger liquid). This is a gapless state of matter, described at the lowest energies by a simple theory: massless, free, relativistic, scalar fields representing density or phase fluctuations. The physical Fermi and bose field operators can be written as exponential "vertex" operators in terms of these massless scalar fields, and this non-trivial connection leads to novel structures in the observable correlation function. Formally, the spectral functions of many observable operators turn out to have a structure similar to that we encountered at the superfluid–insulator quantum critical point in $d = 2$ in Section 11.2.2. However, this superficial similarity should not obscure a crucial physical difference: the $d = 2$ superfluid–insulator quantum critical point has *no* quasiparticle excitations, whereas *all* low–energy states of a Luttinger liquid can be described in terms of the free quasiparticle excitations of the massless scalar field. The novel behavior of a Luttinger liquid arises entirely from the non-trivial connection between the physical observables and the quasiparticle operators, and *not* from the absence of quasiparticles.

We begin in Section 12.1 by considering free fermions in one dimension, and describe their mapping to the theory of relativistic free scalars. We turn to the theory of interacting fermions in Section 12.2, and that of interacting bosons in Section 12.3. In both cases, we find situations in which a finite interaction strength can lead to a phase transition to a gapped insulating phase, with similarities to the superfluid–insulator transition on bosons studied in Chapter 8.

12.1 Non-interacting Fermions

We start with a continuum Fermi field, $\Psi_F(x)$, and expand it in terms of its right- and left-moving components near the two Fermi points:

$$\Psi_F(x) = e^{ik_F x}\Psi_R(x) + e^{-ik_F x}\Psi_L(x). \tag{12.1}$$

The Fermi wavevector, k_F, is related to the fermion density, ρ_0, by $k_F = \pi\rho_0$. We linearize the fermion dispersion about the Fermi points in terms of a Fermi velocity v_F, and then the dynamics of $\Psi_{R,L}$ is described by the simple Hamiltonian

$$H_{FL} = -iv_F \int dx \left(\Psi_R^\dagger \frac{\partial \Psi_R}{\partial x} - \Psi_L^\dagger \frac{\partial \Psi_L}{\partial x} \right), \tag{12.2}$$

which corresponds to the imaginary-time Lagrangian $\mathcal{L}_{FL}$

$$\mathcal{L}_{FL} = \Psi_R^\dagger \left(\frac{\partial}{\partial \tau} - iv_F \frac{\partial}{\partial x} \right) \Psi_R + \Psi_L^\dagger \left(\frac{\partial}{\partial \tau} + iv_F \frac{\partial}{\partial x} \right) \Psi_L. \tag{12.3}$$

We will examine $\mathcal{L}_{FL}$ a bit more carefully and show, somewhat surprisingly, that it can also be interpreted as a theory of free relativistic bosons. The mapping can be rather precisely demonstrated by placing $\mathcal{L}_{FL}$ on a system of finite length L. We choose to place antiperiodic boundary conditions of the Fermi fields $\Psi_{L,R}(x+L) = -\Psi_{L,R}(x)$; this arbitrary choice will not affect the thermodynamic limit $L \to \infty$, which is ultimately all we are interested in. We can expand $\Psi_{L,R}$ in Fourier modes

$$\Psi_R(x) = \frac{1}{\sqrt{L}} \sum_{n=-\infty}^{\infty} \Psi_{Rn} e^{i(2n-1)\pi x/L}, \tag{12.4}$$

and similarly for Ψ_L. The Fourier components obey canonical Fermi commutation relations $\{\Psi_{Rn}, \Psi_{Rn'}^\dagger\} = \delta_{nn'}$ and are described by the simple Hamiltonian

$$\tilde{H}_R = \frac{\pi v_F}{L} \sum_{n=-\infty}^{\infty} (2n-1)\Psi_{Rn}^\dagger \Psi_{Rn} - E_0, \tag{12.5}$$

where E_0 is an arbitrary constant setting the zero of energy, which we adjust to make the ground-state energy of $\tilde{H}_R$ exactly equal to 0; very similar manipulations apply to the left-movers Ψ_L. The ground state of $\tilde{H}_R$ has all fermions states with $n > 0$ empty, while those with $n \leq 0$ are occupied. We also define the total fermion number ("charge"), Q_R, of any state by the expression as

$$Q_R = \sum_n : \Psi_{Rn}^\dagger \Psi_{Rn} :, \tag{12.6}$$

and similarly for Q_L. The colons are the so-called normal-ordering symbol – they simply indicate that the operator enclosed between them should include a c-number subtraction of its expectation value in the ground state of $\tilde{H}_R$, which of course ensures that $Q_R = 0$ in the ground state. Note that Q_R commutes with $\tilde{H}_R$ and so we need only consider states with definite Q_R, which allows us to treat Q_R as simply an integer. The partition function, Z_R, of $\tilde{H}_R$ at a temperature T is then easily computed to be

$$Z_R = \prod_{n=1}^{\infty}(1+q^{2n-1})^2, \tag{12.7}$$

where

$$q \equiv e^{-\pi v_F/TL}. \tag{12.8}$$

The square in (12.7) arises from the precisely equal contributions from the states with n and $-n+1$ in (12.5) after the ground-state energy E_0 has been subtracted out.

We provide an entirely different interpretation of the partition function Z_R. Instead of thinking in terms of occupation numbers of individual fermion states, let us focus instead on particle–hole excitations. We create a particle–hole excitation of "momentum" $n > 0$ above any fermion state by taking a fermion in an occupied state n' and moving it to the unoccupied fermion state $n' + n$. Clearly, the energy change in such a transformation is $2n\pi v_F/L$ and is independent of the value of n'. This independence on n' is a crucial property and is largely responsible for the results that follow. It is a consequence of the linear fermion dispersion in (12.3), and of being in $d = 1$. We interpret the creation of such a particle–hole excitation as being equivalent to the occupation of a state with energy $2n\pi v_F/L$ created by the canonical boson operator $b^{\dagger}_{Rn}$. We can place an arbitrary number of bosons in this state, and we now show how this is compatible with the multiplicity of the particle–hole excitations that can be created in the fermionic language.

The key observation is that there is a precise one-to-one mapping between the fermionic labeling of the states and those specified by the bosons creating particle–hole excitations. Take any fermion state, $|F\rangle$, with an arbitrary set of fermion occupation numbers and charge Q_R. We uniquely associate this state with a set of particle–hole excitations above a particular fermion state that we label $|Q_R\rangle$; this is the state with the lowest possible energy in the sector of states with charge Q_R, that is, $|Q_R\rangle$ has all fermion states with $n \leq Q_R$ occupied and all others unoccupied. The energy of $|Q_R\rangle$ is

$$\frac{\pi v_F}{L}\sum_{n=1}^{|Q_R|}(2n-1) = \frac{\pi v_F Q_R^2}{L}. \tag{12.9}$$

To obtain the arbitrary fermion state, $|F\rangle$, with charge Q_R, we first take the fermion in the "topmost" occupied state in $|Q_R\rangle$, (i.e., the state with $n = Q_R$) and move it to the topmost occupied state in $|F\rangle$ (see Fig. 12.1). We perform the same operation on the fermion in $n = Q_R - 1$ by moving it to the next lowest occupied state in $|F\rangle$. Finally, we repeat until the state $|F\rangle$ is obtained. This order of occupying the boson particle–hole excitations ensures that the $b^{\dagger}_{Rn}$ act in descending order in n. Such an ordering allows one to easily show that the mapping is invertible and one to one. Given any set of occupied boson states $\{n\}$ and a charge Q_R, we start with the state $|Q_R\rangle$ and act on it with the set of Bose operators in the same descending order; their ordering ensures

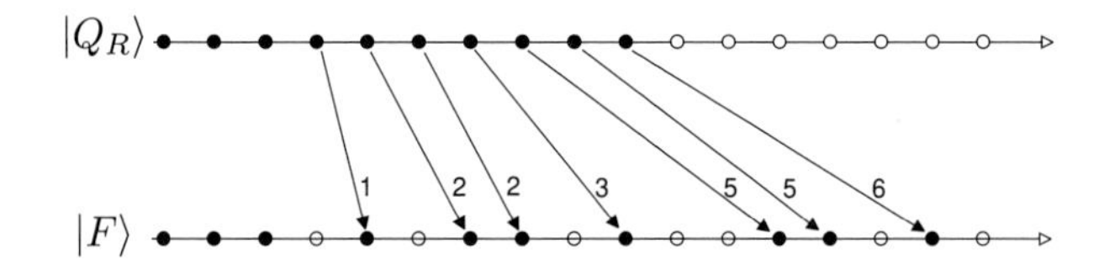

Figure 12.1 Sequence of particle–hole excitations (bosons b_{Rn}) by which one can obtain an arbitrary fermion state $|F\rangle$ from the state $|Q_R\rangle$, which is the lowest energy state with charge Q_R. The filled (open) circles represent occupied (unoccupied) fermion states with energies that increase in units of $2\pi v_F/L$ to the right. The arrows represent bosonic excitations, b_{Rn}, with the integer representing the value of n. Note that the bosons act in descending order in energy upon the descending sequence of occupied states in $|Q_R\rangle$.

that it is always possible to create such particle–hole excitations from the fermionic state, and one is never removing a fermion from an unoccupied state or adding it to an occupied state. The gist of these simple arguments is that the states of the many-fermion Hamiltonian $\tilde{H}_R$ in (12.5) are in one-to-one correspondence with the many-boson Hamiltonian

$$\tilde{H}'_R = \frac{\pi v_F Q_R^2}{L} + \frac{2\pi v_F}{L} \sum_{n=1}^{\infty} n b_{Rn}^\dagger b_{Rn}, \tag{12.10}$$

where Q_R can take an arbitrary integer value. It is straightforward to compute the partition function of $\tilde{H}'_R$ and we find

$$Z'_R = \left[\prod_{n=1}^{\infty} \frac{1}{(1-q^{2n})} \right] \left[\sum_{Q_R=-\infty}^{\infty} q^{Q_R^2} \right]. \tag{12.11}$$

Our pictorial arguments above prove that we must have $Z_R = Z'_R$. That this is the case is an identity from the theory of elliptic functions. (The reader is invited to verify that the expressions (12.7) and (12.11) generate identical power-series expansions in q.)

12.1.1 Tomonaga–Luttinger Liquid Theory

The above gives an appealing picture of bosonization at the level of states and energy levels, but we want to extend it to include operators, and obtain expressions for the bosonized Hamiltonian in a continuum formulation. From the action of the b_{Rn} operator on the fermion states, we can anticipate that it may be proportional to the Fourier components of the fermion density operator. So we consider the operator $\rho_R(x)$ representing the normal-ordered fermion density:

$$\rho_R(x) =: \Psi_R^\dagger(x)\Psi_R(x) := \frac{Q_R}{L} + \frac{1}{L}\sum_{n\neq 0} \rho_{Rn} e^{i2n\pi x/L}, \tag{12.12}$$

where the last step is a Fourier expansion of $\rho_R(x)$; the zero wavevector component is Q_R/L, while non-zero wavevector terms have coefficient ρ_{Rn}. The commutation relations of the ρ_{Rn} are central to our subsequent considerations and require careful evaluation; we have

$$\begin{aligned}[\rho_{Rn}, \rho_{R-n'}] &= \sum_{n_1,n_2} \left[\Psi^\dagger_{Rn_1}\Psi_{Rn_1+n}, \Psi^\dagger_{Rn_2}\Psi_{Rn_2-n'}\right] \\ &= \sum_{n_2} \left(\Psi^\dagger_{Rn_2-n}\Psi_{Rn_2-n'} - \Psi^\dagger_{Rn_2}\Psi_{Rn_2+n-n'}\right). \end{aligned} \tag{12.13}$$

It may appear that a simple change of variables in the summation over the second term in (12.13) ($n_2 \to n_2 - n$) shows that it equals the first, and so the combined expression vanishes. However, this is incorrect because it is dangerous to change variables on expressions that involve the summation over all integer values of n_2 and are therefore individually divergent; rather, we should first decide upon a physically motivated large-momentum cutoff that will make each term finite and then perform the subtraction. We know that the linear spectrum in (12.5) holds only for a limited range of momenta and, for sufficiently large $|n|$, lattice corrections to the dispersion will become important. However, in the low-energy limit of interest here, the high fermionic states at such momenta will be rarely, if ever, excited from their ground-state configurations. We can use this fact to our advantage by explicitly subtracting the ground-state expectation value ("normal-order") from every fermionic bilinear operator we consider; the fluctuations will then be practically zero for the high-energy states in both the linear spectrum model (12.5) and the actual physical systems, and only the low-energy states, where (12.5) is actually a good model, will matter. After such normal ordering, the summation over both terms in (12.12) is well defined and we are free to change the summation variable. As a result, the normal-ordered terms then do indeed cancel, and the expression (12.13) reduces to the subtraction needed to normal order the terms

$$\begin{aligned}[\rho_{Rn}, \rho_{R-n'}] &= \delta_{nn'} \sum_{n_2} \left(\langle \Psi^\dagger_{Rn_2-n}\Psi_{Rn_2-n}\rangle - \langle \Psi^\dagger_{Rn_2}\Psi_{Rn_2}\rangle\right) \\ &= \delta_{nn'} n. \end{aligned} \tag{12.14}$$

This key result shows that the only non-zero commutator is between ρ_{Rn} and ρ_{R-n} and that it is simply the number n. By a suitable rescaling of the ρ_{Rn} it should be evident that we can associate them with canonical bosonic creation and annihilation operators. We do no do this explicitly but simply work directly with the ρ_{Rn} as a set of operators obeying the defining commutation relation (12.14), without making explicit reference to the fermionic relation (12.12). We assert that the Hamiltonians $\tilde{H}_R, \tilde{H}'_R$ are equivalent to

$$\tilde{H}''_R = \frac{\pi v_F Q_R^2}{L} + \frac{2\pi v_F}{L} \sum_{n=1}^{\infty} \rho_{R-n}\rho_{Rn}. \tag{12.15}$$

This assertion is simple to prove. First, it is clear from the commutation relations (12.14) that the eigenvalues and degeneracies of (12.15) are the same as those of (12.10). Second, the definition (12.15) and the commutation relations (12.14) imply that

$$[\tilde{H}''_R, \rho_{R-n}] = \frac{2\pi v_F n}{L} \rho_{R-n}. \tag{12.16}$$

Precisely the same commutation relation follows from the fermionic form (12.5) and the definition (12.12).

We can now perform the same analysis on the left-moving fermions. The expressions corresponding to (12.4), (12.5), (12.12), (12.14), and (12.15) are

$$\Psi_L(x) = \frac{1}{\sqrt{L}} \sum_{n=-\infty}^{\infty} \Psi_{Ln} e^{i(2n-1)\pi x/L}, \tag{12.17}$$

$$\tilde{H}_L = -\frac{\pi v_F}{L} \sum_{n=-\infty}^{\infty} (2n-1) \Psi_{Ln}^\dagger \Psi_{Ln} - E_0, \tag{12.18}$$

$$\rho_L(x) =: \Psi_L^\dagger(x)\Psi_L(x) := \frac{Q_L}{L} + \frac{1}{L} \sum_{n \neq 0} \rho_{Ln} e^{i2n\pi x/L}, \tag{12.19}$$

$$[\rho_{Ln}, \rho_{L-n'}] = -\delta_{nn'} n. \tag{12.20}$$

$$\tilde{H}_L'' = \frac{\pi v_F Q_L^2}{L} - \frac{2\pi v_F}{L} \sum_{n=1}^{\infty} \rho_{Ln} \rho_{L-n}. \tag{12.21}$$

We have now completed a significant part of the bosonization program. We have the "bosonic" Hamiltonian in (12.15) in terms of the operators ρ_{Rn}, which obey (12.14), and we also have the simple explicit relation (12.12) to the fermionic fields (along with the corresponding expressions for the left-movers above). Before proceeding further, we introduce some notation that will allow us to recast the results obtained so far in a compact, local, and physically transparent notation. We combine the operators ρ_{Rn} and ρ_{Ln} (the Fourier components of the left-moving fermions Ψ_L) into two local fields $\phi(x)$ and $\theta(x)$, defined by

$$\begin{aligned} \phi(x) &= -\phi_0 + \frac{\pi Q x}{L} - \frac{i}{2} \sum_{n \neq 0} \frac{e^{i2n\pi x/L}}{n} [\rho_{Rn} + \rho_{Ln}], \\ \theta(x) &= -\theta_0 + \frac{\pi J x}{L} - \frac{i}{2} \sum_{n \neq 0} \frac{e^{i2n\pi x/L}}{n} [\rho_{Rn} - \rho_{Ln}], \end{aligned} \tag{12.22}$$

where $Q = Q_R + Q_L$ is the total charge, $J = Q_R - Q_L$, and ϕ_0 and θ_0 are a pair of angular variables that are canonically conjugate to J and Q, respectively; that is, the only non-vanishing commutation relations between the operators on the right-hand sides of (12.22) are (12.14), $[\phi_0, J] = i$, and $[\theta_0, Q] = i$. For future use, it is also useful define

$$\varphi_R(x) \equiv \phi(x) + \theta(x) \quad , \quad \varphi_L(x) \equiv \phi(x) - \theta(x). \tag{12.23}$$

From (12.22) it is clear that φ_R and φ_L are "chiral" fields, as they only involve operators associated with the right- and left-moving fermions, respectively.

Our objective in introducing these operators is to produce a number of simple and elegant results. First, using (12.22), and the commutators just noted, we have

$$[\phi(x), \nabla\theta(y)] = [\theta(x), \nabla\phi(y)] = -i\pi\delta(x-y), \tag{12.24}$$

implying that $-\nabla\theta/\pi$ is canonically conjugate to ϕ, and $-\nabla\phi/\pi$ is canonically conjugate to θ; alternatively, we can write the unified form

$$[\phi(x), \theta(y)] = i\frac{\pi}{2}\mathrm{sgn}(x-y). \tag{12.25}$$

In terms of the chiral fields, the non-zero commutation relations are

$$[\varphi_R(x), \varphi_R(y)] = i\pi\,\mathrm{sgn}(x-y) \quad , \quad [\varphi_L(x), \varphi_L(y)] = -i\pi\,\mathrm{sgn}(x-y), \tag{12.26}$$

whereas φ_R and φ_L commute with each other. For future applications, it is also useful to express these commutation relations in terms of exponentials of the fields

$$\begin{aligned}
e^{i\alpha\phi(x)}e^{i\beta\theta(y)} &= e^{i\beta\theta(y)}e^{i\alpha\phi(x)}e^{-i\alpha\beta(\pi/2)\mathrm{sgn}(x-y)}, \\
e^{i\alpha\varphi_R(x)}e^{i\beta\varphi_R(y)} &= e^{i\beta\varphi_R(y)}e^{i\alpha\varphi_R(x)}e^{-i\alpha\beta\pi\mathrm{sgn}(x-y)}, \\
e^{i\alpha\varphi_L(x)}e^{i\beta\varphi_L(y)} &= e^{i\beta\varphi_L(y)}e^{i\alpha\varphi_L(x)}e^{i\alpha\beta\pi\mathrm{sgn}(x-y)}.
\end{aligned} \tag{12.27}$$

Second, (12.15) can now be written in the compact, local form

$$\tilde{H}_R'' + \tilde{H}_L'' = \frac{v_F}{2\pi}\int_0^L dx\left[\frac{1}{K}(\nabla\phi)^2 + K(\nabla\theta)^2\right], \tag{12.28}$$

where the dimensionless coupling K has been introduced for future convenience; in the present situation $K=1$, but we will see later that interactions lead to other values of K. The expressions (12.28) and (12.24) can be taken as defining relations, and we could have derived all the properties of the ρ_{Rn}, ρ_{Ln}, θ_0, ϕ_0 as consequences of the mode expansions (12.22), which follow after imposition of the periodic boundary conditions

$$\phi(x+L) = \phi(x) + \pi Q, \qquad \theta(x+L) = \theta(x) + \pi J. \tag{12.29}$$

These conditions show that $\phi(x)$ and $\theta(x)$ are to be interpreted as angular variables of period π. Our final version of the bosonic form of $\tilde{H}_R + \tilde{H}_L$ in (12.5) is contained in Eqns. (12.24), (12.28), and (12.29), and the two formulations are logically exactly equivalent. The Hilbert space splits apart into sectors defined by the integers $Q = Q_R + Q_L$, $J = Q_R - Q_L$, which measure the total charge of the left- and right-moving fermions. Note that

$$(-1)^Q = (-1)^J \tag{12.30}$$

and so the periods of ϕ and θ are together even or odd multiples of π. In terms of the chiral fields, this condition translates into φ_R and φ_L being angular variables with period 2π. All fluctuations in each charge sector are defined by the fluctuations of the local angular bosonic fields $\phi(x)$ and $\theta(x)$, or equivalently by the fermionic fields $\Psi_R(x)$ and $\Psi_L(x)$.

We close this subsection by giving the general form of the effective action for a Tomonaga–Luttinger liquid. The derivation above was limited to the case $K=1$, but we will see later that the generalization to $K \neq 1$ describes a wide class of interacting, compressible, quantum systems in one dimension. From the Hamiltonian (12.28) and the commutation relations (12.24) we can use the standard path-integral approach to write down the imaginary-time action

$$\mathcal{S}_{TL} = \frac{v_F}{2\pi} \int dx d\tau \left[\frac{(\nabla\phi)^2}{K} + K(\nabla\theta)^2 \right] - \frac{i}{\pi} \int dx d\tau \nabla\theta \partial_\tau \phi. \tag{12.31}$$

From this action, we can integrate out θ to obtain an action for the ϕ field alone:

$$\mathcal{S}_{TL} = \frac{1}{2\pi K v_F} \int dx d\tau \left[(\partial_\tau \phi)^2 + v_F^2 (\nabla\phi)^2 \right]. \tag{12.32}$$

This is just the action of a free, massless, relativistic scalar field. Conversely, we also have a "dual" formulation of $\mathcal{S}_{TL}$ in which we integrate out ϕ, and obtain the same action for θ but with $K \to 1/K$

$$\mathcal{S}_{TL} = \frac{K}{2\pi v_F} \int dx d\tau \left[(\partial_\tau \theta)^2 + v_F^2 (\nabla\theta)^2 \right]. \tag{12.33}$$

Finally, it is useful to express (12.31) in terms of the chiral fields φ_R and φ_L using (12.23)

$$\begin{aligned} \mathcal{S}_{TL} = \frac{v_F}{8\pi} \int dx d\tau & \left[\left(\frac{1}{K} + K \right) \left((\nabla\varphi_R)^2 + (\nabla\varphi_L)^2 \right) \right. \\ & \left. + 2 \left(\frac{1}{K} - K \right) \nabla\varphi_R \nabla\varphi_L \right] \\ & - \frac{i}{4\pi} \int dx d\tau \left[\nabla\varphi_R \partial_\tau \varphi_R - \nabla\varphi_L \partial_\tau \varphi_L \right]. \end{aligned} \tag{12.34}$$

The last kinematic "Berry phase" term reflects the commutation relations in (12.26). Note that the left- and right-movers decouple only at $K = 1$, and that is the only case with conformal invariance.

12.1.2 Operator Mappings

We are going to make extensive use of the fields $\phi(x)$, $\theta(x)$ in the following, and so their physical interpretation will be useful. The meaning of ϕ follows from the derivative of (12.22), which with (12.12) gives

$$\nabla\phi(x) = \pi\rho(x) \equiv \pi(\rho_R(x) + \rho_L(x)). \tag{12.35}$$

So the gradient of ϕ measures the total density of particles, and $\phi(x)$ increases by π each time x passes through a particle. The expression (12.35) also shows that we can interpret $\phi(x)$ as the displacement of the particle at position x from a reference state in which the particles are equally spaced as in a crystal; that is, $\phi(x)$ is something like a phonon displacement operator whose divergence is equal to the local change in density. Turning to $\theta(x)$, one interpretation follows from (12.24), which shows that $\Pi_\phi(x) \equiv -\nabla\theta(x)/\pi$ is the canonically conjugate momentum variable to the field $\phi(x)$. So Π_ϕ^2 in the Hamiltonian is the kinetic energy associated with the "phonon" displacement $\phi(x)$.

A physical interpretation of θ is obtained by taking the gradient of (12.22), and we obtain the analog of (12.35):

$$\nabla\theta(x) = \pi(\rho_R(x) - \rho_L(x)); \tag{12.36}$$

hence, gradients of θ measure the difference in density of right- and left-moving particles, that is, the current. Of course, we can combine (12.35) and (12.36) to obtain expressions for the chiral fields separately:

$$\nabla\varphi_R(x) = 2\pi\rho_R(x) \quad , \quad \nabla\varphi_L(x) = 2\pi\rho_L(x). \tag{12.37}$$

Finally, to complete the connection between the fermionic and bosonic theories, we need expressions for the single fermion annihilation and creation operators in terms of the bosons. Here, the precise expressions are dependent upon the short-distance regularization, but these fortunately only affect overall renormalization factors. With the limited aim of neglecting these non-universal renormalizations, the basic result can be obtained by some simple general arguments. First, note that if we annihilate a particle at the position x, from (12.35) the value of $\phi(y)$ at all $y < x$ has to be shifted by π. Such a shift is produced by the exponential of the canonically conjugate momentum operator Π_ϕ:

$$\exp\left(-i\pi\int_{-\infty}^{x}\Pi_\phi(y)dy\right) = \exp\left(i\theta(x)\right). \tag{12.38}$$

However, it is not sufficient to merely create a particle. We are creating a fermion, and the fermionic antisymmetry of the wavefunction can be accounted for if we pick up a minus sign for every particle to the left of x, that is, with a Jordan–Wigner–like factor

$$\exp\left(im\pi\int_{-\infty}^{x}\Psi_F^\dagger(y)\Psi_F(y)dy\right) = \exp\left(imk_F x + im\phi(x)\right), \tag{12.39}$$

where m is any odd integer, and $\Psi_F^\dagger\Psi_F$ measures the *total* density of fermions (see (12.1)), including the contributions well away from the Fermi points. In the second expression in (12.39), the term proportional to k_F represents the density in the ground state, while $\phi(x)$ is the integral of the density fluctuation above that. Combining the arguments leading to (12.38) and (12.39) we can assert the basic operator correspondence

$$\Psi_F(x) = \sum_{m\ \mathrm{odd}} A_m e^{imk_F x + im\phi(x) + i\theta(x)}, \tag{12.40}$$

where the A_m are a series of unknown constants, which depend upon microscopic details. We will see shortly that the leading contribution to (12.40) comes from the terms with $m = \pm 1$, and the remaining terms are subdominant at long distances. Comparison with (12.1) shows clearly that we may make the operator identifications for the right- and left- moving continuum fermion fields

$$\Psi_R \sim e^{i\theta + i\phi}, \qquad \Psi_L \sim e^{i\theta - i\phi}. \tag{12.41}$$

The other terms in (12.40) arise when these basic fermionic excitations are combined with particle–hole excitations at wavevectors that are integer multiples of $2k_F$.

In terms of the chiral fields, the operator correspondences separate simply into left- and right-moving sectors, as they must:

$$\Psi_R \sim e^{i\varphi_R}, \qquad \Psi_L \sim e^{-i\varphi_L}. \tag{12.42}$$

As an alternative to the above derivation, we can also obtain (12.42) by using the commutation relations

$$\begin{aligned}
[\rho_R(x), \Psi_R(y)] &= -\delta(x-y)\Psi_R(y), \\
[\rho_L(x), \Psi_L(y)] &= -\delta(x-y)\Psi_L(y).
\end{aligned} \tag{12.43}$$

It can now be verified that (12.37) and (12.42), combined with the commutation relations (12.26), are consistent with (12.43).

Actually, (12.42) is not precisely correct, but this will not be an issue in our subsequent discussion. From the commutation relations in (12.27) we can verify that $\Psi_R(x)$ and $\Psi_R(x')$ anti-commute with each other for $x \neq x'$, which is precisely the relationship expected for fermion operators (and similarly for Ψ_L). However, upon using (12.42) with (12.27) we find that $\Psi_R(x)$ commutes with $\Psi_L(x')$. This problem can be addressed by introducing the so-called Klein factors

$$\Psi_R \sim F_1 e^{i\varphi_R}, \qquad \Psi_L \sim F_2 e^{-i\varphi_L}, \tag{12.44}$$

which obey the anti-commutation relations $F_i F_j = -F_j F_i$ for $i \neq j$.

It is useful to recall here all the properties of the *chiral* theory, with only right-moving fermions. Such a theory is "anomalous," and cannot be realized in a one-dimensional system on its own. However, it can be realized on the edge of a two-dimensional system, as we will see in Chapter 19 on the integer quantum Hall effect, where we have the "chiral Luttinger" theory described by the following expressions for right-moving fermions and bosons

$$\begin{aligned}
H_{CL} &= -iv_F \int_0^L dx \Psi_R^\dagger \frac{\partial \Psi_R}{\partial x} \\
\mathcal{L}_{CL} &= \Psi_R^\dagger \left(\frac{\partial}{\partial \tau} - iv_F \frac{\partial}{\partial x} \right) \Psi_R \\
\nabla \varphi_R(x) &= 2\pi \rho_R(x) = 2\pi : \Psi_R^\dagger(x) \Psi_R(x) : \\
[\varphi_R(x), \varphi_R(y)] &= i\pi \,\mathrm{sgn}(x-y) \\
H_{CL} &= \frac{v_F}{4\pi} \int_0^L dx \, (\nabla \varphi_R)^2 \\
\mathcal{L}_{CL} &= \frac{1}{4\pi} \left[v_F (\nabla \varphi_R)^2 - i \nabla \varphi_R \partial_\tau \varphi_R \right]. \\
[\rho_R(x), \Psi_R(y)] &= -\delta(x-y)\Psi_R(y) \\
\Psi_R &\sim e^{i\varphi_R}.
\end{aligned} \tag{12.45}$$

12.2 Interacting Fermions

We now add two-body interactions between the Ψ_F fermions. For generic values of the wavevector k_F, the only momentum-conserving interaction for spinless fermions near the Fermi points is

$$H_U = \frac{U}{2}\int dx\left[(\rho_R(x)+\rho_L(x))(\rho_R(x)+\rho_L(x))\right]. \tag{12.46}$$

For special commensurate densities, there can be additional "umklapp" terms, but we defer consideration of such terms to the following section. Using the bosonization formula (12.35), we can write H_U as

$$H_U = \frac{U}{2\pi^2}\int dx(\nabla\phi)^2. \tag{12.47}$$

This can easily be absorbed into the bosonized version of H_{FL} in (12.28) by a redefinition of v_F and K. In this way we have shown that the Hamiltonian $H_{FL}+H_{12}$ is equivalent to (12.28) but with the parameters

$$\begin{aligned} v_F &\to v_F\left[1+\frac{Uv_F}{\pi}\right]^{1/2}, \\ K &= \left[1+\frac{Uv_F}{\pi}\right]^{-1/2}. \end{aligned} \tag{12.48}$$

The values of the parameters only hold for small U; however, the general result of a renormalization of v_F and K, but with no other change, is expected to hold more generally. Notice that now $K \neq 1$, as promised earlier.

We can now evaluate the correlators of the interacting fermion field using the operator mapping in (12.42). These can be obtained by use of the basic identity

$$\langle e^{i\mathcal{O}}\rangle = e^{-\langle\mathcal{O}^2\rangle/2}, \tag{12.49}$$

where $\mathcal{O}$ is an arbitrary linear combination of ϕ and θ fields at different spacetime points; this identity is a simple consequence of the free-field (Gaussian) nature of (12.28). In particular, all results can be reconstructed by combining (12.49) with repeated application of some elementary correlators. The first of these is the two-point correlator of ϕ:

$$\begin{aligned} \frac{1}{2}\langle(\phi(x,\tau)-\phi(0,0))^2\rangle &= \pi v_F K\int\frac{dk}{2\pi}T\sum_{\omega_n}\frac{1-e^{i(kx-\omega_n\tau)}}{\omega_n^2+v_Fk^2} \\ &= \frac{K}{4}\ln\left[\frac{\cosh(2\pi Tx/v_F)-\cos(2\pi T\tau)}{(2\pi T/v_F\Lambda)^2}\right], \end{aligned} \tag{12.50}$$

where Λ is a large-momentum cutoff. Similarly, we have for θ, the correlator

$$
\begin{aligned}
&\frac{1}{2}\langle(\theta(x,\tau)-\theta(0,0))^2\rangle \\
&\qquad = \frac{1}{4K}\ln\left[\frac{\cosh(2\pi Tx/v_F)-\cos(2\pi T\tau)}{(2\pi T/v_F\Lambda)^2}\right]. \qquad (12.51)
\end{aligned}
$$

To obtain the θ, ϕ correlator we use the relation $\Pi_\phi = -\nabla\theta/\pi$ and the equation of motion $i\Pi_\phi = \partial_\tau\phi/(\pi v_F K)$ that follows from the Hamiltonian (12.28); then, by an integration and differentiation of (12.50) we can obtain

$$
\langle\theta(x,\tau)\phi(0,0)\rangle = -\frac{i}{2}\arctan\left[\frac{\tan(\pi T\tau)}{\tanh(\pi Tx/v_F)}\right]. \qquad (12.52)
$$

This expression can also be obtained directly from (12.31). Finally, we can combine these expressions to obtain the fermion correlator (in imaginary time)

$$
\begin{aligned}
&\left\langle\Psi_R^\dagger(x,\tau)\Psi_R(0,0)\right\rangle \sim \\
&\quad \exp\left[-\frac{1}{4}(K+1/K)\ln\left[\frac{\cosh(2\pi Tx/v_F)-\cos(2\pi T\tau)}{(2\pi T/v_F\Lambda)^2}\right]\right. \\
&\qquad\qquad \left. - i\arctan\left[\frac{\tan(\pi T\tau)}{\tanh(\pi Tx/v_F)}\right]\right]. \qquad (12.53)
\end{aligned}
$$

In general, this is a complicated function, but it does have some useful limiting values. At $K=1$ it takes the simple form

$$
\left\langle\Psi_R^\dagger(x,\tau)\Psi_R(0,0)\right\rangle \sim \frac{1}{\sin(\pi T(v_F\tau - ix))} \qquad (12.54)
$$

expected for free fermions. Taking the Fourier transform of (12.53) for general K, and analytically continuing the resulting expressions to real frequencies is, in general, a complicated mathematical challenge; details can be obtained from Refs. [291, 292]. We quote some important results in the limit of $T=0$. The fermion spectral function has the following singularity at small frequencies near $\omega = v_F k$

$$
\begin{aligned}
-\mathrm{Im}G_R^R(k,\omega) &\sim \theta(\omega - v_F k)\,(\omega - v_F k)^{(K+1/K)/2-2} \\
&\qquad \omega > 0, k > 0. \qquad (12.55)
\end{aligned}
$$

At $K=1$, the spectrum function is a delta function $\sim \delta(\omega - v_F k)$ and that is indicative of the presence of quasiparticles in the free-fermion model. However, a key observation is that for $K \neq 1$ the delta function transforms into a branch-cut in the frequency complex plane, and this indicates the absence of fermionic quasiparticles. We can obtain the equal-time fermion Green's function of the original fermion field Ψ_F in (12.1) directly from (12.53):

$$
\langle\Psi_F^\dagger(x)\Psi_F(0)\rangle \sim \frac{\sin(k_F|x|)}{|x|^{(K+1/K)/2}}. \qquad (12.56)
$$

Taking the Fourier transform of this, we conclude that the momentum distribution function of the fermions, $n(k)$, does indeed have a singularity at the Fermi wavevector

$k = k_F$, but that this singularity is not generally a step discontinuity (as it is in Fermi liquids):

$$n(k) \sim -\text{sgn}(k - k_F)|k - k_F|^{(K+1/K)/2-1}. \tag{12.57}$$

Thus, interacting fermions in one dimension realize a new non-Fermi liquid phase, the Tomonaga–Luttinger liquid, whose momentum distribution function has a singularity at the Fermi surface, but the singularity is not the step discontinuity of a Fermi liquid in (2.34), and is instead given by (12.57).

12.2.1 Commensurate Densities

There are conditions under which the Luttinger liquid state is unstable to be a gapped insulator: this requires that the fermion density, ρ_0 is a rational number. The simplest example is when the spinless Fermi gas of Section 12.1 is at half-filling. Then $\rho_0 = 1/2$ and $k_F = \pi/2$. This special value of k_F allows an *Umklapp* process, when two right-moving fermions scatter to become two left-moving fermions: the total momentum transfer is 2π, and this is allowed by the unit periodicity of the underlying lattice. In the continuum formulation, this term is

$$H_U = v \int dx \left[\Psi_R^\dagger \nabla \Psi_R^\dagger \Psi_L \nabla \Psi_L + \Psi_L^\dagger \nabla \Psi_L^\dagger \Psi_R \nabla \Psi_R \right]. \tag{12.58}$$

We can now bosonize this using (12.41), and we obtain the sine-Gordon theory for the Luttinger liquid in the presence of periodic potential:

$$\mathcal{S}_{sG} = \mathcal{S}_{TL} - \lambda \int dx d\tau \cos(4\phi). \tag{12.59}$$

We discuss the properties of such a sine-Gordon theory in some detail in Section 25.2.3, in the context of a more general theory with the action

$$\mathcal{S}_{sG} = \mathcal{S}_{TL} - \lambda \int dx d\tau \cos(p\phi). \tag{12.60}$$

Here, we need the renormalization group (RG) equation for the coupling λ, which follows from (25.44):

$$\frac{d\lambda}{d\ell} = (2 - p^2 K/4)\lambda. \tag{12.61}$$

For the $p = 4$ case of interest to us, there is a critical point at $K = 1/2$, and for $K < 1/2$ there is a flow towards large $|\lambda|$, and we have an instability to a strongly coupled phase. (The full RG flow, shown in Fig. 25.1, has additional complexity which is discussed in Section 25.2.3). This strongly coupled state is expected to be an insulator, but the insulator breaks translational symmetry (so strictly speaking, it is not a Mott insulator). The breaking of translational symmetry can be understood from the fact that $\cos(2\phi)$ and $\sin(2\phi)$ are observables that break translational symmetry. This follows from (12.41),

$$\Psi_R^\dagger \Psi_L \sim e^{-2i\phi}, \tag{12.62}$$

and the fact that

$$\Psi_R^\dagger \Psi_L \to (-1)^n \Psi_R^\dagger \Psi_L \tag{12.63}$$

under translation by n lattice spacings, for $k_F = \pi/2$. When λ flows to $+\infty$ (say), then the values of ϕ will be pinned at $\pi p/2$, where p is an integer. Consequently $\cos(2\phi)$ takes the two possible values $(-1)^p$, and this implies a two-fold breaking of translational symmetry. A possible state is a charge-density wave of fermions with period 2. Similarly, when λ flow to $-\infty$, there are two possible values of $\sin(2\phi)$, and this corresponds to a "valence-bond solid," or a dimerization of the lattice with period 2.

12.3 Bosons in One Dimension

Next, we apply the formalism developed so far to a model of interacting bosons in one dimension

$$\begin{aligned} H_B = &-\frac{\hbar^2}{2m}\int dx \Psi_B^\dagger \nabla^2 \Psi_B \\ &+\frac{1}{2}\int dx dx' \Psi_B^\dagger(x)\Psi_B(x)V(x-x')\Psi_B^\dagger(x')\Psi_B(x'), \end{aligned} \tag{12.64}$$

with a two-body interaction $V(x)$.

For the case of a delta-function interaction,

$$V(x) = V_0 \delta(x), \tag{12.65}$$

we can show that, in the limit $V_0 \to \infty$, the Bose gas is exactly equivalent to a free Fermi gas. This follows from the exact solution of the N-particle Schrödinger equation with the wavefunction

$$\tilde{\Psi}_B(x_1, x_2, \ldots N) = \left[\prod_{i<j} \mathrm{sgn}(x_i - x_j)\right] \tilde{\Psi}_F(x_1, x_2, \ldots N), \tag{12.66}$$

where $\tilde{\Psi}_F$ is the free-fermion Slater determinant, and $\tilde{\Psi}_B$ is the boson wavefunction. The equality (12.66) can be established by examining the nature of the wavefunction as any pair of particles (say x_1 and x_2) approach each other. Then, the fermionic two-particle wavefunction is

$$\tilde{\Psi}_F(x_1, x_2) = e^{i\overline{K}(x_1+x_2)} \sin(k(x_1 - x_2)), \tag{12.67}$$

where $\overline{K}$ and k are the center of mass and relative momenta. It is then easy to check that the boson wavefunction

$$\tilde{\Psi}_B(x_1, x_2) = e^{i\overline{K}(x_1+x_2)} |\sin(k(x_1 - x_2))| \tag{12.68}$$

satisfies the Schrödinger equation in the limit $V_0 \to \infty$. This mapping implies that the Bose and Fermi field operators are related by the Jordan–Wigner transformation we met earlier in (12.39)

$$\Psi_B(x) = \Psi_F(x) \exp\left(i\pi \int_{-\infty}^{x} \Psi_F^\dagger(y)\Psi_F(y)dy \right). \tag{12.69}$$

We will now exploit this mapping, and assume that the effects of moving away from the $V_0 = \infty$ limit, or of having non-delta function interaction, can be absorbed in the resulting Tomonoga–Luttinger liquid theory simply by allowing $K \neq 1$, just as was the case for the Fermi gas (this can be shown more explicitly by regularizing the boson theory on a lattice, and performing a canonical transformation to eliminate all high-energy states that violate the boson hard-core constraint). We can now use (12.69) to express Ψ_B in terms of the continuum fields, ϕ, θ of the Tomonaga–Luttinger theory; using (12.40) we obtain

$$\Psi_B(x) = e^{i\theta} \left[B_0 + B_2 e^{i2\pi\rho_0 x} e^{2i\phi} + B_{-2} e^{-i2\pi\rho_0 x} e^{-2i\phi} + \cdots \right] \tag{12.70}$$

for some constants B_0, $B_{\pm 2}$. In (12.70), we have replaced $k_F = \pi\rho_0$, where ρ_0 is the boson density, because k_F does not have a direct physical interpretation in the Bose gas theory.

Much useful information can now be obtained from (12.70) combined with the correlators of the Tomonaga–Luttinger theory. First we note that (12.70) identifies θ as the *phase* of the Bose–Einstein condensate, and the quantum phase fluctuations are controlled by the simple harmonic theory (12.33). Indeed, up to oscillatory terms associated with the higher-order terms in (12.70), we can compute the two-point equal-time Bose field correlator from (12.51) and obtain

$$\left\langle \Psi_B^\dagger(x)\Psi_B(0) \right\rangle \sim \frac{1}{|x|^{1/2K}}. \tag{12.71}$$

So there is a power-law decay in the superfluid correlations, and no true long-range order. The phase fluctuations have destroyed the Bose–Einstein condensate, but the superfluid stiffness (associated with spatial gradient term in (12.33)) remains finite.

Let us now examine the response of the Bose gas to an external periodic potential under which

$$H_B \to H_B - V_G \int dx \cos(Gx) \Psi_B^\dagger \Psi_B, \tag{12.72}$$

where $2\pi/G$ is the spatial period of the potential. Inserting the expansion (12.70) into (12.72), and assuming that all important fluctuations of the θ and ϕ fields occur at wavelengths much larger than $2\pi/G$, we find that the spatial integral averages to zero *unless* $2\pi\rho_0 = G$. This translates into the condition that there must be exactly one boson per unit cell of the periodic potential. If we allow for omitted higher-order terms in (12.70), we find that a non-zero spatial average is allowed only if there is one boson for an integer number of unit cells. Restricting ourselves to the simplest case of one boson per unit cell, we find that there is modification to the low-energy theory given by

$$\mathcal{S}_{sG} = \mathcal{S}_{TL} - \lambda \int dx d\tau \cos(2\phi) \tag{12.73}$$

where $\lambda \propto V_G$. This is the sine-Gordon field theory.

Before analyzing the sine-Gordon theory, let us note an alternative interpretation of the $e^{2i\phi}$ operator: this operator creates a 2π *vortex* (in spacetime) in the phase of the Bose–Einstein condensate, analogous the spatial vortex considered in Section 7.1. We can conclude this by an argument very similar to that above (12.38). The operator $e^{2i\phi}$ shifts the phase θ by 2π along the spatial line $y < x$, which means it induces a 2π branch-cut in the spacetime configuration of θ. What we also conclude from (12.70) and (12.73) is that each such vortex (which is a tunneling event in spacetime) is accompanied by an oscillating Berry phase factor of $e^{\pm i2\pi\rho_0 x}$. Thus, the background density of bosons endows the vortex with a quantum-mechanical phase factor. This oscillatory Berry phase implies that a spatial average suppresses the matrix element for vortex-tunneling events. So the Tomonaga–Luttinger liquid is generically stable against vortex proliferation. The only exceptions arise for the cases when there is background potential which is commensurate with the boson density, and then there can be a net vortex-tunneling matrix element, as we have illustrated above in $\mathcal{S}_{sG}$.

(Parenthetically, we note that this argument also shows that we can consider the fermion theory at filling $\rho_0 = 1/2$ in (12.59) as a theory of the consequences of *double* vortices in θ, which are the smallest vortices in that case without oscillatory phase factors.)

Finally, let us note the properties of $\mathcal{S}_{sG}$ under a renormalization group analysis. From (12.49) and (12.50), we can compute the equal-time two-point correlator

$$\left\langle e^{2i\phi(x)} e^{-2i\phi(0)} \right\rangle \sim \frac{1}{|x|^{2K}} \tag{12.74}$$

and so conclude that

$$\dim[e^{2i\phi}] = K \tag{12.75}$$

at the Tomonaga–Luttinger liquid fixed point ($\lambda = 0$). So then, for small λ, we have the renormalization group equation

$$\frac{d\lambda}{d\ell} = (2-K)\lambda. \tag{12.76}$$

For $K > 2$, the $\lambda = 0$ fixed point, and so the gapless Tomonaga–Luttinger liquid phase, is *stable* to the introduction of a periodic potential with one boson per unit cell. On the other hand, for $K < 2$, there is a flow towards large $|\lambda|$, and we have an instability to a strongly coupled phase. We cannot predict the strongly coupled quantum state by the present methods, but it is not difficult to make a reasonable surmise. At large $|\lambda|$, the values of ϕ lock to the minima of the $\cos(2\phi)$ term, and hence the fluctuations of the conjugate θ fields are strongly enhanced. So we are then in a gapped phase in which the phase of the Bose–Einstein condensate is ill-defined. This is easy to identify as the Mott insulator of Chapter 8, in which each unit cell has trapped a single boson.

Problems

12.1 Consider a gas of free $S = 1/2$ electrons in one dimension with dispersion $\varepsilon_k = k^2/(2m)$ and density n. Compute the region in the ω, k, plane over which the density–density correlation function (defined as in (9.49)) $\text{Im}\chi_0^R(k,\omega)$ is non-zero. Show that in the limit $k \to 0$ we have the simple result

$$\text{Im}\chi_0^R(k,\omega) = C(k)\left[\delta(\omega - v_F k) - \delta(\omega + v_F k)\right]. \tag{12.77}$$

Thus, there is no particle–hole continuum in the density spectrum in one dimension, only coherent excitations that propagate with a velocity v_F. Compute $C(k)$.

12.2 Impurity in a Luttinger liquid. Consider a single impurity at $x = 0$ in a Luttinger liquid. Its strong effect is back scattering, that is, converting left-moving fermions to right-moving fermions. The impurity action is therefore

$$\mathcal{S}_{\text{imp}} = \lambda \int d\tau \left[\Psi_R^\dagger(x=0,\tau)\Psi_L(x=0,\tau) + \text{H.c.}\right]. \tag{12.78}$$

Obtain the RG equation for λ. For what values of K is the impurity scattering irrelevant?

12.3 Consider a dilute gas of bosons b_i moving on the sites, i of a chain described by the Hamiltonian

$$H = -w\sum_i \left(b_i^\dagger b_{i+1} + b_{i+1}^\dagger b_i - 2b_i^\dagger b_i\right) + \sum_i \left(V n_i(n_i - 1) - \mu n_i\right), \tag{12.79}$$

where $n_i = b_i^\dagger b_i$ is the number operator, w is the hopping matrix element, and V is the on-site repulsion between the bosons. In the limit of large V, states with more than one boson on a site will only occur rarely, and it should pay to restrict the Hilbert space by projecting out such states. However, the elimination will generate a residual interaction of order w^2/V between the states on the restricted space. This interaction can be determined by the effective Hamiltonian method (described in Problem 16.1). Show that, to second order in w, the effective Hamiltonian is

$$\begin{aligned} H_eff &= -w\sum_i \left(b_i^\dagger b_{i+1} + b_{i+1}^\dagger b_i - 2b_i^\dagger b_i\right) - \mu\sum_i n_i \\ &\quad - \frac{2w^2}{V}\sum_i \left(2b_i^\dagger b_{i+1}^\dagger b_{i+1} b_i + b_i^\dagger b_{i-1}^\dagger b_{i+1} b_i + b_i^\dagger b_{i+1}^\dagger b_{i-1} b_i\right), \end{aligned} \tag{12.80}$$

where now the bosons are "hard core", which means that $n_i = 0, 1$ are the only allowed values. Notice now that this reduced Hilbert space is identical to that of spinless fermions. The transformation between the b_i and the spinless fermion operators f_i is the Jordan–Wigner mapping

$$b_i = \prod_{j<i}(1 - 2f_j^\dagger f_j) f_i. \tag{12.81}$$

Verify that (12.81) produces Bose operators that commute between different sites. Insert (12.81) in (12.80) and take the continuum limit with $f_i = \sqrt{a}\Psi_F(x = ia)$, $w = \hbar^2/(2ma^2)$ (a is the lattice spacing) and obtain

$$H_F = \int dx \left[\Psi_F^\dagger \left(-\frac{\hbar^2}{2m}\frac{d^2}{dx^2} - \mu \right) \Psi_F - \frac{8w^2a^3}{V} \frac{d\Psi_F^\dagger}{dx} \Psi_F^\dagger \Psi_F \frac{d\Psi_F}{dx} \right]. \tag{12.82}$$

Finally, decompose the fermion field into left- (Ψ_L) and right- (Ψ_R) moving excitations with a linear dispersion, and obtain the long-wavelength Hamiltonian

$$H_L = \int dx \left[\hbar c \left(\Psi_R^\dagger \frac{d\Psi_R}{dx} - \Psi_L^\dagger \frac{d\Psi_L}{dx} \right) - \frac{32w^2a^3k_F^2}{V} \Psi_R^\dagger \Psi_L^\dagger \Psi_L \Psi_R \right], \tag{12.83}$$

where $c = \hbar k_F/m$ and the Fermi wavevector k_F is given by $\hbar^2 k_F^2/(2m) = \mu$. This is a weakly interacting model of spinless fermions, to which we can apply Luttinger liquid methods.

PART II

FRACTIONALIZATION AND EMERGENT GAUGE FIELDS I

13 Introduction to Gapped Spin Liquids

An intuitive introduction to theory of spin liquids is presented, using the resonating-valence-bond wavefunction. The wavefunctions of the excited spinon and vison states are described, and used to obtain their anyonic properties. An introductory discussion of topological order is also presented.

Part I described quantum phases of matter that were ultimately connected to the free-particle description in a relatively straightforward manner. We began with weakly interacting Fermi gases in Chapter 2, and Bose gases in Chapter 3. For the case of the Bose gas, we introduced the concepts of broken symmetry and long-range order, applied to the U(1) particle number conservation symmetry. These concepts were also useful in the discussion of superconductivity of the Fermi gas in Chapter 4. The subsequent chapters then examined the consequences of fluctuating order, as described by the Landau–Ginzburg theory for a thermally fluctuating superconductor in Chapter 6, and the quantum field theory of a relativistic scalar for the superfluid–insulator quantum phase transition in Chapter 8. Eventually, we did meet situations in which the free-particle description no longer applied: (i) at the superfluid–insulator quantum critical point in 2+1 dimensions in Section 11.2.2, which has no quasiparticle excitations, and (ii) in one spatial dimension in Chapter 12, where the order was only quasi-long range, and the quasiparticles were free phase or density fluctuations.

Parts II and IV will turn to a more radical departure from the free-particle description. The key new idea here will be one of *fractionalization*, in which the lattice fermion or boson turns into a composite of new emergent particles. There is no local operator that can create a single fractionalized particle; consequently, the fractionalized particles must be charged under an emergent gauge field.

Part II will consider fractionalization in a "pure" form, where there are no further topological considerations apart from those arising from the fractionalization of the lattice degrees of freedom. The simplest case of this is the $\mathbb{Z}_2$ spin liquid, which we consider in some detail. Part IV turns to a more intricate realization of fractionalization, where the fractionalized particles ("partons") also have a topological band structure. So, before we can consider these cases, we describe band topology on its own, for unfractionalized particles, in Part III. We survey the many ways band topology can be combined with fractionalization in Chapter 21, and describe them in more detail in the remainder of Part IV. One of the earliest realizations of fractionalization, the fractional quantum Hall effect, does have partons in a non-trivial band topology, and the

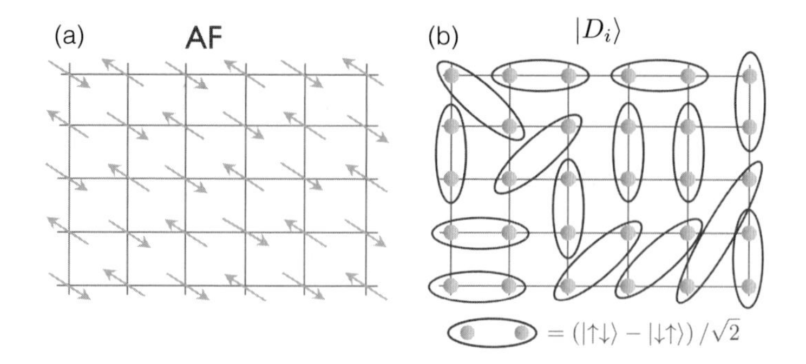

Figure 13.1 (a) The insulating antiferromagnetic (AF) state at $p = 0$. (b) Component of a "resonating-valence-bond (RVB)" wavefunction for the antiferromagnet which preserves spin rotation symmetry; all the $|D_i\rangle$ in Eq. (13.1) have similar pairings of electrons on nearby sites (not necessarily nearest-neighbor).

distinction between the two effects is often obscured in the literature. Our presentation does not describe the fractional quantum Hall effect until Chapter 24.

Much of Parts II and IV focuses on quantum phases associated with the electron Hubbard model in Chapter 9. We restrict our attention to insulating phases here, at doping $p = 0$, where the low-energy degrees of freedom are just spins, described by an effective Hamiltonian such as (9.14). We turn to conducting phases in Part V.

One possible phase on the insulating square lattice is the Néel state in Fig. 9.2, shown here as the antiferromagnet in Fig. 13.1a. The spins are arranged in a checkerboard pattern, so that all the spins in one sublattice are parallel to each other, and anti-parallel to spins on the other sublattice. Two key features of this antiferromagnetic state deserve attention here. Firstly the state breaks a global spin-rotation symmetry, and essentially all of its low-energy properties can be described by well-known quantum field theory methods associated with spontaneously broken symmetries, some of which were described in Chapter 5, Section 9.2.2 and Section 11.2.3. Secondly the wavefunction does not have long-range entanglement, and the exact many-electron wavefunction can be obtained by a series of local unitary transformations on the simple product state sketched in Fig. 13.1a.

Our interest in Parts II and IV is primarily on quantum phases that preserve the symmetries of the underlying Hamiltonian. We can restore the broken spin-rotation symmetry of the antiferromagnetic state by having pairs of spins forming spin singlets, and this led us to the valence-bond solid (VBS) state in Fig. 9.2. However, the VBS state breaks lattice symmetries. To restore lattice symmetries we need a "second level" of entanglement, and take superpositions of the valence-bond configurations themselves. This leads to states with long-range entanglement, as described qualitatively below, and in move detail in Parts II and IV.

Much of Parts II and IV can be interpreted as an answer to a question we posed in Section 8.4: what are the possible ground states of lattice bosons at density 1/2 per site? In Section 16.3 we discuss some general reasons why it is not possible for any ground state to be *trivial*. One possibility is that it can break a translational symmetry so that there is an integer density per unit cell: we have already seen examples of this possibility. Another possibility is that the U(1) boson number symmetry is broken, so that the ground state is a compressible superfluid. The remaining possibility is the focus of our discussion now: there is long-range entanglement, with fractionalized excitations and emergent gauge fields. We study several examples of this alternative.

13.1 The RVB State

We begin with the "resonating-valence-bond" (RVB) state

$$|\Psi\rangle = \sum_i d_i \, |D_i\rangle \,, \tag{13.1}$$

where i extends over all possible pairings of electrons on nearby sites, and a state $|D_i\rangle$ associated with one such pairing is shown in Fig. 13.1b; the d_i are complex coefficients we will leave unspecified here. Note that the electrons in a valence bond need not be nearest neighbors. Each $|D_i\rangle$ is a spin singlet, and so spin-rotation invariance is preserved; the antiferromagnetic exchange interaction is optimized between the electrons within a single valence bond, but not between electrons in separate valence bonds. We also assume that the d_i respect the translational and other symmetries of the square lattice. Such a state was first proposed by Pauling [201] as a description of a simple metal like lithium. We now know that Pauling's proposal is incorrect for such metals. But we will return to a variant of the RVB state in Chapter 29, which does indeed describe a metal. Anderson revived the RVB state many years later [8] as a description of Mott insulators. These are materials with a density of one electron per site, which are driven to be insulators by the Coulomb repulsion between the electrons (contrary to the Bloch theorem for free electrons, which requires metallic behavior at this density).

In a modern theoretical framework, we now realize that the true significance of the Pauling–Anderson RVB proposal was that it was the first ansatz to realize *long-range* quantum entanglement. Similar entanglement appeared subsequently in Laughlin's wavefunction for the fractional quantum Hall state [151], and for RVB states in the absence of time-reversal symmetry [123]. The long-range nature of the entanglement can be made precise by computation of the "topological entanglement entropy" [132, 156, 309]. But here we will be satisfied by a qualitative description of the sensitivity of the spectrum of states to the topology of the manifold on which the square lattice resides. The sensitivity is present irrespective of the size of the manifold (provided it is much larger than the lattice spacing), and so indicates that the information on the quantum entanglement between the electrons is truly long-ranged. A wavefunction that is a product of localized single-particle states would not care about the global topology of the manifold.

13.2 Topological Properties

The basic argument on the long-range quantum information contained in the RVB state is summarized in Fig. 13.2. We place the square lattice on a very large torus (i.e.

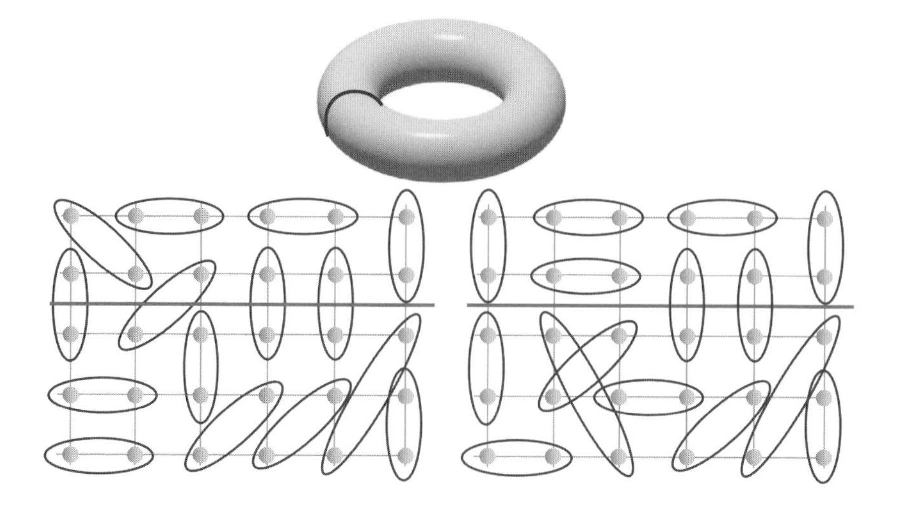

Figure 13.2 Sensitivity of the RVB state to the torus geometry: the number of valence bonds crossing the cut (thick horizontal line) can only differ by an even integer between any two configurations (like those shown), which differ by an arbitrary local arrangement of valence bonds.

impose periodic boundary conditions in both directions), draw an arbitrary imaginary cut across the lattice, indicated by the thick horizontal line, and count the number of valence bonds crossing the cut. It is not difficult to see that any *local* rearrangement of the valence bonds will preserve the number of valence bonds crossing the cut modulo 2. Only very non-local processes can change the parity of the valence bonds crossing the cut; one such process involves breaking a valence bond across the cut into its constituent electrons, and moving the electrons separately around a cycle of the torus crossing the cut, so that they meet on the other side and form a new valence bond that no longer crosses the cut – see Fig. 13.3. Ignoring this very non-local process, we see that the Hilbert space splits into disjoint sectors, containing states with an even or odd number of valence bonds across the cut [140, 277]. Locally, the two sectors are identical, and so we expect them to have ground states (and also excited states) of nearly the same energy for a large-enough torus. The presence of these near-degenerate states is dependent on the global spatial topology, which means it requires periodic boundary conditions around the cycles of the torus, and so can be viewed as a signature of long-range quantum entanglement.

13.3 Emergent Gauge Fields

The above description of topological degeneracy and entanglement relies on a somewhat arbitrary and imprecise trial wavefunction. A precise understanding is provided by a formulation of the physics of the RVB state in terms of an emergent

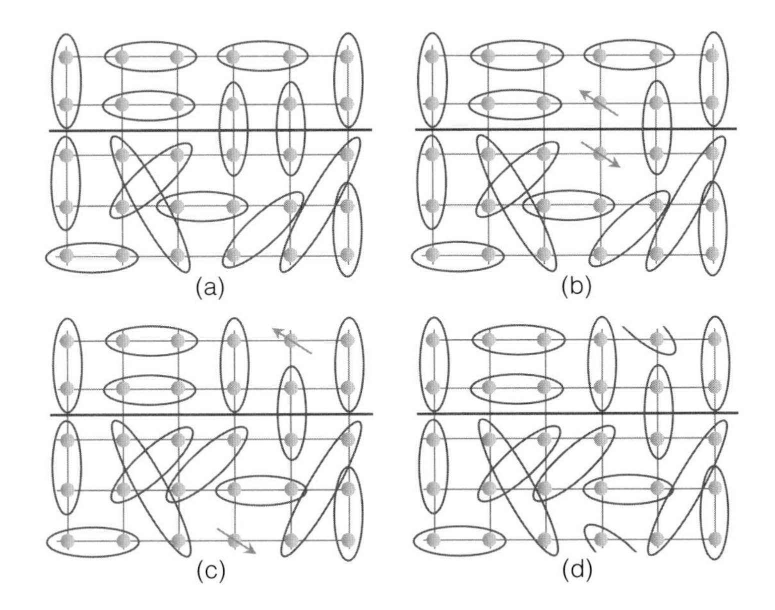

Figure 13.3 Non-local process which changes the parity of the number of valence bonds crossing the cut. A valence bond splits into two spins, which pair up again after going around the torus.

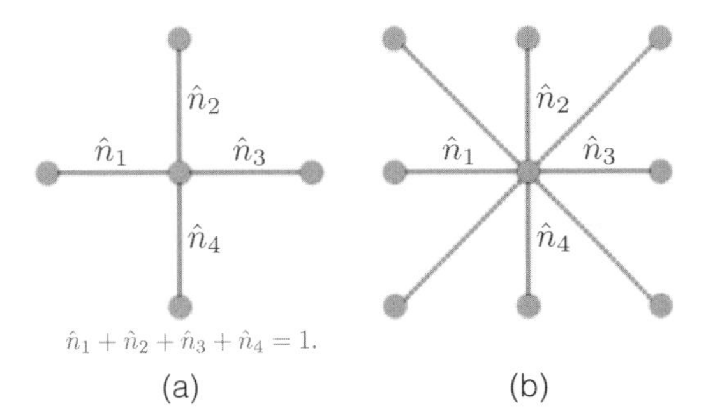

Figure 13.4 (a) Nearest-neighbor valence-bond number operators, proportional to the electric field of a compact U(1) gauge theory. (b) Model with valence bonds connecting the same sublattice; now the constraint on the number operators is modified, and the spin liquid is described by a $\mathbb{Z}_2$ gauge theory.

gauge theory, the first example of which was introduced by Baskaran and Anderson [22]. Such a formulation provides another way to view the nearly degenerate states obtained above on a torus: they are linear combinations of states obtained by inserting fluxes of the emergent gauge fields through the cycles of the torus.

The formulation as a gauge theory [22, 86] becomes evident upon considering a simplified model with valence bonds only between nearest-neighbor sites on the square lattice. We introduce valence-bond number operators $\hat{n}$ on every nearest-neighbor link, and then there is a crucial constraint that there is exactly one valence bond emerging from every site, as illustrated in Fig. 13.4a. After introducing oriented "electric-field"

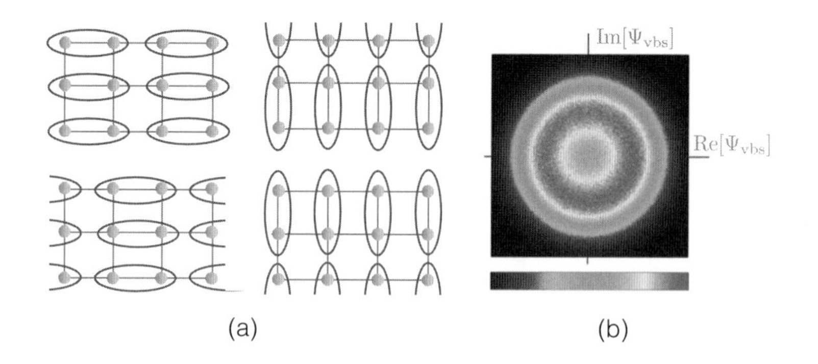

Figure 13.5 (a) The 4 VBS states which break square lattice rotational symmetry. (b) Distribution of the complex VBS order parameter Ψ_{vbs} in the quantum Monte Carlo study by Sandvik [247]; the real and imaginary parts of this order measure the probability of the VBS states in the first and second columns. The near-circular distribution of Ψ_{vbs} reflects an emergent symmetry that is a signature of the existence of a photon. Reprinted with permission from APS.

operators $\hat{e}_{i\mu} = (-1)^{i_x+i_y}\hat{n}_{i\mu}$ (here i labels the sites of the square lattice, and $\mu = x, y$ is a spatial index labeling the two directions), this local constraint can be written in the very suggestive form [86]

$$\Delta_\mu \hat{e}_{i\mu} = \rho_i, \tag{13.2}$$

where Δ_μ is a discrete lattice derivative, and $\rho_i \equiv (-1)^{i_x+i_y}$ is a background "charge" density. The equation (13.2) is analogous to Gauss's law in electrodynamics, and a key indication that the physics of resonating valence bonds is described by an emergent gauge theory. An important difference from Maxwell's U(1) electrodynamics is that the eigenvalues of the electric-field operator $\hat{e}_{i\mu}$ must be integers. In terms of the canonically conjugate gauge field $\hat{a}_{i\mu}$,

$$[\hat{a}_{i\mu}, \hat{e}_{j\nu}] = i\hbar\delta_{ij}\delta_{\mu\nu}, \tag{13.3}$$

the integral constraint translates into the requirement that $\hat{a}_{i\mu}$ is a compact angular variable on a unit circle, and that $\hat{a}_{i\mu}$ and $\hat{a}_{i\mu} + 2\pi$ are equivalent. So there is an equivalence between the quantum theory of nearest-neighbor resonating valence bonds on a square lattice, and compact U(1) electrodynamics in the presence of fixed background charges ρ_i [86]. A non-perturbative analysis of such a theory shows [216, 218] that ultimately there is no gapless "photon" associated with the emergent gauge field $\hat{a}$: compact U(1) electrodynamics is confining in two spatial dimensions, and in the presence of the background charges the confinement leads to the VBS order illustrated in Fig. 13.5. The VBS state breaks square lattice rotation symmetry, and all excitations of the antiferromagnet, including the incipient photon, have an energy gap. In subsequent work, it was realized that the gapless photon can re-emerge at special "deconfined" critical points [84, 261, 290] or phases [108], even in two spatial dimensions. In particular, in certain models with a quantum phase transition between a VBS state and the ordered antiferromagnet in Fig. 13.1a [216, 218, 261], the quantum critical point supports a gapless photon (along with gapless-matter fields). This is illustrated in Fig. 13.5b by numerical results of Sandvik [247]; the circular distribution of valence bonds is evidence for an emergent continuous lattice rotation symmetry, and the associated Goldstone mode is the dual of the photon.

The properties of U(1) gauge theories summarized above are described in more detail in Chapters 25, 26, and 28.

Although U(1) gauge theory does realize spin liquids with long-range entanglement and emergent photons, the gaplessness and "criticality" of the spin liquids indicates the presence of long-range valence bonds, and the Pauling–Anderson trial wavefunctions are poor descriptions of such states. A stable, gapped quantum state with time-reversal symmetry, long-range entanglement and emergent gauge fields was first established in Refs. [119, 219, 230, 231, 303] using a model with short-range valence bonds that also connect sites on the same sublattice (Fig. 13.4b). It was shown [119, 219, 230, 231, 303] that the same-sublattice bonds act like charge ± 2 Higgs fields in the compact U(1) gauge theory, and in such gauge theories there can be [17, 87] a "Higgs" phase. Such a phase realizes a stable, gapped, RVB state preserving all symmetries of the Hamiltonian, including time-reversal symmetry, and is described by an emergent $\mathbb{Z}_2$ gauge theory [119, 230]. The $\mathbb{Z}_2$ gauge theory can be viewed as a discrete analog of the compact U(1) theory in which the gauge field takes only two possible values $\hat{a}_{i\mu} = 0, \pi$. The intimate connection between a spin liquid with a deconfined $\mathbb{Z}_2$ gauge field, and a non-bipartite RVB trial wavefunction like (13.1), was shown convincingly by Wildeboer *et al.* [309]. Upon varying parameters in the underlying Hamiltonian, the $\mathbb{Z}_2$ spin liquid can undergo a confinement transition to a VBS phase, which is described by a dual frustrated Ising model [119, 230]. Since these early works, the $\mathbb{Z}_2$ spin liquid has appeared in a number of other models [88, 101, 133, 180, 257, 304], including the exactly solvable "toric code" [133].

13.4 Excitations of the $\mathbb{Z}_2$ Spin Liquid

A complete study of the $\mathbb{Z}_2$ spin liquid will occupy Chapters 15 and 16 and Section 26.2. Here, we present a simple overview of the structure of its ground state and excitations.

A theory for a stable RVB state with time-reversal symmetry and a gap to all excitations first appeared in Refs. [119, 219, 231, 303], which described a state now called a $\mathbb{Z}_2$ spin liquid. It is helpful to describe the structure of the $\mathbb{Z}_2$ spin liquid in terms of a mean-field ansatz. We write the spin operators on each site, $S_{i\ell}$ ($\ell = x, y, z$), in terms of Schwinger bosons $s_{i\alpha}$ ($\alpha = \uparrow, \downarrow$) [13]

$$S_{i\ell} = \frac{1}{2} s_{i\alpha}^{\dagger} \sigma_{\alpha\beta}^{\ell} s_{i\beta}, \tag{13.4}$$

where σ^{ℓ} are the Pauli matrices, and the bosons obey the local constraint

$$\sum_{\alpha} s_{i\alpha}^{\dagger} s_{i\alpha} = 2S \tag{13.5}$$

on every site i. It is now easy to show that the spin operators obey the required commutation relations, and the $2S+1$ states defined by (13.5) yield the correct matrix elements of all spin operators. Here, we are primarily interested in the case of spin $S=1/2$, but it is useful to also consider the çase of general S. Schwinger fermions can also be used instead, but the description of the $S>1/2$ cases is more cumbersome with them.

At this point, the expression of the spin operators in terms of the $S=1/2$ bosons appears as a formal mathematical trick. However, expressing the spins in terms of spin-1/2 particles naturally predisposes to phases in which these boson are deconfined at long distances, leading to *fractionalization*. Contrast this with the hard-core boson representation in (9.18), where the hard-core bosons carry an integer spin, and we did not obtain any fractionalized state.

The deconfined boson state corresponding to the $\mathbb{Z}_2$ spin liquid is described by an effective Schwinger-boson Hamiltonian [13, 219]

$$\mathcal{H}_b = -\sum_{i<j}\left[P_{ij}s^\dagger_{i\alpha}s_{j\alpha} + Q_{ij}\varepsilon_{\alpha\beta}s^\dagger_{i\alpha}s^\dagger_{j\beta} + \text{H.c.}\right] + \lambda\sum_i s^\dagger_{i\alpha}s_{i\alpha}, \tag{13.6}$$

where $\varepsilon_{\alpha\beta}$ is the antisymmetric unit tensor, λ is chosen to satisfy the constraint in Eq. (13.5) on average, and the $Q_{ij}=-Q_{ji}$ and $P_{ij}=P^*_{ji}$ are a set of variational parameters chosen to optimize the energy of the spin liquid state. Generally, the Q_{ij} and P_{ij} are chosen to be non-zero only between nearby sites, and the "$\mathbb{Z}_2$" character of the spin liquid requires that the links with non-zero Q_{ij} can form closed loops with an odd number of links. The Schwinger-boson parameterization (13.4) is invariant under the U(1) gauge transformation, $s_{i\alpha}\to e^{i\phi_i}s_{i\alpha}$, and odd loops imply that the U(1) is Higgsed down to a $\mathbb{Z}_2$ gauge theory [119, 219, 230, 231, 303]. This Hamiltonian yields a mean-field wavefunction for the spin liquid

$$|\Psi\rangle = \mathcal{P}_{2S}\exp\left(\sum_{i<j} f_{ij}\,\varepsilon_{\alpha\beta}s^\dagger_{i\alpha}s^\dagger_{j\beta}\right)|0\rangle, \tag{13.7}$$

where $|0\rangle$ is the boson vaccum, $\mathcal{P}_{2S}$ is a projection operator that selects only states which obey (13.5), and the boson-pair wavefunction $f_{ij}=-f_{ji}$ is determined by diagonalizing (13.6) by a Bogoliubov transformation. This is closely connected to the Bogoliubov transformation in (3.11) and the wavefunction in (4.7).

Moving to the gapped excited states of the $\mathbb{Z}_2$ spin liquid, we find two distinct types of quasiparticles, illustrated in Fig. 13.6b–d:

(i) A "spinon," shown in Fig. 13.6b, has one unpaired spin and so carries spin $S=1/2$; more specifically, the spinon is the Bogoliubov quasiparticle obtained by diagonalizing $\mathcal{H}_b$ in terms of canonical bosons.

(ii) The second quasiparticle, the "vison," shown in Fig. 13.6c,d, is spinless and it has a more subtle topological character of a vortex in an Ising-like system (hence its name [257]). The vison state is the ground state of a Hamiltonian, $\mathcal{H}^v_b$, obtained from $\mathcal{H}$ by mapping $Q_{ij}\to Q^v_{ij}$, $P_{ij}\to P^v_{ij}$; then the vison state $|\Psi^v\rangle$ has a wavefunction as in (13.7), but with $f_{ij}\to f^v_{ij}$. Far from the center of the vison,

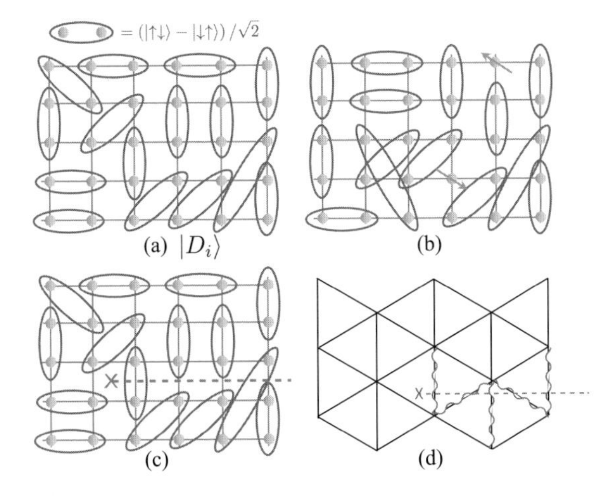

Figure 13.6 (a) Illustration of a component, $|D_i\rangle$, of the RVB wavefunction in (13.1). (b) A pair of $S = 1/2$ spinon excitations. (c) The vison excitation of the $\mathbb{Z}_2$ spin liquid. In terms of (13.1), the co-efficients d_i are modified so that each singlet bond crossing the 'branch-cut' (dashed line) picks up a factor of -1. A similar modification applies to (13.7), and is described in the text. (d) A vison on the triangular lattice for the case of Q_{ij} and P_{ij} non-zero only between nearest-neighbor sites: the wavy lines indicate the Q_{ij} and P_{ij} with a change in their sign in the presence of a vison.

we have $|Q^{\rm v}_{ij}| = |Q_{ij}|$, $|P^{\rm v}_{ij}| = |P_{ij}|$, while closer to the center there are differences in the magnitudes. However, the key difference is in the signs of the link variables, as illustrated in Fig. 13.6c,d: there is a "branch-cut" emerging from the vison core along which $\text{sgn}(Q^{\rm v}_{ij}) = -\text{sgn}(Q_{ij})$ and $\text{sgn}(P^{\rm v}_{ij}) = -\text{sgn}(P_{ij})$. This branch-cut ensures that the $\mathbb{Z}_2$ magnetic flux equals -1 on all loops that encircle the vison core, while other loops do not have non-trivial $\mathbb{Z}_2$ flux.

The spinons and visons have two crucial topological properties:

(i) A spinon and a vison are mutual semions [215]. In other words, adiabatically moving a spinon around a vison (or vice versa) yields a Berry phase of π. This is evident from the structure of the branch-cut in $Q^{\rm v}_{ij}$ and $P^{\rm v}_{ij}$: these $Q^{\rm v}_{ij}$ and $P^{\rm v}_{ij}$ are the hopping amplitudes for the spinon, and they yield an additional phase of π (beyond those provided by P_{ij} and Q_{ij}) every time a spinon crosses the branch-cut.

(ii) A less-well-known and distinct property involves the motion of a single vison without any spinons present; adiabatic motion of a vison around a single lattice site yields a Berry phase of $2\pi S$ [119, 230, 257], as described in Sections 15.4.2 and 16.5.2. The background Berry phase of $2\pi S$ per site for vison motion implies that there are two distinct types of $\mathbb{Z}_2$ spin liquids [119, 183, 186, 230, 257, 258], when there is well-defined spin quantum number, that is, a globally conserved U(1) quantum number. As was first pointed out in Refs. [119, 230], these are "odd-$\mathbb{Z}_2$ spin liquids," which are realized in the present model by half-integer S antiferromagnets, and "even-$\mathbb{Z}_2$ spin liquids," realized here by integer S antiferromagnets. In the $\mathbb{Z}_2$ gauge theory framework (or the related "toric code" [133]), there is a unit-$\mathbb{Z}_2$ electric charge on each lattice site of an odd-$\mathbb{Z}_2$ gauge theory. Further details on the differences between even- and odd-$\mathbb{Z}_2$ spin liquids appear in Chapter 16 and Section 26.2. The RVB state, originally proposed for $S = 1/2$ spins, is an *odd-*$\mathbb{Z}_2$ spin liquid [119].

The modern theory of topological phases focuses on the robust properties of the quantum numbers of the fractionalized excitations (the "anyons"), the Berry phases associated with the motion of these excitations around each other, and the sensitivity to the topological properties of the spatial manifold on which the Hamiltonian resides. We close this chapter by cataloging the properties of $\mathbb{Z}_2$ spin liquid in this language. Many details are described in the following chapters.

- Anyons: $\mathbb{1}$, e, m, ε. The e, m, ε anyons cannot be created from the ground state ($\mathbb{1}$) by any local operator.
- The e and ε are spinons, the m is the 'vison'. The spinons carry spin 1/2, or hard-core boson number $B^\dagger B = 1/2$ (Section 9.2). So the spin symmetry is fractionalized.
- Self-statistics: e and m are bosons, while ε is a fermion.
- Mutual statistics: Any pair of e, m, ε are mutual semions, which means one anyon picks up a (-1) upon encircling any other type of anyon.
- Fusion rules: $e \times m = \varepsilon$, $e \times \varepsilon = m$, $m \times \varepsilon = e$, $e \times e = \varepsilon \times \varepsilon = m \times m = \mathbb{1}$. These describe the possible outcomes when two anyons are brought close to each other.
- Four-fold ground-state degeneracy on a torus.
- Emergent, deconfined $\mathbb{Z}_2$ gauge field.
- No protected edge states in general, but could appear with special symmetries.
- Topological entanglement entropy = $\ln 2$.
- For spin-S antiferromagnets on the square lattice, the single vison states described exhibit "translational symmetry fractionalization" with

$$T_x T_y = T_y T_x e^{2\pi i S}, \tag{13.8}$$

  where T_x, T_y are translation operators by one lattice spacing in the x and y directions.
- Therefore, the RVB state which motivated the discussion of this chapter is an odd-$\mathbb{Z}_2$ spin liquid with $T_x T_y = -T_y T_x$ when acting on vison states.

14 Fractionalization in the XY Model in 2+1 Dimensions

An extension of the familiar classical XY model in three dimenions is used to introduce basic concepts in the theory of fractionalized phases. The phases of the extended XY model are accessed by a representation in terms of an emergent $U(1)$ gauge field, and the Higgs and confining phases of such a gauge theory allow an efficient description of the possible phases, and their anyonic excitations.

This chapter pauses our discussion of two-dimensional antiferromagnets, and introduces many of the key ideas on fractionalization in what I believe is the simplest possible context: the statistical mechanics of the classical XY model on the three-dimensional cubic lattice. We have already met a continuum version of this model in the context of a quantum theory in 2+1 spacetime dimensions – this is the relativistic quantum field theory of a complex scalar Ψ_B in (8.21) with $K_1 = 0$, describing the Mott insulator to superfluid transition in the boson Hubbard model at integer filling. The field Ψ_B (Ψ_B^*) annihilates (creates) excitations in the Mott insulator with boson number -1 ($+1$). After discretizing (8.21) without the K_1 term on a lattice in three-dimensional spacetime lattice, and making a unit magnitude constraint $\Psi_B = e^{i\theta}$, we obtain the partition function of the XY model

$$\begin{aligned}\mathcal{Z}_{XY} &= \prod_i \int_0^{2\pi} \frac{d\theta_i}{2\pi} \exp\left(-H_{XY}\right), \\ H_{XY} &= -J \sum_{\langle ij \rangle} \cos(\theta_i - \theta_j),\end{aligned} \tag{14.1}$$

where the sites i, j reside on the vertices of a cubic lattice with coordinates $\boldsymbol{r}_i, \boldsymbol{r}_j$. As written in (14.1), the statistical mechanics of $\mathcal{Z}_{XY}$ has been thoroughly studied, and is very well understood, and is reviewed in this chapter. There is a large $-J$ "ordered" phase in which the θ_i align in a common direction, corresponding to the superfluid phase of the boson Hubbard model. At small J, we have the "disordered" phase, corresponding to the Mott insulator at integer filling. In the context of $\mathcal{Z}_{XY}$, we refer to the conserved boson-number "charge" as $\mathcal{Q}$. So the disordered phase of (14.1) has gapped excitations with boson number $\mathcal{Q} = \pm 1$. The phase transition between the ordered and disordered phases was described briefly in Chapter 11, and in more detail in the QPT book.

We are interested in this chapter in extensions of the XY model in (14.1), in which the Hamiltonian H_{XY} contains additional short-range interactions consistent with the basic symmetries (described below). I show that it is possible to formulate extensions that feature another "disordered" phase, which has fractionalized excitations with $Q = \pm 1/2$. A convenient way to obtain such extensions is to formulate the partition function $\mathcal{Z}_{XY}$ using the variables of a compact $U(1)$ gauge theory. Finally, I show that the basic phases of the compact $U(1)$ gauge theory are more conveniently realized in a $\mathbb{Z}_2$ gauge theory.

The fractionalized phase realized in this manner in the present chapter has the basic characteristics of a $\mathbb{Z}_2$ spin liquid, outlined in Chapter 13, including excitations with fractionalized charges $Q = \pm 1/2$ and vortex-like vison excitations. It will, however, be an *even*-$\mathbb{Z}_2$ spin liquid, which obeys the relation (13.8) for antiferromagnets with integer spin S, or bosons at integer filling. Strictly speaking, because we are considering an XY order parameter here, the antiferromagnets have to be of the "easy-plane" variety, with spin-anisotropy terms that prefer spin orientation in the XY plane, but this symmetry constraint has no influence on the structure of the spin-liquid phase. We discuss extensions of the $\mathbb{Z}_2$ gauge theory realizing an *odd*-$\mathbb{Z}_2$ spin liquid relevant for $S = 1/2$ easy-plane antiferromagnets and bosons at half-integer filling towards the end of Section 14.2.3, and in Chapter 16. This extension requires the introduction of an intrinsically quantum Berry phase term in (14.1), so that the weights in the partition function are not all positive.

14.1 The Conventional *XY* Model

Let us begin by reviewing the basic characteristics of the XY model in (14.1) in three dimensions. The model is defined in terms of periodic variables θ_i, and so the Hamiltonian is invariant under

$$\theta_i \to \theta_i + 2\pi n_i, \tag{14.2}$$

where the n_i are integers, which can depend upon the sites of index i. There is also a global $U(1)$ symmetry, which requires invariance under

$$\theta_i \to \theta_i + c, \tag{14.3}$$

where c is an arbitrary real number, but independent of i. We discuss extensions of (14.1) below, and these are constrained to also obey (14.2) and (14.3).

The phase diagram of (14.1) is sketched in Fig. 14.1.

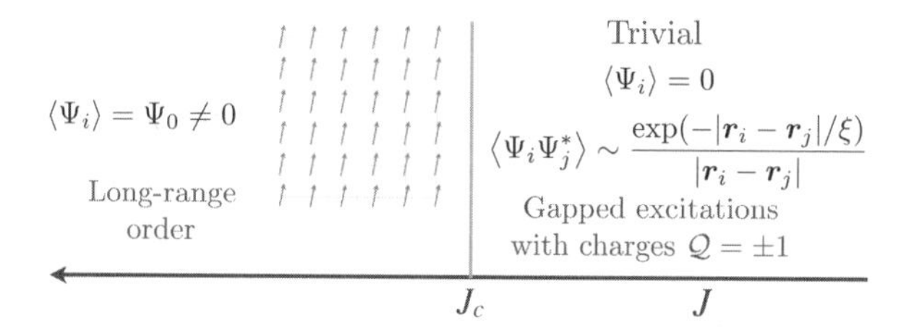

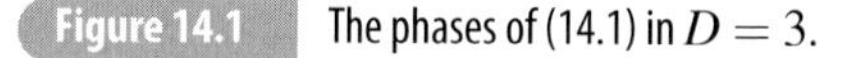
Figure 14.1 The phases of (14.1) in $D = 3$.

At large J, we have phase with long-range order, characterized by the long-range correlation

$$\lim_{|\boldsymbol{r}_i-\boldsymbol{r}_j|\to\infty} \langle \Psi_i \Psi_j^* \rangle = |\Psi_0|^2 \neq 0, \tag{14.4}$$

where we have defined the complex number

$$\Psi_i \equiv e^{i\theta_i}. \tag{14.5}$$

The $U(1)$ symmetry is spontaneously broken, and we can work in an ensemble by choosing an overall phase, so that

$$\langle \Psi_i \rangle = \Psi_0 \neq 0. \tag{14.6}$$

We are not be particularly interested in this ordered phase in the present chapter.

The small-J phase of Fig. 14.1 has no long-range order, and the correlations of Ψ_i decay exponentially. We will often refer to this phase as the "trivial" phase, because of the absence of fractionalization. It will be important for us to also keep track of the power-law prefactor of the exponential decay, which has the form

$$\lim_{|\boldsymbol{r}_i-\boldsymbol{r}_j|\to\infty} \langle \Psi_i \Psi_j^* \rangle \sim \frac{\exp(-|\boldsymbol{r}_i-\boldsymbol{r}_j|/\xi)}{|\boldsymbol{r}_i-\boldsymbol{r}_j|^{(D-1)/2}} \quad , \quad \text{trivial phase} \tag{14.7}$$

in D dimensions with a finite correlation length ξ (we are interested here in $D = 3$). We have written the Ornstein–Zernike form of the decay of two-point correlations in (14.7): this can be obtained most simply from the Fourier transform

$$\int \frac{d^D k}{(2\pi)^D} \frac{e^{i\boldsymbol{k}\cdot\boldsymbol{r}}}{k^2+\xi^{-2}} \sim \frac{e^{-r/\xi}}{r^{(D-1)/2}}, \tag{14.8}$$

but is also known to hold rigorously for small J [35, 42]. An important feature of this "disordered" phase becomes apparent when we view the XY model as a quantum model in 2+1 dimensions (as discussed in the QPT book). This requires analytic continuation to real time and frequency, and we discussed such a continuation in Section 11.2.1. In particular, (11.25) and (11.26) show that the correlator in (14.8) implies the existence of a relativistic particle with mass $\Delta = \xi^{-1}$, with dynamic susceptibility

$$\operatorname{Im}\chi(k,\omega>0) \sim \delta\left(\omega - \sqrt{\Delta^2+k^2}\right). \tag{14.9}$$

This is the Ψ particle, which carries $U(1)$ charge $Q = \pm 1$. An important point for our considerations below is that the existence of this particle with $Q = \pm 1$ is closely tied to the Ornstein–Zernike form of the imaginary-time correlator in (14.8).

14.2 The Extended *XY* Model

The discussion above allows us to outline how we may obtain a distinct disordered phase with fractionalization in an extended XY model. Suppose we can find a model in which the field Ψ fractionalizes into a pair of ϕ particles carrying charge $Q = 1/2$. Then we can write

$$\Psi \sim \phi^2 . \tag{14.10}$$

We imagine that the ϕ particles are nearly free in the fractionalized phase, and so the ϕ correlator can have the Ornstein–Zernike form of (14.7) with correlation length 2ξ. Then, from (14.10), we can conclude that the physically observable Ψ correlator will be the square of the ϕ correlator, and hence, in the fractionalized disordered phase, we have

$$\lim_{|\boldsymbol{r}_i - \boldsymbol{r}_j| \to \infty} \langle \Psi_i \Psi_j^* \rangle \sim \frac{\exp(-|\boldsymbol{r}_i - \boldsymbol{r}_j|/\xi)}{|\boldsymbol{r}_i - \boldsymbol{r}_j|^{(D-1)}} \quad , \quad \text{fractionalized phase.} \tag{14.11}$$

The difference between (14.7) and (14.11) is subtle, and present only in the power-law prefactor. Nevertheless, this difference is sufficient to imply that the fractionalized phase is not smoothly connected to the trivial phase, and there must be a phase transition between them. The difference between the two phases becomes much clearer when we analytically continue (14.11) to the dynamic spin susceptibility $\chi(k, \omega)$. The imaginary part of this susceptibility now does not have a delta function as in (14.9), but a threshold to a continuum for the creation of the two particles each of mass $\Delta/2$. The spectral density above the threshold is controlled by the conservation of energy and momentum to obey

$$\begin{aligned} \operatorname{Im}\chi(k, \omega > 0) &\sim \int d^{D-1}p\, \delta\left(\omega - \frac{\Delta}{2} - \frac{(\boldsymbol{k}/2 + \boldsymbol{p})^2}{2(\Delta/2)} - \frac{\Delta}{2} - \frac{(\boldsymbol{k}/2 - \boldsymbol{p})^2}{2(\Delta/2)}\right) \\ &\sim \left(\omega - \Delta - \frac{k^2}{2\Delta}\right)^{(D-3)/2} , \end{aligned} \tag{14.12}$$

where the last line is non-zero only when it is real, and yields a step function at $\Delta + k^2/(2\Delta)$ in the $D = 3$ of interest here. The difference between the trivial result (14.9) and the fractionalized result (14.12) is the difference between a delta function and a threshold to a continuum, and is easily observable in neutron scattering.

14.2.1 Partition Function

In the remainder of this chapter I will argue that a fractionalized phase with the above structure can indeed be obtained in an extended XY model. We have to move beyond the XY model in (14.1), and consider an extended model with additional terms that obey the symmetries in (14.2) and (14.3). Such extended XY models have been numerically studied in Refs. [186, 258], and here we write models in the general form

$$\begin{aligned} \widetilde{\mathcal{Z}}_{XY} &= \prod_i \int_0^{2\pi} \frac{d\theta_i}{2\pi} \exp\left(-\widetilde{H}_{XY}[\theta]\right) \\ \widetilde{H}_{XY}[\theta] &= -\sum_{ij} J_{ij} \cos(\theta_i - \theta_j) + \sum_{ijk\ell} K_{ijk\ell} \cos(\theta_i + \theta_j - \theta_k - \theta_\ell) + \cdots . \end{aligned} \tag{14.13}$$

Table 14.1 Symmetry charges

Symmetry	$\Psi = e^{i\theta}$	$H = e^{i\vartheta}$	$\phi = e^{i\varphi}$
$U(1)$	1	1	0
$U(1)_{\text{gauge}}$	0	−2	1
$U(1)_{\text{diag}}$	1	0	1/2

We now make a change of variables that will help us expose the possible fractionalization and emergent gauge fields in the large class of models realized by (14.13). We write

$$\Psi_i \equiv H_i \phi_i^2 , \tag{14.14}$$

where

$$H_i \equiv e^{i\vartheta_i} \text{ and } \phi_i \equiv e^{i\varphi_i} , \tag{14.15}$$

so

$$\theta_i = \vartheta_i + 2\varphi_i \bmod(2\pi) . \tag{14.16}$$

The decomposition in (14.14) may seem somewhat arbitrary here, but we have chosen it because it is a simpler version of a procedure that appears more naturally when we consider quantum spin systems in Chapter 15. We will later simplify the decomposition (14.14) to one closer to (14.10) in (14.34), and this simplification also anticipates the analyses in Chapter 15.

Clearly, the decomposition (14.14) and (14.15) is highly redundant. The values of ϑ_i and φ_i are not uniquely fixed by θ_i, and we can we can perform the gauge transformation

$$\vartheta_i \to \vartheta_i + 2\alpha_i \quad , \quad \varphi_i \to \varphi_i - \alpha_i , \tag{14.17}$$

where α_i are arbitrary site-dependent real numbers, without changing θ_i. We refer to the transformation (14.17) as $U(1)_{gauge}$, and we are only interested in models that are invariant under $U(1)_{gauge}$. It is convenient to make a table of the charges, which is shown in Table 14.1. We have some freedom on how to assign the global $U(1)$ charge between H and ϕ, and we have chosen to assign the charge to H; this choice will not modify any gauge-invariant observables. Note that we also have the distinct periodicity constraints analogous to (14.2):

$$\vartheta_i \to \vartheta_i + 2\pi m_i \quad , \quad \varphi_i \to \varphi_i + 2\pi m_i' , \tag{14.18}$$

where m_i, m_i' are arbitrary integers.

We now want to write down $\widetilde{\mathcal{Z}}_{XY}$ using the ϑ_i and φ_i variables, while respecting $U(1)_{gauge}$. A convenient way to do this is to introduce one more auxilliary variable, the emergent gauge field, $a_{i\mu}$. This is a real number that resides on the *links* of the cubic lattice, with $a_{i\mu}$ on the link between sites at $\boldsymbol{r}_i$ and $\boldsymbol{r}_i + \hat{\boldsymbol{e}}_\mu$, where $\hat{\boldsymbol{e}}_\mu$ is a unit vector in the $\mu = x, y, z$ direction. Under the $U(1)$ gauge transformation in (14.17), $a_{i\mu}$ transforms as

$$a_{i\mu} \to a_{i\mu} - \Delta_\mu \alpha_i, \tag{14.19}$$

where Δ_μ denotes a discrete lattice derivative in the μ direction, with $\Delta_\mu f_i \equiv f_{i+\mu} - f_i$ and $i+\mu$ denoting the site at $\boldsymbol{r}_i + \hat{\boldsymbol{e}}_\mu$. It is also important to recognize that the gauge field a_μ is "compact": the action and all observables are periodic functions of a_μ, and invariant under

$$a_{i\mu} \to a_{i\mu} + 2\pi n_{i\mu}, \tag{14.20}$$

where $n_{i\mu}$ are integers. We now write down a $U(1)$ gauge theory, $\mathcal{Z}_U$ consistent the $U(1)$ gauge invariance and the global symmetry

$$\begin{aligned}
\mathcal{Z}_U &= \prod_i \int_0^{2\pi} \frac{d\vartheta_i}{2\pi}\frac{d\varphi_i}{2\pi} \prod_\mu \frac{da_{i\mu}}{2\pi} \exp\left(-H_U[\vartheta, \varphi, a_\mu]\right), \\
H_U[\vartheta, \varphi, a_\mu] &= -J_1 \sum_{i,\mu} \cos(\Delta_\mu \vartheta_i + 2a_{i\mu}) - J_2 \sum_{i,\mu} \cos(\Delta_\mu \varphi_i - a_{i\mu}) \\
&\quad - K \sum_{\square} \cos(\varepsilon_{\mu\nu\lambda} \Delta_\nu a_{i\lambda}),
\end{aligned} \tag{14.21}$$

where the last term is a summation over plaquettes of the cubic lattice, and $\varepsilon_{\mu\nu\lambda}\Delta_\nu a_{i\lambda}$ is the flux though a plaquette. Our claim is that $\mathcal{Z}_U$ is in the class of theories in (14.13), with

$$\prod_{i,\mu} \int_0^{2\pi} \frac{da_{i\mu}}{2\pi} \exp\left(-H_U[\vartheta, \varphi, a_\mu]\right) \approx \exp(-\widetilde{H}_{XY}[\vartheta + 2\varphi]). \tag{14.22}$$

This result follows directly from the requirements of gauge invariance and the global $U(1)$ symmetry. We can make an explicit mapping between the couplings in $\mathcal{Z}_U$ and $\mathcal{Z}_{XY}$ by expanding (14.21) in powers of K, and performing the integrals over $a_{i\mu}$ on each link of the lattice: gauge invariance requires that the results be periodic functions only of $\theta_i = \vartheta_i + 2\varphi_i$. To zeroth order in K we have the following integral on each link

$$\int_0^{2\pi} \frac{da_{i\mu}}{2\pi} \exp\left(J_1 \cos(\Delta_\mu \vartheta_i + 2a_{i\mu}) + J_2 \cos(\Delta_\mu \varphi_i - a_{i\mu})\right), \tag{14.23}$$

which is easily seen to be a function only of $\Delta_\mu \vartheta_i + 2\Delta_\mu \varphi_i = \Delta_\mu \theta_i$; this feature applies to all terms in the K expansion. The remainder of this chapter describes features of the phase diagram of $\mathcal{Z}_U$.

We note the early work of Fradkin and Shenker [87], who studied a model which corresponds to the $J_2 = 0$ limit of (14.21), when φ_i can be dropped and there is no global $U(1)$ symmetry, and some of their results are used below. There are also similarities of the above mappings to early work [25, 58, 311] on emergent $U(1)$ gauge theories for σ-models (see Appendix C).

14.2.2 Phase Diagram

We follow the approach used to successfully analyze the conventional XY model (14.1); we write down a continuum Landau–Ginzburg mean-field theory for the Ψ, and then

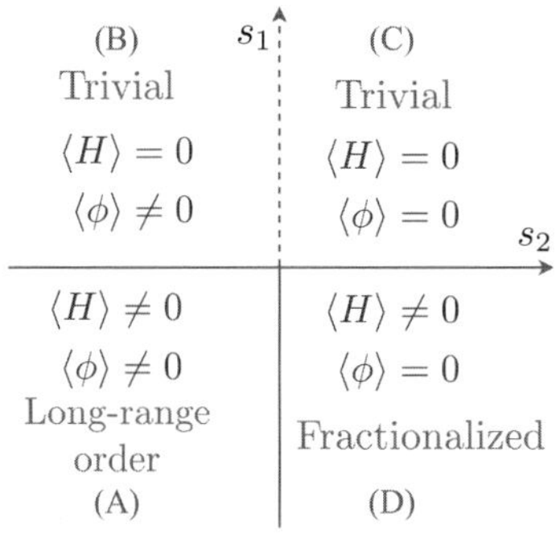

(B) Trivial $\langle H \rangle = 0$ $\langle \phi \rangle \neq 0$

s_1

(C) Trivial $\langle H \rangle = 0$ $\langle \phi \rangle = 0$

s_2

$\langle H \rangle \neq 0$ $\langle \phi \rangle \neq 0$ Long-range order (A)

$\langle H \rangle \neq 0$ $\langle \phi \rangle = 0$ Fractionalized (D)

Figure 14.2 The mean-field phase diagram of (14.24) describing bosons at integer filling, and easy-plane antiferromagnets with integer spin S. The dashed line indicates the absence of a phase transition between phases C and B, both of which are trivial, and only contain excitations with integer $\mathcal{Q}$ charges; this is an example of "Higgs-confinement" continuity. Excitations with half-integer $\mathcal{Q}$ charges are present only in phase D. The full lines indicate phase transitions. The phase transition from A to B is in the XY universality class, as in Fig. 14.1. The transition from A to D is in the XY^* universality class discussed in Section 16.5.2. The transition from D to C is in the Ising* universality class described in Section 16.5.1.

analyze fluctuations about the saddle points of the mean-field theory. For $\mathcal{Z}_U$, we have the continuum fields H, ϕ, and a_μ, and symmetry and gauge invariance yield the following Lagrangian density for the action

$$\mathcal{L}_U = |(\partial_\mu + 2ia_\mu)H|^2 + s_1|H|^2 + u_1|H|^4 + |(\partial_\mu - ia_\mu)\phi|^2 + s_2|\phi|^2 + u_2|\phi|^4 \\ + \nu|H|^2|\phi|^2 + K(\varepsilon_{\mu\nu\lambda}\partial_\nu a_\lambda)^2 + \mathcal{L}_{monopole}. \tag{14.24}$$

The last term $\mathcal{L}_{monopole}$ denotes Dirac monopole configurations of the $U(1)$ gauge field, which require a lattice to properly define at the core of the monopoles; we will not describe this term further here, and defer a full analysis to Section 25.3. Below, we use other arguments to understand the effects of monopoles qualitatively. The analysis is a generalization of that in Chapters 5–7 on the Landau–Ginzburg theory of superconductivity.

Let us obtain the mean-field phase diagram of $\mathcal{L}_U$, ignoring the effects of a_μ. For simplicity, we take $\nu = 0$; the resulting phase diagram is shown in Fig. 14.2. We discuss the nature of the phases, and of the a_μ fluctuations in turn.

A. $\langle H \rangle \neq 0$, $\langle \phi \rangle \neq 0$

It is simplest to begin in the phase with both H and ϕ condensed, in which the effects of a_μ are controlled, and can be analyzed by a direct generalization of the analysis in Chapter 7. Clearly this phase has $\langle \Psi \rangle \neq 0$, and we can therefore identify it with the phase with long-range order in Fig. 14.1. We now see that the properties are nearly identical to the corresponding phase in the conventional XY model, except that the vortices can have somewhat different energetics.

First, we note that condensation of H and/or ϕ makes the a_μ photon massive via the Higgs mechanism; this is closely related to the Meissner effect discussed in Section 5.4.

We can see this from the effective theory for a_μ, once H and/or ϕ are condensed; from (14.24) we obtain a Higgs "mass" term for a_μ:

$$\mathcal{L}_A = a_\mu^2 \left[|\langle \phi \rangle|^2 + |\langle H \rangle|^2 \right] + K(\varepsilon_{\mu\nu\lambda}\partial_\nu a_\lambda)^2 + \mathcal{L}_{monopole} . \tag{14.25}$$

So we can safely ignore the fluctuations of a_μ (and the a_μ monopoles) in considering the long-distance properties of phase A.

To analyze the vortices, as in Chapter 7, let us also allow an external gauge field A_μ, which couples to the global $U(1)$ symmetry of the XY model; from Table 14.1, this will change the spatial gradient term for H in (14.24) to

$$|(\partial_\mu + 2ia_\mu - iA_\mu)H|^2 . \tag{14.26}$$

Now, consider a vortex in which the phase of H winds by $2\pi n_H$, and the phase of ϕ winds by $2\pi n_\phi$, with n_H and n_ϕ integers. In terms of the gauge-invariant XY order parameter, by (14.14), this is a vortex with phase winding $2\pi n_\Psi$ with

$$n_\Psi = n_H + 2n_\phi . \tag{14.27}$$

Let us denote the total A_μ flux in this vortex by Φ_A, and similarly the a_μ flux by Φ_a. Then, generalizing the arguments in Chapter 7, finiteness of the vortex energy in (14.24) modified by (14.26) requires that

$$\Phi_a = 2\pi n_\phi \quad , \quad \Phi_A - 2\phi_a = 2\pi n_H . \tag{14.28}$$

Adding these expressions, we obtain $\Phi_A = 2\pi n_\Psi$, which is exactly the correct expression for the long-range-ordered phase of the conventional XY model. The elementary vortex $n_\Psi = 1$ is obtained by $n_H = 1$, $n_\phi = 0$, $\Phi_a = 0$. However, the double vortex $n_\Psi = 2$ has two possible configurations: $n_H = 2$, $n_\phi = 0$, $\Phi_a = 0$ and $n_H = 0$, $n_\phi = 1$, $\Phi_a = 2\pi$. These two vortex configurations differ by an a_μ flux of 2π, and so a Dirac monopole in $\mathcal{L}_{monopole}$ can induce tunneling between them, and we need only keep the lower energy combination. In a parameter regime with $|s_1| \gg |s_2|$, phase winding with n_H non-zero can become expensive; so the extended XY model can lead to a novel situation where a double vortex with $n_\Psi = 2$, $n_\phi = 1$ can become less expensive than a vortex with $n_\Psi = 1$. But the physical quantum numbers of all the allowed vortices remain identical to those of the conventional XY model.

B. $\langle H \rangle = 0$, $\langle \phi \rangle \neq 0$

Here, ϕ is condensed, and so let us set $\phi = \phi_0$, with a gauge chosen so that ϕ_0 is real. This condensation of ϕ makes the a_μ photon massive via the Higgs mechanism, as in (14.25). Moreover, with ϕ condensed, then we can reduce (14.14) to

$$\Psi \sim H \quad \text{in phase B.} \tag{14.29}$$

With this identification, and the suppression of a_μ, the remaining theory for H is just the XY model for Ψ, and so $\langle H \rangle = 0$ implies that we are in the trivial phase of the XY model, as indicated in Fig. 14.2. The massive H excitations have integer charges $Q = \pm 1$.

With ϕ condensed in phase B, the theory $\mathcal{L}_U$ does have vortex saddle points, and the reader may wonder why this does not make a difference to the identification as a trivial phase. Because H is not condensed, it is not sensible to consider the winding n_H, and also not the winding in the XY order parameter n_Ψ. However, the winding n_ϕ is well defined. A vortex with $n_\phi = 1$ has $\Phi_a = 2\pi$. But the monopoles render such a flux invisible, given the periodicity property of a_μ discussed near (14.20), that is, monopoles are tunneling events that change global flux by 2π, and this is possible because the associated Dirac string is invisible due to (14.20) these issues will be discussed in more detail in Section 25.3.

C. $\langle H \rangle = 0$, $\langle \phi \rangle = 0$

At large positive s_1 and s_2, both H and ϕ are restricted to very small magnitudes, and so we can initially set them to zero. Then the remaining term in $\mathcal{L}_U$ describes the statistical mechanics of a gapless photon a_μ in three spacetime dimensions. Such a gauge theory is confining, which means no free gauge-charged particles are allowed. Here, this is easy to see in the small-K expansion noted above (14.23); we argued there that such an expansion generates the XY model for $\Psi \sim H\phi^2$, and so only the Ψ particle survives as a free excitation. The confining property also holds for large K, but establishing this is more subtle: we have to couple the gapless photon to monopole configurations decribed by $L_{monopole}$. We discuss the large-K confining phase of the $U(1)$ gauge theory using a duality transformation in Section 25.3. The resulting magnitude for Ψ is small in the confining phase, and so we conclude that phase C coincides with the trivial phase in Fig. 14.1. This is confirmed by the fact that the only gauge-neutral combinations of H and ϕ have integer values of global $U(1)$ charge $\mathcal{Q}$.

Phase B is fundamentally the same as phase C, and there is no phase transition between them: this is an example of "Higgs-confinement continuity," and is linked to the fact that ϕ carries a unit gauge charge, as we shall see shortly.

D. $\langle H \rangle \neq 0$, $\langle \phi \rangle = 0$

Finally, we turn to the novel fractionalized phase of interest, whose creation has been the objective of our model building. The analysis of fluctuations about the mean field in phase D is similar to that of phase B, except that it is H that is condensed, and not ϕ. As in phase B, let us now set $H = H_0$, with a gauge chosen so that H_0 is real. However, there is the crucial difference that H carries gauge charge 2, while ϕ carries gauge charge 1. Related to this is the fact that (14.29) is now replaced by

$$\Psi \sim \phi^2 \quad \text{in phase D,} \tag{14.30}$$

coinciding with our fractionalization ansatz (14.10). Because of the Higgs mass for a_μ, the ϕ excitations behave like nearly free massive particles, and so the gauge fluctuations won't modify the long-range part of the Ψ correlator obtained from (14.30), and it will have the needed fractionalized form in (14.11).

In our charge assignments in Table 14.1, we defined the field H to be charged both under $U(1)$ and $U(1)_{gauge}$. It is now convenient to redefine the $U(1)_{gauge}$ symmetry so that the H field is neutral under the global charge, and we can work with a global symmetry that has not been broken. To this end, we shift $a_\mu \to a_\mu + A_\mu/2$, and then the global charge associated with A_μ is $U(1)_{diag} = U(1) + U(1)_{gauge}/2$. We now see from Table 14.1 that the ϕ excitations indeed have a global $U(1)$ charge $Q = 1/2$.

We now note a crucial emergent property of phase D, which was not on our mind when we designed $\mathcal{Z}_U$. With the condensation of H, we can consider saddle points of $\mathcal{L}_U$ that are $n_H = \pm 1$ vortices in H with winding $2\pi n_H$. The winding numbers n_ϕ and n_Ψ are not meaningful in phase D, but the second relation in (14.28) implies that $\Phi_a = \pm\pi$ (after the shift in a_μ noted above). Such an a_μ flux is physically observable, and so this vortex is a physical object. Note that because of the periodicity (14.20), the flux of π coincides with the flux of $-\pi$. So this vortex line can be interpreted as the wordline of a real emergent particle present in the fractionalized phase, which is its own anti-particle: this is nothing but the *vison* of Section 13.4. Moreover, the ϕ excitations in phase D are the analog of the *spinon* of Section 13.4: the ϕ field picks up an Aharonov–Bohm phase factor of -1 when encircling the $\pm\pi$ flux of this vortex. Thus, phase D has essentially all of the structure of the $\mathbb{Z}_2$ spin liquid presented at the end of Section 13.4. One important point is that phase D here realizes the "even"-$\mathbb{Z}_2$ spin liquid, with S an integer in (13.8). The more interesting, and physically relevant, case of the odd-$\mathbb{Z}_2$ gauge theory is discussed in Chapter 16.

A key point worth noting here is that the distinction between the fractionalized phase D and the remaining phases in Fig. 14.2 is "topological" and robust, and does *not* rely upon the presence of the global $U(1)$ symmetry in the extended XY model. The distinction remains even when there are terms in the action that explicitly break the global $U(1)$ symmetry, so that the charge Q is not defined: then the distinction between phases A, B, and C disappears, but phase D remains distinct. This topological distinction relies upon the existence of the stable vison excitations in phase D, which survive the breaking of the global $U(1)$ symmetry, and are not present in the other phases of Fig. 14.2. We will see in Section 14.2.3 that the spinons carry $\mathbb{Z}_2$ gauge charges, and in Section 16.3 that the existence of the visons translates into a topological degeneracy of ground states on a torus, which is only present in phase D; both are robust features of phase D, which are independent of the presence or not of the global $U(1)$ symmetry.

14.2.3 Emergent $\mathbb{Z}_2$ Gauge Theory

While the theory of the phases of Fig. 14.2 in terms of the compact $U(1)$ gauge theory in $\mathcal{Z}_U$ is internally consistent and satisfactory, it is possible to obtain the same phases in a simpler theory. An important point is that the gapless photon a_μ does not make an appearance in any of the phases, which is an indication that we could rewrite the theory without a $U(1)$ gauge field.

The reuqired theory is obtained by taking the large J_1 limit of $\mathcal{Z}_U$ in (14.21). We choose a gauge in which $\vartheta_i = 0$. Then we see that the J_1 is optimized when

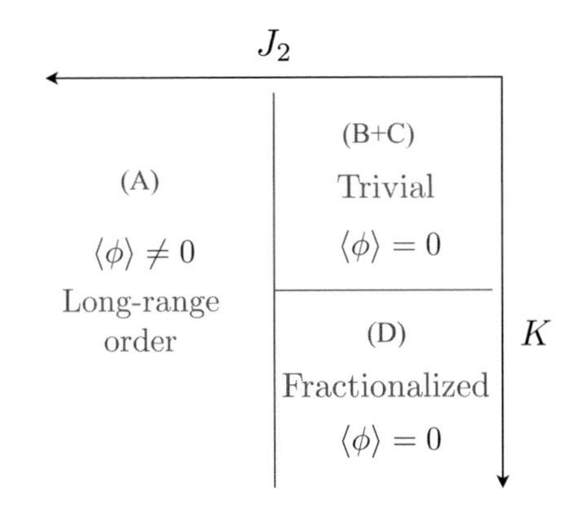

Figure 14.3 Schematic phase diagram of the $\mathbb{Z}_2$ gauge theory $\mathcal{Z}_{\mathbb{Z}_2}$ in (14.32) in $D = 3$ describing bosons at integer filling and easy-plane antiferromagnets with integer spin S in $d = 2$. The labels of the phases are the same as those in Fig. 14.2 for $\mathcal{Z}_U$.

$$a_{i\mu} = 0, \pi \quad , \quad e^{ia_{i\mu}} \equiv Z_{i,i+\mu} = \pm 1 , \tag{14.31}$$

and we therefore restrict a_μ to these discrete values. This has effectively eliminated the H field, and reduced the $U(1)$ gauge field $a_{i\mu}$ to a $\mathbb{Z}_2$ gauge field $Z_{i,i+\mu}$. The partition function (14.21) now becomes

$$\begin{aligned} \mathcal{Z}_{\mathbb{Z}_2} &= \sum_{Z_{ij}=\pm 1} \prod_i \int_0^{2\pi} \frac{d\varphi_i}{2\pi} \exp\left(-H_{\mathbb{Z}_2}[\varphi, Z]\right) , \\ H_{\mathbb{Z}_2}[\varphi, Z] &= -J_2 \sum_{\langle ij \rangle} Z_{ij} \cos(\varphi_i - \varphi_j) - K \sum_{\square} \prod_{ij \in \square} Z_{ij} . \end{aligned} \tag{14.32}$$

The action of this partition function is not invariant under a $U(1)$ gauge transformation, but is invariant under a remnant $\mathbb{Z}_2$ gauge transformation

$$e^{i\varphi_i} \rightarrow \eta_i e^{i\varphi_i} \quad , \quad Z_{ij} \rightarrow \eta_i Z_{ij} \eta_j , \tag{14.33}$$

where $\eta_i = \pm 1$ is the $\mathbb{Z}_2$ gauge remnant of α_i in (14.17) and (14.19). The partition function (14.32) describes an extended XY model, with the identification of the gauge-invariant order parameter

$$\Psi_i = \phi_i^2 = e^{2i\varphi_i} \tag{14.34}$$

obtained from (14.14), and so (14.10) has now become an equality. It is now much easier, than for $\mathcal{Z}_U$, to perform the summation over Z_{ij} in a small-K expansion of (14.32), and hence obtain the terms of the extended XY model in (14.13). So $\mathcal{Z}_{\mathbb{Z}_2}$ is a better minimal choice for a theory of the extended XY model, involving only the field $\phi = e^{i\varphi}$ with charge $Q = 1/2$, and a $\mathbb{Z}_2$ gauge field. We note that a $\mathbb{Z}_2$ gauge theory similar to (14.32) was used by Lammert et al. [149, 150, 281] to propose fractionalized phases in the classical phase diagram of a nematic liquid crystal.

The phases of $\mathcal{Z}_{\mathbb{Z}_2}$ can be analyzed in a manner similar to $\mathcal{Z}_U$ and a schematic phase diagram is shown in Fig. 14.3. The analysis requires some knowledge of the structure of the $\mathbb{Z}_2$ gauge theory, which we turn to in Chapter 16. The phases in Fig. 14.3 have the same basic structure as the correspondingly labeled phases in Fig. 14.2, and the fractionalized ϕ excitations with $Q = \pm 1/2$ are deconfined only in phase D, where they carry a $\mathbb{Z}_2$ gauge charge (by (14.33)). The vison excitations of phase D are now simpler: they correspond to configurations over Z_{ij} with $\prod_{ij\in\square} Z_{ij} = -1$, in a background of $\prod_{ij\in\square} Z_{ij} = 1$, as illustrated in Fig. 16.2. The presence of the visons, and spinon excitations with $\mathbb{Z}_2$ charges, are robust features of phase D, which survive even when the global $U(1)$ symmetry is explicitly broken.

As was the case for $\mathcal{Z}_U$, the partition function $\mathcal{Z}_{\mathbb{Z}_2}$ in (14.32) describes the even-$\mathbb{Z}_2$ spin liquid case, corresponding to integer spin-S antiferromagnets, or bosons at integer filling. The extension of the $\mathbb{Z}_2$ gauge theory to the *odd* case corresponding to half-integer spin-S antiferromagnets, or bosons at half-integer filling is considered in Chapter 16. In the spacetime lattice discretization being considered in the present chapter, this odd extension corresponds to adding an additional Berry phase term to (14.32), which depends on the $\mathbb{Z}_2$ gauge field on the temporal links:

$$H_{\mathbb{Z}_2}[\varphi,\sigma] \to H_{\mathbb{Z}_2}[\varphi,\sigma] - i\pi S \sum_i (1 - Z_{i,i+\tau}) \,. \tag{14.35}$$

In terms of the partition function in (14.32), the additional term in (14.35) corresponds to an overall factor so that (14.32) is modified to

$$\mathcal{Z}_{\mathbb{Z}_2} = \sum_{Z_{ij}=\pm 1} \prod_i \int_0^{2\pi} \frac{d\varphi_i}{2\pi} \left[\prod_i Z_{i,i+\tau}\right]^{2S} \exp\left(-H_{\mathbb{Z}_2}[\varphi, Z]\right) \,, \tag{14.36}$$

where $H_{\mathbb{Z}_2}$ is as in (14.32); the factor in square brackets is gauge invariant after we impose periodic boundary conditions in the temporal direction. Such a factor corresponds to including "Polyakov loops" on each spatial site of the lattice. A derivation [229, 239, 257] of (14.35) from spin Berry phases in the $U(1)$ gauge theory parent is presented in Appendix C and (26.4), and its consequence in the $\mathbb{Z}_2$ gauge theory below (26.37). See also Problem 16.2 connecting (14.35) to the corresponding $\mathbb{Z}_2$ gauge theory Hamiltonian formulation in Chapter 16, from which we will see that the Berry phase in (14.35) and (14.36) is intimately connected to the phase factor in (13.8) on the action of translations on visons.

A numerical study of the odd extension of (14.32) in (14.35) was performed in Ref. [198], and the phase diagram shown in Fig. 14.4 was obtained. Because the Berry phase factor in (14.36) can be either positive or negative for half-integer S, direct simulation of the partition function (14.36) is not possible; instead, the simulation was performed after a duality mapping [198] similar to those discussed in Chapters 25 and 26. The main change from Fig. 14.3 is that the "trivial" phase has been replaced by a valence-bond solid (shown in Fig. 9.2), a phenomenon which is treated theoretically in Section 16.5.2. Indeed, a notable feature of Fig 14.4 is that there is no completely trivial phase: every phase has either a broken symmetry or fractionalization. We will see in Chapter 16 that this is a fundamental feature of the odd case, describing half-integer spin-S antiferromagnets, or bosons at half-integer filling. The absence of a trivial phase implies that fractionalized phases are *more* likely in quantum systems with Berry phases

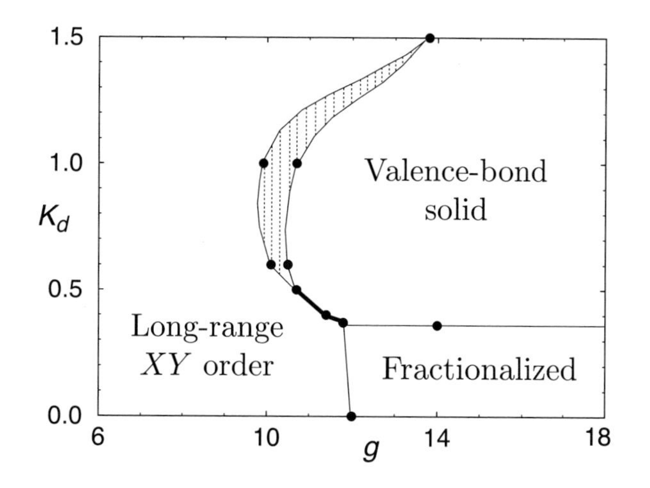

Figure 14.4 Numerical phase diagram of a Villain formulation of the "odd"-extension of (14.32), applying to half-integer spin-S easy-plane antiferromagnets or bosons at half-integer filling in $D = 3, d = 2$, obtained in Ref. [198]. The coupling K_d is related to K via $\tanh(K_d) = e^{-2K}$, and $J_2 = 4/g$. The valence-bond solid has the same order as in Fig. 9.2. The ordered phase at small g has long-range order in Ψ. The phases could not be precisely identified within the dashed region. Note the similarity of these phases to those in the experimental survey in Chapter 1 in Fig. 1.8. Reprinted with permission from APS.

(as in (14.35) and (14.36)) than in classical systems with only positive weights in the partition function. We will see another example of Berry phase induced deconfinement in the discussion of deconfined quantum critical points in Section 28.1.

Problems

14.1 Compute the two-point correlator of ϕ^2 using the relativistic field theory (10.2) in its gapped and symmetric phase, at order u^0. Take the imaginary part, and obtain a more precise version of (14.12). Argue that the form of the threshold singularity is not modified at higher orders in u.

14.2 Expand (14.32) to order K^4, and sum over the Z_{ij}. So obtain an explicit form of the extended XY model (14.13).

15 Theory of Gapped $\mathbb{Z}_2$ Spin Liquids

A large-N expansion of square- and triangular-lattice antiferromagnets is obtained using a representation of the spins in terms of Schwinger bosons. This yields magnetically ordered phases, $\mathbb{Z}_2$ spin liquids, and theories for the transitions between these phases. Topological excitations of the $\mathbb{Z}_2$ spin liquid are used to systematically obtain the properties of visons.

This chapter moves from the classical XY models of Chapter 14 to realistic quantum spin models, which realize phases with fractionalization. We begin our formal study of spin-liquid phases of the Hamiltonian in (9.14), which we write in a more general form

$$\mathcal{H} = \sum_{i,j} J_{ij} \boldsymbol{S}_i \cdot \boldsymbol{S}_j . \tag{15.1}$$

We consider the general case of $\boldsymbol{S}_i$ being spin-S quantum spin operators on the sites, i, of a two-dimensional lattice. The J_{ij} are short-ranged antiferromagnetic exchange interactions. We will mainly consider here the square and triangular lattices with nearest-neighbor interactions, but the methods generalize to a wide class of lattices and interaction ranges. The results on the triangular lattice obtained in this chapter apply to observations on the insulator $\mathrm{KYbSe_2}$ [250].

As discussed in Section 9.2, for spin $S = 1/2$, such models can be mapped onto theories of hard-core bosons B_i, with the operator correspondence

$$B_i = S_{i-} \quad , \quad B_i^\dagger B_i - 1/2 = S_{iz} , \tag{15.2}$$

so that an occupied (empty) boson state is a spin-up (-down) state. In this chapter we study quantum states in which the hard-core boson number is *fractionalized*, and we have excitations with $B^\dagger B = \pm 1/2$. As discussed briefly in Chapter 13, this is achieved most naturally in a formalism that uses a "parton" construction, in which the spin operators are expressed in terms of half-charged particles at the outset. Such constructions invariably lead to emergent gauge symmetries, and analyzing the structure of such emergent gauge theories is a major objective of Parts II and IV.

As in the classical XY analysis of Chapter 14, we find that our analysis initially leads to a compact $U(1)$ gauge theory, which can be simplified to a $\mathbb{Z}_2$ gauge theory under suitable conditions.

15.1 Parton Formulation

A careful examination of the non-magnetic "spin-liquid" phases requires an approach that is designed explicitly to be valid in a region well separated from Néel long-range order, and preserves $SU(2)$ symmetry at all stages. It should also be designed to naturally allow for neutral $S = 1/2$ excitations. To this end, we introduce the Schwinger boson description [13], in terms of elementary $S = 1/2$ bosons. For the group $SU(2)$ the complete set of $(2S+1)$ states on site i are represented as follows

$$|S,m\rangle \equiv \frac{1}{\sqrt{(S+m)!(S-m)!}}(s^{\dagger}_{i\uparrow})^{S+m}(s^{\dagger}_{i\downarrow})^{S-m}|0\rangle, \tag{15.3}$$

where $m = -S,\ldots,S$ is the z component of the spin ($2m$ is an integer). We have introduced two flavors of Schwinger bosons on each site, created by the canonical operator $s^{\dagger}_{i\alpha}$, with $\alpha = \uparrow,\downarrow$, and $|0\rangle$ is the vacuum with no Schwinger bosons. The total number of Schwinger bosons, n_s, is the same for all the states; therefore

$$s^{\dagger}_{i\alpha}s^{\alpha}_{i} = n_s \tag{15.4}$$

with

$$n_s = 2S \tag{15.5}$$

(we will henceforth assume an implied summation over repeated upper and lower indices). It is not difficult to see that the above representation of the states is completely equivalent to the operator identity in (13.4) between the spin and Schwinger boson operators

$$\boldsymbol{S}_i = \frac{1}{2}s^{\dagger}_{i\alpha}\,\boldsymbol{\sigma}^{\alpha}_{\beta}\,s^{\beta}_{i}, \tag{15.6}$$

where $\ell = x,y,z$ and the σ^{ℓ} are the usual 2×2 Pauli matrices. In the present chapter, we distinguish between upper and lower spin indices $\alpha,\beta,\ldots$ because this clarifies the large M limit to be discussed shortly.

Note that the Schwinger bosons s_{α} are (roughly) the "square root" of the hard-core boson B noted above. In particular, for $S = 1/2$, $S_+ = s^{\dagger}_{\uparrow}s_{\downarrow} \sim B^{\dagger}$. So the Schwinger boson $s_{\uparrow}$ ($s_{\downarrow}$) carries boson number $1/2$ ($-1/2$).

The spin states on the two sites i, j can combine to form a singlet in a unique manner – the wavefunction of the singlet state is particularly simple in the boson formulation:

$$\left(\varepsilon^{\alpha\beta}s^{\dagger}_{i\alpha}s^{\dagger}_{j\beta}\right)^{2S}|0\rangle. \tag{15.7}$$

Finally we note that, using the constraint (15.4), the following Fierz-type identity generalizing (9.30) can be established

$$\left(\varepsilon^{\alpha\beta}s^{\dagger}_{i\alpha}s^{\dagger}_{j\beta}\right)\left(\varepsilon_{\gamma\delta}s^{\gamma}_{i}s^{\delta}_{j}\right) = -2\boldsymbol{S}_i\cdot\boldsymbol{S}_j + n_s^2/2 + \delta_{ij}n_s, \tag{15.8}$$

where ε is the totally antisymmetric 2×2 tensor

$$\varepsilon = \begin{pmatrix} 0 & 1 \\ -1 & 0 \end{pmatrix}. \tag{15.9}$$

This implies that $\mathcal{H}$ can be rewritten in the form (apart from an additive constant)

$$\mathcal{H} = -\frac{1}{2} \sum_{<ij>} J_{ij} \left(\varepsilon^{\alpha\beta} s^{\dagger}_{i\alpha} s^{\dagger}_{j\beta} \right) \left(\varepsilon_{\gamma\delta} s^{\gamma}_{i} s^{\delta}_{j} \right). \tag{15.10}$$

This form makes it clear that $\mathcal{H}$ counts the number of singlet bonds.

We have so far defined a one-parameter (n_s) family of models $\mathcal{H}$ for a fixed realization of the J_{ij}. Increasing n_s makes the system more classical and a large-n_s expansion is therefore not suitable for studying the quantum-disordered phase. For this reason we introduce a second parameter – the flavor index α on the bosons is allowed to run from $1, \ldots, 2M$ with M an arbitrary integer. This therefore allows the bosons to transform under $SU(2M)$ rotations. However the $SU(2M)$ symmetry turns out to be too large. We want to impose the additional restriction that the spins on a pair of sites are able to combine to form a singlet state, thus generalizing the valence-bond structure of $SU(2)$ – this valence-bond formation is clearly a crucial feature determining the structure of the states that do not have long-range magnetic order. It is well known that this is impossible for $SU(2M)$ for $M > 1$ – there is no generalization of the second-rank, antisymmetric, invariant tensor ε to general $SU(2M)$.

The proper generalization turns out to be the group $USp(2M)$ [219]. This group is defined by the set of $2M \times 2M$ unitary matrices U such that

$$U^T \mathcal{J} U = \mathcal{J} \tag{15.11}$$

where

$$\mathcal{J}_{\alpha\beta} = \mathcal{J}^{\alpha\beta} = \begin{pmatrix} & 1 & & & & \\ -1 & & & & & \\ & & & 1 & & \\ & & -1 & & & \\ & & & & \ddots & \\ & & & & & \ddots \end{pmatrix} \tag{15.12}$$

is the generalization of the ε tensor to $M > 1$. It is clear that $USp(2M) \subset SU(2M)$ for $M > 1$, while $USp(2) \cong SU(2)$. The s^{α}_{i} bosons transform as the fundamental representation of $USp(2M)$; the "spins" on the lattice therefore belong to the symmetric product of n_s fundamentals, which is also an irreducible representation. Valence bonds

$$\mathcal{J}^{\alpha\beta} s^{\dagger}_{i\alpha} s^{\dagger}_{j\alpha} \tag{15.13}$$

can be formed between any two sites; this operator is a singlet under $USp(2M)$ because of (15.11). The form (15.10) of $\mathcal{H}$ has a natural generalization to general $USp(2M)$:

$$\mathcal{H} = -\sum_{i>j} \frac{J_{ij}}{2M} \left(\mathcal{J}^{\alpha\beta} s^{\dagger}_{i\alpha} s^{\dagger}_{j,\beta} \right) \left(\mathcal{J}_{\gamma\delta} s^{\gamma}_{i} s^{\delta}_{j} \right), \tag{15.14}$$

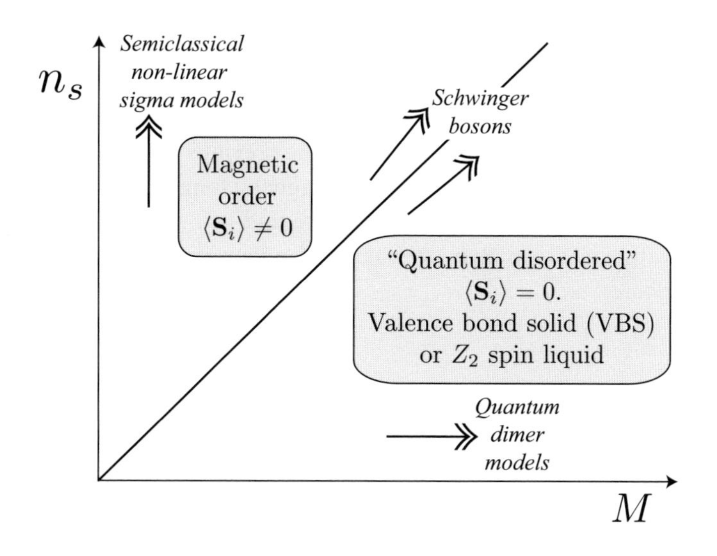

Figure 15.1 Phase diagram of the two-dimensional USp$(2M)$ antiferromagnet $\mathcal{H}$ as a function of the "spin" n_s. The "quantum-disordered" region preserves $Sp(M)$ spin-rotation invariance, and there is no magnetic long-range order; however, the ground states here have new types of emergent order (VBS, or Z_2 topological order), that are described in the text.

where the indices $\alpha, \beta, \gamma, \delta$ now run over $1, \ldots, 2M$. We recall also that the constraint (15.4) must be imposed on every site of the lattice.

We now have a two-parameter (n_s, M) family of models $\mathcal{H}$ for a fixed realization of the J_{ij}. It is very instructive to consider the phase diagram of $\mathcal{H}$ as a function of these two parameters (Fig. 15.1).

The limit of large n_s, with M fixed leads to the semi-classical theory. For the special case of $SU(2)$ antiferromagnets with a two-sublattice collinear Néel ground state, the semi-classical fluctuations are described by the $O(3)$ non-linear sigma model, which is clearly related to the $O(3)$ relativistic scalar field theory discussed in Chapter 10. For other models, the structure of the non-linear sigma models is rather more complicated and will not be considered here.

A second limit in which the problem simplifies is M large at fixed n_s [4, 216], which is taken using the Schwinger fermion representation. It can be shown that in this limit the ground state is "quantum disordered." Further, the low-energy dynamics of $\mathcal{H}$ is described by an effective quantum dimer model [216, 222], with each dimer configuration representing a particular pairing of the sites into valence bonds. There have been extensive studies of such quantum dimer models, and we note some of them later in Sections 16.4.2 and 26.1.2. Such quantum dimer model studies in the "quantum-disordered" region of Fig. 15.1 have yielded phases that were obtained earlier [219] by the methods described below.

The most interesting solvable limit, studied in the present chapter, is obtained by fixing the ratio of n_s and M

$$\kappa = \frac{n_s}{M} \tag{15.15}$$

and subsequently taking the limit of large M. The implementation of $\mathcal{H}$ in terms of bosonic operators also turns out to be naturally suited for studying this limit. The parameter κ is arbitrary; tuning κ modifies the slope of the line in Fig. 15.1 along which the large M limit is taken. From the previous limits discussed above, one might expect that the ground state of $\mathcal{H}$ has magnetic long-range order for large κ and is quantum disordered for small κ. Indeed, we find below that for any set of J_{ij} there is a critical value of $\kappa = \kappa_c$ that separates the magnetically ordered and the quantum-disordered phase.

15.2 Mean-Field Theory

We begin by analyzing $\mathcal{H}$ at $M = \infty$ with $n_s = \kappa M$. As noted above, this limit is most conveniently taken using the bosonic operators. We may represent the partition function of $\mathcal{H}$ by

$$Z = \int \mathcal{D}\mathcal{Q}\mathcal{D}s\mathcal{D}\lambda \exp\left(-\int_0^\beta \mathcal{L}d\tau\right), \tag{15.16}$$

where

$$\begin{aligned}\mathcal{L} = &\sum_i \left[s_{i\alpha}^\dagger \left(\frac{d}{d\tau} + i\lambda_i\right) s_i^\alpha - i\lambda_i n_s \right] \\ &+ \sum_{<i,j>} \left[M\frac{J_{ij}|\mathcal{Q}_{i,j}|^2}{2} - \frac{J_{ij}\mathcal{Q}_{i,j}^*}{2} \mathcal{J}_{\alpha\beta} s_i^\alpha s_j^\beta + \text{H.c.} \right].\end{aligned} \tag{15.17}$$

Here, the λ_i fix the boson number of n_s at each site; τ-dependence of all fields is implicit; $\mathcal{Q}$ was introduced by a Hubbard–Stratonovich decoupling of $\mathcal{H}$. Notice the close similarity of this procedure to the transformations needed for the Landau–Ginzburg theory of superconductivity in Chapter 6.

However, the present theory has a crucial additional feature that was not present in Chapter 6. The Lagrangian $\mathcal{L}$ has a $U(1)$ gauge invariance under which

$$\begin{aligned} s_{i\alpha}^\dagger &\to s_{i\alpha}^\dagger \exp\left(i\rho_i(\tau)\right), \\ \mathcal{Q}_{ij} &\to \mathcal{Q}_{ij} \exp\left(-i\rho_i(\tau) - i\rho_j(\tau)\right), \\ \lambda_i &\to \lambda_i + \frac{\partial \rho_i}{\partial \tau}(\tau), \end{aligned} \tag{15.18}$$

which is the analog of (14.17) and (14.19) in the classical XY model: we will see below that the $\mathcal{Q}$ play the role of the H field on the triangular lattice, while s_α play the role of ϕ. The functional integral over $\mathcal{L}$ faithfully represents the partition function, but does require gauge fixing. This gauge invariance leads to emergent gauge-field degrees of freedom, as we see below.

The $1/M$ expansion of the free energy can be obtained by integrating out of $\mathcal{L}$ the $2M$-component $s,\bar{s}$ fields to leave an effective action for $\mathcal{Q}$, λ having coefficient M (because $n_s \propto M$). Thus, the $M \to \infty$ limit is given by minimizing the effective action with respect to "mean-field" values of $\mathcal{Q} = \bar{\mathcal{Q}}$, $i\lambda = \bar{\lambda}$ (we are ignoring here the possibility of magnetic long-range order, which requires an additional condensate $x^\alpha = \langle b^\alpha \rangle$). This is in turn equivalent to solving the mean-field Hamiltonian

$$\mathcal{H}_{MF} = \sum_{<i,j>} \left(M \frac{J_{ij}|\bar{\mathcal{Q}}_{ij}|^2}{2} - \frac{J_{ij}\bar{\mathcal{Q}}^*_{i,j}}{2} \mathcal{J}_{\alpha\beta} s_i^\alpha s_j^\beta + \text{H.c.} \right) + \sum_i \bar{\lambda}_i (s^\dagger_{i\alpha} s_i^\alpha - n_s), \quad (15.19)$$

a more general version of which appeared in (13.6). This Hamiltonian is quadratic in the boson operators and all its eigenvalues can be determined by a Bogoluibov transformation, closely connected to that in (3.11). This leads in general to an expression of the form

$$\mathcal{H}_{MF} = E_{MF}[\bar{\mathcal{Q}}, \bar{\lambda}] + \sum_\mu \omega_\mu [\bar{\mathcal{Q}}, \bar{\lambda}] \gamma^\dagger_{\mu\alpha} \gamma^\alpha_\mu. \quad (15.20)$$

The eigenstates are labeled by the index μ, and the number of eigenstates equals the number of sites in the system. E_{MF} is the ground-state energy and is a functional of $\bar{\mathcal{Q}}$, $\bar{\lambda}$; ω_μ is the eigenspectrum of excitation energies, which is also a function of $\bar{\mathcal{Q}}$, $\bar{\lambda}$; and the γ^α_μ represent the bosonic eigenoperators. The excitation spectrum thus consists of non-interacting spinor bosons. The ground state is determined by minimizing E_{MF} with respect to the $\bar{\mathcal{Q}}_{ij}$ subject to the constraints

$$\frac{\partial E_{MF}}{\partial \bar{\lambda}_i} = 0. \quad (15.21)$$

The saddle-point value of the $\bar{\mathcal{Q}}$ satisfies

$$\bar{\mathcal{Q}}_{ij} = \langle \mathcal{J}_{\alpha\beta} s_i^\alpha s_j^\beta \rangle. \quad (15.22)$$

Note that $\bar{\mathcal{Q}}_{ij} = -\bar{\mathcal{Q}}_{ji}$, indicating that $\bar{\mathcal{Q}}_{ij}$ is a directed field – an orientation has to be chosen on every link.

These saddle-point equations have been solved for the square and triangular lattices with nearest-neighbor exchange J, and they lead to stable and translationally invariant solutions for $\bar{\lambda}_i$ and $\bar{\mathcal{Q}}_{ij}$. The only saddle-point quantity that does not have the full symmetry of the lattice is the orientation of the $\bar{\mathcal{Q}}_{ij}$. Note that although it appears that such a choice of orientation appears to break inversion or reflection symmetries, such symmetries are actually preserved: the $\bar{\mathcal{Q}}_{ij}$ are not gauge-invariant, and all gauge-invariant observables do preserve all symmetries of the underlying Hamiltonian. For the square lattice, we have $\bar{\lambda}_i = \bar{\lambda}$, $\bar{\mathcal{Q}}_{i,i+\hat{x}} = \bar{\mathcal{Q}}_{i,i+\hat{y}} = \bar{\mathcal{Q}}$. Similarly, on the triangular lattice we have $\bar{\mathcal{Q}}_{i,i+\hat{e}_p} = \bar{\mathcal{Q}}$ for $p = 1,2,3$, where the unit vectors

$$\begin{aligned} \hat{e}_1 &= (1/2, \sqrt{3}/2), \\ \hat{e}_2 &= (1/2, -\sqrt{3}/2), \\ \hat{e}_3 &= (-1, 0), \end{aligned} \quad (15.23)$$

point between nearest-neighbor sites of the triangular lattice. The orientation of the $\bar{\mathcal{Q}}_{ij}$ on the triangular lattice is sketched in Fig. 15.2.

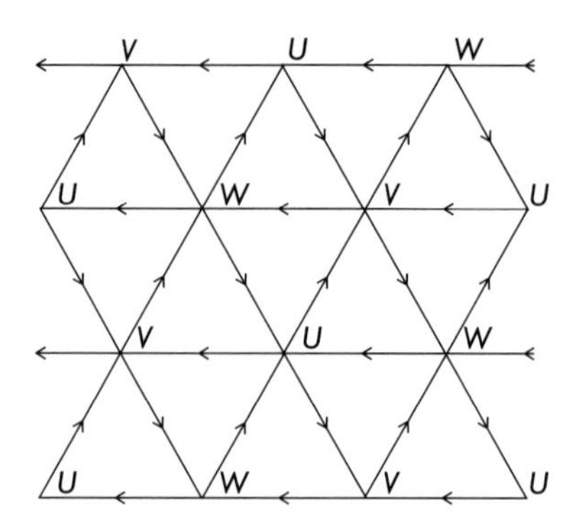

Figure 15.2 Orientation of the nearest neighbor $\bar{Q}_{ij}$ on the triangular lattice. Also shown are the labels of the three sublattices.

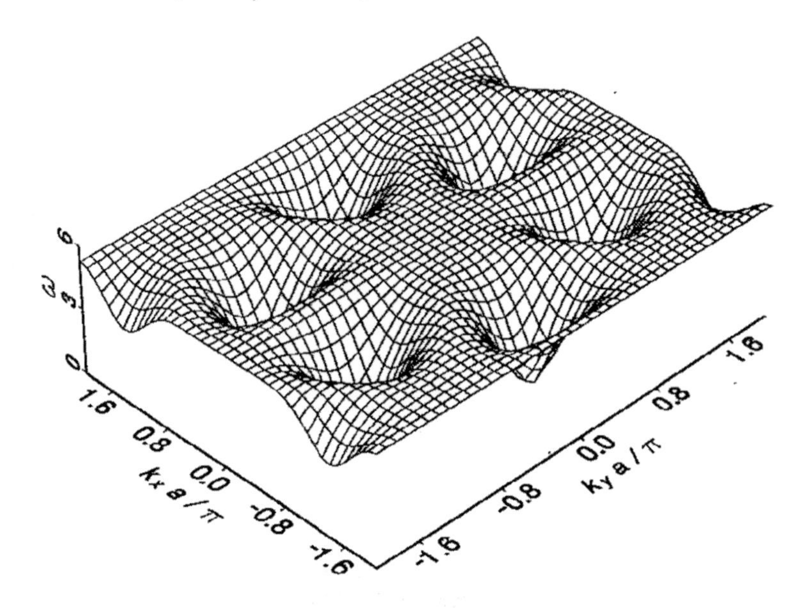

Figure 15.3 Spinon dispersion on the triangular lattice [231]. Reprinted with permission from APS.

We can also compute the dispersion ω_k of the γ_k excitations. These are bosonic particles that carry spin $S = 1/2$ ("spinons"), and so they carry fractionalized boson-number charge $Q = \pm 1/2$. The dispersion on the square lattice is

$$\omega_k = \left(\bar{\lambda}^2 - J^2 \bar{Q}^2 (\sin k_x + \sin k_y)^2\right)^{1/2} \tag{15.24}$$

while that on the triangular lattice is [231]

$$\omega_k = \left(\bar{\lambda}^2 - J^2 \bar{Q}^2 (\sin k_1 + \sin k_2 + \sin k_3)^2\right)^{1/2}, \tag{15.25}$$

with $k_p = \boldsymbol{k} \cdot \hat{e}_p$. These are the spinons, and the spinon dispersion on the triangular lattice is plotted in Fig. 15.3.

Notice that the spinons have minima at two degenerate points in the Brillouin zone for both lattices. For the square lattice, the minima are at $\boldsymbol{k} = \pm(\pi/2, \pi/2)$, while for the triangular lattice they are at $\boldsymbol{k} = \pm(4\pi/3, 0)$ (and at wavevectors separated from these by reciprocal lattice vectors). So there are a total of four spinon excitations in both cases: two associated with the spin degeneracy of $S_z = \pm 1/2$, and two associated with the degeneracy in the Brillouin-zone spectrum.

15.3 Excitation Spectrum

We have already described the four-fold degenerate low-energy spinon excitations above. Here, we address the nature of the spin-singlet excitations.

In the context of the large-M expansion, this question reduces to understanding the nature of the spectrum of the Q_{ij} and λ_i fluctuations about the large-M saddle point described above. At the outset, we can view such fluctuations as composites of two-spinon excitations, as both Q_{ij} and λ_i couple to spinon-pair operators, and so conclude that such excitations should not be viewed as the "elementary" excitations of the quantum state found so far. Furthermore, the saddle point has not broken any global symmetries of the Hamiltonian, and so it would appear that no such composite excitation has any reason to be low energy without fine-tuning.

However, it does turn out that there are separate elementary excitations in the singlet sector, and these arise from two distinct causes: (i) the gauge invariance in (15.18) leads to a gapless "photon" excitation; and (ii) there are topologically non-trivial configurations of Q_{ij}, which lead to excitations that would be not be evident in a naive $1/M$ expansion. Excitations in the class (i) arise in the square-lattice case, while those in class (ii) appear on the triangular lattice, and these will be considered separately in the following subsections.

15.3.1 Gauge Excitations

The gauge transformations in (15.18) act on the phases of the Q_{ij}, and so it is appropriate to just focus on the fluctuations of the phases of the Q_{ij} that are non-zero in the large-M limit. We will separate the discussions for the square and triangular lattices, because the results are very different.

Square Lattice

We define

$$\begin{aligned} Q_{i,i+\hat{x}} &= \bar{Q} \exp\left(i\Theta_{ix}\right), \\ Q_{i,i+\hat{y}} &= \bar{Q} \exp\left(i\Theta_{iy}\right). \end{aligned} \tag{15.26}$$

Then, the gauge transformations in (15.18) can be written as

$$\begin{aligned} \Theta_{ix}(\tau) &\to \Theta_{ix}(\tau) - \rho_i(\tau) - \rho_{i+x}(\tau), \\ \Theta_{iy}(\tau) &\to \Theta_{iy}(\tau) - \rho_i(\tau) - \rho_{i+y}(\tau), \\ \lambda_i &\to \lambda_i + \frac{\partial \rho_i}{\partial \tau}(\tau). \end{aligned} \tag{15.27}$$

The question before us is whether (15.18) imposes on us the presence of a gapless photon in the low-energy and long-wavelength limit. The answer is affirmative, and the required result is obtained by parameterizing such fluctuations as follows

$$\begin{aligned} \Theta_{ix}(\tau) &= \eta_i a_x(\boldsymbol{r}, \tau), \\ \Theta_{iy}(\tau) &= \eta_i a_y(\boldsymbol{r}, \tau), \\ \lambda_i &= -i\bar{\lambda} - \eta_i a_\tau(\boldsymbol{r}, \tau), \end{aligned} \tag{15.28}$$

where the a_μ are assumed to be smooth functions of spacetime parameterized by the continuum spatial coordinate r, and imaginary time τ; the factor $\eta_i = \pm 1$ on the two checkerboard sublattices of the square lattice, so that η_i has opposite signs on any pair of nearest-neighbor sites. Then, taking the continuum limit of (15.27) with $\rho_i(\tau) = \eta_i \rho(r, \tau)$, we deduce from (15.28) that

$$\begin{aligned} a_x &\to a_x - \partial_x \rho, \\ a_y &\to a_y - \partial_y \rho, \\ a_\tau &\to a_\tau - \partial_\tau \rho. \end{aligned} \tag{15.29}$$

So we reach the very important conclusion that a_μ transforms just like a continuum $U(1)$ gauge field! This a_μ gauge field is essentially the same as the a_μ gauge field discussed in Section 13.3, and which we introduced in (14.21) for the classical XY model of Chapter 14. Note that the factor η_i in this $U(1)$ gauge transformation implies from (15.18) that the spinons s_i carry opposite gauge charges on the two sublattices – these spinons are the analog of ϕ in the classical XY model of Chapter 14.

As in traditional field-theoretic analyses, (15.29) imposes the requirement that the long-wavelength action of the a_μ fluctuations must have the form

$$\mathcal{S}_b = \int d^3x \frac{1}{2K'} (\varepsilon_{\mu\nu\lambda} \partial_\nu a_\lambda)^2, \tag{15.30}$$

and this describes a gapless a_μ photon excitation, with a suitable velocity of "light." So, on the square lattice, the spectrum of spin-singlet states includes a linearly dispersing photon mode. Such a state is a $U(1)$ spin liquid, and not a $\mathbb{Z}_2$ spin liquid. Actually, the gapless photon of this $U(1)$ spin liquid is ultimately not stable because of monopole tunneling events, just as in phase A of Fig. 14.2; this involves a long and interesting story [217, 218, 256, 261], which we will turn to in Chapter 28. The remainder of the discussion in this chapter is restricted to the triangular lattice where, as we show below, the $U(1)$ photon is gapped by the Higgs mechanism, just as in phase D of Fig. 14.2.

Triangular Lattice

Now we have to consider three separate values of $\mathcal{Q}_{ij}$ per site, and so we replace (15.26) by

$$\mathcal{Q}_{i,i+\hat{e}_p} = \bar{\mathcal{Q}} \exp\left(i\Theta_{p,i}\right), \tag{15.31}$$

where $p = 1,2,3$, the vectors $\hat{e}_p$ were defined (15.23), $\bar{Q}$ is the mean-field value, and Θ_p is a real phase. The effective action for the $\Theta_{p,i}$ must be invariant under

$$\Theta_{p,i} \to \Theta_{p,i} - \rho_i - \rho_{i+\hat{e}_p}. \tag{15.32}$$

Upon performing a Fourier transform, with the link variables Θ_p placed on the center of the links, the gauge invariance takes the form

$$\Theta_p(\boldsymbol{k}) \to \Theta_p(\boldsymbol{k}) - 2\rho(\boldsymbol{k})\cos(k_p/2). \tag{15.33}$$

The momentum $\boldsymbol{k}$ takes values in the first Brillouin zone of the triangular lattice. This invariance implies that the effective action for the Θ_p can only be a function of the following gauge-invariant combinations:

$$I_{pq}(\boldsymbol{k}) = 2\cos(k_q/2)\Theta_p(\boldsymbol{k}) - 2\cos(k_p/2)\Theta_q(\boldsymbol{k}). \tag{15.34}$$

We now wish to take the continuum limit at points in the Brillouin zone where the action involves only gradients of the Θ_p fields and thus has the possibility of gapless excitations. The same analysis could have been applied to the square lattice, in which case there is only one invariant I_{xy}. In this case, we choose $\boldsymbol{k} = \boldsymbol{g} + \boldsymbol{q}$, with $\boldsymbol{g} = (\pi,\pi)$ (this corresponds to the choice of η_i above) and $\boldsymbol{q}$ small; then $I_{xy} = q_x\Theta_y - q_y\Theta_x$, which is clearly the $U(1)$ flux invariant under (15.29).

The situation is more complex for the case of the triangular lattice [231]. Now there are three independent I_{pq} invariants, and it is not difficult to see that only two of the three values of $\cos(k_p/2)$ can vanish at any point of the Brillouin zone. One such point is the wavevector

$$\boldsymbol{g} = \frac{2\pi}{\sqrt{3}a}(0,1), \tag{15.35}$$

where

$$\begin{aligned} \boldsymbol{g}\cdot\hat{e}_1 &= \pi, \\ \boldsymbol{g}\cdot\hat{e}_2 &= -\pi, \\ \boldsymbol{g}\cdot\hat{e}_3 &= 0. \end{aligned} \tag{15.36}$$

Taking the continuum limit with the fields varying with momenta close to $\boldsymbol{g}$, we find that the I_{pq} depend only upon gradients of Θ_1 and Θ_2. It is also helpful to parametrize the Θ_p in the following manner (analogous to (15.28))

$$\begin{aligned} \Theta_1(\boldsymbol{r}) &= ia_1(\boldsymbol{r})e^{i\boldsymbol{g}\cdot\boldsymbol{r}}, \\ \Theta_2(\boldsymbol{r}) &= -ia_2(\boldsymbol{r})e^{i\boldsymbol{g}\cdot\boldsymbol{r}}, \\ \Theta_3(\boldsymbol{r}) &= H(\boldsymbol{r})e^{i\boldsymbol{g}\cdot\boldsymbol{r}}. \end{aligned} \tag{15.37}$$

It can be verified that the condition for the reality of Θ_p is equivalent to demanding that a_1, a_2, H be real. We will now take the continuum limit with a_1, a_2, H varying slowly on the scale of the lattice spacing. It is then not difficult to show that the invariants I_{pq} reduce to (after a Fourier transformation)

$$
\begin{aligned}
I_{12} &= \partial_2 a_1 - \partial_1 a_2, \\
I_{31} &= \partial_1 H - 2a_1, \\
I_{32} &= \partial_2 H - 2a_2,
\end{aligned} \tag{15.38}
$$

where ∂_i is the spatial gradient along the direction $\hat{e}_i$. Thus, the a_1, a_2 are the components of a $U(1)$ gauge field, where the components are taken along an "oblique" coordinate system defined by the axes $\hat{e}_1, \hat{e}_2$; this is just as in the square lattice. However, in addition to I_{12}, we also have the invariants I_{31} and I_{32} in the triangular lattice; we observe that this involves the field H, which transforms like the phase of a charge-± 2 Higgs field under the $U(1)$ gauge invariance: indeed H is the analog of the field H introduced in the classical XY model in Chapter 14, and the present triangular-lattice spin liquid is the analog of phase D of Fig. 14.2. So the fluctuations of an isotropic triangular lattice will be characterized by an action of the form

$$
\mathcal{S}_b = \int d^3x \frac{1}{2K'} \left[I_{12}^2 + I_{31}^2 + I_{32}^2 \right], \tag{15.39}
$$

which replaces (15.30). This is the action expected in the *Higgs phase* of a $U(1)$ gauge theory. The Higgs condensate gaps out the $U(1)$ photon, and so there are no gapless singlet excitations on the triangular lattice. This is a necessary condition for mapping the present state onto a $\mathbb{Z}_2$ spin liquid.

The presentation so far of the gauge fluctuations described by a charge-± 2 Higgs field coupled to a $U(1)$ gauge field would be appropriate for an anisotropic triangular lattice in which the couplings along the $\hat{e}_3$ direction are different from those along $\hat{e}_1$ and $\hat{e}_2$. For an isotropic triangular lattice, all three directions must be treated equivalently, and then there is no simple way to take the continuum limit in the gauge sector; we have to work with the action in (15.39), but with the invariants specified as in (15.34). Such an action does not have a gapless photon anywhere in the Brillouin zone, and all gauge excitations remain gapped. There are other choices for the wavevector $\boldsymbol{g}$ in (15.35) at which the other pairs of values of $\cos(k_p/2)$ vanish; these are the points

$$
\frac{2\pi}{\sqrt{3}} \left(\frac{\sqrt{3}}{2}, -\frac{1}{2} \right) \quad , \quad \frac{2\pi}{\sqrt{3}} \left(\frac{-\sqrt{3}}{2}, -\frac{1}{2} \right), \tag{15.40}
$$

which are related to the analysis above by the rotational symmetry of the triangular lattice.

15.3.2 Topological Excitations

The analysis in Section 15.3.1 described small fluctuations in the phases of the $\mathcal{Q}_{ij}$ about their saddle-point values $\bar{\mathcal{Q}}$. On the triangular lattice, we found that such fluctuations led only to gapped excitations, which at higher energies become part of the two-spinon continuum.

Now we consider excitations that involve large deviations from the spatially uniform saddle-point values, and that turn out to be topologically protected. These excitations are closely connected to the vortices in charged superfluids that were described in

Section 7.2. There, we considered a scalar field Ψ with charge q coupled to the electromagnetic $U(1)$ gauge field $\boldsymbol{A}$. We found stable vortex-like saddle points with flux $n\Phi_0$, with $\Phi_0 = hc/q$, for all integer n. We have seen above that, on the triangular lattice, the Schwinger boson state has fluctuations described by a charge-2 Higgs field H coupled to a $U(1)$ gauge field a_μ. In this case, we are normalizing the gauge field so that $\hbar c \Rightarrow 1$, and so we can expect vortex solutions with a_μ flux $n(2\pi)/2$, for all integers n. However, this is not quite correct. A crucial difference between the present theory and the electromagnetic gauge field is that the a_μ gauge field is "compact": this means that a gauge field $a_{x,y}$ is identical to $a_{x,y} + 2\pi$, and tunneling events that change the total flux by 2π are allowed (these are "monopoles," which will be considered further in Section 25.3 and Chapter 26). This means that all vortex solutions with even n are identical to each other, as are those with odd n. The $n = 0$ case corresponds to no vortex at all, and so there is only a single non-trivial vortex with $n = 1$ and flux π. This is the sought-after *vison*. Note that because flux π and $-\pi$ are identical, the vison saddle point preserves time-reversal symmetry. In contrast, the vortices in Section 7.2 break time reversal, with $\pm n$ vortices time-reversal partners of each other. We note that this vison is closely related to that discussed in Section 14.2.2 for phase D of the classical XY model.

In this section, we obtain the vison saddle-point solution by working with a lattice effective action: this is essential to account for the influence of the monopoles. So we look for spatially non-uniform solutions of the saddle-point equations (15.21) and (15.22). In general, solving such equations is a demanding numerical task, and so we are satisfied with a simplified analysis that is valid when the spin gap is large. In the large-spin-gap limit, we can integrate out the Schwinger bosons, and write the energy as a local functional of the $\mathcal{Q}_{ij}$. This functional is strongly constrained by the gauge transformations in (15.18): for time-independent $\mathcal{Q}_{ij}$, this functional takes the form

$$E[\{\mathcal{Q}_{ij}\}] = -\sum_{i<j}\left(\alpha|\mathcal{Q}_{ij}|^2 + \frac{\beta}{2}|\mathcal{Q}_{ij}|^4\right) - K \sum_{\text{even loops}} \mathcal{Q}_{ij}\mathcal{Q}^*_{jk}\cdots\mathcal{Q}^*_{\ell i}. \tag{15.41}$$

Here α, β, and K are coupling constants determined by the parameters in the Hamiltonian of the antiferromagnet. We have shown them to be site-independent, because we have only displayed terms in which all links/loops are equivalent; they can depend upon links/loops for longer-range couplings provided the full lattice symmetry is preserved.

We can now search for saddle points of the energy functional in (15.41). Far from the center of the vison, we have $|\mathcal{Q}^v_{ij}| = \bar{\mathcal{Q}}$, so that the energy differs from the ground-state energy only by a finite amount. Closer to the center, there are differences in the magnitudes. However, the key difference is in the signs of the link variables, as illustrated in Fig. 15.4: there is a "branch-cut" emerging from the vison core along which $\text{sgn}(\mathcal{Q}^v_{ij}) = -\text{sgn}(\bar{\mathcal{Q}}_{ij})$, as in Fig. 13.6d. The results of a numerical minimization [114] of $E[\{\mathcal{Q}_{ij}\}]$ on the the triangular lattice are shown in Fig. 15.4. The magnitudes of $\mathcal{Q}^v_{ij}$ are suppressed close to the vison, and converge to $\bar{\mathcal{Q}}_{ij}$ as we move away from the vison (modulo the sign change associated with the branch cut), analogous to the Abrikosov vortices examined in Chapter 7. Despite the branch-cut breaking the three-fold rotation symmetry, the gauge-invariant fluxes of $\mathcal{Q}^v_{ij}$ preserve the rotation symmetry.

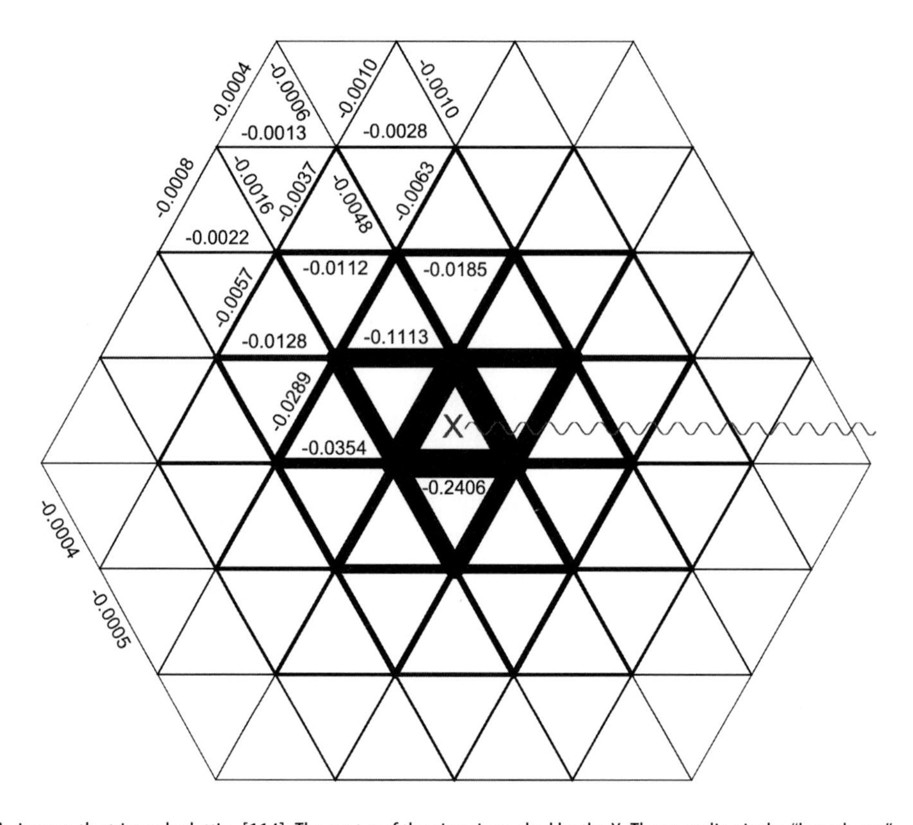

Figure 15.4 A vison on the triangular lattice [114]. The center of the vison is marked by the X. The wavy line is the "branch-cut" where we have $\text{sgn}(\mathcal{Q}^{\text{v}}_{ij}) = -\text{sgn}(\bar{\mathcal{Q}}_{ij})$ only on the links crossed by the line. Plotted is the minimization result of $E[\{\bar{\mathcal{Q}}_{ij}\}]$ with $\alpha = 1, \beta = -2, K = 0.5$. Minimization is done with the cluster embedded in a vison-free lattice with all nearest-neighbor links equal to $\bar{\mathcal{Q}}_{ij}$. The numbers are $(\bar{\mathcal{Q}}_{ij} - \mathcal{Q}^{\text{v}}_{ij})$ and the thickness of the links are proportional to $(\mathcal{Q}^{\text{v}}_{ij} - \bar{\mathcal{Q}}_{ij})^{1/2}$. Reprinted with permission from APS.

So we have found a stable real-vortex solution that preserves time reversal, and has a finite excitation energy. We have also anticipated that this vortex is identified with the vison particle of the $\mathbb{Z}_2$ spin liquid: more evidence for this identification is given in Sections 15.4.2 and 15.4.3.

15.4 Dynamics of Excitations

For the case of the triangular lattice, Section 15.3 has identified two types of elementary excitations: bosonic spinons with a two-fold spin and a two-fold lattice degeneracy, and a topological excitation that we have anticipated will become the vison particle of a $\mathbb{Z}_2$ spin liquid. I now describe the dynamics of the interactions between these excitations, and indeed verify that they reproduce the general structure associated with the $\mathbb{Z}_2$ spin liquid.

A similar analysis can also be carried for the $U(1)$ spin liquid on the square lattice. However, let us defer consideration of this case to Section 28.1.

15.4.1 Bosonic Spinons

The general structure of the theory controlling the low-energy spectrum becomes clearer upon taking a suitable continuum limit of the Lagrangian in (15.17), while replacing $\mathcal{Q}_{ij} = \bar{\mathcal{Q}}_{ij}$ and $i\lambda_i = \bar{\lambda}$. We take the continuum limit after separating three sites, u, v, w, in each unit cell (see Fig. 15.2). We write the boson operators on these sites as $s_u^\alpha = u_\alpha$, $s_v^\alpha = v_\alpha$ etc. Then, to the needed order in spatial gradients, the Lagrangian density becomes [113]

$$
\begin{aligned}
\mathcal{L} = & \, u_\alpha^* \frac{\partial u_\alpha}{\partial \tau} + v_\alpha^* \frac{\partial v_\alpha}{\partial \tau} + w_\alpha^* \frac{\partial w_\alpha}{\partial \tau} + \bar{\lambda}\left(|u_\alpha|^2 + |v_\alpha|^2 + |w_\alpha|^2\right) \\
& - \frac{3J\bar{\mathcal{Q}}}{2} \mathcal{J}_{\alpha\beta}\left(u_\alpha v_\beta + v_\alpha w_\beta + w_\alpha u_\beta\right) + \text{c.c.} \\
& + \frac{3J\bar{\mathcal{Q}}}{8} \mathcal{J}_{\alpha\beta}\left(\nabla u_\alpha \cdot \nabla v_\beta + \nabla v_\alpha \cdot \nabla w_\beta + \nabla w_\alpha \cdot \nabla u_\beta\right] + \text{c.c.}
\end{aligned} \tag{15.42}
$$

We now perform a unitary transformation to new variables $x_\alpha, y_\alpha, z_\alpha$. These are chosen to diagonalize only the non-gradient terms in $\mathcal{L}$.

$$
\begin{aligned}
\begin{pmatrix} u_\alpha \\ v_\alpha \\ w_\alpha \end{pmatrix} = & \frac{z_\alpha}{\sqrt{6}} \begin{pmatrix} 1 \\ \zeta \\ \zeta^2 \end{pmatrix} + \mathcal{J}_{\alpha\beta} \frac{z_\beta^*}{\sqrt{6}} \begin{pmatrix} -i \\ -i\zeta^2 \\ -i\zeta \end{pmatrix} + \frac{y_\alpha}{\sqrt{6}} \begin{pmatrix} 1 \\ \zeta \\ \zeta^2 \end{pmatrix} + \mathcal{J}_{\alpha\beta} \frac{y_\beta^*}{\sqrt{6}} \begin{pmatrix} i \\ i\zeta^2 \\ i\zeta \end{pmatrix} \\
& + \frac{x_\alpha}{\sqrt{3}} \begin{pmatrix} 1 \\ 1 \\ 1 \end{pmatrix},
\end{aligned} \tag{15.43}
$$

where $\zeta \equiv e^{2\pi i/3}$. The tensor structure above makes it clear that this transformation is rotationally invariant, and that $x_\alpha, y_\alpha, z_\alpha$ transform as spinors under $SU(2)$ spin rotations (for convenience, we consider the case $USp(2) \equiv SU(2)$ in this section). Inserting Eq. (15.43) into $\mathcal{L}$ we find

$$
\begin{aligned}
\mathcal{L} = & \, x_\alpha^* \frac{\partial x_\alpha}{\partial \tau} + y_\alpha^* \frac{\partial z_\alpha}{\partial \tau} + z_\alpha^* \frac{\partial y_\alpha}{\partial \tau} + (\bar{\lambda} - 3\sqrt{3}J\bar{\mathcal{Q}}/2)|z_\alpha|^2 \\
& + (\bar{\lambda} + 3\sqrt{3}J\bar{\mathcal{Q}}/2)|y_\alpha|^2 + \bar{\lambda}|x_\alpha|^2 + \frac{3J\bar{\mathcal{Q}}\sqrt{3}}{8}\left(|\partial_x z_\alpha|^2 + |\partial_y z_\alpha|^2\right) + \cdots .
\end{aligned} \tag{15.44}
$$

The ellipsis indicates omitted terms involving spatial gradients in the x_α and y_α, which we will not keep track of. This is because the fields y_α and x_α are massive relative to z_α, and so can be integrated out. This yields the effective Lagrangian

$$
\begin{aligned}
\mathcal{L}_z = & \frac{1}{(\bar{\lambda} + 3\sqrt{3}J\bar{\mathcal{Q}}/2)}|\partial_\tau z_\alpha|^2 + \frac{3J\bar{\mathcal{Q}}\sqrt{3}}{8}\left(|\partial_x z_\alpha|^2 + |\partial_y z_\alpha|^2\right) \\
& + (\bar{\lambda} - 3\sqrt{3}J\bar{\mathcal{Q}}/2)|z_\alpha|^2 + \cdots .
\end{aligned} \tag{15.45}
$$

Note that the omitted spatial gradient terms in x_α, y_α do contribute a correction to the spatial gradient term in (15.45), and we have not accounted for this.

So we reach the important conclusion that the spinons are described by a relativistic complex scalar field z_α. Counting the two values of α, and the particle and anti-pariticle excitations, we have a total of four spinons, as expected.

Next, we consider the higher-order terms in (15.45), which arise from including the fluctuations of the gapped fields $\mathcal{Q}$ and λ. Rather than computing these from the microscopic Lagrangian, it is more efficient to deduce their structure from symmetry considerations. The representation in (15.43), and the connection of the u_α, v_α, w_α to the lattice degrees of freedom, allow us to deduce the following symmetry transformations of the x_α, y_α, z_α:

- Under a global spin rotation by the $SU(2)$ matrix $g_{\alpha\beta}$, we have $z_\alpha \to g_{\alpha\beta} z_\beta$, and similarly for x_α, and y_α.
- Under a 120° lattice rotation, we have $u_\alpha \to v_\alpha$, $v_\alpha \to w_\alpha$, $w_\alpha \to u_\alpha$. From (15.43), we see that this symmetry is realized by

$$z_\alpha \to \zeta z_\alpha \;,\; y_\alpha \to \zeta y_\alpha \;,\; x_\alpha \to x_\alpha . \tag{15.46}$$

  Note that this is distinct from the $SU(2)$ rotation because $\det(\zeta) \neq 1$.

It is easy to verify that Eq. (15.44) is invariant under all the symmetry operations above. These symmetry operators make it clear that the only allowed quartic term for the Heisenberg Hamiltonian is $\left(\sum_\alpha |z_\alpha|^2\right)^2$: this quartic term added to $\mathcal{L}_z$ yields a theory with $O(4)$ symmetry, corresponding to rotations between the four real fields that can be extracted from the two complex fields z_α. So the theory reduces to the $M = 4$ case of the relativistic scalar field theories of Chapters 10 and 11.

We also observe from (15.45) that the z_α field will condense when

$$r = (\bar{\lambda} - 3\sqrt{3} J \bar{\mathcal{Q}}/2) \tag{15.47}$$

becomes negative as κ is varied across κ_c. This condensation breaks the spin-rotation symmetry, and leads to a quantum phase transition to a phase with coplanar antiferromgnetic long-range order, as illustrated in Fig. 15.5. The order parameter of this coplanar antiferromagnet is related to $\mathbb{Z}_2$ gauge-invariant bilinears of z_α by (generalizing the relation (9.15) for the bipartite lattice)

$$\boldsymbol{S}_i \propto \operatorname{Im}\left[\exp\left(i\boldsymbol{Q}\cdot\boldsymbol{r}\right) \varepsilon_{\alpha\gamma} z_\gamma \boldsymbol{\sigma}_{\alpha\beta} z_\beta\right] , \tag{15.48}$$

where the wavevector $\boldsymbol{Q} = (4\pi/a)(1/3, 1/\sqrt{3})$. This the $\mathbb{Z}_2$ spin liquid to coplanar antiferromagnet phase transition, and is a close analog of the transition from phase D to phase C of the classical XY model in Fig. 14.2, or a corresponding transition in Fig. 14.4: the z_α field is the analog of the ϕ field, and the main difference is that the global spin-rotation symmetry here is larger than the $U(1)$ symmetry of the XY model, and the action for the z_α has an $O(4)$ symmetry (corresponding to rotations among the four real and imaginary components of z_α).

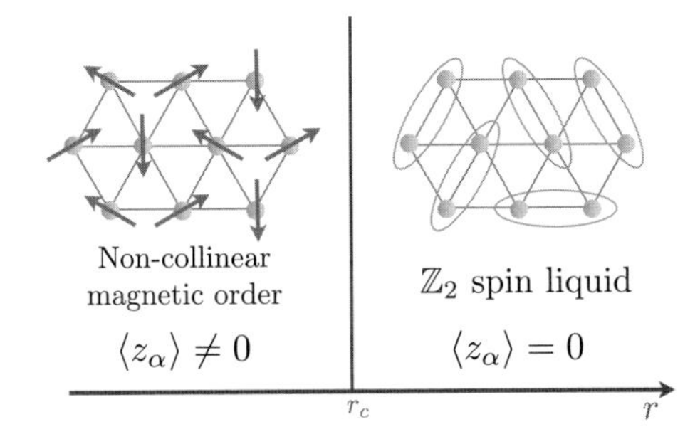

Figure 15.5 Magnetic ordering transition driven by tuning r in (15.47). Fractionalized anyonic excitations are present only for $r > r_c$, and so there is a "confinement" transition at $r = r_c$. The critical theory is expressed in terms of bosonic spinons z_α, and is an example of deconfined critical theory. This transition is the analog of the transition between phases C and D in Figs. 14.2 and 14.3, and a similar transition in Fig. 14.4. There is evidence for such a transition in $KYbSe_2$ [250].

We can extend the connection to the models of Chapter 14 to the remaining phases in the phase diagrams in Figs. 14.3 and 14.4 simply by extending the theory in (14.36) to $O(4)$ symmetry. So now we have

$$\mathcal{Z}_{\mathbb{Z}_2} = \sum_{Z_{ij}=\pm 1} \prod_i \int dz_{i\alpha} \delta\left(\sum_\alpha |z_{i\alpha}|^2 - 1\right) \left[\prod_i Z_{i,i+\tau}\right]^{2S} \exp\left(-H_{\mathbb{Z}_2}[z_\alpha, Z]\right),$$
$$H_{\mathbb{Z}_2}[z_\alpha, Z] = -J_2 \sum_{\langle ij \rangle} Z_{ij}\left(z^*_{i\alpha} z_{j\alpha} + \text{c.c.}\right) - K \sum_{\triangle,\square} \prod_{ij \in \triangle,\square} Z_{ij}, \tag{15.49}$$

where the $Z_{i\alpha}$ reside on the sites i of a three-dimensional hexagonal lattice (i.e., stacked triangular lattices), and the K term acts on the triangular plaquettes of the spatial plane, and the rectangular plaquettes along the temporal direction. The prefactor in the square brackets is the Berry phase term for half-integer-spin S, which is important for replacing the trivial phase by the valence-bond solid, as in Fig. 14.4. This Berry phase is directly linked to the vison Berry phase discussed in Section 15.4.2, as we will see in Section 16.5.2. As in Fig. 14.4, the theory (15.49) is expected to also display valence-bond solid phases, along with the phases in Fig. 15.5.

The transition from the ordered antiferromagnet to the $\mathbb{Z}_2$ spin liquid is not influenced by the Berry phase term (because the visons remains gapped at this transition), and is described by the $O(4)$ Wilson–Fisher critical theory [50, 51] considered in Section 11.2. However, there is an important difference in the structure of the observable order parameter. Note that z_α was obtained from the continuum limit of the spinon s_α, and so it is a fractionalized degree of freedom, carrying a unit $\mathbb{Z}_2$ charge. Correlators of z_α are therefore not observable, only those of gauge-invariant bilinear combinations. This is denoted by stating the universality class of the transition is actually $O(4)^*$. This critical theory has the same exponents as the $O(4)$ theory, but some observables in a finite geometry are different [308]. As the critical fields of the theory are fractionalized spinons, this is an example of a "deconfined critical point," and we will meet others in Section 16.5.2 and Chapters 26 and 28.

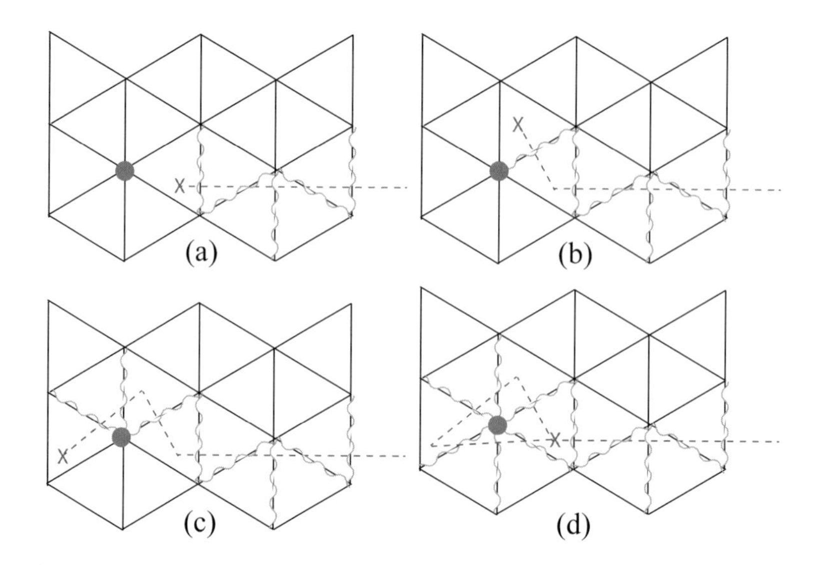

Figure 15.6 Adiabatic motion of a vison (denoted by the X) around a single site of the triangular lattice (denoted by the filled circle). The initial state is in (a), and the final state is in (d), and these differ by a gauge transformation under which $s_{i\alpha} \to -s_{i\alpha}$ only on the filled-circle site.

15.4.2 Motion of Visons

Let us now consider the motion of the vison elementary excitation, which is illustrated in Fig. 15.6. The vison is located at the center of a triangle, and so can tunnel between neighboring triangular cells. We are interested here in any possible Berry phases the vison could pick up upon tunneling around a closed path.

In Section 15.3.2, we characterized the vison by the saddle-point configuration $\mathcal{Q}^{\mathrm{v}}_{ij}$ of the bond variables in the Hamiltonian (15.19). By diagonalizing this Hamiltonian [114, 231], we can show that the wavefunction of the vison can be written as (compare to (4.7) and (13.7))

$$|\Psi^{\mathrm{v}}\rangle = \mathcal{P}\exp\left(\sum_{i<j} f^{\mathrm{v}}_{ij}\,\mathcal{J}_{\alpha\beta}s^{\dagger}_{i\alpha}s^{\dagger}_{j\beta}\right)|0\rangle, \tag{15.50}$$

where $|0\rangle$ is the boson vaccum, $\mathcal{P}$ is a projection operator that selects only states that obey (15.4), and the boson-pair wavefunction $f^{\mathrm{v}}_{ij} = -f^{\mathrm{v}}_{ji}$ is determined from (15.19) by a Bogoliubov transformation.

Let us now consider the motion of a single vison [114]. The gauge-invariant Berry phases are those associated with a periodic motion, and so let us consider the motion of a vison along a general closed loop $\mathcal{C}$. We illustrate the simple case where $\mathcal{C}$ encloses a single site of the triangular lattice in Fig. 15.6. The wavy lines indicate $\mathrm{sgn}(\mathcal{Q}^{\mathrm{v}}_{ij}) = -\mathrm{sgn}(\bar{\mathcal{Q}}_{ij})$, as in Fig. 15.4. The last state is gauge-equivalent to the first state, after the gauge transformation $s^{\alpha}_i \to -s^{\alpha}_i$ only for the site i marked by the filled circle. As long as the vison wavefunction can be chosen to be purely real, it is clear that no Berry phase is accumulated from the time evolution of the wavefunction as the vison tunnels around the path $\mathcal{C}$. However, there can still be a non-zero Berry phase because a gauge transformation is required to map the final state to the initial state. The analysis in Fig. 15.6 shows that the required gauge transformation is

$$\begin{aligned} s_i^\alpha &\to -s_i^\alpha, \quad \text{for } i \text{ inside } \mathcal{C}, \\ s_i^\alpha &\to s_i^\alpha, \quad \text{for } i \text{ outside } \mathcal{C}. \end{aligned} \tag{15.51}$$

By (15.4), each site has $n_s = 2S$ bosons, and so the total Berry phase accumulated by $|\Psi^v\rangle$ is

$$\pi n_s \times (\text{number of sites enclosed by } \mathcal{C}). \tag{15.52}$$

This Berry phase is equivalent to the relationship (13.8) between the translation operators in the x and y directions acting on the vison states for the square lattice. For the important case of $S = 1/2$, the vison experiences a flux of π for every site of the triangular lattice. This phase factor of π is related to an "anomaly" associated with the global $U(1)$ boson-number symmetry, and translational symmetry [32, 68], and was first noted in Refs. [119, 230] as a feature of $\mathbb{Z}_2$ spin liquids with half-integer spin. In particular, this result implies the resonating-valence-bond state is an *odd*-$\mathbb{Z}_2$ spin liquid. It is the vison Berry phase in (15.52) that leads to the Berry phase term in (14.35) for the XY model.

A notable features of (15.52) is that the quantized integer value of $n_s = 2S$ is important. Memory of this quantization was lost in the mean-field theory of Section 15.1, which was sensible also for non-integer values of n_s. So inclusion of the vison fluctuations restores the quantization of spin. A more complete theory for the vison fluctuations is given in Chapter 16.

15.4.3 Semions and Fermions

Before the identification of the present Schwinger boson spin liquid with the $\mathbb{Z}_2$ spin liquid, we need to establish that the spinons and visons are mutual semions. This is immediately apparent from a glance at Figs. 15.4 and 15.7.

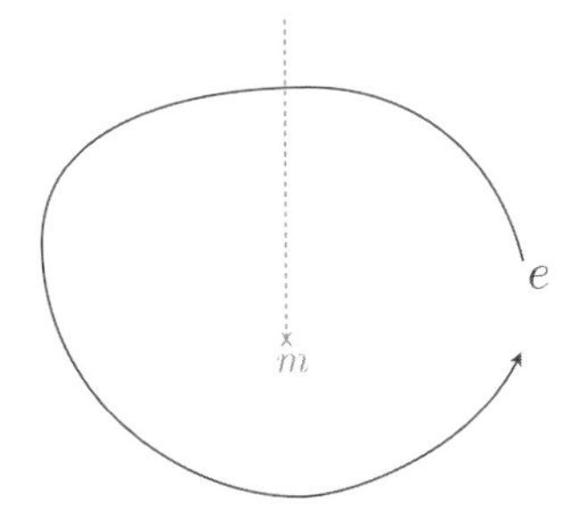

Figure 15.7 Mutual statistics of e and m particles. This process leads to a Berry phase of -1, when the e particle crosses the branch-cut of the m particle.

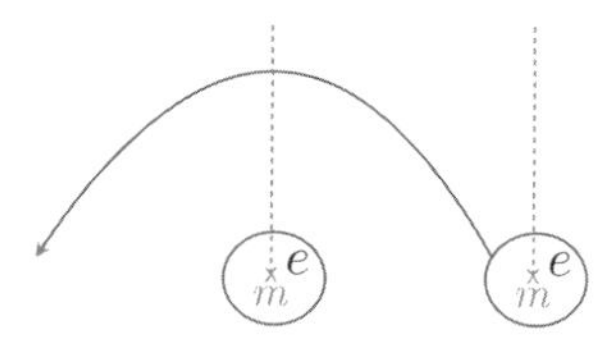

Figure 15.8 Two ε particles undergoing an exchange: after traversing the path shown, a translation returns the ε particles to the original state. Each ε particle is a bound state of a vison (the m particle) and the s_α bosonic spinon (the e particle). This process leads to a Berry phase of -1, when the moving e particle crosses the branch-cut of the stationary m particle.

The Q^v_{ij} transport the spinons from site to site, and for spinon encircling a vison in a large circuit, the only difference between the cases with and without the vison is the branch-cut. This branch-cut yields an additional phase of π in the vison amplitude, and provides the needed phase for mutual semion statistics [219, 303].

We can now identify the e, ε, and m anyons, in the abstract topological characterization of the $\mathbb{Z}_2$ spin liquid at the end of Section 13.4. The e anyon is the Schwinger boson itself, s_α. This is a mutual semion with respect to the vison, and so we identify the vison with the m particle. Finally, the ε anyon is obtained by the fusion $\varepsilon = e \times m$, and so the ε anyon is a bound state of e and m. The ε anyon is a *fermion* as can be deduced by computing the Berry phase associated with exchanging one bound state of e and m with another bound state, as shown in Fig. 15.8. (Note that this "long-distance" Berry phase is multiplied by the "short-distance" vison motion phases discussed in Section 15.4.2.) It is quite remarkable that a microscopic theory of bosonic spins $\boldsymbol{S}$, expressed in terms of fractionalized bosons s_α, yields an excitation that is a fermion; this is one indication of the presence of long-range entanglement and topological order.

An alternative formulation of the $\mathbb{Z}_2$ spin liquid on the triangular lattice proceeds by expressing the spins $\boldsymbol{S}$ in terms of Schwinger fermions f_α; we consider such formulations for other spin liquids in Chapters 22, 23, and 28. For the $\mathbb{Z}_2$ spin liquid, the f_α spinons would become the ε particles, and the bosonic e would be the bound state of the f_α and the m vison. So ultimately, independent of whether we choose to fractionalize the $\boldsymbol{S}$ spins in terms of bosonic or fermionic partons, we obtain the same characterization of the observable excitations in the resulting $\mathbb{Z}_2$ spin liquid. This identity also extends to the symmetry transformations of the anyons, as has been shown in some detail on the kagome lattice [163].

Problem

15.1 Insert the anstaz (15.31) into the mean-field ansatz (15.19) for the triangular lattice. Expand the resulting ansatz to order Θ_p^2, and then integrate out the Schwinger bosons to obtain the effective action $\mathcal{S}[\Theta_p(\boldsymbol{k})]$. The gauge invariance (15.34) requires that this be of the form

$$\mathcal{S}[\Theta_p(\boldsymbol{k})] = \frac{1}{2}\int \frac{d^2k}{(2\pi)^2} \sum_{p\neq q} K(\boldsymbol{k}) I_{pq}(\boldsymbol{k}) I_{pq}(-\boldsymbol{k}). \tag{15.53}$$

Show that it is indeed of this form, and obtain an expression for $K(\boldsymbol{k})$.

16 $\mathbb{Z}_2$ Gauge Theory

An effective $\mathbb{Z}_2$ gauge theory of spin liquids is obtained from the Schwinger-boson theory. The weak and strong coupling expansions of the $\mathbb{Z}_2$ gauge theory enable a unified description of both the spin liquid and confining phases. Half-integer spin antiferromagnets are described by an "odd"-$\mathbb{Z}_2$ gauge theory, which is used to describe the deconfined criticality of the transition from the $\mathbb{Z}_2$ spin liquid to the confining valence-bond solid. The odd-$\mathbb{Z}_2$ gauge theory is also connected to the quantum dimer model in the strong-coupling limit. Finally, a connection is made between $\mathbb{Z}_2$ gauge theory and arrays of laser-pumped Rydberg atoms.

Chapter 15 gave an essentially complete presentation of the properties of a $\mathbb{Z}_2$ spin liquid, derived from a theory in which the spins $\boldsymbol{S}$ were fractionalized into bosonic partons. The purpose of this chapter is to (i) establish that the $\mathbb{Z}_2$ spin liquid is stable for a finite range of parameters by accounting for non-perturbative corrections to the large-N expansion, and (ii) address the *confinement* transitions of this spin liquid into conventional states without fractionalization, similar to those in Figs. 9.2, 14.3, and 14.4. Such confinement transitions proceed by the condensation of one of the bosonic anyons of the $\mathbb{Z}_2$ spin liquid. We have already discussed the confinement transition associated with the condensation of the e particle in Section 15.4.1; the condensation of the z_α particle leads to a magnetic state with non-collinear order via a deconfined critical point with emergent $O(4)$ symmetry. This chapter addresses the other possible confinement transition, driven by the condensation of the m anyon, the vison. As was first shown in Refs. [119, 230], the condensation of visons is described by an effective $\mathbb{Z}_2$ gauge theory of the visons. For half-integer spins (or bosons at half-filling), the confining state turns out to be the valence-bond solid (VBS) of Fig. 9.2.

16.1 From the Large-N Path Integral to a $\mathbb{Z}_2$ Gauge Theory

The theory for the $\mathbb{Z}_2$ spin liquid in Chapter 15 began by fractionalizing the spin operator into Schwinger bosons $s_{i\alpha}$, and then writing down the path integral (15.16) in terms of auxilliary variables Q_{ij} that reside on the links of the lattice, and a Lagrange multiplier on each site λ_i. This path integral had the fundamental property of invariance under the $U(1)$ gauge transformation in (15.18). Subsequently, as dictated by the large-N limit, we integrated out the $s_{i\alpha}$ spinons, and examined the structure of effective action for Q_{ij} and λ_i. We found a saddle point for the ground state $\bar{Q}_{ij}$, and also a saddle point for the vison Q^v_{ij} in Section 15.3.2 and Fig. 15.4.

In this chapter, I would like to examine the path integral over the Q_{ij} and λ_i in more detail, going well beyond the large-N expansion employed in Chapter 15. In particular, we want to sum over the vison saddle points in the path integral, along with "instantons" between the saddle points, and even allow the vison density to become large, so that we move across the transition associated with vison condensation. I now argue that all of this physics can be mapped onto a $\mathbb{Z}_2$ gauge theory. This mapping was established in Refs. [119, 230] by employing a duality mapping to an Ising model on the dual lattice. Here we shall proceed in a direct approach, focusing on a Hamiltonian description. The analysis here is the analog of the mapping to the $\mathbb{Z}_2$ gauge theory in (14.32) from the $U(1)$ gauge theory in (14.21) for the classical XY model.

Working directly with the path integrals defined in Chapter 15 is far too complicated, and it pays to obtain a simpler effective theory that retains the essential physics. For this purpose, an important observation is that the saddle-point values for the ground state $\bar{Q}_{ij}$, and for the vison Q^v_{ij}, are both real. Moreover, far from the core of the vison in Fig. 15.4, Q^v_{ij} differs from $\bar{Q}_{ij}$ only by a change in sign along the branch-cut. This indicates that, to include the visons, we can just perform the path integral over real Q_{ij}.

In the path integral over the real Q_{ij}, we need to include terms that endow the system with a quantum time dependence. After integrating out the gapped spinons, we will generate terms in the effective action for Q_{ij} of the form

$$\sum_{\langle ij \rangle} \int d\tau \left[(\partial_\tau Q_{ij})^2 + V(Q_{ij}) \right] . \tag{16.1}$$

Here, we are using a gauge in which $\lambda_i(\tau)$ has been made τ independent after a suitable gauge transformation from (15.18). On its own, (16.1) implies that there is a "particle" on each link of the lattice, with a discrete spectrum of non-degenerate energy levels. Let us assume that the potential V, which is symmetric under a sign change of Q_{ij}, has deep minima at values $Q_{ij} = \pm 1$, after a suitable rescaling of the Q_{ij}. There will be instanton events in the path integral, which tunnel between these minima, and so we obtain two closely spaced lowest-energy eigenstates, with an energy spacing we equate to $2g$. We focus our attention on these two states: so we have a "qubit" on each link of the lattice, and we introduce Pauli matrices X_{ij}, Y_{ij}, Z_{ij}, which act on each qubit. From this reasoning, we have argued for the replacement

$$Q_{ij} \Rightarrow Z_{ij} \tag{16.2}$$

in the large-N path integral; this is the precise analog of the mapping (14.31) for the XY model. This is the converse of the reasoning that related the Ising model to the theory of a real scalar field, and we have restricted the Q_{ij} to only two possible values, ± 1. Moreover, the energies of the qubit on each link can be accounted for by a term in the Hamiltonian

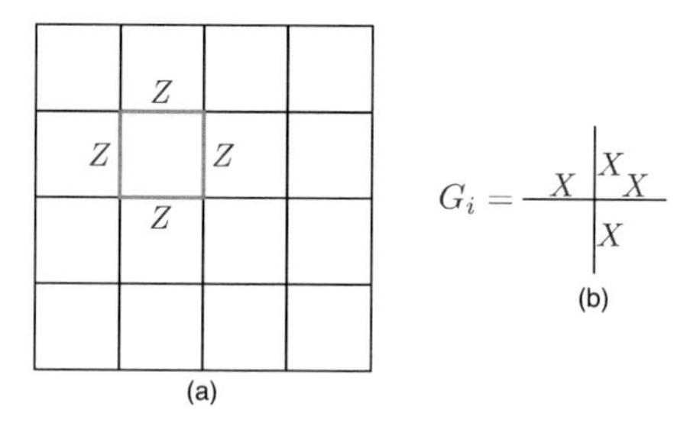

Figure 16.1 (a) The plaquette term of the $\mathbb{Z}_2$ lattice gauge theory. (b) The operators G_i that generate $\mathbb{Z}_2$ gauge transformations.

$$-g \sum_{\langle ij \rangle} X_{ij}. \tag{16.3}$$

It is worth noting that the mapping between (16.1) and (16.3) is very similar to the mapping from the transverse-field Ising model and the path integral of the ϕ^4 field theory, which is discussed at length in the QPT book [234].

As always, we have to play close attention to gauge invariance. With the restriction of the Q_{ij} to real values, the $U(1)$ gauge transformation in (15.18) has effectively been "Higgsed" to a residual $\mathbb{Z}_2$ gauge transformation

$$\begin{aligned}
Z_{ij} &\to Z_{ij}\rho_i\rho_j, \\
X_{ij} &\to X_{ij}, \\
s^\dagger_{i\alpha} &\to s^\dagger_{i\alpha}\rho_i,
\end{aligned} \tag{16.4}$$

where $\rho_i = \pm 1$ generates the $\mathbb{Z}_2$ gauge transformation. Note the close analogy to the $\mathbb{Z}_2$ gauge transformation in (14.33) for the classical XY model in (14.32).

Let us now turn to spatial fluctuations in Q_{ij}, for which the gauge transformation (16.4) will play a crucial role. An important point is that the replacement (16.2) is sufficient to capture the vison saddle points we encountered in Section 15.3.2. From Fig. 15.4, we observe that a vison is simply a state with $\mathbb{Z}_2$ gauge flux equal to -1, that is,

$$\prod_{\langle ij \rangle \in C} Z_{ij} = -1, \tag{16.5}$$

where C is any contour that encloses the vison, similar to the discussion on phase D in Section 14.2.3 for the classical XY model. Note that the product (16.5) is invariant under the $\mathbb{Z}_2$ gauge transformation (16.4). Moreover, the energy of the vison saddle point can be accounted for by a gauge-invariant term in the Hamiltonian, which is

$$-K \sum_{\square} \prod_{\ell \in \square} Z_\ell, \tag{16.6}$$

where $\square$ indicates the elementary plaquettes on the square lattice, as indicated in Fig. 16.1a. We are now using the symbol ℓ to represent a link between the square-lattice sites i, j. The term in (16.6) measures the $\mathbb{Z}_2$ "flux" in each square plaquette. A configuration of Z_{ij} with two visons is sketched in Fig. 16.2.

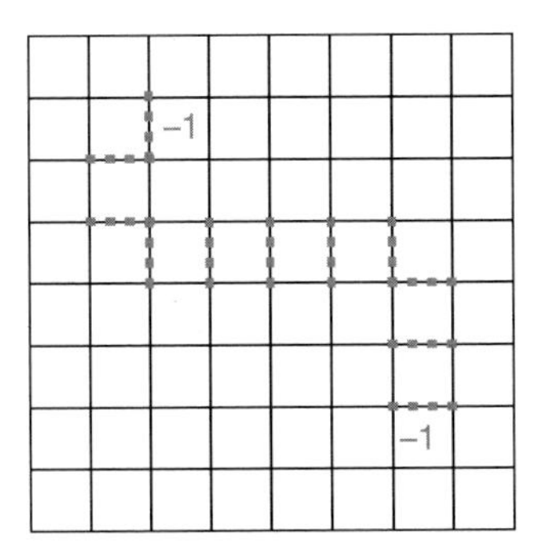

Figure 16.2 Two visons (indicated by the -1 in the plaquettes) connected by an invisible string. The dashed lines indicate the links, ℓ, on which $Z_\ell = -1$. The X_ℓ operators on these links act on $|0\rangle$ in (16.15) to create a pair of separated visons. The plaquettes with an even number of dashed lines on their edges carry no $\mathbb{Z}_2$ fluxes, and so are "invisible."

Note that we are presenting the effective $\mathbb{Z}_2$ gauge theory here on a nearest-neighbor square lattice. This is mainly for notational simplicity, and it is straightforward to extend the $\mathbb{Z}_2$ gauge theory to any lattice. The necessary ingredient is that the large-N saddle point we begin with must have the lattice $U(1)$ gauge symmetry broken down to $\mathbb{Z}_2$, as in Section 15.3.1 for the triangular lattice. For the case of the square lattice, the large-N theory requires that we include non-nearest-neighbor exchange interactions, which yield saddle-point values for the $\mathcal{Q}_{ij}$ without a bipartite structure. It is not difficult to extend our analysis here to include such longer-range links in (16.6), but we avoid it in the interests of keeping the presentation readable.

The energy cost of the vison configuration in (16.5) is exactly $2K$, as only a single elementary plaquette at the vison center has a change in the sign of the $\mathbb{Z}_2$ flux. So we can equate $2K$ to the energy of the vison saddle point in Fig. 15.4; in the large-N limit, we obtain the scaling $K \sim N$. The tunneling associated with g in (16.3) leads to a change in sign of $\mathcal{Q}_{ij}$ on a link; if this link is next to the center of a vison, we see from Fig. 15.4 or (16.5) that the result is the motion of the vison by a single lattice spacing across the link. Thus, the visons are now mobile, and we have included an instanton event between different vison saddle points of the path integral. From this reasoning, we can conclude that the large-N limit has $g \sim e^{-aN}$, for some constant a. So the vison tunneling events are rare in the large-N limit, and the main purpose of our present analysis is to introduce a framework in which we can understand the physics at larger g.

16.2 Hamiltonian of the $\mathbb{Z}_2$ Gauge Theory

We now collect all the ingredients assembled in Section 16.1 to specify the Hamiltonian of the $\mathbb{Z}_2$ gauge theory investigated in this chapter. As noted above, the discussion is presented entirely in terms of a $\mathbb{Z}_2$ gauge theory on the square lattice, and generalizations to other lattices is straightforward. The degrees of freedom of the quantum model are qubits on the links, ℓ, of a square lattice. The Pauli operators X_ℓ, Y_ℓ, Z_ℓ act on these qubits. The Hamiltonian combining (16.6) and (16.3) is

$$\mathcal{H}_{\mathbb{Z}_2} = -K \sum_{\square} \prod_{\ell \in \square} Z_\ell - g \sum_{\ell} X_\ell \,, \tag{16.7}$$

where $\square$ indicates the elementary plaquettes on the square lattice, as indicated in Fig. 16.1a. The model (16.7) is just the Hamiltonian version of the classical $\mathbb{Z}_2$ gauge theory defined by the K term in (14.32), in the "temporal" gauge, which sets all $Z_{i,i+\tau}$ to unity, as discussed further in Section 16.2.1.

The key property constraining the structure of possible terms in (16.7) is the requirement of invariance under the $\mathbb{Z}_2$ gauge transformations in (16.4); this allows the "flux" term proportional to K, and also the "kinetic" term in (16.3). For completeness, we note that we could also include the spinon terms in (15.19), so that the complete $\mathbb{Z}_2$ gauge theory for visons *and* spinons is

$$\mathcal{H}_{\mathbb{Z}_2}^{vs} = \mathcal{H}_{\mathbb{Z}_2} + \sum_{\langle ij \rangle} \left(-\widetilde{J}_{ij} Z_{ij} \mathcal{J}_{\alpha\beta} s_i^\alpha s_j^\beta + \text{H.c.} \right) + \sum_i \bar{\lambda}_i (s_{i\alpha}^\dagger s_i^\alpha - n_s) \,, \tag{16.8}$$

where we recall from (15.4) that the integer $n_s = 2S$ for a spin-S antiferromagnet. The spinon-hopping term in (16.8) is the analog of the J_2 term in (14.32) for the classical XY model. All terms in (16.8) are invariant under (16.4). We assume that the spinons have a large energy gap in this chapter, and so work mainly with the vison-only Hamiltonian in (16.7). However, the presence of a background density of n_s spinons has an important influence even in the approach based on the Hamiltonian (16.7), as we now discuss.

On the infinite square lattice, we can define operators on each site, i, of the lattice, which commute with the pure gauge theory $\mathcal{H}_{\mathbb{Z}_2}$ (see Fig. 16.1b)

$$G_i = \prod_{\ell \in +} X_\ell \,, \tag{16.9}$$

which clearly obey $G_i^2 = 1$. We have $G_i Z_\ell G_i = \rho_i Z_\ell$, where $\rho_i = -1$ only if the site i is at the end of link ℓ, and $\rho_i = 1$ otherwise: the G_i generates a space-dependent $\mathbb{Z}_2$ gauge transformation on the site i, equivalent to the operation of ρ_i in (16.4). There are an even number of Z_ℓ emanating from each site in the K term in $\mathcal{H}_{\mathbb{Z}_2}$, and so

$$[\mathcal{H}_{\mathbb{Z}_2}, G_i] = 0 \,. \tag{16.10}$$

The key ingredients in our analysis of $\mathcal{H}_{\mathbb{Z}_2}$ are the values assigned to the conserved $\mathcal{Z}_2$ gauge charges G_i. Before we specify these values, let us note that it is possible to extend the definition of G_i in (16.9) so that it commutes with the vison–spinon Hamiltonian in (16.8), which is the $\mathbb{Z}_2$ gauge theory *with* matter. It is easy to check that the required term is

$$G_i^{vs} = G_i \exp\left(i\pi s_{i\alpha}^\dagger s_i^\alpha \right) . \tag{16.11}$$

This ensures that each $s_{i\alpha}$ spinon carries a unit $\mathbb{Z}_2$ gauge charge, as required by (16.4). The full system of visons and spinons should be gauge neutral, and so we have

$$G_i^{vs} = 1, \quad \mathbb{Z}_2 \text{ gauge theory with matter}. \tag{16.12}$$

But recall from (15.4) that the spinon number appearing in (16.11) is exactly constrained to equal $n_s = 2S$, and the exponential factor in (16.11) evaluates to $(-1)^{2S}$. In the remainder of this chapter we will work with the $\mathbb{Z}_2$ gauge theory (16.7), in which the spinon matter fields have been integrated out; the constraint on the $s_{i\alpha}$ number is realized in such a theory by placing the $s_{i\alpha}$ in a valence bond whose number is measured by the $\mathbb{Z}_2$ gauge fields: this is clear from (16.2), which identifies Z_{ij} with the singlet bond annihilation operator $\varepsilon_{\alpha\beta} s_{i\alpha} s_{j\beta}$. From this analysis, we conclude that the vison-only theory $\mathcal{H}_{\mathbb{Z}_2}$ in (16.7), with no dynamic $\mathbb{Z}_2$ electric charges, should be examined in the Hilbert-space sector in which the $\mathbb{Z}_2$ gauge charges take the values

$$G_i = (-1)^{2S}, \quad \text{“pure” } \mathbb{Z}_2 \text{ gauge theory without matter}. \tag{16.13}$$

(We note that $\mathbb{Z}_2$ gauge theories with “relativistic matter” without Berry phases, as in (14.36) and (15.49), belong in the “without-matter” category in (16.13).) We will see below that this gauge charge constraint is precisely that needed to reproduce the Berry phase of the vison motion computed in Section 15.4.2. For half-integer S, the constraint $G_i = -1$ corresponds to the additional term in (14.35) for the classical XY model coupled to a $\mathbb{Z}_2$ gauge theory in (14.32): see Problem 16.2 for further details on this mapping.

In the language of the $\mathbb{Z}_2$ gauge theory, integer S antiferromagnets correspond to a “pure” $\mathbb{Z}_2$ gauge theory with no electrically charged matter fields. This is the “even”-$\mathbb{Z}_2$ gauge theory, where the spinons in (16.8) are gapped, and n_s is even. The case of half-odd-integer S is sometimes called an “odd”-$\mathbb{Z}_2$ gauge theory, and this case is of interest for the resonating-valence-bond theory of spin $S = 1/2$ antiferromagnets. We will see below that the properties of the even- and odd-$\mathbb{Z}_2$ gauge theories are very different. The odd gauge theory corresponds to placing a static background $\mathbb{Z}_2$ electric charge on each lattice site. The system has to be globally neutral, and so, on a torus of size $L_x \times L_y$, the number of sites, $L_x L_y$, has to be even for there to be any states which satisfy $G_i = -1$.

Finally, we note that the connection outlined above between the $U(1)$ gauge theory of Chapter 15 and the $\mathbb{Z}_2$ explored in the present chapter is studied in more detail in Section 26.2.

16.2.1 Wegner's $\mathbb{Z}_2$ gauge theory

As a historical aside, we note that the $\mathbb{Z}_2$ theory in (16.7) first appeared in Wegner's pioneering lattice gauge theory paper [301] for the case of $G_i = 1$, that is, the even-$\mathbb{Z}_2$ gauge theory. Wegner, and the subsequent lattice gauge theory literature, has not considered the odd-$\mathbb{Z}_2$ gauge theory that is important for our purposes.

Wegner defined the $\mathbb{Z}_2$ gauge theory as a classical statistical mechanics partition function on the cubic lattice. He considered the partition function [301]

$$\begin{aligned} \widetilde{\mathcal{Z}}_{\mathbb{Z}_2} &= \sum_{\{Z_{ij}\}=\pm 1} \exp\left(-\widetilde{\mathcal{H}}_{\mathbb{Z}_2}/T\right), \\ \widetilde{\mathcal{H}}_{\mathbb{Z}_2} &= -K \sum_{\square} \prod_{(ij)\in\square} Z_{ij}, \end{aligned} \tag{16.14}$$

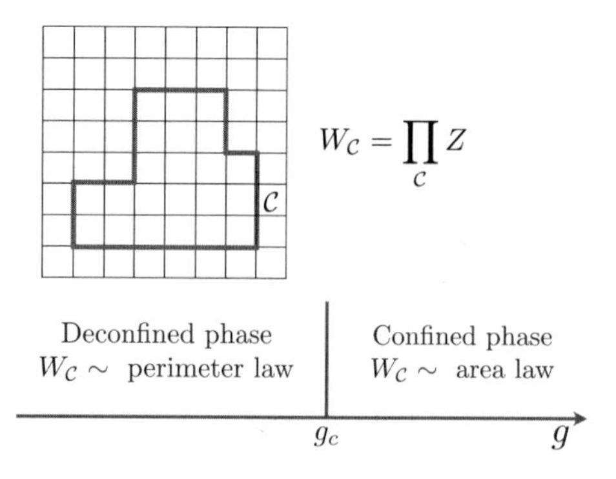

Figure 16.3 The Wegner-Wilson loop operator W_C on the closed loop C. Shown above is a schematic ground-state phase diagram of $\mathcal{H}_{\mathbb{Z}_2}$ for integer S, with the distinct behaviors of W_C in the deconfined and confined phases.

which is just the K term in (14.32). The degrees of freedom in this partition function are the binary variables $Z_{ij} = \pm 1$ on the links $\ell \equiv (ij)$ of the cubic lattice. The $\square$ indicates the elementary plaquettes of the cubic lattice. The quantum Hamiltonian is obtained from the classical theory in (16.14) by the usual quantum-to-classical mapping discussed in the QPT book [234], similar to that relating the quantum Ising model in a transverse field to a classical Ising model in one additional dimension. By such an analysis it can be shown [146] that the three-dimensional classical model in (16.14) is equivalent to the two-dimensional quantum model in (16.7) for $G_i = 1$ with the Z_{ij} on the spatial links mapping to the Z_{ij} operator of the quantum model. The extension of the three-dimensional classical model in (16.14) to the case $G_i = -1$ was described in Refs. [119, 230], and leads to the additional term in (14.35) in spacetime lattice formulation.

Wegner [301] showed that there were two phases with exponentially decaying correlations in the theory, which are necessarily separated by a phase transition. One of these phases (the "deconfined" phase below) corresponds to the $\mathbb{Z}_2$ spin liquid, and so the connection to Wegner's work establishes the stability of the $\mathbb{Z}_2$ spin liquid for a finite range of parameters, beyond the large-N expansion of Chapter 15. Remarkably, unlike all previously known cases, Wegner's phase transition was not required by the presence of a broken symmetry in one of the phases; there was no local order parameter characterizing the phase transition. Instead, Wegner argued for the presence of a phase transition using the behavior of the Wegner–Wilson loop operator W_C, which is the product of Z_α on the links of any closed contour C on the direct square lattice, as illustrated in Fig. 16.3. (W_C is usually, and improperly, referred to just as a Wilson loop.) The two phases are:

(i) At $g \gg K$ we have the "confining" phase. In this phase W_C obeys the area law: $\langle W_C \rangle \sim \exp(-\alpha A_C)$ for large contours C, where A_C is the area enclosed by the contour C and α is a constant. This behavior can easily be seen by a small-K expansion of $\langle W_C \rangle$: one power of K is needed for every plaquette enclosed by C for the first non-vanishing contribution to W_C. The rapid decay of $\langle W_C \rangle$ is a consequence of the large fluctuations in the $\mathbb{Z}_2$ flux, $\prod_{\ell \in \square} Z_\ell$, through each plaquette, that is, the proliferation of visons in

the quantum model. This proliferation of visons implies that particles carrying a $\mathbb{Z}_2$ electric charge, that is, spinons, will be confined in this phase.
(ii) At $K \gg g$ we have the "deconfined" phase. In this phase, the $\mathbb{Z}_2$ flux is expelled, the visons have a large gap, and $\prod_{\ell\in\Box} Z_\ell$ usually equals $+1$ in all plaquettes. We see later in Section 26.2 that the flux expulsion is analogous to the Meissner effect in superconductors. The small residual fluctuations of the flux lead to a perimeter-law decay, $\langle W_C\rangle \sim \exp(-\alpha' P_C)$ for large contours C, where P_C is the perimeter of the contour C and α' is a constant.

16.3 Topological Order at Small g

While Wegner's analysis yields a satisfactory description of the pure $\mathbb{Z}_2$ gauge theory, the Wegner–Wilson loop is, in general, not a useful diagnostic for the existence of a phase transition. Once we add dynamic matter fields, W_C invariably has a perimeter-law decay, although the confinement–deconfinement phase transition can persist.

The modern interpretation of the small-g phase of $\mathbb{Z}_2$ lattice gauge theory is that it is characterized by the presence of $\mathbb{Z}_2$ "topological" order [17, 88, 101, 133, 171, 219, 303]. The stability of the small-g phase implies that the $\mathbb{Z}_2$ spin-liquid phase obtained in Chapter 15 by large-N methods is stable, and we have already described its topological characteristics. We now describe two characteristics of this topological order in the context of the $\mathbb{Z}_2$ gauge theory. Both characteristics can survive the introduction of additional degrees of freedom; but we will see in Section 28.4 that the first is more robust, and is present even in cases with gapless excitations carrying $\mathbb{Z}_2$ charges.

The first characteristic is that there are stable low-lying excitations of the small-g phase in the infinite lattice model, which cannot be created by the action of any local operator on the ground state (i.e., there are "superselection" sectors [133]). This excitation is, of course, the "vison," which carries $\mathbb{Z}_2$ flux of -1 [138, 215, 257]. Recall that the ground state of the deconfined phase expelled the $\mathbb{Z}_2$ flux: at $g = 0$ the state with all qubits up, $|\Uparrow\rangle$, (i.e., eigenstates of Z_ℓ with eigenvalue $+1$) is a ground state, and this has no $\mathbb{Z}_2$ flux. This state is not an eigenstate of the G_i, but this is easily remedied by a gauge transformation:

$$|0\rangle = \prod_i (1 + (-1)^{2S} G_i) |\Uparrow\rangle \,. \tag{16.15}$$

It is easily shown that this state is an eigenstate of all the G_i with the eigenvalues obeying (16.13). When we apply the X_ℓ operator on a link ℓ, the neighboring plaquettes acquire a $\mathbb{Z}_2$ flux of -1. We need a non-local "string" of X_ℓ operators to separate these $\mathbb{Z}_2$ fluxes so that we obtain two well-separated vison excitations; see Fig. 16.2. Each vison is stable in its own region, and the motion of visons is described in Section 16.5.

The second topological characteristic emerges upon considering the low-lying states of $\mathcal{H}_{\mathbb{Z}_2}$ on a topologically non-trivial geometry, like the torus. A key observation in such geometries is that the G_i (and their products) do not exhaust the set of operators

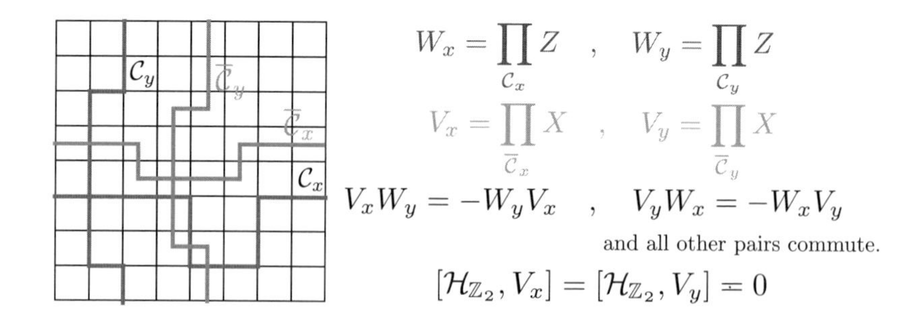

Figure 16.4 Operators in a torus geometry; periodic boundary conditions are implied on the lattice.

that commute with $\mathcal{H}_{\mathbb{Z}_2}$. On a torus, there are two additional independent operators that commute with $\mathcal{H}_{\mathbb{Z}_2}$: these operators, V_x, V_y, are illustrated in Fig. 16.4 (these are analogs of 'tHooft loops). The operators are defined on contours, $\overline{\mathcal{C}}_{x,y}$, which reside on the dual square lattice, and encircle the two independent cycles of the torus. The specific contours do not matter, because we can deform the contours locally by multiplying them with the G_i.

It is also useful to define Wegner–Wilson loop operators $W_{x,y}$ on direct lattice contours $\mathcal{C}_{x,y}$, which encircle the cycles of the torus; note that the $W_{x,y}$ do not commute with $\mathcal{H}_{\mathbb{Z}_2}$, while the $V_{x,y}$ do commute. Because the contour $\mathcal{C}_x$ intersects the contour $\overline{\mathcal{C}}_y$ an odd number of times (and similarly with $\mathcal{C}_y$ and $\overline{\mathcal{C}}_x$) we obtain the anti-commutation relations

$$W_x V_y = -V_y W_x \quad , \quad W_y V_x = -V_x W_y \,, \tag{16.16}$$

while all other pairs commute. Note that $W_{x,y}$ and the $V_{x,y}$ commute with all the G_i.

With this algebra of topologically non-trivial operators at hand, we can now identify the second distinct signature of the small-g phase with topological order. All eigenstates of $\mathcal{H}_{\mathbb{Z}_2}$ on the torus can also be made eigenstates of V_x and V_y. The ground state $|0\rangle$ is not an eigenstate of $V_{x,y}$, but is instead an eigenstate of $W_{x,y}$ with $W_x = W_y = 1$. The state $V_x |0\rangle$ is easily seen to be an eigenstate of $W_{x,y}$ with $W_x = 1$ and $W_y = -1$; so this state has $\mathbb{Z}_2$ flux of -1 through one of the holes of the torus, as is illustrated in Fig. 16.5. At $g = 0$, the state $V_x |0\rangle$ is also a ground state of $\mathcal{H}_{\mathbb{Z}_2}$, degenerate with $|0\rangle$. Similarly, we can create two other ground states, $V_y |0\rangle$ and $V_y V_x |0\rangle$, which are also eigenstates of $W_{x,y}$ with distinct eigenvalues. So, at $g = 0$, we have a four-fold degeneracy in the ground state, and all other states are separated by an energy gap.

When we turn on a non-zero g, the ground states will no longer be eigenstates of $W_{x,y}$ because these operators do not commute with $\mathcal{H}_{\mathbb{Z}_2}$. Instead, the ground states will become eigenstates of $V_{x,y}$; at $g = 0$ we can take the linear combinations $(1 \pm V_x)(1 \pm V_y)|0\rangle$ to obtain degenerate states with eigenvalues $V_x = \pm 1$ and $V_y = \pm 1$. At non-zero g, these four states will no longer be degenerate, but only acquire an exponentially small splitting of order $gL(g/K)^L$, where L is a linear dimension of the torus. This is because the four states differ from each other only by global topological operators, and any non-zero matrix element between them requires the application of $-g\sum_\ell X_\ell$ on loops that encircle the lattice. Alternatively stated, the tunneling amplitude between states

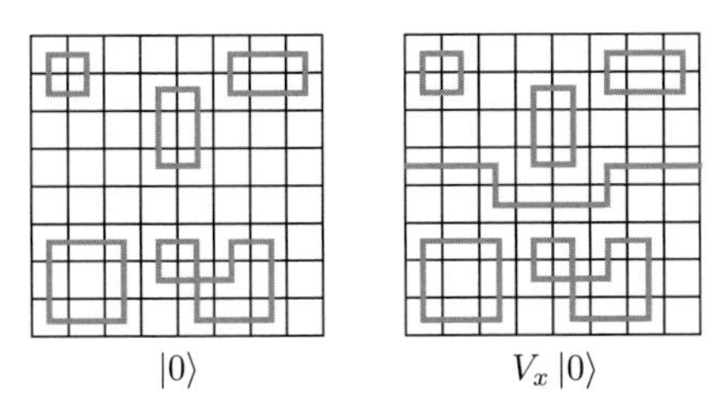

Figure 16.5 A typical term in the expansion for the state $|0\rangle$ in (16.15), and for the state $V_x|0\rangle$ on a torus (periodic boundary conditions are implied). All links of the direct lattice have $|\uparrow\rangle$ state, apart from those crossing a thick line which have $|\downarrow\rangle$. Each term has a prefactor of $(-1)^{2S}$ for each site of the direct lattice enclosed by the loops of thick lines. The state $V_x|0\rangle$ always has a single thick line that encircles the cycle of the torus, unlike the state $|0\rangle$. Notice that every plaquette of the direct lattice cuts an even number of thick lines, and so has $\mathbb{Z}_2$ flux $+1$ and both states are ground states at $g = 0$. The state $|0\rangle$ has $W_x = W_y = 1$, while the state $V_x|0\rangle$ has $W_x = 1$ and $W_y = -1$. At small non-zero g, there is a non-zero tunneling amplitude between $|0\rangle$ and $V_x\,|0\rangle$ of order g^{L_x}, where L_x is the length of $\overline{\mathcal{C}}_x$.

with distinct $\mathbb{Z}_2$ fluxes through the holes of the torus is exponentially small in the size of the system.

We can encapsulate these tunneling terms in an effective Hamiltonian H_{eff} as a 4×4 matrix which acts on this space of lowest-energy states. We can also project the operators $V_{x,y}$ and $W_{x,y}$ onto this space of four states, which then become a set of 4×4 matrices that obey the algebraic relations in Fig. 16.4. After this projection, the $W_{x,y}$ will differ from the $g=0$ expressions in Fig. 16.4 by a canonical transformation that can be computed order by order in g; see Problem 16.3. The operators $V_{x,y}$ must commute with the effective Hamiltonian, and so we can write the most general form

$$H_{eff} = c_1 V_x + c_2 V_y + c_3 V_x V_y\,, \tag{16.17}$$

where $c_{1,2,3}$ are constants of order $gL(g/K)^L$.

The presence of these four lowest energy states, which are separated by an energy splitting that vanishes exponentially with the linear size of the torus, is one of the defining characteristics of $\mathbb{Z}_2$ topological order. We can take linear combinations of these four states to obtain distinct states with eigenvalues $W_x = \pm1$, $W_y = \pm1$ of the $\mathbb{Z}_2$ flux through the holes of the torus; or we can take energy eigenvalues, which are also eigenstates of $V_{x,y}$ with $V_x = \pm1$, $V_y = \pm1$. These features are present throughout the entire deconfined phase.

Finally, we turn to the subtle role played by translational symmetry. The considerations below apply for any g, and play a crucial role in our discussion of the large-g limit in the subsequent sections. Let T_x (T_y) be the operator that translates the system by one lattice spacing along the x (y) direction. Clearly, the operators $T_{x,y}$ commute with the Hamiltonian in (16.7). Now, consider the operators V_x and V_y, defined as in Fig. 16.4 on a $L_x \times L_y$ torus, for convenience on contours $\overline{\mathcal{C}}_x$ and $\overline{\mathcal{C}}_y$ that are straight, of lengths L_x and L_y respectively. These operators V_x and V_y also commute with the Hamiltonian. But, as illustrated in Fig. 16.6, $V_{x,y}$ and $T_{x,y}$ don't always commute with each other:

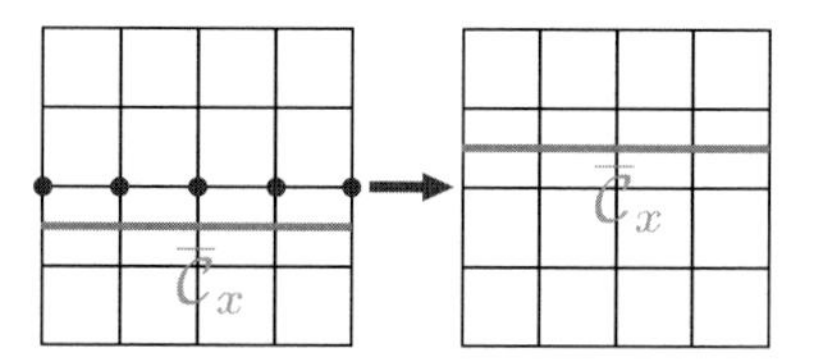

Figure 16.6 The operator V_x on the contour $\overline{\mathcal{C}}_x$ is translated by T_y upon the action of G_i on the encircled sites. This yields (16.18).

$$T_x V_y = (-1)^{2SL_y} V_y T_x \quad , \quad T_y V_x = (-1)^{2SL_x} V_x T_y \, . \tag{16.18}$$

The relations in (16.18) are valid on *any* state and apply for all g. They have particularly strong consequences for half-odd-integer S, when T_x, V_y or T_y, V_x can anti-commute with each other for certain system sizes. As $T_{x,y}$ and $V_{x,y}$ all commute with the Hamiltonian, all the eigenstates of the Hamiltonian must realize various representations of the algebra in (16.18). This immediately implies that the ground state of $\mathcal{H}_{\mathbb{Z}_2}$ *cannot be trivial for any g and half-odd-integer S*. Here, by a trivial ground state, we mean one which is insensitive to global features such as the size of the lattice or its topology. This is a version of the Oshikawa–Hastings theorem [105, 194] of the impossibility of trivial states in boson systems at half filling. In the context of $\mathcal{H}_{\mathbb{Z}_2}$ with half-odd-integer S our analysis in this section leads us to conclude that (i) the ground state has topological order at small g, and (ii) at large g, either the topological order survives or there is broken translational symmetry.

We now comment on the nature of the T_x, T_y operators within the four-dimensional space of (near) ground states on a torus in the small-g topological phase. For integer S, the relation (16.18) is trivial, and so T_x and T_y both reduce to unit operators on this space. However, for half-integer S, by comparing (16.18) with (16.16) we deduce that the nature of these operators depends upon the parities of L_x and L_y:

$$\begin{aligned}
L_x \text{ even and } L_y \text{ even} &\Rightarrow T_x = 1, T_y = 1, \\
L_x \text{ even and } L_y \text{ odd} &\Rightarrow T_x = W_x, T_y = 1, \\
L_x \text{ odd and } L_y \text{ even} &\Rightarrow T_x = 1, T_y = W_y.
\end{aligned} \tag{16.19}$$

Note that T_x and T_y commute with each other in the ground-states subspace in all cases.

We close this section by noting that the $\mathbb{Z}_2$ topological order described above can also be realized in a $U(1) \times U(1)$ Chern–Simons gauge theory [88, 171]. This will be described in Section 17.1.1, where (17.14) contains alternative expressions for the W_i and V_i which obey the same commutation relations as in (16.16).

16.4 Large-g Limit

Unlike the small-g limit, the large-g limit is dramatically different for integer and half-odd-integer S. We will consider these cases in the separate subsections.

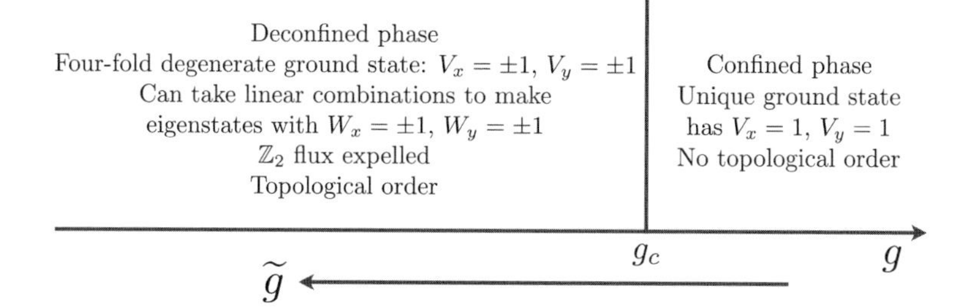

Figure 16.7 An updated version of the phase diagram of $\mathcal{H}_{\mathbb{Z}_2}$ in Fig. 16.3 for integer S. The confinement–deconfinement phase transition is described by the Ising* Wilson–Fisher CFT. This is the same transition as that between phases A+B and D in Figs. 14.2 and 14.3.

16.4.1 Integer S: Trivial Ground State

With $G_i = 1$, we immediately obtain a trivial, unique ground state at large g, with a large energy gap to all excitations. At $g = \infty$, the ground state, $|\Rightarrow\rangle$, has all qubits pointing to the right (i.e., all qubits are eigenstates of X_ℓ with eigenvalue $+1$). This state clearly has eigenvalues $V_x = V_y = +1$. States with $V_x = -1$ or $V_y = -1$ must have at least one qubit pointing to the left, and so cost a large-energy g: such states cannot be degenerate with the ground state, even in the limit of an infinite volume for the torus. See Fig. 16.7.

16.4.2 Half-Odd-Integer S: Quantum-dimer Model

A notable feature of the confining phase obtained in Fig. 16.7 for integer S is that it is completely "trivial." It has no broken symmetry, no degeneracy on the torus, and no fractionalized excitations. We motivated the $\mathbb{Z}_2$ gauge theory in the introduction to this chapter as the low-energy theory of an antiferromagnet of $\boldsymbol{S}$ spins, and so it is natural to ask to what state of the antiferromagnet this confining state corresponds. A glance at Fig. 9.2 does *not* yield any suitable candidates; all states there break either spin-rotation or lattice symmetries. However, recall that Fig. 9.2 corresponds to $S = 1/2$ antiferromagnets. And, precisely for this case, we derived a Berry phase in the motion of visons in Section 15.4.2, corresponding to the projective realization of translational symmetries in (13.8). Clearly, we need to include this Berry phase in our effective gauge theory here, and we argued in Section 16.2 that this corresponds to the case $G_i = -1$ on each site. As we will see below, this is sufficient to remove the trivial confining state.

The value $G_i = -1$ implies that there must be an odd number of $|\leftarrow\rangle$ qubits on the links ending on each site. Every one of these $|\leftarrow\rangle$ qubits costs an energy $2g$, and so at large g we need to minimize the total number of $|\leftarrow\rangle$ qubits. As illustrated in Fig. 16.8, there is a very large number of possible $|\leftarrow\rangle$ qubit configurations, and these are in one-to-one correspondence with the dimer close-packings of the lattice. So the ground state at $g = \infty$ is highly degenerate, and indeed there is an extensive ground-state entropy.

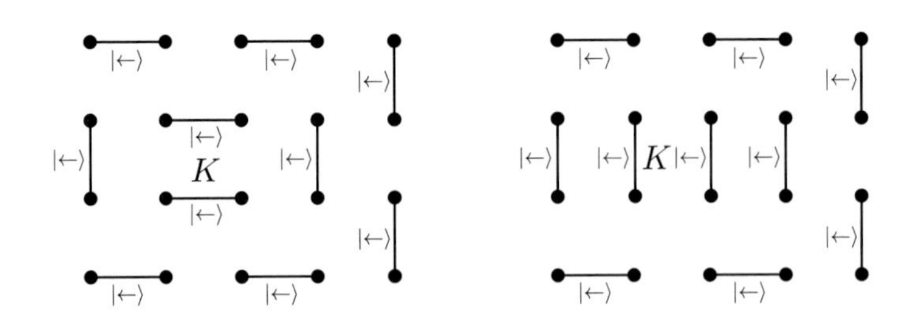

Figure 16.8 Connection between the large-g ground states of the $\mathbb{Z}_2$ gauge theory $\mathcal{H}_{\mathbb{Z}_2}$ for half-odd-integer S and quantum-dimer packings. All links with a dimer have a qubit in the $|\leftarrow\rangle$ state (i.e., the eigenstate of X_ℓ with eigenvalue -1), while the remaining links are in the orthogonal $|\rightarrow\rangle$ state. Each lattice site neighbors exactly one $|\leftarrow\rangle$ state, and so $G_i = -1$ on all sites. The left and right states are connected by the K term in (16.7) applied to the plaquette shown. This mapping applies on any lattice. Compare to the mapping to the interface model in Fig. 26.3, which applies only bipartite lattices.

Tracing our mappings back to the spin models of Chapter 15, it is clear that we can identify the dimers with singlet bonds of the underlying spins of the antiferromagnet. Recall that we identified $\mathcal{Q}_{ij}$, the singlet annihilation operator, with Z_{ij} in (16.2). It is therefore natural to identify the conjugate X_{ij} operator with the dimer-number operator.

For a proper description of the antiferromagnet, it is clear that the $g = \infty$ limit is singular, and we have to include corrections to this limit. At first order in K/g, it is simple to see that the K term in (16.7) induces a dimer resonance term: that is, it interchanges the $|\leftarrow\rangle$ and $|\rightarrow\rangle$ qubits, as illustrated in Fig. 16.8. This off-diagonal term leads to a quantum-dimer model: the Hilbert space of this model is in one-to-one correspondence with dimer packings, and the effective Hamiltonian on this dimer space is [222]

$$\mathcal{H}_{qd} = -K\sum_{\square}\Big(|\rangle\langle| + |\rangle\langle|\Big) + V\sum_{\square}\Big(|\rangle\langle| + |\rangle\langle|\Big). \tag{16.20}$$

The first-order correction in K/g yields only the K term in (16.20), but we have included the commonly considered V term, which appears at higher orders in the K/g expansion. Although we have presented $\mathcal{H}_{qd}$ on the square lattice, there is a natural generalization of such models to the large-g limit of $\mathcal{H}_{\mathbb{Z}_2}$ on any lattice.

The next step is the determination of the spectrum of $\mathcal{H}_{qd}$. This is a difficult problem, which has been addressed by a variety of numerical and analytic methods. The case of the square lattice is addressed in Section 26.1.2; we present general arguments that, on bipartite lattices, the quantum dimer model only has confining states that break the lattice translational symmetry with the appearance of VBS order, as shown in Fig. 26.4. In the following section, we address the nature of the translational symmetry breaking by developing a theory of the condensation of the m anyons, and obtain the same result; this analysis starts from the spectrum of the m visons in the deconfined topological phase, but after the condensation of the anyons it is also possible to describe the spectrum in the confining phase.

The large-g limit of the $G_i = -1$ $\mathbb{Z}_2$ gauge theory on the triangular lattice also maps to a quantum dimer model with one dimer on each triangular lattice site. This quantum-dimer model has been studied numerically and, for the $V = 0$ case relevant for the large-g expansion, it has a confining ground state with broken translational symmetry [179, 180, 320]. This can also be described by the analogous vison condensation theory of Section 16.5.

16.5 Visons and Anyon Condensation

This section describes in more detail the spectrum of the excited states of the $\mathbb{Z}_2$ gauge theory in (16.7). We begin in the small-g topological phase and compute the spectrum of the vison excitations. General symmetry arguments will then allow us to write down a field theory for the condensation of the visons. We will then also be able to use this field theory to understand the structure of the large-g confining phase, and of the phase transition to it. As in Section 16.4, the even- and odd-$\mathbb{Z}_2$ gauge theories are addressed separately.

16.5.1 Integer S: Ising* Criticality

A pair of well-separated vison states are obtained by applying a string of X_ℓ operators to the ground state in (16.15), as shown in Fig. 16.2. For integer S, the action of $-gX_\ell$ on the four links surrounding the vison plaquette move the vison by one lattice spacing in either direction, and so the single vison dispersion is

$$\varepsilon_{\boldsymbol{k}}^{v} = 2K - 2g(\cos(k_x) + \cos(k_y)) + \mathcal{O}(g^2/K). \tag{16.21}$$

So we obtain a single real particle with a dispersion minimum at $\boldsymbol{k} = 0$, and a gap of $2(K - g)$.

As g increases, this vison gap will decrease, until it vanishes at the confinement transition already shown in Figs. 16.3 and 16.7. This is also the transition from phase D to phase A+B in the classical XY model in Figs. 14.2 and 14.3. Wegner also determined the critical properties of the transition. He performed a Kramers–Wannier duality transformation, and showed that the $\mathbb{Z}_2$ gauge theory was equivalent to the classical Ising model in $D = 3$ dimensions. This establishes that the confinement–deconfinement transition is in the universality class of the the Ising Wilson–Fisher [310] conformal field theory (CFT) in three spacetime dimensions (a CFT3). The phase with the dual Ising order is the confining phase, and the phase with Ising "disorder" is the deconfined phase.

Indeed, we can identify the real Ising field ϕ with the vison creation and annihilation operator. This is clear from the structure of the duality transformation from the $\mathbb{Z}_2$ gauge theory: the plaquette flux operator is dual to the Ising spin operator, which resides at the center of the plaquette. We can now derive the field theory obeyed by this Ising field by starting in the deconfined phase, when the vison excitations are gapped;

we write down a free field theory on a lattice whose dispersion reproduces (16.21) at small g. A suitable Hamiltonian is

$$\mathcal{H}_\phi = \sum_j \left[\frac{\pi_j^2}{2} + \frac{\omega_0^2}{2} \phi_j^2 \right] - \sum_{\langle jj' \rangle} t_{jj'} \phi_j \phi_{j'}, \tag{16.22}$$

where j, j' represent sites on the *dual* square lattice (on which the vison is centered), and π_j is the canonically conjugate momentum to ϕ_j:

$$[\phi_j, \pi_{j'}] = i\delta_{jj'} . \tag{16.23}$$

Matching the dispersion of (16.22) to (16.21) at small g, we obtain $\omega_0 = 2K$ and a nearest-neighbor hopping $t = g\omega_0$.

The confinement transition occurs when the Ising field ϕ condenses, near which point we need to include higher-order corrections to $\mathcal{H}_\phi$. The simplest allowed term is $\sum_j \phi_j^4$, representing the scattering of a pair of visons. We take the continuum limit near the critical point, and therefore obtain the relativistic field theory of Chapters 10 and 11 for $N = 1$. This has the (2+1)-dimensional Lagrangian density

$$\mathcal{L}_\phi = (\partial_\mu \phi)^2 + \widetilde{g}\phi^2 + \widetilde{u}\phi^4, \tag{16.24}$$

where the coupling $\widetilde{g}$ increases as g decreases (see Fig. 16.7). So, for $g < g_c$, the ϕ field is gapped, and the ϕ excitations correspond to the visons. The confining phase for $g > g_c$ is obtained by the condensation of the visons. A another derivation of this Ising critical theory is presented in Section 26.2.2.

We also note that the critical theory is not precisely the Ising Wilson–Fisher CFT, but what is often called the Ising* theory. In the Ising* theory, the only allowed operators are those which are invariant under $\phi \to -\phi$, where ϕ is the Ising primary field [252, 308].

16.5.2 Half-Odd-Integer S: VBS Order and XY^* Criticality

We compute the vison motion for $G_i = -1$ in a perturbation theory in g applied to the vison state shown in Fig. 16.2. T_x and T_y do not commute when acting on a vison state $|v\rangle$:

$$T_x T_y |v\rangle = -T_y T_x |v\rangle . \tag{16.25}$$

The proof of this relation is presented in Fig. 16.9. This implies that the vison accumulates a Berry phase of π when transported around a single square-lattice site. This is precisely the phase factor illustrated in Fig. 15.6 in Section 15.4.2 for the underlying spin system on the triangular lattice, and also discussed in Section 13.4.

Following the computation in Section 16.5.1, we determine the dispersion of the vison excitations for $G_i = -1$. The explicit computation is presented in Ref. [230], but here we obtain the result by a general argument similar to that used to obtain (16.24) for the even-$\mathbb{Z}_2$ gauge theory. As indicated in Fig. 16.9, each vison moves in a background π flux per plaquette of the dual lattice due to the presence of the electric charges on the sites of the direct lattice. So we account for this flux by modifying the hopping $t_{jj'}$ in

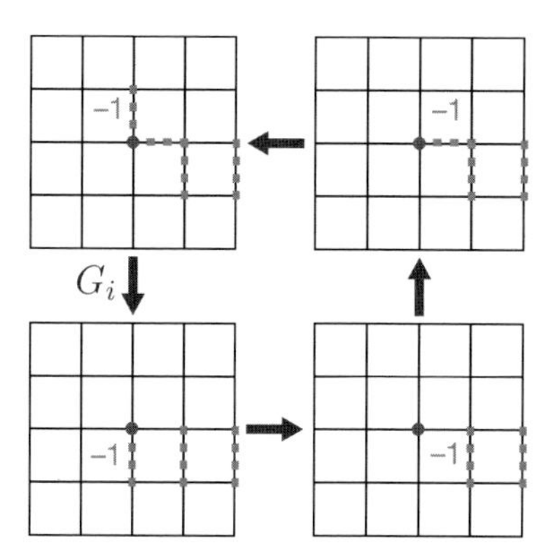

Figure 16.9 Starting from the lower left, we illustrate a vison undergoing the operations $T_x, T_y, T_x^{-1}, T_y^{-1}$. The final state differs from the initial state by the action of G_i on the single encircled site. Using $G_i = -1$, we then obtain (16.25), the π Berry phase of a vison moving on the path shown.

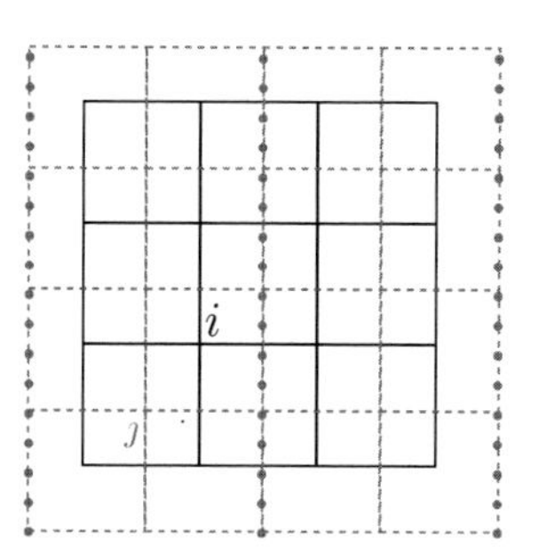

Figure 16.10 The $\mathbb{Z}_2$ gauge theory resides on the sites of the direct lattice of sites i (connected by full lines), while the Ising theory for the visons resides on the dual lattice of sites j (connected by dashed lines). The visons hop with amplitude t to nearest neighbors on the dual lattice, except for the dotted links with amplitude $-t$. This ensures that the vison experiences π flux upon encircling any site of the direct lattice, as required for half-odd-integer S in Section 15.4.2.

(16.22) to that shown in Fig. 16.10. In momentum space, the hopping matrix is a 2×2 matrix

$$\begin{pmatrix} -2t\cos(k_y) & -2t\cos(k_x) \\ -2t\cos(k_x) & 2t\cos(k_y) \end{pmatrix}. \tag{16.26}$$

After diagonalizing this matrix, from (16.22) we find that the vison dispersion is

$$\varepsilon_k^v = \left[\omega_0^2 \pm 4t\left(\cos^2(k_x) + \cos^2(k_y)\right)^{1/2}\right]^{1/2}. \tag{16.27}$$

Now we turn to the confinement transition where the vison gap vanishes. This transition was not considered by Wegner [301] for the odd-$\mathbb{Z}_2$ gauge theory. This transition is that from the fractionalized phase to the bond-order phase in Fig. 14.4 for the XY model with the Berry phase term in (14.35). As with the even gauge theory in Section 16.5.1, performing the Kramers–Wannier duality with the condition $G_i = -1$ leads to an Ising model in a transverse field on the dual lattice. However, in the odd gauge theory, the signs of the couplings in each spatial plaquette are frustrated [119, 257]. Such a fully frustrated Ising model has been investigated in experiments on superconducting qubits [131].

Here, we will derive the low-energy theory of this fully frustrated Ising model by simple symmetry arguments applied to the vison disperson in (16.27). Near the confinement transition, we can focus energy on the lowest-energy vison states. From (16.27), we observe that the minimum of the vison dispersion is at *two* distinct points in the reduced Brillouin zone of the lattice in Fig. 16.10; these are the points $\boldsymbol{k} = (0,0)$ and $\boldsymbol{k} = (0,\pi)$. So the continuum theory will be expressed in terms of *two* real fields, in contrast to the single real field for the even case in Section 16.5.1. We label these states as

$$\begin{aligned} \varphi_1 &= \phi(\boldsymbol{k} = (0,0)) \text{ on the first sublattice,} \\ \varphi_2 &= \phi(\boldsymbol{k} = (0,\pi)) \text{ on the second sublattice.} \end{aligned} \tag{16.28}$$

Now we need a continuum theory for $\varphi_{1,2}$ in the vicinity of the confinement transition, the analog of (16.24) for the even case. In principle, this can be obtained by a g/K perturbation theory of the underlying $\mathbb{Z}_2$ gauge theory Hamiltonian. However, it is more illuminating to derive the general answer from symmetry principles. The simplest symmetry is translations in the y directions, which follow directly from (16.28):

$$T_y : \varphi_1 \to \varphi_1 \quad ; \quad \varphi_2 \to -\varphi_2\,. \tag{16.29}$$

The operation of T_x is more subtle, because the pattern in Fig. 16.10 is not invariant under translation by one lattice spacing. However, we can restore translational invariance by a $\mathbb{Z}_2$ gauge transformation, and this gauge action yields the needed transformation [18, 114, 199]

$$T_x : \varphi_1 \to \varphi_2 \quad ; \quad \varphi_2 \to \varphi_1\,. \tag{16.30}$$

We can now immediately verify from (16.29) and (16.30) that

$$T_x T_y = -T_y T_x, \tag{16.31}$$

as needed for the vison states from Section 15.4.2 and (16.25). The relation (16.31) also explains the double degeneracy of the low-energy vison states, as 2×2 matrices are the smallest realization of this algebra. We emphasize that the anti-commutation relation (16.31) applies only for single vison states for half-odd-integer S. Contrast this with the relations (16.19) in the ground-state subspace, where T_x and T_y commute with each other.

In a similar manner, we can also obtain the action of rotations [18, 114, 199]

$$R_{\pi/2} : \varphi_1 \to \frac{1}{\sqrt{2}}(\varphi_1 + \varphi_2) \quad ; \quad \varphi_2 \to \frac{1}{\sqrt{2}}(\varphi_1 - \varphi_2), \tag{16.32}$$

where $R_{\pi/2}$ is the symmetry of rotations about a dual lattice point. The transformations in (16.29), (16.30), and (16.32), and their compositions, form the projective symmetry group, which constrains the theory of the topological phase and of its phase transitions. Direct computation shows that the group generated by (16.29), (16.30), and (16.32) is

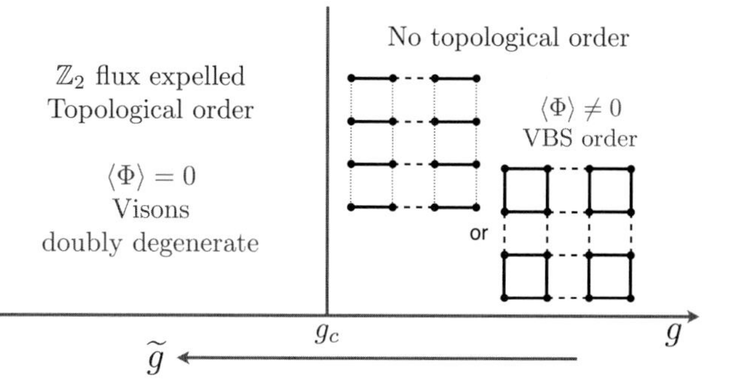

Figure 16.11 Phase diagram of the odd $\mathbb{Z}_2$ gauge theory defined by (16.7) and $G_i = -1$. Compare to Fig. 16.7 for the even-$\mathbb{Z}_2$ gauge theory. The transition above is also the transition between the fractionalized phase and bond order in Fig. 14.4. This phase diagram is an example of deconfined criticality, and a numerical study appeared early on in Ref. [119].

the 16-element non-abelian dihedral group D_8 [114]. We combine these real particles into a single complex field

$$\Phi = e^{-i\pi/8}(\varphi_1 + i\varphi_2). \tag{16.33}$$

With these phase factors, Φ transforms under D_8 as

$$T_x : \Phi \to e^{i\pi/4}\Phi^* \quad ; \quad T_y : \Phi \to e^{-i\pi/4}\Phi^* \quad ; \quad R_{\pi/2} : \Phi \to \Phi^* . \tag{16.34}$$

Note again that under the vison D_8 operations in (16.34), T_x and T_y anti-commute (as required by (16.31)). Then the effective theory for Φ is the simplest Lagrangian invariant under the D_8 symmetry:

$$\mathcal{L}_\Phi = |\partial_\mu \Phi|^2 + \widetilde{g}|\Phi|^2 + \widetilde{u}|\Phi|^4 - \overline{\lambda}\left(\Phi^8 + \Phi^{*8}\right). \tag{16.35}$$

This is the generalization of the Ising Lagrangian for a real field in (16.24) with $\mathbb{Z}_2$ symmetry for the even case. For the odd gauge theory we have a complex field Φ with D_8 symmetry.

The phase diagram of $\mathcal{L}_\Phi$ is modified from Fig. 16.7 to Fig. 16.11. The topological phase has a gapped Φ excitation. A crucial difference from the even-$\mathbb{Z}_2$ gauge theory is that this excitation is doubly degenerate; $\mathcal{L}_\Phi$ is sufficiently high order that the degeneracy between the real and imaginary parts of Φ is not broken. So the vison is a complex relativistic particle, unlike the real particle in Section 16.5.1. This double degeneracy in the vison states is a feature of the symmetry-enriched topological order [45, 69], and is intimately linked to the D_8 symmetry and to the anti-commutation relation [18] in (16.25); it is not possible to obtain vison states that form a representation of the algebra of T_x and T_y without this degeneracy.

Turning to the confined phase where Φ is condensed, the non-trivial transformations in (16.34) imply that lattice symmetries must be broken. This is because we can define a $\mathbb{Z}_2$ gauge-invariant order parameter, which is non-zero when Φ condenses, and which is not invariant under lattice symmetries. This is the VBS order parameter

$$\mathcal{O}_{VBS} = e^{-i\pi/4}\Phi^{*2}. \tag{16.36}$$

Using (16.34), we obtain the transformations of the VBS order parameter [119]:

$$T_x : \mathcal{O}_{VBS} \to -\mathcal{O}^*_{VBS} \quad ; \quad T_y : \mathcal{O}_{VBS} \to \mathcal{O}^*_{VBS} \quad ; \quad R_{\pi/2} : \mathcal{O}_{VBS} \to -i\mathcal{O}^*_{VBS}. \tag{16.37}$$

Note that T_x and T_y commute when acting on the VBS order, as they must on any gauge-invariant observable. The precise pattern of the broken symmetry depends upon the sign of $\overline{\lambda}$, and the two possibilities are shown in Fig. 16.11. The columnar ordering pattern is four-fold degenerate and corresponds to

$$\langle \mathcal{O}_{VBS} \rangle \sim 1, i, -1, -i, \tag{16.38}$$

where the plaquette ordering pattern is also four-fold degenerate with

$$\langle \mathcal{O}_{VBS} \rangle \sim e^{i\pi/4}, e^{i3\pi/4}, e^{i5\pi/4}, e^{i7\pi/4}. \tag{16.39}$$

So an important feature of the odd Ising gauge theory is that there is no "trivial" phase without symmetry breaking and fractionalization; the background gauge charges induce a breaking of translational symmetry in the confining phase. When extended to the boson model at filling ν, this implies the absence of trivial phases at half filling: this is consistent with the requirements of various rigorous arguments on the ground states of such models.

Finally, we address the confinement–deconfinement transition in Fig. 16.11. In (16.35), the $\overline{\lambda}$ term is an irrelevant perturbation to $\mathcal{L}_\Phi$, and the critical point of $\mathcal{L}_\Phi$ is the XY^* Wilson–Fisher CFT [119, 230, 256, 261] (contrast this with the Ising* Wilson–Fisher CFT in Fig. 16.7). This is an example of a "deconfined critical point" (hence the asterisk on XY) because the field theory is expressed in terms of the gauge-dependent field Φ, which is not an observable; only the "square" of the field is the VBS order parameter in (16.36). So the order parameter has fractionalized at the critical point. We present another formulation of this critical point in Section 26.2.3, where we obtain a dual formulation in which the deconfined criticality has an emergent $U(1)$ gauge field.

We note that the above phase diagram also applies to quantum dimer models on the square lattice [119, 230]. The extension to quantum dimer models on other lattices have also been considered [84, 114, 180, 181, 182, 290].

16.6 Models of Rydberg Atoms

This section is motivated by a recent experiment [254] on rubidium (Rb) atoms individually trapped in an array of optical tweezers, and pumped by optical tweezers to Rydberg states (see Fig. 16.12). Each Rydberg atom effectively becomes a two-state system (a qubit), and the whole system can be described by the Fendley–Sengupta–Sachdev (FSS) model [76] with the Hamiltonian

$$\mathcal{H}_{FSS} = \sum_\ell \left[\frac{\Omega}{2}\left(B_\ell + B_\ell^\dagger\right) - \Delta N_\ell \right] + \frac{1}{2}\sum_{\ell \neq \ell'} V(\boldsymbol{r}_\ell - \boldsymbol{r}_{\ell'}) N_\ell N_{\ell'}. \tag{16.40}$$

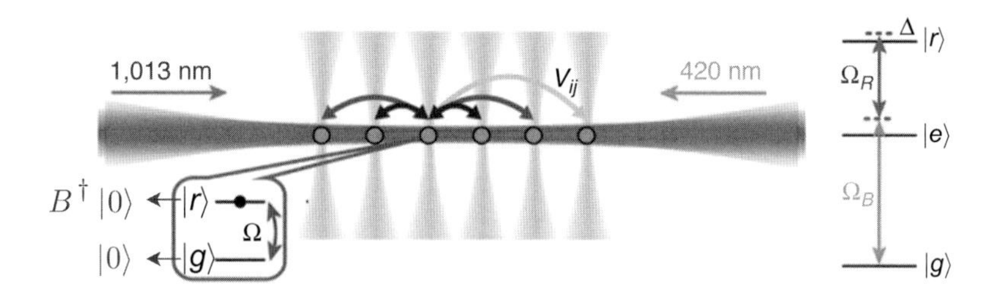

Figure 16.12 An array of Rb atoms trapped in tweezers and pumped by lasers with wavelengths 1013 nm and 420 nm to the Rydberg state $|r\rangle$ [29]. The quantum dynamics of these levels is described by $\mathcal{H}_{FSS}$ in (16.40) after identifying the $|r\rangle$ state with the occupied state of a hard-core boson B. Reprinted with permission from Springer Nature.

Here ℓ labels a set of lattice points with position $\boldsymbol{r}_\ell$; in the notation of the present chapter, the symbol ℓ has been reserved for the links of a lattice, and our notation here anticipates the connection to $\mathbb{Z}_2$ gauge theory to be made below. At the moment, though, $\mathcal{H}_{FSS}$ is not a lattice gauge theory, and B_ℓ is the annihilation operator of a boson that does not carry gauge charges. This is a "hard core" boson, and so the boson-number operator N_ℓ obeys

$$N_\ell \equiv B_\ell^\dagger B_\ell \quad , \quad N_\ell = 0, 1\,. \tag{16.41}$$

The function $V(\boldsymbol{r})$ represents repulsive interactions between the bosons on different sites, and is specified later.

The FSS model was originally motivated by a different connection to experiments on ultracold atoms [241]. The connection to the recent Rydberg atom experiment [254] is illustrated in Fig. 16.12. We identify the ground state of each atom, $|g\rangle$, with the empty boson state $|0\rangle$, and the Rydberg state, $|r\rangle$, with the filled boson state $B^\dagger|0\rangle$. These are coupled by the external lasers with a Rabi frequency Ω. The frequency of the external laser is adjusted so that the detuning away from resonance of the $|g\rangle$ to $|r\rangle$ transition is Δ. The potential $V(\boldsymbol{r})$ represents the van der Waals interaction when both atoms are in their Rydberg states; there is no appreciable van der Waals interaction of an atom in its ground state.

The basic physics of this system is that of the "Rydberg blockade"; the interaction $V(\boldsymbol{r})$ can be large at short distances, so that there is a large energy cost for two nearby atoms to both be in the $|r\rangle$ state, that is, for $N_\ell = 1$ on both sites. This induces quantum correlations between the atomic states, and this effect has been exploited to realize a number of interesting phases of quantum matter in recent experiments [29, 65, 129, 254].

We are interested here in configurations of the FSS model that can realize a $\mathbb{Z}_2$ spin liquid. Our strategy in Section 9.2 and Chapter 15 has been to identify hard-core bosons with spin operators $B \to S_-$, fractionalize the spin into spinons, and then study if the spinon theory can have a deconfined phase. However, a different strategy has so far proved useful for the FSS model [19, 224, 245, 246, 288]: we identify the two boson

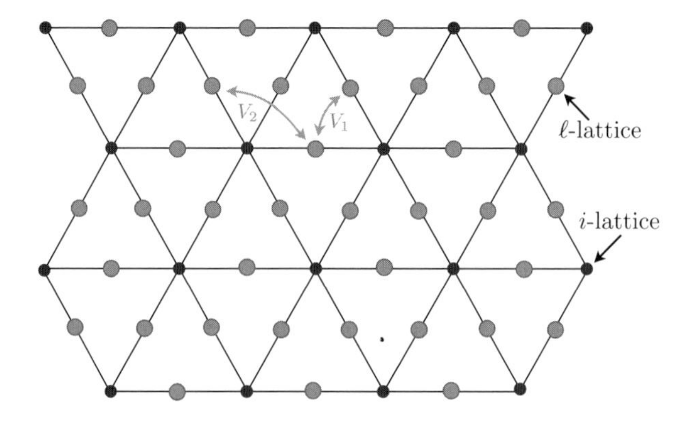

Figure 16.13 The ℓ-lattice is a kagome lattice, and the i-lattice is a triangular lattice. Rydberg atoms are on the ℓ-lattice. This model was numerically studied in Ref. [245].

states on each site with the qubits of the $\mathbb{Z}_2$ gauge theory of Section 16.2. Specifically, we make the identification

$$\begin{aligned} B_\ell + B_\ell^\dagger &\Leftrightarrow Z_\ell, \\ N_\ell &\Leftrightarrow (1 - X_\ell)/2. \end{aligned} \tag{16.42}$$

In the large-g limit of the odd-$\mathbb{Z}_2$ gauge theory of Section 16.4.2 and Fig. 16.8, the B boson is therefore identified with a *dimer*. Then, without approximation, we can write the FSS model as a model of interacting qubits

$$\mathcal{H}_{FSS} = \frac{1}{2}\sum_\ell [\Omega Z_\ell + \Delta X_\ell] + \frac{1}{2}\sum_{\ell \neq \ell'} \frac{V(\boldsymbol{r}_\ell - \boldsymbol{r}_{\ell'})}{4}(1 - X_\ell)(1 - X_{\ell'}). \tag{16.43}$$

In contrast to the dimer model considered in Section 16.2, the dimer model of (16.43) does not satisfy a dimer close-packing constraint. The Z_ℓ term can annihilate or create a dimer on site ℓ independent of the occupation of neighboring sites. In the language of the $\mathbb{Z}_2$ gauge theory, this is related to the fact that (16.43) is not invariant under the $\mathbb{Z}_2$ gauge transformation in (16.4). But notice that a Z_ℓ term does appear in the theory that includes spinons in (16.8); it is made $\mathbb{Z}_2$ gauge invariant by the presence of spinons that carry $\mathbb{Z}_2$ gauge charges. In a similar manner, to study possible $\mathbb{Z}_2$ spin-liquid states, we explore making (16.43) gauge invariant by introducing zero-energy matter fields, which carry a $\mathbb{Z}_2$ gauge charge. We introduce an "i-lattice" of sites $i, j, \ldots$, so that the center of the (i, j) link on the i-lattice coincides with the ℓ sites in (16.43); the latter sites belong to the "ℓ-lattice." We want to introduce the i-lattice in a manner that does not break any symmetries of the ℓ-lattice as illustrated in two cases in Figs. 16.13 and 16.14. Note that for a given ℓ-lattice, it is not always possible to define an i-lattice that does not break some symmetries of the ℓ-lattice, for example, square and honeycomb ℓ-lattices do not have a corresponding i-lattice that preserves all symmetries of the ℓ-lattice. Also, the ℓ-lattice is sometimes called the medial lattice of the i-lattice.

Having found suitable ℓ- and i-lattices, we place the Rydberg atoms on the ℓ-lattice, and introduce a new set of qubits on the i-lattice. The i-lattice qubits are acted on by

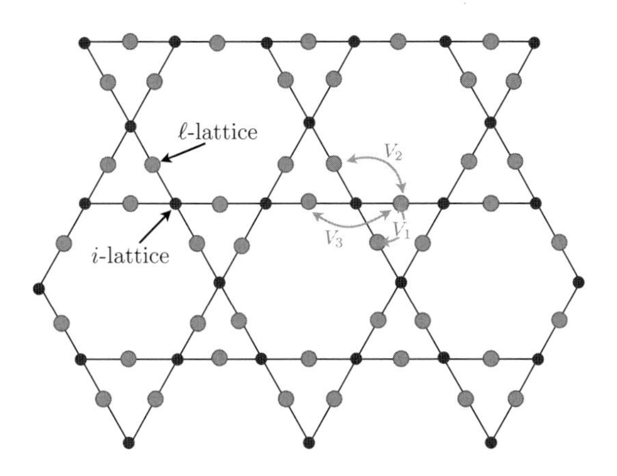

Figure 16.14 The ℓ-lattice is a ruby lattice, and the i-lattice is a kagome lattice. Rydberg atoms are on the ℓ-lattice. The model with $V_1 = V_2 = V_3 = \infty$, and the remaining $V(\boldsymbol{r}) = 0$, displays a $\mathbb{Z}_2$ spin liquid [288].

Pauli matrices $\tau_i^{x,y,z}$, and these transform under the $\mathbb{Z}_2$ lattice gauge transformations in (16.4) by

$$\begin{aligned} \tau_i^z &\to \tau_i^z \rho_i, \\ \tau_i^x &\to \tau_i^x. \end{aligned} \tag{16.44}$$

Then, an explicitly $\mathbb{Z}_2$ gauge-invariant form of the FSS Hamiltonian is

$$\mathcal{H}_{FSS} = \frac{\Omega}{2} \sum_{\langle ij \rangle} \tau_i^z Z_{ij} \tau_j^z + \frac{\Delta}{2} \sum_{\ell} X_\ell + \frac{1}{2} \sum_{\ell \neq \ell'} \frac{V(\boldsymbol{r}_\ell - \boldsymbol{r}_{\ell'})}{4} (1 - X_\ell)(1 - X_{\ell'}), \tag{16.45}$$

where $Z_{ij} \equiv Z_\ell$ on the ℓ-lattice site between the i and j sites on the i-lattice. This form of $\mathcal{H}_{FSS}$ is gauge invariant under (16.4) and (16.44). The exact equivalence of (16.45) to (16.43), and hence to (16.40), can is established by fixing the gauge, by choosing $\rho_i = \tau_i^z$.

With the introduction of the τ^z Ising matter fields, we introduce an infinite number of gauge charges G_i that commute with $\mathcal{H}_{FSS}$. These generalize (16.9) to

$$G_i = \tau_i^x \prod_{\ell \text{ ends on } i} X_\ell, \tag{16.46}$$

and we impose the Gauss law constraint:

$$G_i = 1, \tag{16.47}$$

which is the analog of $G_i^{vs} = 1$ in (16.12). With this constraint, it is easy to see that the Ising matter fields do not introduce any additional states, and the Hilbert space of (16.45) is identical to that of the original Rydberg model (16.40). Given this identification between (16.45) and (16.40), we can view the i-lattice $\tau^{x,y,z}$ spins as "ancilla qubits" introduced only to enable the description of entangled states on the physical qubits of the ℓ-lattice Rydberg atoms.

Unlike the conventional Hamiltonian of a $\mathbb{Z}_2$ gauge theory with Ising matter, (16.45) does not contain an on-site term such as $\sum_i \tau_i^x$ that can gap the matter field. In the absence of such a term, one might worry that τ^z will condense, and this will lead to a Higgs/confining phase of the $\mathbb{Z}_2$ gauge theory, as in the spinon condensation transitions discussed in Section 15.4.1. Now the task is to find choices for the ℓ- and i-lattices, and for $V(\boldsymbol{r})$, so that the $\mathbb{Z}_2$ gauge theory (16.45) has a $\mathbb{Z}_2$ spin-liquid phase, that is the $\mathbb{Z}_2$ gauge charges are deconfined. We need the τ^x gauge-charged matter fields to be gapped, similar to the bosonic spinons in Section 15.2. This can be achieved here; from the first term in (16.45), we see that the motion of the Ising matter τ^z requires a Z operation, and by (16.42) this can add a B_ℓ boson, which leads to a large energy cost from the Rydberg interaction $V(\mathbf{r})$. So τ^x gauge-charge fluctuations are expensive, and this could help stabilize a deconfined phase of the $\mathbb{Z}_2$ gauge theory (16.45). Then we can eliminate the τ^z matter fields in an Ω expansion, and this will induce terms involving the gauge-invariant product of Z_ℓ around closed loops, similar to those in (16.6) and Fig. 16.1; these are the terms in (16.48) below. Such terms suppress fluctuations in $\mathbb{Z}_2$ flux, and so stabilize a deconfined phase.

A proposal for a possible $\mathbb{Z}_2$ spin liquid phase for Rydberg atoms on the kagome ℓ-lattice in Fig. 16.13 was made in Refs. [245, 246, 320], and supported by some numerical evidence. Reference [288] proposed the ruby ℓ-lattice configuration shown in Fig. 16.14, and provided convincing numerical evidence for a $\mathbb{Z}_2$ spin-liquid phase for a specific $V(\boldsymbol{r})$; they included a perfect blockade on the first, second, and third neighbor sites, with $V_1 = V_2 = V_3 = \infty$ in Fig. 16.14, and the remaining $V(\boldsymbol{r}) = 0$. This blockaded model is a version of the "PXP model" [160, 241, 283, 322] on the ruby lattice (although, with our choice of X, Y, Z axes in (16.42), the terminology "PZP model" would be more appropriate here). With this blockade, at large positive Δ, we have to maximize the number of ℓ sites with $X_\ell = -1$, that is, atoms in the $|r\rangle$ state. It is not difficult to see that the set of all such configurations are precisely the dimer coverings of the kagome lattice, and the reasoning is similar to that in Section 16.4.2. Consequently, quantum fluctuations of the dimers in an Ω expansion can be mapped onto a quantum dimer model on the kagome lattice [288], similar to that in (16.20). The dimer flipping terms in this dimer model are the $\mathbb{Z}_2$ flux terms in (16.7) on the ℓ-lattice

$$\mathcal{H}_{loop} = -\sum_{loops} K_{loop} \prod_{\ell_1,\ell_2,\ell_3,\ldots,\in loop} Z_{\ell_1} Z_{\ell_2} Z_{\ell_3}, \ldots, \tag{16.48}$$

which commute with the contraints in (16.46); recall that it was in the large-K limit we found the deconfined phase. Similar comments apply to the model of Fig. 16.13, for $V_1 = V_2 = \infty$ [245].

Recent experiments [254] on Rydberg atoms in the configuration in Fig. 16.14 have yielded evidence for topological correlations.

Problems

16.1 We will work out the excitation spectrum of the $\mathbb{Z}_2$ gauge theory in both the confining and deconfined phases. We examine the Hamiltonian

$$H = -K\sum_{\square}\prod_{\ell\in\square} Z_\ell - g\sum_{\ell} X_\ell. \tag{16.49}$$

(a) First, consider the easy case of the deconfined phase $K \gg g$. Starting with the state with all spins up, compute the dispersion of a π-flux particle to order g.

(b) Now, consider the confined case $g \gg K$. Compute the ground-state energy to order K^2.

(c) We now consider the dispersion of the lowest quasiparticle exctiation for $g \gg K$. The lowest mobile excitation is the analog of a "glueball" in quantum chromodynamics; it is a set of four spins which have flipped to $|\leftarrow\rangle$ from $|\rightarrow\rangle$ around a plaquette (verify that any smaller set of flipped spins is not mobile). We will use an effective Hamiltonian method to compute the dispersion of this excitation.

We describe this method for a general Hamiltonian. Consider a particular Hamiltonian, H, whose spectrum consists of two sets of states, labeled a and b, which are separated by a large energy gap. Let us label these states $|a,i\rangle$ and $|b,j\rangle$. The Hamiltonian has non-vanishing matrix elements between these two sets of states. We would like to derive an effective Hamiltonian that acts only on the a states, so that we no longer have to consider the b states. This is done by performing a canonical transformation, so that the transformed Hamiltonian H' has vanishing matrix elements between the a and b states. The Hamiltonian H' acts only within the a subspace, and its matrix elements can be computed in an expansion in the inverse energy gap. To leading order, the matrix elements of H' are:

$$\begin{aligned}\langle a,i|H'|a,i'\rangle &= \langle a,i|H|a,i'\rangle \\ &+\sum_j \frac{\langle a,i|H|b,j\rangle\langle b,j|H|a,i'\rangle}{2}\left(\frac{1}{E_{a,i}-E_{b,j}}+\frac{1}{E_{a,i'}-E_{b,j}}\right).\end{aligned} \tag{16.50}$$

In our case, a labels all single quasiparticle states, labeled by the location i of the quasiparticle. And b labels other states connected to the single quasiparticle states by the Hamiltonian. Note that b includes states that have an energy both above and below that of a states.

Using the method just described, obtain the dispersion of the glueball to order K^2/g. Note that there are two sets of intermediate states: one in which the number of flipped spins increases, and the other in which the number of flipped spins decreases. And there is an exact cancellation between these states for processes which involve an apparent non-local hopping of the glueball.

16.2 Start from the Hamiltonian (16.7), along with the constraint (16.13), and express its partition function as a discrete-time path integral. This starts from the Trotter product representation $\exp\left(-\beta \mathcal{H}_{\mathbb{Z}_2}\right) = \prod_{i=1}^{N} \exp\left(-\Delta\tau \mathcal{H}_{\mathbb{Z}_2}\right)$, with $N\Delta\tau = \beta$, and $N \to 0$, $\Delta\tau \to 0$. Insert complete sets of Z_ℓ eigenstates between each exponential, and impose the constraint (16.13) by a discrete Lagrange multiplier $Z_{i,i+\tau}$, which will become the time component of the gauge field. In this manner, establish the equivalence of (16.7) and (16.13) to the cubic-lattice gauge theory of Chapter 14 in (14.32) and (14.36) with $J_2 = 0$. This establishes the S-dependent Berry phase factor in (14.36). A similar analysis has been performed in Appendix A of the arXiv version of Ref. [90]; the reader should set the fermion number $n_r = 1$ rigidly in their analysis.

16.3 Compute the wavefunctions of the four ground states of the $\mathbb{Z}_2$ gauge theory on a torus to order g^2. These states will be exactly degenerate at this order, and can be chosen to be eigenstates of V_x and V_y. Use these wavefunctions to compute the corrections to the W_x and W_y operators in Fig. 16.4 at order g^2. The operators $W_{x,y}$ should be chosen to satisfy the anti-commutation relations in (16.16) in the ground-state subspace, that is, they should be the operators that change the sign of the eigenvalues of $V_{x,y}$.

17 Chern–Simons Gauge Theories

The $U(1)$ Chern–Simons gauge theory is introduced, along with an exact description of its properties. The topological degeneracy on a torus, the quasiparticle statistics, and the Hall conductance are obtained. A connection is also established between the $\mathbb{Z}_2$ spin liquid and a $U(1) \times U(1)$ Chern–Simons theory. The bulk–boundary correspondence between the bulk Chern–Simons theory and the boundary chiral Luttinger liquid is established.

Chapters 13, 15, and 16 describe a gapped "topological" state: the $\mathbb{Z}_2$ spin liquid. Such states have quasiparticle excitations that are "anyons", which means they pick up non-trivial phase factors upon encircling each other, even while they are separated by large distances. It turns out that the $\mathbb{Z}_2$ spin liquid, and other topological states we will consider, can be described in a common formalism: that of the Chern–Simons gauge theory.

We consider in this chapter the abelian Chern–Simons theory with the imaginary-time action

$$\mathcal{S}_{CS} = \int d^3x \left[\frac{i}{4\pi} \varepsilon_{\mu\nu\lambda} a^I_\mu K_{IJ} \partial_\nu a^J_\lambda \right], \tag{17.1}$$

where I,J are indices extending over N values $1,\ldots,N$, a^I_μ are N $U(1)$ gauge fields, and the K_{IJ} are integers in a symmetric $N \times N$ matrix.

Different choices of K lead to different topological phases. We argue that the $\mathbb{Z}_2$ spin liquid has $N = 2$ and the symmetric K matrix

$$K = \begin{pmatrix} 0 & 2 \\ 2 & 0 \end{pmatrix}. \tag{17.2}$$

Much of our analysis is carried out for the simplest case with $N = 1$. We will see later in Chapter 24 that this describes the Laughlin fractional quantum Hall states, which have

$$K = m, \tag{17.3}$$

with m an odd (even) integer for fermions (bosons).

In addition to the structure described by $\mathcal{S}_{CS}$, specification of a particular state of condensed matter often requires the quasiparticle quantum numbers, and the transformations of the gauge fields and the quasiparticles under various global symmetries of the Hamiltonian. We consider such issues in Section 17.3.

17.1 Chern–Simons Theory on a Torus

We now discuss the quantization of (17.1) on a spatial torus of size $L_x \times L_y$. One important property of (17.1) is that it is exactly invariant under the gauge transformations $a^I_\mu \to a^I_\mu - \partial_\mu \zeta^I$, where ζ^I generates the gauge transformation; there is no surface term upon integration by parts on a torus, and the variation in the action vanishes exactly.

For simplicity, we consider first the case $N = 1$, with $K = m$; the methods below can be generalized to other values of N and K.

We work in the gauge $a_\tau = 0$, where τ is imaginary time. However, we cannot just set $a_\tau = 0$ in (17.1). We have to examine the equation of motion obtained by varying a_τ, which for the pure Chern–Simons theory is simply the zero-flux condition

$$\varepsilon_{ij}\partial_i a_j = 0. \tag{17.4}$$

But this does not imply the theory is pure gauge and trivial. We still have to consider fluxes around the cycles of the torus. So, up to a gauge transformation, we can choose the solutions of (17.4) as constants we parameterize as

$$a_x = \frac{\theta_x}{L_x} \quad , \quad a_y = \frac{\theta_y}{L_y}, \tag{17.5}$$

in terms of new variables θ_x and θ_y. Now let us consider the influence of a "large" gauge transformation on (17.5), generated by

$$\zeta = \frac{2\pi\ell x}{L_x}, \tag{17.6}$$

where ℓ is an integer. Such a gauge transformation is permitted because $e^{i\zeta}$ is single-valued on the torus, and it is always $e^{i\zeta}$ that appears as a gauge-transformation factor on any underlying particles. Under the action of (17.6) we have

$$\theta_x \to \theta_x - 2\pi\ell. \tag{17.7}$$

So only the value of θ_x modulo 2π can be treated as a gauge-invariant quantity, and θ_x is an "angular" variable. A similar argument applies to θ_y. We therefore introduce the Wilson-loop operators

$$W_x \equiv e^{i\theta_x} \quad , \quad W_y \equiv e^{i\theta_y}. \tag{17.8}$$

These are the gauge-invariant observables that characterize Chern–Simons theory on a torus.

Inserting (17.5) into (17.1), we find that the dynamics of $\theta_{x,y}$ is described by the simple action

$$\mathcal{S}_\theta = \frac{im}{2\pi}\int d\tau\, \theta_y \frac{d\theta_x}{d\tau}. \tag{17.9}$$

This is a purely kinematic action, and it shows that $(m/(2\pi))\theta_y$ is the canonically conjugate momentum to θ_x. There is no Hamiltonian, and so the energy of all states is zero. Upon promoting $\theta_{x,y}$ to operators, this action implies the commutation relation

$$[\hat{\theta}_x, \hat{\theta}_y] = \frac{2\pi i}{m}. \tag{17.10}$$

In terms of the gauge-invariant Wilson loop operators, this commutation relation is equivalent to

$$\hat{W}_x \hat{W}_y = e^{-2\pi i/m} \hat{W}_y \hat{W}_x. \tag{17.11}$$

This is the fundamental operator relation that controls the quantum Chern–Simons theory on a torus.

For the simplest non-trivial case of $m = 2$, we see that $\hat{W}_x$ and $\hat{W}_y$ anti-commute. So they must act on a Hilbert space that is at-least two-fold degenerate, because the smallest matrices that anti-commute are the Pauli matrices; we can choose $\hat{W}_x = \sigma^x$ and $\hat{W}_y = \sigma^z$. So the $U(1)$ Chern–Simons theory on the torus at level $m = 2$ has a two-dimensional Hilbert space at zero energy.

It is not difficult to generalize the above argument to the general integer m. As $(\hat{W}_y)^m$ commutes with all other Wilson loop operators, we can demand that it equal the unit matrix. Then, the eigenvalues of $\hat{W}_y$ can only be $e^{2\pi i\ell/m}$, with $\ell = 0, 1, \ldots, m-1$. So we introduce the m states $|\ell\rangle$ obeying

$$\hat{W}_y |\ell\rangle = e^{2\pi i\ell/m} |\ell\rangle. \tag{17.12}$$

The relationship (17.11) can be satisfied by demanding that $\hat{W}_x$ is a cyclic "raising" operator on these states:

$$\hat{W}_x |\ell\rangle = |(\ell+1)(\text{mod } m)\rangle. \tag{17.13}$$

So, the $U(1)$ Chern–Simons theory on the torus at level m has an m-fold ground-state degeneracy.

17.1.1 $\mathbb{Z}_2$ Spin Liquid

It is easy to extend this analysis to the $N = 2$ case with the K matrix given by (17.2). We introduce the operators

$$W_i = \exp\left(i \int_{C_i} a^1_\mu dx_\mu\right) \quad , \quad V_i = \exp\left(i \int_{C_i} a^2_\mu dx_\mu\right), \tag{17.14}$$

where $C_{x,y}$, with $i = x, y$, are contours that encircle the contours of the torus. These correspond to the identically named operators in Section 16.3. Then, by a parallel analysis, it is easy to see that these operators obey the relations

$$W_x V_y = -V_y W_x \quad , \quad W_y V_x = -V_x W_y, \tag{17.15}$$

while remaining pairs commute. These were precisely the relations found in the $\mathbb{Z}_2$ gauge theory in (16.16), and they imply a four-fold degeneracy on the torus.

More generally, it can be shown that the torus degeneracy is $|\det K|$.

17.1.2 Path-Integral Quantization

Returning to the $N = 1$ case, it is instructive to obtain the above results by regularizing the action (17.9) by adding higher-derivative terms, so that the Hamiltonian does not vanish, and all states are not exactly at zero energy. By adding a bare Maxwell term to the Chern–Simons theory, we can extend (17.9) to

$$\mathcal{S}_\theta = \int d\tau \left[\frac{\mathcal{M}}{2}\left(\frac{d\theta_x}{d\tau}\right)^2 + \frac{\mathcal{M}}{2}\left(\frac{d\theta_y}{d\tau}\right)^2 + i\mathcal{A}_x \frac{d\theta_x}{d\tau} + i\mathcal{A}_y \frac{d\theta_y}{d\tau} \right], \tag{17.16}$$

with

$$(\mathcal{A}_x, \mathcal{A}_y) = (m\theta_y/(2\pi), 0). \tag{17.17}$$

But this is precisely the (imaginary-time) Lagrangian of a fictitious particle with coordinates (θ_x, θ_y) and mass $\mathcal{M}$ moving in the presence of "magnetic field" specified by a vector potential $(\mathcal{A}_x, \mathcal{A}_y)$. We are interested in the spectrum in the limit $\mathcal{M} \to 0$, when (17.16) reduces to (17.9). The strength of the magnetic field is $\mathcal{B} = \partial_{\theta_x}\mathcal{A}_y - \partial_{\theta_y}\mathcal{A}_x = -m/(2\pi)$. We can now introduce a wavefunction $\psi(\theta_x, \theta_y)$ obeying the Schrödinger equation

$$\mathcal{H}\,\psi(\theta_x, \theta_y) = \mathcal{E}\,\psi(\theta_x, \theta_y), \tag{17.18}$$

where the Hamiltonian is

$$\mathcal{H} = \frac{1}{2\mathcal{M}}\left(\frac{1}{i}\frac{\partial}{\partial\theta_x} - \mathcal{A}_x\right)^2 + \frac{1}{2\mathcal{M}}\left(\frac{1}{i}\frac{\partial}{\partial\theta_y} - \mathcal{A}_y\right)^2. \tag{17.19}$$

A subtle feature in the solution of this familiar Hamiltonian is the nature of the periodic boundary conditions on θ_x and θ_y. This fictitious particle moves on a torus of size $(2\pi) \times (2\pi)$, not to be confused by the torus of size $L_x \times L_y$ for the original Chern–Simons theory. The total "magnetic" flux is therefore $4\pi^2\mathcal{B}$, and the total number of "magnetic"-flux quanta is $4\pi^2|\mathcal{B}|/(2\pi) = m$. So, from the Landau-level analysis to be carried out in Chapter 19, we expect that the eigenstates of $\mathcal{H}$ are m-fold degenerate, just as we concluded from the arguments above using the Wilson loop operators. In computing the eigenstates of $\mathcal{H}$, we run into the difficulty that the vector potential in (17.17) is not explicitly a periodic function of θ_y, and instead obeys

$$\begin{aligned} \mathcal{A}_x(\theta_x, \theta_y + 2\pi) &= \mathcal{A}_x(\theta_x, \theta_y) + m, \\ \mathcal{A}_y(\theta_x, \theta_y + 2\pi) &= \mathcal{A}_y(\theta_x, \theta_y). \end{aligned} \tag{17.20}$$

But we can make the vector potential periodic by using the gauge transformation

$$\mathcal{A}_i \to \mathcal{A}_i - \partial_i \zeta, \tag{17.21}$$

with $\zeta = m\theta_x$. So we need to solve (17.18) and (17.19) subject to the boundary conditions

$$\psi(\theta_x + 2\pi, \theta_y) = \psi(\theta_x, \theta_y), \tag{17.22}$$

$$\psi(\theta_x, \theta_y + 2\pi) = e^{im\theta_x}\psi(\theta_x, \theta_y). \tag{17.23}$$

The Landau-level eigenstates of (17.19) in an infinite plane are obtained in Chapter 19. We focus only on the lowest Landau-level states, as these are the only ones that will survive the $\mathcal{M} \to 0$ limit. Imposing only the boundary condition (17.22), we obtain the unnormalized eigenstates

$$\phi_\ell(\theta_x, \theta_y) = \exp\left(i\ell\theta_x - \frac{m}{4\pi}\left(\theta_y - \frac{2\pi\ell}{m}\right)^2\right), \tag{17.24}$$

where ℓ is any integer. Notice that these states obey

$$\phi_\ell(\theta_x, \theta_y + 2\pi) = e^{im\theta_x}\phi_{\ell-m}(\theta_x, \theta_y). \tag{17.25}$$

Now it is evident that we can also satisfy the second boundary condition (17.23) with m different orthogonal wavefunctions $\psi_\ell(\theta_x, \theta_y)$, with $\ell = 0, 1, \ldots, m-1$, which are given by

$$\psi_\ell(\theta_x, \theta_y) = \sum_{p=-\infty}^{\infty} \phi_{\ell+mp}(\theta_x, \theta_y). \tag{17.26}$$

These are related to Jacobi theta functions. We have again reached the conclusion that the $U(1)$ Chern–Simons theory at level m has an m-fold degenerate ground state on the torus.

17.2 Quasiparticles and Their Statistics

Let us now introduce a set of gapped quasiparticle excitations to the Chern–Simons theory, in infinite two-dimensional space. While the Chern–Simons theory is "trivial" on its own, without any dynamic excitations, it does induce non-trivial statistical interactions between the quasiparticles. In other words, the quasiparticles acquire non-trivial Berry phases along their trajectories due to the presence of the other particles. A quasiparticle excitation is labeled by its ℓ-vector, ℓ_I, which is a set of N integers representing its charges under the N gauge fields. The Aharonov–Bohm–Berry phase acquired by this quasiparticle along its trajectory C is

$$\exp\left(i\int_C dx_\mu \ell_I a^I_\mu\right). \tag{17.27}$$

In terms of the action (17.1), this extends the action to

$$\mathcal{S}_{CS} = \int d^3x \left[\frac{i}{4\pi}\varepsilon_{\mu\nu\lambda} a^I_\mu K_{IJ}\,\partial_\nu a^J_\lambda + i j_\mu\,\ell_I a^I_\mu\right], \tag{17.28}$$

where j_μ is the current of the quasiparticle.

Now consider a quasiparticle ℓ_I stationary at the origin of space. The saddle-point equations of (17.28) imply that this particle produces a flux tube of a^I obeying

$$K_{IJ}\boldsymbol{\nabla} \times \boldsymbol{a}^J = 2\pi\ell_I\,\delta^2(\vec{r})\hat{z}. \tag{17.29}$$

If we take a second identical quasiparticle, also with charges ℓ_I, and have it encircle the origin halfway, it will acquire a phase factor θ_ℓ, associated with the evaluation of (17.27) for the solution of (17.29), given by

$$\theta_\ell = \pi \ell^T K^{-1} \ell . \tag{17.30}$$

This value of θ_ℓ determines the *self statistics* of this type of quasiparticle/anyon, as a half circle can be turned into an exchange without any additional Berry phases. Note $\theta = 0 \bmod (2\pi)$ corresponds to bosons, and $\theta = \pi \bmod (2\pi)$ corresponds to fermions.

Next, we can take a quasiparticle with a charge vector ℓ'_I all the way around the ℓ_I quasiparticle and determine the angle controlling their *mutual statistics*:

$$\theta_{\ell,\ell'} = 2\pi \ell^T K^{-1} \ell' . \tag{17.31}$$

17.2.1 $\mathbb{Z}_2$ spin liquid

For the $\mathbb{Z}_2$ spin liquid, with the K matrix (17.2), the ℓ vectors of the e, m, and ε particles are:

$$\ell^e = \begin{pmatrix} 1 \\ 0 \end{pmatrix} \quad , \quad \ell^m = \begin{pmatrix} 0 \\ 1 \end{pmatrix} \quad , \quad \ell^\varepsilon = \begin{pmatrix} 1 \\ 1 \end{pmatrix} . \tag{17.32}$$

It is now instructive to verify that the self and mutual statistics of these particles are exactly those obtained in the $\mathbb{Z}_2$ gauge theory.

17.3 Coupling to an External Gauge Field

In many physically important cases, topological phases have a globally conserved $U(1)$ quantum number, such as the electrical charge, boson number, or spin component. Associated with this globally conserved $U(1)$, we can create a fixed background $U(1)$ gauge field A_μ, and describe its coupling to the Chern–Simons theory. It is conventional to write this coupling by a "mutual" Chern–Simons term between the A_μ and a^I_μ, so that (17.28) is extended to

$$\mathcal{S}_{CS} = \int d^3x \left[\frac{i}{4\pi} \varepsilon_{\mu\nu\lambda} a^I_\mu K_{IJ} \partial_\nu a^J_\lambda + \frac{i}{2\pi} t_I A_\mu \varepsilon_{\mu\nu\lambda} \partial_\nu a^I_\lambda + i j_\mu \ell_I a^I_\mu \right] , \tag{17.33}$$

where the t_I are a set of integers that determine the $U(1)$ charges of the quasiparticle. Again using (17.29), we see that the $U(1)$ charge of the quasiparticle ℓ is

$$Q = \ell^T K^{-1} t . \tag{17.34}$$

In the absence of quasiparticles, it is possible to "integrate out" the a^I_μ from (17.33), and obtain an effective action for A_μ (e.g., by adding Maxwell terms for a^I_μ to regularize the integral)

$$\mathcal{S}_A = (t^T K^{-1} t) \int d^3x \left[\frac{i}{4\pi} \varepsilon_{\mu\nu\lambda} A_\mu \partial_\nu A_\lambda \right]. \tag{17.35}$$

For the case where the $U(1)$ charge is the electrical charge, the value of $t^T K^{-1} t$ is the Hall conductance, in units of e^2/h.

17.3.1 $\mathbb{Z}_2$ Spin Liquid

For the $\mathbb{Z}_2$ spin liquid realized as a phase of a boson (spin) model with a conserved boson number Q, we have the t_I vector

$$t = \begin{pmatrix} 0 \\ 1 \end{pmatrix}. \tag{17.36}$$

This associates an A_μ charge with the a^2_μ flux that is carried by each spinon. It is now easy to show that the e and ε particles (the spinons) have boson number $Q = 1/2$, while the m particle has $Q = 0$. Also, the Hall conductance vanishes, as expected from the time-reversal symmetry of the $\mathbb{Z}_2$ spin liquid.

17.4 Physics at the Edge

We return to the Chern–Simons theory in (17.1), and describe its quantization in the geometry of Fig. 19.1. To begin with, we will just consider the $U(1)$ theory with $N = 1$, and with $K = m$. The first important property of $\mathcal{S}_{CS}$ is that it is not invariant under a gauge transformation $a_\mu \to a_\mu - \partial_\mu \zeta$ in the presence of an edge. Instead, we obtain a surface term

$$\mathcal{S}_{CS} \to \mathcal{S}_{CS} - \frac{im}{2\pi} \int dx d\tau \, \zeta \left(\partial_\tau a_x - \partial_x a_\tau \right) \Big|_{y=0}. \tag{17.37}$$

The proper way to understand this lack of gauge invariance is to regard the Chern–Simons theory as an effective theory for microscopic degrees of freedom that are gauge invariant. So, while the Chern–Simons theory properly describes the low-energy physics in the bulk, it evidently fails to do so on the boundary. There must be additional degrees of freedom on the boundary, which restore gauge invariance. Ultimately, we have to return to a suitable microscopic model to directly determine the degrees of freedom on the edge. We will do so in Chapters 18 and 19, when we consider specific situations that give rise to an effective Chern–Simons gauge theories.

For now, we employ a somewhat ad hoc procedure, which defines a gauge-invariant theory in the bulk, and introduces new degrees of freedom on the boundary designed to restore gauge invariance. We work directly with the Chern–Simons action in the geometry of Fig. 19.1. The variation of the action is [67]:

$$\delta \mathcal{S}_{CS} = \frac{im}{2\pi} \int d^3x \left[\delta a_\mu \left(\varepsilon_{\mu\nu\lambda} \partial_\nu a_\lambda \right) \right] + \frac{im}{4\pi} \int dx d\tau \left(a_x \delta a_\tau - a_\tau \delta a_x \right) \Big|_{y=0}. \tag{17.38}$$

To make the variation vanish, we require the usual zero-flux condition, $\varepsilon_{\mu\nu\lambda}\partial_\nu a_\lambda = 0$, in the bulk. But, on the boundary, we must also impose a secondary condition to define the theory under the stationary action principle; a convenient choice is to set $a_\tau = 0$ (and hence also $\delta a_\tau = 0$) at $y = 0$. We now find that the fluctuations of the gauge field near the boundary are no longer pure gauge, in contrast to the situation in the bulk.

Let us quantize the system by choosing the gauge $a_\tau = 0$ in the bulk. Then, (17.4) continues to hold for the spatial components of the gauge field, and so we can solve this constraint by the choice

$$a_i = \partial_i \varphi \tag{17.39}$$

in terms of a scalar field φ. As in (17.7), we can use large gauge transformations to argue that φ should be physically equivalent to $\varphi + 2\pi$, and so φ takes values on a unit circle. Inserting (17.39) into $\mathcal{S}_{CS}$, and integrating over y, we obtain the edge action

$$\mathcal{S}_e = -\frac{im}{4\pi}\int dx d\tau\, \partial_\tau \varphi \partial_x \varphi, \tag{17.40}$$

where the fields are now implicitly evaluated at $y = 0$. Now we notice that at $m = 1$ this is precisely the kinematic term in the bosonic representation of a free chiral fermion obtained in the last line of (12.34). For general m, following the arguments in Chapter 12 on Luttinger liquids, we can write (17.40) as a commutation relation

$$[\varphi(x_1), \varphi(x_2)] = -i\frac{\pi}{m}\operatorname{sgn}(x_1 - x_2). \tag{17.41}$$

In addition to the kinematic term in (17.40), non-zero energetic terms are also permitted at the boundary, provided they are consistent with the residual shift symmetry $\varphi \to \varphi$+constant; these would arise from higher-order terms in the bulk, like the Maxwell terms considered earlier in (17.16). In an operator language, including the lowest-order spatial gradient, we obtain the Hamiltonian

$$\mathcal{H}_\varphi = \frac{mv}{4\pi}\int dx\,(\partial_x \varphi)^2, \tag{17.42}$$

where v is a coupling constant with units of velocity, and the prefactor of m has been chosen so that v is the actual velocity of φ excitations in (17.43). This interpretation of v becomes clearer in the action for the path integral, which is the final form of the edge theory [302]:

$$\mathcal{S}_e = \frac{m}{4\pi}\int dx d\tau \left[-i\partial_\tau \varphi \partial_x \varphi + v(\partial_x \varphi)^2\right]. \tag{17.43}$$

This is a theory of left-moving chiral bosons at velocity v, and is also known as the $U(1)$ Kac–Moody theory at level m. At $m = 1$, we can conclude from our previous analysis of Luttinger liquids in Chapter 12 that (17.43) is precisely the bosonized version of the free chiral fermion theory.

At other values of m, $\mathcal{S}_e$ remains a Gaussian theory, and so it is possible to compute all correlators on the edge using the methods developed in Chapter 12. In particular, a useful result that can be obtained by such methods is

$$\langle \varphi(x,\tau)\varphi(0,0)\rangle = -\frac{1}{m}\ln(x - iv\tau) + \cdots. \tag{17.44}$$

We will use this result below.

Quantum Hall systems also have a conserved $U(1)$ charge in the bulk and, as discussed below, this is important for the stability of the chiral boson theory in (17.43) towards external perturbations on the edge. From (17.33), the external electromagnetic potential A_μ couples via the term

$$\mathcal{S}_{Aa} = \int d^3x \left[\frac{i}{2\pi} a_\mu \varepsilon_{\mu\nu\lambda} \partial_\nu A_\lambda \right]. \tag{17.45}$$

Using a non-zero electrostatic potential A_τ, which is independent of y, and integrating over y in (17.45) reduces $\mathcal{S}_{Aa}$ to $(i/(2\pi)) \int dx d\tau A_\tau \partial_x \varphi$, and so we may identify the charge density as

$$\rho(x) = \frac{1}{2\pi} \partial_x \varphi, \tag{17.46}$$

which is precisely the relation obtained in (12.37) in our discussion of Luttinger liquids.

We can also identify the fate of the quasiparticle operators on the boundary by considering the adiabatic transport of the quasiparticles via the bulk between two points, x_1 and x_2, on the boundary. The present $N = 1$ Chern–Simons theory is characterized by m species of quasiparticles with the $\ell = 1, 2, \ldots, m$. Such a process for a quasiparticle ℓ would be accompanied by the Berry phase

$$\exp\left(i\ell \int_{(x_1,0)}^{(x_2,0)} d\vec{x} \cdot \vec{a} \right) = e^{i\ell(\varphi(x_2) - \varphi(x_1))}, \tag{17.47}$$

where the integral on the left-hand side is along a path in the bulk of the sample, and the right-hand side follows from (17.39). So we find a bulk–boundary correpondence between quasiparticles in the bulk and operators

$$\psi_\ell = e^{i\ell\varphi} \tag{17.48}$$

in the gapless theory on the boundary. Using (17.41) and (17.46), we can verify the commutation relation

$$[\rho(x_1), \psi_\ell(x_2)] = \frac{\ell}{m} \delta(x_1 - x_2) \psi_\ell(x_2), \tag{17.49}$$

which confirms that the quasiparticle ψ_ℓ carries charge ℓ/m. When we realize this topological state as a fractional quantum Hall state of fermions or bosons (with m odd and even, respectively), the operator ψ_ℓ has unit charge, and corresponds to adding or removing the underlying fermion or boson. We also note the commutation relation (by generalizing the identities in (12.27))

$$\psi_\ell(x_1) \psi_{\ell'}(x_2) = \exp\left(i \frac{\pi \ell \ell'}{m} \right) \psi_{\ell'}(x_2) \psi_\ell(x_1). \tag{17.50}$$

This is the boundary manifestation of the self/mutual statistics of the corresponding quasiparticles in the bulk; compare (17.50) with (17.30) and (17.31). We can also compute the two-point correlator of the quasiparticle operators from (17.44):

$$\langle \psi_\ell(x_1) \psi_{\ell'}(x_2) \rangle \sim (x_1 - x_2)^{-\ell\ell'/m}. \tag{17.51}$$

17.4.1 $\mathbb{Z}_2$ Spin Liquids

First, we note that the generalization of the edge theory (17.43) to arbitrary N in the presence of an external field A_μ follows from (17.33)

$$\mathcal{S}_e = \frac{1}{4\pi}\int dx d\tau \left[-iK_{IJ}\partial_\tau \varphi^I \partial_x \varphi^J + v_{IJ}(\partial_x \varphi^I)(\partial_x \varphi^J) + \frac{i}{2\pi} t_I A_\mu \varepsilon_{\mu\nu}\partial_\nu \varphi^I \right], \quad (17.52)$$

where the K matrix is as in (17.1), and v_{IJ} is a positive–definite matrix determining a set of velocities.

For the $\mathbb{Z}_2$ spin liquid, with the K matrix in (17.2), we choose to label the boundary scalars by θ and ϕ (rather than φ^1 and φ^2), both defined modulo 2π:

$$a_i^1 = \partial_i \theta \quad , \quad a_i^2 = \partial_i \phi. \quad (17.53)$$

Then the boundary kinematic action is (dropping the coupling to the external field A_μ)

$$\mathcal{S}_e = -\frac{i}{\pi}\int dx d\tau \, \partial_x \theta \partial_\tau \phi, \quad (17.54)$$

which implies the commutation relation

$$[\phi(x_1), \theta(x_2)] = i\frac{\pi}{2}\mathrm{sgn}(x_1 - x_2). \quad (17.55)$$

Following the arguments above, and the ℓ values in (17.32), we can identify $e^{i\theta}$ as the e particle on the boundary, $e^{i\phi}$ as the m particle on the boundary, and $e^{i\theta + i\phi}$ as the ε particle on the boundary.

Remarkably, (17.54) is precisely the kinematics of the Luttinger liquid of spinless fermions in (12.25). This theory is non-chiral, and has equal numbers of left- and right-moving excitations. The simplest terms in the Hamiltonian for these edge excitations are

$$\mathcal{H}_e = \int dx \left[\frac{K_1}{2}(\partial_x \phi)^2 + \frac{K_2}{2}(\partial_x \theta)^2 \right], \quad (17.56)$$

and this also coincides with the Hamiltonian for the Luttinger liquid.

However, unlike the quantum Hall cases described above for $N = 1$, the gapless non-chiral edge states described by $\mathcal{H}_e$ for the K matrix (17.2) are generally not stable. For the $\mathbb{Z}_2$ spin liquid, both $e^{2i\phi}$ and $e^{2i\theta}$ are trivial bosonic excitations; this corresponds to the fact that, in the bulk, two visons or two spinons can fuse into trivial excitations. Consequently, the general edge Hamiltonian is [20]

$$\mathcal{H}_e = \int dx \left[\frac{K_1}{2}(\partial_x \phi)^2 + \frac{K_2}{2}(\partial_x \theta)^2 - \lambda_1 \cos(2\phi) - \lambda_2 \cos(2\theta) \right]. \quad (17.57)$$

We have the scaling dimensions $\dim[\lambda_1] = 2 - K$ and $\dim[\lambda_2] = 2 - 1/K$, where K is the Luttinger parameter (see Problem 17.2): so the scaling dimension of either λ_1 or λ_2 is positive for all values of K, one of the coupling constants is always relevant, and the edge spectrum is always gapped. However, it is possible to choose couplings so that the edge gap is much smaller than the bulk gap, and so the concept of a separate edge theory of the $\mathbb{Z}_2$ spin liquid makes sense even in the gapped case.

In the presence of additional global symmetries in the underlying lattice model, it is possible that the cosine terms conspire to leave at least one mode gapless. The symmetry constraints for the existence of $\mathbb{Z}_2$ spin liquids with gapless edge states have been explored in the literature [115, 148, 162].

Problems

17.1 Quantize the Chern–Simons theory for the $\mathbb{Z}_2$ spin liquid on a torus using the K matrix in (17.2) by the method of Section 17.1. In this manner, obtain (17.15).

17.2 Compute the scaling dimensions of $\lambda_{1,2}$ in (17.57) using the correlators in (25.43).

PART III

BAND TOPOLOGY

18 Berry Phases and Chern Numbers

The quantum-mechanical Berry phase is introduced, and applied to a single spin in a magnetic field, and to an electron moving in Bloch bands. The Chern numbers of Bloch bands are shown to lead to chiral edge states, and a quantized Hall conductivity.

Part III pauses our discussion of fractionalized phases and moves to a different set of topics that will eventually be important for the continuation of our discussion of fractionalized phases in Part IV. Our focus here is on free fermion models. It might seem that there is little more to say about such systems beyond that found in undergraduate text books. But there is a surprising richness in this subject, some of which was only uncovered in recent years.

From a broader perspective, we are concerned with the geometry of the space of eigenstates of a Hamiltonian, as the Hamiltonian is changed over a space of "coupling constants." This geometry is characterized by a Berry connection, and a Berry flux, which have properties similar to the vector potential and magnetic field in electromagnetism. The Berry phase is a line integral over the Berry connection. This was originally discovered as an important contribution to the evolution of the wavefunction under a Hamiltonian, which varies adiabatically in time. But it is important to keep in mind that the Berry connection is a geometric characterization that has applications in numerous physical contexts, quite apart from the adiabatic process.

As we will see in Section 18.3, in applications to crystalline materials, the Bloch crystal momentum provides the space of coupling constants over which the Berry phase can be computed. This leads to a classification of the band structure of crystals in distinct topological classes, some of which are described in Chapter 20.

18.1 Berry Phases

Consider a general Hamiltonian $\mathcal{H}(R_\mu)$, which depends on a finite set of parameters R_μ, $\mu = 1,2,\ldots$. Note that R_μ are coupling constants in the Hamiltonian, and are not connected to the quantum degrees of freedom. Assume that we can determine the eigenstates $|n,R_\mu\rangle$ of this Hamiltonian for each R_μ:

$$\mathcal{H}(R_\mu)\,|n,R_\mu\rangle = \varepsilon_n(R_\mu)\,|n,R_\mu\rangle\,, \qquad (18.1)$$

with energy eigenvalues $\varepsilon_n(R_\mu)$, $n = 1,2,3,\ldots$, which we assume are non-degenerate. The eigenstates in (18.1) are determined only up to an overall phase, and we can pick another set of eigenstates by a transformation

$$\left|n,R_\mu\right\rangle \to e^{i\phi_n(R_\mu)}\left|n,R_\mu\right\rangle, \tag{18.2}$$

with arbitrary $\phi_n(R_\mu)$.

We now define the "Berry connection" as a characterization of the geometry of the space of quantum states defined by varying the parameter R_μ; we have

$$A^n_\mu(R_\mu) = i\left\langle n,R_\mu\right|\frac{\partial}{\partial R_\mu}\left|n,R_\mu\right\rangle \tag{18.3}$$

at each point in the R_μ parameter space. We write (18.3) more compactly as

$$A^n_\mu = i\left\langle n|\partial_\mu n\right\rangle, \tag{18.4}$$

where the dependence on R_μ of A^n_μ and the eigenstate $|n\rangle$ is implicit, and the ∂_μ is understood to apply to the R_μ dependence. Under the transformation in (18.2), we see that A_μ transforms as

$$A^n_\mu \to A^n_\mu - \partial_\mu\phi_n, \tag{18.5}$$

that is, A^n_μ behaves like a vector potential of a $U(1)$ gauge symmetry defined by the transformation in (18.2).

The geometric characterization of the Berry connection can be made gauge invariant by computing the $U(1)$ gauge flux, which in this context is known as the "Berry curvature." First, let us note a few simple identities following from the normalization of the quantum states, and integration by parts:

$$\begin{aligned}\langle n|n\rangle &= 1,\\ \left\langle\partial_\mu n|n\right\rangle + \left\langle n|\partial_\mu n\right\rangle &= 0,\\ \left\langle\partial_\mu n|n\right\rangle &= \left\langle n|\partial_\mu n\right\rangle^*.\end{aligned} \tag{18.6}$$

These identities imply that A^n_μ is real, and can be written more symmetrically as

$$A^n_\mu = \frac{i}{2}\left(\left\langle n|\partial_\mu n\right\rangle - \left\langle\partial_\mu n|n\right\rangle\right). \tag{18.7}$$

Now we can compute the gauge-invariant Berry curvature of the state $|n\rangle$:

$$F^n_{\mu\nu} = \partial_\mu A^n_\nu - \partial_\nu A^n_\mu = i\left(\left\langle\partial_\mu n|\partial_\nu n\right\rangle - \left\langle\partial_\nu n|\partial_\mu n\right\rangle\right). \tag{18.8}$$

The above analysis shows that $F^n_{\mu\nu}$ is gauge invariant, but the expression in (18.8) involves gauge-dependent quantities at all the intermediate steps. It is useful, and possible, to write down $F^n_{\mu\nu}$ in a manner that is explicitly gauge invariant. First, we take the derivative of (18.1)

$$\partial_\mu\mathcal{H}|n\rangle + \mathcal{H}|\partial_\mu n\rangle = \partial_\mu\varepsilon_n|n\rangle + \varepsilon_n|\partial_\mu n\rangle, \tag{18.9}$$

and then take the overlap with $\langle n'|$, where $|n'\rangle$ is any other eigenstate of $\mathcal{H}$, not the same as $|n\rangle$:

$$\left\langle n'\right|\partial_\mu\mathcal{H}\left|n\right\rangle = \left(\varepsilon_n - \varepsilon_{n'}\right)\left\langle n'|\partial_\mu n\right\rangle. \tag{18.10}$$

Then, we insert the identity $\sum_{n'} |n'\rangle \langle n'| = 1$ into (18.8) and obtain

$$F^n_{\mu\nu} = i \sum_{n' \neq n} \frac{\langle n| \partial_\mu \mathcal{H} |n'\rangle \langle n'| \partial_\nu \mathcal{H} |n\rangle - \langle n| \partial_\nu \mathcal{H} |n'\rangle \langle n'| \partial_\mu \mathcal{H} |n\rangle}{(\varepsilon_n - \varepsilon_{n'})^2}, \tag{18.11}$$

where the $n' = n$ contribution to (18.8) vanishes. It is now easy to show that (18.11) is explicitly invariant under (18.2).

This geometric structure of the eigenstates of $\mathcal{H}(R_\mu)$ is often stated in terms of a "Berry phase" $\gamma_n(\mathcal{C})$ of the state $|n\rangle$ associated with a closed curve $\mathcal{C}$ in the R_μ parameter space. This Berry phase is given by

$$\begin{aligned} \gamma_n(\mathcal{C}) &= \oint_{\mathcal{C}} A^n_\mu(R_\mu) dR_\mu \\ &= \int_{\mathcal{S}} dS_{\mu\nu} F^n_{\mu\nu}. \end{aligned} \tag{18.12}$$

The second line uses the generalized Stokes theorem to convert the integral to one over a surface $\mathcal{S}$ enclosed by the closed curve $\mathcal{C}$.

The Berry phase $\gamma_n(\mathcal{C})$ is also called an "adiabatic phase," because it is an extra phase accumulated by the wavefunction under the time-dependent Schrödinger equation upon an adiabatic time-dependent variation of the parameters R_μ around the closed curve $\mathcal{C}$. More precisely, let us introduce a time-dependent Hamiltonian

$$\mathcal{H}(t) \equiv \mathcal{H}(R_\mu(t)), \tag{18.13}$$

where the parameters R_μ have now become time dependent, and we are interested in solving the equation

$$i\hbar \frac{d}{dt} |\Psi(t)\rangle = \mathcal{H}(t) |\Psi(t)\rangle \tag{18.14}$$

for the time evolution of the wavefunction $|\Psi(t)\rangle$. Consider the case in which the initial state $|\Psi(0)\rangle$ is the eigenstate $|n\rangle$, and we evolve smoothly and slowly along the curve $\mathcal{C}$ in parameter space so that $R_\mu(0) = R_\mu(\mathcal{T})$. Then, the adiabatic theorem states that as $\mathcal{T} \to \infty$

$$|\Psi(\mathcal{T})\rangle = \exp\left(i\gamma_n(\mathcal{C}) - \frac{i}{\hbar} \int_0^{\mathcal{T}} \varepsilon_n(R_\mu(t)) dt \right) |\Psi(0)\rangle, \tag{18.15}$$

where $\varepsilon_n(R_\mu)$ are the instantaneous energy eigenvalues in (18.1). We leave the proof of this theorem as an exercise.

18.2 Berry Phase of a Spin

An important case of a non-trivial phase is a single spin $\boldsymbol{S}$ in a Zeeman field $\boldsymbol{h}$ with Hamiltonian

$$\mathcal{H} = \boldsymbol{h} \cdot \boldsymbol{S}. \tag{18.16}$$

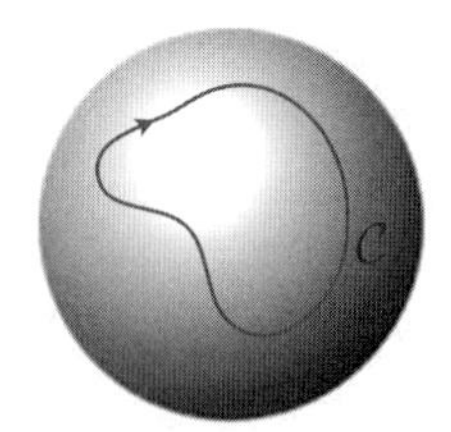

Figure 18.1 Berry phase of a spin moving along the contour C is S times the spherical area enclosed by C.

We consider Zeeman fields with fixed magnitude $|\boldsymbol{h}| = h$ but variable direction, so the parameter space R_μ is two-dimensional and has the topology of a sphere. We can use spherical coordinates with

$$\boldsymbol{h} = h(\sin\theta\cos\phi, \sin\theta\sin\phi, \cos\theta) \tag{18.17}$$

and so now $R_\mu \Rightarrow (\theta, \phi)$. For spin $S = 1/2$, the Hilbert space is also two-dimensional, with $|n\rangle \Rightarrow |\pm\rangle$, and we focus on the state

$$|-\rangle = \begin{pmatrix} e^{-i\phi}\sin(\theta/2) \\ -\cos(\theta/2) \end{pmatrix}. \tag{18.18}$$

This yields the Berry connection

$$A_\theta^- = \langle -|\frac{\partial}{\partial\theta}|-\rangle = 0, \quad A_\phi^- = \langle -|\frac{\partial}{\partial\phi}|-\rangle = \sin^2(\theta/2). \tag{18.19}$$

From this we obtain the Berry curvature defined in (18.8):

$$F_{\theta\phi}^- = S\sin\theta, \tag{18.20}$$

for $S = 1/2$. A similar computation can be performed for general S (see Appendix A.2), and yields (18.20). From (18.12), and the area element on the sphere, we obtain the Berry phase illustrated in Fig. 18.1

$$\gamma_-(C) = S \times (\text{Oriented area enclosed by } C \text{ on a unit sphere}). \tag{18.21}$$

The oriented area is undetermined modulo 4π, but the phase factor $\exp(i\gamma_-(C))$ is single-valued for an S integer or a half-integer.

It is also possible to express the spin Berry phase (18.20) in a way that makes the spin-rotation invariance manifest. One such expression is (A.38) of Appendix A, involving an emergent dimension and integral over a two-dimensional spacetime, as in a Wess–Zumino term. Alternatively, we can use the vector potential of a Dirac monopole at the center of the unit sphere, $\boldsymbol{A}(\boldsymbol{N})$, where $\boldsymbol{N}$ is a unit vector representing the orientation of the spin (as in (A.38)), and

$$\boldsymbol{\nabla}_N \times \boldsymbol{A} = \boldsymbol{N}. \tag{18.22}$$

Then we can write (18.21) as

$$\gamma_-(C) = S\oint_C \boldsymbol{A}(\boldsymbol{N}) \cdot d\boldsymbol{N}. \tag{18.23}$$

An interesting special case is when the field $\boldsymbol{h}$ is restricted to lie in the x–z plane. Then, the Hamiltonian in (18.16) is real, and so is invariant under a suitable action of time-reversal symmetry. Nevertheless, the Berry phase can be non-zero for a path $\mathcal{C}$ that maps out a great circle on the sphere; in this case $\gamma_-(\mathcal{C}) = -2\pi S$, and $\exp(i\gamma_-(\mathcal{C})) = -1$ for half-integer S.

18.3 Berry Curvature of Bloch Bands

We consider a general model of fermions c (assumed spinless for simplicity) moving in a potential $V(\boldsymbol{r})$ that is periodic under translations by a Bravais lattice of vectors $\boldsymbol{R}$

$$V(\boldsymbol{r}+\boldsymbol{R}) = V(\boldsymbol{r}). \tag{18.24}$$

We are interested in the Berry curvature associated with the eigenfunctions of the Schrödinger equation:

$$\left(-\frac{\hbar^2\nabla_r^2}{2m} + V(\boldsymbol{r})\right)\Psi(\boldsymbol{r}) = E\Psi(\boldsymbol{r}). \tag{18.25}$$

Let us first review some basic aspects of band theory. Bloch's theorem states that all eigenfunctions of (18.25) are of the form

$$\Psi_{n,\boldsymbol{k}}(\boldsymbol{r}) = e^{i\boldsymbol{k}\cdot\boldsymbol{r}}u_{n,\boldsymbol{k}}(\boldsymbol{r}), \tag{18.26}$$

where $\boldsymbol{k}$ is the "crystal momentum," n is the "band index," and the Bloch wavefunction $u_{n,\boldsymbol{k}}(\boldsymbol{r})$ has the same periodicity as the potential $V(\boldsymbol{r})$:

$$u_{n,\boldsymbol{k}}(\boldsymbol{r}+\boldsymbol{R}) = u_{n,\boldsymbol{k}}(\boldsymbol{r}). \tag{18.27}$$

The periodicity is usefully expressed in terms of the reciprocal lattice of wavevectors $\boldsymbol{G}$ that obey

$$e^{i\boldsymbol{G}\cdot\boldsymbol{R}} = 1. \tag{18.28}$$

Then, we can write both $V(\boldsymbol{r})$ and $u_{n,\boldsymbol{k}}(\boldsymbol{r})$ in the analog of a Fourier series expansion

$$\begin{aligned} V(\boldsymbol{r}) &= \sum_{\boldsymbol{G}} V_{\boldsymbol{G}} e^{i\boldsymbol{G}\cdot\boldsymbol{r}}, \\ u_{n,\boldsymbol{k}}(\boldsymbol{r}) &= \sum_{\boldsymbol{G}} u_{n,\boldsymbol{k},\boldsymbol{G}} e^{i\boldsymbol{G}\cdot\boldsymbol{r}}. \end{aligned} \tag{18.29}$$

Inserting (18.29) into (18.26), we can see that eigenstates at crystal momenta $\boldsymbol{k}$ and $\boldsymbol{k}+\boldsymbol{G}$ are related to each other by a rearrangement of the Fourier series expansion. Assuming the eigenstates at each n and $\boldsymbol{k}$ are non-degenerate, the wavefunctions $\Psi_{n,\boldsymbol{k}+\boldsymbol{G}}(\boldsymbol{r})$ and $\Psi_{n,\boldsymbol{k}}(\boldsymbol{r})$ must be the same functions of $\boldsymbol{r}$, apart from an $\boldsymbol{r}$-independent phase factor. So, there exists a function $\theta_n(\boldsymbol{k})$ with which the function

$$\overline{\Psi}_{n,\boldsymbol{k}}(\boldsymbol{r}) \equiv \exp\left(i\theta_n(\boldsymbol{k})\right)\Psi_{n,\boldsymbol{k}}(\boldsymbol{r}) \tag{18.30}$$

is a periodic function of $\boldsymbol{k}$ over the extended Brillouin zone:

$$\overline{\Psi}_{n,\boldsymbol{k}+\boldsymbol{G}}(\boldsymbol{r}) = \overline{\Psi}_{n,\boldsymbol{k}}(\boldsymbol{r}) \quad , \quad \varepsilon_n(\boldsymbol{k}+\boldsymbol{G}) = \varepsilon_n(\boldsymbol{k}). \tag{18.31}$$

The distinct values of the crystal momentum $\boldsymbol{k}$ extend over the first Brillouin zone of the Bravais lattice, which has the topology of a torus.

We now consider the Berry curvature of the Bloch wavefunction $u_{n,\boldsymbol{k}}(\boldsymbol{r})$ while treating $\boldsymbol{k}$ as the realization of parameter space R_μ. The parameter space is therefore a d-dimensional torus, in contrast to Section 18.2, where the parameter space was a sphere. From (18.25), the analog of (18.1) is now

$$\mathcal{H}(\boldsymbol{k})u_{n,\boldsymbol{k}}(\boldsymbol{r}) = \varepsilon_n(\boldsymbol{k})u_{n,\boldsymbol{k}}(\boldsymbol{r}), \tag{18.32}$$

where

$$\mathcal{H}(\boldsymbol{k}) = \frac{1}{2m}(-i\hbar\boldsymbol{\nabla}_{\boldsymbol{r}} + \hbar\boldsymbol{k})^2 + V(\boldsymbol{r}). \tag{18.33}$$

Then the Berry connection in (18.4) is a vector-valued function of crystal momentum in each band n:

$$\boldsymbol{A}_n(\boldsymbol{k}) = i\left\langle u_{n,\boldsymbol{k}}\right| \boldsymbol{\nabla}_{\boldsymbol{k}} \left|u_{n,\boldsymbol{k}}\right\rangle. \tag{18.34}$$

We write the Berry curvature in (18.8) in terms of a "magnetic" field $\boldsymbol{b}_n(\boldsymbol{k})$ in crystal momentum space:

$$\boldsymbol{b}_n(\boldsymbol{k}) = i\left\langle \boldsymbol{\nabla}_{\boldsymbol{k}} u_{n,\boldsymbol{k}}\right| \times \left|\boldsymbol{\nabla}_{\boldsymbol{k}} u_{n,\boldsymbol{k}}\right\rangle. \tag{18.35}$$

Finally, we are in a position to obtain the topological invariant characterizing the Bloch wavefunction $u_{n,\boldsymbol{k}}(\boldsymbol{r})$. This is obtained by computing the integral of the Berry curvature over the first Brillouin zone. We restrict ourselves to the spatial dimension $d = 2$, where the Berry curvature $b_n(\boldsymbol{k})$ is a scalar (obtained from (18.35) as a vector in the z direction):

$$\begin{aligned}\frac{1}{2\pi}\int_{\text{1st B.Z.}} dk_x dk_y\, b_n(\boldsymbol{k}) &= \oint d\boldsymbol{k}\cdot \boldsymbol{A}_n(\boldsymbol{k}) \\ &= i\oint d\boldsymbol{k}\cdot \left\langle u_{n,\boldsymbol{k}}\right| \boldsymbol{\nabla}_{\boldsymbol{k}} \left|u_{n,\boldsymbol{k}}\right\rangle. \end{aligned} \tag{18.36}$$

This is an integral on the boundary of the first Brillouin zone. We now express the integrand in terms of the periodic function in (18.30). Using the fact that for any point on the boundary $\boldsymbol{k}$ there is a corresponding point $\boldsymbol{k}+\boldsymbol{G}$, the contributions to the integral in (18.36) from $\overline{\Psi}_{n,\boldsymbol{k}}$ vanish, and the only non-vanishing contribution is that from the phase $\theta_n(\boldsymbol{k})$; so we have

$$C_n \equiv \frac{1}{2\pi}\int_{\text{1st B.Z.}} dk_x dk_y\, b_n(\boldsymbol{k}) = \frac{1}{2\pi}\oint d\boldsymbol{k}\cdot \boldsymbol{\nabla}_{\boldsymbol{k}}\theta_n(\boldsymbol{k}). \tag{18.37}$$

The phase $\theta_n(\boldsymbol{k})$ is only defined modulo 2π, and this establishes that (18.37) evaluates to an integer. We can identified this integer with the *Chern number* C_n. Every Bloch band in two dimensions is characterized by a topological invariant, the value of C_n. In systems with time-reversal symmetry, and without spin–orbit coupling, we can choose the Bloch wavefunctions to be real, and then $C_n = 0$. However, without time-reversal

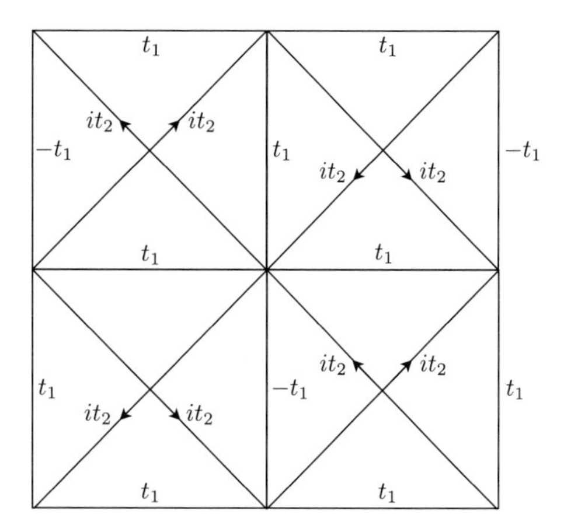

Figure 18.2 Tight-binding model with flux $\pi/2$ in each elementary right triangle.

symmetry, C_n can be non-zero, and its value is of great physical importance, as we see in Section 18.4. We can use (18.11) to present an expression for the value of C_n that is manifestly gauge invariant:

$$C_n = \frac{i}{2\pi}\int_{\text{1st B.Z.}} dk_x dk_y \sum_{n' \neq n} \frac{\langle u_{n,\boldsymbol{k}} | \boldsymbol{\nabla}_{\boldsymbol{k}} \mathcal{H}(\boldsymbol{k}) | u_{n',\boldsymbol{k}} \rangle \times \langle u_{n',\boldsymbol{k}} | \boldsymbol{\nabla}_{\boldsymbol{k}} \mathcal{H}(\boldsymbol{k}) | u_{n,\boldsymbol{k}} \rangle}{(\varepsilon_n(\boldsymbol{k}) - \varepsilon_{n'}(\boldsymbol{k}))^2} . \tag{18.38}$$

We now describe a simple and important example of a band structure that leads to non-zero Chern numbers; this example appears in our discussion of chiral spin liquids in Chapter 22, and also in subsequent chapters in Part IV. We consider a tight-binding model on the square lattice that has a magnetic flux of a half-flux quantum per unit cell; this the simplest case of a Hofstadter model. The tight-binding model is illustrated in Fig. 18.2. The nearest-neighbor hoppings t_1 have been chosen so that there is a π flux in each square plaquette of the square lattice. The t_1 hoppings are purely real, and so time-reversal symmetry is not yet broken. To break time reversal, and also to induce a band gap, we need to include imaginary second-neighbor hoppings of magnitude t_2; their phases are chosen so that each elementary right triangle has a spatially uniform flux of $\pi/2$. We can obtain the band structure of Fig. 18.2 by using a two-site unit cell of the A and B sublattices (as in Fig. 8.2), where the momentum-space Hamiltonian is

$$H_f = \sum_{a,b} f_a^\dagger(\boldsymbol{k}) M_{ab}(\boldsymbol{k}) f_b(\boldsymbol{k}) \tag{18.39}$$

where a, b represent A, B and

$$M(\boldsymbol{k}) = \begin{pmatrix} -2t_2\sin(k_x+k_y) - 2t_2\sin(k_x-k_y) & -2t_1\cos(k_x) - 2it_1\sin(k_y) \\ -2t_1\cos(k_x) + 2it_1\sin(k_y) & 2t_2\sin(k_x+k_y) + 2t_2\sin(k_x-k_y) \end{pmatrix} . \tag{18.40}$$

The band structure obtained by diagonalizing M is shown in Fig. 18.3. There is a band gap between the upper and lower bands of $8|t_2|$ at $\boldsymbol{k} = (\pm\pi/2, 0)$ and, as t_2 vanishes, the

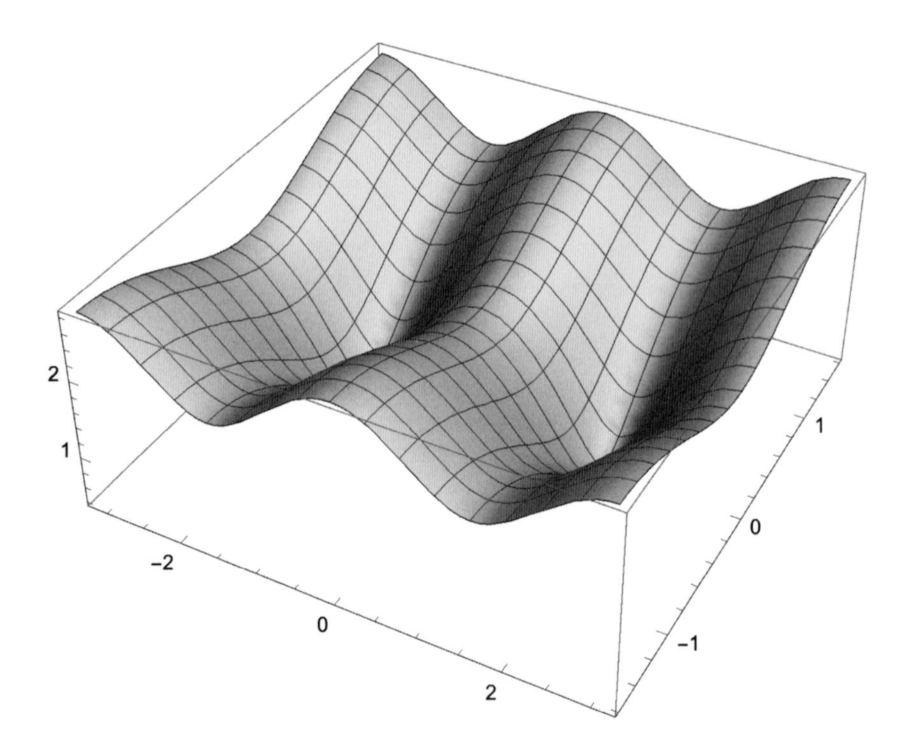

Figure 18.3 The upper band of the Hamiltonian in (18.39) for $t_1 = 1, t_2 = 0.1$. The lower band has energies that are the negative of the upper band.

bands meet at Dirac points, as discussed further in Chapter 22. Computing the Chern number via (18.38) shows that the two bands have $C_n = \pm 1$, with the sign determined by the sign of t_2. The Chern number is ill-defined at $t_2 = 0$ because of the vanishing band gap.

An important and remarkable property of bands with a non-zero Chern number is the existence of gapless edge states within the band gap. Imagine a situation where the value of t_2 slowly changes its value along some spatial direction until it ultimately flips its sign. Then we have regions of the system with opposite Chern numbers, and they must be connected via a gapless region, for that is the only way the Chern number of any band can change. In this case, there will be two gapless modes, corresponding to the two Dirac nodes at $\boldsymbol{k} = (\pm\pi/2, 0)$. Consider, next, an infinite strip of finite width along the x direction. Across each edge of the strip there is a transition from Chern number ± 1 to the empty state, which can be considered the gap of a trivial insulator; this situations leads a single edge state on each edge. We numerically verify this by exact diagonalization of a strip of width 200, whose spectrum is shown in Fig. 18.4. We see two bands of states within the gap. These correspond to left- and right-moving states on opposite edges of the strip.

There is an explicit analytic argument for the existence of such chiral edge states in Chapter 19.

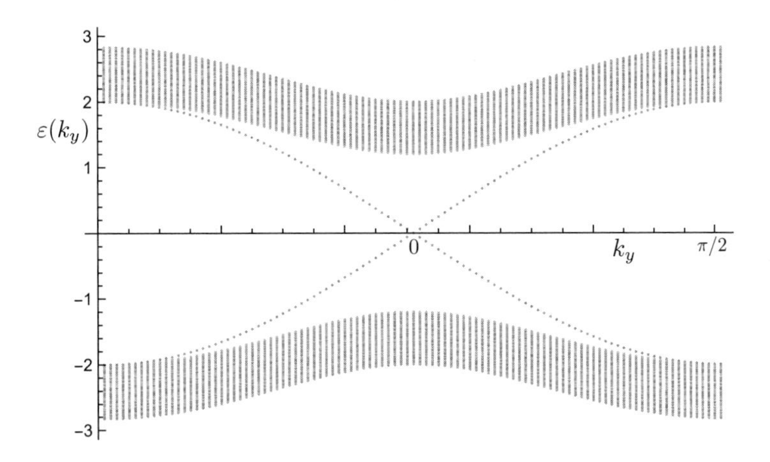

Figure 18.4 Spectrum of a strip of width 200 for the tight-binding model of Fig. 18.2. All eigenstates have definite momentum k_y.

18.4 Chern Insulators

Chern insulators are two-dimensional crystalline insulators with occupied bands having non-zero Chern numbers. They have the remarkable property that their Hall conductivity is quantized at zero temperature, and is related to the sum of the Chern numbers of the occupied bands. This was established in a classic paper by Thouless et al. [278], and we follow their approach.

We use the Kubo formula to compute the Hall conductivity. In the presence of an external vector potential $\boldsymbol{A}(\boldsymbol{r})$, the Hamiltonian in (18.38) is modified to

$$\mathcal{H}(\boldsymbol{k}) = \frac{1}{2m}\left(-i\hbar\boldsymbol{\nabla}_{\boldsymbol{r}} + \hbar\boldsymbol{k} - e\boldsymbol{A}(\boldsymbol{r})\right)^2 + V(\boldsymbol{r}). \tag{18.41}$$

This implies that the electron current operator is $(e/\hbar)\boldsymbol{\nabla}_{\boldsymbol{k}}\mathcal{H}(\boldsymbol{k})$. Inserting this current operator into the Kubo formula, and evaluating the current–current correlator for free fermions, we obtain an expression for the Hall conductivity at a Matsubara frequency $i\omega_m$:

$$\sigma_{xy}(i\omega_m) = \frac{e^2}{i\hbar^2\omega_m}\sum_{n\neq n'}\int\frac{d^2k}{4\pi^2}\left\langle u_{n,\boldsymbol{k}}\right|\partial_{k_x}\mathcal{H}(\boldsymbol{k})\left|u_{n',\boldsymbol{k}}\right\rangle\left\langle u_{n',\boldsymbol{k}}\right|\partial_{k_y}\mathcal{H}(\boldsymbol{k})\left|u_{n,\boldsymbol{k}}\right\rangle$$
$$\times\frac{f(\varepsilon_{n'}(\boldsymbol{k})) - f(\varepsilon_n(\boldsymbol{k}))}{\varepsilon_n(\boldsymbol{k}) - \varepsilon_{n'}(\boldsymbol{k}) + i\hbar\omega_m}, \tag{18.42}$$

where $f(\varepsilon)$ is the Fermi function. At $T = 0$, with the chemical potential in a band gap, we can expand in powers of ω_n and using (18.38) obtain the d.c. Hall conductivity:

$$\sigma_{xy} = \frac{e^2}{h}\sum_{\text{occupied } n} C_n. \tag{18.43}$$

Thus, the Hall conductivity of a Chern insulator is an integer multiple of the e^2/h. Another argument for the quantization of the Hall conductivity is given in Chapter 19.

Problem

18.1 Consider a tight-binding model on the square lattice in the presence of a magnetic field with flux $\Phi/\Phi_0 = 1/5$ per unit cell. The site i is at $\boldsymbol{r}_i = (x_i, y_i)$, with x_i, y_i integers. Begin by using a model with only nearest-neighbor hopping matrix elements given by

$$t_{i,i+\hat{x}} = -t_x \quad , \quad t_{i,i+\hat{y}} = -t_y\, \omega^{x_i} \tag{18.44}$$

(and their Hermitian conjugates), with $\omega = e^{2\pi i/5}$, $\boldsymbol{r}_{i+\hat{x}} = (x_i+1, y_i)$, $\boldsymbol{r}_{i+\hat{y}} = (x_i, y_i+1)$.

(a) These hopping elements have a unit cell of five sites, and so the momentum-space Hamiltonian $H(k_x, k_y)$ is a 5×5 matrix. Write down this 5×5 matrix. Its eigenvalues are $\varepsilon(k_x, k_y)$.

(b) Choose $t_x = 1.0$ and $t_y = 1.0$ (In Mathematica, it is important to put the decimal point – otherwise it will attempt an algebraic evaluation of the eigenvalues.) Plot the energy eigenvalues when the lattice is placed on a cylinder of circumference $L = 50$. So we have periodic boundary conditions along the x direction, and $x_i = 51$ is the same position as $x_i = 1$. The momentum k_x is quantized to the values $k_x = 2\pi j/(50)$, $j = 1, \ldots, 10$ (recall that the first Brillouin zone is $0 < k_x \leq 2\pi/5$, $0 < k_y \leq 2\pi$). Plot all the eigenvalues $\varepsilon(k_x, k_y)$ as a function of k_y alone, with k_x taking all the possible quantized values listed. In Mathematica, this is accomplished by the command

```
ListPlot[Transpose[Table[Flatten[Table[Eigenvalues[
hof[2 Pi j/50, ky]], {j, 1, 10}]], {ky, 0, 2 Pi, 2 Pi/n}]]],
```

where `hof[`k_x`,`k_y`]` yields $H(k_x, k_y)$, and n is some large integer fixing the number of points you want to take along the k_y axis (I chose $n = 500$). You should find that the eigenvalues separate into five bands.

(c) What are the Hall conductivities when the Fermi level is in each of the band gaps? Using Mathematica, compute the Chern numbers from (18.38).

(d) Compute and plot the eigenvalues in a strip of width $L = 50$ in the x direction (with open boundary conditions) and infinite in the y direction. Now k_y remains a good quantum number, but there is no periodicity in the x direction. So you can think of the system as a one-dimensional crystal along the y direction, with a unit cell of width 50 sites along the x direction. The Hamiltonian `H1D[ky]` is now a 50×50 tridiagonal matrix (i.e., the only non-zero matrix elements of $\texttt{H1D}_{i,j}$ have $|i - j| \leq 1$), which is a function of k_y. The near-diagonal elements are all $-t_x$, and the 50 diagonal matrix elements are 10 copies of the diagonal matrix elements of `hof[`k_x`,`k_y`]`. Write a program to generate this matrix.

(e) Plot all the 50 eigenvalues $\varepsilon_{1D}(k_y)$ as a function of k_y. In Mathematica, this is accomplished by

```
ListPlot[Transpose[Table[
Eigenvalues[H1D[ky]], {ky, 0, 2 Pi, 2 Pi/n}]]].
```

Comparing with the eigenvalues in (b), you should see additional dispersing states within the band gaps. These are the edge states. Relate the number of edge states within each band gap to the Hall conductivities computed in (c).

(f) Repeat (b)–(e) after adding an additional second-neighbor hopping in the x direction only:

$$t_{i,i+2\hat{x}} = -t_2 \tag{18.45}$$

(and its Hermitian conjugate). Choose $t_2 = 0.5$.

19 Integer Quantum Hall States

A theory of non-interacting electrons in the Landau levels of a strong magnetic field is presented. The theory yields chiral edge states and a quantized Hall conductivity. A connection is made to anomaly inflow arguments in quantum field theory.

This chapter considers electrons moving in two dimensions in the presence of an applied magnetic field. This situation is closely related to the Chern insulators studied in Section 18.4; we imagine focusing on the bottom of the band, and reducing the strength of the magnetic field so that the flux per unit cell is much smaller than the flux quantum. In this limit, we can ignore the periodic potential of the lattice, and work directly with the single-particle Hamiltonian in (19.1) below. In this limit, we will be able to study many physical properties in explicit detail.

19.1 Non-relativistic Particles

19.1.1 Landau Levels

We consider the single-particle Hamiltonian

$$H_0 = -\frac{1}{2M}(\boldsymbol{\nabla} - ie\boldsymbol{A})^2 \tag{19.1}$$

of a particle of mass M moving in two dimensions in the presence of a magnetic field $B\hat{z} = \boldsymbol{\nabla} \times \boldsymbol{A}$. We work in a gauge for the vector potential that preserves the translational symmetry along the x direction:

$$\boldsymbol{A} = (-By, 0). \tag{19.2}$$

Then the eigenstates of H_0 are of the form

$$\psi_{n,k}(x,y) = \frac{1}{\sqrt{L_x}} e^{ikx} \phi_{n,k}(y), \tag{19.3}$$

where $n = 0, 1, 2, \ldots$ labels the energy eigenvalues, and $\phi_{n,k}(y)$ obeys

$$-\frac{1}{2M}\frac{d^2\phi_{n,k}}{dy^2} + \frac{1}{2M}(k + eBy)^2\,\phi_{n,k}(y) = E_n \phi_{n,k}(y). \tag{19.4}$$

This is just the Schrödinger equation of a shifted harmonic oscillator, allowing us to determine the eigenstates and energy eigenvalues. The eigenvalues are

$$E_n = (n+1/2)\omega_c \quad , \quad n = 0,1,2,\ldots \quad , \quad \omega_c = eB/M\,, \tag{19.5}$$

where ω_c is the cyclotron frequency. These are the dispersionless (independent of k) Landau levels; the independence on k follows from the independence of the eigenenergies on the shift of the harmonic-oscillator position. We also introduce the dimensionless coordinate $\bar{y} = y/\ell$, where

$$\ell = \frac{1}{\sqrt{M\omega_c}} = \frac{1}{\sqrt{eB}}\,. \tag{19.6}$$

Then the eigenvalue equation becomes

$$-\frac{1}{2}\frac{d^2\phi_{n,k}}{d\bar{y}^2} + \frac{1}{2}(k\ell+\bar{y})^2\,\phi_{n,k}(\bar{y}) = (n+1/2)\,\phi_{n,k}(\bar{y})\,. \tag{19.7}$$

The eigenfunctions are the harmonic-oscillator eigenstates

$$\phi_{n,k}(y) = \frac{\pi^{-1/4}}{\sqrt{2^n n!}} H_n(y+k\ell)\exp\left(-\frac{(y+k\ell)^2}{2}\right), \tag{19.8}$$

where $H_n(y)$ are the Hermite polynomials.

In a sample of size $L_x \times L_y$, k is quantized in integer multiples of $2\pi/L_x$. And the shift in the harmonic-oscillator eigenstates implies that the range of allowed values of k is $-L_y/(2\ell^2)$ to $L_y/(2\ell^2)$. So the degeneracy of each Landau level is (after recalling Planck's constant)

$$\frac{L_x}{2\pi}\frac{L_y}{\ell^2} = \frac{AB}{h/e}, \tag{19.9}$$

where A is the area, and h/e is the flux quantum, that is, the degeneracy is the number of flux quanta in the sample.

An important observation is that none of these states carry any current:

$$\begin{aligned} I_x &= -\frac{e}{Mi}\langle n,k|\partial_x - ieA_x|n,k\rangle \\ &= -\frac{e}{ML_x}\int dy|\phi_{n,k}(y)|^2(k+eBy) \\ &= 0\,. \end{aligned} \tag{19.10}$$

For the computation of the conductivity later, we need the matrix elements:

$$\begin{aligned} \int dy\,\phi_{n,k}(y)\,\partial_y\phi_{m,k}(y) &= \sqrt{\frac{M}{2}}\,(\delta_{m,n+1} - \delta_{m,n-1})\,, \\ \int dy\,\phi_{n,k}(y)(y+k\ell)\,\phi_{m,k}(y) &= \sqrt{\frac{M}{2}}\,(\delta_{m,n+1} + \delta_{m,n-1})\,. \end{aligned} \tag{19.11}$$

19.1.2 Symmetric Gauge

For future application to the fractional case, it is useful to write down the wavefunction of the fully filled lowest Landau level in the "symmetric" gauge, where

$$\boldsymbol{A} = -\frac{1}{2}\boldsymbol{r} \times \boldsymbol{B} \tag{19.12}$$

and $\boldsymbol{B} = -B\hat{z}$. Then, the single-particle wavefunctions in the lowest Landau level are analytic functions of $z = (x+iy)/\ell$ times a Gaussian

$$\varphi_m(x,y) = \frac{1}{\sqrt{2\pi\ell^2 2^m m!}} z^m e^{-|z|^2/4}. \tag{19.13}$$

The integer $m \geq 0$ is the eigenvalue of the angular momentum. From this wavefunction, we can write the wavefunction of the fully filled lowest Landau level as

$$\Phi(\boldsymbol{r}_1, \boldsymbol{r}_2, \dots, \boldsymbol{r}_N) \propto P(z_1, z_2, \dots, z_N) \prod_{i=1}^{N} e^{-|z_i|^2/4}, \tag{19.14}$$

where $P(z_1, z_2, \dots, z_N)$ is a polynomial in all the z_i given by the Slater determinant

$$\begin{aligned} P(z_1, z_2, \dots, z_N) &= \begin{vmatrix} 1 & z_1 & z_1^2 & \dots & z_1^{N-1} \\ 1 & z_2 & z_2^2 & \dots & z_2^{N-1} \\ \dots & \dots & \dots & \dots & \dots \\ \dots & \dots & \dots & \dots & \dots \\ 1 & z_N & z_N^2 & \dots & z_N^{N-1} \end{vmatrix} \\ &= \prod_{i<j} (z_j - z_i). \end{aligned} \tag{19.15}$$

This particular Slater determinant is the Vandermonde determinant, and the last expression follows from the requirement that the P vanish whenever any $z_i = z_j$.

19.1.3 Hall Conductivity

We compute the conductivity σ_{xy} by the Kubo formula.

We introduce an operator $c_{n,k}$, which annihilates the electron in the Landau-level state $\psi_{n,k}(x,y)$. Then, the electron field operator is

$$\Psi(x,y) = \int \frac{dk}{2\pi} \sum_n \psi_{n,k}(x,y) c_{n,k} \tag{19.16}$$

and the electron current operator is

$$\boldsymbol{J}(x,y) = \frac{1}{2Mi} \left[\Psi^\dagger (\boldsymbol{\nabla} - i\boldsymbol{A}) \Psi - (\boldsymbol{\nabla} + i\boldsymbol{A}) \Psi^\dagger \Psi \right]. \tag{19.17}$$

Then, applying the Kubo formula at a frequency ω_n, we obtain

$$\sigma_{xy}(\omega_n) = \frac{1}{\omega_n}\frac{1}{L_x L_y}\sum_k \sum_{n,m}\left[\frac{f(E_n)-f(E_m)}{i\omega_n - E_n + E_m}\right] \tag{19.18}$$
$$\times \left[\int dy\, \phi_{n,k}(y)\, \partial_y \phi_{m,k}(y)\right]\left[\int dy\, \phi_{n,k}(y)(y+k\ell)\phi_{m,k}(y)\right],$$

where $f(E)$ is the Fermi function. In the limit $T \to 0$, we find that only two levels contribute to the sum: the ones just above and below the Fermi level. Carefully evaluating this expression using the results above, we find

$$\sigma_{xy} = \frac{\nu e^2}{h}, \tag{19.19}$$

where ν is the number of filled Landau levels.

19.1.4 Chern–Simons term

We can express the result of Section 19.1.3 in terms of an effective action for the external electromagnetic field A_μ. The non-zero Hall conductivity implies a term $\sim \sigma_{xy} A_x \partial_\tau A_y$ upon integrating out the electrons. Making this gauge invariant (or explicitly computing the one-loop diagram in the presence of a general A_μ), we find the famous Chern–Simons term in the external field A_μ for a single filled Landau level:

$$S_A = \frac{i}{4\pi}\int d^3x\, \varepsilon_{\mu\nu\lambda} A_\mu \partial_\nu A_\lambda . \tag{19.20}$$

Here, we have absorbed the factor of e in the definition of A_μ, and use units with $\hbar = 1$. For now, A_μ is a fixed, background, external field, and is not fluctuating. We will meet Chern–Simons terms of dynamic gauge fields when we consider the fractional quantum Hall effect.

19.2 Relativistic Particles (Graphene)

Now we consider a Dirac fermion, with unit Fermi velocity, which describes the low-energy spectrum of graphene with Hamiltonian

$$H = -i\boldsymbol{\sigma}\cdot(\boldsymbol{\nabla} - i\boldsymbol{A}). \tag{19.21}$$

The eigenvalue equation can be written as

$$\begin{pmatrix} 0 & -i(\partial_x - iA_x) - (\partial_y - iA_y) \\ -i(\partial_x - iA_x) + (\partial_y - iA_y) & 0 \end{pmatrix}\begin{pmatrix} \psi_1 \\ \psi_2 \end{pmatrix}$$
$$= E\begin{pmatrix} \psi_1 \\ \psi_2 \end{pmatrix}. \tag{19.22}$$

This translates into a Schroedinger-like equation for each component:

$$\left[-(\boldsymbol{\nabla} - i\boldsymbol{A})^2 + (\boldsymbol{\nabla}\times\boldsymbol{A})\right]\psi_{1,2} = E^2\,\psi_{1,2}. \tag{19.23}$$

The eigenvalues and eigenfunctions are now easily obtained from the solution in Section 19.1.1:

$$E_n = \mathrm{sgn}(n)\sqrt{2B|n|} \quad , \quad n = \ldots -2, -1, 0, 1, 2\ldots. \tag{19.24}$$

Note that the Landau levels now have both positive and negative energies, as does the Dirac dispersion in a zero magnetic field. Also note the special Landau level at exactly zero energy. The degeneracy of each Landau level is still given by (19.9).

The Hall-conductivity computation can be carried out in a manner similar to Section 19.1.3. A similar computation yields (see Problem 19.1)

$$\sigma_{xy} = (n + 1/2)\frac{e^2}{2\pi\hbar} \tag{19.25}$$

when the chemical potential is just above the Landau level with energy E_n (we have reinserted $\hbar$ in the last step). So, we find here too that the Hall conductivity increases by an integer multiple of e^2/h every time the Fermi level crosses a Landau level. We can understand the offset of $1/2$ by using particle–hole symmetry. We expect σ_{xy} to flip its sign between the cases where the Fermi level is just above and below the zeroth Landau level; so we expect that just above and below the zero-energy Landau level, the Hall conductivity $= \pm e^2/(2h)$, as is the case. Because of the additional valley degeneracy of the Dirac fermions, the observed Hall conductivity yields integer values (except on the surface of a topological insulator).

19.3 Edge states

We return to consideration of the non-relativistic Hamiltonian, although very similar results apply also to the relativistic case. Let us look at the situation in which the sample is only present for $y < 0$, and so has a edge at $y = 0$ (see Fig. 19.1). We consider the single-particle Hamiltonian

$$H_0 = -\frac{1}{2M}(\boldsymbol{\nabla} - i\boldsymbol{A})^2 + V(y). \tag{19.26}$$

As translational symmetry is preserved along the x direction, we can continue to work with the vector potential in (19.2). The eigenstates are as in (19.3)

$$\psi_{n,k_x}(x,y) = \frac{1}{\sqrt{L_x}}e^{ik_x x}\phi_n(y) \tag{19.27}$$

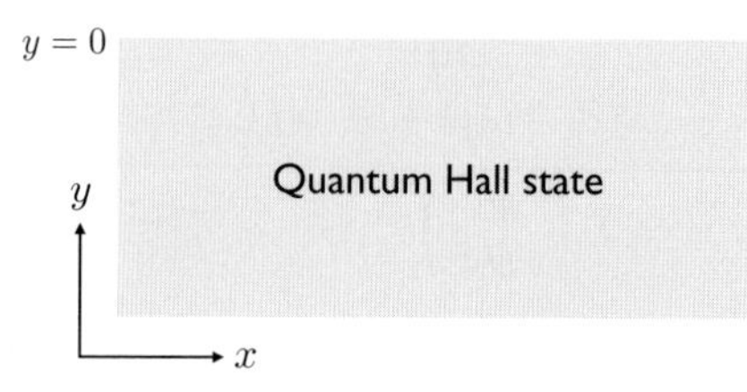

Figure 19.1 Edge of a semi-infinite quantum Hall state at $y = 0$.

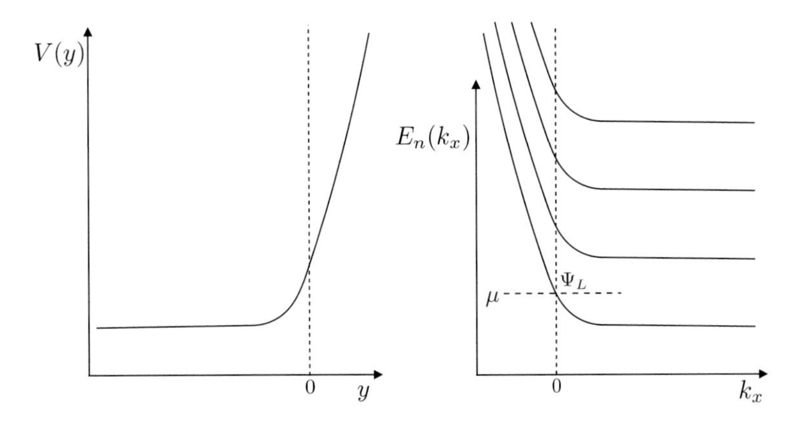

Figure 19.2 Energy levels, $E_n(k_x)$ of electrons in a potential $V(y)$ and a magnetic field. The chemical potential μ is chosen so that only the $n=0$ level is occupied. The left-moving chiral fermions, Ψ_L, describe the excitations on the $y=0$ edge.

and $\phi_n(y)$ obeys

$$-\frac{1}{2M}\frac{d^2\phi_n}{dy^2}+\left[\frac{1}{2M}(k_x+By)^2+V(y)\right]\phi_n(y)=E_n(k_x)\phi_n(y), \tag{19.28}$$

with $E_n(k_x)$ the energy eigenvalue that disperses as a function of k_x.

More generally, we can take $V(y)=0$ in the bulk of the sample, far from the edge, without loss of generality. Here, the eigenstates in (19.28) are harmonic-oscillator states centered at $y=-k_x/B$. So we expect the eigenstates at large positive k_x to be within the sample, and insensitive to the edge. But, near the edge of the sample at $y=0$, we expect $V(y)$ to increase rapidly to confine the electrons within the sample. Therefore, as k_x decreases through 0, we expect the eigenstates to approach the edge of the sample, and for $E_n(k_x)$ to increase. We can estimate the change in E_n by perturbation theory

$$\begin{aligned}E_n(k_x)&=(n+1/2)\hbar\omega_c+\int dy|\phi_n^0(y)|^2V(y)+\cdots,\\&=(n+1/2)\hbar\omega_c+V(k_x\ell^2)+\cdots,\end{aligned} \tag{19.29}$$

where $\phi_n^0(y)$ are the eigenstates for $V=0$. See the sketch of energy levels in Fig. 19.2. A crucial feature of the states distorted by the potential $V(y)$ is that they now carry a non-zero current. We have

$$\begin{aligned}I_x&=-\frac{e}{ML_x}\int dy|\phi_n(y)|^2(k_x+eBy)\\&=-\frac{e}{L_x}\frac{\partial E_n(k_x)}{\partial k_x}\quad\text{by the Feynman–Hellman theorem,}\\&\approx -e\frac{\bar{v}}{L_x},\end{aligned} \tag{19.30}$$

where

$$\bar{v} = \frac{c}{B}\frac{\partial V}{\partial y}\bigg|_{y=k_x\ell^2} \tag{19.31}$$

is the drift velocity in the presence of a potential gradient and a magnetic field. This is the velocity of "skipping orbits" at the edge of the sample.

Now, we consider the situation for the $n=1$ case, where only the lowest, $n=0$, Landau level is fully occupied in the bulk. We see from Fig. 19.2 that in such a situation $E_0(k_x)$ will necessarily cross the chemical potential once as a function of decreasing k_x. This implies the existence of gapless, one-dimensional, fermionic excitations on the edge of the sample. The fermions all move with velocity $\bar{v} = dE_0(k_x)/dk_x$, and so from (19.29) are left-moving chiral fermions. Notice that there is no right-moving counterpart, at least on the edge near $y=0$. If the sample had another edge far away at some $y<0$, that edge would support a right-moving chiral fermion.

It is interesting to note that a single left-moving chiral fermion cannot appear by itself in any strictly one-dimensional system. In the presence of a lattice, the fermion dispersion $E(k_x)$ of such a system must be a periodic function of k_x, and no periodic function can cross the Fermi level only once. But, on the edge of a two-dimensional system, it is possible for $E(k_x)$ to cross the Fermi level just once, as we have shown above.

It is simple to write down a low-energy effective theory for the left-moving chiral fermion at the edge of the simple. Using the notation and methods of Chapter 12 on Luttinger liquids, we have the imaginary action

$$\mathcal{S}_L = \int dx d\tau \Psi_L^\dagger \left(\frac{\partial}{\partial \tau} + i\bar{v}\frac{\partial}{\partial x}\right)\Psi_L. \tag{19.32}$$

This is the universal low-energy theory of the edge of a quantum Hall sample at $n=1$. Note that, unlike non-chiral Luttinger liquids, there are no marginal interaction corrections to the free theory. All such interaction corrections involve right-moving fermions too, which are absent in the present system.

Let us also recall the bosonized version of this chiral theory from (12.34):

$$\mathcal{L}_{CL} = \frac{1}{4\pi}\left[\bar{v}(\partial_x\varphi_L)^2 + i\partial_x\varphi_L\partial_\tau\varphi_L\right], \tag{19.33}$$

in terms of the chiral boson field φ_L. This formulation will be more useful for the fractional quantum Hall effect, when the edge states are described by theory similar to (19.33), but the free-fermion theory in (19.32) will no longer apply.

19.3.1 Quantized Hall Conductance

Let us now compute the Hall conductance in a manner that highlights the crucial role played by the edge states. We apply a voltage V across the sample. Then, in a current-carrying steady state, the chemical potential μ_T on the top edge, and the chemical potential μ_B on the bottom edge obey

$$-eV = \mu_T - \mu_B\,. \tag{19.34}$$

We compute the total current in this situation from (19.30)

$$I_x = -\frac{e}{L_x} L_x \int \frac{dk_x}{2\pi} \sum_n \frac{\partial E_n(k_x)}{\partial k_x} f(E_n(k_x)), \tag{19.35}$$

where $f(E_n(k_x))$ is the mean occupation number of the state $\phi_n(k_x)$. At $T = 0$, only the levels with $n < \nu$ contribute, where ν Landau levels are occupied, and the integral over k can be evaluated exactly to yield

$$\begin{aligned} I_x &= -\nu \frac{e}{2\pi}(\mu_T - \mu_B) \\ &= \nu \frac{e^2}{2\pi\hbar} V. \end{aligned} \tag{19.36}$$

This computation also highlights the connection between the Hall conductance of the two-dimensional sample and the ordinary longitudinal conductance of the edge state [40]. In the Landauer approach, we compute the conductance by connecting the one-dimensional conductor to two reservoirs maintained at a chemical potential difference eV. Then, the current is carried by the electrons in this energy interval; so we obtain from (19.32)

$$I = e \int \frac{dk}{2\pi} \bar{v} \left[\theta(\bar{v}k) - \theta(\bar{v}k - eV)\right] = \frac{e^2}{2\pi\hbar} V, \tag{19.37}$$

again reinserting $\hbar$ in the last step. Notice that $\bar{v}$ cancels out, and the conductance is just e^2/h. This argument also shows that impurities on the edge cannot make a difference; because the edge fermions are chiral, there is no back scattering, and the transmission coefficient remains unity.

It is useful to redo the computation of the edge conductance using the bosonized theory (19.33); this will allow later us to easily extend the result to the fractionalized case [124]. We already know that the density operator on the edge is $\partial_x \varphi_L/(2\pi)$, and so, by the continuity equation, we conclude that the current operator is $I = -ie\partial_\tau \varphi_L/(2\pi)$. In the Landauer–Buttiker picture, we need a computation of the linear response of the Gaussian theory in (19.33) to an external voltage $V(x)$, which reaches values that differ by V between the two reservoirs. So we write

$$I(x) = \int dx' D^R(x - x', \omega \to 0) V(x'), \tag{19.38}$$

where D^R is the retarded response function of the chiral edge theory. It is straightforward to compute this function for a Matsubara frequency ω_n from the Gaussian action

$$D^R(x - x', i\omega_n) = \frac{e^2}{2\pi} \int \frac{dk}{2\pi} e^{ik(x-x')} \frac{k\omega_n}{ik\omega_n - \bar{v}k^2}. \tag{19.39}$$

Analytically, continuing $i\omega_n \to \omega + i\eta$, and performing the momentum integral, we obtain

$$D^R(x - x', \omega) = \frac{e^2}{2\pi} \theta(x - x') \frac{i\omega}{\bar{v}} e^{i(\omega + i\eta)(x-x')/\bar{v}}. \tag{19.40}$$

We observe a one-sided response, with the current at x depending only upon the voltage at $x' < x$, a signature of the chiral nature of the edge theory. Performing the integral of x' in (19.38), we obtain the expected quantized result:

$$I = \frac{e^2}{2\pi\hbar} V. \tag{19.41}$$

19.4 Anomaly Inflow Arguments

We have shown how edge states appear by considering a specific potential $V(y)$ acting on free fermions in a magnetic field. It seems clear that the appearance of the edge states is a robust phenomenon, which will also apply to more general potentials, and also survive the introduction of interactions. We now describe "anomaly inflow arguments" [41], which establish this "topological" stability.

Consider the partition function generalizing (19.26) to a generic Hamiltonian of electrons c, in the presence of a generic fixed external gauge field A_μ and potential $V(x,y)$:

$$Z[A_\mu] = \int \mathcal{D}c \exp\left[-\mathcal{S}(c, A_\mu)\right]. \tag{19.42}$$

We are considering a well-defined problem of electrons confined to a fixed region of space, and so the partition function should be gauge invariant:

$$Z[A_\mu + \partial_\mu \lambda] = Z[A_\mu]. \tag{19.43}$$

For the bulk of the sample, we have already computed the partition function from a single filled Landau level in (19.20), and so we write

$$\begin{aligned} Z[A_\mu] &= Z_{bulk}[A_\mu] Z_{boundary}[A_\mu], \\ Z_{bulk}[A_\mu] &= \exp\left(-\frac{i}{4\pi}\int d^3x\, \varepsilon_{\mu\nu\lambda} A_\mu \partial_\nu A_\lambda\right). \end{aligned} \tag{19.44}$$

The key point is that bulk is not gauge invariant up to a boundary term, and it is easy to show that

$$Z_{bulk}[A_\mu + \partial_\mu \lambda] = Z_{bulk}[A_\mu] \exp\left(-\frac{i}{2\pi}\int dx d\tau \lambda \left(\partial_\tau A_x - \partial_x A_\tau\right)\right), \tag{19.45}$$

where we have retained only the boundary term at $y = 0$. Consistency with (19.43) now demands that there be an "anomaly" in the boundary theory at $y = 0$, $Z_{boundary}[A_\mu]$, which cancels the bulk anomaly in (19.45). We now show that the chiral Luttinger-liquid theory defined by (19.32) and (19.33) satisfies this requirement.

The expression in (19.45) shows that the anomaly is associated with the application of an electric field $E = \partial_t A_x - \partial_x A_t$ to the boundary theory. In addition there is indeed something anomalous about the chiral boundary theory in the presence of E: the density of fermions is not conserved. In the case where E is independent of x, the filled fermion states just move up in momentum k via (setting $e = 1$)

$$\frac{dk}{dt} = E. \tag{19.46}$$

So the change in density $\rho_L = \Psi_L^\dagger \Psi_L$ is

$$\frac{d\rho_L}{dt} = \frac{1}{2\pi}\frac{dk_F}{dt} = \frac{E}{2\pi}. \tag{19.47}$$

For the spatially non-uniform case, this generalizes to an anomalous term in the equation for the conservation of the fermion current $J^\mu = (\rho_L, \bar{v}\Psi_L^\dagger \Psi_L)$:

$$\partial_\mu J^\mu = \frac{E}{2\pi}. \tag{19.48}$$

The expression (19.48) can be obtained from the equations of motion for the Fourier components of ρ_L in the presence of an electric field, after carefully accounting for the Schwinger term in the commutator of ρ_L with itself.

Now let us connect (19.48) to (19.45). We can write the current as

$$\langle J^\mu \rangle = -i\frac{\delta \ln Z_{boundary}(A_\mu)}{\delta A_\mu}, \tag{19.49}$$

and consequently

$$\left\langle \partial_\mu J^\mu \right\rangle = i\frac{\delta \ln Z_{boundary}(A_\mu + \partial_\mu \lambda)}{\delta \lambda}. \tag{19.50}$$

Using (19.48), we verify that the λ dependence of $Z_{boundary}(A_\mu + \partial_\mu \lambda)$ is precisely that needed to cancel the bulk anomaly of (19.45).

It is useful to also present this result in the language of the chiral boson. The density is $\rho_L = \partial_x \varphi_L/(2\pi)$, and so the chiral-boson commutator implies the density commutator

$$[\rho_L(x), \rho_L(y)] = \frac{i}{2\pi}\delta'(x-y). \tag{19.51}$$

Now, we use the Hamiltonian in the presence of an external potential V (so that $E = -\partial_x V$),

$$H = \int dx \left[\pi v_F \rho_L^2 + V\rho_L\right], \tag{19.52}$$

to the equation of motion for ρ_L which coincides with (19.48).

Problem

19.1 Obtain the Hall conductivity in (19.25) of Dirac fermions by applying the Kubo formula, as in (19.18) for non-relativistic electrons. See Ref. [204] for expressions for the wavefunctions and the matrix elements of the current operator.

20 Topological Insulators and Superconductors

Topological band structures with time-reversal symmetry are described: the Su–Schrieffer–Heeger model in one dimension, and the Kane–Mele model in two dimensions. The theory of topological superconductors in one and two dimensions is presented, describing their edge states with Majorana fermions.

Chapters 18 and 19 have described free-fermion systems in two spatial dimensions, which contain protected edge states. These are now understood to be early examples of distinct types of band topology, which fall under under precisely defined classes of topological insulators and superconductors by their properties under certain discrete symmetries. In this classification one considers the topological properties of most general Hamiltonians consistent with some symmetry constraints. The examples in Chapters 18 and 19 have *no* symmetry constraints, and this feature partially obscures the role of symmetry in their properties. The most general Hamiltonian with no symmetry will, in general, break time-reversal symmetry, and this was the case with the models considered in Chapters 18 and 19. Indeed, we found that time-reversal symmetry breaking was needed to obtain non-zero Chern numbers, and hence protected edge states. Somewhat unexpectedly, in the modern classification this last feature is disregarded, and non-zero Chern numbers are assumed possible whenever a Hamiltonian is in a class without time-reversal symmetry.

I do not present the complete band topology classification here, and refer the reader to a number of other reviews [6, 15, 28, 135, 226, 248]. I will only discuss a few examples that are important for the subsequent discussion of correlated phases. With the restricted purpose of connecting to the notation of the classification, Table 20.1 presents the 10-fold classification of free-fermion topological insulators and superconductors in spatial dimension d. This table specifies symmetries of the single-particle Hamiltonian, which are anti-unitary time-reversal $\mathcal{T}$, anti-unitary particle-hole $\mathcal{C}$, and unitary sublattice or chiral $\mathcal{S}$; examples of these symmetries are given in the following sections. The columns under $\mathcal{T}$, $\mathcal{C}$, $\mathcal{S}$ specify the squares of these symmetries when non-zero, and 0 when the symmetries are absent. The models considered in Chapters 18 and 19 belong to class A in $d = 2$ without any discrete symmetries; from Table 20.1 we see that they are classified as $\mathbb{Z}$ – this corresponds to the integer Chern number, and the integer number of edge states characterizing such systems. Note that for $d \leq 3$, $d = 2$ is the only dimension in which class A has topological states.

Table 20.1 10-fold way

Class	$\mathcal{T}$	$\mathcal{C}$	$\mathcal{S}$	$d=1$	$d=2$	$d=3$
A	0	0	0	0	$\mathbb{Z}$	0
AIII	0	0	1	$\mathbb{Z}$	0	$\mathbb{Z}$
AI	1	0	0	0	0	0
BDI	1	1	1	$\mathbb{Z}$	0	0
D	0	1	0	$\mathbb{Z}_2$	$\mathbb{Z}$	0
DIII	−1	1	1	$\mathbb{Z}_2$	$\mathbb{Z}_2$	$\mathbb{Z}$
AII	−1	0	0	0	$\mathbb{Z}_2$	$\mathbb{Z}_2$
CII	−1	−1	1	$\mathbb{Z}$	0	$\mathbb{Z}_2$
C	0	−1	0	0	$\mathbb{Z}$	0
CI	1	−1	1	0	0	$\mathbb{Z}$

20.1 Su–Schrieffer–Heeger Model

A simple example of band topology with symmetry is provided by the Su–Schrieffer–Heeger (SSH) model illustrated in Fig. 20.1. We consider a tight-binding model in one dimension, with a two-sublattice structure. The fermion operators c_{iA} and c_{iB} annihilate fermions in unit cell i on the A and B sublattice, respectively. So we have the Hamiltonian

$$H = -\sum_i \left[w c_{iA}^\dagger c_{iB} + v c_{iB}^\dagger c_{i+1,A} + \text{H.c.} \right], \tag{20.1}$$

where w and v are the real hopping matrix elements shown in Fig. 20.1. We transform to one-dimensional momentum space, and introduce the two-component fermion

$$C_p = \begin{pmatrix} c_{Ap} \\ c_{Bp} \end{pmatrix}. \tag{20.2}$$

Then, we have

$$H = \sum_p C_p^\dagger H(p) C_p, \tag{20.3}$$

where the momentum-space Hamiltonian is

$$H(p) = d_x(p)\tau^x + d_y(p)\tau^y + d_z(p)\tau^z, \tag{20.4}$$

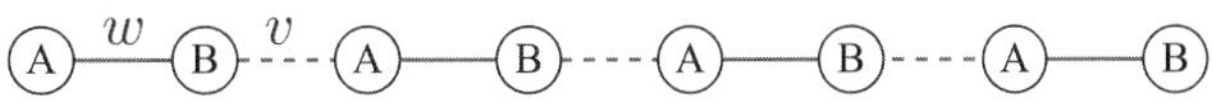

Figure 20.1 The Su–Schrieffer–Heeger model: a one-dimensional lattice of A and B "molecules." This realizes the $\mathbb{Z}$ invariant of class BDI in $d=1$, with protected zero energy edge states for $|v| > |w|$.

with $\tau^{x,y,z}$ Pauli matrices acting on the sublattice space of (20.2), and the "***d*** vector"

$$\boldsymbol{d}(p) = -(w + v\cos(p), v\sin(p), 0). \tag{20.5}$$

One of the symmetry properties important for topological considerations is a unitary "sublattice" or "chiral" symmetry $\mathcal{S}$, which anti-commutes with the Hamiltonian:

$$\begin{aligned} \mathcal{S} &= \tau^z, \\ \mathcal{S}H(p)\mathcal{S} &= -H(p), \end{aligned} \tag{20.6}$$

because $d_z(p) = 0$. Another symmetry property is the anti-unitary symmetry $\mathcal{C}$, under which

$$\begin{aligned} \mathcal{C} &= \tau^z, \\ \mathcal{C}H(p)\mathcal{C} &= -H^*(-p). \end{aligned} \tag{20.7}$$

It is easy to show that $\mathcal{S}^2 = 1$ and $\mathcal{C}^2 = 1$. The Hamiltonian also has the product symmetry $\mathcal{T} = \mathcal{C}\mathcal{S}$, under which $H(p)* = H(-p)$. By Table 20.1, these symmetries place the SSH model in class BDI in $d = 1$, which has a $\mathbb{Z}$ topological invariant. Indeed, this topological invariant is easy to find by mapping the Hamiltonian to that of $S = 1/2$ spin in a Zeeman field in (18.16). The vector $\boldsymbol{d}(p)$ lies in the x–y plane, and as p extends in the Brillouin zone $0 \leq p \leq 2\pi$, the $\boldsymbol{d}$ vector maps out a closed path in the plane. Provided there is a band gap, that is, $|\boldsymbol{d}(p)| \neq 0$, the topological invariant is the number of times this path encircles the origin. Alternatively, we can define the unit vector

$$\hat{d}(p) = \frac{\boldsymbol{d}(p)}{|\boldsymbol{d}(p)|}, \tag{20.8}$$

and then $\hat{d}(p)$ is a path on the equator of the unit sphere of the spin in Section 18.2 – it is constrained to lie on the equator by $\mathcal{S}$ symmetry. The $\mathbb{Z}$ topological invariant is the number of times the equator is encircled. For an analytic expression similar to the Chern number in (18.37), we define

$$|\boldsymbol{d}(p)|e^{i\theta(p)} = d_x(p) + id_y(p), \tag{20.9}$$

and then the integer invariant is

$$C = \frac{1}{2\pi}\int_0^{2\pi} dp\,\partial_p\theta(p). \tag{20.10}$$

The two eigenstates of $H(p)$ have invariants $\pm C$, as in (18.21). For the expression in (20.5), C vanishes when $|w| < |v|$, and then the band structure is not topological.

Another important consequence of the symmetry in (20.6) is that the eigenvalues of $H(p)$ come in pairs with opposite signs. For every eigenstate $|E\rangle$ with non-zero energy E, there is an eigenstate $\mathcal{S}|E\rangle$ with energy $-E$, as follows easily from (20.6). Moreover, the states $|E\rangle$ and $\mathcal{S}|E\rangle$ have to be orthogonal to each other because they are eigenstates of $H(p)$ with distinct eigenvalues.

Similar to the arguments in Section 18.3, the existence of this invariant implies that there are protected edge states at the ends of a finite SSH chain. Near the edges, we have to cut the chain so that the A and B molecules remain intact, and this ensures that there is $\mathcal{S}$ symmetry even for a finite chain. We can see the presence of these edge states in the limit $v \to \infty$ from Fig. 20.1. Then, the bulk eigenstates are odd and even linear combinations c_{iB} and $c_{i+1,A}$, with eigenvalues $\pm v$. However, there are isolated zero-energy eigenstates on the edges, c_{1A} and c_{LB}, on a chain of L unit cells with $2L$ sites. It is now easy to see that these zero-energy eigenstates remain exponentially localized on the edges of the chain as v is reduced. Gapping these states would require terms such as $c_{1A}c_{1A}$, which violate the $\mathcal{S}$ symmetry. There will be some mixing between the edge states for a finite chain of order $\exp(-\alpha L)$, and so the eigenvalues will be $\sim \pm\exp(-\alpha L)$. As in Section 18.3, these edge states remain near zero energy as long as the invariant in (20.10) is non-zero. For an infinite chain, it is actually possible to determine the wavefunction of the zero-energy eigenstate exactly. Denoting the eigenstate by the vector (A_i, B_i), $i = 1, 2, \ldots$ we obtain the zero eigenvalue equations for the left edge:

$$\begin{aligned} wB_1 &= 0, \\ vB_{i-1} + wB_i &= 0, \quad i \geq 2, \\ wA_i + vA_{i+1} &= 0, \quad i \geq 1. \end{aligned} \tag{20.11}$$

These equations have the solution

$$A_i = (-w/v)^{i-1} \quad , \quad B_i = 0. \tag{20.12}$$

This is an exponentially decaying solution provided $|v| > |w|$, which is indeed the condition for protected edge states, as we saw in the discussion below (20.10). Note also that the zero-energy state is polarized exclusively on the A sublattice. Similarly, there will be a zero-energy state on the right edge, which is polarized exclusively on the B sublattice.

We noted above that the SSH chain belongs to the class BDI, with a $\mathbb{Z}$ invariant, and so a generalized model can have an arbitrary integer number of edge states. We can realize such states by coupling an arbitrary number of SSH chains in a manner in which it preserves $\mathcal{S}$ symmetry. We now describe a modification of the SSH chain so that it is no longer in the class BDI in Table 20.1, but moves to class D. The topological invariant for class D is $\mathbb{Z}_2$, and this leads to a novel structure, the analog of which we have not encountered in Chapters 18 and 19. We obtain a model in class D by adding, in the same-sublattice, purely imaginary hopping it shown in Fig. 20.2 [287]. The Hamiltonian H is now the sum of (20.1) with

$$H_1 = -it \sum_i \left[c_{iA}^\dagger c_{i+1,A} - c_{iB}^\dagger c_{i+1,B} + \text{H.c.} \right]. \tag{20.13}$$

In momentum space, this retains the form in (20.4), but changes the $\boldsymbol{d}$ vector from (20.5) to

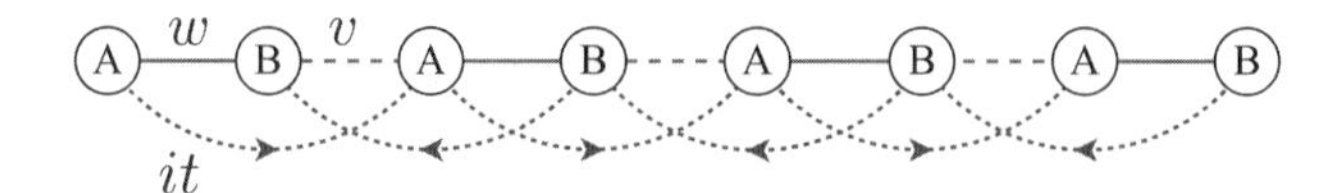

Figure 20.2 An extended SSH model, with a purely imaginary hopping it. This realizes the $\mathbb{Z}_2$ invariant of class D in $d = 1$ for non-zero t.

$$\boldsymbol{d}(p) = -(w + v\cos(p), v\sin(p), 2t\sin(p)). \tag{20.14}$$

The chiral symmetry in (20.6) now does not apply, but the symmetry in (20.7) continues to do so; by Table 20.1, this puts the system in class D, as time-reversal symmetry is also not present.

The unit $\boldsymbol{d}$ vector now no longer lies in the equator, and so the winding number is not a topological invariant. Nevertheless, one can see from an analysis similar to the $v \gg w$ analysis of class BDI in Fig 20.1, that the zero-energy edge states survive in the model of Fig. 20.2. These edge states are protected by a $\mathbb{Z}_2$ invariant defined as follows. We notice that the mapping (20.7) sends p to $-p$, and so it is helpful to focus on the time-reversal invariant momenta (TRIMs) $p = 0, \pi$. The Hamiltonian takes a very simple form at these TRIMs:

$$H(p=0) = -(w+v)\sigma^x \quad , \quad H(p=\pi) = -(w-v)\sigma^x. \tag{20.15}$$

Let us assume that $w > 0$ and $v > 0$. Then, for $w > v$, we notice that the occupied lower-energy bulk band is the state with $\sigma^x = 1$ for both $p = 0$ and $p = \pi$. On the other hand, in the regime where there are edge states, the lower-energy state has $\sigma^x = 1$ for $p = 0$ and $\sigma^x = -1$ for $p = \pi$. This allows us to deduce the $\mathbb{Z}_2$ invariant:

$$v_2 = \hat{d}_x(p=0)\,\hat{d}_x(p=\pi), \tag{20.16}$$

and the protected edge states are present for $v_2 = -1$. The condition (20.7) requires the $\boldsymbol{d}$ vector orient along $(\pm 1, 0, 0)$ for $p = 0$ and $p = \pi$; the invariant (20.16) informs us whether changing p from 0 to π leads to a $\hat{d}(p)$, which connects opposite poles of the sphere, or returns to the same pole.

Finally, it is interesting to consider an explicit example that differentiates the $\mathbb{Z}$ invariant of Fig. 20.1 from the $\mathbb{Z}_2$ invariant of Fig. 20.2. We consider two copies of an SSH chain as shown in Fig. 20.3 [287]. When the chains are decoupled with $t = 0$, we obtain two protected edge states, one each from each chain. However, when we introduce a non-zero t, the model falls into class D, and this case has no protected edge states; this can easily be seen in the expansion from large v. In other words, the number of edge states is only protected modulo 2 in class D in $d = 1$.

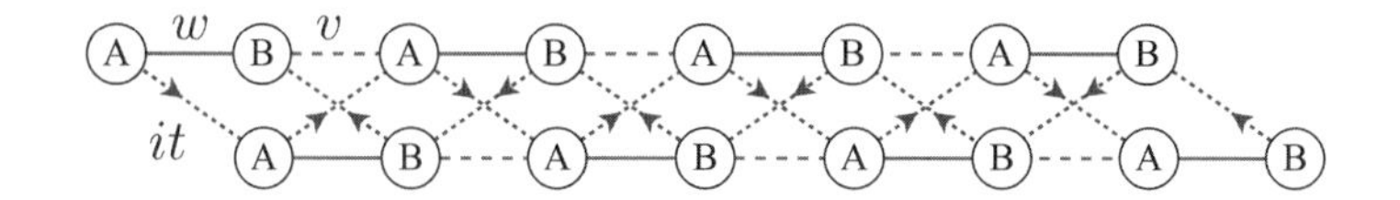

Figure 20.3 Two SSH chains coupled with an imaginary hopping it. This case has no protected edge state for $t \neq 0$.

20.2 Kane–Mele Insulators

We now turn to topological insulators in dimension $d = 2$. We have already discussed Chern insulators with a $\mathbb{Z}$ invariant in Section 18.4, which are in class A in Table 20.1, and require broken time-reversal symmetry for a non-zero Chern number. It was pointed out by Kane and Mele [125, 126] that topological insulators with a $\mathbb{Z}_2$ invariant are possible with time-reversal symmetry for spinful fermions with $\mathcal{T}^2 = -1$; this is class AII in Table 20.1.

Here, I present a brief discussion of the structure of this insulator [164] using the Chern–Simons theory of Chapter 17. First, let us consider a system in which the number of spin-up and spin-down electrons are independently conserved (there is no spin-flip scattering) and there is a bulk $U(1) \times U(1)$ symmetry. We imagine we can apply two independent external gauge fields $A_{\uparrow\mu}$ and $A_{\downarrow\mu}$, which can couple to these electrons. We also allow for time-reversal symmetry. Naively, this would imply that the Chern numbers of the bands of the electrons are zero. However, this is not the case, because time reversal flips the particle spin and also the Chern number of a band. So, we could have a situation in which the spin-up particles occupy a band with Chern number $C_\uparrow = 1$, while the spin-down particles occupy a band with Chern number $C_\downarrow = -1$. Then, integrating out the electrons, we obtain a Chern–Simons theory for the gauge fields $A_{\uparrow\mu}$ and $A_{\downarrow\mu}$ as in (17.1) with the K matrix

$$K = \begin{pmatrix} 1 & 0 \\ 0 & -1 \end{pmatrix}. \tag{20.17}$$

From the results in Chapter 17, we can see that such a K matrix only allows fermionic excitations in the bulk, or their composites; therefore, the bulk is trivial. Note also that $|\det K| = 1$, so there is no degeneracy on the torus. This is just as expected, given the construction of such a state using free electrons in Chern bands.

The arguments in Chapter 17 and 19 show that the edge of such a system has two copies of the integer quantum Hall edge, one left-moving with spin up, and the other right-moving with spin down. So, we can write the edge theory in terms of a left-moving fermion $\Psi_\uparrow$ and a right-moving fermion $\Psi_\downarrow$:

$$\mathcal{S}_e = \int dx d\tau \left[\Psi_\uparrow^\dagger \left(\frac{\partial}{\partial \tau} + iv \frac{\partial}{\partial x} \right) \Psi_\uparrow + \Psi_\downarrow^\dagger \left(\frac{\partial}{\partial \tau} - iv \frac{\partial}{\partial x} \right) \Psi_\downarrow \right]. \tag{20.18}$$

An important observation made by Kane and Mele [125, 126] was that the $U(1) \times U(1)$ symmetry is actually not required, and the edge states described by (20.18) survive even in the presence of a single $U(1)$ symmetry, provided time reversal is preserved. This is crucial for realizing this insulator because, with spin–orbit interactions, the number of spin-up and spin-down electrons is not separately conserved, and there is only a single $U(1)$ symmetry associated with the conservation of the total number of electrons. In addition spin–orbit interactions do conserve time-reversal symmetry.

One way to see this is to check for allowed edge operators that can gap out the pair of edge states in (20.18). In the absence of any symmetries, we could imagine a back-scattering term such as

$$\mathcal{H}'_e = \int dx \left[v(x)\Psi^\dagger_\uparrow(x)\Psi_\downarrow(x) + v^*(x)\Psi^\dagger_\downarrow(x)\Psi_\uparrow(x) \right], \tag{20.19}$$

where we have even allowed for the breaking of translational symmetry on the edge by a disordered coupling $v(x)$. Such a term, if present, would gap out the edge states of (20.18). However, the key observation is that a term such as (20.19) is forbidden in systems with time-reversal symmetry. The time-reversal operation is anti-unitary and so

$$\mathcal{T}v(x)\mathcal{T}^{-1} \to v^*(x), \tag{20.20}$$

while the electron operator transforms as

$$\mathcal{T}\Psi_\uparrow\mathcal{T}^{-1} \to \Psi_\downarrow \quad , \quad \mathcal{T}\Psi_\downarrow\mathcal{T}^{-1} \to -\Psi_\uparrow. \tag{20.21}$$

Note that $\mathcal{T}^2 = -1$ on the electron operator. Now we can see that (20.19) is not invariant under time reversal. A single-fermion backscattering term is therefore forbidden, and the edge states remain gapless. In particular, the zero-energy states at momentum $k_x = 0$ are protected by a Kramers degeneracy.

To see that this insulator has a $\mathbb{Z}_2$ invariant, we use an argument similar to that in Fig. 20.3 for the SSH chain. We take two copies of the insulator just described, with edge modes $\Psi_{1\uparrow}, \Psi_{1\downarrow}$ and $\Psi_{2\uparrow}, \Psi_{2\downarrow}$. Then, it is possible to have a term which gaps out the edge states and preserves time-reversal symmetry:

$$\begin{aligned}\mathcal{H}''_e = \int dx \Big[& v(x)\Psi^\dagger_{1\uparrow}(x)\Psi_{2\downarrow}(x) + v^*(x)\Psi^\dagger_{2\downarrow}(x)\Psi_{1\uparrow}(x) \\ & -v^*(x)\Psi^\dagger_{1\downarrow}(x)\Psi_{2\uparrow}(x) - v(x)\Psi^\dagger_{2\uparrow}(x)\Psi_{1\downarrow}(x) \Big].\end{aligned} \tag{20.22}$$

More generally, this argument shows that only the number of edge modes modulo 2 is protected, and hence this topological insulator is of $\mathbb{Z}_2$ nature.

There have also been studies of the effect of electron–electron interactions on the edge states of the Kane–Mele insulator [314, 316]. These can lead to back scattering in the presence of time reversal, but are only relevant for sufficiently strong interactions.

20.3 Odd-Parity Superconductors

A second important class of topological band structure arises in superconductors, when we consider the Bogoliubov Hamiltonian of the fermionic quasiparticles, similar to those studied in Sections 4.3 and 9.3. We consider here the case of triplet pairing, when the spatial wavefunction of the Cooper pair has to have odd parity. In the simplest situation, we can just ignore the spin of the electrons (assuming all electrons are spin-polarized in the same direction), and so we consider here the Bogoliubov theory of pairing of spinless fermions f_i on a lattice of sites i. Then, a general Hamiltonian is

$$H_f = -\sum_i \mu f_i^\dagger f_i - \sum_{\langle ij \rangle} \left(t f_i^\dagger f_j + t^* f_j^\dagger f_i \right) - \sum_{\langle ij \rangle} \left(\Delta_{ij} f_i^\dagger f_j^\dagger + \Delta_{ij}^* f_j f_i \right), \tag{20.23}$$

which has chemical potential μ, nearest-neighbor hopping t, and nearest-neighbor pairing $\Delta_{ij} = -\Delta_{ji}$. We will consider the cases of spatial dimensions $d = 1$ and $d = 2$ separately in the following subsections.

20.3.1 Dimension $d = 1$

We can always choose a gauge in which t is real and positive, but let us take the pairing to be complex

$$\Delta_{i,i+\hat{x}} = \Delta_1 + i\Delta_2, \tag{20.24}$$

with $\Delta_{1,2}$ real. Considering an infinite chain, we can write the Hamiltonian in momentum space using a Nambu spinor

$$\Psi_p = \begin{pmatrix} f_p \\ f_{-p}^\dagger \end{pmatrix}. \tag{20.25}$$

Note that the spinless nature of the fermions introduces a redundancy in the Nambu notation, with

$$\Psi_p^\dagger = \Psi_{-p}^T \tau^x, \tag{20.26}$$

where $\tau^{x,y,z}$ are Pauli matrices acting on the Nambu space, and T-superscript is a transpose. So Ψ_p and Ψ_{-p} describe the same fermions, and this will play an important role in our analysis of edge states. The Hamiltonian is

$$H_f = \frac{1}{2} \sum_p \Psi_p^\dagger H_{\mathrm{BdG}}(p) \Psi_p, \tag{20.27}$$

where the factor of $1/2$ is associated with (20.26), and

$$H_{BdG}(p) = d_x(p)\tau^x + d_y(p)\tau^y + d_z(p)\tau^z, \tag{20.28}$$

with the $\boldsymbol{d}$ vector

$$\boldsymbol{d}(p) = (2\Delta_2 \sin(p), 2\Delta_1 \sin(p), -\mu - 2t\cos(p)). \tag{20.29}$$

This has the same form as the SSH Hamiltonian in (20.3) with the $\boldsymbol{d}$ vector in (20.14). From this, we can easily conclude that the topological class is D with a $\mathbb{Z}_2$ invariant for $\Delta_{1,2}$ both non-zero, and BDI with a $\mathbb{Z}$ invariant when only one of Δ_1 and Δ_2 are non-zero. We can also see from the topological invariants of these classes that a topologically non-trivial phase is obtained for $|\mu| < 2t$.

Upon consideration of the edge states, the reality constraint in (20.26) makes an important difference; rather than obtaining a zero mode of a complex fermion f at each edge, we obtain a real "Majorana" fermion zero mode at each edge. We can define two

Majorana fermions on each site by decomposing the operators f_i and $f_i^\dagger$ into Majorana fermions γ_{Ai} and γ_{Bi}

$$f_i = \frac{1}{2}(\gamma_{Bi} + i\gamma_{Ai}) \quad , \quad f_i^\dagger = \frac{1}{2}(\gamma_{Bi} - i\gamma_{Ai}) . \tag{20.30}$$

The Majorana fermions anti-commute with distinct labels, and square to unity

$$\gamma_{Ai}^2 = \gamma_{Bi}^2 = 1 . \tag{20.31}$$

The operators γ_{Ai} and γ_{Bi} act upon the two-dimensional Hilbert space of empty and occupied states of the fermion f_i, and they are equivalent to the Pauli σ_i^x and σ_i^y operators on this subspace. The Pauli σ_i^z operator is identified with fermion parity:

$$i\gamma_{Bi}\gamma_{Ai} = 2f_i^\dagger f_i - 1 = \pm 1 . \tag{20.32}$$

Note, however, that this does not imply that the set of fermions f_i map to a set of Pauli matrices with such a relation; the fermions anti-commute for different i, while the Pauli matrices commute.

We now insert (20.30) into (20.23), and write the lattice Hamiltonian in terms of the Majorana fermions, working in the case $\Delta_2 = 0$ for simplicity. For a chain of length L, we obtain

$$H_f = -\frac{\mu}{2}\sum_{i=1}^{L}(1 + i\gamma_{Bi}\gamma_{Ai}) + \frac{i}{2}\sum_{i=1}^{L-1}\left[(\Delta_1 + t)\gamma_{Ai}\gamma_{B,i+1} + (\Delta_1 - t)\gamma_{B,i}\gamma_{A,i+1}\right] . \tag{20.33}$$

We now observe the similarity between the inter-site term in (20.33) and the SSH Hamiltonian in (20.1), and highlight this is Fig. 20.4. We can compute the edge modes of H_f using an analysis that parallels that of the SSH Hamiltonian (20.1) in Section 20.1. The main difference is that the complex fermion eigenmodes of a real symmetric matrix are now replaced by the Majorana fermion eigenmodes of an imaginary antisymmetric matrix. We will not enter into details, apart from pointing out the solvable model with a Majorana zero mode localized exactly on the edge of the chain noted in Fig. 20.4, which is similar to the $w = 0$ edge mode in Fig. 20.1.

20.3.2 Dimension $d = 2$

Here, we consider superconductors described by (20.23) on the square lattice, which break time-reversal symmetry in "$p + ip$" pairing, with

$$\Delta_{i,i+\hat{x}} = \Delta \quad , \quad \Delta_{i,i+\hat{y}} = i\Delta \tag{20.34}$$

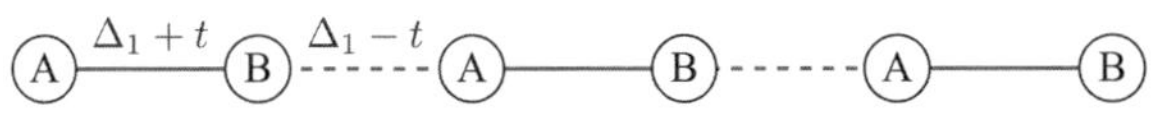

A $\Delta_1 + t$ B $\Delta_1 - t$ A B A B

Figure 20.4 Pictorial represenation of the Majorana Hamiltonian in (20.33). There is an exact Majorana zero mode localized on the A site on the left edge for $\mu = 0$ and $\Delta_1 = -t$. Compare Fig. 20.1 for the SSH model.

where real $\Delta > 0$. The Nambu spinor representation and Fourier transform of H_f proceeds just as in (20.25), (20.27), and (20.28), apart from the replacement of p by a vector $\boldsymbol{p}$. The reality condition in (20.26) also continues to apply. The $\boldsymbol{d}$ vector in (20.29) is now given by

$$\boldsymbol{d}(\boldsymbol{p}) = (2\Delta \sin p_y, 2\Delta \sin p_x, -\mu - 2t\cos p_x - 2t\cos p_y)\,. \tag{20.35}$$

The first two components of $\boldsymbol{d}(\boldsymbol{p})$ describe odd-parity pairing, which is $p_x + ip_y$ pairing in the continuum. Note

$$H^T_{BdG}(-\boldsymbol{p}) = -\tau^x H_{BdG}(\boldsymbol{p})\,\tau^x, \tag{20.36}$$

which establishes consistency with (20.26), and shows the presence of particle–hole $\mathcal{C}$ symmetry of class D in $d = 2$ in Table 20.1.

The eigenvalues of $H_{BdG}(\boldsymbol{p})$ are $\pm|\boldsymbol{d}(\boldsymbol{p})|$ and these describe two particle–hole symmetric fermion bands; only the lower one is occupied in the ground state. We can compute the Chern number of this band by the usual methods of band theory for fermions without pairing, as discussed in Section 18.3. Alternatively, we can think of $H_{BdG}(\boldsymbol{p})$ of fictitious spin 1/2 in the presence of a $\boldsymbol{p}$-dependent "Zeeman" field $\boldsymbol{d}(\boldsymbol{p})$. From the expression for the Berry phase of spin 1/2 in Section 18.2, we then obtain the Chern number as a measure of the number of times $\boldsymbol{d}(\boldsymbol{p})$ wraps the sphere in spin space. Using either method, we obtain the Chern number

$$\nu = \frac{1}{4\pi}\int d^2p\, \frac{\boldsymbol{d}(\boldsymbol{p})}{|\boldsymbol{d}(\boldsymbol{p})|^3}\cdot\left(\frac{\partial \boldsymbol{d}(\boldsymbol{p})}{\partial p_x}\times\frac{\partial \boldsymbol{d}(\boldsymbol{p})}{\partial p_y}\right). \tag{20.37}$$

This Chern number is non-zero for $|\mu| < 4t$, when the third component of (20.35) changes sign in the Brillouin zone. We focus on the regime with $\mu - 4t > 0$ and small, where a continuum limit is possible.

As in Section 18.4, we can now conclude that there is a single edge state associated with the lower occupied band. However, there is a crucial difference in the present situation with paired fermions arising from the redundancy in (20.26). This redundancy implies that the edge mode contains only *half* the degrees of freedom of the edge mode of a Chern insulator.

An elegant way to extract half the fermion degrees is to express the physics in terms of Majorana fermions. For an ordinary Chern insulator, we can write the edge theory as

$$H_{CI,edge} = \sum_p v_F\, p\, \eta_p^\dagger \eta_p, \tag{20.38}$$

where η_p is a canonical "complex" fermion of a one-dimensional momentum p. Let us write this Hamiltonian using Majorana fermions γ_{1p} and γ_{2p}, which obey the momentum-space generalization of relations near (20.31)

$$\begin{aligned}\gamma^\dagger_{ap} &= \gamma_{a,-p},\\ \gamma_{a,-p}\gamma_{b,p'} + \gamma_{b,p'}\gamma_{a,-p} &= 2\delta_{ab}\delta_{pp'},\end{aligned} \tag{20.39}$$

with $a, b = 1, 2$. Then, using

$$\eta_p = \frac{1}{2}(\gamma_{1p} + i\gamma_{2p}), \tag{20.40}$$

the edge theory of the Chern insulator can be written as

$$H_{CI,edge} = \frac{1}{4}\sum_p v_F p\,(\gamma_{1,-p}\gamma_{1p} + \gamma_{2,-p}\gamma_{2p}). \tag{20.41}$$

Now we return to the edge theory associated with the H_{BdG} in (20.27) with the $\boldsymbol{d}$ vector in (20.35). If we ignore the constraint in (20.26), the analysis would proceed just as in the Chern insulator, and we would end up with the theory in (20.41). A full analysis can proceed by writing (20.27), (20.35) in terms of Majorana fermions, and then analyzing the edge structure in that approach. However, we can quickly obtain the answer by noting that (20.26) projects out half the degrees of freedom, and that should also be the case on edge. And it is easy to pull out half the degrees of freedom in (20.41) simply by dropping one species of the Majorana fermion; so we obtain the theory of the edge state of H_{BdG}:

$$H_{edge} = \frac{1}{4}\sum_p v_F p\,\gamma_{-p}\gamma_p, \tag{20.42}$$

expressed in terms of a single chiral Majorana fermion γ_p in one spatial dimension. In a more complete analysis, the dropped Majorana fermions would appear on the opposite edge of the sample.

In real space and imaginary time, the action of the chiral Majorana theory is

$$\mathcal{S}_{edge} = \frac{1}{4}\int dx d\tau\, \gamma\left(\frac{\partial}{\partial \tau} - i v_F \frac{\partial}{\partial x}\right)\gamma, \tag{20.43}$$

where $\gamma(x, \tau)$ is a Grassman field.

PART IV

FRACTIONALIZATION AND EMERGENT GAUGE FIELDS II

21 Parton Theories

This chapter provides a unified bird's-eye view of the fractionalized phases studied in this book. The all-in-one tool used to construct such phases is the parton method. All the important phases with anyonic excitations are obtained by placing the partons in topological bands.

Part IV continues the discussion of phases with fractionalized excitations and emergent gauge fields that were studied earlier in Part II. This part will expand the library of such phases by applying the results of Part III on free-fermion band topology.

This chapter provides a unified bird's-eye view of the fractionalized phases studied in this book. The all-in-one tool used to construct such phases is the parton method. We first met this in Chapter 14 when we fractionalized the XY order parameter Ψ into two bosonic partons ϕ, and then in Chapter 15, where we fractionalized the spin operators into bosonic partons s_i^α (recalled in Appendix D); these fractionalizations led to the theory of the $\mathbb{Z}_2$ spin liquid. The fractionalization methods have similarities to early work in the particle-theory literature [25, 58, 311], although the latter did not include the spin Berry phases (see Appendix C). It is also possible to fractionalize spin operators into fermionic partons, and we will use this method in Part IV. A significant advantage of fermionic partons is that it is possible for the partons to acquire one of the non-trivial band topologies we described in Part III – we will see that this band topology of partons has a profound and interesting feedback on the structure of the fractionalized phase itself.

It should be noted that the existence of a particular parton construction is not a guarantee that the associated fractionalization occurs in a particular microscopic model. But experience has shown that the parton method is usually the simplest and most direct way of completely understanding the structure of a phase once one has some other evidence of the nature of the fractionalization.

21.1 Spin Fractionalization into Bosonic Partons

Let us now summarize the phases studied by fractionalizing the spin into bosonic partons in Part II; see Fig. 21.1. We discuss the three possibilities in Fig. 21.1 in turn:

- The simplest state formed by bosons is a superfluid, realized by Bose condensation, as discussed in Chapter 3. When bosonic partons condense, we obtain a magnetically ordered state, as described in Section 15.4.1 for the triangular lattice. The bosonic partons carry both a gauge charge and a spin quantum number; the Bose

Bosons	Bose partons $S = b_\alpha^\dagger \frac{\sigma_{\alpha\beta}}{2} b_\beta$
Superfluid	Antiferromagnetic order
Paired superfluid	Square lattice: $\mathbb{CP}^1$ model or $SO(5)_1$ WZW model Triangular lattice: $\mathbb{Z}_2$ spin liquid
SPT	Chiral spin liquid

Figure 21.1 The correspondence between the phases of bosons, and the phases of spin systems obtained by expressing the spins in terms of bosonic partons. The Wess–Zumino–Witten (WZW) model is a theory of a five-component field representing the Néel and valence bond solid orders.

condensation fully higgses the gauge symmetry, and the spin index leads to broken spin-rotation symmetry.

- It is possible for pairs of bosons to form molecules, and for the molecular bosons to condense. The Q_{ij} variables in Chapter 15 are molecules of bosonic partons, and we found two distinct possibilities upon their condensation. If the condensation pattern was such that the $U(1)$ gauge symmetry was broken to $\mathbb{Z}_2$, then we obtained the $\mathbb{Z}_2$ spin liquid of Chapter 15. On the other hand, if the paired-parton condensate left a "staggered" $U(1)$ gauge theory unbroken, we obtain gapless spin liquids described by the $\mathbb{CP}^1$ model, is be discussed in Chapter 28.
- The last possibility in Fig. 21.1 is a route that is not studied in this book. If the bosonic partons form a suitable symmetry-protected topological (SPT) state, then the spin system realizes a chiral spin liquid [106]. We study the chiral spin liquid using fermionic partons in Chapter 22, as noted below in Section 21.2.

21.2 Spin Fractionalization into Fermionic Partons

Next, we turn to phases obtained by decomposing spins into fermionic partons. Here, the possibilities turn out to be quite rich, thanks to the band topologies studied in Part III, and are listed in Fig. 21.2.

- The simplest state for fermions is the Fermi liquid studied in Chapter 2. With fermionic partons, we obtain a spinon Fermi surface. The spinon quasiparticles on the Fermi surface experience strong gauge fluctuations, and this leads to a Fermi surface where the excitations are not quasiparticles. We examine the spinon Fermi surface state in Chapter 34.
- Fermions can also form paired Bardeen–Cooper–Schrieffer (BCS) states, studied in Chapter 4. With fermionic partons, we obtain a $\mathbb{Z}_2$ spin liquid, which can be matched to $\mathbb{Z}_2$ spin liquids obtained from bosonic partons, as we noted in Section 15.4.3.

Electrons	Fermion partons $S = f_\alpha^\dagger \frac{\sigma_{\alpha\beta}}{2} f_\beta$
Fermi liquid	Bose metal Spinon Fermi surface
BCS superconductor	$\mathbb{Z}_2$ spin liquid
Chern insulator (class A)	Chiral spin liquid
$p_x + ip_y$ superfluid (class D)	ITO-spin liquid with non-abelian Ising anyons (Kitaev honeycomb)
Semi-metals with gapless Dirac spectrum (graphene)	Gapless spin liquid Square lattice: $N_f = 2$ $SU(2)$ QCD or $SO(5)_1$ WZW model Triangular lattice: $N_f = 4$ QED
Complex SYK non-Fermi liquid	SY spin liquid

Figure 21.2 The correspondence between the phases of fermions and the phases of spin systems obtained by expressing the spins in terms of fermionic partons. The $SU(2$ quantum chromodynamics (QCD) and $U(1)$ quantum electrodynamics (QED) theories have spinons with a massless Dirac spectrum coupled to emergent gauge fields.

- An interesting possibility is for fermions is to form a band insulator with band topology in class A of Table 20.1, and this leads to the Chern insulator of Section 18.4. We see in Chapter 22 that fermionic partons in a Chern insulator realize a chiral spin liquid.
- Next, we consider fermions forming a paired superfluid in class D of Table 20.1 so that the Bogoliubov quasiparticles are in bands with a non-zero Chern number. With fermionic partons, this possibility leads to a remarkable state discussed in Chapter 23, which has "Ising" anyons with non-abelian statistics. Such a state has Ising topological order (ITO), and is realized in spin systems without spin-rotation symmetry (e.g., in the presence of spin–orbit interactions), as also discussed in Chapter 23.
- Fermions can form semi-metals with the spectrum of massless Dirac particles, as in graphene. When fermionic partons have a massless Dirac dispersion, the coupling to the gauge field is important, and we obtain strongly coupled relativistic gauge theories. The nature of such gauge theories for the square lattice are discussed in Chapter 28. There has also been interesting work on such gauge theories on the triangular and kagome lattices [268, 270], which could well be the low-energy description of models like (15.49) near the point where the phases in Fig. 1.8 and Fig. 14.4 meet.
- Finally, fermions with random interactions in a Sachdev–Ye–Kitaev (SYK) model can form a non-Fermi liquid without quasiparticle excitations, as discussed in Chapter 32. When fermionic partons form a SYK liquid, a gapless spin liquid is obtained, which is examined in Chapter 33.

Fermions	Fermion partons $c = \psi_1\psi_2\psi_3$
ψ_1, ψ_2, ψ_3: IQH	Laughlin and Jain FQH
ψ_3: Fermi liquid ψ_1, ψ_2: IQH	Fermi surface in the half-filled Landau level (HLR)
ψ_3: $p_x + ip_y$ superfluid (class D) ψ_1, ψ_2: IQH	Moore-Read non-abelian FQH

Figure 21.3 Fractional quantum Hall (FQH) states obtained by the decomposition of the electron into three fermionic partons.

21.3 Quantum Hall States

Let us now turn to quantum Hall states. The integer quantum Hall state was discussed in Chapter 19, and some fractional quantum Hall states are described in Chapter 24. We summarize in Fig. 21.3 how these fractional states can be obtained by the parton method. We assume the electron is fully spin polarized, and ignore its spin. Then the electron c is fractionalized into a product of three fermionic partons, ψ_1, ψ_2, and ψ_3. The possibilities noted in Fig. 21.3 are:

- When all three fermionic partons occupy fully filled Landau levels in integer quantum Hall (IQH) states, we obtain the most commonly observed Laughlin and Jain fractional quantum Hall states, as described in Chapter 24.
- When the ψ_1, ψ_2 fermionic partons occupy fully filled Landau levels, but the parton ψ_3 moves in a vanishing average magnetic field and forms a Fermi liquid, we obtain the Halperin–Lee–Read (HLR) compressible Hall state. The interaction of the gauge field with the Fermi surface excitations is important, as discussed in Section 34.3.
- An interesting possibility is that the ψ_3 fermion of the HLR state pairs to form a superfluid in class D of Table 20.1. This leads to the non-abelian Moore–Read state, and is discussed in Section 24.4.

21.4 Correlated Metals

Finally, we turn to correlated metallic phases, which will be the focus of Part V. A commonly used approach here is to fractionalize the electron into the product of a bosonic parton and a fermionic parton. Possible phases are listed in Fig. 21.4.

Fermions and bosons	Fermi and Bose partons $c = f\, b^\dagger$
f_α: Fermi liquid b: Superfluid	Fermi liquid with heavy quasiparticles
f_α: Spinon band b: Normal liquid	Model for pseudogap metal ?
f: Fermi liquid b_α: Paired superfluid	$\mathbb{Z}_2$ holon metal

Figure 21.4 Fractionalization of the electron into a boson b and fermion f. The spin of the electron can reside either on the fermion (denoted f_α) or on the boson (denoted b_α). We do not use these popular electron fractionalizations in this book, and instead advocate the paramagnon fractionalization in Section 31.4.

- Placing the spin of the electron on the fermionic parton, and condensing the bosonic parton, leads to a Fermi liquid. This is one of the rationales behind the "vanilla" state [10] discussed in Section 9.3.
- Rather than condensing the bosonic parton, we can consider the possibility that the bosons form a normal liquid at intermediate temperatures. This has been used as a starting point for a model for the pseudogap metal phase of the cuprate superconductors [153]. We argue in Section 31.4 that writing the electron as a product of a fermion and a boson is not a well-behaved fractionalization for the t–J model with $t \gg J$. Section 31.4 also presents arguments that a better approach is one which fractionalizes the paramagnon of Section 9.4. The paramagnon fractionalization theory leads directly to a fractionalized Fermi liquid (FL*) model for the pseudogap metal (see Fig. 31.9b). We see in Section 31.4 that the paramagnon fractionalization theory has a resemblance to the successful three-fermionic parton method for fractional quantum Hall states noted in Section 21.2.
- Another possibility is that the electronic spin resides on the bosonic partons, and these partons pair-condense to form a $\mathbb{Z}_2$ spin liquid, as noted in Section 21.1 and Chapter 15. The fermionic partons are spinless, and can form a metallic Fermi surface state. The resulting state is called a "holon metal" in the context of the underdoped cuprates, and is noted briefly in Section 31.4 (see Fig. 31.8).

The discussion above of metallic phases with electron fractionalization is in the context of the single-band t–J model. The discussion of the metallic phases of Kondo lattice models in Chapters 30 and 31 also involves a "hybridization" or "slave" boson, which we denote P. From the perspective of this chapter (and this book) the P boson should not be viewed as a parton, but as a molecular bound state of an electron c_α in one band, and a fermionic spinon f_α in a *separate* band.

22 The Chiral Spin Liquid

The chiral spin-liquid state of quantum antiferromagnets is obtained by expressing the spin operator in terms of Schwinger fermions, and placing these fermions in a band structure with a non-zero Chern number.

This chapter describes a case of parton band topology from Section 21.2.

We return to the $S = 1/2$ antiferromagnet on the square lattice, considered earlier in Chapter 15 for the theory of the $\mathbb{Z}_2$ spin liquid:

$$\mathcal{H} = \sum_{i,j} J_{ij} \boldsymbol{S}_i \cdot \boldsymbol{S}_j, \tag{22.1}$$

where $\boldsymbol{S}_i$ are spin-1/2 quantum spin operators on the sites i of the square lattice. As discussed near (9.18), we can, equivalently, consider this as a theory of hard-core bosons B_i, with the operator correspondence

$$B_i = S_{i-} \quad , \quad B_i^\dagger B_i - 1/2 = S_{iz}, \tag{22.2}$$

so that an occupied (empty) boson state is a spin-up (-down) state.

In contrast to the Schwinger-boson representation used in Chapter 15, we will now employ the Schwinger-*fermion* representation

$$\boldsymbol{S}_i = \frac{1}{2} f_{i\alpha}^\dagger \boldsymbol{\sigma}_{\alpha\beta} f_{i\beta}, \tag{22.3}$$

where $f_{i\alpha}$ are canonical fermions and the σ^a, $a = x, y, z$, are the usual 2×2 Pauli matrices. The fermions obey the constraint

$$\sum_\alpha f_{i\alpha}^\dagger f_{i\alpha} = 1 \quad , \quad \text{for all } i. \tag{22.4}$$

The subsequent steps closely parallel those of Chapter 15; we decouple the fermion quartic terms to obtain a free-fermion mean-field Hamiltonian, and then examine the gauge structure of the fluctuations. We find here that the fermions can choose to acquire a mean-field Hamiltonian with non-trivial band topology (class A from Table 20.1), and this has strong consequences for the anyon structure of the resulting spin-liquid state: specifically, a non-zero Chern number in the spinon bands leads to a Chern–Simons term in the gauge theory of the spin liquid [306].

22.1 Mean-Field theory

Rather than explicitly carrying out the mean-field computation, we outline the basic steps:

- We use simple identities to write the exchange interaction on each link as

$$-\frac{J_{ij}}{2}\left(f^{\dagger}_{i\alpha}f_{j\alpha}\right)\left(f^{\dagger}_{j\beta}f_{i\beta}\right). \tag{22.5}$$

- We decouple this interaction with a Hubbard–Stratonovich field Q_{ij}:

$$\frac{J_{ij}}{2}|Q_{ij}|^2 - Q_{ij}\left(f^{\dagger}_{i\alpha}f_{j\alpha}\right) - \text{H.c.} \tag{22.6}$$

- We examine a saddle point in which we set $Q_{ij} = t_{ij}$, and examine fluctuations in the phases of the Q_{ij}.

Computations of this type lead to mean-field Hamiltonians for the f_α spinons in the following form:

$$H_f = \varepsilon_0 \sum_i f^{\dagger}_{i\alpha}f_{i\alpha} - \sum_{i<j}\left(t_{ij}f^{\dagger}_{i\alpha}f_{j\alpha} + t^{*}_{ij}f^{\dagger}_{j\alpha}f_{i\alpha}\right). \tag{22.7}$$

Here, t_{ij} has become the spinon hopping, and ε_0 is the saddle point of the Lagrange multiplier imposing the constraint (22.4); we will find $\varepsilon_0 = 0$ for the cases we consider here.

For most choices of the hopping matrix elements in (22.7), the f fermions acquire a Fermi surface, and we obtain a spin liquid with a "spinon Fermi surface." The coupling between the emergent gauge field and the spinon Fermi surface leads to many singular effects, which are discussed further in Chapter 34. Here, we consider the case with the hopping matrix elements t_{ij} shown in Fig. 18.2, with nearest-neighbor hopping of magnitude t_1 and purely imaginary second-neighbor hopping of magnitude t_2, for which there is no Fermi surface and the spinon excitations are gapped. The imaginary diagonal hopping matrix elements imply that time-reversal symmetry has been spontaneously broken.

For $t_2 = \varepsilon_0 = 0$, the Hamiltonian in (18.39) has Dirac nodes at $\boldsymbol{k} = (\pm\pi/2, 0)$. We focus on the vicinities of these points by writing $\boldsymbol{k} = (\pm\pi/2 + q_x, q_y)$ and expand for small q_x, q_y. We also introduce Pauli matrices τ^x, τ^y, τ^z in the sublattice space. Then we can write the Hamiltonian as

$$H_f = f^{\dagger}\left[\pm 2t_1 q_x \tau^x + 2t_1 q_y \tau^y \mp 4t_2 \tau^z\right] f, \tag{22.8}$$

where we don't write out the spin (α,β), sublattice (a,b), or valley (momenta near $(\pm\pi/2,0)$) indices on the spinons f. (Note that because of the spin label in (22.7), the present case has 2 copies of the model considered in Section 18.3.) This is the Hamiltonian for two-component Dirac fermions with masses $\pm 4t_2$. We introduce the relativistic notation with fermion fields defined by $\psi = f$, $\bar{\psi} \equiv \psi^{\dagger}\tau^z$, and $\gamma^{\mu} = (\tau^z, -\tau^y, \tau^x)$, and then the imaginary time Lagrangian corresponding to the Dirac fermion near one of the valleys in H_f is

$$\mathcal{L}_f = \bar{\psi}\gamma^\mu \partial_\mu \psi + M\bar{\psi}\psi, \tag{22.9}$$

where we have absorbed the Fermi velocity $2t_1$ by rescaling time, and $M = \pm 4t_2$.

We make a number of important observation from $\mathcal{L}_f$:

- For $t_2 = 0$, in the π-flux state, the spinon spectrum is gapless, and the low-energy excitations are described by four species of massless two-component Dirac fermions: two species from the valleys at $(\pm\pi/2, 0)$, and two species from the spin index α.
- For $t_2 \neq 0$, the Dirac fermions acquire a "mass" (a "gap") by breaking time-reversal symmetry.
- The lower band is fully occupied, while the upper band is empty; this ensures the half-filling condition at $\varepsilon_0 = 0$. As we showed in Section 18.3, for $t_2 \neq 0$, the occupied band has a non-zero Chern number ± 1 (sign determined by the sign of t_2). The unoccupied has the opposite Chern number.

22.2 Gauge Fluctuations

As in Chapter 15, the most important feature determining the structure of fluctuations is the gauge symmetry under which

$$\begin{aligned} f_{i\alpha}^\dagger &\to f_{i\alpha}^\dagger \exp\left(i\rho_i(\tau)\right), \\ Q_{ij} &\to Q_{ij} \exp\left(-i\rho_i(\tau) + i\rho_j(\tau)\right). \end{aligned} \tag{22.10}$$

However, the nature of the long-wavelength limit in the theory of the gauge fluctuations is now simpler than the bosonic spinon case in Chapter 15. Because of the opposite signs in the transformation of Q_{ij} in (22.10), we can write the fluctuations as

$$Q_{ij} = t_{ij} \exp\left(i \int_{\boldsymbol{r}_i}^{\boldsymbol{r}_j} d\boldsymbol{r} \cdot \boldsymbol{a}(\boldsymbol{r})\right), \tag{22.11}$$

and then the vector field $\boldsymbol{a}$ transforms as $\boldsymbol{a} \to \boldsymbol{a} + \boldsymbol{\nabla}\rho$, just like the vector potential of a $U(1)$ gauge field. As in Chapter 15, we obtain the time component of the gauge field from the fluctuations of the Lagrange multiplier imposing the constraint in (22.4). Using gauge invariance, we can now simply describe the consequences of long-wavelength gauge fluctuations on the spinons: the Lagrangian $\mathcal{L}_f$ is modified to

$$\mathcal{L}_f = \bar{\psi}\gamma^\mu (\partial_\mu - ia_\mu)\psi + M\bar{\psi}\psi. \tag{22.12}$$

So, we have a $U(1)$ gauge field a_μ coupled to four species of Dirac fermions with mass M.

The massless case $M = 0$ of the π-flux state has a number of subtleties which we consider in Chapter 28, and we limit ourselves to the case where $M \neq 0$. In this case, we can integrate spinons out, and by a close parallel of the computations in Section 18.4 we find that the unit Chern number of the filled spinon band leads to a $U(1)$ Chern–Simons gauge theory for a_μ at level $m = 2$:

$$\mathcal{L}_a = \frac{2i}{4\pi} \int d^3x \varepsilon_{\mu\nu\lambda} a_\mu \partial_\nu a_\lambda . \tag{22.13}$$

The $m = 2$ arises from the sum over the spin indices of the spinons.

The Lagrangian $\mathcal{L}_a$ describes the *chiral spin liquid*, and we can deduce its properties from Chapter 17: the f_α spinons are "semions" with statistical angle $\pi/2$, and there is a two-fold degeneracy on the torus.

22.3 Edge States

The chiral spin liquid has protected edge states. We can deduce the edge theory from the general Chern–Simons theory considerations in Section 17.4, but let us employ a more direct approach here.

The filled spinon band has Chern number 1, and so prior to imposing gauge constraints, this band has free-fermion edge states, as we described in Section 18.3 and Fig. 18.4. Because of the additional spin index here, there are two copies of the edge states in Fig. 18.4. By gauge invariance, the gauge field couples minimally to the edge fermions just as in the bulk; so in its bosonized form analogous to (19.33), we can write the edge theory as

$$\mathcal{L}_{edge} = \frac{1}{4\pi} \sum_\alpha \left[\bar{v} (\partial_x \varphi_\alpha)^2 + i \partial_x \varphi_\alpha \partial_\tau \varphi_\alpha \right] + \frac{i}{2\pi} \sum_{\mu,\nu=x,\tau} \sum_\alpha \varepsilon_{\mu\nu} a_\mu \partial_\nu \varphi_\alpha , \tag{22.14}$$

where $\alpha = \uparrow, \downarrow$ is a spin index, and the last term couples the gauge field to the fermion current on the edge. The integral over the a_μ gauge field now yields the constraint $\varphi_\uparrow + \varphi_\downarrow =$ constant. So we write $\varphi_\uparrow = -\varphi_\downarrow = \varphi$, and obtain the final form of the edge theory:

$$\mathcal{L}_{edge} = \frac{2}{4\pi} \left[\bar{v} (\partial_x \varphi)^2 + i \partial_x \varphi \partial_\tau \varphi \right] . \tag{22.15}$$

This is exactly the edge theory obtained in Section 17.4 on the Chern–Simons theories for $m = 2$.

The spinon operators on the edge are clearly $e^{\pm i\varphi}$. From these spinon operators, we can obtain the spin-flip (or boson B creation) operator

$$S_+ = f_\uparrow^\dagger f_\downarrow \sim e^{i\varphi_\uparrow} e^{-i\varphi_\downarrow} = e^{2i\varphi} . \tag{22.16}$$

It is also interesting to consider the spin density in the z direction (or the boson-number operator):

$$\frac{1}{2} \left(f_\uparrow^\dagger f_\uparrow - f_\downarrow^\dagger f_\downarrow \right) = \frac{1}{4\pi} \left(\partial_x \varphi_\uparrow - \partial_x \varphi_\downarrow \right) = \frac{1}{2\pi} \partial_x \varphi . \tag{22.17}$$

The underlying antiferromagnet has $SU(2)$ symmetry, and so we expect the correlators of $S_\pm$ and S_z to be equal to each other. And from the correlators of the $m = 2$ edge theory discussed earlier, we can indeed verify that their two-point correlators both decay as $(x_1 - x_2)^{-2}$.

22.4 *SU*(2) Gauge Theory

The Schwinger-fermion representation of spin operators in (22.3) actually has a larger gauge $SU(2)$ gauge symmetry. For the present chiral spin liquid, the $SU(2)$ gauge theory leads to the same state as the $U(1)$ gauge theory, as I now show. However, we will be able to exploit the $SU(2)$ gauge approach to obtain new spin-liquid phases in Chapter 28.

First, let us recall (22.3), and let us also define the Nambu pseudospin operators

$$\boldsymbol{T}_i = \frac{1}{2}\left(f_{i\downarrow}^\dagger f_{i\uparrow}^\dagger + f_{i\uparrow} f_{i\downarrow}, i\left(f_{i\downarrow}^\dagger f_{i\uparrow}^\dagger - f_{i\uparrow} f_{i\downarrow}\right), f_{i\uparrow}^\dagger f_{i\uparrow} + f_{i\downarrow}^\dagger f_{i\downarrow} - 1\right). \tag{22.18}$$

For a more transparent presentation of the symmetries, it is useful to write the fermions as 2×2 matrices:

$$\boldsymbol{f}_i = \begin{pmatrix} f_{i\uparrow} & -f_{i\downarrow}^\dagger \\ f_{i\downarrow} & f_{i\uparrow}^\dagger \end{pmatrix}. \tag{22.19}$$

This matrix obeys the relation

$$\boldsymbol{f}_i^\dagger = \sigma^y \boldsymbol{f}_i^T \sigma^y. \tag{22.20}$$

We can now write the spin and Nambu pseudospin operators as

$$\boldsymbol{S}_i = \frac{1}{4}\mathrm{Tr}(\boldsymbol{f}_i^\dagger \boldsymbol{\sigma} \boldsymbol{f}_i) \quad , \quad \boldsymbol{T}_i = \frac{1}{4}\mathrm{Tr}(\boldsymbol{f}_i^\dagger \boldsymbol{f}_i \boldsymbol{\sigma}). \tag{22.21}$$

The unit $\boldsymbol{f}$ occupancy constraint in (22.4) can be stated as the vanishing of the pseudospins on each site:

$$\boldsymbol{T}_i = 0. \tag{22.22}$$

Now we can see that the full gauge symmetry generated by the constraints is $SU(2)$ [153], because the constraint (22.22) and $\boldsymbol{S}_i$ are invariant under

$$SU(2)_g : \quad \boldsymbol{f}_i \to \boldsymbol{f}_i U_g(i), \tag{22.23}$$

where $U_g(i)$ is a $SU(2)$ matrix ($U_g^\dagger U_g = 1$, $\det(U_g) = 1$), which can depend upon space and time. This is distinct from the global $SU(2)$ spin-rotation symmetry, which acts as

$$SU(2)_s : \quad \boldsymbol{f}_i \to U_s \boldsymbol{f}_i, \tag{22.24}$$

with U_s space and time independent.

Turning to the spin model, we now ask whether it is possible to obtain a mean-field Hamiltonian similar to (22.7) that is invariant under $SU(2)$ gauge transformations. We begin by writing down a general Hamiltonian invariant under the global $SU(2)_s$ spin rotation in (22.24):

$$H_f = -\sum_{i<j} \mathrm{Tr}\left(\boldsymbol{f}_i^\dagger \boldsymbol{f}_j u_{ij}\right), \tag{22.25}$$

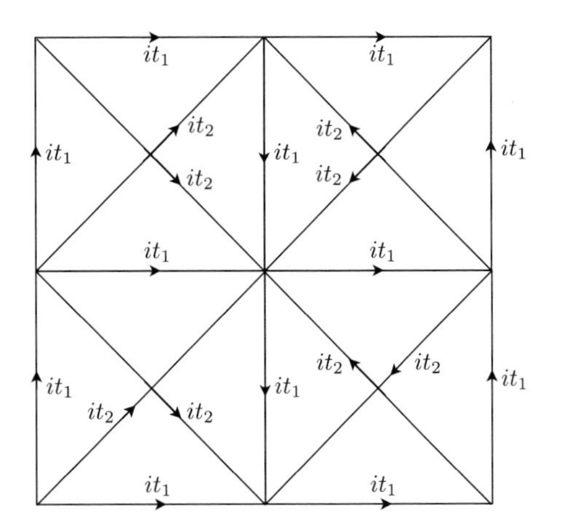

Figure 22.1 Unit cell of the saddle-point Hamiltonian, H_f in (22.27) with all imaginary hoppings. All fluxes are the same as those in Fig. 18.2.

where u_{ij} is a 2×2 matrix. Requiring (22.25) to be Hermitian using (22.20) we obtain the restriction

$$u_{ij} = -\sigma^y u_{ij}^* \sigma^y . \tag{22.26}$$

Finally, we also require that the $SU(2)_g$ gauge symmetry is unbroken in the mean-field Hamiltonian (22.25), and this constrains u_{ij} to be proportional to the unit matrix. Then, the only solution to (22.26) is that u_{ij} is a purely imaginary diagonal matrix, and (22.25) becomes

$$\begin{aligned} H_f &= -i \sum_{i<j} \bar{t}_{ij} \mathrm{Tr} \left(f_i^\dagger f_j \right) \\ &= -i \sum_{i<j} \bar{t}_{ij} \left(f_{i\alpha}^\dagger f_{j\alpha} - f_{j\alpha}^\dagger f_{i\alpha} \right), \end{aligned} \tag{22.27}$$

where the $\bar{t}_{ij}$ are real numbers. This is of the form of our earlier mean-field Hamiltonian in (22.7), but the t_{ij} in Fig. 18.2 are not purely imaginary. However, it is not difficult to perform a $U(1)$ gauge transformation to the t_{ij} in Fig. 18.2 so that all hopping terms are purely imaginary. The result is shown in Fig. 22.1, and the reader can verify that all the $U(1)$ fluxes in Fig. 22.1 are identical to those in Fig. 18.2.

The subsequent analysis of fluctuations proceeds as for the $U(1)$ gauge theory. We consider fluctuations in which $\bar{t}_{ij} \to \bar{t}_{ij} \exp(i a_{ij})$, where a_{ij} is now the spatial component of a $SU(2)$ gauge field. Integrating out the fermions yields a Chern–Simons term for the $SU(2)$ gauge field, but it is now at level $m = 1$, unlike the level $m = 2$ for the $U(1)$ gauge theory; this is because there is now only a single flavor of an $SU(2)$ gauge-charged fermion, and the α index on f_α has been absorbed into an $SU(2)$ color index. I now refer the reader to analyses of non-abelian Chern–Simons theories [172], from which it is known that the present Chern–Simons theory is the same as that obtained in Section 22.2, that is, the $SU(2)$ theory at Chern–Simons level 1 is the same as the $U(1)$ theory at Chern–Simons level 2.

23 Non-Abelian Ising Anyons

We begin with a fermionic spinon theory of the $\mathbb{Z}_2$ spin liquid, and show that it transforms into a new phase when the fermionic spinons have a non-trivial band topology. The band topology leads to Majorana zero modes bound to visons, and hence to non-abelian topological order.

The non-abelian Ising topological order (ITO) phase was discovered by Kitaev [134] using an exactly solvable model on the honeycomb lattice. This is a model of spin-1/2 electrons on the sites of a honeycomb lattice, interacting via Ising-type spin–spin couplings with a bond-dependent orientation. Strong spin–orbit couplings are required to obtain such interactions in practice, and there are materials, such as α-$RuCl_3$, whose spins are described by a Hamiltonian that contains the terms of the Kitaev Hamiltonian.

Rather than considering the exactly solved model, we consider a general and simple theory that can have an ITO phase. We proceed by decomposing the spin-1/2 electrons into fermionic spinons, as in Chapter 22 on chiral spin liquids. We assume there is a spinon-pair condensate, and so the $U(1)$ gauge symmetry has been broken down to $\mathbb{Z}_2$, as was the case for bosonic spinons in Chapter 15. Finally, because of the strong spin–orbit interactions, which do not conserve spin, we drop the spin indices on the spinons entirely. So, as in the Chapters 15 and 16 on $\mathbb{Z}_2$ spin liquids, we end up with a $\mathbb{Z}_2$ gauge theory coupled to spinons f_i on some lattice of sites i, similar to (16.8). We therefore consider the following Hamiltonian, on the square lattice for simplicity

$$\begin{aligned} H &= H_f + H_{\mathbb{Z}_2}, \\ H_f &= -\sum_i \mu f_i^\dagger f_i \\ &\quad -t\sum_{\langle ij\rangle} Z_{ij}\left(f_i^\dagger f_j + f_j^\dagger f_i\right) - \sum_{\langle ij\rangle} Z_{ij}\left(\Delta_{ij} f_i^\dagger f_j^\dagger + \Delta_{ij}^* f_j f_i\right), \\ H_{\mathbb{Z}_2} &= -K\sum_{\square}\prod_{\langle ij\rangle\in\square} Z_{ij} - g\sum_\ell X_\ell. \end{aligned} \tag{23.1}$$

Here i, j denotes the sites of a square lattice, $\langle ij\rangle$ denotes a nearest-neighbor pair of sites, and ℓ denotes a link of the square lattice. Note that the fermionic matter Hamiltonian in (23.1) is the same as (20.23) in Section 20.3 on topological superconductors. The $\mathbb{Z}_2$ gauge theory has operators Z_ℓ, X_ℓ acting on qubits on every link ℓ. The spinons have a nearest-neighbor hopping t, and a nearest-neighbor pairing $\Delta_{ij} = -\Delta_{ji}$, which we specify shortly. Because of the pairing, the $U(1)$ gauge invariance has broken down to $\mathbb{Z}_2$, and H is invariant under a $\mathbb{Z}_2$ gauge transformation similar to (16.4)

$$f_i \to \rho_i f_i \quad , \quad Z_{ij} \to \rho_i Z_{ij} \rho_j , \tag{23.2}$$

with $\rho_i = \pm 1$ and X_ℓ invariant.

Generally, because of the spinon pairing, the spinon Hamiltonian H_f (with $Z_{ij} = 1$) is gapped. Then, for $g \ll K$, the $\mathbb{Z}_2$ gauge fluctuations are weak, and we can assume that we can safely integrate the spinons out. In this manner, we recover the $\mathbb{Z}_2$ spin liquid we have already studied in Chapters 15 and 16, with the f spinons realizing the ε quasiparticle. And, as before, the m particles are excitations carrying $\mathbb{Z}_2$ flux of -1.

However, the above conclusion is too facile. The band topology of the f spinons can be in a non-trivial class D for the case of $p + ip$ pairing:

$$\Delta_{i,i+\hat{x}} = \Delta \quad , \quad \Delta_{i,i+\hat{y}} = i\Delta , \tag{23.3}$$

just as in (20.34) and discussed in Section 20.3.2, and also noted in Fig. 21.2. This requires the absence of time-reversal symmetry, which can arise either from spontaneous or explicit breaking in the underlying spin Hamiltonian. We showed in Section 20.3.2 that the class D band topology led to protected chiral Majorana edge modes. Such edge modes are also present here, and indeed are a distinctive property of the ITO phase. However, the band topology also has an important influence on the properties of the visons in the $\mathbb{Z}_2$ gauge theory, and completely changes their topological properties; this is described below in Section 23.1.

Note also that for $g \gg K$ the $\mathbb{Z}_2$ gauge theory is confining, and we obtain a "trivial" phase with the f spinons confined.

23.1 Visons and Majorana Zero Modes

The most dramatic consequence of the Chern number of the spinon bands is the resulting structure of the vison excitations of the $\mathbb{Z}_2$ gauge theory. I show here that each well-separated vison harbors a Majorana zero mode. This has the remarkable consequence of inducing non-abelian statistics among multiple visons, as discussed in Section 23.2.

We consider the eigenmodes of the spinon Hamiltonian H_f in a fixed background of the $\mathbb{Z}_2$ gauge field Z_{ij}. Kitaev presented a rigorous treatment of a general class of such H_f in Appendix C of Ref. [134]. Employing a Pfaffian invariant similar to that appearing in the one-dimensional case in Section 20.3.1, he established the existence of a Majorana mode with each vison, which approaches zero energy as the visons are well separated from each other. Below, a less rigorous argument is presented, which employs the continuum limit near the bottom of the fermion band at $\mu = -4t$.

First, we note some general properties of H_f. Consider a lattice of M sites, with periodic boundary conditions (so there are no edges). We have a general Z_{ij}, so there is no translational symmetry. We can diagonalize H_f by a unitary transformation $\mathcal{U}$ among the spinon operators of the form

$$\begin{pmatrix} f_1 \\ f_2 \\ \vdots \\ f_M \\ f_1^\dagger \\ f_2^\dagger \\ \vdots \\ f_M^\dagger \end{pmatrix} = \mathcal{U} \begin{pmatrix} \eta_1 \\ \eta_2 \\ \vdots \\ \eta_M \\ \eta_1^\dagger \\ \eta_2^\dagger \\ \vdots \\ \eta_M^\dagger \end{pmatrix}, \tag{23.4}$$

where η_s are a new set of canonical fermions, obeying $\eta_s \eta_{s'}^\dagger + \eta_{s'}^\dagger \eta_s = \delta_{ss'}$. In terms of these new fermions, H_f takes the simple form

$$H_f = \sum_{s=1}^{M} \varepsilon_s \left(2\eta_s^\dagger \eta_s - 1\right) \tag{23.5}$$

for some eigenvalues ε_s. The eigenmodes H_f come in pairs with opposite signs, as can easily be shown to be the case from its general form. There are a total of 2^M states in the Hilbert space described by (23.5), as there are in the underlying lattice with M sites.

Now consider H_f for the case where $Z_{ij} = Z_{ij,v}$ describes a *pair* of well-separated visons as in Fig. 16.2: that is, $Z_{ij,v}$ has $\prod_{\ell \in \square} Z_\ell = -1$ only for the plaquette $\square$ containing a vison, and is unity otherwise. As discussed in Chapter 16, this flux configuration requires a branch-cut connecting the two visons. Now, the claim is that, provided the bulk spinon band structure has a non-zero Chern number, one of the ε_s will approach zero exponentially rapidly as the separation between the visons becomes large. Let us call this special, very small, eigenvalue ε_0. Then, it is convenient to split η_0 into its Majorana components, as in (20.30):

$$\gamma_1 = \eta_0 + \eta_0^\dagger \tag{23.6}$$

$$\gamma_2 = -i(\eta_0 - \eta_0^\dagger), \tag{23.7}$$

so that H_f can be written as

$$H_f = i\varepsilon_0 \gamma_1 \gamma_2 \; + \text{ high-energy terms.} \tag{23.8}$$

The further claim of the vison capturing a Majorana zero mode is that the operators γ_1 and γ_2 are *separately localized* around the two visons, which means γ_1 is a linear combination of the spinon f_i and $f_i^\dagger$ operators around the first vison, while γ_2 is a linear combination of the spinon operators around the second vison.

We now establish these claims in the continuum limit by focusing on the vicinity of the first vison (say). We place this vison at the origin of coordinates in a plane, and use polar coordinates r, θ. We see below that the the radial equation is a continuum generalization of the zero-mode equation for the Su–Schrieffer–Heeger (SSH) model in (20.11), and also has a simple zero-mode solution as a decaying exponential.

We assume that the presence of a vison can be accounted for by a radial dependence in $\mu(r)$ and $\Delta(r)$ (we shift μ by $4t$ so the condition to be in the bulk topological phase is $\mu > 0$ and small). We take a $\mu(r) < 0$ for small r, so that the "core" region is non-topological, while we take $\mu(r) > 0$ for large r so that the bulk is topological. In the continuum limit, valid for small μ, we can write H_{BdG} in (20.27), (20.28), and (20.35) in polar coordinates (r, θ) as

$$H_{BdG} = \frac{1}{2} \begin{pmatrix} -\mu(r) & 2\Delta(r)e^{i\theta}\left(\frac{\partial}{\partial r} + \frac{i}{r}\frac{\partial}{\partial \theta}\right) \\ -2\Delta(r)e^{-i\theta}\left(\frac{\partial}{\partial r} - \frac{i}{r}\frac{\partial}{\partial \theta}\right) & \mu(r) \end{pmatrix}. \tag{23.9}$$

This acts on a wavefunction $\Psi(r, \theta)$. Apart from the vison core, accounted for above by $\mu(r)$ and $\Delta(r)$, we also have to account for the "branch-cut" of $Z_{ij} = -1$ in $Z_{ij,v}$ emanating from the core of the vortex. This branch-cut imposes anti-periodic boundary conditions on Ψ in the angular direction:

$$\Psi(r, \theta + 2\pi) = -\Psi(r, \theta). \tag{23.10}$$

Remarkably, (23.9) turns out to have a zero-energy eigenmode $\Psi_0(r, \theta)$ for precisely such boundary conditions. Let us write an eigenfunction of H_{BdG} in the form

$$\Psi(r) = \frac{1}{\sqrt{r}} \begin{pmatrix} -e^{i\theta/2} g_1(r) \\ e^{-i\theta/2} g_2(r) \end{pmatrix}. \tag{23.11}$$

The boundary conditions (23.10) are now satisfied, and the eigenvalue equation at energy E becomes

$$\begin{aligned} \mu(r) g_1(r) + 2\Delta(r)\frac{dg_2}{dr} &= -E g_1(r), \\ 2\Delta(r)\frac{dg_1}{dr} + \mu(r) g_2(r) &= E g_2(r). \end{aligned} \tag{23.12}$$

In general, the eigenmodes of (23.12) come in pairs with energy $\pm E$, associated with the particle–hole symmetry $g_1(r) \leftrightarrow g_2(r)$. However, there is a special lone eigenvalue at $E = 0$ that has

$$g_1(r) = g_2(r) \propto \exp\left(-\frac{1}{2}\int_0^r \frac{\mu(r')}{\Delta(r')} dr'\right), \tag{23.13}$$

which is the continuum analog of the zero-mode solution of the SSH model in (23.12).

We can now write down a Majorana zero-mode operator γ associated with this eigenmode:

$$\gamma = \int d^2 r i g_1(r) \left[-e^{i\theta/2} f(r, \theta) + e^{-i\theta/2} f^\dagger(r, \theta)\right], \tag{23.14}$$

where $f(r, \theta)$ is the continuum limit of the spinon operator f_i. Note that γ changes the fermion parity; it either increases or decreases the fermion number by unity. As the fermion number is only conserved modulo 2 in the paired phase in the bulk, it is only the fermion parity that is significant. Note also that the zero-mode Majorana in (23.14) obeys $\gamma^\dagger = \gamma$.

Finally, the most important property of the Majorana fermion γ is that it is localized exclusively near the single vison we focused on. It does not have a component on the distant vison at the other end of the branch-cut.

23.2 Non-Abelian Statistics

We now consider a situation with $2N$ well-separated visons. By the arguments above, we expect that the low-energy subspace described by N complex fermions η_s, $s = 1, \ldots, N$, has 2^N states. Let us write these states as

$$|n_1, n_2, \ldots, n_N\rangle\,, \tag{23.15}$$

with $n_s = 0, 1$ measuring the number of complex η_s fermions. Associated with these N complex fermions are $2N$ Majorana (near-)zero modes localized separately near each of the visons. Let us assume that the N complex fermions have been chosen so that the $2N$ Majoranas γ_s localized near the $2N$ visons are given by

$$\begin{aligned}\gamma_{2s-1} &= \eta_s + \eta_s^\dagger\,,\\ \gamma_{2s} &= -i\left(\eta_s - \eta_s^\dagger\right).\end{aligned} \tag{23.16}$$

For the anyonic wavefunctions we have met so far, exchange (or braiding) of a pair of anyons lead to a statistical phase factor acquired by the single-component wavefunction. Here, we show that the exchange of a pair of visons leads to a non-trivial unitary transformation in the 2^N-dimensional Hilbert space in (23.15). Moreover, upon performing successive braiding operations, the final transformation depends upon the order of the operations. In other words, a non-abelian representation of the braid group is realized by the states in (23.15).

We do compute the complete unitary transformation here, but focus only its non-abelian part. There is an additional abelian phase factor associated with the "topological spin," which we ignore – see Ref. [134] for more details. But the non-abelian part of the unitary transformation can be computed by a simple argument due to Ivanov [117]. Consider the simplest non-trivial case with four visons, shown in Fig. 23.1. The low-energy subspace now consists of the four states $|n_1, n_2\rangle$. Let us braid the visons γ_1 and γ_2 as shown in Fig. 23.1. For the choice of Z_{ij} branch-cuts associated with each

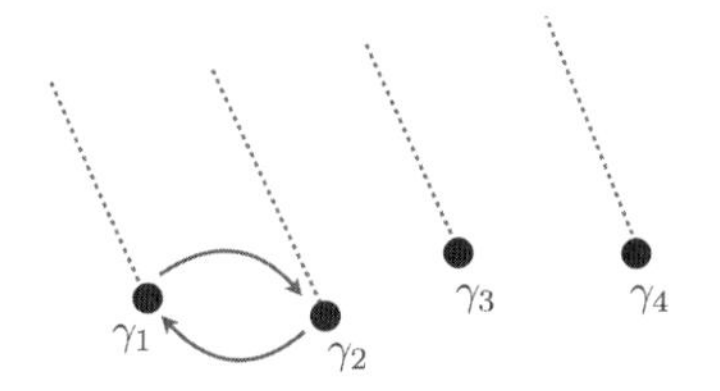

Figure 23.1 Four visons and their Majorana zero modes γ_s, and branch-cuts in Z_{ij}.

vison shown, we observe that γ_1 intersects a branch-cut, while γ_2 does not. This implies that under this braiding operation

$$\begin{aligned} \gamma_1 &\to -\gamma_2, \\ \gamma_2 &\to \gamma_1 . \end{aligned} \tag{23.17}$$

The unitary operator implementing this operation is

$$U_{12} = \frac{1}{\sqrt{2}}(1 + \gamma_1 \gamma_2); \tag{23.18}$$

that is, $U_{12}\gamma_1 U_{12}^\dagger = -\gamma_2$, $U_{12}\gamma_2 U_{12}^\dagger = \gamma_1$. Similarly, for clockwise exchanges of neighboring vortices we have

$$U_{s,s+1} = \frac{1}{\sqrt{2}}(1 + \gamma_s \gamma_{s+1}). \tag{23.19}$$

We can now work out the actions of $U_{s,s+1}$ on the four-dimensional Hilbert space $|n_1, n_2\rangle$. We find [6]

$$\begin{aligned} U_{12}\,|n_1,n_2\rangle &= e^{i\pi(1-2n_1)/4}\,|n_1,n_2\rangle, \\ U_{23}\,|n_1,n_2\rangle &= \frac{1}{\sqrt{2}}\,|n_1,n_2\rangle + i\frac{(-1)^{n_1}}{\sqrt{2}}\,|1-n_1, 1-n_2\rangle, \\ U_{34}\,|n_1,n_2\rangle &= e^{i\pi(1-2n_2)/4}\,|n_1,n_2\rangle. \end{aligned} \tag{23.20}$$

The novel feature is the action of U_{23}; this leads to a 2×2 braiding matrix, which will not commute with other braiding operations.

Ultimately, the key features leading to non-abelian statistics are the “halfing” of the spinon f into Majoranas γ_s bound at the vortices and the familiar -1 picked up by a spinon when crossing the vison branch-cut.

It is useful to also work out the fusion rules implied by the above, and compare to the $\mathbb{Z}_2$ spin liquid. In the present model, the $\mathbb{Z}_2$ spin liquid is obtained when the spinon bands have a zero Chern number. Then, the spinons realize the ε particle, the visons are the m particle, and a bound state of m and ε is the bosonic e spinon.

For the ITO phase, the vison binds *half* a ε spinon, that is, a Majorana, to form a quasiparticle usually referred to as the σ particle. So, the quasiparticle content of the ITO is 1, ε, and σ, and there is no e particle. The fusion rules are:

$$\varepsilon \times \varepsilon = 1, \tag{23.21}$$

$$\varepsilon \times \sigma = \sigma, \tag{23.22}$$

$$\sigma \times \sigma = 1 + \varepsilon. \tag{23.23}$$

The first rule is as in a $\mathbb{Z}_2$ spin liquid. The second rule shows that bringing a spinon close to the σ particle associated with γ_1 just flips the fermion parity with $n_1 \to 1 - n_1$. The third rule shows that the vison branch-cuts annihilate each other, and for the fusion of σ particles $\gamma_{1,2}$ we are left with the two-dimensional Hilbert space $|n_1\rangle$.

The expressions (23.21)–(23.23) are also rules for the operator product expansion for the $c = 1/2$ Ising conformal field theory (CFT) in 1+1 dimensions, and hence the name ITO. This is an example of the close connection between 2+1-dimensional anyon phases, and 1+1-dimensional CFTs, and we also saw examples of this connection in the chapter on abelian Chern–Simons gauge theory. Indeed, these are the simplest cases of a deep and general connection established by Witten [312].

23.3 Connections to Odd-Parity Superconductors

We have presented the Majorana zero modes above as features of the vison excitations of $\mathbb{Z}_2$ gauge theory. But as we saw in Chapter 15, there is an intimate connection between Abrikosov vortices of superconductors and visons in $\mathbb{Z}_2$ spin liquids. And, indeed, Majorana zero modes can also be present in the vortices of odd-parity superconductors with spin–orbit coupling [6].

To see this explicitly here, we consider a vortex in a superconductor in which the pairing field is changed from Δ_0 to $\Delta_0 e^{i\theta}$ around a vortex, with θ the polar angle around the vortex center. We have to discretize this on a lattice, and a natural choice is

$$\Delta_{ij} = \Delta_{0,ij} e^{i(\tilde{\theta}_i + \tilde{\theta}_j)/2}. \tag{23.24}$$

A crucial point here is that we have to choose $\tilde{\theta}_i = \theta_i$ modulo 2π so that there is no branch-cut between any pair of sites i and j for a given Δ_{ij}; this ensures a smooth continuum limit to a single-valued $\Delta_0 e^{i\theta}$. In other words, $\tilde{\theta}_i$ needs to be multi-valued to ensure that the physical pairing field Δ_{ij} is single-valued and smooth. Now we perform a gauge transformation of the fermions:

$$f_i \to f_i e^{i\theta_i/2}, \tag{23.25}$$

where we make a single-valued gauge choice for the value of θ_i at each lattice site i. Then we see from (23.1) that $\Delta_{ij} = \Delta_{0,ij}$ apart from a branch-cut of -1 across every bond in which there is a jump in the value of θ_i by 2π. This can be absorbed into a Z_{ij}, which is precisely that of a vison at the core of the vortex.

24 Fractional Quantum Hall States

The theories of the fractionalized quantum Hall states are developed by fractionalizing the electron into three fermionic partons, and placing the fermionic partons into Landau levels. This is shown to yield a description of the abelian Laughlin and Jain states, and also of the non-abelian Moore–Read state.

We return to the problem considered in Chapter 19 of electrons of density ρ moving in two dimensions in a large magnetic field of strength B. Each Landau level can accommodate a density of electrons $\rho = B/\Phi_0$, where $\Phi_0 = h/e$ is the flux quantum, and so we define the filling fraction as

$$\nu = \frac{\rho}{B/\Phi_0}. \tag{24.1}$$

We considered integer values of ν in Chapter 19, and here we describe a class of fractional quantum Hall states that appear at certain rational values of $\nu < 1$.

In keeping with our discussion of $\mathbb{Z}_2$ spin liquids in Chapter 15, and of chiral spin liquids in Chapter 22, and as promised in Chapter 21, we describe the fractional quantum Hall states by fractionalizing the electron into "partons," which are fermionic. We describe the parton construction and its mean-field theory in Section 24.1. Then we turn to a description of fluctuations using the Chern–Simons gauge theory framework of Chapter 17. However, before we embark on this, we first revisit the integer quantum Hall states, and place them in the context of Chern–Simons theories with an emergent gauge field in Section 24.3.1.

Historically, the Chern-Simons theory of the fractional quantum Hall state was first developed using a flux-attachment transformation which converted the electron into a composite boson [326]. We will not describe this here, and obtain equivalent results by the parton method.

24.1 Partons

We consider here states in which the electron c (assumed spinless) fractionalizes into three fermionic "partons," ψ_p, $p = 1, 2, 3$ [118]:

$$c(\boldsymbol{r}) = \psi_1(\boldsymbol{r})\,\psi_2(\boldsymbol{r})\,\psi_3(\boldsymbol{r}). \tag{24.2}$$

This parton construction is best understood by discretizing continuous space to a lattice of sites i:

$$c_i = \psi_{1i}\psi_{2i}\psi_{3i}. \tag{24.3}$$

Then on each site i there are two possible states, either zero or one electron; we identify the empty (full) electron state with the empty (full) ψ_i so that

$$c_i^\dagger |0\rangle = \psi_{1i}^\dagger \psi_{2i}^\dagger \psi_{3i}^\dagger |0\rangle . \tag{24.4}$$

This correspondence is achieved by imposing two constraints, which we choose as

$$\psi_{1i}^\dagger \psi_{1i} = \psi_{3i}^\dagger \psi_{3i} \quad , \quad \psi_{2i}^\dagger \psi_{2i} = \psi_{3i}^\dagger \psi_{3i} . \tag{24.5}$$

The constraints in (24.5) reduce the eight possible states of the fermions on each site to the two physical states.

As in our study of spin liquids, this decomposition introduces a gauge redundancy. There are two $U(1)$ gauge symmetries corresponding to (24.5) under which

$$\psi_1 \to e^{i\rho_1} \psi_1 \quad , \quad \psi_3 \to e^{-i\rho_1} \psi_3 \tag{24.6}$$

and

$$\psi_2 \to e^{i\rho_2} \psi_2 \quad , \quad \psi_3 \to e^{-i\rho_2} \psi_3 . \tag{24.7}$$

(More properly, there is actually a $SU(3)$ gauge redundancy, but we will ignore this, because the $U(1) \times U(1)$ theory turns out to be sufficient to capture the topological order properly.) We can now proceed with the usual Hubbard–Stratonovich decompositions of the electron Hamiltonian and obtain an effective theory of the ψ partons coupled to two $U(1)$ gauge fields $b_{1\mu}$ and $b_{2\mu}$. Also including a vector potential A_μ for the applied magnetic field, which we choose to apply on the ψ_3 fermion, the spatial gradient terms of the parton theory have the form

$$\begin{aligned} &\frac{1}{2m_1} \psi_1^\dagger (\boldsymbol{\nabla} - i\boldsymbol{b}_1)^2 \psi_1 + \frac{1}{2m_2} \psi_2^\dagger (\boldsymbol{\nabla} - i\boldsymbol{b}_2)^2 \psi_2 \\ &+ \frac{1}{2m_3} \psi_3^\dagger (\boldsymbol{\nabla} + i\boldsymbol{b}_1 + i\boldsymbol{b}_2 - ie\boldsymbol{A})^2 \psi_3 . \end{aligned} \tag{24.8}$$

Depending the structure of the saddle point we are examining, the fermion masses $m_{1,2,3}$ can be different from each other. Let us define the net fields experienced by the three partons:

$$\begin{aligned} B_1 \hat{z} &= \langle \boldsymbol{\nabla} \times \boldsymbol{b}_1 \rangle , \\ B_2 \hat{z} &= \langle \boldsymbol{\nabla} \times \boldsymbol{b}_2 \rangle , \\ B_3 \hat{z} &= - \langle \boldsymbol{\nabla} \times \boldsymbol{b}_1 \rangle - \langle \boldsymbol{\nabla} \times \boldsymbol{b}_2 \rangle + e \boldsymbol{\nabla} \times \boldsymbol{A} . \end{aligned} \tag{24.9}$$

Clearly, we have

$$eB = B_1 + B_2 + B_3 , \tag{24.10}$$

where $B\hat{z} = \boldsymbol{\nabla} \times \boldsymbol{A}$. Note also that the densities of all partons are equal to the electron density $\rho_{1,2,3} = \rho$.

The main idea here is to choose the values of $B_{1,2,3}$ so that each parton fully occupies an integer number of Landau levels, and so there is gap in the parton excitation spectrum; this ensures that fluctuations about this mean-field state are not too strong. If the partons fully occupy N_p Landau levels, then, using (24.1) and (24.10), we obtain

$$\frac{1}{\nu} = \frac{1}{N_1} + \frac{1}{N_2} + \frac{1}{N_3}. \tag{24.11}$$

The filling of the Landau levels indicates there is likely a gap to all excitations at such values of ν. However, we do have to ensure that gauge fluctuations about the saddle point are not singular and do not destabilize the mean-field saddle point above.

24.1.1 Laughlin State

The simplest case arises when all partons fully occupy a single Landau level. Then we must have

$$\rho = \frac{B_1}{e\Phi_0} = \frac{B_2}{e\Phi_0} = \frac{B_3}{e\Phi_0}. \tag{24.12}$$

From (24.1), (24.10), and (24.12) we therefore deduce

$$\nu = \frac{1}{3}. \tag{24.13}$$

From (24.2), and the Vandermonde determinant wavefunction in (19.15) for the integer quantum Hall state, we obtain the celebrated Laughlin wavefunction at $\nu = 1/3$:

$$\Psi(z_1, z_2 \dots z_N) = \left[\prod_{j>i}(z_j - z_i)^3\right] \prod_i \exp\left(-|z_i|^2/4\right). \tag{24.14}$$

24.1.2 Jain States

The other important set of fractional quantum Hall states are the Jain states; the Laughlin and Jain states account for the vast majority of the states observed in experiments. The Jain states are obtained when not all the integers $N_{1,2,3}$ in (24.11) are set equal to unity. One set of Jain fractional states are obtained when partons ψ_1 and ψ_2 occupy a single Landau level, while the parton ψ_3 occupies N Landau levels. Then (24.12) is replaced by

$$\rho = \frac{B_1}{e\Phi_0} = \frac{B_2}{e\Phi_0} = \frac{NB_3}{e\Phi_0}. \tag{24.15}$$

and (24.1), (24.10), and (24.15) yield the filling fraction

$$\nu = \frac{N}{2N+1}. \tag{24.16}$$

The wavefunction is now

$$\Psi(z_1, z_2, \dots, z_N) = \left[\prod_{j>i}(z_j - z_i)^2\right] \left[\chi_N(z_1, z_2 \dots, z_N)\right] \prod_i \exp\left(-|z_i|^2/4\right), \tag{24.17}$$

where χ_N is the wavefunction of the state with N Landau levels occupied. Note that χ_N is not an analytic function of the z_i, and also contains dependence on z_i^* from the higher

Landau levels. In numerical studies, these non-analytic terms are usually dropped by projecting the wavefunction to the lowest Landau level.

24.2 Edge Theory of the Fractional Quantum Hall States

We now use the parton construction to obtain the theory of gapless excitations on the edge. We have already described the edge theory of the integer quantum Hall state in Chapter 19 using the free-electron theory, and we recall its bosonized version in Section 24.2.1. This bosonized approach generalizes readily to the fractional quantum Hall states, along the lines of the chiral spin liquid discussed in Chapter 22. We will consider the bulk theory of the fractional quantum Hall states in Section 24.3.

24.2.1 Integer Quantum Hall States

We consider an integer quantum Hall state with N occupied Landau levels. From Chapter 19, we know that each Landau level has a chiral fermion edge state. We can bosonize each edge state into a chiral boson φ^I. Then, the edge state theory will have the canonical form of (17.52), with the action

$$\mathcal{S}_e = \frac{1}{4\pi}\int dx d\tau \sum_{I,J=1}^{N}\left[-iK_{IJ}\partial_\tau\varphi^I\partial_x\varphi^J + v_{IJ}(\partial_x\varphi^I)(\partial_x\varphi^J) + \frac{i}{2\pi}t_I A_\mu \varepsilon_{\mu\nu}\partial_\nu\varphi^I\right], \quad (24.18)$$

and

$$K_{IJ} = \delta_{IJ} \quad , \quad t_I = 1 \text{ for all } I, \quad (24.19)$$

and some velocities v_{IJ}.

The following discussion does not explicitly consider the coupling to the external gauge field A_μ for simplicity. The gauge field re-instated in Section 24.3, where we also specify the t_I.

24.2.2 Laughlin State

We have three fully occupied lowest Landau levels of the three partons, and each has a chiral fermion edge state, which we bosonize into an edge chiral boson $\varphi_{1,2,3}$, as in (19.33). As in the case of the chiral spin liquid in Chapter 22, we have the coupling to the gauge fields associated with the gauge invariances in (24.6) and (24.7), which we denoted $b_{1\mu}$ and $b_{2\mu}$, respectively, below (24.7). In this manner, we obtain the edge Lagrangian

$$\begin{aligned}\mathcal{L}_{edge} = &\frac{1}{4\pi}\sum_{p=1}^{3}\left[\bar{v}_p(\partial_x\varphi_p)^2 + i\partial_x\varphi_p\partial_\tau\varphi_p\right] \\ &+ \frac{i}{2\pi}\sum_{\mu,\nu=x,\tau}\varepsilon_{\mu\nu}b_{1\mu}\left[\partial_\nu\varphi_1 - \partial_\nu\varphi_3\right] + \frac{i}{2\pi}\sum_{\mu,\nu=x,\tau}\varepsilon_{\mu\nu}b_{2\mu}\left[\partial_\nu\varphi_2 - \partial_\nu\varphi_3\right]. \quad (24.20)\end{aligned}$$

After performing the integral over $b_{1\mu}$ and $b_{2\mu}$ we can set $\varphi_1 = \varphi_2 = \varphi_3 = \varphi$, and we obtain the $N = 1$ edge theory with $m = 3$:

$$\mathcal{L}_{edge} = \frac{3}{4\pi}\left[\bar{v}(\partial_x\varphi)^2 + i\partial_x\varphi\partial_\tau\varphi\right]. \tag{24.21}$$

The quasiparticle operators on the edge are $e^{i\varphi}$ and $e^{2i\varphi}$, carrying $U(1)$ electromagnetic charges $1/3$ and $2/3$, respectively. From (24.2), we see that the electron creation/annihilation operator is $e^{\pm 3i\varphi}$. And, finally, the electron density equals the density of any of the partons, and so the electron-density operator is $(\partial_x\varphi)/(2\pi)$.

24.2.3 Jain States

The edge states of the partons ψ_1 and ψ_2 remain as in the Laughlin state. But for the parton ψ_3, we now have N edge states from the N filled Landau levels; we represent these by chiral bosons $\varphi_{3,s}$, where $s = 1,\ldots,N$. So we have the edge Lagrangian

$$\begin{aligned}
\mathcal{L}_{edge} = {} & \frac{1}{4\pi}\sum_{p=1}^{2}\left[\bar{v}_p(\partial_x\varphi_p)^2 + i\partial_x\varphi_p\partial_\tau\varphi_p\right] \\
& + \frac{1}{4\pi}\sum_{s=1}^{N}\left[\bar{v}_{3,s}(\partial_x\varphi_{3,s})^2 + i\partial_x\varphi_{3s}\partial_\tau\varphi_{3,s}\right] \\
& + \frac{i}{2\pi}\sum_{\mu,\nu=x,\tau}\varepsilon_{\mu\nu}b_{1\mu}\left[\partial_\nu\varphi_1 - \sum_{s=1}^{N}\partial_\nu\varphi_{3,s}\right] \\
& + \frac{i}{2\pi}\sum_{\mu,\nu=x,\tau}\varepsilon_{\mu\nu}b_{2\mu}\left[\partial_\nu\varphi_2 - \sum_{s=1}^{N}\partial_\nu\varphi_{3,s}\right].
\end{aligned} \tag{24.22}$$

After performing the integral over $b_{1\mu}$ and $b_{2\mu}$, we solve the constraints by introducing N chiral bosons φ^I, with $I = 1,\ldots,N$ and

$$\begin{aligned}
\varphi_1 &= \varphi^1 \\
\varphi_2 &= \varphi^1 \\
\varphi_{3,1} &= \varphi^1 - \sum_{s=2}^{N}\varphi^s \\
\varphi_{3,s} &= \varphi^s \quad , \quad s = 2,\ldots,N
\end{aligned} \tag{24.23}$$

Then the edge action takes the canonical form in (24.18) for some velocity v_{IJ} and the K matrix given by

$$\begin{aligned}
& K_{11} = 3; \\
& K_{1I} = K_{I1} = -1, \quad I = 2,\ldots,N; \\
& K_{II} = 2, \quad I = 2,\ldots,N; \\
& K_{IJ} = 1, \quad I = 2,\ldots,N,\ J = 2,\ldots,N,\ I \neq J.
\end{aligned} \tag{24.24}$$

We write the explicit forms for $N = 2, 3, 4$ with $\nu = 2/5, 3/7, 4/9$ as

$$K = \begin{pmatrix} 3 & -1 \\ -1 & 2 \end{pmatrix}, \quad K = \begin{pmatrix} 3 & -1 & -1 \\ -1 & 2 & 1 \\ -1 & 1 & 2 \end{pmatrix},$$

$$K = \begin{pmatrix} 3 & -1 & -1 & -1 \\ -1 & 2 & 1 & 1 \\ -1 & 1 & 2 & 1 \\ -1 & 1 & 1 & 2 \end{pmatrix}. \tag{24.25}$$

It can be verified that $\det(K) = 2N+1$. Now the quasiparticle operators are $e^{i\varphi^I}$ and the electron-density operator is $(\partial_x \varphi^1)/(2\pi)$. The electron operators can be constructed from (24.2) and (24.23), and this leads to the operators $e^{2i\varphi^1 + i\varphi^s}$ with $s \geq 2$, and $e^{3i\varphi^1 - i\sum_{s=2}^{N} \varphi^2}$; it can be verified that all these operators have the same scaling dimension.

24.3 Bulk Gauge Theory of the Fractional Quantum Hall States

Now we follow the same strategy as on the boundary, and derive the Chern–Simons gauge theory in 2+1 dimensions. We will find that it is in accord with the bulk–edge correspondence postulated in Section 17.4.

24.3.1 Integer Quantum Hall States

As in Section 24.2, we pause in our analysis of fractional states to recast the discussion of the integer quantum Hall states in Chapter 19 in the framework of the Chern–Simons theories of Chapter 17.

In our original treatment in Chapter 19, we considered electrons without any emergent gauge fields, and coupled the electrons directly to the external gauge field A_μ. However, in Section 17.3, we found it convenient to couple the external gauge field A_μ to the *fluxes* of an emergent gauge field a^I_μ as determined by the vector t^I. We will now reformulate the low-energy theory of Chapter 19 so that A_μ couples instead to an emergent gauge field.

We do this by coupling the electrons in the Ith Landau level to a separate emergent gauge field a^I_μ. Then, if we integrate out the electrons in the Landau levels, we obtain a Chern–Simons term $(i/(4\pi))K_{IJ}\varepsilon_{\mu\nu\lambda}a^I_\mu \partial_\nu a^J_\lambda$ with same K matrix as in (24.19). Then, taking the variation of the effective action with respect to the Ith gauge field, we obtain the current of the electrons in the Ith Landau level:

$$j^I_\mu = \frac{1}{2\pi}\varepsilon_{\mu\nu\lambda}\partial_\nu a^I_\lambda. \tag{24.26}$$

So we can regard a^I_μ as an emergent gauge field accounting for the conservation of the current within each Landau level $\partial_\mu j^I_\mu = 0$. Furthermore, (24.26) allows us to write the coupling to the external gauge $A_\mu \sum_I j^I_\mu$ in terms of the coupling of A_μ to a^I_μ. Putting this together, we obtain precisely the bulk Chern–Simons theory in (17.33) with N emergent gauge fields, and the K matrix and t_I exactly as in (24.19). We can also verify from the expressions in Chapter 17 that the Hall conductivity and statistical angles have the expected values, and there is no ground-state degeneracy on the torus because $\det K = 1$.

24.3.2 Laughlin State

As in Section 24.3.1, we couple each parton Landau level to a $U(1)$ gauge field $a_{p\mu}$, $p = 1,\ldots,3$. Upon integrating out the partons, we obtain a Chern–Simons term at level 1 for each gauge field. The current $j_{p\mu}$ of the parton ψ_p is then expressed in terms of $a_{p\mu}$ just as in (24.26).

Next we (i) impose the constraints following from (24.6) and (24.7), $j_{1\mu} = j_{3\mu}$, and $j_{2\mu} = j_{3\mu}$ by gauge fields $b_{1\mu}$ and $b_{2\mu}$; and (ii) couple the current $j_{1\mu}$, which equals the electron current, to the external gauge field A_μ. In this manner, we obtain the Lagrangian

$$\begin{aligned}\mathcal{L}_{CS} &= \frac{i}{4\pi}\sum_{p=1}^{3} \varepsilon_{\mu\nu\lambda} a_{p\mu}\partial_\nu a_{p\lambda} + \frac{i}{2\pi}\varepsilon_{\mu\nu\lambda} A_\mu \partial_\nu a_{1\lambda} \\ &+ \frac{i}{2\pi} b_{1\mu}\varepsilon_{\mu\nu\lambda}\partial_\nu (a_{1\lambda} - a_{3\lambda}) + \frac{i}{2\pi} b_{1\mu}\varepsilon_{\mu\nu\lambda}\partial_\nu (a_{2\lambda} - a_{3\lambda}). \end{aligned} \tag{24.27}$$

Performing the integral over $b_{1\mu}$ and $b_{2\mu}$, we can set $a_{p\mu} = a_\mu$, and obtain the Chern–Simons theory of the Laughlin state with $N = 1$ and $m = 3$:

$$\mathcal{L}_{CS} = \frac{3}{4\pi}\varepsilon_{\mu\nu\lambda} a_\mu \partial_\nu a_\lambda + \frac{i}{2\pi}\varepsilon_{\mu\nu\lambda} A_\mu \partial_\nu a_\lambda .$$

Using (17.35), this yields the Hall conductivity $\sigma_{xy} = e^2/(3h)$.

24.3.3 Jain States

The procedure parallels that implemented for the Laughlin state above, and maps to the corresponding edge-state theory.

Now we introduce bulk gauge fields $a_{1\mu}$, $a_{2\mu}$, and $a_{3,s\mu}$ with $s = 1,\ldots,N$. Then, as in (24.27), we have

$$\begin{aligned}\mathcal{L}_{CS} = &\frac{i}{4\pi}\sum_{p=1}^{2} \varepsilon_{\mu\nu\lambda} a_{p\mu}\partial_\nu a_{p\lambda} + \frac{i}{4\pi}\sum_{s=1}^{N} \varepsilon_{\mu\nu\lambda} a_{3,s\mu}\partial_\nu a_{3,s\lambda} \\ &+ \frac{i}{2\pi}\varepsilon_{\mu\nu\lambda} A_\mu \partial_\nu a_{1\lambda} \\ &+ \frac{i}{2\pi} b_{1\mu}\varepsilon_{\mu\nu\lambda}\partial_\nu \left(a_{1\lambda} - \sum_{s=1}^{N} a_{3,s\lambda} \right) \\ &+ \frac{i}{2\pi} b_{2\mu}\varepsilon_{\mu\nu\lambda}\partial_\nu \left(a_{2\lambda} - \sum_{s=1}^{N} a_{3,s\lambda} \right). \end{aligned} \tag{24.28}$$

After performing the integral over $b_{1\mu}$ and $b_{2\mu}$, we solve the constraints by introducing N gauge fields a^I_μ, with $I = 1,\ldots,N$, which are related to $a_{1\mu}$, $a_{2\mu}$, and $a_{3,s\mu}$ just as for φ in (24.23). Then we obtain the Chern–Simons theory in its conventional form:

$$\mathcal{L}_{CS} = \frac{i}{4\pi}\varepsilon_{\mu\nu\lambda}a^I_\mu K_{IJ}\,\partial_\nu a^J_\lambda + \frac{i}{2\pi}t_I A_\mu \varepsilon_{\mu\nu\lambda}\partial_\nu a^I_\lambda\,, \tag{24.29}$$

with the same K matrix as in (24.24) and $t_I = \delta_{I1}$. Now computing $t^T K^{-1} t$ we obtain from (17.35) the Hall conductivity $\sigma_{xy} = (N/(2N+1))e^2/h$.

24.4 Moore–Read State

Here, we consider a novel state that forms at the filling fraction $\nu = 1/2$. We approach this filling by taking the $N \to \infty$ limit of the Jain state in (24.16). From (24.12), we obtain $B_3 = 0$, and so the ψ_3 parton is in a zero effective magnetic field. As in all the Jain states, the ψ_1 and ψ_2 partons fully occupy single Landau levels. At the mean-field level, we expect the ψ_3 partons to form a Fermi surface. Such a mean-field state has gapless excitations, and does not display a quantized Hall effect. The fluctuations of the emergent gauge fields have to be included, and they have singular effects on the fermions at the Fermi surface. This novel Halperin–Lee–Read state is discussed further in Section 34.3.

However, another interesting possibility is the Moore–Read state, which is obtained when the ψ_3 fermions that move in a net zero field form an odd-parity superfluid in class D, as in Section 20.3.2 and in Chapter 23. Indeed, the state formed by the ψ_3 fermions is closely analogous to the state formed by the f spinons in Chapter 23. The condensate of ψ_3 pairs will higgs the $b_{1\mu} + b_{2\mu}$ gauge field down to $\mathbb{Z}_2$, and the vortices in the ψ_3-pair condensate will display non-abelian statistics. And, as described in Section 20.3.2, there will be chiral Majorana edge states. We also have to consider the edge theory associated with the ψ_1 and ψ_2 integer quantum Hall states. This is the same as in the chiral spin-liquid state of Chapter 22, with the $b_{1\mu} - b_{2\mu}$ $U(1)$ gauge field playing the same role as the a_μ field. To summarize, the edge theory of the Moore–Read state is the sum of the edge theories in Chapters 22 and 23.

25 Dualities of *XY* Models and *U*(1) Gauge Theories

We describe the duality of the XY model in two dimensions to a Coulomb gas of vortices, and the analogous duality of the $U(1)$ gauge theory in three dimensions to a Coulomb gas of monopoles. This leads to the Dasgupta–Halperin particle–vortex duality of the critical point of the XY model in three dimensions.

In our discussion of bosons at integer filling on a d-dimensional lattice in Chapter 8, we showed that their superfluid–insulator transition was described by the field theory of a complex scalar field ψ in $D = d+1$ spacetime dimensions with the action

$$\mathcal{S}_\psi = \int d^{d+1}x \left[|\partial_\mu \psi|^2 + r|\psi|^2 + u|\psi|^4 \right]. \tag{25.1}$$

The same theory also describes the thermal phase transition of the classical XY model on a D-dimensional lattice with sites i and the partition function

$$\mathcal{Z} = \prod_i \int_0^{2\pi} d\theta_i \exp\left(\frac{K}{\pi} \sum_{\langle ij \rangle} \cos(\theta_i - \theta_j) \right). \tag{25.2}$$

We also discussed extensions of (25.2) in Chapter 14 as a model for fractionalization in $D = 3$.

For the case $d = 1$, we applied Luttinger liquid theory to a quantum gas of bosons at integer filling in one dimension in Section 12.3. We obtained a dual description in terms of a sine-Gordon field theory for the field ϕ, where $e^{2i\phi}$ was the vortex operator. Here, we present a different derivation of this duality, starting from (25.2), and then generalize this duality to higher dimensions.

25.1 *XY* model in $D = 1$

First, we consider the simplest case of the classical XY model in one dimension, when there is no ordered phase at any temperature.

The lattice partition function is

$$\mathcal{Z} = \prod_i \int_0^{2\pi} d\theta_i \exp\left(\frac{K}{\pi} \sum_i \cos(\theta_i - \theta_{i+1}) \right), \tag{25.3}$$

where

$$K \equiv \frac{\pi J}{T} \tag{25.4}$$

in terms of the "exchange interaction" J and the temperature of the classical model T. We also define the complex order parameter

$$\psi = e^{i\theta}. \tag{25.5}$$

Anticipating that there is no ordered phase at any T, let us work in the low-T limit, $T \ll J$. Then, we expect that θ_i varies slowly with i, and we can take the continuum limit to write

$$\mathcal{Z} = \int \mathcal{D}\theta(x) \exp\left(-\frac{K}{2\pi}\int dx \left(\frac{d\theta}{dx}\right)^2\right). \tag{25.6}$$

As this is a Gaussian action, we can easily evaluate the correlation functions of the order parameter

$$\begin{aligned}\langle \psi(x)\psi^*(0)\rangle &= \exp\left(-\frac{\pi}{K}\int \frac{dk}{2\pi}\frac{(1-\cos(kx))}{k^2}\right) \\ &= \exp\left(-\frac{\pi|x|}{2K}\right).\end{aligned} \tag{25.7}$$

So the correlation length is

$$\xi = \frac{2K}{\pi} = \frac{2J}{T}, \tag{25.8}$$

which diverges only at $T = 0$. We also expect exponential decay of correlations at very high T, and so the low-T and high-T limits are smoothly connected without an intervening phase transition.

25.1.1 Quantum Interpretation

We can also interpret (25.6) as the Feynman path integral of a particle with mass K/π and coordinate θ, moving on a unit circle with imaginary time x. The Hamiltonian of this quantum particle is ($\hbar = 1$)

$$H = -\frac{\pi}{2K}\frac{d^2}{d\theta^2}. \tag{25.9}$$

So the eigenenergies are

$$E_n = \frac{\pi n^2}{2K} \quad , \quad n = 0, \pm 1, \pm 2, \ldots. \tag{25.10}$$

In terms of these eigenstates $|n\rangle$, the correlation function in (25.7) can we written as (for $x > 0$)

$$
\begin{aligned}
\langle \psi(x)\psi^*(0)\rangle &= \langle 0|\,\psi \exp(-Hx)\psi^*\,|0\rangle \\
&= \langle -1|\exp(-Hx)\,|-1\rangle \\
&= \exp(-E_{-1}x),
\end{aligned}
\tag{25.11}
$$

which agrees with Eq. (25.7).

This quantum interpretation is also easily extended to the D-dimensional XY model in (25.2). We introduce an integer-valued boson-number operator n_i on each site i of a $d = D - 1$-dimensional lattice. This is conjugate to the XY "rotor" variable θ_i:

$$
[\theta_i, n_j] = i\delta_{ij}\,. \tag{25.12}
$$

Then, the Hamiltonian of the quantum rotor model is

$$
\mathcal{H}_r = h\sum_i n_i^2 - \frac{K}{\pi}\sum_{\langle ij\rangle}\cos(\theta_i - \theta_j). \tag{25.13}
$$

The imaginary Feynman path integral of $\mathcal{H}_r$, after discretizing the time direction, leads back to (25.2).

25.2 Vortices in the *XY* Model in $D = 2$

Now, we turn to the quantum Bose gas in $d = 1$, considered in Section 12.3. At integer filling, this is described by the quantum rotor Hamiltonian in (25.13) in $d = 1$, and hence the classical XY model in $D = 2$.

First, we carry out precisely the same analysis as that carried out above in $D = 1$. We will ultimately show that such an analysis yields the correct results as $T \to 0$ even in $D = 2$. However, unlike $D = 1$, the results apply only below a critical temperature T_{KT}.

The continuum theory analog of (25.6) is

$$
\mathcal{Z} = \int \mathcal{D}\theta(x)\exp\left(-\frac{K}{2\pi}\int d^2x\,(\nabla_x\theta)^2\right). \tag{25.14}
$$

The correlator in (25.7) now maps to

$$
\begin{aligned}
\langle \psi(x)\psi^*(0)\rangle &= \exp\left(-\frac{\pi}{K}\int^{\Lambda}\frac{d^2k}{4\pi^2}\frac{(1-\cos(kx))}{k^2}\right) \\
&\approx \exp\left(-\frac{1}{2K}\ln(\Lambda|x|)\right) \quad , \quad |x| \to \infty \\
&= 1/(\Lambda x)^{1/(2K)}.
\end{aligned}
\tag{25.15}
$$

So, at low T, the two-point correlator decays only as a power law. On the other hand, we know from the high-temperature expansion that, at sufficiently high T, the two-point correlator must decay exponentially. These differences are resolved by a vortex unbinding transition at T_{KT}.

25.2.1 Duality Transform in $D = 2$

We recall the partition function

$$\mathcal{Z} = \prod_i \int_0^{2\pi} d\theta_i e^{-\mathcal{S}}, \tag{25.16}$$

where the action is the XY model

$$\mathcal{S} = -\frac{K}{\pi} \sum_{\langle i,j \rangle} \cos(\theta_i - \theta_j). \tag{25.17}$$

Our treatment below follows the classic paper José et al. [121].

It is convenient to write this in a lattice gauge theory notation:

$$\mathcal{S} = -\frac{K}{\pi} \sum_{i,\mu} \cos(\Delta_\mu \theta_i), \tag{25.18}$$

where μ extends over x, τ, the two directions of spacetime. Here, Δ_μ defines a discrete lattice derivative with $\Delta_\mu f(x_i) \equiv f(x_i + \hat{\mu}) - f(x_i)$, with $\hat{\mu}$ a vector of unit length.

Now we introduce the Villain representation

$$e^{-K(1-\cos(\theta))/\pi} \approx \sum_{n=-\infty}^{\infty} e^{-K(\theta - 2\pi n)^2/(2\pi)}, \tag{25.19}$$

which is clearly valid for large K. We will use it for all values of K; this is permitted, because the right-hand side preserves an essential feature for all K – periodicity in θ. Then we can write the partition function as

$$\mathcal{Z} = \sum_{m_{i\mu}} \prod_i \int_0^{2\pi} \frac{d\theta_i}{2\pi} e^{-\mathcal{S}}, \tag{25.20}$$

with

$$\mathcal{S} = \frac{K}{2\pi} \sum_{i,\mu} (\Delta_\mu \theta_i - 2\pi m_{i\mu})^2, \tag{25.21}$$

where the $m_{i\mu}$ are independent integers on all the links of the square lattice. Now we need the exact Fourier series representation of a periodic function of θ:

$$\sum_{n=-\infty}^{\infty} e^{-K(\theta - 2\pi n)^2/(2\pi)} = \frac{1}{\sqrt{2K}} \sum_{p=-\infty}^{\infty} e^{-\pi p^2/(2K) - ip\theta}. \tag{25.22}$$

Note that both sides of the equation are invariant under $\theta \to \theta + 2\pi$. Then, (25.20) can be rewritten as (ignoring overall normalization constants)

$$\mathcal{Z} = \sum_{p_{i\mu}} \prod_i \int_0^{2\pi} \frac{d\theta_i}{2\pi} e^{-\mathcal{S}}, \tag{25.23}$$

with

$$\mathcal{S} = \frac{\pi}{2K} \sum_{i,\mu} p_{i\mu}^2 + i p_{i\mu} \Delta_\mu \theta_i. \tag{25.24}$$

Again, the $p_{i\mu}$ are an independent set of integers on the links of the square lattice. The advantage of (25.24) is that all the integrals over the θ_i factorize, and each θ_i integral can be performed exactly. Each integral leads to a divergence-free constraint on the $p_{i\mu}$ integers:

$$\Delta_\mu p_{i\mu} = 0. \tag{25.25}$$

We can view $p_{i\mu}$ as the number current of a boson (particle) moving in $D = 2$ lattice spacetime. The divergence-free constraint shows that the number of these particles is conserved. We can solve this constraint by writing $p_{i\mu}$ as the "curl" of another integer-valued field $h_{\bar{j}}$, which resides on the sites $\bar{j}$ of the dual lattice:

$$p_{i\mu} = \varepsilon_{\mu\nu}\Delta_\nu h_{\bar{j}}. \tag{25.26}$$

Then, the partition function becomes that of a "height" or "solid-on-solid" (SOS) model:

$$\mathcal{Z} = \sum_{h_{\bar{j}}} e^{-\mathcal{S}}, \tag{25.27}$$

with

$$\mathcal{S} = \frac{\pi}{2K}\sum_{\bar{j},\mu}\left(\Delta_\mu h_{\bar{j}}\right)^2. \tag{25.28}$$

This can also describe the statistical mechanics of a two-dimensional surface upon which atoms are being added discretely on the sites $\bar{j}$, and $h_{\bar{j}}$ is the height of the atomic surface.

We are now almost at the final, dual, form of the original XY model in (25.18). We simply have to approximate the discrete height field $h_{\bar{j}}$ by a continuous field $\phi_{\bar{j}}$. Formally, we can do this by writing, for any function $f(h)$,

$$\begin{aligned}
\sum_{h=-\infty}^{\infty} f(h) &= \int_{-\infty}^{\infty} d\phi f(\phi/\pi) \sum_{h=-\infty}^{\infty} \delta(\phi - \pi h) \\
&= \frac{1}{\pi}\sum_{p=-\infty}^{\infty}\int_{-\infty}^{\infty} d\phi f(\phi/\pi) e^{2ip\phi} \\
&= \frac{1}{\pi}\sum_{p=-\infty}^{\infty}\int_{-\infty}^{\infty} d\phi f(\phi/\pi) e^{(\ln y)p^2 + 2ip\phi} \\
&= \frac{1}{\pi}\int_{-\infty}^{\infty} d\phi f(\phi/\pi)\left[1 + 2\sum_{p=1}^{\infty} y^{p^2}\cos(2p\phi)\right]
\end{aligned} \tag{25.29}$$

for $y = 1$. We now substitute (25.29) into (25.28), and then examine the situation for $y \to 0$; this is the only "unjustified" approximation we make here. It is justified more carefully by José et al. [121] using renormalization-group arguments, who show that all of the basic physics is apparent already at small y. In such a limit we can write the exact result (25.29) as approximately

$$\sum_{h=-\infty}^{\infty} f(h) \approx \frac{1}{\pi}\int_{-\infty}^{\infty} d\phi f(\phi/\pi) \exp\left(2y\cos(2\phi)\right). \tag{25.30}$$

Substituting (25.30) into (25.28), we find that our final dual theory of the XY model is the sine-Gordon theory, as obtained earlier in Section 12.3 by different methods:

$$\mathcal{Z} = \prod_{\bar{j}} \int d\phi_{\bar{j}} e^{-\mathcal{S}_{sG}}, \tag{25.31}$$

with

$$\mathcal{S}_{sG} = \frac{1}{2\pi K} \sum_{\bar{j},\mu} (\Delta_\mu \phi_{\bar{j}})^2 - 2y \sum_{\bar{j}} \cos(2\phi_{\bar{j}}). \tag{25.32}$$

If we now expand in powers of y, and integrate out ϕ, we obtain the partition function of a plasma of vortices and anti-vortices:

$$\mathcal{Z} = \sum_{N=0}^{\infty} \frac{y^{2N}}{(N!)^2} \prod_{i=1}^{N} \int dx_{+i} dx_{-i} \exp\left(K \sum_{j \neq k=1}^{2N} p_j p_k \ln |x_j - x_k| \right), \tag{25.33}$$

where $x_j = x_{+j}$ and $p_j = 1$ for $j = 1, \ldots, N$ (representing the vortices) and $x_j = x_{-,j-N}$ and $p_j = -1$ for $j = N+1, \ldots, 2N$ (representing the anti-vortices). So we can identify

$$V_\pm = e^{\pm 2i\phi} \tag{25.34}$$

as the vortex/anti-vortex operators. These vortices are the same objects as those discussed in Section 7.1.

25.2.2 Mappings of Observables

We begin with the boson current J_μ. This can obtained by coupling the XY model to a fixed external gauge field A_μ by replacing (25.18) with

$$\mathcal{S} = -\frac{K}{\pi} \sum_{i,\mu} \cos(\Delta_\mu \theta_i - A_{i\mu}), \tag{25.35}$$

and defining the current

$$J_{i\mu} = \frac{\delta \mathcal{S}}{\delta A_{i\mu}}. \tag{25.36}$$

Note that because the chemical potential of the bosons is i times the time component of the vector potential (in Euclidean time), the boson-density operator is i times J_τ (this factor of i must be kept in mind in all Euclidean path integrals). Then, carrying through the mappings above we find that the sine-Gordon action is replaced by

$$\mathcal{S}_{sG} = \frac{1}{2\pi K} \sum_{\bar{j},\mu} (\Delta_\mu \phi_{\bar{j}})^2 - 2y \sum_{\bar{j}} \cos(2\phi_{\bar{j}}) + \frac{i}{\pi} A_\mu \varepsilon_{\mu\nu} \Delta_\nu \phi_{\bar{j}}. \tag{25.37}$$

So now we can identify the current operator

$$J_{i\mu} = \frac{i}{\pi} \varepsilon_{\mu\nu} \Delta_\nu \phi_{\bar{j}}. \tag{25.38}$$

Note that this current is automatically conserved: that is, $\Delta_\mu J_\mu = 0$.

Another useful observable is the *vorticity*. A little thought using the action (25.21) shows that we can identify

$$v_{\bar{j}} = \varepsilon_{\mu\nu}\Delta_{\mu}m_{i\nu} \tag{25.39}$$

as the integer vorticity on site $\bar{j}$. So we extend the action (25.21) with a fixed source field $\lambda_{\bar{j}}$ to

$$\mathcal{S} = \frac{K}{2\pi}\sum_{i,\mu}(\Delta_{\mu}\theta_i - 2\pi m_{i\mu})^2 + i2\pi\lambda_{\bar{j}}\varepsilon_{\mu\nu}\Delta_{\mu}m_{i\nu}. \tag{25.40}$$

Now, every vortex/anti-vortex in the partition function at site $\bar{j}$ appears with a factor of $e^{\pm i2\pi\lambda_{\bar{j}}}$. For the duality mapping we need an extended version of the identify (25.22)

$$\sum_{n=-\infty}^{\infty} e^{-K(\theta-2\pi n)^2/(2\pi)+i2\pi n\lambda} = \frac{1}{\sqrt{2K}}\sum_{p=-\infty}^{\infty} e^{-\pi(p-\lambda)^2/(2K)-i(p-\lambda)\theta}, \tag{25.41}$$

which holds for any real λ, θ, and K. Then we find that (25.32) maps to

$$\mathcal{S}_{sG} = \frac{1}{2\pi K}\sum_{\bar{j},\mu}\left(\Delta_{\mu}\phi_{\bar{j}}\right)^2 - 2y\sum_{\bar{j}}\cos(2\phi_{\bar{j}} - 2\pi\lambda_{\bar{j}}). \tag{25.42}$$

We therefore observe that each factor $e^{\pm i2\pi\lambda_{\bar{j}}}$ appears with a factor of $ye^{\pm 2i\phi}$. So we identify y with the vortex "fugacity," and confirm that $V_{\pm} = e^{\pm 2i\phi}$ is the vortex/anti-vortex operator.

25.2.3 Renormalization-Group Analysis

We now discuss some important properties of the sine–Gordon field theory $\mathcal{S}_{sG}$ in (25.42) as a function of the dimensionless coupling K and the vortex fugacity y.

First, we note that $\mathcal{S}_{sG}$ at $y=0$ is precisely the dual theory of the low-temperature theory in (25.14). We can compute correlators of the operators $e^{ip\theta}$ and $e^{ip\phi}$ in such a phase as before and obtain

$$\left\langle e^{ip\theta(x)}e^{-ip'\theta(0)}\right\rangle \sim \delta_{pp'}/x^{p^2/2K} \quad , \quad \left\langle e^{ip\phi(x)}e^{-ip'\phi(0)}\right\rangle \sim \delta_{pp'}/x^{p^2K/2}; \tag{25.43}$$

similar results were obtained earlier in Sections 12.2 and 12.3. Note that the correlator of the vortex operator $e^{2i\phi}$ is $\sim \exp(-2K\ln(|x|))$, which is precisely the exponential of the vortex/anti-vortex interaction energy.

Note that these correlators are both power laws, indicating that the theory is scale invariant along the line $y=0$ (indeed it is conformally invariant). From (25.43) we see that this is a line of critical points along which the exponents vary continuously as a function of the dimensionless parameter K. The technology of renormalization-group scale transformations can therefore be applied freely at any point along this line. We can talk of scaling dimensions of operators, and the results (25.43) show that

$$\dim[e^{ip\theta}] = \frac{p^2}{4K} \quad , \quad \dim[e^{ip\phi}] = \frac{p^2K}{4}. \tag{25.44}$$

Using this, and the scaling dimensions (25.44) for $p=2$, we immediately obtain the scaling dimension $\dim[y] = 2-K$ along the $y=0$ line. This can be written as a renormalization-group flow equation under the rescaling $\Lambda \to \Lambda e^{\ell}$:

$$\frac{dy}{d\ell} = (2-K)y. \tag{25.45}$$

So, the critical fixed line $y=0$ is stable for $K<2$. However, this flow equation is not the complete story, especially when K approaches 2. For $|K-2| \sim |y|$ we see that the term on the right-hand side is not linear in the small parameter y, but quadratic. To be consistent, then, we also have to consider other terms of order y^2 that might arise in the flow equations. As we see below, there is a renormalization of K that appears at this order.

The flow equations at order y^2 are generated by decomposing the field $\phi(x)$ into a background slowly varying component $\phi_<(x)$ and a rapidly varying component $\phi_>(x)$, which are integrated out to order y^2:

$$\phi(x) = \phi_<(x) + \phi_>(x), \tag{25.46}$$

where $\phi_<$ has spatial Fourier components at momenta smaller than $\Lambda e^{-\ell}$, while $\phi_>$ has components between $\Lambda e^{-\ell}$ and Λ. Inserting (25.46) into (25.42), to linear order in y, we generate the following effective coupling for $\phi_<$:

$$\begin{aligned} y\int d^2x \langle\cos(2\phi_<(x)+2\phi_>(x))\rangle_0 \\ = y\int d^2x\cos(2\phi_<(x))\left\langle e^{i2\phi_>(x)}\right\rangle_0 \\ = y\int d^2x\cos(2\phi_<(x))e^{-2\langle\phi_>^2\rangle_0} \\ \approx y\left(1-K\frac{d\Lambda}{\Lambda}\right)\int d^2x\cos(2\phi_<(x)), \end{aligned} \tag{25.47}$$

where the subscript 0 indicates an average with respect to the free $y=0$ Gaussian action of $\phi_>$, and $d\Lambda = \Lambda(1-e^{-\ell})$. When combined with a rescaling of coordinates $x \to xe^{-\ell}$ to restore the cutoff to its original value, it is clear that (25.47) leads to the flow equation (25.45). The same procedure applies to quadratic order in y. As the algebra is a bit cumbersome, I only schematically indicate the steps. We generate terms such as

$$\begin{aligned} y^2\int d^2x_1d^2x_2\cos(2\phi_<(x_1)\pm 2\phi_<(x_2))\exp\left(\mp 4\langle\phi_>(x_1)\phi_>(x_2)\rangle_0\right) \\ = y^2\int d^2x_1d^2x_2\cos(2\phi_<(x_1)\pm 2\phi_<(x_2))\exp\left(\mp f(x_1-x_2)d\Lambda\right), \end{aligned} \tag{25.48}$$

where $f(x_1-x_2)$ is some regularization-dependent function that decays on the spatial scale $\sim\Lambda^{-1}$. For this last reason we may expand the other terms in (25.48) in powers of x_1-x_2. The terms with the $+$ sign then generate a $\cos(4\phi)$ interaction; we ignore this term as the analog of the arguments used to obtain (25.45) shows that this

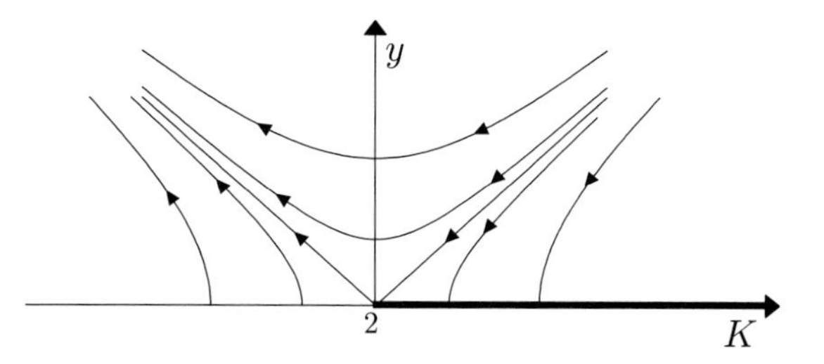

Figure 25.1 RG flows of the sine-Gordon theory $\mathcal{S}_{sG}$. The thick fixed line corresponds to the low-temperature phase of the XY model, $T < T_{KT}$.

term is strongly irrelevant for $K \sim 2$. The terms with the $-$ sign generate gradients on $\phi_<$ and therefore lead to a renormalization of K. In this manner we obtain the flow equation

$$\frac{dK}{d\ell} = -\delta y^2, \tag{25.49}$$

where δ is a positive, regularization-dependent constant (it also depends upon K, but we can ignore this by setting $K = 2$ in δ at this order).

A fairly complete understanding of the properties of $\mathcal{S}_{sG}$ follows from an analysis of (25.45) and (25.49). The flow trajectories are shown in Fig. 25.1. They lie along the hyperbolae $4\delta y^2 - (2-K)^2 = \text{constant}$.

To facilitate the integration of the flow equations (25.45) and (25.49) we change variables to

$$y_{1,2} = \sqrt{\delta} y \mp (K/2 - 1). \tag{25.50}$$

Then (25.45) and (25.49) become

$$\begin{aligned} \frac{dy_1}{d\ell} &= y_1(y_1 + y_2), \\ \frac{dy_2}{d\ell} &= -y_2(y_1 + y_2). \end{aligned}$$

It is clear from these equations that one integral is simply $y_1 y_2 = C$, where C is a constant determined by the initial conditions; the first equation is then easily integrated to give

$$\tan^{-1}\frac{y_1(\ell)}{\sqrt{C}} - \tan^{-1}\frac{y_1(0)}{\sqrt{C}} = \sqrt{C}\ell. \tag{25.51}$$

By the usual scaling argument, the characteristic inverse correlation length ξ^{-1} in the disordered phase is of order $e^{-\ell^*}$, where ℓ^* is the value of ℓ over which y_1 grows from an initial value of order

$$\varepsilon \sim \frac{T - T_{KT}}{T_{KT}} \ll 1 \tag{25.52}$$

to a value of order unity. From the initial conditions, we expect the constant C to also be of order ε, and so let us choose $C = \varepsilon$; then a straightforward analysis of (25.51) gives us

$$\xi^{-1} \sim \exp\left(-\frac{\pi}{2\sqrt{\varepsilon}}\right). \tag{25.53}$$

This singularity, and the flow analysis above, are characteristic of a "Kosterlitz–Thouless" transition.

25.3 $U(1)$ Gauge Theory with Monopoles in $D = 3$

A $U(1)$ gauge theory with a Maxwell action is a Gaussian theory, and so its spectrum is easily determined: we have a gapless photon with one polarization. This single photon can also be written as a gapless scalar field in $D = 3$, as we see below.

However, in applications related to an emergent gauge field a_μ arising from fractionalization of lattice degrees of freedom, we have to allow for monopole configurations of a_μ in the spacetime path integral. We can write this schematically as an action

$$\mathcal{S}_a = \int d^3x \left[\frac{K}{2}\left(\varepsilon_{\mu\nu\lambda}\partial_\nu a_\lambda\right)^2 - y\left(\mathcal{M}_a + \mathcal{M}_a^\dagger\right)\right], \tag{25.54}$$

where y is the monopole fugacity, analogous to the vortex fugacity in (25.42).

We perform duality transforms closely analogous to those carried out above by placing the continuum theory (25.54) on a cubic lattice in spacetime with an action that is 2π periodic in the lattice a_μ. Then, just in the XY model, monopoles are automatically included in the lattice integral. So we consider

$$\mathcal{Z}_a = \prod_{i,\mu} \int_0^{2\pi} da_{i\mu} \exp\left(K \sum_{\square} \cos\left(\varepsilon_{\mu\nu\lambda}\Delta_\nu a_{i\lambda}\right)\right), \tag{25.55}$$

It is useful to first write down the lattice Hamiltonian form of (25.55), which we do in Section 25.3.1. Then, we sproceed to the duality transform of (25.55) in Section 25.3.2.

25.3.1 Quantum Hamiltonian

The mapping of (25.55) to a quantum Hamiltonian parallels that of the mapping from the XY model in (25.2) to the quantum rotor model in (25.13). Now, we have rotor angular variables $a_{i\alpha}$, $\alpha = x, y$, on the links of a square lattice. Its canonically conjugate integer-valued field is $e_{i\alpha}$, the electric field:

$$\left[a_{i\alpha}, e_{j\beta}\right] = i\delta_{ij}\delta_{\alpha\beta}. \tag{25.56}$$

Then, the needed quantum Hamiltonian is

$$\mathcal{H}_a = h\sum_{i\alpha} e_{i\alpha}^2 - K\sum_{\square} \cos\left(\varepsilon_{\alpha\beta}\Delta_\alpha a_{i\beta}\right). \tag{25.57}$$

It is easy to show that the divergence of the electric field commutes with $\mathcal{H}_a$ at all sites i. We impose the Gauss law constraint

$$\Delta_\alpha e_{i\alpha} = 0. \tag{25.58}$$

Now we can obtain a path integral for $\mathcal{H}_a$ and obtain the theory (25.55) (see Problem 25.1).

25.3.2 Duality

As in the XY model, we proceed by replacing the exponent of the cosine in (25.55) by a periodic Gaussian, and then perform an exact integral over the a_μ. Now the divergence-free condition in (25.25) is replaced by a curl-free condition

$$\varepsilon_{\lambda\nu\mu}\Delta_\lambda p_{\bar{j}\mu} = 0, \tag{25.59}$$

where $p_{\bar{j}\mu}$ is an integer-valued field on the links of the dual lattice. The constraint is solved by writing $p_{\bar{j}\mu}$ as the gradient of a height field $h_{\bar{j}}$ now on the sites of the dual lattice:

$$p_{\bar{j}\mu} = \Delta_\mu h_{\bar{j}}. \tag{25.60}$$

This is very similar to the situation for the $D=2$ XY model in (25.26), and the remaining analysis is then the same as in this model. We replace the height field $h_{\bar{j}}$ by a real-valued field $\phi \sim 2\pi h$, where the convention differs by a factor of 2 from $D=2$. Then we obtain finally the dual version of (25.54): a sine-Gordon theory in $D=3$

$$\mathcal{S}_{sG} = \frac{1}{8\pi^2 K}\sum_{\bar{j},\mu}\left(\Delta_\mu \phi_{\bar{j}}\right)^2 - 2y\sum_{\bar{j}}\cos(\phi_{\bar{j}}), \tag{25.61}$$

where y is the monopole fugacity. We can also obtain, using the same arguments as for the XY model, the operator correspondence

$$e^{i\phi} \sim \mathcal{M}_a. \tag{25.62}$$

So monopoles play the same role in the $D=3$ $U(1)$ gauge theory as vortices in the $D=2$ XY model.

We note that at $y=0$, (25.61) is the theory of a gapless scalar field, which is the promised dual form of the $U(1)$ gauge theory without monopoles.

25.3.3 Confinement

We can now perform a renormalization-group analysis of the $D=3$ theory (25.61) along the lines of Section 25.2.3 for $D=2$. The main important observation here is that the free-photon theory at $y=0$ is *always* unstable, which means there is no range of values of K for which there is a fixed line at $y=0$ as in $D=2$. The renormalization-group flow predicts that y is unstable towards large values, implying a dense plasma of monopoles (similar to the vortex plasma for $K<2$ in $D=2$). Such a phase will not have a gapless photon excitation. Instead, it is analogous to the confining phase of the $\mathbb{Z}_2$

gauge theory discussed earlier, where the proliferation of visons lead to confinement of electric charges; here, the proliferation of monopoles leads to electric confinement.

25.4 Particle–Vortex Duality of the *XY* model in $D = 3$

The initial analysis in $D = 3$ tracks that in $D = 2$; everything in Section 25.2.1 until (25.25) also applies in $D = 3$. However, the solution of the divergence-free condition on the boson-number current, $p_{i\mu}$, now takes the form

$$p_{i\mu} = \varepsilon_{\mu\nu\lambda}\Delta_\nu h_{\bar{j}\lambda}, \tag{25.63}$$

where $h_{\bar{j}\mu}$ is now an integer-valued field on the links of the dual lattice. Then, promoting $h_{\bar{j}\mu}$ to a continuous field $a_{\bar{j}\mu}/(2\pi)$ (which replaces ϕ/π in (25.30)), the sine-Gordon theory in (25.31) and (25.32) is replaced by

$$\mathcal{Z} = \prod_{\bar{j},\mu} \int da_{\bar{j}\mu}\, e^{-\mathcal{S}}, \tag{25.64}$$

with

$$\mathcal{S} = \frac{1}{2K}\sum_{\bar{j},\mu} \left(\varepsilon_{\mu\nu\lambda}\Delta_\nu a_{\bar{j}\lambda}\right)^2 - 2y\sum_{\bar{j}} \cos(a_{\bar{j}\mu})). \tag{25.65}$$

Notice that the first term has the form of a Maxwell term in electrodynamics, and is invariant under gauge transformations. We can also make the second term gauge invariant by introducing an angular scalar field $\vartheta_{\bar{j}}$ on the links of the dual lattice. Then, we obtain the action of scalar electrodynamics on a lattice, which is the the final form of the lattice dual theory:

$$\mathcal{Z} = \prod_{\bar{j}} \int d\vartheta_{\bar{j}} \prod_{\bar{j},\mu} \int da_{\bar{j}\mu}\, e^{-\mathcal{S}_{qed}}, \tag{25.66}$$

with

$$\mathcal{S}_{qed} = \frac{1}{2K}\sum_{\bar{j},\mu} \left(\varepsilon_{\mu\nu\lambda}\Delta_\nu a_{\bar{j}\lambda}\right)^2 - 2y\sum_{\bar{j}} \cos(\Delta_\mu \vartheta_{\bar{j}} - a_{\bar{j}\mu}). \tag{25.67}$$

Clearly, we have not changed anything, apart from an overall constant in the path integral; we can absorb the ϑ by a gauge transformation of $a_{\bar{j}\mu}$, and then the $\vartheta_{\bar{j}}$ integrals just yield a constant prefactor.

25.4.1 Mapping of Observables

The mappings in Section 25.2.2 have direct generalizations to $D = 3$.

By coupling the XY model to an external vector potential A_μ, we find that the boson-current operator is given by (replacing (25.38))

$$J_{i\mu} = \frac{i}{2\pi}\varepsilon_{\mu\nu\lambda}\Delta_\nu a_{\bar{j}\lambda}. \tag{25.68}$$

So the boson current maps to the electromagnetic flux of the dual scalar quantum electrodynamics (QED) theory.

It is also useful to consider the two-point correlator of the boson field ψ by performing the duality mapping on

$$\langle \psi(R_1)\psi^*(R_2)\rangle = \frac{1}{\mathcal{Z}}\prod_i \int_0^{2\pi} d\theta_i \exp\left(\frac{K}{\pi}\sum_i \cos(\theta_i - \theta_{i+1}) + i\sum_i m_i\theta_i\right). \tag{25.69}$$

We have introduced the fixed-integer-valued field m_i, which is given above by

$$m_i = \delta_{i,R_1} - \delta_{i,R_2}. \tag{25.70}$$

From the duality mapping, we can now see that m_i corresponds to the charge of *Dirac monopole* insertions in the $U(1)$ gauge field a_μ; the total a_μ flux carried by the Dirac monopole at R_1 (R_2) is 2π (-2π). This follows from the generalization of (25.25) here to

$$\Delta_\mu p_{i\mu} = m_i \tag{25.71}$$

and the expression in (25.63), $p_{i\mu} = \varepsilon_{\mu\nu\lambda}\Delta_\nu a_{\bar{j}\lambda}/(2\pi)$. So $\psi(R)$ is the monopole operator, $\mathcal{M}_a(R)$, at the spacetime point R, which is analogous to the statement in $D = 2$ that $e^{2i\phi}$ was the vortex operator.

25.4.2 Universal Continuum Theory

The connection described so far may appear specialized to particular lattice XY models. However, it is possible to state the particle–vortex mapping in rather precise and universal times as an exact correspondence between two different field theories, as was first argued by Dasgupta and Halperin [60].

In direct boson perspective, we have already seen that the vicinity of the superfluid–insulator transition is described by a field theory for the complex field $\psi \sim e^{i\theta}$:

$$\begin{aligned}\mathcal{Z}_\psi &= \int \mathcal{D}\psi\, e^{-\mathcal{S}_\psi},\\ \mathcal{S}_\psi &= \int d^3x \left[|\partial_\mu \psi|^2 + r|\psi|^2 + u|\psi|^4\right].\end{aligned} \tag{25.72}$$

In the dual-vortex formulation, we can deduce a field theory from (25.67) for the complex field $\phi \sim e^{i\vartheta}$:

$$\begin{aligned}\mathcal{Z}_\phi &= \int \mathcal{D}\phi\mathcal{D}a_\mu\, e^{-\mathcal{S}_\phi},\\ \mathcal{S}_\phi &= \int d^3x \left[|(\partial_\mu - ia_\mu)\phi|^2 + s|\phi|^2 + v|\phi|^4 + \frac{1}{2K}\left(\varepsilon_{\mu\nu\lambda}\partial_\nu a_\lambda\right)^2\right].\end{aligned} \tag{25.73}$$

Note that there is no monopole fugacity term in (25.73) analogous to that in (25.54). We showed that $\mathcal{M}_a$ is the operator ψ in the XY model $\mathcal{Z}_\psi$, and so monopole sources are forbidden by the global $U(1)$ symmetry $\psi \to \psi e^{i\theta}$ of the XY model.

The precise claim is that the universal theory describing the phase transition in $\mathcal{S}_\psi$, as the parameter r is tuned across a symmetry-breaking transition at $r = r_c$, is identical to the theory describing the phase transition in $\mathcal{S}_\phi$ as a function of the tuning parameter s. A key observation is that the duality reverses the phases in which the fields are condensed; in particular the phases are

- ***XY* order**: In $\mathcal{S}_\psi$: $\langle\psi\rangle \neq 0$ and $r < r_c$. However, in $\mathcal{S}_\phi$: $\langle\phi\rangle = 0$ and $s > s_c$.
- ***XY* disorder**: In $\mathcal{S}_\psi$: $\langle\psi\rangle = 0$ and $r > r_c$. However, in $\mathcal{S}_\phi$: $\langle\phi\rangle \neq 0$ and $s < s_c$.

We can also extend this precise duality to include the presence of an arbitrary spacetime-dependent external gauge field A_μ. In the particle theory, this couples minimally to ψ, as expected:

$$\mathcal{Z}_\psi[A_\mu] = \int \mathcal{D}\psi\, e^{-\mathcal{S}_\psi},$$
$$\mathcal{S}_\psi = \int d^3x \left[|(\partial_\mu - iA_\mu)\psi|^2 + r|\psi|^2 + \frac{u}{2}|\psi|^4 \right], \tag{25.74}$$

while in the vortex theory, the coupling follows from (25.68):

$$\mathcal{Z}_\phi[A_\mu] = \int \mathcal{D}\phi\mathcal{D}a_\mu\, e^{-\mathcal{S}_\phi},$$
$$\mathcal{S}_\phi = \int d^3x \left[|(\partial_\mu - ia_\mu)\phi|^2 + s|\phi|^2 + \frac{v}{2}|\phi|^4 + \frac{1}{2K}\left(\varepsilon_{\mu\nu\lambda}\partial_\nu a_\lambda\right)^2 \right.$$
$$\left. + \frac{i}{2\pi}\varepsilon_{\mu\nu\lambda}A_\mu\partial_\nu a_\lambda \right]. \tag{25.75}$$

The last term is a "mutual" Chern–Simons term. The equivalence between (25.74) and (25.75) is reflected in the equality of their partition functions as functionals of A_μ

$$\mathcal{Z}_\psi[A_\mu] = \mathcal{Z}_\phi[A_\mu], \tag{25.76}$$

after a suitable normalization, and mappings between renormalized couplings away from the quantum critical point. This is a powerful non-perturbative connection between two strongly interacting field theories, and holds even for large and spacetime-dependent A_μ. It maps arbitrary multi-point correlators of the particle current to associated correlators of the electromagnetic flux in the vortex theory. In particular, by taking one derivative of both theories with respect to A_μ, we have the operator identification

$$\psi^*\partial_\mu\psi - \psi\partial_\mu\psi^* = \frac{1}{2\pi}\varepsilon_{\mu\nu\lambda}\partial_\nu a_\lambda \tag{25.77}$$

between the particle and the vortex theories.

Also, as established in Section 25.4.1 the field operator ψ in the theory $\mathcal{Z}_\psi$ corresponds to a Dirac monopole "background insertion" $\mathcal{M}_a$ in the field a_μ in $\mathcal{Z}_\phi$:

$$\psi \sim \mathcal{M}_a\,. \tag{25.78}$$

Let us now consider the nature of the excitations in the XY-ordered and-disordered phases in turn.

XY Order

In the particle theory, $r < r_c$, the ψ field is condensed. So the only low-energy excitation is the Nambu–Goldstone mode associated with the phase of ψ. We write $\psi \sim e^{i\theta}$, and the effective theory for θ is

$$\mathcal{S} = \frac{\rho_s}{2} \int d^3x \left(\partial_\mu \theta - A_\mu\right)^2, \tag{25.79}$$

where ρ_s is the helicity modulus, and we have included the form of the coupling to the external field A_μ by gauge invariance.

In the vortex theory, $s > s_c$, and so ϕ is uncondensed and gapped. Let us ignore the ϕ field to begin with. Then, the only gapless fluctuations are associated with the photon a_μ, and its low-energy effective action is

$$\mathcal{S} = \int d^3x \left[\frac{1}{8\pi^2 \rho_s} \left(\varepsilon_{\mu\nu\lambda} \partial_\nu a_\lambda\right)^2 + \frac{i}{2\pi} \varepsilon_{\mu\nu\lambda} A_\mu \partial_\nu a_\lambda \right]. \tag{25.80}$$

In 2+1 dimensions, there is only one polarization of a gapless photon, and this corresponds precisely to the gapless θ scalar in the boson theory. Indeed, it is not difficult to prove that (25.79) and (25.80) are exactly equivalent, using a Hubbard–Stratanovich transformation, which is essentially a continuous version of the discrete angle–integer transforms we have used in our duality analysis so far.

Turning to gapped excitations, in the vortex theory we have ϕ particles and anti-particles. They interact via a long-range force mediated by the exchange of the gapless photon. For static vortices, this interaction has the form of a Coulomb interaction $\sim \ln(r)$ in 2+1 dimensions. In the boson theory, this logarithmic interaction precisely computes the interaction between vortices in the XY ordered phase.

XY disorder

This phase is simplest in the ψ theory: there are gapped particle and anti-particle excitations, quanta of ψ, which carry total A_μ charge $Q = 1$ and $Q = -1$, respectively. These excitations only have short-range interactions (associated with u).

The situation in the vortex theory is subtle, but ultimately yields the same set of excitations. The ϕ field is condensed. Consequently, by the Higgs mechanism, the a_μ gauge field has a non-zero "mass" and has a gap – so there is no gapless photon mode, as expected. But where are the excitations with quantized charges $Q = \pm 1$? These are *vortices in vortices*. In particular, $\mathcal{S}_\phi$ has solutions of its saddle-point equations that are Abrikosov vortices, as discussed in Section 7.2. Because ϕ is condensed, any finite energy solution of ϕ must have the phase winding of ϕ exactly match the line integral of a_μ. In particular, we can look for time-independent vortex saddle-point solutions centered at the origin in which

$$\phi(x) = f(|x|) e^{i\vartheta(x)}, \tag{25.81}$$

where the angle ϑ winds by 2π upon encircling the origin. The saddle-point equations show that $f(|x| \to 0) \sim |x|$, while

$$f(|x| \to \infty) = \sqrt{\frac{-s}{v}}. \tag{25.82}$$

Under these conditions, it is not difficult to show that finiteness of the energy requires

$$\oint dx_i \partial_i \vartheta = \oint dx_i a_i = \int d^2x \varepsilon_{ij} \partial_i a_j \tag{25.83}$$

on any contour far from the center of the vortex. As the phase ϑ must be single-valued, we have from (25.77) and (25.83) that

$$Q = \frac{1}{2\pi} \int d^2x \varepsilon_{ij} \partial_i a_j = \pm 1 \tag{25.84}$$

in the Abrikosov vortex/anti-vortex, as required. So the important conclusion is that the $Q = \pm 1$ particle and anti-particle excitations of the Mott insulator are the vortices and anti-vortices of the dual-vortex theory.

There is another way to run through the above vortices-in-vortices argument. Let us apply the mapping from (25.74) to (25.75) to the vortex theory. In other words, let us momentarily think of ϕ as the boson and a_μ as an external source field. Then the particle-to-vortex mapping from (25.74) to (25.75) applied to (25.75) yields a theory for a new dual scalar $\widetilde{\psi}$, and a new gauge field b_μ controlled by the action

$$\begin{aligned} \mathcal{S}_{\widetilde{\psi}} = \int d^3x \Bigg[& |(\partial_\mu - i b_\mu)\widetilde{\psi}|^2 + \widetilde{r}|\widetilde{\psi}|^2 + \frac{\widetilde{u}}{2}|\widetilde{\psi}|^4 + \frac{1}{2\widetilde{K}} (\varepsilon_{\mu\nu\lambda}\partial_\nu b_\lambda)^2 \\ & + \frac{i}{2\pi}\varepsilon_{\mu\nu\lambda} a_\mu \partial_\nu b_\lambda \frac{1}{2K} (\varepsilon_{\mu\nu\lambda}\partial_\nu a_\lambda)^2 + \frac{i}{2\pi}\varepsilon_{\mu\nu\lambda} A_\mu \partial_\nu a_\lambda \Bigg]. \end{aligned} \tag{25.85}$$

Now we can exactly perform the Gaussian integral over a_μ; to keep issues of gauge invariance transparent, it is convenient to first decouple the Maxwell term using an auxilliary field P_μ:

$$\begin{aligned} \mathcal{S}_{\widetilde{\psi}} = \int d^3x \Bigg[& |(\partial_\mu - i b_\mu)\widetilde{\psi}|^2 + \widetilde{r}|\widetilde{\psi}|^2 + \frac{\widetilde{u}}{2}|\widetilde{\psi}|^4 + \frac{1}{2\widetilde{K}} (\varepsilon_{\mu\nu\lambda}\partial_\nu b_\lambda)^2 \\ & + \frac{i}{2\pi}\varepsilon_{\mu\nu\lambda} a_\mu \partial_\nu b_\lambda \frac{K}{2} P_\mu^2 - i\varepsilon_{\mu\nu\lambda} P_\mu \partial_\nu a_\lambda + \frac{i}{2\pi}\varepsilon_{\mu\nu\lambda} A_\mu \partial_\nu a_\lambda \Bigg]. \end{aligned} \tag{25.86}$$

We can now perform the integral over a_μ, and obtain a delta-function constraint that sets

$$P_\mu = b_\mu + A_\mu - \partial_\mu \alpha, \tag{25.87}$$

where α is an arbitrary scalar corresponding to a gauge choice; so we have

$$\begin{aligned} \mathcal{S}_{\widetilde{\psi}} = \int d^3x \Bigg[& |(\partial_\mu - i b_\mu)\widetilde{\psi}|^2 + \widetilde{r}|\widetilde{\psi}|^2 + \frac{\widetilde{u}}{2}|\widetilde{\psi}|^4 + \frac{1}{2\widetilde{K}} (\varepsilon_{\mu\nu\lambda}\partial_\nu b_\lambda)^2 \\ & + \frac{K}{2}(b_\mu + A_\mu - \partial_\mu \alpha)^2 \Bigg]. \end{aligned} \tag{25.88}$$

The last term in (25.88) implies that the b_μ gauge field has been "Higgsed" to the value $b_\mu = -A_\mu + \partial_\mu \alpha$. Setting b_μ to this value, we observe that α can be gauged away, and then $\mathcal{S}_{\widetilde{\psi}}$ reduces to the original particle theory in (25.74). So applying the particle-to-vortex duality to the vortex theory yields back the particle theory.

Problems

25.1 1 Write down the phase-space path integral for $\mathcal{H}_a$ in (25.57), treating the $a_{i\alpha}$ as coordinates, and the $e_{i\alpha}$ as momenta. Impose the Gauss law constraint (25.58) by a Lagrange multiplier, which is the time component of the $U(1)$ gauge field $a_{i\tau}$. Then, integrate out the $e_{i\alpha}$, discretize the resulting path integral in time, and obtain the theory (25.55).

2 Modify the Gauss law constraint to (26.3), and obtain the partition function in (26.4).

26 Applications of Dualities to Spin Liquids

The gauge theory of the gapped $U(1)$ spin liquid is dualized to a height model, where the monopole Berry phases lead to offsets in the heights. The height model is always in a flat phase, and this describes confining valence-bond solid states for half-integer spin antiferromagnets. The confinement transitions of even and odd $\mathbb{Z}_2$ spin liquids are described by mappings generalizing the Dasgupta–Halperin duality.

In our bosonic parton treatment of spin liquids on the square lattice in Chapter 15, we found in Section 15.3.1 a mean-field saddle point with gapped bosonic spinons and an emergent $U(1)$ gauge field. The $U(1)$ gauge field had a photon excitation, and we deferred consideration of its consequences. We now return to this model, and apply the lessons of duality from Chapter 25.

We also re-examine the $\mathbb{Z}_2$ spin liquids in Chapter 15 and 16 by embedding them in a $U(1)$ gauge theory (we did the reverse in Chapters 14 and 15), which will allow us to apply the dualities of Chapter 25. This enables us to obtain a deeper understanding of $\mathbb{Z}_2$ spin liquids and their phase transitions, while reproducing some of the results of Chapter 16 by different methods. We also connect to the Chern–Simons formulation of $\mathbb{Z}_2$ spin liquids in Section 17.2.1.

26.1 *U*(1) Spin Liquids

Let us consider the theory of the $U(1)$ gauge fluctuations about the square-lattice saddle point of Section 15.3.1 after the gapped spinons have been integrated out. We expect that monopole insertions are allowed in the path integral, and so, by the arguments in the previous chapter, we conclude that the $U(1)$ gauge field should be confining. So we appear to have obtained a gapped, trivial state of the antiferromagnet with no broken symmetries. This finding runs afoul with the Hastings–Oshikawa theorem [105, 194] that such states are not possible, at least for spin $S = 1/2$. So it is too simplistic to use the naive action for the $U(1)$ gauge theory, and it must be modified in some way.

The required modification becomes apparent when we reason in a manner similar to the $\mathbb{Z}_2$ gauge theory, where we had run into a similar problem in Section 16.5.2. We found there that the solution was to introduce the "odd" gauge theory, with the Gauss law contraint $G_i = -1$ on each site. This -1 arises from the fact that the spinons in the ground state also carry a unit $\mathbb{Z}_2$ gauge charge. The same is true for the $U(1)$ spin liquid, where we had found below (15.28) that the bosonic spinons carry charges $+1$ and -1 on the two sublattices, and there are $2S$ spinons in the ground state on each site, as in (15.4).

So the required $U(1)$ gauge theory has the same Hamiltonian we met in Section 25.3.1

$$\mathcal{H}_a = h \sum_{i\alpha} e_{i\alpha}^2 - K \sum_{\square} \cos\left(\varepsilon_{\alpha\beta} \Delta_\alpha a_{i\beta}\right), \tag{26.1}$$

with the commutation relations

$$\left[a_{i\alpha}, e_{j\beta}\right] = i\delta_{ij}\delta_{\alpha\beta}. \tag{26.2}$$

The gauge field takes values on a unit circle $a_{i\alpha} \equiv a_{i\alpha} + 2\pi$, and so the eigenvalues of the electric field $e_{i\alpha}$ are the integers. The modified Gauss law from the background spinon charges is

$$\Delta_\alpha e_{i\alpha} = 2S\eta_i, \tag{26.3}$$

where $\eta_i = 1$ ($\eta_i = -1$) on sublattice A (B).

The path-integral formulation of the partition function of $\mathcal{H}_a$ yields the following on a cubic lattice in spacetime (see Problem 25.1)

$$\widetilde{\mathcal{Z}}_a = \prod_{j\mu} \int_0^{2\pi} \frac{da_{j\mu}}{2\pi} \exp\left(K \sum_{\square} \cos\left(\varepsilon_{\mu\nu\lambda} \Delta_\nu a_{j\lambda}\right) - i2S \sum_j \eta_j a_{j\tau} \right). \tag{26.4}$$

The new term is the Berry phase proportional to $2S$. An explicit derivation [229, 239] of the from (26.4), starting from the individual spin Berry phase in Section 18.2, is presented in Appendix C. We now apply the duality transformations of the previous chapter to (26.4), and obtain a theory for the monopoles $\mathcal{M}_a$. We find that the monopoles also acquire a Berry phase, and this leads to non-trivial transformation of the monopoles under the symmetries of the lattice. And consequently, in the confining phase where the monopoles proliferate, we show that there has to be a broken lattice symmetry for half-integer S, leading to consistency with the Hastings–Oshikawa theorem.

26.1.1 Duality Mapping

We now proceed with a duality mapping that parallels that in Section 25.3.2. We first rewrite the partition function in $2+1$ spacetime dimensions by replacing the cosine interaction in (26.4) by a Villain sum [121, 289] over periodic Gaussians:

$$\mathcal{Z}_a = \sum_{\{q_{\bar{j}\mu}\}} \prod_{j\mu} \int_0^{2\pi} \frac{da_{j\mu}}{2\pi} \exp\left(-\frac{K}{2} \sum_{\square} \left(\varepsilon_{\mu\nu\lambda} \Delta_\nu a_{j\lambda} - 2\pi q_{\bar{j}\mu}\right)^2 - i2S \sum_j \eta_j a_{j\tau} \right), \tag{26.5}$$

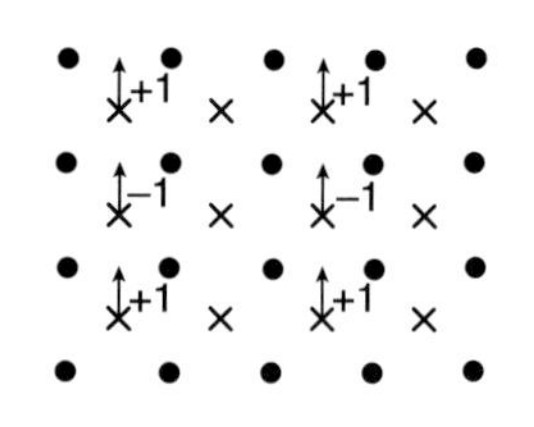

Figure 26.1 Specification of the non-zero values of the fixed field $b^0_{\bar{j}\mu}$. The circles are the sites of the direct lattice, j, while the crosses are the sites of the dual lattice, $\bar{j}$; the latter are also offset by half a lattice spacing in the direction out of the paper (the $\mu = \tau$ direction). The $b^0_{\bar{j}\mu}$ are all zero for $\mu = \tau, x$, while the only non-zero values of $b^0_{\bar{j}y}$ are shown above. Notice that the b^0 flux obeys (26.7).

where the $q_{\bar{j}\mu}$ are integers on the links of the *dual* cubic lattice, which pierce the plaquettes of the direct lattice. Throughout this section we use the index $\bar{j}$ to refer to sites of this dual lattice, while j refers to the direct lattice on sites on which the spins are located.

We will now perform a series of exact manipulations on (26.5), which lead to a dual *interface* model [86, 217, 218]. This dual model has only positive weights – this fact, of course, makes it much more amenable to a standard statistical analysis. This first step in the duality transformation is to rewrite (26.5) by the Poisson summation formula:

$$\sum_{\{q_{\bar{j}\mu}\}} \exp\left(-\frac{K}{2}\sum_{\square}\left(\varepsilon_{\mu\nu\lambda}\Delta_\nu a_{j\lambda} - 2\pi q_{\bar{j}\mu}\right)^2\right)$$
$$= \sum_{\{b_{\bar{j}\mu}\}} \exp\left(-\frac{1}{2K}\sum_{\bar{j}} b^2_{\bar{j}\mu} - i\sum_{\square}\varepsilon_{\mu\nu\lambda} b_{\bar{j}\mu}\Delta_\nu a_{j\lambda}\right), \qquad (26.6)$$

where $b_{\bar{j}\mu}$ (like $q_{\bar{j}\mu}$) is an integer-valued vector field on the links of the dual lattice (here, and below, we drop overall normalization factors in front of the partition function). Next, we write the Berry phase in a form more amenable to duality transformations. We choose a "background" $b_{\bar{j}\mu} = b^0_{\bar{j}\mu}$ flux that satisfies

$$\varepsilon_{\mu\nu\lambda}\Delta_\nu b^0_{\bar{j}\lambda} = \eta_j \delta_{\mu\tau}, \qquad (26.7)$$

where j is the direct lattice site in the center of the plaquette defined by the curl on the left-hand side. Any integer-valued solution of (26.7) is an acceptable choice for $b^0_{\bar{j}\mu}$, and a convenient choice is shown in Fig 26.1. Using (26.7) to rewrite the Berry phase in (26.5), applying (26.6), and shifting $b_{\bar{j}\mu}$ by the integer $2Sb^0_{\bar{j}\mu}$, we obtain a new exact representation of $\mathcal{Z}_a$ in (26.5):

$$\mathcal{Z}_a = \sum_{\{b_{\bar{j}\mu}\}} \prod_{j\mu} \int_0^{2\pi} \frac{da_{j\mu}}{2\pi} \exp\left(-\frac{1}{2K}\sum_{\bar{j},\mu}\left(b_{\bar{j}\mu} - 2Sb^0_{\bar{j}\mu}\right)^2 - i\sum_{\square}\varepsilon_{\mu\nu\lambda} b_{\bar{j}\mu}\Delta_\nu a_{j\lambda}\right). \qquad (26.8)$$

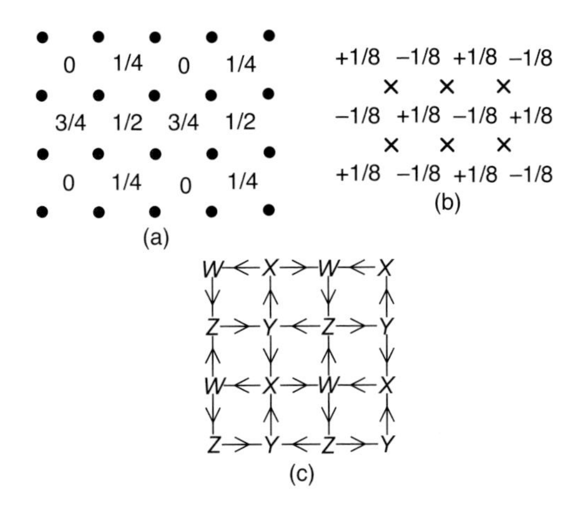

Figure 26.2 Specification of the non-zero values of the fixed fields (a) $\mathcal{X}_{\bar{j}}$, (b) $\mathcal{Y}_{j\mu}$, and (c) $\varepsilon_{\mu\nu\lambda}\Delta_\nu\mathcal{Y}_{j\lambda}$ introduced in (26.10). The notational conventions are as in Fig. 26.1. Only the $\mu = \tau$ components of $\mathcal{Y}_{j\mu}$ are non-zero, and these are shown in (b). Only the spatial components of $\varepsilon_{\mu\nu\lambda}\Delta_\nu\mathcal{Y}_{j\lambda}$ are non-zero, and these are oriented as in (c) with magnitude 1/4. The four dual sublattices, W, X, Y, Z, are also indicated in (c). Note that $\mathcal{X}_W = 0$, $\mathcal{X}_X = 1/4$, $\mathcal{X}_Y = 1/2$, and $\mathcal{X}_Z = 3/4$.

The integral over the $a_{j\mu}$ can be performed independently on each link, and its only consequence is the imposition of the constraint $\varepsilon_{\mu\nu\lambda}\Delta_\nu b_{\bar{j}\lambda} = 0$. We solve this constraint by writing $b_{\bar{j}\mu}$ as the gradient of a integer-valued "height" $h_{\bar{j}}$ on the sites of the dual lattice, and so obtain

$$\mathcal{Z}_h = \sum_{\{h_{\bar{j}}\}} \exp\left(-\frac{1}{2K}\sum_{\bar{j},\mu}(\Delta_\mu h_{\bar{j}} - 2Sb^0_{\bar{j}\mu})^2\right). \tag{26.9}$$

I emphasize that, apart from an overall normalization, we have $\mathcal{Z}_h = \mathcal{Z}_a$ exactly. This is the promised 2+1-dimensional interface, or height, model in almost its final form. The same height model was obtained in Section 25.3.2, but without the $b^0_{\bar{j}\mu}$ term, that is, with $S = 0$.

The physical properties of (26.9) become clearer by converting the "frustration" $b^0_{\bar{j}\mu}$ in (26.9) into offsets for the allowed height values. This is done by decomposing $b^0_{\bar{j}\mu}$ into curl and divergence free parts and writing it in terms of new fixed fields $\mathcal{X}_{\bar{j}}$ and $\mathcal{Y}_{j\mu}$, as follows:

$$b^0_{\bar{j}\mu} = \Delta_\mu \mathcal{X}_{\bar{j}} + \varepsilon_{\mu\nu\lambda}\Delta_\nu \mathcal{Y}_{j\lambda}. \tag{26.10}$$

The values of these new fields are shown in Fig 26.2. Inserting (26.10) into (26.9), we can now write the height model as [218]

$$\mathcal{Z}_h = \sum_{\{h_{\bar{j}}\}} \exp\left(-\frac{1}{2K}\sum_{\bar{j}}\left(\Delta_\mu h_{\bar{j}} - 2S\Delta_\mu \mathcal{X}_{\bar{j}}\right)^2\right). \tag{26.11}$$

Finally, as in Section 25.3.2, we replace the height field $h_{\bar{j}}$ by a real-valued field $\phi \sim 2\pi(h - 2S\mathcal{X})$ to obtain our modified sine-Gordon theory

$$\mathcal{S}_{sG} = \frac{1}{8\pi^2 K}\sum_{\bar{j},\mu}\left(\Delta_\mu \phi_{\bar{j}}\right)^2 - 2y\sum_{\bar{j}}\cos\left(\phi_{\bar{j}} - 4\pi S\mathcal{X}_{\bar{j}}\right), \tag{26.12}$$

where y is the monopole fugacity. From the operator correspondence

$$\mathcal{M}_a \sim e^{i\phi}, \tag{26.13}$$

we conclude that the monopole

$$\boxed{\mathcal{M}_{a\bar{j}} \text{ has Berry phase } e^{-i4\pi S\mathcal{X}_{\bar{j}}}.} \tag{26.14}$$

This is the main result of this subsection. For $S = 1/2$, we see from Fig. 26.2a that monopoles have Berry phases 1, i, -1, $-i$ on the four dual sublattices of the square lattice. From the symmetry transformations of this Berry phase for $S = 1/2$ (shown later in (26.39)), we can see that it has the same symmetries as the valence-bond solid (VBS) order parameter $\mathcal{O}_{VBS}$ that we met in Section 16.5.2 on odd-$\mathbb{Z}_2$ gauge theories. This implies the operator correspondence

$$\mathcal{M}_a \sim \mathcal{O}_{VBS} \quad , \quad \text{for half-integer } S \tag{26.15}$$

and that the monopole plasma phase has VBS order for such S.

26.1.2 Phases of the Height Model

Rather than working with the sine-Gordon theory (26.12), it is instructive to work out the phases of the theory by returning to the height model (26.11), and making contact with the height-model literature.

We define a new height variable

$$H_{\bar{j}} \equiv h_{\bar{j}} - 2S\mathcal{X}_{\bar{j}} \tag{26.16}$$

and then the partition function is simply

$$\mathcal{Z}_h = \sum_{\{h_{\bar{j}}\}} \exp\left(-\frac{1}{2K}\sum_{\bar{j}}\left(\Delta_\mu H_{\bar{j}}\right)^2\right). \tag{26.17}$$

The $\mathcal{Y}_{j\mu}$ have dropped out, while the $\mathcal{X}_{\bar{j}}$ act only as fractional offsets (for S not an even integer) to the integer heights. From (26.16) we see that for half-odd-integer S the height is restricted to be an integer on one of the four sublattices, an integer plus 1/4 on the second, an integer plus 1/2 on the third, and an integer plus 3/4 on the fourth; the fractional parts of these heights are as shown in Fig 26.2a; the steps between neighboring heights are always an integer plus 1/4, or an integer plus 3/4. For S an odd integer, the heights are integers on one square sublattice, and half-odd integers on the second sublattice. Finally, for even-integer S, the offset has no effect and the height is an integer on all sites. We discuss these classes of S values in turn in the following subsections.

S-Even Integer

In this case, the offsets $2S\mathcal{X}_{\bar{j}}$ are all integers, and (26.11) is just an ordinary three-dimensional height model that we already discussed in Section 25.3.3, which has been much studied in the literature [82, 85, 121]. Unlike the two-dimensional case, three-dimensional height models generically have no roughening transition, and the interface is always smooth [82, 85]. With all heights integers, the smooth phase breaks no lattice symmetries. So square-lattice antiferromagnets with an S-even integer can have a paramagnetic ground state with a spin gap and no broken symmetries. The smooth interface corresponds to confinement in the dual compact $U(1)$ gauge theory [207]: consequently, the z_a of $\mathcal{Z}$ are confined, and the elementary excitations are $S = 1$ quasiparticles, similar to the φ_α of $\mathcal{S}_\varphi$. This is in accord with the exact ground state for an $S = 2$ antiferromagnet on the square lattice found by Affleck et al., the Affleck–Kennedy–Lieb–Tasaki (AKLT) state [3].

S-Half-Odd Integer

Now, the heights of the interface model can take four possible values, which are integers, plus the offsets on the four square sublattices shown in Fig 26.2a. As for the S-even integer case above, the interface is always smooth, which means that any state of (26.11) has a fixed average interface height

$$\overline{H} \equiv \frac{1}{N_d} \sum_{\bar{j}=1}^{N_d} \langle H_{\bar{j}} \rangle, \tag{26.18}$$

where the sum is over a large set of N_d dual lattice points that respect the square-lattice symmetry. *Any* well-defined value for $\overline{H}$ breaks the uniform shift symmetry of the height model under which $H_{\bar{j}} \to H_{\bar{j}} \pm 1$. In the present context, only the value of $\overline{H}$ modulo integers is physically significant, and so the breaking of the shift symmetry is not important by itself. However, after accounting for the height offsets, we now prove that *any smooth interface must also break a lattice symmetry with the development of VBS order*; this means that $\mathcal{Z}_a$ in (26.5) describes spin-gap ground states of the lattice antiferromagnet, which necessarily have spontaneous VBS order.

The proof of this central result becomes clear upon a careful study of the manner in which the height model in (26.11) and (26.16) implements the 90°-rotation symmetry about a direct square-lattice point. Consider such a rotation under which the dual sublattice points in Fig 26.2c interchange as

$$W \to X, \;\; X \to Y, \;\; Y \to Z, \;\; Z \to W. \tag{26.19}$$

The terms in the action in (26.16) will undergo a 90° rotation under this transformation provided the integer heights $h_{\bar{j}}$ transform as

$$h_W \to h_X, \;\; h_X \to h_Y, \;\; h_Y \to h_Z, \;\; h_Z \to h_W - 1. \tag{26.20}$$

Notice the all-important -1 in the last term – this compensates for the "branch-cut" in the values of the offsets $\mathcal{X}_{\bar{j}}$ as one goes around a plaquette in Fig 26.2c. From (26.20), it

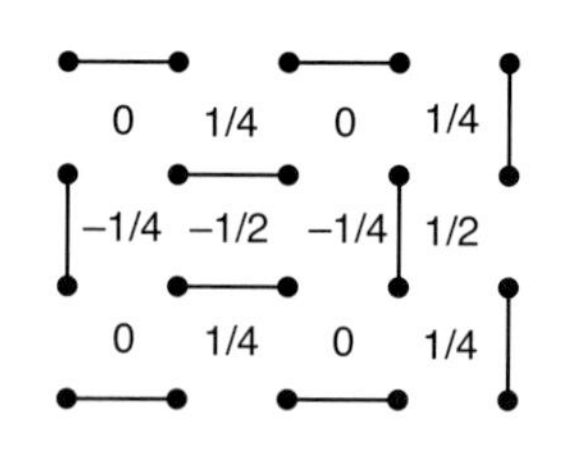

Figure 26.3 Mapping between the quantum dimer model and the interface model $\mathcal{Z}_h$ in (26.11). Each dimer on the direct lattice is associated with a step in height of $\pm 3/4$ on the link of the dual lattice that crosses it. All other height steps are $\pm 1/4$. Each dimer represents a singlet valence bond between the sites. Such a mapping of the quantum dimer model to the interface model only works on bipartite lattice, in contrast to the mapping to the odd-$\mathbb{Z}_2$ gauge theory in Fig. 16.8, which applies on any lattice.

is evident that the average height $\overline{H} \to \overline{H} - 1/4$ under the 90°-rotation symmetry under consideration here. Hence, a smooth interface with a well-defined value of $\overline{H}$ *always* breaks this symmetry.

We now make this somewhat abstract discussion more physical by presenting a simple interpretation of the interface model in the language of the $S = 1/2$ antiferromagnet [327]. From Fig 26.2a it is clear that nearest-neighbor heights can differ either by 1/4 or 3/4 (modulo integers). To minimize the action in (26.11), we should choose the interface with the largest possible number of steps of $\pm 1/4$. However, the interface is frustrated, and it is not possible to make all steps $\pm 1/4$ and at least a quarter of the steps must be $\pm 3/4$. Indeed, there is a precise one-to-one mapping between interfaces with the minimal number of $\pm 3/4$ steps (we regard interfaces differing by a uniform integer shift in all heights as equivalent) and the dimer coverings of the square lattice; the proof of this claim is illustrated in Fig 26.3. We identify each dimer with a singlet valence bond between the spins, and so each interface corresponds to a quantum state with each spin locked in a singlet valence bond with a particular nearest neighbor. Fluctuations of the interface in imaginary time between such configurations correspond to quantum-tunneling events between such dimer states, and an effective Hamiltonian for this is provided by the quantum dimer model [222].

The nature of the possible smooth phases of the interface model are easy to determine from the above picture and by standard techniques from statistical theory [218, 327]. As a simple example, the above mapping between interface heights and dimer coverings allows one to deduce that interfaces with average height $\overline{H} = 1/8, 3/8, 5/8, 7/8$ (modulo integers) correspond to the four-fold degenerate bond-ordered states in Fig 26.4a. To see this, select the interface with $h_{\bar{j}} = 0$ for all $\bar{j}$: this interface has the same symmetry as Fig 26.4a, and a simple computation summing over sites from (26.16) shows that this state has an average height $\overline{H} = -(0 + 1/4 + 1/2 + 3/4)/4 = -3/8$ for $S = 1/2$. The remaining three values of $\overline{H}$ correspond to the three other states obtained by successive 90° rotations of Fig 26.4a. In a similar manner, interfaces with $\overline{H} = 0, 1/4, 1/2, 3/4$ (modulo integers) correspond to the four-fold degenerate plaquette bond-ordered states in Fig 26.4b. A simple example of such an interface is the "disordered-flat" state [223] in which $h_{\bar{j}} = 0$ on all sites $\bar{j}$, except for the W

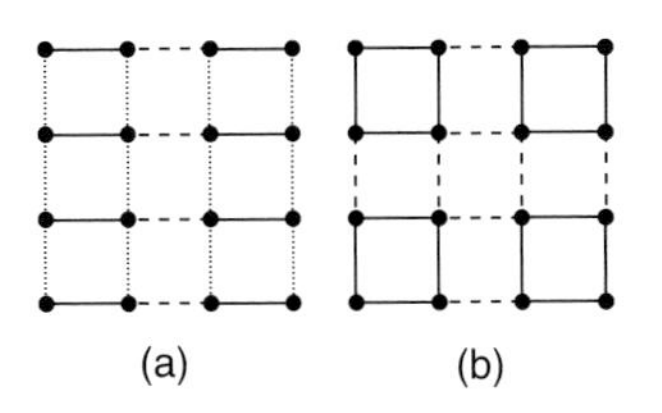

Figure 26.4 Sketch of the two simplest possible states with bond order for $S = 1/2$ on the square lattice: (a) the columnar and (b) plaquette VBS states. Here, the distinct line styles encode the different values of $\langle \vec{S}_i \cdot \vec{S}_j \rangle$ on the links.

sublattices, which have $\mathcal{X}_{\bar{j}} = 0$; for these sites we have $h_{\bar{j}}$ fluctuate randomly between $h_{\bar{j}} = 0$ and $h_{\bar{j}} = 1$, and independently for different $\bar{j}$. The average height of such an interface is $\overline{H} = -((0+1)/2 + 1/4 + 1/2 + 3/4)/4 = -1/2$ for $S = 1/2$, and the mapping to dimer coverings in Fig 26.3 shows easily that such an interface corresponds to the state in Fig 26.4b. All values of $\overline{H}$ other than those quoted above are associated with eight-fold degenerate bond-ordered states with a superposition of the orders in Fig 26.4a and b.

All these phases are expected to support non-zero spin quasiparticle excitations that carry spin $S = 1$, but not $S = 1/2$. Despite the local corrugation in the interface configuration introduced by the offsets, the interface remains smooth on the average, and this continues to correspond to confinement in the dual compact $U(1)$ gauge theory [207]. Consequently the spinons are confined in pairs.

S-Odd Integer

This case is similar to that of the S-half-odd integer, and we will not consider it in detail. The Berry phases again induce bond order in the spin-gap state, but this order only leads to a two-fold degeneracy with *nematic order*. This has possible relevance to the pnictide superconductors [295].

26.2 Gapped $\mathbb{Z}_2$ Spin Liquids

I have already presented a fairly complete theory of $\mathbb{Z}_2$ spin liquids in Chapter 15, and of their confinement transitions, in Chapter 16.

The deconfined topological phase, with an emergent $\mathbb{Z}_2$ gauge field, and gapped e, ε, m, anyons is stable. When we include lattice symmetries, $\mathbb{Z}_2$ spin liquids come in two varieties, odd and even, depending upon the spinon occupations of the ground state: they have a Gauss law constraint $G_i = -1$ and $G_i = 1$, respectively.

We also discussed the confinement transitions of the $\mathbb{Z}_2$ spin liquid, driven by the condensation of visons (the m particle). For even-$\mathbb{Z}_2$ spin liquids, the confinement transition is in the universality class of 2+1 Ising model (more precisely, it is the Ising* theory), and the confining phase is trivial with no broken symmetries. On the other hand, for odd-$\mathbb{Z}_2$ spin liquids, the confining phase has VBS order, and the critical theory is in the XY* class.

Here, we will re-examine these properties of the $\mathbb{Z}_2$ spin liquid from the perspective of $U(1)$ gauge theories and their duality transforms discussed in Section 26.1. This yields (i) an explicit derivation of the Chern–Simons gauge theory formulation of the $\mathbb{Z}_2$ spin liquid that we wrote down in (17.2), and (ii) a new perspective on the critical theory of the confinement transition of the odd-$\mathbb{Z}_2$ spin as a paradigm of *deconfined criticality* that we discussed at the end of Section 16.5.2.

We already saw in our discussion of the bosonic parton theory of spin systems in Chapter 15 that the $\mathbb{Z}_2$ gauge theory arises by the condensation of a charge-2 Higgs field in a $U(1)$ gauge theory; we also saw a simpler version of this connection in Chapter 14 on a classical XY model. So a natural candidate for our discussion of $\mathbb{Z}_2$ spin liquids is to extend the $U(1)$ gauge theory above by introducing a lattice Higgs field with $U(1)$ charge 2:

$$H_i = e^{i\Theta_i} \tag{26.21}$$

and extending the Hamiltonian theory in (26.1) and (26.2) to

$$\begin{aligned} \mathcal{H}_H = & -K \sum_{\square} \cos\left(\varepsilon_{\alpha\beta} \Delta_\alpha a_{i\beta}\right) + h \sum_{i,\alpha} e_{i\alpha}^2 \\ & -L \sum_{i\alpha} \cos(\Delta_\alpha \Theta_i - 2a_{i\alpha}) + \widetilde{h} \sum_i \hat{N}_i^2 \,. \end{aligned} \tag{26.22}$$

Here, $\hat{N}_i$ is the conjugate integer-valued number operator to Θ_i:

$$[\Theta_i, \hat{N}_j] = i\delta_{ij}\,. \tag{26.23}$$

Note that (26.22) is invariant under the gauge transformation

$$a_{i\alpha} \to a_{i\alpha} + \Delta_\alpha f_i \quad , \quad \Theta_i \to \Theta_i + 2f_i\,. \tag{26.24}$$

The extension of the Gauss law constraint in (26.3) is

$$\Delta_\alpha e_{i\alpha} - 2\hat{N}_i = 2S\eta_i. \tag{26.25}$$

The above theory is simply the Hamiltonian version of the theory (14.21) of the classical XY model, in the limit of small J_2 where we ignore the gapped ϕ excitations, and after we set the right-hand side of the Gauss law (26.25) to zero.

We describe the properties of the theory in (26.22) in the remainder of this section. First, in Section 26.2.1 we will show that, in the limit of large L, the $U(1)$ gauge theory in (26.22) reduces to the previously considered $\mathbb{Z}_2$ gauge theories in Chapter 16, with the mapping of the Gauss law constraint

$$G_i = \exp\left(i2\pi S\eta_i\right)\,. \tag{26.26}$$

Then, we apply the duality transforms to (26.22) for the even and odd cases in Sections 26.2.2 and 26.2.3, respectively.

26.2.1 Mapping to $\mathbb{Z}_2$ Gauge Theory

We now present an argument that is the close analog of that in Section 14.2.3. Let us choose the gauge $\Theta_i = 0$, and consider the L term in the Hamiltonian on a single link; dropping site and link indices, we have the Hamiltonian

$$\mathcal{H}_1 = he^2 - L\cos(2a) \quad , \quad [a,e] = i. \tag{26.27}$$

This is the Hamiltonian for a rotor with angular coordinate a, in the presence of a potential $-L\cos(2a)$. At large L, the rotor will be localized near $a = 0, \pi$. So there be a doublet of low-energy levels with a small spacing $2g$, corresponding to even and odd combinations of the localized states near $a = 0, \pi$. As this doublet is well separated from the higher excited states for large L, we project the Hamiltonian on this doublet. We introduce Pauli operators X, Z, which act on this doublet, and the low-energy Hamiltonian is then

$$\mathcal{H}_1 \approx -gX. \tag{26.28}$$

The ground state of $\mathcal{H}_1$ consists of linear combinations of states $|e\rangle$ with e even, while the first excited state has e odd. So, in the projected space, we have the correspondence

$$X \leftrightarrow \exp\left(i\pi\eta\, e\right), \tag{26.29}$$

where $\eta = \pm 1$. Similarly, for the Pauli Z operator we have

$$Z \leftrightarrow \exp\left(i\pi\tilde{\eta}\, a\right). \tag{26.30}$$

Now we apply these mappings to the full lattice theory in (26.22). We choose $\eta = \eta_i$ and $\tilde{\eta} = \eta_i$ on the x-directed links, and $\tilde{\eta} = -\eta_i$ on the y-directed links. Then $\mathcal{H}_H$ maps directly to the $\mathbb{Z}_2$ gauge theory of Chapter 16, with the K term corresponding to products of the Z operators on plaquettes, and (26.28) to the transverse field term. Moreover, the constraint (26.25) maps onto the value of G_i in (26.26).

The discrete-time path-integral version of this theory is (14.36) without the matter field φ at $J_2 = 0$.

26.2.2 Even-$\mathbb{Z}_2$ Gauge Theory

Let us analyze the properties of the $U(1)$ gauge theory (26.22) in a continuum theory for the case $G_i = 1$.

We impose the constraint in (26.25) by a Lagrange multiplier $a_{i\tau}$, which serves as a time component of the gauge field. The continuum limit is expressed in terms of a $U(1)$ gauge field a_μ ($\mu = x, y, \tau$) and the Higgs field H, and takes the form of a standard relativistic theory of the Higgs field with the Lagrangian density

$$\begin{aligned}
\mathcal{L}_{U(1)} &= \mathcal{L}_H + \mathcal{L}_{monopole}, \\
\mathcal{L}_H &= |(\partial_\mu - 2ia_\mu)H|^2 + g|H|^2 + u|H|^4 + K(\varepsilon_{\mu\nu\lambda}\partial_\nu a_\lambda)^2, \\
\mathcal{L}_{monopole} &= -y\left(\mathcal{M}_a + \mathcal{M}_a^\dagger\right).
\end{aligned} \tag{26.31}$$

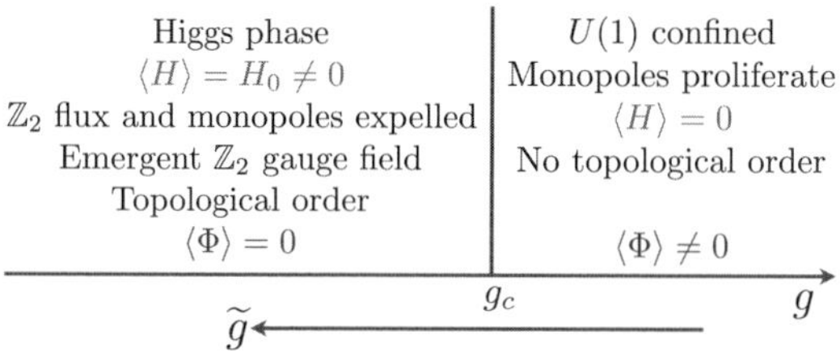

Figure 26.5 Phase diagram of the $U(1)$ gauge theory in (26.31), and the correspondence to the phases of the even-$\mathbb{Z}_2$ gauge theory (compare Fig. 16.7). This is also the transition between phases A+B and D in Figs. 14.2 and 14.3 for the classical XY model. The vison field Φ represents a 2π vortex in H, corresponding to $p = \pm 1$ in Fig. 26.6. The above is also the phase diagram of the theory for the visons in (26.34), as a function of $\widetilde{g}$; this vison theory shows that the critical point is described by the Ising* Wilson–Fisher conformal field theory.

This is precisely the theory (14.24) of the classical XY model, in the limit of large s_2 when ϕ can be ignored. The gauge invariance in (26.24) has now been lifted to the continuum

$$a_\mu \to a_\mu + \partial_\mu f \quad , \quad H \to He^{2if} . \tag{26.32}$$

We allow for Dirac monopole instantons in which the $U(1)$ gauge flux changes by 2π. These are represented schematically by the source term $\mathcal{L}_{monopole}$, and such instantons are present because of the periodicity of the gauge field on the lattice.

The two phases of $\mathcal{L}_{U(1)}$ correspond to the two phases of the even-$\mathbb{Z}_2$ gauge theory as sketched in Fig. 26.5; compare to Fig. 16.7 obtained directly for the $\mathbb{Z}_2$ gauge theory, and to the transition between phases A+B and D in Figs. 14.2 and 14.3 for the classical XY model. For $g > g_c$, we have no Higgs condensate, $\langle H \rangle = 0$, and then $\mathcal{L}_{U(1)}$ reduces to a pure $U(1)$ gauge theory with monopole sources in the action; we know from Section 25.3 that this theory is confining, and this corresponds to the confining phase of the $\mathbb{Z}_2$ gauge theory. For $g < g_c$, we realize the Higgs phase with $\langle H \rangle = H_0 \neq 0$, which corresponds to the deconfined phase of the $\mathbb{Z}_2$ gauge theory. Because of the presence of a gauge field, such a condensate does not correspond to a broken symmetry. But the Higgs phase is topological because there is a stable point-like topological defect, realizing the vison of the deconfined phase of the $\mathbb{Z}_2$ gauge theory. This defect is similar to the finite-energy Abrikosov vortex of the Landau–Ginzburg theory, and is sketched in Fig. 26.6; the phase of H winds by $2\pi p$ around the core of the defect (p is an integer), and this traps a $U(1)$ gauge flux of πp. However, because of the presence of monopoles, the flux is conserved only modulo 2π, and so there is only a single $\pm\pi$ flux defect, which preserves time-reversal symmetry. This π flux is clearly the analog of the $\mathbb{Z}_2$ flux of -1 for the vison.

Let us now apply the Dasgupta–Halperin duality to (26.31). While applying this duality, we view a_μ temporarily as a background gauge field, and obtain the dual theory of the scalar H. From the results in the previous chapter, we obtain a dual theory in which H is replaced by a dual scalar Φ and a dual gauge field b_μ. Remarkably, demonstrating the power of this approach, we can also represent $\mathcal{L}_{monopole}$ in terms of Φ. It is

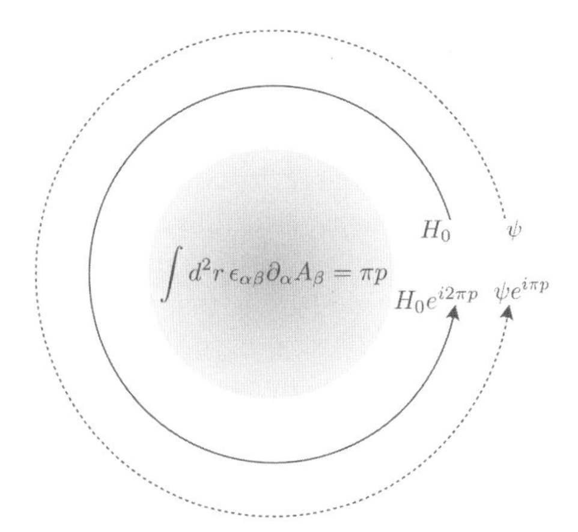

Figure 26.6 Structure of an Abrikosov vortex saddle point of (26.31). The Higgs field magnitude $|\langle H\rangle| \to 0$ as $r \to 0$. Far from the vortex core, $|\langle H\rangle| \to |H_0| \neq 0$, and the phase of H_0 winds by $2\pi p$, where p is an integer. There is a gauge flux trapped in the vortex core; far from the core, the gauge field screens the Higgs field gradients, and so the energy of the vortex is finite. The trapped flux is defined only modulo 2π because of the monopole source term, and so ultimately all odd values of p, map to the same vortex (the vison). The dashed line indicates the Berry phase picked up by a particle ψ with unit a_μ $U(1)$ gauge charge; in the $\mathbb{Z}_2$ gauge theory, ψ can be either the e or ε anyon.

clear that the field Φ represents the π flux vortex (vison) illustrated in Fig. 26.6, because it is the vortex around which H winds by 2π. A monopole insertion carries flux 2π, and so turns out to correspond here to the operator Φ^2:

$$\mathcal{M}_a \sim \Phi^2 . \tag{26.33}$$

In this manner, we obtain the following theory, which is the particle-vortex dual of (26.31), including the monopole insertion [119, 230]

$$\begin{aligned} \mathcal{L}_{d,U(1)} &= \mathcal{L}_\Phi + \mathcal{L}_{\text{cs}} + \mathcal{L}_{monopole}, \\ \mathcal{L}_\Phi &= |(\partial_\mu - ib_\mu)\Phi|^2 + \widetilde{g}|\Phi|^2 + \widetilde{u}|\Phi|^4, \\ \mathcal{L}_{cs} &= \frac{i}{\pi}\varepsilon_{\mu\nu\lambda} a_\mu \partial_\nu b_\lambda, \\ \mathcal{L}_{monopole} &= -y\left(\Phi^2 + \Phi^{*2}\right). \end{aligned} \tag{26.34}$$

Note here that we have retained the a_μ gauge field explicitly because the gapped spinon excitations (the e and ε particles) carry a unit a_μ charge. It is also possible [230] to explicitly derive (26.34) by carrying out the duality transformation on the lattice using a Villain form of the original lattice gauge theory in (26.22).

Now we can obtain one of our promised results. In the topological phase, $g < g_c$, the field Φ is gapped. So we can drop the Φ fluctuations, and this phase described by the Chern–Simons theory

$$\mathcal{L}_{cs} = \frac{i}{4\pi}\varepsilon_{\mu\nu\lambda} a^I_\mu K_{IJ} \partial_\nu a^J_\lambda \quad , \quad K = \begin{pmatrix} 0 & 2 \\ 2 & 0 \end{pmatrix}, \tag{26.35}$$

with $a^1_\mu = a_\mu$ and $a^2_\mu = b_\mu$. This is precisely the K matrix asserted in (17.2).

As long as the spinons (the e and ε particles) are gapped, we can integrate out the a_μ gauge field, and obtain a field theory for the transition in Fig. 26.5. The integral over a_μ fixes the b_μ gauge field to fluxes of 0 and π, that is, b_μ is a $\mathbb{Z}_2$ gauge field, and Φ carries a b_μ $\mathbb{Z}_2$ gauge charge. Reassuringly, the terms in $\mathcal{L}_{monopole}$ are invariant under $\mathbb{Z}_2$ gauge transformations. Near the critical point, we can neglect the gapped b_μ $\mathbb{Z}_2$ gauge field, and the critical theory is simply $\mathcal{L}_\Phi + \mathcal{L}_{monopole}$ at $b_\mu = 0$.

The important feature here is that the monopole term is strongly relevant at the critical point. For $y > 0$ (say), $\mathcal{L}_{monopole}$ prefers the real part of Φ over the imaginary part of Φ: so effectively, at low energies, $\mathcal{L}_{d,U(1)}$ is actually the theory of a real (and not complex) scalar. The phase where Φ is condensed, corresponding to the proliferation of $\mathbb{Z}_2$ flux in the $\mathbb{Z}_2$ gauge theory, is the confining phase, as illustrated in Fig. 26.5. And the phase where Φ is gapped is the deconfined phase; this is a gapped real particle carrying $\mathbb{Z}_2$ flux, the vison.

A result that can be obtained from (26.34) is the universality class of the confinement–deconfinement transition. We integrate out the always gapped imaginary part of Φ, and then $\mathcal{L}_{d,U(1)}$ becomes the Wilson–Fisher theory of the Ising transition in 2+1 dimensions, as already noted in Section 16.5 by other methods. So the phase transition is in the Ising universality class. Strictly speaking, as we already noted in Section 16.5, the transition is actually is in the Ising* universality class [252, 308]. This differs from the Ising universality by dropping operators that are odd under $\Phi \to -\Phi$, because the topological order prohibits the creation of single visons.

26.2.3 Odd $\mathbb{Z}_2$ gauge theory

We now extend the analysis of Section 26.2.2 to the odd-$\mathbb{Z}_2$ gauge theory, in which G_i defined in (26.26) obeys

$$G_i = -1 \tag{26.36}$$

on all sites i.

In the path-integral formulation, the extension of the theory in (26.31) to the odd case is [119, 230]

$$\begin{aligned}
\mathcal{S}_{o,U(1)} &= \int d^3x \left(\mathcal{L}_H + \mathcal{L}_{monopole}\right) + \mathcal{S}_B, \\
\mathcal{L}_H &= |(\partial_\mu - 2ia_\mu)H|^2 + g|H|^2 + u|H|^4 + K(\varepsilon_{\mu\nu\lambda}\partial_\nu a_\lambda)^2, \\
\mathcal{L}_{monopole} &= -y\left(\mathcal{M}_a + \mathcal{M}_a^\dagger\right), \\
\mathcal{S}_B &= i\sum_i \eta_i \int d\tau A_{i\tau}.
\end{aligned} \tag{26.37}$$

The key new feature here (compared to the even case in (26.31)) is the spatially oscillating Berry phase term in $\mathcal{S}_B$ (see Appendix C), which requires us to take the continuum limit with great care. As in (26.31), we have the theory (14.24) of the XY model, in the limit of large s_2 when ϕ can be ignored; the Berry phase $\mathcal{S}_B$ is the $U(1)$ parent of the

$\mathbb{Z}_2$ Berry phase in (14.35), and the latter follows from the former by setting $A_{i\tau} = 0, \pi$, and using (14.31).

A full computation of the dual from of (26.22) was presented in Refs. [119, 230]. That computation is not presented here, but the dual theory is deduced by using some general arguments similar to those in Section 26.2.2. First, as before, we perform the Dasgupta–Halperin transformation to replace H by a dual vison field Φ and a $U(1)$ gauge field b_μ. We then ask about the proper continuum action for Φ. This was actually already worked out in Section 16.5.2, where we showed that Φ acquired non-trivial transformations under lattice symmetries. In particular, we found that

$$T_x : \Phi \to e^{i\pi/4}\Phi^* \quad ; \quad T_y : \Phi \to e^{-i\pi/4}\Phi^* \quad ; \quad R_{\pi/2} : \Phi \to \Phi^* , \tag{26.38}$$

where $T_{x,y}$ are the translations by a unit lattice spacing, and $R_{\pi/2}$ is the symmetry of rotations about a dual lattice point. The transformations in (26.38), and their compositions, realize a projective symmetry group, which is the 16-element non-abelian dihedral group D_8 [114]. As in Section 26.2.2, it is still the case that the monopole operator $\mathcal{M}_a \sim \Phi^2$ as in (26.33). So, from (26.38), we deduce the monopole transformations

$$T_x : \mathcal{M}_a \to i\mathcal{M}_a^* \quad ; \quad T_y : \mathcal{M}_a \to -i\mathcal{M}_a^* \quad ; \quad R_{\pi/2} : \mathcal{M}_a \to \mathcal{M}_a^* . \tag{26.39}$$

It is reassuring to note that these are also the transformations implied by the monopole Berry phases and the identification in (26.14) for the $U(1)$ gauge theory [99, 218, 256, 261]. Note that under the vison D_8 operations in (26.38), T_x and T_y anti-commute:

$$T_x T_y = -T_y T_x \quad , \quad \text{acting on } \Phi \tag{26.40}$$

(as we also saw in (16.25) in our earlier discussion of the odd-$\mathbb{Z}_2$ gauge theory), while they commute under the monopole operations in (26.39).

$$T_x T_y = T_y T_x \quad , \quad \text{acting on } \mathcal{M}_a . \tag{26.41}$$

Using these symmetries, we can now obtain the generalization of the dual theory in (26.34) to the odd case. Only allowing terms that are invariant under the above symmetries, we obtain

$$\begin{aligned}
\mathcal{L}_{od,U(1)} &= \mathcal{L}_\Phi + \mathcal{L}_{cs} + \mathcal{L}_{monopole}, \\
\mathcal{L}_\Phi &= |(\partial_\mu - i b_\mu)\Phi|^2 + \widetilde{g}|\Phi|^2 + \widetilde{u}|\Phi|^4, \\
\mathcal{L}_{cs} &= \frac{i}{\pi}\varepsilon_{\mu\nu\lambda} a_\mu \partial_\nu b_\lambda, \\
\mathcal{L}_{monopole} &= -y_4\left(\Phi^8 + \Phi^{*8}\right).
\end{aligned} \tag{26.42}$$

We recall here that the gapped spinon excitations (the e and ε particles) carry a unit a_μ charge. The important new feature of (26.42) is that $\mathcal{L}_{monopole}$ now involves eight powers of the vison field operator (compared to two powers for the even case in (26.34)). This implies that only quadrupled monopoles are permitted in the action, with fugacity y_4, in contrast to single monopoles in (26.34). All smaller monopoles cancel out of the action due to quantum interference arising from Berry phases from $\mathcal{S}_B$ in (26.37).

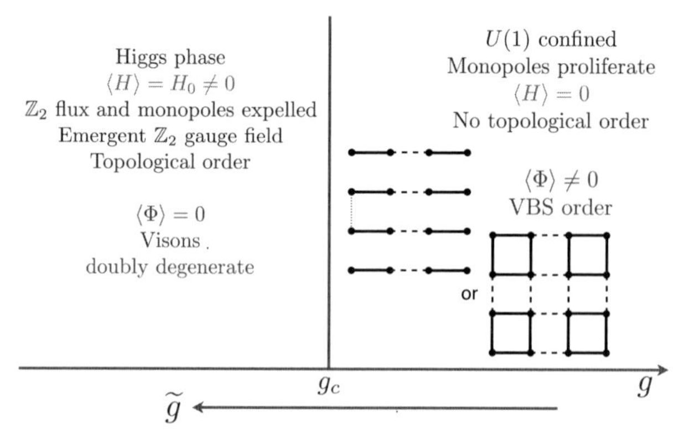

Figure 26.7 Phase diagram of the $U(1)$ gauge theory in (26.37), which describes the physics of the odd-$\mathbb{Z}_2$ gauge theory. Compare to Fig. 26.5 for the even-$\mathbb{Z}_2$ gauge theory, and Fig. 16.11 obtained directly for the odd-$\mathbb{Z}_2$ gauge theory. The transition above is also that between the fractionalized phase and bond order in Fig. 14.4 for the XY model with the Berry phase in (14.35). The vison field Φ represents a 2π vortex in H. The theory for the visons is (26.42), with the tuning parameter $\widetilde{g}$. Monopoles are suppressed at the deconfined critical point at $g = g_c$, and consequently there is an emergent critical $U(1)$ photon described by the deconfined critical theory $\mathcal{L}_H$ in (26.37); in the dual representation of the doubly degenerate visons, the critical theory is the XY^* Wilson–Fisher CFT, described by $\mathcal{L}_\Phi$ in (26.42). In contrast, monopoles are not suppressed at $g = g_c$ in the even-$\mathbb{Z}_2$ gauge theory phase diagram of Fig. 26.5. This phase diagram is the earliest example of deconfined criticality, and a numerical study appeared in Ref. [119].

The phase diagram of $\mathcal{L}_{od,U(1)}$ is modified from Fig. 26.5 to Fig. 26.7 (the latter should be compared to Fig. 16.11 of the odd-$\mathbb{Z}_2$ gauge theory). The topological phase has a gapped Φ excitation. A crucial difference from the even-$\mathbb{Z}_2$ gauge theory is that this excitation is doubly degenerate: $\mathcal{L}_{monopole}$ is sufficiently high order that the degeneracy between the real and imaginary parts of Φ is no longer broken (unlike in (26.34)). So the vison is a complex relativistic particle, unlike the real particle in Section 26.2.2. This double degeneracy in the vison states is a feature linked to the D_8 symmetry and the anti-commutation of T_x and T_y: it is not possible to obtain vison states that form a representation of the algebra of T_x and T_y without this degeneracy.

Turning to the confined phase where Φ is condensed, the non-trivial transformations in (26.38) imply that lattice symmetries must be broken. The precise pattern of the broken symmetry depends upon the sign of y_4, and the two possibilities are shown in Fig. 26.7.

Finally, we address the confinement–deconfinement transition in Fig. 26.7, which is also the transition between the bond-ordered and fractionalized phases in Fig. 14.4 for the XY model with the Berry phase in (14.35). In (26.42), $\mathcal{L}_{\text{monopole}}$ is an irrelevant perturbation to $\mathcal{L}_\Phi$, and the critical point of $\mathcal{L}_\Phi$ is the XY* Wilson–Fisher CFT [119, 230, 256, 261], a conclusion we also reached in Section 16.5.2 (contrast this with the Ising* Wilson–Fisher CFT in Fig. 26.5 for the even-$\mathbb{Z}_2$ gauge theory). Undoing the duality mapping back to (26.37), we note that the XY^* Wilson–Fisher CFT undualizes precisely to $\mathcal{L}_H$. So, $\mathcal{S}_B$ and $\mathcal{L}_{monopole}$ in (26.37) combine to render each other irrelevant in the critical theory; the Berry phases in $\mathcal{S}_B$ suppress the monopole tunneling events. Consequently, the resulting $U(1)$ gauge theory, $\mathcal{L}_H$, retains a critical photon; this is the phenomenon of *deconfined criticality* [119, 230, 256, 261], as the monopoles do turn relevant once we are in the VBS phase with $g > g_c$. The embedding of the $\mathbb{Z}_2$ gauge theory into the $U(1)$ gauge theory is now not optional; it is necessary to obtain a complete description of the critical theory of the phase transition in Fig. 26.7. And the critical theory is $\mathcal{L}_H$ in (26.37), the abelian Higgs model in 2+1 dimensions, which describes a critical scalar coupled to a $U(1)$ gauge field, that is, the naive continuum limit of the lattice $U(1)$ gauge theory yields the correct answer for the critical theory, and monopoles and Berry phases can be ignored. This should be contrasted with the even-$\mathbb{Z}_2$ gauge theory case in Section 26.2.2, where monopoles were relevant.

In closing, we note that the above phase diagram also applies to quantum dimer models on the square lattice [119, 230]. The extension to quantum dimer models on other lattices have also been considered [84, 114, 180, 181, 182, 290].

27 Boson–Fermion and Fermion–Fermion Dualities

The duality between relativistic Dirac fermions and relativistic bosons in 2+1 dimensions is derived by using different parton constructions for the superfluid–insulator transition of lattice bosons. A fermion–fermion duality is obtained by describing the transition from a Chern insulator to a trivial insulator using a decomposition of the electron into three partons. The fermion–fermion duality is applied to obtain the Dirac composite-fermion theory of the quantum Hall states near a half-filled Landau level.

This chapter describes new classes of dualities in three dimensions, which generalize the particle–vortex Dasgupta–Halperin duality we obtained initially in Section 25.4.2. This is a duality between relativistic bosons on both sides of the critical point of the XY model, and hence is a boson–boson duality.

Here, we obtain dualities between relativistic bosons and fermions, and also a fermion–fermion duality. The strategy starts with a lattice model that exhibits a continuous quantum phase transition. Then, by applying different parton constructions to the lattice model, we obtain apparently different continuum field theory descriptions of the transition. By an appeal to universality, we are able to deduce dualities between the continuum theories.

We use these dualities in Section 27.4 to shed additional light on the fractional quantum Hall states studied in Chapter 24.

27.1 Fermion–Boson Duality I

We map between the Wilson–Fisher theory of the superfluid–insulator transition of bosons described by the $D = 3$ XY model (obtained in Section 8.3), and a dual model of Dirac fermions. This mapping was originally obtained from a lattice model for bosons in Ref. [44], but we follow the parton approach of Ref. [21].

27.1.1 Superfluid–Insulator Transition

Let us consider bosons, B_i, on the square lattice at an average density of 1/2 per site. However, we apply a staggered on-site potential so that the density of bosons is one per unit cell:

$$H_b = \varepsilon_0 \sum_i \eta_i B_i^\dagger B_i - \sum_{i<j} t_{ij} \left(B_i^\dagger B_j + B_j^\dagger B_i \right) + \frac{U}{2} \sum_i n_i (n_i - 1), \tag{27.1}$$

where $n_i = B_i^\dagger B_i$ and $\eta_i = +1$ ($\eta_i = -1$) on sublattice A (sublattice B) of the square lattice. Note that $\varepsilon_0 \neq 0$ breaks the sublattice symmetry, and there are two sites per unit cell. We work at an average density $\mathcal{Q} = \langle B_i^\dagger B_i \rangle = 1/2$. At small U the ground state is a superfluid and at large U the ground state is a non-degenerate ("trivial") insulator with all bosons on the B sublattice. The quantum phase transition can be shown, using the same methods as those used in Chapter 8 for the Hubbard model with $\varepsilon_0 = 0$ and $\mathcal{Q} = 1$, to be in the universality class of the Wilson–Fisher conformal filed theory (CFT) with $N = 2$ component real fields (see Problem 8.1). We can write the partition function for the vicinity of the critical point as a field theory for the complex superfluid order parameter Φ:

$$\mathcal{Z}_b[A] = \int \mathcal{D}\Phi \exp\left(- \int d^3x \mathcal{L}_b \right),$$

$$\boxed{\mathcal{L}_b = |(\partial_\mu - iA_\mu)\Phi|^2 + s|\Phi|^2 + u|\Phi|^4 .} \tag{27.2}$$

For future convenience, we have introduced an external $U(1)$ gauge field A_μ which couples minimally to the current associated with the boson number $\mathcal{Q}$.

27.1.2 Fermionic Partons

Let us now describe the same phase transition by decomposing the boson B into two fermionic partons:

$$B = f_1 f. \tag{27.3}$$

We arrange the mean-field Hamiltonian of these partons so that the low-energy excitations reside on the f fermion, while the auxiliary fermion f_1 is a spectator in a gapped state. In particular we place f_1 in a filled band that has Chern number $C_1 = -1$, while f undergoes a transition from a band with Chern number $C = 1$ to Chern number $C = 0$.

Let us discuss the Hamiltonian for f for such a Chern-number-changing transition. We choose

$$H_f = \varepsilon_0 \sum_i \eta_i f_i^\dagger f_i - \sum_{i<j} \left(t_{ij} f_i^\dagger f_j + t_{ij}^* f_j^\dagger f_i \right). \tag{27.4}$$

Whereas the t_{ij} in H_b were real, now the t_{ij} acquire additional phase factors to account for the average flux. We take first and second neighbor hopping t_1 and t_2 as shown in Fig. 27.1 (compare Fig. 18.2 for the chiral spin liquid). We employ a two-site unit cell, and then the momentum-space Hamiltonian is

$$H_f = \sum_{\boldsymbol{k}} f_\alpha^\dagger(\boldsymbol{k}) M_{\alpha\beta}(\boldsymbol{k}) f_\beta(\boldsymbol{k}), \tag{27.5}$$

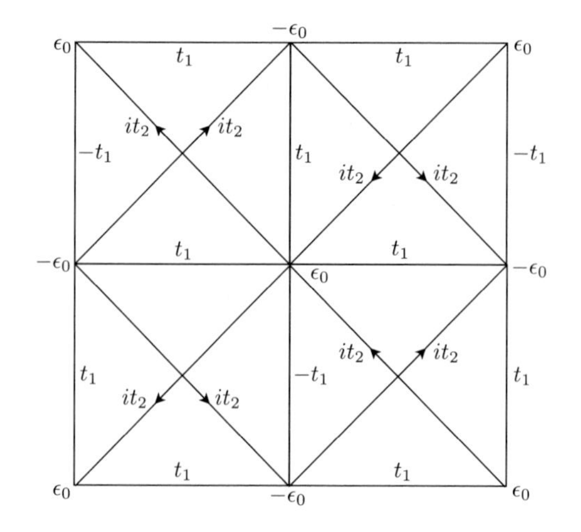

Unit cell of the saddle-point Hamiltonian H_f for the fermions

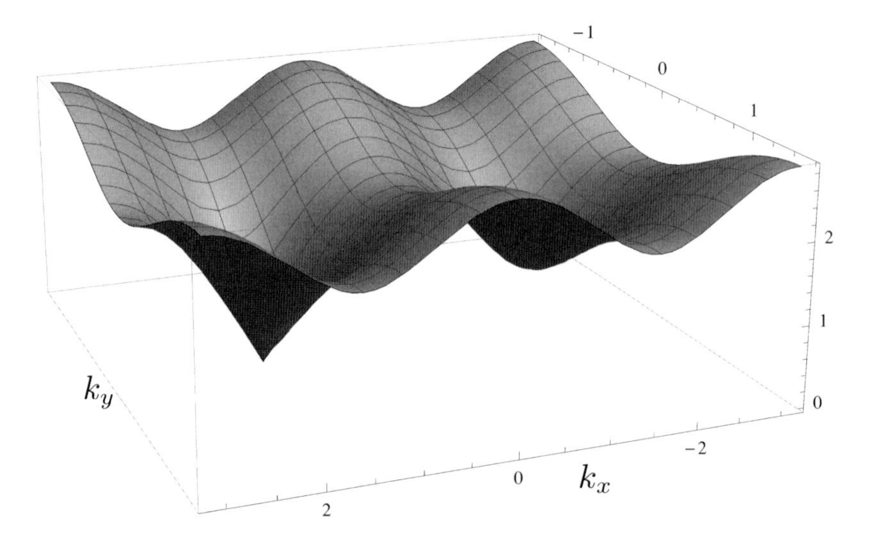

The upper band of the Hamiltonian in (27.5) for $t_1 = 1, t_2 = 0.1, \varepsilon_0 = 0.4$. There is a massless Dirac node at $(-\pi/2, 0)$, while that at $(\pi/2, 0)$ remains massive.

where α, β represent sublattices A and B and

$$M(\boldsymbol{k}) = \begin{pmatrix} \varepsilon_0 - 2t_2\sin(k_x+k_y) - 2t_2\sin(k_x-k_y) & -2t_1\cos(k_x) - 2it_1\sin(k_y) \\ -2t_1\cos(k_x) + 2it_1\sin(k_y) & -\varepsilon_0 + 2t_2\sin(k_x+k_y) + 2t_2\sin(k_x-k_y) \end{pmatrix}. \tag{27.6}$$

The band structure obtained from $M(\boldsymbol{k})$ is shown in Fig. 27.2. Note that there is Dirac node at $\boldsymbol{k} = (-\pi/2, 0)$ for $\varepsilon_0 = 4t_2$; this is our quantum critical point of interest, when the band gap vanishes. For $\varepsilon_0 > 4t_2$, we have a trivial insulator in which the lower band is fully occupied and $C = 0$, while for $\varepsilon_0 < 4t_2$ we have a Chern insulator with $C = 1$ (as can be computed from the methods of Chapter 18).

We arrange a similar Hamiltonian for f_1, which is always a gapped Chern insulator with $C_1 = -1$.

27.1.3 Gauge Theory

Next let us consider the gauge fluctuations about these mean-field states. We will show that the state with $C = 0$, $C_1 = -1$ is actually a trivial insulator of the gauge-invariant bosons B, while the $C = 1$, $C_1 = -1$ state is a superfluid of B.

The gauge invariance associated with the parton decomposition (27.3) requires the introduction of an emergent $U(1)$ gauge field b_μ. We couple the external $U(1)$ gauge field A_μ to the parton f_1. So the parton f is coupled to the gauge field b_μ, while the parton f_1 is coupled to the gauge field $-b_\mu + A_\mu$. This corresponds to the charge assignments

Gauge field	f	f_1
b_μ	1	-1
A_μ	0	1

(27.7)

Integrating out both fermionic partons in the gapped phases, we obtain the effective Lagrangian

$$\mathcal{L}_b = \frac{iC}{4\pi}\varepsilon_{\mu\nu\lambda} b_\mu \partial_\nu b_\lambda + \frac{iC_1}{4\pi}\varepsilon_{\mu\nu\lambda}(b_\mu - A_\mu)\partial_\nu(b_\lambda - A_\lambda). \tag{27.8}$$

Chern Numbers $C = 0, C_1 = -1$

First, let us consider the structure of (27.8) for $C = 0$, $C_1 = -1$. As $\mathcal{L}_b$ depends only upon $b_\mu - A_\mu$, the integral over b_μ yields a result independent of A_μ, and so there is no Hall conductivity. Let us also verify that there are no edge states following the methods of Chapter 17.4, because counter-propagating edge states could also lead to vanishing Hall conductivity. Only the f_1 band has a possible edge state, and an associated edge chiral boson φ_1; coupling this edge state to the gauge fields in the standard manner according to (27.7), we obtain

$$\mathcal{L}_{edge} = -\frac{i}{4\pi}\partial_x\varphi_1\partial_\tau\varphi_1 + \frac{i}{2\pi}\varepsilon_{\mu\nu}\left(-b_\mu + A_\mu\right)\partial_\nu\varphi_1. \tag{27.9}$$

Note that we have only included the important kinematic terms in the edge theory, and dropped the spatial-gradient terms. Now the integral over b_μ yields the condition $\varphi_1 =$ constant, and so there is no edge mode. This verifies that we obtain a trivial insulator for $C = 0$, $C_1 = -1$.

Chern Numbers $C = 1, C_1 = -1$

Next, let us consider the structure of (27.8) for $C = 1$, $C_1 = -1$. Now, (27.8) reduces to

$$\mathcal{L}_b = \frac{i}{2\pi}\varepsilon_{\mu\nu\lambda} b_\mu \partial_\nu A_\lambda - \frac{i}{4\pi}\varepsilon_{\mu\nu\lambda} A_\mu \partial_\nu A_\lambda . \tag{27.10}$$

The integral over b_μ yields a zero flux condition $\sim \delta\left(\varepsilon_{\mu\nu\lambda}\partial_\nu A_\lambda\right)$. This is just the Meissner effect. If we had included Maxwell terms in the effective gauge theory (27.8), we would have obtained a Higgs mass term $\sim A_\mu^2$. Let us also verify there are no active edge states. Now we have chiral bosons φ, φ_1 associated with the two bands with non-zero Chern number, and coupled to the gauge fields according to (27.7); so the edge theory is

$$\mathcal{L}_{edge} = \frac{i}{4\pi}\partial_x\varphi\partial_\tau\varphi - \frac{i}{4\pi}\partial_x\varphi_1\partial_\tau\varphi_1 + \frac{i}{2\pi}\varepsilon_{\mu\nu} b_\mu \left(\partial_\nu\varphi - \partial_\nu\varphi_1\right) + \frac{i}{2\pi}\varepsilon_{\mu\nu} A_\mu \partial_\nu \varphi_1 . \tag{27.11}$$

Now, the integral over b_μ yields $\varphi = \varphi_1$ (up to a constant), and then the first two terms in (27.11) cancel: there is no edge mode.

We have now verified our central claim: the two phases of the parton theory across the gapless point at $\varepsilon_0 = 4t_2$ precisely match the low-energy properties of the original boson model undergoing a superfluid–insulator transition.

27.1.4 Quantum Critical Theory

Now let us consider the neighborhood of the point $\varepsilon_0 = 4t_2$. We can still safely integrate out the f_1 fermions, but have to keep the low-energy f fermions active.

For $t_2 = \varepsilon_0 = 0$, this Hamiltonian has Dirac nodes at $\boldsymbol{k} = (\pm\pi/2, 0)$, and it is instructive to keep track of the vicinities of both nodes. We write $\boldsymbol{k} = (\pm\pi/2 + q_x, q_y)$ and expand for small q_x, q_y. We also introduce Pauli matrices τ^x, τ^y, τ^z in the sublattice space. Then we can write the Hamiltonian as

$$H_f = f^\dagger\left[\pm 2t_1 q_x \tau^x + 2t_2 q_y \tau^y + (\varepsilon_0 \mp 4t_2)\tau^z\right] f . \tag{27.12}$$

This is the Hamiltonian of two species of two-component Dirac fermions, with one species carrying a light mass $m = \varepsilon_0 - 4t_2$, and other a heavy mass $M = \varepsilon_0 + 4t_2$ (see Fig. 27.2).

Let us focus on the light-mass fermion. We employ relativistic notation by defining $\psi = c$, $\bar\psi \equiv \psi^\dagger\tau^z$, and $\gamma^\mu = (\tau^z, -\tau^y, \tau^x)$, introduce a coupling to the gauge field b_μ by gauge invariance, and then the imaginary-time Lagrangian corresponding to the light-mass Dirac fermion in H_f is

$$\mathcal{L}_f^0 = \bar\psi\gamma^\mu\left(\partial_\mu - i b_\mu\right)\psi + m\bar\psi\psi , \tag{27.13}$$

where we have absorbed the Fermi velocity $2t_1$ by rescaling time. If we use this continuum theory to naively integrate out the f fermion, we obtain a Chern–Simons term for b_μ with a half-integral coefficient

$$\frac{i\,\mathrm{sgn}(m)}{8\pi}\varepsilon_{\mu\nu\lambda} b_\mu \partial_\nu b_\lambda . \tag{27.14}$$

Clearly, this has to be compensated by a contribution from the heavy fermion, so that the total Chern number is either 0 or 1 depending upon the sign of m. The proper gauge-invariant way to account for this additional contribution is to use the η-invariant to represent the path integral over ψ [260, 313]. However, it is common to denote this additional contribution by an explicit $C = 1/2$ contribution from the heavy fermion; more properly, this $C = 1/2$ contribution is from "the rest of the band" after the Dirac node has been separately accounted for. As the remaining band remains gapped, we do not expect any critical corrections from this contribution. Consequently, the complete low-energy theory for the f bands is then

$$\mathcal{L}_f = \bar{\psi}\gamma^\mu(\partial_\mu - ib_\mu)\psi + m\bar{\psi}\psi + \frac{i}{8\pi}\varepsilon_{\mu\nu\lambda}b_\mu\partial_\nu b_\lambda \,. \tag{27.15}$$

Finally, we add the $C_1 = -1$ contribution of the f_1 band from (27.8) to obtain the complete low-energy theory for the quantum critical point:

$$\boxed{\mathcal{L}_\psi = \bar{\psi}\gamma^\mu(\partial_\mu - ib_\mu)\psi + m\bar{\psi}\psi - \frac{i}{8\pi}\varepsilon_{\mu\nu\lambda}b_\mu\partial_\nu b_\lambda + \frac{i}{2\pi}\varepsilon_{\mu\nu\lambda}A_\mu\partial_\nu b_\lambda - \frac{i}{4\pi}\varepsilon_{\mu\nu\lambda}A_\mu\partial_\nu A_\lambda \,.} \tag{27.16}$$

We can now present the final statement of the boson–fermion duality. The fermionic partition function with the Lagrangian in (27.16) equals the bosonic partition function $\mathcal{Z}_b[A]$ in (27.2):

$$\begin{aligned} \mathcal{Z}_\psi[A] &= \int \mathcal{D}\psi\mathcal{D}b_\mu \exp\left(-\int d^3x\,\mathcal{L}_\psi\right) \\ &= \mathcal{Z}_b[A] \,. \end{aligned} \tag{27.17}$$

In the bosonic theory, we tune across the superfluid–insulator quantum critical point by changing the parameter s. In the fermionic theory we tune across the critical point where the sign of m changes.

It is now interesting to note an important operator correspondence between (27.16) and (27.2). At the lattice level we have in (27.8) that $B \sim ff_1$. Because of the Chern number $C_1 = -1$ on f_1, we note that a 2π flux in b_μ corresponds to an f_1 particle. So we can conclude that a monopole in b, $\mathcal{M}_b$, corresponds to an f_1 particle. However, in the presence of gapless Dirac fermions at the critical point $m = 0$, the monopole operator has a single fermionic zero mode [34, 127]. Gauge invariance fixes the occupation number of this mode, and this turns out to be precisely the same as the requirement of gauge invariance of $B \sim ff_1$. So, choosing the appropriately gauge-invariant monopole [34, 127] associated with (27.16), we have the operator correspondence $B \sim \mathcal{M}_b$.

27.2 Fermion–Boson Duality II

We obtain our second exact duality by introducing another external gauge field C_μ, and adding the term

$$-\frac{i}{2\pi}\varepsilon_{\mu\nu\lambda}C_\mu\partial_\nu A_\lambda+\frac{i}{4\pi}\varepsilon_{\mu\nu\lambda}A_\mu\partial_\nu A_\lambda \tag{27.18}$$

to both sides of the duality in Section 27.1. Then, we promote the gauge field A_μ to a dynamic gauge field, and integrate over it. In keeping with our notational convention, we map $A_\mu \to a_\mu$, and then perform the integration.

On the bosonic side of the duality we obtain the partition function

$$\mathcal{Z}_b[C]=\int \mathcal{D}\Phi\mathcal{D}a_\mu \exp\left(-\int d^3x\,\mathcal{L}_b\right), \tag{27.19}$$

$$\boxed{\mathcal{L}_b=|(\partial_\mu-ia_\mu)\Phi|^2+s|\Phi|^2+u|\Phi|^4+\frac{i}{4\pi}\varepsilon_{\mu\nu\lambda}a_\mu\partial_\nu a_\lambda-\frac{i}{2\pi}\varepsilon_{\mu\nu\lambda}C_\mu\partial_\nu a_\lambda\,.} \tag{27.20}$$

On the fermionic side we obtain

$$\mathcal{Z}_\psi[C]=\int \mathcal{D}\psi\mathcal{D}b_\mu\mathcal{D}a_\mu \exp\left(-\int d^3x\,\mathcal{L}_\psi\right), \tag{27.21}$$

where

$$\mathcal{L}_\psi=\bar{\psi}\gamma^\mu(\partial_\mu-ib_\mu)\psi+m\bar{\psi}\psi-\frac{i}{8\pi}\varepsilon_{\mu\nu\lambda}b_\mu\partial_\nu b_\lambda+\frac{i}{2\pi}\varepsilon_{\mu\nu\lambda}a_\mu\partial_\nu b_\lambda-\frac{i}{2\pi}\varepsilon_{\mu\nu\lambda}C_\mu\partial_\nu a_\lambda\,. \tag{27.22}$$

From the path integral over a_μ, we obtain the constraint $b_\mu=C_\mu$. So the final form of the fermionic theory equals the bosonic partition function in (27.19) with

$$\begin{aligned}\mathcal{Z}_\psi[C]&=\int \mathcal{D}\psi \exp\left(-\int d^3x\,\mathcal{L}_\psi\right)\\&=\mathcal{Z}_b[C],\end{aligned} \tag{27.23}$$

$$\boxed{\mathcal{L}_\psi=\bar{\psi}\gamma^\mu(\partial_\mu-iC_\mu)\psi+m\bar{\psi}\psi-\frac{i}{8\pi}\varepsilon_{\mu\nu\lambda}C_\mu\partial_\nu C_\lambda\,.} \tag{27.24}$$

Similar to the case in Section 27.1, the gauge-invariant monopole $\mathcal{M}_a$ [34, 127] corresponds to the fermion ψ.

27.3 Fermion–Fermion Duality

I now present a fermion–fermion duality between fermions c, and "composite" fermions f_c. This has useful applications to the fractional quantum Hall effect, where c represents the physical electron, and f_c realizes Jain's composite fermions, as we see in Section 27.4.

Let us consider the situation when the fermion c undergoes an integer quantum Hall transition from a state with $\sigma_{xy}=0$ to $\sigma_{xy}=e^2/h$. This is easily described by placing c in a band whose Chern number changes from 0 to 1. Indeed, we considered just such a band structure in Section 27.1 for the fermion f. The transition is described at low energies by a single Dirac fermion ψ, whose mass changes sign. So the first low-energy theory for this transition is simply that in (27.15), after replacing the emergent gauge field b_μ by the external gauge field A_μ,

$$\boxed{\mathcal{L}_\psi = \bar{\psi}\gamma^\mu(\partial_\mu - iA_\mu)\psi + m\bar{\psi}\psi + \frac{i}{8\pi}\varepsilon_{\mu\nu\lambda}A_\mu\partial_\nu A_\lambda .} \tag{27.25}$$

The partition function is

$$Z_\psi[A] = \int \mathcal{D}\psi \exp\left(-\int d^3x\,\mathcal{L}_\psi\right). \tag{27.26}$$

The transition is driven by a change in the sign of the fermion mass at $m = 0$.

Now, let us describe this integer quantum Hall transition by the parton construction. We use the same parton decomposition as that used in Chapter 24 in the description of the Jain fractional quantum Hall states. So we write the fermion c in terms of the three fermionic partons f_c, f_1, f_2:

$$c = f_c f_1 f_2. \tag{27.27}$$

As usual, we place the partons in filled bands characterized by an integer Chern number. We label these Chern numbers as C_c, C_1, and C_2, respectively. We now show that the Chern numbers $C_c = 0$, $C_1 = 1$, $C_2 = 1$ describe an insulator with $\sigma_{xy} = 0$, while the Chern numbers $C_c = -1$, $C_1 = 1$, $C_2 = 1$ describe a Chern insulator with $\sigma_{xy} = e^2/h$. This matches the integer quantum Hall transition of the parent fermion c. And the transition is driven by the change in Chern number of the composite fermion f_c from 0 to -1. So our low-energy theory is expressed in terms of the corresponding Dirac composite fermion ψ_c.

27.3.1 Gapped Phases

The parton decomposition (27.27) requires us to introduce two gauge fields to impose the equalities of the three parton currents at all spacetime points. We choose to couple f_c to an emergent gauge field a_μ, f_1 to the emergent gauge field $-a_\mu - b_\mu$, and f_2 to the gauge field $b_\mu + A_\mu$. We represent these charge assignments in the following table.

Gauge field	f_c	f_1	f_2
a_μ	1	-1	0
b_μ	0	-1	1
A_μ	0	0	1

(27.28)

Assuming we are in the gapped phases, we can integrate out the fermions and obtain an effective action for gauge fields analogous to (27.8)

$$\mathcal{L}_{a,b} = \frac{iC_c}{4\pi}\varepsilon_{\mu\nu\lambda}a_\mu\partial_\nu a_\lambda + \frac{iC_1}{4\pi}\varepsilon_{\mu\nu\lambda}(a_\mu + b_\mu)\partial_\nu(a_\lambda + b_\lambda) + \frac{iC_2}{4\pi}\varepsilon_{\mu\nu\lambda}(b_\mu + A_\mu)\partial_\nu(b_\lambda + A_\lambda). \tag{27.29}$$

Let us now examine this theory for the two cases noted above.

Chern numbers $C_c = 0, C_1 = 1, C_2 = 1$

For this case, in (27.29), we can integrate out a_μ by shifting $a_\mu \to a_\mu - b_\mu$, and then integrate over b_μ after shifting $b_\mu \to b_\mu - A_\mu$, yielding trivial results. So there is no Chern–Simons term for A_μ.

We also need to verify that there no edge states, because we have not ruled out counter-propagating edge states that yield a net-zero σ_{xy}. We obtain the edge theory by introducing chiral bosons $\varphi_{1,2}$ for the fermions $f_{1,2}$, and there is no edge state associated with f_c. Coupling these bosons to the gauge fields minimally according to the charge assignments in (27.28), the analog of (27.11) is

$$\mathcal{L}_{edge} = \frac{i}{4\pi}\partial_x\varphi_1\partial_\tau\varphi_1 + \frac{i}{4\pi}\partial_x\varphi_2\partial_\tau\varphi_2 + \frac{i}{2\pi}\varepsilon_{\mu\nu}a_\mu\left(-\partial_\nu\varphi_1\right)$$
$$+\frac{i}{2\pi}\varepsilon_{\mu\nu}b_\mu\left(-\partial_\nu\varphi_1 + \partial_\nu\varphi_2\right) + \frac{i}{2\pi}\varepsilon_{\mu\nu}A_\mu\partial_\nu\varphi_2. \tag{27.30}$$

The integral over a_μ sets $\varphi_1 = \text{constant}$, and then the integral of b_μ sets $\varphi_2 = \text{constant}$. So there is no edge state, and we indeed have a trivial insulator.

Chern Numbers $C_c = -1, C_1 = 1, C_2 = 1$

Now, (27.29) becomes

$$\mathcal{L}_{a,b} = \frac{i}{4\pi}\varepsilon_{\mu\nu\lambda}b_\mu\partial_\nu b_\lambda + \frac{i}{2\pi}\varepsilon_{\mu\nu\lambda}a_\mu\partial_\nu b_\lambda + \frac{i}{4\pi}\varepsilon_{\mu\nu\lambda}(b_\mu + A_\mu)\partial_\nu(b_\lambda + A_\lambda). \tag{27.31}$$

The integral over a_μ sets $b_\mu = 0$, up to a gauge transformation. Then we are left with a Chern–Simons term for A_μ with a unit coefficient, implying $\sigma_{xy} = e^2/h$.

As above, we have to verify the nature of the edge states. Now we have three chiral bosons $\varphi_{c,1,2}$, and the analog of (27.30) from (27.28) is

$$\mathcal{L}_{edge} = -\frac{i}{4\pi}\partial_x\varphi_c\partial_\tau\varphi_c + \frac{i}{4\pi}\partial_x\varphi_1\partial_\tau\varphi_1 + \frac{i}{4\pi}\partial_x\varphi_2\partial_\tau\varphi_2 + \frac{i}{2\pi}\varepsilon_{\mu\nu}a_\mu\left(\partial_\nu\varphi_c - \partial_\nu\varphi_1\right)$$
$$+\frac{i}{2\pi}\varepsilon_{\mu\nu}b_\mu\left(-\partial_\nu\varphi_1 + \partial_\nu\varphi_2\right) + \frac{i}{2\pi}\varepsilon_{\mu\nu}A_\mu\partial_\nu\varphi_2. \tag{27.32}$$

Now, the integrals over a_μ and b_μ set $\varphi_c = \varphi_1 = \varphi_2$ (up to constants). Then (27.32) reduces to precisely the edge theory of a Chern insulator in a Chern band with $C = 1$.

27.3.2 Quantum Critical Theory

It is now easy to obtain the quantum critical theory, following the methods of the sections above. We take the low-energy limit of the active composite fermion f_c undergoing a transition from Chern number $C_c = 0$ to $C_c = -1$, represent it by a low-energy Dirac fermion ψ_c as in (27.15), and add the contributions of the $f_{1,2}$ fermions from (27.29) to obtain

$$
\begin{aligned}
\mathcal{L}_c &= \bar{\psi}_c \gamma^\mu (\partial_\mu - i a_\mu)\psi_c - m\bar{\psi}_c\psi_c - \frac{i}{8\pi}\varepsilon_{\mu\nu\lambda} a_\mu \partial_\nu a_\lambda \\
&\quad + \frac{i}{4\pi}\varepsilon_{\mu\nu\lambda}(a_\mu + b_\mu)\partial_\nu (a_\lambda + b_\lambda) + \frac{i}{4\pi}\varepsilon_{\mu\nu\lambda}(b_\mu + A_\mu)\partial_\nu (b_\lambda + A_\lambda) \\
&= \bar{\psi}_c \gamma^\mu (\partial_\mu - i a_\mu)\psi_c - m\bar{\psi}_c\psi_c + \frac{i}{8\pi}\varepsilon_{\mu\nu\lambda} a_\mu \partial_\nu a_\lambda + \frac{i}{2\pi}\varepsilon_{\mu\nu\lambda} a_\mu \partial_\nu b_\lambda \\
&\quad + \frac{2i}{4\pi}\varepsilon_{\mu\nu\lambda} b_\mu \partial_\nu b_\lambda + \frac{i}{2\pi}\varepsilon_{\mu\nu\lambda} b_\mu \partial_\nu A_\lambda + \frac{i}{4\pi}\varepsilon_{\mu\nu\lambda} A_\mu \partial_\nu A_\lambda \,. \qquad (27.33)
\end{aligned}
$$

It is now tempting to integrate out the b_μ gauge field above, and to obtain a theory with only one emergent gauge field a_μ. However, as emphasized in Ref. [253], this is dangerous, and does not properly capture all the topological properties of the theory. But if we are only interested in the bulk properties of critical theory in a perturbative expansion (say in an inverse expansion in the number of fermion flavors), we can go ahead and do so. So, with this caution, we can perform the integral over the b_μ gauge field and obtain our promised fermion–fermion duality

$$
\begin{aligned}
\mathcal{Z}_c[A] &= \int \mathcal{D}\psi_c \mathcal{D}a \exp\left(-\int d^3x\,\mathcal{L}_c\right) \\
&= \mathcal{Z}_\psi[A], \qquad (27.34)
\end{aligned}
$$

where $\mathcal{Z}_\psi[A]$ is defined in (27.26) and

$$
\boxed{\mathcal{L}_c = \bar{\psi}_c \gamma^\mu (\partial_\mu - i a_\mu)\psi_c - m\bar{\psi}_c\psi_c - \frac{i}{4\pi}\varepsilon_{\mu\nu\lambda} a_\mu \partial_\nu A_\lambda + \frac{i}{8\pi}\varepsilon_{\mu\nu\lambda} A_\mu \partial_\nu A_\lambda \,.} \qquad (27.35)
$$

I reiterate that (27.33) is a more complete representation of the field theory on the composite-fermion side of the duality [253].

Similar to the previous cases, the assignment of the gauge charges in (27.27) can be used to deduce the representation of the fermion operator ψ in terms of theory for ψ_c; we find that $\psi_c \sim \mathcal{M}_a^2 \mathcal{M}_b^\dagger$.

Unlike the previous dualities considered in this chapter, the fermion–fermion duality has no Chern–Simons term in an internal gauge field on either side of the duality, at least in the form in (27.35). In this respect, it is similar to the Dasgupta–Halperin boson–boson duality studied in Chapter 23.

27.4 Fractional Quantum Hall Effect: Dirac Composite Fermions

We have previously described the Jain states in Chapter 24 by using a parton theory where we assumed that the fermions occupied Landau levels. We can now use the Dirac composite-fermion approach above to obtain an alternative theory of the Jain states [266, 267]. The Dirac approach has the advantage of allowing one to preserve the particle–hole symmetry of the half-filled Landau level.

A useful starting point to introducing Dirac composite fermions is to consider graphene at its charge-neutrality point in the presence of a magnetic field, as in Section 19.2. Because of the particle–hole symmetry of graphene, the $n = 0$ Landau level in graphene is exactly half filled, and we are at $\nu = 1/2$. By particle–hole symmetry, this state has $\sigma_{xy} = 0$. If we are considering a non-relativistic fermion, as in GaAs, then quantum Hall states would have their conductivity shifted by $e^2/(2h)$, which means the Hall conductivity of the half-filled Landau level would be $e^2/(2h)$.

So, starting from graphene at its charge-neutrality point, we consider the theory

$$\mathcal{L}_{graphene} = \overline{\psi}\gamma_\mu \left(\partial_\mu - iA_\mu\right)\psi + \cdots, \tag{27.36}$$

where ψ creates an electron in the form of a two-component Dirac fermion. The spatial components of A_μ represent the applied magnetic field with $\boldsymbol{\nabla} \times \boldsymbol{A} = B$. The time component, iA_τ, is the applied chemical potential, which allows us to consider electron densities away from the charge-neutrality point of graphene. If we were considering the non-relativistic quantum Hall states, as in GaAs, then we would shift the Hall conductivity (and the density) by adding a Chern–Simons term in the external gauge field

$$\mathcal{L}_{GaAs} = \overline{\psi}\gamma_\mu \left(\partial_\mu - iA_\mu\right)\psi + \frac{i}{8\pi}\varepsilon_{\mu\nu\lambda}A_\mu\partial_\nu A_\lambda + \cdots. \tag{27.37}$$

The remainder of the discussion here is carried out using $\mathcal{L}_{GaAs}$, as that is the case considered in most of the literature. It is easy to translate back to graphene, by appropriate shifts in the Hall conductivity and the density.

To describe the $\nu = 1/2$ state and its vicinity, we perform the fermion–fermion duality on $\mathcal{L}_{GaAs}$, to obtain the theory of Dirac composite fermions, ψ_c:

$$\mathcal{L}_c = \overline{\psi}_c\gamma_\mu \left(\partial_\mu - ia_\mu\right)\psi_c - \frac{i}{4\pi}\varepsilon_{\mu\nu\lambda}A_\mu\partial_\nu a_\lambda + \frac{i}{8\pi}\varepsilon_{\mu\nu\lambda}A_\mu\partial_\nu A_\lambda + \cdots. \tag{27.38}$$

As has been our convention, the field a_μ is a dynamical $U(1)$ gauge field that has to be integrated over, while A_μ is the external, "background," electromagnetic gauge field.

We now introduce some basic notation, and obtain important relations by comparing the saddle-point equations (27.37) and (27.38). Let ρ be the density of electrons. Then, the filling factor of the Landau level ν is (in units with $\hbar = e = 1$)

$$\nu = \frac{2\pi\rho}{B}. \tag{27.39}$$

Taking the derivative of $\mathcal{L}_{GaAs}$ with respect to A_τ, we have

$$\rho = -\left\langle\overline{\psi}\gamma_\tau\psi\right\rangle + \frac{B}{4\pi}. \tag{27.40}$$

So, at $\nu = 1/2$, the density of Dirac electrons in graphene, $-\left\langle\overline{\psi}\gamma_\tau\psi\right\rangle$, vanishes. Let us also represent the average internal gauge field on the composite fermions by $b = \boldsymbol{\nabla} \times \boldsymbol{a}$. Now we take the derivative of $\mathcal{L}_c$ with respect to A_τ. This yields

$$\rho = -\frac{b}{4\pi} + \frac{B}{4\pi}, \tag{27.41}$$

which we can also write as

$$b = -B(2\nu - 1)\,. \tag{27.42}$$

So, at $\nu = 1/2$, the average magnetic field on the Dirac composite fermions b vanishes. This is the primary advantage of the composite-fermion formulation: we have mapped a problem at high magnetic field to one at vanishing magnetic field.

We also need some relations for the density of the Dirac composite fermions. Let us denote the density of the Dirac composite fermions by

$$\rho_c = -\left\langle \overline{\Psi}_c \gamma_\tau \Psi_c \right\rangle\,. \tag{27.43}$$

Then, taking the derivative of $\mathcal{L}_c$ with respect to a_τ, we obtain

$$\rho_c = \frac{B}{4\pi}\,. \tag{27.44}$$

So the density of Dirac composite fermions is determined by the applied magnetic field.

In general, the theory of composite fermions therefore has a non-zero density ρ_c and an applied average field b. So, if anything, it is more complicated than the original problem of electrons, as we also have to consider a fluctuating gauge field a_μ. However, the composite-fermion theory simplifies under two conditions: (i) if the average field b vanishes, or (ii) if exactly an integer number of its Landau levels are filled, so that there is an energy gap to composite-fermion excitations.

In case (i), which happens exactly at $\nu = 1/2$, we have a finite density of composite fermions in a vanishing magnetic field, which will form a Fermi surface. We do have to consider the influence of the a_μ gauge fluctuations on this Fermi surface, and this we defer to Section 34.3. Note that the present formulation of the half-filled Landau-level problem is dual to that presented in Section 24.4, but leads to the same low-energy effective theory.

Let us consider here case (ii), when the n-th Landau level of Dirac fermions is fully filled. Recall that the allowed values of n are $\ldots, -2, -1, 0, 1, 2, \ldots$, and that the corresponding Hall conductivity of the Dirac fermions is $(n + 1/2)e^2/h$. In this situation, the relationship between ρ_c and b is

$$\rho_c = \left(n + \frac{1}{2}\right)\frac{b}{2\pi}\,. \tag{27.45}$$

From (27.42), (27.44), and (27.45), we obtain the Jain filling fractions

$$\nu = \frac{n}{2n+1}\,. \tag{27.46}$$

The simplest cases of the fractional quantum Hall states are $n = 1$, $\nu = 1/3$ and $n = -2$, $\nu = 2/3$. Notice that for the Dirac composite fermions, the cases $n = 1$ and $n = -2$ are particle–hole symmetric: the chemical potential is between the adjacent Landau levels with $|n| = 1$ and $|n| = 2$, and the composite-fermion Hall conductivity is $\pm(3/2)e^2/h$. This particle–hole symmetry is a crucial feature of the Dirac composite-fermion theory.

This theory can also describe integer quantum Hall states at the special values $n = 0, -1$ corresponding to $\nu = 0, 1$; in these cases the zero-th Dirac Landau level is either fully filled or empty, and these cases are also particle–hole symmetric.

28 Gapless Spin Liquids

The phase transition from the Néel ordered phase to the valence-bond solid leads to a deconfined critical theory of a gapless spin liquid. The description using bosonic partons leads to a $\mathbb{CP}^1$ field theory, and the description using fermionic partons leads to an $SU(2)$ gauge theory with massless Dirac spinons. A stable gapless $\mathbb{Z}_2$ spin-liquid phase is obtained by condensing Higgs fields in the latter theory.

This chapter returns to the basic problem that historically was the motivation for much of the analyses in Parts II and IV: the nature of spin-liquid states of spin $S = 1/2$ antiferromagnets on the square lattice. In our bosonic parton analysis in Chapter 15, the basic phase diagram in Fig. 15.1 shows that the magnetically ordered Néel state can have a quantum phase transition to a "quantum-disordered" state, and we examine here the nature of such a transition.

For the case of the triangular lattice, the quantum-disordered state is a $\mathbb{Z}_2$ spin liquid, as was described in Chapter 15; this sets up the phase diagram in Fig. 15.5, and the quantum phase transition was described by a well-understood $O(4)^*$ field theory in Section 15.4.1.

For the case of the square lattice with a bipartite Néel ordered state, we found in Section 15.3.1 that the corresponding quantum-disordered state was described by a $U(1)$ gauge theory. We studied the nature of the $U(1)$ gauge theory more completely using a duality mapping in Section 26.1, and found that the quantum-disordered state was a confining state of a $U(1)$ spin liquid with valence-bond solid (VBS) order. This sets up the phase diagram for the square-lattice antiferromagnet in Fig. 28.1, which is the analog of Fig. 15.5 for the triangular or other non-bipartite lattices. We recall Fig. 1.7, which showed an example of a quantum transition between two such phases in $SrCu_2(BO_3)_2$, albeit for a case in which the VBS state is only two-fold degenerate.

This chapter investigates the nature of the Néel–VBS transition in Fig. 28.1. Both phases are confining, and break distinct symmetries. In the conventional Landau–Ginzburg–Wilson framework, such a transition is expected to be first-order, or to have an intermediate coexistence phase. Here we describe the possibility of deconfined criticality: that there is a second-order transition, with the critical theory a gapless spin liquid with fractionalized spinon degrees of freedom. Evidence for deconfined criticality has emerged in experiments on $SrCu_2(BO_3)_2$ [56]. I will discuss two distinct approaches to this transition, using bosonic spinons in Section 28.1, and fermionic spinons in Sections 28.2 and 28.3. Finally, in Section 28.4, we consider the possibility

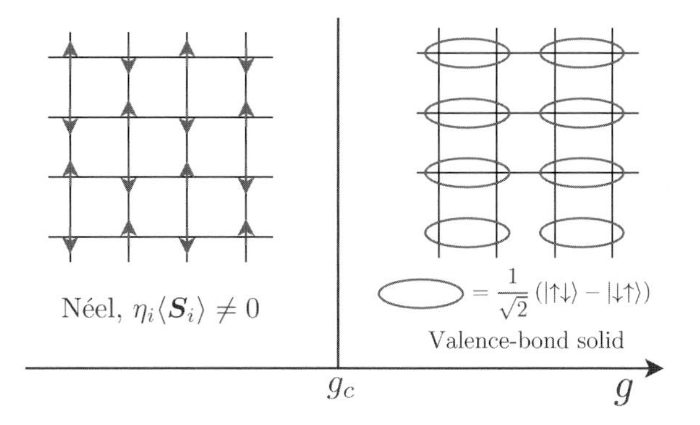

Figure 28.1 Possible phase diagram of the $S = 1/2$ square-lattice antiferromagnet tuned by further neighbor exchange interactions.

of an intermediate gapless $\mathbb{Z}_2$ spin-liquid phase, a possibility supported by recent numerical studies.

28.1 $U(1)$ Spin Liquids on the Square Lattice: Bosonic Spinons

We return to our treatment in Chapter 15 of square-lattice antiferromagnets using Schwinger bosons $s_{i\alpha}$. After decoupling the spin exchange interactions to boson pairing terms, and diagonalizing the boson Hamiltonian by a Bogoliubov transformation, we find bosonic spinons with dispersion

$$\omega_{\boldsymbol{k}} = \left(\bar{\lambda}^2 - J^2 \bar{\mathcal{Q}}^2 (\sin k_x + \sin k_y)^2\right)^{1/2}, \tag{28.1}$$

where $\bar{\lambda}$ is the mean value of the Lagrange multiplier imposing the boson-number constraint, and $\bar{\mathcal{Q}}$ is the mean value of the boson pairing field. We now notice that this dispersion has a minimum at $\boldsymbol{k} = \pm(\pi/2, \pi/2)$, and the minimum spinon gap is $\sqrt{\bar{\lambda}^2 - 4J^2\bar{\mathcal{Q}}^2}$. We are interested here in the situation when this gap vanishes. The point where the gap is exactly zero is our candidate for a gapless quantum critical theory. Beyond this point, the bosonic spinons condense, and we obtain the Néel state, with the same symmetry as the classical antiferromagnet with spins polarized in opposite orientations on the two checkerboard sublattices.

In the vicinity of this gap-vanishing transition, we can write down a continuum theory for the spinons, coupled to the emergent $U(1)$ gauge field a_μ. Recall that we expressed the phases of the bond fields in (15.26)

$$\begin{aligned} \mathcal{Q}_{i,i+\hat{x}} &= \bar{\mathcal{Q}} \exp\left(i\Theta_{ix}\right), \\ \mathcal{Q}_{i,i+\hat{y}} &= \bar{\mathcal{Q}} \exp\left(i\Theta_{iy}\right), \end{aligned} \tag{28.2}$$

and the gauge fields are obtained by the parameterization in (15.28):

$$\begin{aligned}
\Theta_{ix}(\tau) &= \eta_i a_x(\boldsymbol{r},\tau), \\
\Theta_{iy}(\tau) &= \eta_i a_y(\boldsymbol{r},\tau), \\
\lambda_i &= -i\bar{\lambda} - \eta_i a_\tau(\boldsymbol{r},\tau),
\end{aligned} \tag{28.3}$$

where $\eta_i = \pm 1$ on the checkerboard sublattices. For the spinons, we introduce the wavevector at the minimum spinon gap $\boldsymbol{k}_0 = (\pi/2, \pi/2)$ and parameterize on the A and B sublattices:

$$\begin{aligned}
s^\alpha_{Ai} &= \psi^\alpha_1(\boldsymbol{r}_i) e^{i\boldsymbol{k}_0 \cdot \boldsymbol{r}_i}, \\
s^\alpha_{Bi} &= -i\mathcal{J}^{\alpha\beta}\psi_{2\beta}(\boldsymbol{r}_i) e^{i\boldsymbol{k}_0 \cdot \boldsymbol{r}_i},
\end{aligned} \tag{28.4}$$

where $\mathcal{J}^{\alpha\beta} = \varepsilon^{\alpha\beta}$ for the $SU(2)$ spin case ($\mathcal{J}^{\alpha\beta}$ was defined in (15.12)). Next, we insert these parameterizations into the spinon action, perform a gradient expansion, and transform the Lagrangian $\mathcal{L}$ into

$$\begin{aligned}
\mathcal{L} = \int \frac{d^2r}{a^2} \Bigg[& \psi^*_{1\alpha}\left(\frac{d}{d\tau} + ia_\tau\right)\psi^\alpha_1 + \psi^{\alpha *}_2\left(\frac{d}{d\tau} - ia_\tau\right)\psi_{2\alpha} \\
& + \bar{\lambda}\left(|\psi^\alpha_1|^2 + |\psi_{2\alpha}|^2\right) - 4J\bar{Q}_1\left(\psi^\alpha_1\psi_{2\alpha} + \psi^*_{1\alpha}\psi^{\alpha *}_2\right) \\
& + J\bar{Q}_1 a^2 \left[(\boldsymbol{\nabla} + i\boldsymbol{a})\psi^\alpha_1 (\boldsymbol{\nabla} - i\boldsymbol{a})\psi_{2\alpha} \right. \\
& \quad \left. + (\boldsymbol{\nabla} - i\boldsymbol{a})\psi^*_{1\alpha}(\boldsymbol{\nabla} + i\boldsymbol{a})\psi^{\alpha *}_2\right] \Bigg].
\end{aligned} \tag{28.5}$$

We now introduce the fields

$$\begin{aligned}
z^\alpha &= (\psi^\alpha_1 + \psi^{\alpha *}_2)/\sqrt{2}, \\
\pi^\alpha &= (\psi^\alpha_1 - \psi^{\alpha *}_2)/\sqrt{2}.
\end{aligned}$$

Following the definitions of the underlying spin operators, it is not difficult to show that the Néel order parameter φ_a is related to the z^α by

$$\varphi_a = z^*_\alpha \sigma^{a\alpha}{}_\beta z^\beta. \tag{28.6}$$

From Eq. (28.5), it is clear that the the π fields turn out to have mass $\bar{\lambda} + 4J\bar{Q}_1$, while the z fields have a mass $\bar{\lambda} - 4J\bar{Q}_1$, which vanishes at the transition to the long-range-order phase. The π fields can therefore be safely integrated out, and $\mathcal{L}$ yields the following effective action, valid at distances much larger than the lattice spacing [217, 218]:

$$S_{eff} = \int \frac{d^2r}{\sqrt{8}a} \int_0^{c\beta} d\tau \left\{ |(\partial_\mu - ia_\mu)z^\alpha|^2 + \frac{\Delta^2}{c^2}|z^\alpha|^2 \right\}. \tag{28.7}$$

Here, μ extends over x, y, τ; $c = \sqrt{8}J\bar{Q}_1 a$; is the spin-wave velocity; and $\Delta = (\bar{\lambda}^2 - 16J^2\bar{Q}_1^2)^{1/2}$ is the gap towards spinon excitations. Thus the long-wavelength theory consists of a massive, spin-1/2, relativistic, boson z^α (spinon) coupled to a $U(1)$ gauge field a_μ.

On general symmetry grounds, and by analogy with the arguments in Chapter 26, we extend (28.7) to a theory for the vicinity of the quantum critical point at which the spinon gap vanishes [239]:

$$\begin{aligned} \mathcal{S}_{U(1)} &= \int d^3x \left(\mathcal{L}_z + \mathcal{L}_{monopole}\right) + \mathcal{S}_B, \\ \mathcal{L}_z &= |(\partial_\mu - i a_\mu) z_\alpha|^2 + g|z_\alpha|^2 + u\left(|z_\alpha|^2\right)^2 + K(\varepsilon_{\mu\nu\lambda}\partial_\nu a_\lambda)^2, \\ \mathcal{L}_{monopole} &= -y\left(\mathcal{M}_a + \mathcal{M}_a^\dagger\right), \\ \mathcal{S}_B &= i2S\sum_i \eta_i \int d\tau\, a_{i\tau}. \end{aligned} \tag{28.8}$$

The theory $\mathcal{L}_z$ is also known as the $\mathbb{CP}^1$ model. For a complete description, we have to include monopoles $\mathcal{M}_a$ in the gauge field a_μ, and also the Berry phase of the spinons in the ground state, as in Chapter 26. The Néel order parameter in (9.15) is the $U(1)$ gauge-invariant bilinear of the spinon field:

$$\mathcal{N} = z_\alpha^* \boldsymbol{\sigma}_{\alpha\beta} z_\beta, \tag{28.9}$$

where $\boldsymbol{\sigma}$ are the Pauli matrices (compare to (15.48) for the coplanar order in $\mathbb{Z}_2$ gauge theories).

As we tune the coupling g in (28.8), we can expect the two phases shown in Fig. 28.1:

(i) Néel phase, $g < g_c$: the spinon z_α condenses in a Higgs phase with $\langle z_\alpha \rangle \neq 0$. The a_μ gauge field is higgsed, and spin-rotation symmetry is broken by the opposite polarization of the spins on the two sublattices.
(ii) VBS, $g > g_c$: the spinons are gapped, and then we apply the effective theory for the $U(1)$ gauge field in Section 26.1. For half-integer-spin S, we conclude there is VBS order.

We now obtain a potential gapless spin liquid if there is a continuous quantum phase transition at $g = g_c$. As we showed in Chapter 26, for half-integer-spin S, the single monopole terms in (28.8) average to zero at long wavelengths from the Berry phases, and only quadrupoled monopole terms survive. So we can simplify the continuum theory for the vicinity of the quantum critical point to [256, 261]:

$$\mathcal{L}_z = |(\partial_\mu - i a_\mu) z_\alpha|^2 + g|z_\alpha|^2 + u\left(|z_\alpha|^2\right)^2 + K(\varepsilon_{\mu\nu\lambda}\partial_\nu a_\lambda)^2 - y_4\left(\mathcal{M}_a^4 + \mathcal{M}_a^{\dagger 4}\right), \tag{28.10}$$

where y_4 is the quadrupoled monopole fugacity. There is ample numerical evidence that y_4 is irrelevant near a possible critical point, and so the question reduces to whether the theory $\mathcal{L}_z$ at $y_4 = 0$ exhibits a critical point that realizes a conformal field theory in 2+1 dimensions. This is a question that has been studied extensively in numerics, with no firm conclusion. But the weight of the current evidence points to a "complex" critical point, that is, there are very large correlation lengths due the proximity to a renormalization-group fixed point, but the fixed point itself resides in a regime where it is necessary to analytically continue the theory to complex values of the couplings [107,

167, 188, 273, 294, 299]. In any case, it is clear that a "deconfined critical" description is suitable over a substantial length scale, with fractionalized spinons interacting with a $U(1)$ gauge field in the absence of monopoles.

28.2 *U*(1) Spin Liquids on the Square Lattice: Fermionic Spinons

We now examine the same antiferromagnet in Section 28.1, but using fermionic spinons. An exact computation should, of course, give the same results from the two approaches. But if an approximate computation gives the same phases, then we would expect the field theories of the critical points in between them are dual to each other; indeed, this was the strategy of Chapter 27.

We return to the $U(1)$ spin liquid that we described in Chapter 22 on the chiral spin liquid. We already showed there that in the absence of the time-reversal symmetry breaking term (i.e., by setting $t_2 = 0$ in Fig. 18.2), we obtained a theory of four species of massless, two-component Dirac fermions. Here, we generalize this theory to include a "staggered flux" (as in Fig. 28.2; see Section 28.3), and couple it to a $U(1)$ gauge field and monopoles to obtain

$$\mathcal{L}_\psi = \sum_{i=1}^{4} \bar{\psi}_i \gamma^\mu (\partial_\mu - i a_\mu) \psi_i - \left(y_i \mathcal{M}_{ai} + y_i^* \mathcal{M}_{ai}^\dagger \right), \tag{28.11}$$

where we have allowed for the possible spatial dependence of the monopole fugacity, as was the case in Chapter 24. A full analysis of this theory requires a complete understanding of the monopole terms, and their signatures under various symmetry transformations, and there has been important progress in this direction recently [268, 270]. While there is still some theoretical uncertainty, the most likely outcome is that (28.11) is unstable to either the Néel or VBS state, and can also realize a dual description [294] of the critical fluctuations associated with the bosonic theory (28.10).

There have also been studies of similar $U(1)$ spin liquids of gapless Dirac spinons on the triangular and kagome lattices [268, 270]. The possibility of a $U(1)$ gapless spin-liquid phase remains open, and the instabilities to various confining phases with broken symmetries have also been described. It would be interesting to elucidate the connection between this $U(1)$ gauge field + fermionic spinon approach to the phase diagram of the triangular or kagome lattice antiferromagnet, and the $\mathbb{Z}_2$ gauge theory + bosonic spinon approach of (15.49).

28.3 Gapless *SU*(2) Spin Liquids

We can obtain another perspective on the spin liquid discussed in Section 28.2 by employing the $SU(2)$ gauge theory used for the chiral spin liquid in Section 22.4. To restore time-reversal symmetry, and obtain a gapless spin liquid, we set $t_2 = 0$ in

Fig. 22.1. Then, the Hamiltonian in (22.27) describes massless Dirac fermions, which will be coupled to a $SU(2)$ gauge field. The continuum formulation of this theory can be obtained by following the same procedure as below (18.39), but we have to carefully account for the $SU(2)$ gauge symmetry. For this purpose, it is convenient to express the lattice fermion $\boldsymbol{f}_i$ in (22.19) in terms of four Majorana fermions $\chi_{0i}, \chi_{xi}, \chi_{yi}, \chi_{zi}$ by

$$f = \frac{1}{\sqrt{2}}(\chi_0 + i\chi_a\sigma^a). \tag{28.12}$$

To work out the dispersion relation of the Hamiltonian in Fig. 22.1, we increase our unit cell by one lattice site in the x direction and so χ acquires an additional sublattice index $m = A, B$. In momentum space, we then have

$$\begin{aligned} H &= \sum_{\boldsymbol{k}} \chi_{-\boldsymbol{k}}^{\mathrm{T}} H(\boldsymbol{k}) \chi_{\boldsymbol{k}}, \\ H(\boldsymbol{k}) &= -2t_1 \left[\sin(k_y)\rho^z + \sin(k_x)\rho^x\right]. \end{aligned} \tag{28.13}$$

Here, ρ^i are Pauli operators acting on the sublattice space, $m = A, B$. This Hamiltonian is diagonal in the $0, a$ indices in (28.12), and the gauge was chosen to have this feature. The Hamiltonian in (28.13) has Dirac points at $k_y = 0, \pi$, $k_x = 0$. Labelling these Dirac points by another index $v = 1, 2$, and expanding around these two points, we decompose our Majorana operator as

$$\chi_{m,i} \sim \rho^x \chi_{m,v=1}(\boldsymbol{x}) + (-1)^{i_y} \chi_{m,v=2}(\boldsymbol{x}). \tag{28.14}$$

With this, the Hamiltonian reduces to

$$H \approx 2it_1 \sum_{v=1,2} \chi_v^T (\rho^x \partial_x - \rho^z \partial_y) \chi_v, \tag{28.15}$$

with the sublattice and $0, a$ indices implicit. This gives the continuum Lagrangian

$$\mathcal{L} = 2it_1 \bar{\chi}_v \gamma^\mu \partial_\mu \chi_v, \tag{28.16}$$

where $\gamma^0 = \rho^y$, $\gamma^x = i\rho^z$, $\gamma^y = i\rho^x$, and $\bar{\chi} \equiv \chi^T \gamma^0$. Here, we have chosen to express $\mathcal{L}$ in the Minkowski metric $(+,-,-)$; we ultimately move to the Euclidean metric below to perform calculations.

We now define the 4×2 matrix operator

$$X_{\alpha,v;\beta} = \frac{1}{\sqrt{2}}\left(\chi_{0,v}\delta_{\alpha\beta} + i\chi_{a,v}\sigma^a_{\alpha\beta}\right) \tag{28.17}$$

and $\bar{X} = X^\dagger \gamma^0$, where the sublattice/Dirac index m is left implicit. This allows us write our Lagrangian as

$$\mathcal{L} = 2it_1 \operatorname{Tr}\left(\bar{X}\gamma^\mu \partial_\mu X\right). \tag{28.18}$$

In this form, the Hamiltonian describes eight massless Majorana fermions (these are two-component "relativistic" Majorana fermions with an additional sublattice index). The $SU(2)$ gauge symmetry acts on the right index (β in (28.17)) of X, and the gradient in $\mathcal{L}$ must be replaced by the appropriate covariant gradient when the gauge field is included. Specifically, we have

$$D_\mu X = \partial_\mu X + iX a_\mu \tag{28.19}$$

and the $SU(2)$ gauge symmetry in (22.23) acts as

$$X \to X U_g, \quad a_\mu \to U_g^\dagger a_\mu U_g - i\partial_\mu U_g^\dagger U_g, \tag{28.20}$$

and the full Lagrangian is

$$\mathcal{L} = 2it_1 \operatorname{Tr}\left(\bar{X}\gamma^\mu D_\mu X\right). \tag{28.21}$$

Global spin rotations act on the left index (α in (28.17)) of X, and global valley rotations act on the v index. These global rotations combine to yield an emergent, low-energy $Sp(4)/\mathbb{Z}_2 \equiv SO(5)$ global symmetry in this spin liquid [214, 294].

To summarize, we have now obtained a continuum theory of the π-flux phase of the square-lattice antiferromagnet; this is an $SU(2)$ gauge theory with $N_f = 2$ massless Dirac fermions. This theory has an emergent global $SO(5)$ symmetry, which combines the $SO(3)$ global spin rotations with various lattice symmetries. Note that, unlike (28.11), there are no monopole insertions, because they are not present in an $SU(2)$ gauge theory in 2+1 dimensions. Nevertheless, (28.21) is a strongly coupled gauge theory, and has been the focus of a number of numerical studies. While the existence of a conformal critical theory has not been established, the numerics display critical correlations over very large length scales.

A possible ultimate fate of the theory is confinement into a phase where the $SO(5)$ symmetry is broken by the condensation of a Lorentz-invariant fermion bilinear, which transforms as a vector under $SO(5)$ symmetry. Remarkably, the five components of this $SO(5)$ vector turn out to be precisely the Néel ($N_{x,y,z}$) and VBS ($V_{1,2}$, see (16.36)) order parameters that appeared in the bosonic-spinon theory in Section 28.1; we have

$$(V_1, V_2, N_x, N_y, N_z) = \operatorname{Tr}\left(\bar{X}\boldsymbol{\Gamma}X\right), \tag{28.22}$$

with

$$\boldsymbol{\Gamma} = (\mu^z, -\mu^x, \sigma^x\mu^y, \sigma^y\mu^y, \sigma^z\mu^y), \tag{28.23}$$

where μ^a are Pauli matrices acting on the valley index.

28.4 Gapless $\mathbb{Z}_2$ Spin Liquid on the Square Lattice

Recent numerical studies of frustrated square lattice Hamiltonians [43, 78, 79, 112, 161, 192, 296] have provided evidence for an extended intermediate gapless spin liquid between the Néel and VBS states of Fig. 28.1. An attractive candidate for this phase is a gapless $\mathbb{Z}_2$ spin liquid [257]. Unlike the gapless $U(1)$ spin liquids examined above, the gapless $\mathbb{Z}_2$ spin liquid is stable to gauge fluctuations over at least some regime of parameters because the gauge sector is gapped, that is, even though the spinons are gapless,

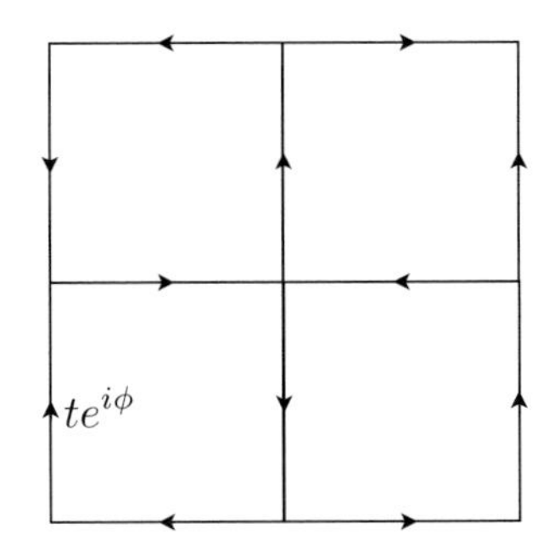

Figure 28.2 Staggered flux hopping.

the visons remain gapped. The gapped visons are sufficient to retain $\mathbb{Z}_2$ topological order.

I describe here a gapless $\mathbb{Z}_2$ spin liquid obtained by turning on a d_{xy} pairing between the spinons in a $U(1)$ staggered flux phase. This pairing amplitude acts like a charge-2 Higgs field, and higgses the $U(1)$ gauge symmetry down to $\mathbb{Z}_2$. This parallels the situation found for bosonic spinons in Chapter 15.

We employ the staggered flux hopping for the fermionic spinons $f_{i\alpha}$ shown in Fig 28.2. After turning on pairing in the d_{xy} channel, and introducing fermions $f_{A\alpha}$, $f_{B\alpha}$ on the two sublattices, we find the Hamiltonian acting on $(f_{A,\boldsymbol{k}\uparrow}, f_{B,\boldsymbol{k}\uparrow}, f^{\dagger}_{A,-\boldsymbol{k},\downarrow}, f^{\dagger}_{B,-\boldsymbol{k},\downarrow})^T$:

$$H(\boldsymbol{k}) = \begin{pmatrix} 0 & A_{\boldsymbol{k}} & B_{\boldsymbol{k}} & 0 \\ A^*_{\boldsymbol{k}} & 0 & 0 & -B_{\boldsymbol{k}} \\ B^*_{\boldsymbol{k}} & 0 & 0 & -A^*_{\boldsymbol{k}} \\ 0 & -B^*_{\boldsymbol{k}} & -A_{\boldsymbol{k}} & 0 \end{pmatrix}, \tag{28.24}$$

where

$$A_{\boldsymbol{k}} = -2t(e^{-i\phi}\cos(k_x) + e^{i\phi}\cos(k_y)) \quad , \quad B_{\boldsymbol{k}} = 4\Delta e^{i\theta}\sin(k_x)\sin(k_y)\,. \tag{28.25}$$

Now we have taken a general complex d_{xy} order parameter $\Delta e^{i\theta}$; for the choice of gauge in Fig. 28.2, it has opposite signs on the two sublattices:

$$\Delta_{A,xy} = \Delta e^{i\theta} \quad , \quad \Delta_{B,xy} = -\Delta e^{i\theta}\,. \tag{28.26}$$

The eigenvalues of (28.24) are

$$\varepsilon^2_{\boldsymbol{k}} = [\mathrm{Im}(A_{\boldsymbol{k}})]^2 + [\mathrm{Re}(A_{\boldsymbol{k}}) \pm |B_{\boldsymbol{k}}|]^2\,, \tag{28.27}$$

which is independent of θ. This dispersion is plotted in Fig. (28.3). Note the presence of Dirac nodal points, representing gapless spinon excitations. The dispersion does not have full square-lattice symmetries. However, spinons are not individually observable, and so this does not necessarily imply that square-lattice symmetry has been broken. We have to examine gauge-invariant products of the bond variables, and confirm that all such products do indeed preserve square-lattice symmetry [306].

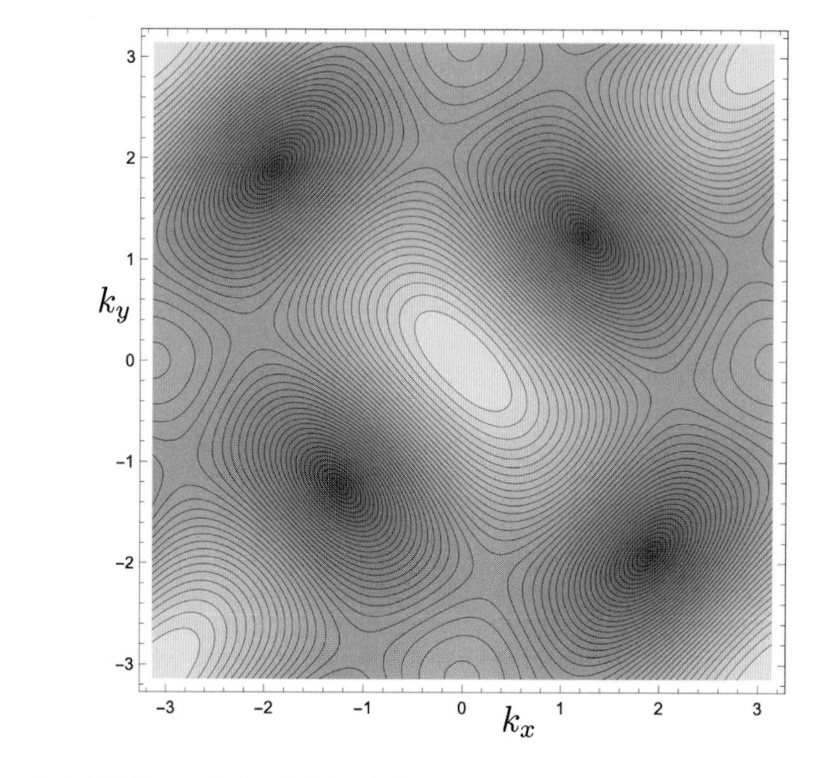

Figure 28.3 Dispersion in (28.27) for $t = 1, \phi = 0.6, \Delta = 0.3$.

The $U(1)$ gauge transformation in the staggered flux phase acts in a uniform manner, $f_{i\alpha} \to f_{i\alpha}e^{i\rho_i}$. Consequently, the d_{xy} pairing $\Delta e^{i\theta}$ does indeed act like a charge-2 Higgs field, which higgs the photon, and leaves a phase with $\mathbb{Z}_2$ topological order.

PART V

CORRELATED METALS

29 Kondo Impurity Model

The Kondo model describes a magnetic impurity in a metal. The renormalization-group and a large-N analysis show that the mobile electrons screen a spin $S = 1/2$ impurity so that it resembles a non-magnetic impurity at temperatures below the Kondo temperature. The Bose Kondo model of an impurity in an insulating quantum-critical magnet is also described.

The Kondo impurity model was introduced to describe the behavior of impurities of transition metal ions (e.g., Mn) in simple metals (e.g., Cu). It was observed that the impurity ion acquires a local magnetic moment that interacts non-trivially with the host conduction electrons. More recently, similar models have also been used to describe small quantum dots coupled to mobile electrons in leads.

The importance of the Kondo model in condensed-matter physics rests on its central role in the development of our understanding of the consequences of strong interactions in metals. In Chapter 2, we found that strong interactions don't do much in metals, apart from renormalizing the effective mass and residue of the quasiparticle excitations. That ultimately continues to be the case in the simplest Kondo model, but the strong renormalizations occur in an interesting, non-trivial, but nevertheless computable manner.

We also consider the "Bose Kondo" model in Section 29.5; in this case, the local magnetic moment interacts with gapless bosonic excitations in the bulk, such as those that may be found near a magnetic quantum phase transition. Strictly speaking, this model does not apply to a correlated metal; however, we see a close connection in Chapter 33 to the Sachdev–Ye–Kitaev model of a correlated metal studied in Chapter 32.

29.1 Resonant-Level Model

We begin by a simple model of non-interacting electrons on a lattice with a single impurity (see Fig. 29.1)

$$H_{RLM} = \sum_{k,\alpha} \varepsilon_k c_{k\alpha}^{\dagger} c_{k\alpha} + \sum_{\alpha} \left[\varepsilon_d d_\alpha^{\dagger} d_\alpha - w \left(d_\alpha^{\dagger} c_{0\alpha} + c_{0\alpha}^{\dagger} d_\alpha \right) \right]. \tag{29.1}$$

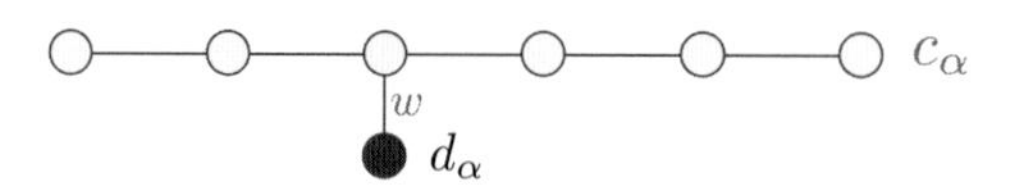

Figure 29.1 Resonant-level model of free electrons: conduction electrons c_α hybridize with a localized d_α state with amplitude w. The c_α move on a $d > 1$-dimensional lattice, although only one dimension is shown.

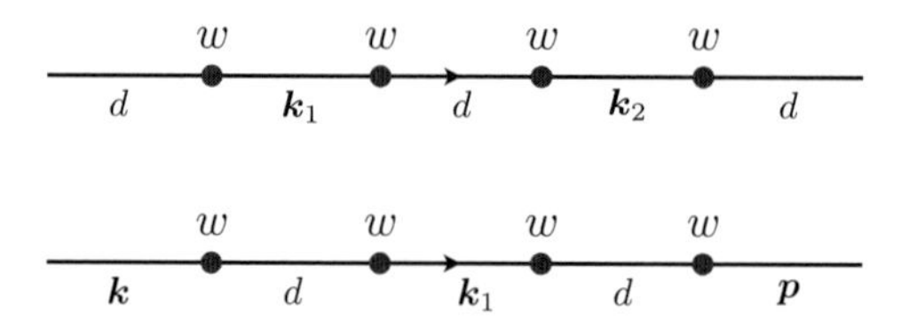

Figure 29.2 Feynman diagrams for G_{dd} and the conduction electrons.

The $c_{k\alpha}$ are the conduction electrons with dispersion ε_k in a perfect lattice. They scatter off an additional impurity atom at the origin of spatial coordinates, and the electron on the impurity atom is represented by d_α. The scattering is represented by the tunneling matrix element w between the impurity atom and the lattice site at the origin, with

$$c_{0\alpha} \equiv \frac{1}{\sqrt{V}} \sum_k c_{k\alpha} \,. \tag{29.2}$$

This is a model of non-interacting electrons, and despite the lack of translational invariance, it is not difficult to solve it exactly. We can sum all the Feynman diagrams in powers of w, as shown in Fig. 29.2, and obtain the Green's function of the d electron

$$\begin{aligned} [G_{dd}(i\omega_n)]^{-1} &= i\omega_n - \varepsilon_d - \frac{1}{V}\sum_k \frac{w^2}{i\omega_n - \varepsilon_k} \\ &= i\omega_n - \varepsilon_d - \int d\varepsilon \frac{d(\varepsilon) w^2}{i\omega_n - \varepsilon} \,, \end{aligned} \tag{29.3}$$

where $d(\varepsilon)$ is the single spin density of states of the conduction electrons. We approximate the density of states by its value of the Fermi level, $d(0)$, and absorb the real part of the integral over ε into a renormalization of ε_d. Then we obtain the final answer

$$G_{dd}(i\omega_n) = \frac{1}{i\omega_n - \varepsilon_d + i\Gamma \mathrm{sgn}(\omega_n)} \,, \tag{29.4}$$

where

$$\Gamma = \pi w^2 d(0) \,. \tag{29.5}$$

The density of electronic states on the d site is now seen to be Lorentzian of width Γ:

$$\rho_d(\omega) = -\frac{1}{\pi}\text{Im}\, G_{dd}(\omega + i\eta)$$
$$= \frac{1}{\pi}\frac{\Gamma}{(\omega - \varepsilon_d)^2 + \Gamma^2}. \tag{29.6}$$

This is the resonant level: the d electron is mostly on the d site with energy ε_d, but it has a lifetime of $1/\Gamma$, as it can "decay" by *coherent* tunneling into the conduction band. The quotes around decay indicate that this is not an incoherent decay involving exchange of energy with other electrons, and the single-particle eigenstates of H_{RLM} are infinitely long lived, albeit not plane waves. We can also obtain the conduction-electron Green's function, which is now not diagonal in $\boldsymbol{k}$:

$$G(\boldsymbol{k},\boldsymbol{p},i\omega_n) = \frac{\delta_{\boldsymbol{k},\boldsymbol{p}}}{i\omega_n - \varepsilon_{\boldsymbol{k}}} + \frac{w^2}{V}\frac{G_{dd}(i\omega_n)}{(i\omega_n - \varepsilon_{\boldsymbol{k}})(i\omega_n - \varepsilon_{\boldsymbol{p}})}. \tag{29.7}$$

Unlike the case of the Fermi liquid, notice now that $\text{Im}\, G_{dd}^{-1}(z)$ does not vanish as $z \to 0$ (as in (2.38) and (2.39)), but equals the non-zero constant Γ. This is a consequence of the lack of translational invariance, not the breakdown of the quasiparticle concept. The width of the resonant level Γ signifies a coherent mixing of the localized d-level state with the continuum of conduction-electron states. It is not a measure of the scattering of quasiparticles that exchange energy, which we considered in Chapter 2. So, in disordered systems, we have to use more complicated correlators to deduce the lifetime of the true quasiparticles, which are not momentum eigenstates (as we see in Section 30.2.2).

For subsequent considerations when we include the effect of interactions, it is useful to characterize the system by its response to a uniform applied Zeeman field coupling as $-h\sum_i S_{zi}$, where $\vec{S}_i$ is the electron spin operator on the site i. We characterize the response by the local spin susceptibility $\chi_i = \langle S_{zi}\rangle/h$. For a system without an impurity (or far from the impurity, when present), this spin susceptibility can be computed just like the compressibility, and we obtain the Pauli susceptibility of a metal χ_P (which is the spin susceptibility per unit volume):

$$\chi_P = \frac{d(0)}{2}. \tag{29.8}$$

With an impurity, we are interested in the behavior of χ_i in the vicinity of the impurity. In the "flat density of states" approximation, which was made between (29.3) and (29.4), the response on the d site is dominated entirely by the response to the local field; then we can easily compute the impurity susceptibility on the d site as

$$\chi_{imp} = -\frac{T}{2}\sum_{\omega_n}[G_{dd}(i\omega_n)]^2$$
$$= \frac{\Gamma}{2\pi(\varepsilon_d^2 + \Gamma^2)}. \tag{29.9}$$

At resonance, that is, for a d level that is at the Fermi level ($\varepsilon_d = 0$) and Γ small, the susceptibility $\chi_{imp} \sim 1/\Gamma$ is greatly enhanced over other sites.

29.2 Adding Interactions

Now we add interactions to H_{RLM}. It is not difficult to argue that interactions in the conduction band will not do much, and merely renormalize the bare conducting electrons to electron-like quasiparticles with a modified dispersion. However, interactions on the impurity site can have a strong effect, and we only include those. In the applications to quantum dots, this is the analog of including the "Coulomb blockade" on the quantum dot. In this manner, we obtain the Hamiltonian of the Anderson impurity model:

$$H_A = H_{RLM} + U_d \, d_\uparrow^\dagger d_\uparrow d_\downarrow^\dagger d_\downarrow \,. \tag{29.10}$$

In the context of the perturbation theory in U, it is easy to see from the diagrammatic perturbation expansion in Fig. 29.2 that the relationship between the conduction-electron and d-electron Green's function in (29.7) still applies – we simply have to replace the d Green's function by a renormalized Green's function including all interactions. These interactions modify the form of the d-electron Green's function from (29.3) to

$$[G_{dd}(i\omega_n)]^{-1} = i\omega_n - \varepsilon_d - \Sigma_{dd}(i\omega_n) - \frac{1}{V}\sum_k \frac{|w|^2}{i\omega_n - \varepsilon_k} \,. \tag{29.11}$$

Here, $\Sigma_{dd}(i\omega_n)$ is the only change from the presence of the interaction U_d.

The task for the remainder of this chapter is to understand the behavior of Σ_{dd} as a function of temperature and frequency for large U_d, and deduce consequences for other physical observables. A first guess would be to simply compute $\Sigma_{dd}(\omega)$ in a perturbative expansion in powers of U_d, and hope that the result applies also for large U_d, as it did for Fermi liquid theory, albeit with possibly large renormalizations of various parameters. In fact, this hope is realized. However, it took some time for the condensed-matter community to appreciate this, and much new physics emerged from understanding the intricate structure of the crossover to large U_d in this analysis. An important part of the new physics is the emergence of a new energy scale, the Kondo temperature T_K, which can be much smaller than all other energy scales when U_d is large. For $T \ll T_K$, we do indeed realize the Fermi liquid behavior of a non-interacting resonant-level model, but with strong renormalizations. And for $T_K \ll T \ll E_F$, we have "local-moment" physics, which we describe shortly.

The analysis for large U_d proceeds most naturally, as it did for the Hubbard model in Section 9.1, by performing a canonical transformation to an effective Hamiltonian acting on the low-energy subspace. This turns out to be the Kondo Hamiltonian. Let us examine the spectrum on the d level, in the limit that $|\varepsilon_d|$ and U_d are much larger than w. Then there are four possible states on the d level, with energies

$$\begin{aligned} |0\rangle &\quad\Rightarrow\quad E = 0, \\ d_\alpha^\dagger |0\rangle &\quad\Rightarrow\quad E = \varepsilon_d, \\ d_\uparrow^\dagger d_\downarrow^\dagger |0\rangle &\quad\Rightarrow\quad E = 2\varepsilon_d + U_d \,. \end{aligned} \tag{29.12}$$

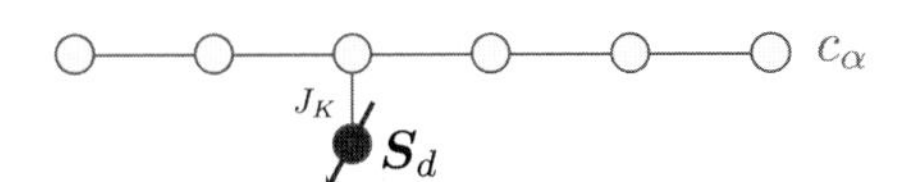

Figure 29.3 Kondo impurity model: conduction electrons c_α with a spin $\boldsymbol{S}_d$ and exchange interactions J_K.

The non-trivial situation arises when $\varepsilon_d \ll 0, 2\varepsilon_d + U_d$. In this case, the d level is doubly degenerate, and has either a spin-up or a spin-down electron, and other states on the d site have a much larger energy. This is precisely the analog of the situation in the square-lattice Hubbard model in Section 9.1, where we replaced each site by an $S = 1/2$ spin, and derived an effective superexchange interaction between them. In the present situation, we only replace the d site by an $S = 1/2$ spin and, by a very similar Schrieffer–Wolff transformation, we obtain the celebrated Kondo impurity model, sketched in Fig. 29.3:

$$H_K = \sum_{\boldsymbol{k}} \varepsilon_{\boldsymbol{k}} c^\dagger_{\boldsymbol{k}\alpha} c_{\boldsymbol{k}\alpha} + J_K \boldsymbol{S}_d \cdot c^\dagger_{0\alpha} \frac{\boldsymbol{\sigma}_{\alpha\beta}}{2} c_{0\beta}. \tag{29.13}$$

Here, $\boldsymbol{S}_d$ is an $S = 1/2$ spin operator acting on the two states $d^\dagger_\uparrow |0\rangle$, $d^\dagger_\downarrow |0\rangle$ on the d site, and we recall (29.2) for the conduction-electron operator on the impurity site $c_{0\alpha}$. So H_K describes the conduction electrons interacting with an $S = 1/2$ spin with an exchange interaction J_K, which is antiferromagnetic; as in the Hubbard model, the Kondo exchange interaction is antiferromagnetic. Its value is

$$J_K = 2w^2 \left(\frac{1}{-\varepsilon_d} + \frac{1}{\varepsilon_d + U_d} \right). \tag{29.14}$$

Note that for $\varepsilon_d \ll 0, 2\varepsilon_d + U_d$, both energy denominators are large and positive. If we take the limit of two sites, which we considered for the Hubbard model, this reduces to the familiar expression $J = 4t^2/U$. The Schrieffer–Wolff transformation also generates an additional potential scattering term $\sim w^2/U_d$ for the conduction electrons, which we have dropped in (29.13).

29.3 Renormalization Theory

As a first step towards understanding the large-U_d limit, we can perform a perturbation expansion of H_K in powers of J_K. As we see below, such an expansion leads to correlators that have a logarithmic dependence upon external frequency (at $T = 0$). This naturally suggests an application of the renormalization group, which allows us to resum the logarithmically singular terms. From such an analysis we find that the system becomes strongly coupled below an energy scale T_K, which is non-zero even for very small J_K; an estimate of T_K is given below. The perturbative expansion in J_K fails for temperatures $T < T_K$, and this strong coupling regime is addressed systematically in Section 29.4 by another method.

Generating a perturbation expansion in powers of J_K is not entirely straightforward because the d spins are now no longer free fermions, and there is no Wick's theorem for the correlations of the spin operators. However, it is nevertheless possible to use diagrammatic methods by using the Schwinger fermion decomposition of the spin operator in (22.3) along with the unit fermion constraint in (22.4); in our case there is no site index, and we perform this decomposition of the d spin:

$$\boldsymbol{S}_d = \frac{1}{2} f_\alpha^\dagger \boldsymbol{\sigma}_{\alpha\beta} f_\beta \quad , \quad f_\alpha^\dagger f_\alpha = 1. \tag{29.15}$$

Remarkably, for the case of single spin, it is possible to impose the constraint in (29.15) exactly, diagram by diagram, using the Abrikosov method. This method involves imposing a chemical potential $-\lambda$ on the Schwinger fermions, and taking the $\lambda \to \infty$ limit to impose the single fermion constraint. So we consider the Hamiltonian, generalizing (29.13),

$$H_K = \sum_k \varepsilon_k c_{k\alpha}^\dagger c_{k\alpha} + \lambda f_\alpha^\dagger f_\alpha + J_K \left(f_\gamma^\dagger \frac{\boldsymbol{\sigma}_{\gamma\delta}}{2} f_\delta \right) \cdot \left(c_{0\alpha}^\dagger \frac{\boldsymbol{\sigma}_{\alpha\beta}}{2} c_{0\beta} \right). \tag{29.16}$$

The constraint in (29.15) is implemented by computing $\left\langle \mathcal{O} f_\alpha^\dagger f_\alpha \right\rangle_\lambda / \left\langle f_\alpha^\dagger f_\alpha \right\rangle_\lambda$, where $\mathcal{O}$ is any observable, and taking the limit $\lambda \to \infty$. In practice, this is straightforward to implement, and usually involves omitting graphs in which the f_α flow both forward and backward in time. Note also that we are now representing the impurity spin by a single fermionic spinon f_α. This is not the same as the original d fermion because it obeys the constraint in (29.15).

The key physics of the Kondo model becomes evident upon considering the renormalization of the J_K coupling to second order in J_K^2. This is given by the two graphs in Fig. 29.4. The first graph in Fig. 29.4 evaluates to

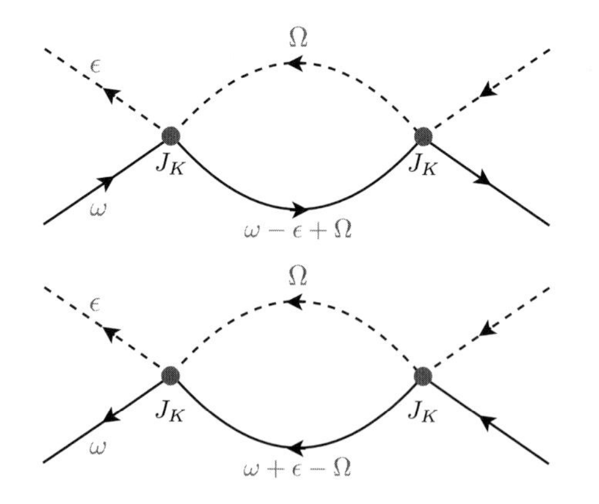

Figure 29.4 Renormalization of the Kondo exchange coupling J_K. The full line is the spinon f_α, while the dashed line is the conduction electron.

$$\frac{J_K^2}{2}\frac{1}{V}\sum_k \int \frac{d\Omega}{2\pi} \frac{1}{(i\Omega - \varepsilon_k)(i\omega - i\varepsilon + i\Omega - \lambda)}$$
$$= \frac{J_K^2}{2}\frac{1}{V}\sum_k \frac{\theta(-\varepsilon_k)}{i\omega - i\varepsilon + \varepsilon_k - \lambda}, \tag{29.17}$$

while the second evaluates to

$$\frac{J_K^2}{2}\frac{1}{V}\sum_k \int \frac{d\Omega}{2\pi} \frac{1}{(i\Omega - \varepsilon_k)(i\omega + i\varepsilon - i\Omega - \lambda)}$$
$$= \frac{J_K^2}{2}\frac{1}{V}\sum_k \frac{-\theta(-\varepsilon_k)}{i\omega + i\varepsilon - \varepsilon_k - \lambda}. \tag{29.18}$$

We should take the frequency of the incoming spinon to be just above threshhold, with $i\omega - \lambda$ small. We can also set the frequency of the external conduction electron to zero, ε. Then, the integral over $\boldsymbol{k}$ leads to a logarithmic dependence upon the external frequency of the spinon. As in Section 25.2.3, we can treat this using the renormalization group, by only integrating out high-energy conduction electrons. In the simplest model, we assume a flat band density of states that extends for $|\varepsilon_k| < D$, where $2D$ is the bandwith. In the renormalization-group computation, we only integrate out the highest-energy electrons with $D - \delta D < |\varepsilon_k| < D$. Then, the sum of (29.17) and (29.18) evaluates to a renormalization of the exchange coupling:

$$J_K \to J_K + \frac{J_K^2}{V} \sum_{D-\delta D < -\varepsilon_k < D} \frac{1}{\varepsilon_k}. \tag{29.19}$$

Performing the $\boldsymbol{k}$ integral, and writing $\delta D = D\delta\ell$, we obtain the "poor person" renormalization-group flow

$$\frac{dJ_K}{d\ell} = d(0) J_K^2 + \mathcal{O}(J_K^3). \tag{29.20}$$

This flow is sketched in Fig. 29.5. For the ferromagnetic Kondo problem, $J_K < 0$, the renormalization-group flow is towards $J_K = 0$: in this case the impurity spin is essentially decoupled from the conduction electrons, and its coupling to the conduction electrons can be treated perturbatively. However, our interest here is the antiferromagnetic case, with $J_K > 0$, in which case (29.20) informs us that J_K increases without bound under the renormalization-group flow, no matter how small the initial positive value of J_K. Specifically, the integral of (29.20) is

$$J_K(\ell) = \frac{1}{1/J_K(0) - d(0)\ell}. \tag{29.21}$$

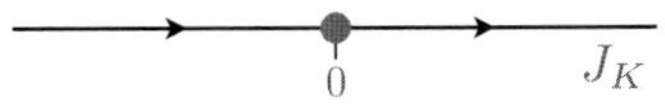

Figure 29.5 Flow of the renormalization-group equation (29.20). For ferromagnetic exchange, $J_K < 0$, the flow is to the fixed point at $J_K = 0$. For antiferromagnetic exchange, $J_K > 0$, the flow is towards strong coupling, $J_K \to \infty$.

If we now start with a very small positive bare value $J_K(\ell=0)$, we see from (29.21) that the renormalized exchange is of order unity at $\ell=\ell^* \sim 1/(d(0)J_K)$; equivalently, when $De^{-\ell^*} \sim T_K$, where T_K is the Kondo temperature,

$$T_K \sim D\exp\left(-\frac{1}{J_K d(0)}\right). \tag{29.22}$$

From the point of view of the Kondo Hamiltonian, the expression for T_K is non-perturbative, given its singular dependence upon J_K. However, one should note that this expression is ultimately non-singular at small U_d as $U_d \sim 1/J_K$: this is a hint that the low-energy physics is actually adiabatically connected to the free resonant-level model, albeit with strong renormalizations, as we now discuss. But first, it should be noted that the flow of J_K to infinity predicted by (29.20) is not a reliable prediction of the present analysis because we cannot trust (29.20) when J_K becomes large – it was obtained in a perturbative expansion in J_K. Computations by Wilson using a numerical renormalization-group scheme showed that the flow is indeed to $J_K \to \infty$, and the predictions of (29.20) are qualitatively correct.

Given this flow to large J_K, we can understand the qualitative fate of the model by examining the ground state of H_K in the limit of large J_K. In this limit, the energy is minimized if the impurity spin $\boldsymbol{S}_d$ locks into a spin singlet with a *single* conduction electron at the site 0. No other electron can occupy this site, and therefore we can describe the remaining electrons by the renormalized free-electron Hamiltonian

$$H_R = \sum_{\boldsymbol{k},\alpha} \varepsilon_{\boldsymbol{k}} c_{\boldsymbol{k}\alpha}^\dagger c_{\boldsymbol{k}\alpha} + V_0 c_{0\alpha}^\dagger c_{0\alpha} \tag{29.23}$$

and take the limit $V_0 \to \infty$ to prevent any other electrons from occupying the impurity site. As H_R is free-electron-like, and there are no dynamical degrees of freedom at the impurity, it is not difficult to take this limit using scattering theory, and we obtain an effective Fermi-liquid description of the scattering states. Indeed, the remarkable conclusion is that in the strong coupling limit at $T \ll T_K$, the Kondo model reduces to a model of non-interacting electrons qualitatively similar to the resonant-level model of Section 29.1.

We can use this interpretation to deduce some important features of the T dependence of the impurity spin susceptibility χ_{imp}, introduced towards the end of Section 29.1. At temperatures $T \gg T_K$, the perturbation theory in J_K is reliable, and so, at leading order, the impurity susceptibility is given by the Curie susceptibility of an isolated spin-$1/2$ electron:

$$\chi_{imp} = \frac{1}{4T} \quad , \quad T_K \ll T \ll U_d\,. \tag{29.24}$$

For $T \ll T_K$, we expect a mapping to the resonant-level model, which has a finite impurity susceptibility $\sim 1/\Gamma$, as obtained in (29.9). This χ_{imp} is determined by the width of the resonant level Γ, and a natural guess is that the width of the Kondo "resonance" (as it is known) should be T_K. So we have

$$\chi_{imp} \sim \frac{1}{T_K} \quad , \quad T \ll T_K\,, \tag{29.25}$$

where the coefficient depends upon the precise definition of T_K. We can combine (29.24) and (29.25) into a crossover function between the two limiting regimes:

$$\chi_{imp} = \frac{1}{4T}\Phi(T/T_K). \tag{29.26}$$

An important implication of the Kondo renormalization-group flow is that the crossover function $\Phi(\bar{T})$ is a *universal* function for $J_K \ll D$, determined by the renormalization-group flow of the Kondo model from $J_K = 0$ to $J_K = \infty$. At $\bar{T} \gg 1$, $\Phi \to 1$ so that we have the susceptibility of a free moment. For $\bar{T} \ll 1$ we have $\Phi(\bar{T}) \sim \bar{T}$ so that χ_{imp} is finite.

Similar universal crossovers apply to other observables of a Kondo impurity in a metal.

29.4 Large-M Theory

This section describes a method [109] that can yield explicit results for crossover functions such as those in (29.26). This is obtained by generalizing the $SU(2)$ spin symmetry of the Kondo model to $SU(M)$, and examining the large-M limit, similar to those employed for spin liquids in Chapters 15 and 22. It yields the correct qualitative behavior both at low and high T, and the crossover between these limits. And, as we see in Chapter 30, the large M is also a powerful tool in examining the Kondo lattice model.

First, we realize the spin $\boldsymbol{S}_d$ by the spinon f_α in (29.15). In this section, we implement the constraint in (29.15) by a Lagrange multiplier in the path integral, in contrast to the Abrikosov method in Section 29.3.

To enable the generalization to $SU(M)$, we first write the $SU(2)$ model in a manner which does not involve the Pauli matrices. We use the identity

$$\boldsymbol{\sigma}_{\alpha\beta} \cdot \boldsymbol{\sigma}_{\gamma\delta} = 2\delta_{\alpha\delta}\delta_{\beta\gamma} - \delta_{\alpha\beta}\delta_{\gamma\delta} \tag{29.27}$$

to write the Kondo interaction as

$$\frac{J_K}{2}\boldsymbol{S}_d \cdot c_{0\gamma}^\dagger \boldsymbol{\sigma}_{\gamma\delta} c_{0\delta} = -\frac{J_K}{2}\left(f_\alpha^\dagger c_{0\alpha}\right)\left(c_{0\beta}^\dagger f_\beta\right) - \frac{J_K}{4}\left(f_\alpha^\dagger f_\alpha\right)\left(c_{0\beta}^\dagger c_{0\beta}\right). \tag{29.28}$$

After using the constraint in (29.15), the second term in (29.28) is just a shift in the local chemical potential, and we will ignore it from now on. The generalization to $SU(M)$ is now straightforward; the indices $\alpha, \beta = 1, \ldots, M$, and the constraint is

$$f_\alpha^\dagger f_\alpha = \frac{M}{2}. \tag{29.29}$$

As we see below, to obtain a suitable large-M saddle point, we also need to replace $J_K/2$ by J_K/M.

We can now write the path integral for the Kondo model:

$$\mathcal{Z}_K = \int \mathcal{D}f_\alpha \mathcal{D}c_{\mathbf{k}\alpha}\mathcal{D}\lambda \exp\left(-\int_0^\beta d\tau\left[\mathcal{L}_0 + \mathcal{L}_1\right]\right),$$

$$\mathcal{L}_0 = \sum_k c_{k\alpha}^\dagger \left(\frac{\partial}{\partial \tau} + \varepsilon_k \right) c_{k\alpha} + f_\alpha^\dagger \left(\frac{\partial}{\partial \tau} + i\lambda \right) f_\alpha - i\lambda \frac{M}{2},$$

$$\mathcal{L}_1 = -\frac{J_K}{M} \left(f_\alpha^\dagger c_{0\alpha} \right) \left(c_{0\beta}^\dagger f_\beta \right). \quad (29.30)$$

The large-M theory is obtained by a method parallel to that followed in obtaining the Landau–Ginzburg theory from the Bardeen–Cooper–Schrieffer (BCS) theory in Chapter 6. We decouple the Kondo exchange term by a Hubbard–Stratonovich field $P(\tau)$. Then we obtain

$$\mathcal{Z}_K = \int \mathcal{D}\lambda \mathcal{D}P \mathcal{D}f_\alpha \mathcal{D}c_{\mathbf{k}\alpha} \exp\left(-\int_0^\beta d\tau \left[\mathcal{L}_0 + \mathcal{L}_Q \right] \right),$$

$$\mathcal{L}_Q \doteq \frac{M|P|^2}{J_K} - P f_\alpha^\dagger c_{0\alpha} - P^* c_{0\alpha}^\dagger f_\alpha. \quad (29.31)$$

Before proceeding, we notice an important property of $\mathcal{Z}_K$: it is invariant under an emergent $U(1)$ gauge symmetry, under which

$$f_\alpha \to f_\alpha e^{i\phi(\tau)},$$

$$P \to P e^{i\phi(\tau)},$$

$$\lambda \to \lambda - \frac{\partial \phi}{\partial \tau}. \quad (29.32)$$

This gauge symmetry is not so crucial in the Kondo model, as it is always possible to work in a convenient fixed gauge, but it will play a crucial role when we consider the Kondo lattice in Chapter 30. Clearly, this gauge symmetry is closely connected to that encountered in (22.10) in the study of spin liquids in Part IV.

We return to taking the large-M limit of $\mathcal{Z}_K$. In the form (29.31), the action is quadratic in the fermions, and so we can integrate them out. As all the fermions have M components, this yields an effective action for P and λ, which has an M prefactor. Consequently, the large-M limit is obtained by replacing P and λ by their saddle-point values. We go ahead and do this and replace P by $\overline{P}$, and $i\lambda$ by $\overline{\lambda}$ (we are anticipating here that the saddle-point value of λ is purely imaginary). Then the problem reduces to the following free-fermion Hamiltonian

$$\overline{H}_K = \frac{M|\overline{P}|^2}{J_K} - \overline{P} f_\alpha^\dagger c_{0\alpha} - \overline{P}^* c_{0\alpha}^\dagger f_\alpha + \sum_{\mathbf{k}} \varepsilon_{\mathbf{k}} c_{\mathbf{k}\alpha}^\dagger c_{\mathbf{k}\alpha}$$

$$+ \overline{\lambda} f_\alpha^\dagger f_\alpha - \overline{\lambda} \frac{M}{2}. \quad (29.33)$$

Our remaining task is to find the ground-state energy of $\overline{H}_K$, and demand that it is stationary with respect to variations in $\overline{P}$ and $\overline{\lambda}$. The latter task is simplified by the Feynman–Hellman theorem: in any eigenstate $|G\rangle$ of $\overline{H}_K$ we have

$$\frac{\partial}{\partial \overline{P}^*} \langle G | \overline{H}_K | G \rangle = \langle G | \frac{\partial \overline{H}_K}{\partial \overline{P}^*} | G \rangle \quad (29.34)$$

and so

$$\overline{P} = \frac{J_K}{M} \left\langle c_{0\alpha}^\dagger f_\alpha \right\rangle. \quad (29.35)$$

Similarly, the corresponding equation for $\overline{\lambda}$ is

$$\frac{1}{M}\langle f_\alpha^\dagger f_\alpha\rangle = \frac{1}{2}. \tag{29.36}$$

We can evaluate these expectation values from the Green's functions of $\overline{H}_K$, which are the same as those of the resonant-level model, H_{RLM} in (29.1). As in (29.4), we have

$$G_{ff}(i\omega_n) = \frac{1}{i\omega_n - \overline{\lambda} + i\Gamma_P \mathrm{sgn}(\omega_n)}, \tag{29.37}$$

where

$$\Gamma_P = \pi|\overline{P}|^2 d(0), \tag{29.38}$$

and

$$G_{fc_0} = -G_{ff}(i\omega_n)\frac{\overline{P}}{V}\sum_k \frac{1}{i\omega_n - \varepsilon_k}. \tag{29.39}$$

For the frequency summations required for (29.35) and (29.36), we employ the spectral representation

$$G_{ff}(i\omega_n) = \int_{-\infty}^{\infty}\frac{d\Omega}{\pi}\frac{A_f(\Omega)}{i\omega_n - \Omega}, \tag{29.40}$$

where

$$A_f(\Omega) = \frac{\Gamma_P}{(\Omega-\overline{\lambda})^2 + \Gamma_P^2} \tag{29.41}$$

is the Lorentzian spectral density of the f level. Then, (29.35) becomes

$$\begin{aligned}\overline{P} &= \frac{J_K\overline{P}}{V}\int_{-\infty}^{\infty}\frac{d\Omega}{\pi}A_f(\Omega)\sum_k T\sum_{\omega_n}\frac{-1}{(i\omega_n-\varepsilon_k)(i\omega_n-\Omega)} \\ &= J_K\overline{P}\int_{-\infty}^{\infty}\frac{d\Omega}{\pi}A_f(\Omega)\int d\varepsilon\, d(\varepsilon)\frac{f(\varepsilon)-f(\Omega)}{\Omega-\varepsilon}.\end{aligned} \tag{29.42}$$

Similarly, (29.36) is

$$\int_{-\infty}^{\infty}\frac{d\Omega}{\pi}A_f(\Omega)f(\Omega) = \frac{1}{2}. \tag{29.43}$$

We now have to determine the saddle-point values of $\overline{P}$ and $\overline{\lambda}$ by solving (29.42) and (29.43). Fortunately, the solution of (29.43) is simple:

$$\overline{\lambda} = 0, \tag{29.44}$$

because then we pick up exactly half of the Lorentzian spectral density. So the renormalized f level is exactly at the Fermi level, which means it is exactly resonant (this is sometimes called the Abrikosov–Suhl or Kondo resonance). The value of $\overline{P}$ is determined by (29.42), which is the analog here of the BCS equation that determined the gap parameter. We first evaluate the integral of ε in the limit of a flat density of states at $T = 0$

$$\int d\varepsilon\, d(\varepsilon)\frac{f(\varepsilon)-f(\Omega)}{\Omega-\varepsilon} \approx d(0)\int_{-D}^{D} d\varepsilon\frac{\theta(-\varepsilon)-\theta(-\Omega)}{\Omega-\varepsilon}$$

$$
\begin{aligned}
&= d(0)\left[\ln\left|\frac{D+\Omega}{\Omega}\right| - \theta(-\Omega)\ln\left|\frac{D+\Omega}{D-\Omega}\right|\right] \\
&\approx d(0)\ln\left|\frac{D}{\Omega}\right|,
\end{aligned}
\tag{29.45}
$$

when $|\Omega| \sim \Gamma_P \ll D$. So (29.42) yields

$$
\begin{aligned}
\frac{1}{J_K d(0)} &= \int_{-\infty}^{\infty} \frac{d\Omega}{\pi} A_f(\Omega) \ln\left|\frac{D}{\Omega}\right| \\
&= \ln\left(\frac{D}{\Gamma_P}\right).
\end{aligned}
\tag{29.46}
$$

So, we have our main result: the value of $\overline{P}$ is determined by (29.38) from the width of the Kondo resonance, which is

$$
\Gamma_P = D\exp\left(-\frac{1}{J_K d(0)}\right), \tag{29.47}
$$

consistent with the estimate of the Kondo temperature in (29.22). We can now compute other physical properties in the large-M expansion, and it is natural that they will be determined by the exponentially low-energy scale Γ_P in (29.47).

29.5 Bose Kondo Model

This section briefly considers a model in which the impurity spin is in an insulator, and the free-fermion environment is replaced by low-energy bosonic spin fluctuations. Such a model arises in the presence of a vacancy in an antiferromagnet, as illustrated in Fig. 29.6. The coupling between the spin and the bulk spin excitations is particularly important when the bulk undergoes a quantum phase transition; in Fig. 29.6, such a transition can be realized by tuning the value of J_2/J_1, when there is a transition from a trivial gapped paramagnet to a magnetically ordered Néel state. The Bose Kondo

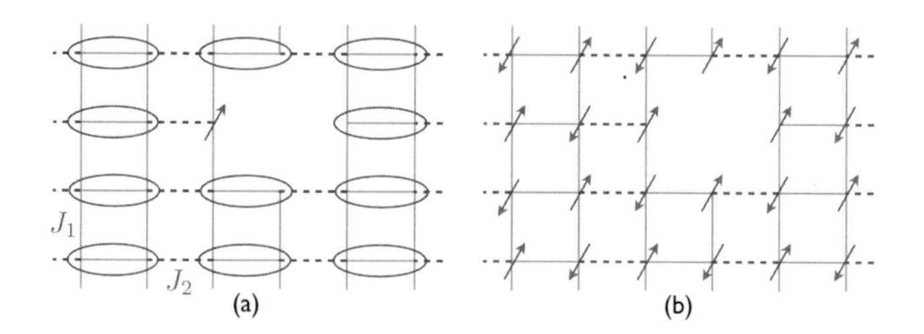

Figure 29.6 A coupled ladder antiferromagnet with antiferromagnetic bonds J_1 and weaker antiferromagnetic bonds J_2, with a single vacancy. We are interested in the behavior of the impurity spin while the bulk undergoes a quantum phase transition from a trivial gapped paramagnet in (a), to a Néel state in (b), with increasing J_2/J_1. The ellipses in (a) respresent valence bonds, as in Fig. 13.1.

problem we consider here will also be relevant for analyzing certain random quantum spin liquids in Section 33.3.2.

The bulk transition in Fig. 29.6 is described by the relativistic field theory of a three-component real scalar field ϕ_a in 2+1 dimensions in (10.2), as has been discussed at length in the QPT book. Here, in the interests of simplicity, we neglect the self-interaction u in (10.2), which is actually important to describ the physical situation in Fig. 29.6. So the field ϕ_a will be Gaussian, and the coupling of such a field to an impurity spin was considered by Sengupta [255] and others [23, 57, 189, 238, 293, 300]. The effects of u were described in Ref. [57, 238, 293], but will not be considered here.

We consider the model that generalizes the $SU(2)$ spin-rotation symmetry to $SU(M)$, as this will be important for the application in Section 33.3.2. In this case, the impurity spin S^a, and the bulk scalar field ϕ_a both have $a = 1,\ldots,M^2-1$. So we examine the Hamiltonian

$$H_{imp} = \gamma_0 S^a \phi_a(0) + \frac{1}{2}\int d^d x \left[\pi_a^2 + (\partial_x \phi_a)^2\right]. \tag{29.48}$$

Here, x is the d-dimensional bulk spatial coordinate, $\phi_a(x)$ is the scalar field, and $\pi_a(x)$ is its canonically conjugate momentum. The "Kondo" coupling between the spin and the bulk is γ_0, and we are interested in its renormalization-group flow.

The $SU(M)$ spin S^a acts on the antisymmetric, self-conjugate representation of $SU(M)$. Such a spin can be realized by fermions ("spinons") f_α, $\alpha = 1,\ldots,M$, obeying the constraint

$$\sum_\alpha f_\alpha^\dagger f_\alpha = \frac{M}{2}, \tag{29.49}$$

and the operator representation

$$S^a = f_\alpha^\dagger T^a_{\alpha\beta} f_\beta, \tag{29.50}$$

where the matrices T^a (which are 1/2 times the Pauli matrices for $M = 2$) obey

$$\mathrm{Tr}(T^a T^b) = \frac{1}{2}\delta^{ab}, \quad T^a T^a = \frac{M^2-1}{2M}\cdot\mathbf{1}, \quad T^a_{\alpha\beta}T^a_{\gamma\delta} = \frac{1}{2}\left(\delta_{\alpha\delta}\delta_{\beta\gamma} - \frac{1}{M}\delta_{\alpha\beta}\delta_{\gamma\delta}\right). \tag{29.51}$$

Far from the impurity, the time-dependent correlators of the scalar field are

$$\langle \phi(x,\tau)\phi(x,0)\rangle \sim \frac{1}{|\tau|^{2-\varepsilon}}, \quad |x| \to \infty, \tag{29.52}$$

where $\varepsilon = 3-d$; this can be deduced from the $x \ll \xi$ case of (10.26). Our task here is to determine the exponent, α, characterizing the autocorrelation function of the impurity spin

$$\langle S^a(\tau) S^a(0)\rangle \sim \frac{1}{|\tau|^\alpha}. \tag{29.53}$$

We restrict our attention to $d < 3$ (i.e., $\varepsilon > 0$) when the coupling γ_0 is relevant, and the spin-autocorrelation function cannot be determined by bare perturbation theory in γ_0. I present in Section 29.5.1 a renormalization-group analysis in powers of ε, and the main result is that

$$\alpha = \varepsilon \tag{29.54}$$

to *all* orders in ε and γ_0 [293], and for all M.

Note that a result of the form in (29.53) implies that the impurity spin is *not* screened, unlike the fermion Kondo problem considered in earlier sections of this chapter. The Bose Kondo coupling γ_0 does not flow to infinity under renormalization, as does the fermion Kondo coupling J_K. Instead, as we see below, γ_0 is attracted to a finite coupling fixed point, which leads to the critical scaling in (29.53).

29.5.1 Renormalization-Group Analysis

The main input to the renormalization-group analysis is a perturbative evaluation of the impurity spin autocorrelator in powers of γ_0. I only present the one-loop results here, although a two-loop evaluation has been given in Refs. [122, 293]. We follow the strategy of Ref. [293] and use time-ordered perturbation theory to expand the correlator in powers of γ_0, insert the two-point correlators of the bulk fields, and then explicitly evaluate the traces over the S^a. We write the correlator as

$$\langle S^a(\tau) S^a(0) \rangle = \frac{N}{D}, \tag{29.55}$$

and the perturbative expansions of the numerator and denominator are represented by the diagrams shown in Figs. 29.7 and 29.8. Note that these are not Feynman diagrams, and there is no Wick's theorem. The oriented line represents the worldline of the spin, and the diagrams indicate the ordering of the operators whose traces are to be evaluated. The numerator and denominator have to be evaluated separately, and there is no automatic cancellation of disconnected contributions. The diagrams in Figs. 29.7 and 29.8 yield

$$D = 1 + \gamma_0^2 L_0 \left(D_{1\phi} + D_{2\phi} + D_{3\phi} \right), \tag{29.56}$$

$$N = L_0 + \gamma_0^2 \left(L_1 D_{1\phi} + L_2 D_{2\phi} + L_3 D_{3\phi} \right), \tag{29.57}$$

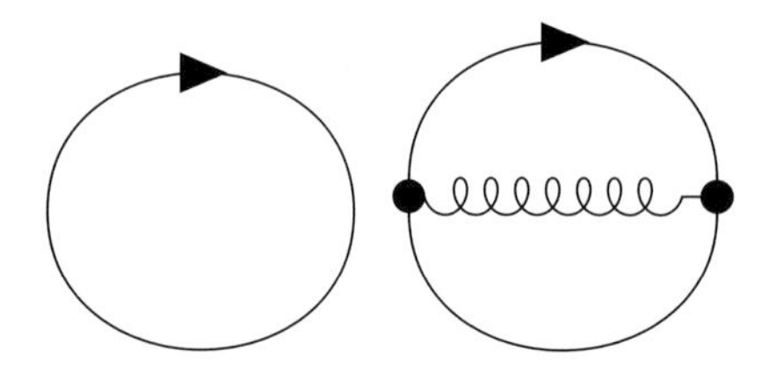

Figure 29.7 Diagrams contributing to the denominator D in (29.56), of (29.55). The oriented line denotes the trajectory of the $SU(M)$ spin in imaginary time, a filled circle is a γ_0 vertex, and the spiral curve denotes the ϕ propagator.

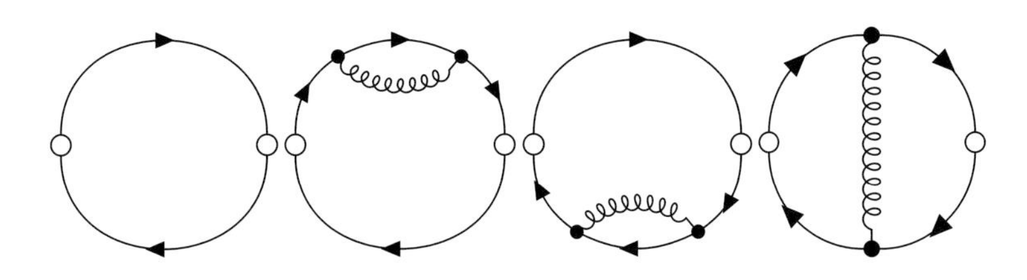

Figure 29.8 Diagrams contributing to the numerator N in (29.57), of (29.55). Conventions as in Fig. 29.7, and an open circle denotes the external S^a operator.

where the average over the group representation $\langle \mathcal{O} \rangle \equiv (\text{Tr}\mathcal{O}) / (\text{Tr}\mathbb{1})$ is carried out by the expressions

$$
\begin{aligned}
L_0 &= \langle S^a S^a \rangle = M(M+1)/8, \\
L_1 &= \left\langle S^a S^b S^b S^a \right\rangle = L_0^2, \\
L_2 &= \left\langle S^a S^a S^b S^b \right\rangle = L_0^2, \\
L_3 &= \left\langle S^a S^b S^a S^b \right\rangle = M^2(M+1)(M-3)/64.
\end{aligned}
\tag{29.58}
$$

Also,

$$
\begin{aligned}
D_{1\phi} &= \int_0^{\tau} d\tau_1 \int_{\tau_1}^{\tau} d\tau_2 G_\phi(\tau_1 - \tau_2) = -\frac{\widetilde{S}_{d+1}\tau^{\varepsilon}}{\varepsilon(1-\varepsilon)}, \\
D_{2\phi} &= \int_{\tau}^{\beta} d\tau_1 \int_{\tau_1}^{\beta} d\tau_2 G_\phi(\tau_1 - \tau_2) = -\frac{\widetilde{S}_{d+1}\tau^{\varepsilon}}{\varepsilon(1-\varepsilon)}, \\
D_{3\phi} &= \int_0^{\tau} d\tau_1 \int_{\tau}^{\beta} d\tau_2 G_\phi(\tau_1 - \tau_2) = \frac{2\widetilde{S}_{d+1}\tau^{\varepsilon}}{\varepsilon(1-\varepsilon)}.
\end{aligned}
\tag{29.59}
$$

Note we evaluate the above integrals at $T = 0$, by extending the integrals appropriately as explained in Ref. [293]. Here,

$$
G_\phi(\tau) = \int \frac{d^d k}{(2\pi)^d} \frac{d\omega}{2\pi} \frac{e^{-i\omega\tau}}{k^2 + \omega^2} = \frac{\widetilde{S}_{d+1}}{|\tau|^{d-1}},
\tag{29.60}
$$

with $\widetilde{S}_d = \Gamma(d/2-1)/(4\pi^{d/2})$.

We can now apply these perturbation-theory results to compute the renormalization-group equations. In the usual field-theoretic renormalization group, the perturbative results for the S^a correlator can be used to fix its renormalization constant

$$
S^a = \sqrt{Z_s} S_R^a.
\tag{29.61}
$$

Similarly, we define a renormalized couplings constant by

$$
\gamma_0 = \frac{\mu^{\varepsilon/2} Z_\gamma}{\sqrt{\widetilde{S}_{d+1}}} \gamma,
\tag{29.62}
$$

where μ is a renormalization scale. Because of the simple bilinear structure of the γ_0 term in (29.48), it is easy to see that the same graphs contribute to the renormalization constants Z_s and Z_γ, and we obtain an exact relation to all orders in γ:

$$Z_S = \frac{1}{Z_\gamma^2}. \tag{29.63}$$

It is this relation that leads to the main exponent identity in (29.54). We can see this by computing the β function of γ and the anomalous spin exponent η_S:

$$\beta(\gamma) = \mu \frac{d\gamma}{d\mu} \quad , \quad \eta_S(\gamma) = \frac{d \ln Z_S}{d \ln \mu}. \tag{29.64}$$

Then a direct evaluation using (29.61), (29.62), and (29.63) shows that

$$\beta(\gamma) = \gamma(\eta_S(\gamma) - \varepsilon). \tag{29.65}$$

In other words, at any fixed point of the β function with a non-zero γ, we must have $\eta_S(\gamma) = \varepsilon$, and hence (29.54) holds.

It remains to show that a such a fixed point actually exists, and is attractive in the flow to low energies. This we can only do order by order in γ, and the function $\beta(\gamma)$ is not known exactly. To the order we have computed results above, we have

$$Z_S = 1 - \frac{\gamma^2}{\varepsilon} L_\gamma, \tag{29.66}$$

where

$$L_\gamma = \frac{L_1 + L_2 - 2L_3}{L_0} = M. \tag{29.67}$$

This yields the β function

$$\beta(\gamma) = -\frac{\varepsilon}{2}\gamma + \frac{M}{2}\gamma^3 + \mathcal{O}(\gamma^5). \tag{29.68}$$

So the needed attractive fixed point is indeed present in the flow to low energies at $\gamma^* = (\varepsilon/M)^{1/2}$, at least at this order in perturbation theory. I emphasize that although there are higher-order corrections in (29.68), there are no such corrections to (29.54). This fact is important later in Section 33.3.2.

Problems

29.1 (a) Consider a generic free-electron Hamiltonian of the form

$$H = \sum_{a,b} c_a^\dagger h_{ab} c_b \tag{29.69}$$

and let ϕ_a^μ be the μ-th eigenvector and ε_μ the corresponding eigenvalue of the matrix h_{ab}. Show that the Green's function is

$$G_{ab}(i\omega_n) = \sum_\mu \frac{\phi_a^\mu \phi_b^{\mu *}}{i\omega_n - \varepsilon_\mu}. \tag{29.70}$$

Hence, argue that the free energy of H at a temperature T can be written as

$$\begin{aligned} F &= -T \sum_\mu \ln\left(1 + e^{-\varepsilon_\mu/T}\right) \\ &= T \int_{-\infty}^{\infty} \frac{d\Omega}{\pi} \ln\left(1 + e^{-\Omega/T}\right) \sum_a \operatorname{Im} G_{aa}(\Omega + i\eta). \end{aligned} \tag{29.71}$$

(b) Now consider the resonant-level model in (29.1). By thinking of H_{RLM} as acting on the space of $\boldsymbol{k}$ and d orbitals, show that in the limit $V \to \infty$ the change in the free energy due to the presence of the d state is

$$\Delta F = 2T \int_{-\infty}^{\infty} \frac{d\Omega}{\pi} \ln\left(1 + e^{-\Omega/T}\right) \operatorname{Im} G_{dd}(\Omega + i\eta), \tag{29.72}$$

along with an additional contribution from $G(\boldsymbol{k}, \boldsymbol{k}, i\omega_n)$. Show that this additional contribution vanishes for the flat density of states near the Fermi level.

(c) Now let us apply these formulae for the free energy to the Kondo impurity model. Take $\overline{\lambda} = 0$ at the outset. Show that the free energy of the impurity spin in the large-N theory is given by (at $N = 2$)

$$F_K = \frac{2|\overline{Q}|^2}{J_K} - 2T \int_{-\infty}^{\infty} \frac{d\Omega}{\pi} \ln\left(1 + e^{-\Omega/T}\right) A_f(\Omega). \tag{29.73}$$

(d) It now remains to evaluate the integral over Ω to obtain F_K as a function of $\overline{Q}$, and then minimize F_K to find the optimum value of $\overline{Q}$. We consider the case $T \to 0$. After cutting off the integral in (29.73) at $|\Omega| = D$, show that

$$F_K = \frac{2|\overline{Q}|^2}{J_K} - \frac{\Gamma}{\pi} \ln\left(1 + \frac{D^2}{\Gamma^2}\right), \tag{29.74}$$

where $\Gamma = \pi |\overline{Q}|^2 d(0)$. Show that the minimum of (29.74) is always at $\overline{Q} \neq 0$, and so find that the optimum value of $\overline{Q}$ for $J_K d(0) \ll 1$ agrees with that obtained in this chapter.

30 The Heavy Fermi Liquid

The Kondo lattice model describes a lattice of $S = 1/2$ spins coupled to a separate band of mobile electrons. The lattice manifestation of the Kondo effect leads to a heavy Fermi liquid state, with a large Fermi surface, and quasiparticles with a large effective mass. The Luttinger relation of Fermi liquid theory on the volume enclosed by the Fermi surface is described, and applied to the Kondo lattice model.

The Kondo lattice is the preferred model to describe the physics of a number of intermetallic compounds. These compounds contain a transition metal or a rare-earth metal with a localized orbital with strong local Coulomb interactions that prefer a net magnetic moment on each site. This moment then interacts with the mobile conduction electrons arising from the lighter elements. The key difference from the previous chapter is that the moments are not isolated impurity sites, but arranged periodically in a perfect lattice. So the Bloch crystal momentum is a good quantum number to describe the electronic states, including those associated with the moments on the electronic sites. There are also interesting applications of Kondo lattice models to twisted bilayer graphene [271].

We begin by generalizing the interacting resonant-level model, that is, the Anderson model in (29.10), to the Anderson lattice model sketched in Fig. 30.1. The resonant d site is now replaced by a lattice of d sites, each of which mix with the conduction electrons c_α with the hybridization w, and there is an on-site repulsion U_d on every d site:

$$H_{AL} = \sum_k \left[\varepsilon_k c_{k\alpha}^\dagger c_{k\alpha} + \varepsilon_k^d d_{k\alpha}^\dagger d_{k\alpha}\right] + \sum_i \left[-w\left(d_{i\alpha}^\dagger c_{i\alpha} + c_{i\alpha}^\dagger d_{i\alpha}\right) + U_d d_{i\uparrow}^\dagger d_{i\uparrow} d_{i\downarrow}^\dagger d_{i\downarrow}\right]. \tag{30.1}$$

Note that we neglect the weaker interactions on the c sites. As the two bands mix, only the total number of electrons is conserved, and we denote the total density per unit cell as $1 + \rho_c$, with $0 < \rho_c < 1$.

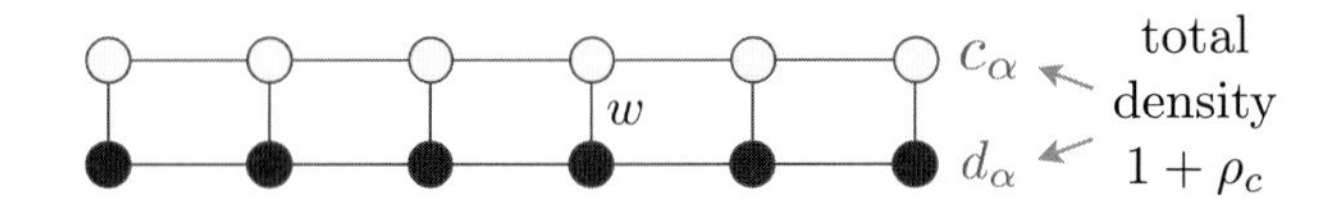

Figure 30.1 Anderson lattice model: conduction-band electrons c_α and d-band electrons d_α with band-mixing hybridization w. Two electrons on the same site of the d band repel with energy U_d.

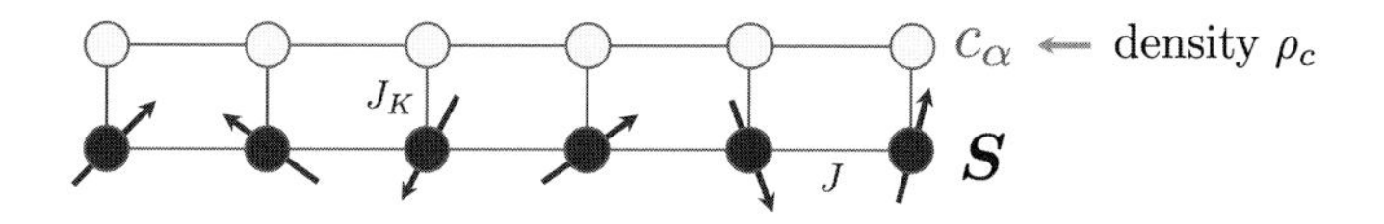

Figure 30.2 Kondo lattice model: conduction electrons c_α coupled to $S = 1/2$ spins $\boldsymbol{S}$.

We are interested in the large-U_d limit, with the chemical potential chosen so that there is exactly one electron in every d site, just as in Chapter 29. Then, we can perform the Schrieffer–Wolff transformation on (30.1), following the same procedure as Section 29.2, and hence obtain the lattice generalization of the Kondo impurity model in (29.13). This Kondo lattice model is sketched in Fig. 30.2. The Kondo lattice Hamiltonian is expressed in terms of $S = 1/2$ spins $\boldsymbol{S}_i$ on each d site:

$$H_{KL} = \sum_{\boldsymbol{k}} \varepsilon_{\boldsymbol{k}} c^\dagger_{\boldsymbol{k}\alpha} c_{\boldsymbol{k}\alpha} + \frac{J_K}{2} \sum_i \boldsymbol{S}_i \cdot c^\dagger_{i\alpha} \boldsymbol{\sigma}_{\alpha\beta} c_{i\beta} \,. \tag{30.2}$$

The Schrieffer–Wolff transformation also generates an exchange interaction between the d sites, but we defer consideration of the resulting Kondo–Heisenberg model until Section 31.1. As the d sites have electron density that is exactly unity, the density of conduction electrons is now ρ_c, as shown in Fig. 30.2.

In Section 30.1, we apply the large-M method [52, 109] to the Kondo lattice model in (30.2), and find that the Kondo impurity model has a natural and simple generalization to the lattice. Kondo screening applies also to the lattice model, and we obtain a "heavy Fermi liquid" (HFL) state involving both the conduction electrons and the local moments. This state has a Fermi surface, and the volume enclosed by the Fermi surfaces counts *all* electrons: the conduction electrons and the local moments, for a total density of $1 + \rho_c$. The narrow Kondo resonance width translates, as we shall see, to a large renormalized mass at this large Fermi surface.

In Section 30.2 we turn to a more general consideration of the Luttinger relation, which constrains the volume enclosed by the Fermi surface in Fermi liquids. We connect this analysis to the Kondo impurity model in Section 30.2.3, and to the Kondo lattice model in Section 30.2.4.

30.1 The Kondo Lattice Heavy Fermi Liquid

We proceed with an analysis of (30.2) using the large-M approach of Section 29.4. The initial steps are exactly the same: we represent the spin by constrained fermionic spinons $f_{i\alpha}$, now with an additional site label. We impose the constraint by a Lagrange multiplier $\lambda_i(\tau)$ on each site, and decouple the Kondo interaction by a Hubbard–Stratonovich field $P_i(\tau)$ on each site. Finally, we reduce the theory to its large-M saddle point, where the Lagrange multiplier is replaced by a site-independent value $i\lambda_i(\tau) \Rightarrow \overline{\lambda}$, and also $P_i(\tau) \Rightarrow \overline{P}$. Then, the large-$M$ saddle point is replaced by a saddle-point Kondo lattice Hamiltonian of free fermions generalizing (29.33) (V is now the number of lattice sites):

$$\overline{H}_{KL} = \frac{MV|\overline{P}|^2}{J_K} + \sum_k \left[-\overline{P} f^\dagger_{k\alpha} c_{k\alpha} - \overline{P}^* c^\dagger_{k\alpha} f_{k\alpha} + \varepsilon_k c^\dagger_{k\alpha} c_{k\alpha} \right] + \overline{\lambda} \sum_k f^\dagger_{k\alpha} f_{k\alpha} - \overline{\lambda} \frac{MV}{2}. \tag{30.3}$$

The most important difference from the impurity model is that the f spinons have now acquired a momentum label, and the Hamiltonian is diagonal in momentum. The saddle-point equations determining the values of $\overline{\lambda}$ and $\overline{P}$ are now (replacing (29.35) and (29.36))

$$\overline{P} = \frac{J_K}{MV} \sum_k \left\langle c^\dagger_{k\alpha} f_{k\alpha} \right\rangle, \tag{30.4}$$

$$\frac{1}{2} = \frac{1}{MV} \sum_k \left\langle f^\dagger_{k\alpha} f_{k\alpha} \right\rangle. \tag{30.5}$$

We also introduce the density of conduction electrons ρ_c, which is important for the following:

$$\frac{\rho_c}{2} = \frac{1}{MV} \sum_k \left\langle c^\dagger_{k\alpha} c_{k\alpha} \right\rangle. \tag{30.6}$$

It is easy to compute the Green's functions of $\overline{H}_{KL}$ by summing diagrams order by order in $\overline{P}$, as shown in Fig. 30.3 (compare Fig. 29.2 for the resonant-level model). It is convenient to write them in the following form:

$$[G_{cc}(\boldsymbol{k}, i\omega_n)]^{-1} = i\omega_n - \varepsilon_k - \frac{|\overline{P}|^2}{i\omega_n - \overline{\lambda}}, \tag{30.7}$$

$$G_{fc}(\boldsymbol{k}, i\omega_n) = \frac{-\overline{P}}{(i\omega_n - \overline{\lambda})} G_{cc}(\boldsymbol{k}, i\omega_n), \tag{30.8}$$

$$G_{ff}(\boldsymbol{k}, i\omega_n) = \frac{1}{i\omega_n - \overline{\lambda}} + \frac{|\overline{P}|^2}{(i\omega_n - \overline{\lambda})^2} G_{cc}(\boldsymbol{k}, i\omega_n). \tag{30.9}$$

These equations correspond to (29.37)–(29.39) for the resonant-level model. We now insert G_{fc} in the saddle-point equation (30.4), and obtain

$$\frac{\overline{P}}{J_K} = \frac{T}{V} \sum_{k,\omega_n} \frac{-\overline{P}}{(i\omega_n - \varepsilon_k)(i\omega_n - \overline{\lambda}) - |\overline{P}|^2}. \tag{30.10}$$

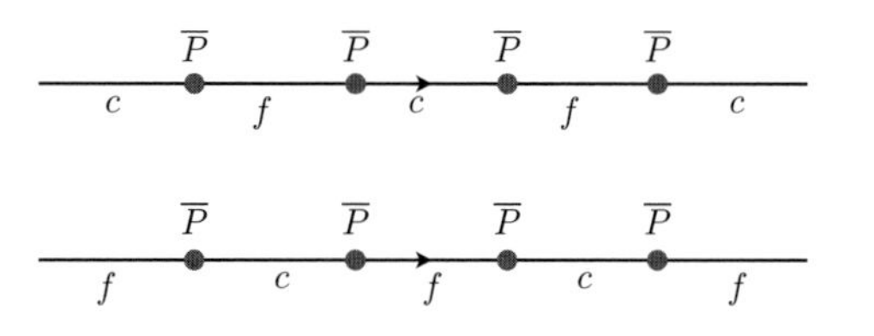

Figure 30.3 Feynman diagrams for G_{cc} and G_{ff}. All lines carry the same momentum and frequency.

To evaluate the frequency summation, we notice that the denominator in (30.10) has poles at the energies $z = E_{\boldsymbol{k}}^{\pm}$ where

$$2E_{\boldsymbol{k}}^{\pm} = \varepsilon_{\boldsymbol{k}} + \overline{\lambda} \pm \left[(\varepsilon_{\boldsymbol{k}} - \overline{\lambda})^2 + 4|\overline{P}|^2 \right]^{1/2}. \tag{30.11}$$

These are, of course, the single-particle eigenenergies of $\overline{H}_{KL}$. Evaluating the frequency summation in (30.10), we obtain (compare to (29.42))

$$\frac{\overline{P}}{J_K} = \frac{\overline{P}}{V} \sum_{\boldsymbol{k}} \frac{f(E_{\boldsymbol{k}}^{-}) - f(E_{\boldsymbol{k}}^{+})}{E_{\boldsymbol{k}}^{+} - E_{\boldsymbol{k}}^{-}}. \tag{30.12}$$

Before solving (30.12) we need to constrain the chemical potentials acting on the c and f fermions. These are fixed by the density ρ_c of the c fermions in (30.6), and the constraint (30.5) on the f fermions. It is easier to first fix the total density of fermions, which leads to the relation

$$1 + \rho_c = \frac{2}{V} \sum_{\boldsymbol{k}} \left[f(E_{\boldsymbol{k}}^{+}) + f(E_{\boldsymbol{k}}^{-}) \right]. \tag{30.13}$$

We work under conditions in which the total density of the conduction electrons per site $\rho_c < 1$. Then, at $T = 0$, (30.13) shows that we can have $E_{\boldsymbol{k}}^{+} > 0$ for all $\boldsymbol{k}$, while $E_{\boldsymbol{k}}^{-} < 0$ for some finite domain of $\boldsymbol{k}$; this is the portion of the Brillouin zone inside the Fermi surface (see Fig. 30.4). Similarly, from (30.5), we obtain, at $T = 0$ (compare to (29.43)),

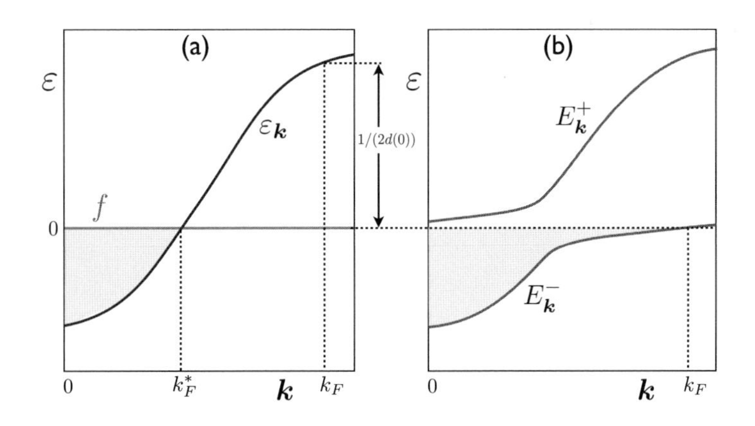

Figure 30.4 Schematic band structure in the HFL phase. In (a), we show the conduction electron band $\varepsilon_{\boldsymbol{k}}$, and the decoupled f band at zero energy; the conduction-electron states with energy $\varepsilon_{\boldsymbol{k}} < 0$ are occupied and have density ρ_c (this determines the value of k_F^*), while the f band is halffilled. The HFL state is obtained when the mixing between the bands is non-zero (given by $\overline{P}$ in (30.19)), and the f band is shifted by $\overline{\lambda}$ in (30.17). This results in the bands shown in (b). Only the lower band in (b) is occupied, up to the wavevector k_F, for a total density of $1 + \rho_c$. The values of k_F in (a) and (b) are the same, with the value of ε_{k_F} specified by (30.20). See Fig. 31.2 for a description of how (a) is modified in the FL* phase.

$$\frac{1}{2} = \frac{1}{V}\sum_{\boldsymbol{k}} \frac{1}{2}\left(1 + \frac{\varepsilon_k - \overline{\lambda}}{\left[(\varepsilon_k - \overline{\lambda})^2 + 4|\overline{P}|^2\right]^{1/2}}\right)\theta(-E_{\boldsymbol{k}}^-). \tag{30.14}$$

Returning to (30.12), determining $\overline{P}$ at $T = 0$, we obtain

$$\frac{\overline{P}}{J_K} = \frac{\overline{P}}{V}\sum_k \frac{\theta(-E_{\boldsymbol{k}}^-)}{\left[(\varepsilon_k - \overline{\lambda})^2 + 4|\overline{P}|^2\right]^{1/2}}. \tag{30.15}$$

We now have to solve (30.15) and (30.14) to obtain the values of the saddle-point parameters, $\overline{\lambda}$ and $\overline{P}$. We will examine the nature of the solution more carefully below after consideration of the Luttinger theorem. However, we can already notice an important point: the Kondo logarithmic divergence as $\overline{P} \to 0$, found in (29.45) for the Kondo impurity model, is also present in the Kondo lattice model. This is clear from (30.15), which has a logarithmic divergence at $\overline{P} = 0$ when the conduction band crosses the f level, that is, when $\varepsilon_k = \overline{\lambda}$, provided the density of conduction-electron states is finite at the Fermi level. This means that no matter how small we make J_K, we can have a solution with a non-zero $\overline{P}$; this is also the lowest-energy solution, and so we obtain a heavy Fermi liquid. We see in the next section that a minimum-energy solution with $\overline{P} = 0$ becomes possible once we allow the f moments to interact directly with each other.

30.1.1 Solution of Saddle-Point Equations

We now present an analytic solution of the saddle-point equations (30.13)–(30.15) in the limit of small $\overline{P}$, for the case of a flat band density of states $d(\varepsilon) = (1/V)\sum_k \delta(\varepsilon - \varepsilon_k) \approx d(0)$; the structure of the solution is sketched in Fig. 30.4.

The Fermi surface is present at $E_{\boldsymbol{k}}^- = 0$, and from (30.10) this translates to $\varepsilon_k = |\overline{P}|^2/\overline{\lambda}$. From (30.13), assuming a constant density of states, we therefore deduce that the limits of the summation over $\boldsymbol{k}$ in (30.13), (30.14), and (30.15) translate into bounds on ε_k:

$$-\frac{\rho_c + 1}{2d(0)} + \frac{|\overline{P}|^2}{\overline{\lambda}} < \varepsilon_k < \frac{|\overline{P}|^2}{\overline{\lambda}}. \tag{30.16}$$

We can now evaluate (30.14) in the limit $\overline{P} \to 0$, and obtain the value of $\overline{\lambda}$:

$$\overline{\lambda} = 2d(0)|\overline{P}|^2. \tag{30.17}$$

(In contrast, recall that for the Kondo impurity model, we had $\overline{\lambda} = 0$.) We now see from the upper limit in (30.16) that the Fermi surface is at $\varepsilon_k = 1/(2d(0))$ when $\overline{P}$ is small, but non-zero, as shown in Fig. 30.4. This should be contrasted from the Fermi surface location $\varepsilon_k = 0$ for a decoupled conduction-electron band. This increase in ε_k is precisely that needed to accomodate the 1/2 electron per site per spin component associated with the f fermions, that is, we have a large Fermi surface.

We can now obtain the value of $\overline{P}$ by evaluating (30.15) to logarithmic accuracy

$$\frac{1}{J_K} = 2d(0)\ln(D/|\overline{P}|), \tag{30.18}$$

where D is an energy of order of the bandwidth. So $\overline{P}$ is exponentially small,

$$|\overline{P}| \sim D\exp\left(-\frac{1}{2d(0)J_K}\right), \tag{30.19}$$

the same as the estimate for the Kondo impurity model in the argument of the exponent.

Finally, let us examine the structure of the conduction-electron Green's function near the Fermi level. We expand the expression for G_{cc} in (30.7) for small ω_n, and obtain

$$G_{cc}(\boldsymbol{k}, i\omega_n) \approx \frac{Z}{i\omega_n - Z\left[\varepsilon_{\boldsymbol{k}} - \dfrac{1}{2d(0)}\right]}, \tag{30.20}$$

where

$$Z = |2d(0)\overline{P}|^2. \tag{30.21}$$

So there is an exponentially small quasiparticle residue Z, and a quasiparticle effective mass $m^* = m/Z$, which is exponentially large (here m is the band mass associated with the dispersion $\varepsilon_{\boldsymbol{k}}$). Notice also the shift in the Fermi energy in (30.20), illustrated in Fig. 30.4. These are characteristic properties of the heavy Fermi liquid.

30.2 The Luttinger Relation

In Chapter 2, we alluded to one of the most remarkable features of Fermi liquid theory: the momentum-space volume enclosed by the Fermi surface defined by (2.39) is independent of the interactions, and depends only on the total electron density. Actually, this result is more general than Fermi liquid theory, and holds also in non-Fermi liquids without quasiparticle excitations, as discussed in Chapter 34. Moreover, there are deep connections of this "classic" result to key ideas in the modern theories of phases with fractionalization and anomalies; we encounter these connections in Section 31.3.2. In Section 30.2.1, we discuss this result in the simplest context of one-band model considered in Chapter 2, extend to disordered systems in Section 30.2.2, and consider applications to the Kondo models in Sections 30.2.3 and 30.2.4.

30.2.1 One-Band Model

I present a proof of the Luttinger relation following the classic textbook treatments, but use an approach that highlights its connections to the modern developments. Specifically, there is a fundamental connection between the Luttinger relation and $U(1)$

symmetries [53, 209]: any many-body quantum system has a Luttinger relation associated with each $U(1)$ symmetry, and this connects the density of the $U(1)$ charge in the ground state to the volume enclosed by its Fermi surfaces. This relation applies both to systems of fermions and bosons, or mixtures of fermions and bosons. However, the relation does not apply if the $U(1)$ symmetry is "broken" or "higgsed" by the condensation of a boson carrying the $U(1)$ charge. As bosons are usually condensed at low temperatures (see Chapter 3), the Luttinger relation is not often mentioned in the context of bosons. However, there can be situations when bosons do not condense, for example, if the bosons bind with fermions to form a fermionic molecule, and then the molecules form a Fermi surface; then we have to apply the Luttinger relation to the boson density [209].

We begin by noting a simple argument on why there could even be a relation between a short-time correlator (the density, given by an "ultraviolet" (UV) equal-time correlator) and a long-time correlator (the Fermi surface is the locus of zero short-time (correlator) energy excitations in a Fermi liquid, an "infrared" (IR) property). In the fermion path integral (see Appendix B), the free-particle term in the Lagrangian is

$$\mathcal{L}_c^0 = \sum_p c_p^\dagger \left(\frac{\partial}{\partial \tau} + \varepsilon_p^0 - \mu \right) c_p, \tag{30.22}$$

where we have now chosen to extract the chemical potential μ explicitly from the bare dispersion ε_p^0. The expression in (30.22) is invariant under global $U(1)$ symmetry:

$$c_p \to c_p e^{i\theta} \quad , \quad c_p^\dagger \to c_p^\dagger e^{-i\theta} \tag{30.23}$$

as are the rest of the terms in the Lagrangian describing the interactions between the electrons. However, let us now 'gauge' this global symmetry by allowing θ to have a *linear* dependence on imaginary time τ:

$$c_p \to c_p e^{\mu\tau} \quad , \quad c_p^\dagger \to c_p^\dagger e^{-\mu\tau}. \tag{30.24}$$

Note that in the Grassman path integral, c_p and $c_p^\dagger$ are independent Grassman numbers and so the two transformations in (30.24) are not inconsistent with each other. The interaction terms in the Lagrangian are explicitly invariant under the time-dependent $U(1)$ transformation in (30.24). The free-particle Lagrangian in (30.22) is not invariant under (30.24) because of the presence of the time derivative term; however, application of (30.24) shows that μ cancels out of the transformed $\mathcal{L}_c^0$, and so has completely dropped out of the path integral. We seem to have reached the absurd conclusion that the properties of the electron system are independent of μ: this is explicitly incorrect even for free particles.

What is wrong with the above argument that "gauges away" μ by the transformation in (30.24)? The answer becomes clear from the expression for the total electron density:

$$\rho_e = \frac{1}{V} \sum_p \int_{-\infty}^{\infty} \frac{d\omega}{2\pi} G(\boldsymbol{p}, i\omega) e^{i\omega 0^+}. \tag{30.25}$$

The transformation in (30.24) corresponds to a shift in frequency $\omega \to \omega + i\mu$ of the contour of integration, and this is not permitted because of singularities in $G(\boldsymbol{p}, i\omega)$.

However, as shown below, it is possible to manipulate (30.25) into a part that contains the full answer, and a remainder that vanishes because manipulations similar to the failed frequency shift in (30.24) become legal.

The key step to extracting the non-zero part is to use the following simple identity, which follows directly from Dyson's equation (2.26):

$$
\begin{aligned}
G(\boldsymbol{p}, i\omega) &= G_{ff}(\boldsymbol{p}, i\omega) + G_{LW}(\boldsymbol{p}, i\omega), \\
G_{ff}(\boldsymbol{p}, i\omega) &\equiv i\frac{\partial}{\partial \omega} \ln\left[G(\boldsymbol{p}, i\omega)\right], \\
G_{LW}(\boldsymbol{p}, i\omega) &\equiv -iG(\boldsymbol{p}, i\omega)\frac{\partial}{\partial \omega}\Sigma(\boldsymbol{p}, i\omega).
\end{aligned} \tag{30.26}
$$

The non-zero part is G_{ff}: it is a frequency derivative, and so its frequency integral in (30.25) is not difficult to evaluate exactly after carefully using the $e^{i\omega 0^+}$ convergence factor. The subscript of G_{ff} denotes that this the only term that is non-vanishing for free fermions; indeed, we will see below that the frequency integral of G_{ff} has the same value for interacting fermions as for free fermions with the same Fermi surface. The remaining contribution from G_{LW} vanishes for free particles (which have vanishing Σ). Therefore, establishing the Luttinger relation, that is, the invariance of the volume enclosed by the Fermi surface, reduces then to establishing that the contribution of G_{LW} to (30.25) vanishes.

We consider the latter important step first. We would like to show that

$$
\sum_{\boldsymbol{p}} \int_{-\infty}^{\infty} \frac{d\omega}{2\pi} G_{LW}(\boldsymbol{p}, i\omega) = 0. \tag{30.27}
$$

We now show that (30.27) follows from the transformations of G_{LW} under the gauge transformation in (30.24) for an imaginary chemical potential:

$$
c_{\boldsymbol{p}} \to c_{\boldsymbol{p}} e^{+i\omega_0 \tau} \quad , \quad c_{\boldsymbol{p}}^\dagger \to c_{\boldsymbol{p}}^\dagger e^{-i\omega_0 \tau}. \tag{30.28}
$$

The argument relies on the existence of a functional, $\Phi_{LW}\left[G(\boldsymbol{p}, i\omega)\right]$, of the Green's function, called the Luttinger–Ward functional, so that the self-energy is its functional derivative

$$
\Sigma(\boldsymbol{p}, i\omega) = \frac{\delta \Phi_{LW}}{\delta G(\boldsymbol{p}, i\omega)}. \tag{30.29}
$$

The existence of such a functional can be seen diagrammatically, in which the Luttinger–Ward functional equals the interaction-dependent terms for the free energy written in a "skeleton" graph expansion in terms of the fully renormalized Green's function. Taking the functional derivative with respect to $G(\boldsymbol{p}, \omega)$ is equivalent to cutting a single G from all such graphs in all possible ways, and these are just the graphs for the self interaction dependent energy. For a more formal argument, see Ref. [208]. An important property of the Luttinger–Ward functional is its invariance under frequency shifts:

$$
\Phi\left[G(\boldsymbol{p}, i\omega + i\omega_0)\right] = \Phi\left[G(\boldsymbol{p}, i\omega)\right], \tag{30.30}
$$

for any fixed ω_0. Here, we are regarding Φ as a functional of two distinct functions $f_{1,2}(\omega)$, with $f_1(\omega) \equiv G(\boldsymbol{p}, i\omega + i\omega_0)$ and $f_2(\omega) = G(\boldsymbol{p}, i\omega)$, and Φ evaluates to the same value for these two functions. Now note that this frequency shift is nothing but the gauge transformation in (30.28); therefore, (30.30) follows from the fact that such frequency shifts are allowed in Φ_{LW}. The singularity on the real frequency axis is sufficiently weak so that the frequency shifts are legal in a Fermi liquid; but we note that in the non-Fermi liquid Sachdev–Ye–Kitaev model considered in Chapter 32, the Green's functions are significantly more singular at $\omega = 0$, and the analogs of (30.27) and (30.30) do not apply; this is described in Section 32.2.2. For the Fermi liquid, we can now expand (30.30) to first order in ω_0, using (30.29), and integrating by parts we establish (30.27).

Now that we have disposed of the offending term in (30.26), we can return to (30.25) and evaluate

$$\rho_e = \frac{i}{V} \sum_{\boldsymbol{p}} \int_{-\infty}^{\infty} \frac{d\omega}{2\pi} \frac{\partial}{\partial \omega} \ln \left[G(\boldsymbol{p}, i\omega) \right] e^{i\omega 0^+} . \tag{30.31}$$

We evaluate the ω integral by distorting the contour in the frequency plane. For this, we need to carefully understand the analytic structure of the integrand. This is subtle, because there are two types of branch-cuts. One arises from the Green's function: $G(\boldsymbol{p}, z)$ has a branch-cut along the real axis $\text{Im}(z) = 0$, with $\text{Im}G(\boldsymbol{p}, z) \leq 0$ for $\text{Im}(z) = 0^+$, $\text{Im}G(\boldsymbol{p}, z) \geq 0$ for $\text{Im}(z) = 0^-$ and $\text{Im}G(\boldsymbol{p}, z) = 0$ for $z = 0$. The other branch-cut is from the familiar $\ln(z)$ function: we take this on the positive real axis, with a discontinuity of $2i\pi$. First, we account for the branch-cut in $G(\boldsymbol{p}, z)$, by distorting the contour of integration in (30.31) to pick up the discontinuity $\text{Im}G(\boldsymbol{p}, z)$:

$$\rho_e = \frac{-i}{V} \sum_{\boldsymbol{p}} \int_{-\infty}^{0} \frac{dz}{2\pi} \frac{\partial}{\partial z} \ln \left[\frac{G(\boldsymbol{p}, z + i0^+)}{G(\boldsymbol{p}, z + i0^-)} \right] . \tag{30.32}$$

Note from (2.26) and (2.38) that on the real frequency axis $\text{Im}\, G(\boldsymbol{p}, z + i0^\pm) \to 0$ as $z \to 0$ or $-\infty$. Consequently, the only possible values of $\ln[G(\boldsymbol{p}, z + i0^+)/G(\boldsymbol{p}, z + i0^-)]$ are $0, \pm 2\pi i$ as $z \to 0$ or $-\infty$, from the branch-cut of the logarithm. So we obtain from (30.32)

$$\begin{aligned} \rho_e &= \frac{-i}{2\pi V} \sum_{\boldsymbol{p}} \ln \left[\frac{G(\boldsymbol{p}, i0^+)}{G(\boldsymbol{p}, i0^-)} \right] \\ &= \frac{1}{V} \sum_{\boldsymbol{p}} \theta \left(-\varepsilon_{\boldsymbol{p}}^0 + \mu - \Sigma(\boldsymbol{p}, i0^+) \right) \\ &= \frac{1}{V} \sum_{\boldsymbol{p}} \theta \left(-\varepsilon_{\boldsymbol{p}} \right) , \end{aligned} \tag{30.33}$$

where we have used (2.28) and (2.38). Because the branch-cut of the logarithm extends to $z = +\infty$, only negative values of $\varepsilon_{\boldsymbol{p}}$ contribute to the z integral extending from $z = -\infty$ to $z = 0$. The equation (30.33) is the celebrated Luttinger relation, equating the electron density to the volume enclosed by the Fermi surface of the quasiparticles $\varepsilon_{\boldsymbol{p}} = 0$. In the presence of a crystalline lattice, there can be additional bands that are either fully filled or unoccupied; such bands yield a contribution of unity or zero, respectively to (30.33).

To summarize, the Luttinger relation is intimately connected to the $U(1)$ symmetry of electron-number conservation. Indeed, we can obtain a Luttinger relation for each $U(1)$ symmetry of any system consisting of fermions or bosons. The result follows from the invariance of the Luttinger–Ward functional under the transformation in (30.28), in which we gauge the global symmetry to a linear time dependence: in this respect, there is a resemblance to 'tHooft anomalies in quantum field theories. If the $U(1)$ symmetry is "broken" by the condensation of a boson that carries $U(1)$ charge, the Luttinger relation no longer applies. We will find this point of view very useful when we consider systems that have a modified Luttinger relation due to the presence of emergent gauge symmetries in Section 31.2.

30.2.2 Disordered Systems

Our discussion of the Luttinger relation has so far assumed perfect crystalline symmetry, so the quasiparticles energies $\varepsilon_{\boldsymbol{p}}$ are functions of the crystal momentum $\boldsymbol{p}$. The Luttinger relation applies also to systems without crystalline symmetry, although it is expressed in a form involving quantities that are not easy to observe.

Let us consider a lattice of sites i, with a bare electron hopping t_{ij}, which has no particular symmetry. Then, the electron Green's function $G_{ij}(i\omega)$ is a matrix indexed by the lattice sites, as is the self-energy $\Sigma_{ij}(i\omega)$. These are related by Dyson's equation, which now has a matrix form

$$[(i\omega+\mu)\delta_{ij}+t_{ij}-\Sigma_{ij}(i\omega)]\,G_{jk}=\delta_{ik}\,. \tag{30.34}$$

The low-lying quasiparticles are no longer plane-wave eigenstates, but the arguments leading to (2.38) still apply, and we have

$$\operatorname{Im}\left[\Sigma_{ij}(\Omega+i0^+)\right]\to 0 \text{ as } \Omega\to 0 \text{ at } T=0. \tag{30.35}$$

We now proceed with the computation of the average density, which generalizes (30.25) to

$$\rho_e=\frac{1}{N}\int_{-\infty}^{\infty}\frac{d\omega}{2\pi}\operatorname{Tr}\left[G(i\omega)\right]e^{i\omega 0^+}, \tag{30.36}$$

where N is the number of sites, and the trace is over the site indices. The analysis is then a close parallel of that carried out for clean systems. The existence of the Luttinger–Ward functional now replaces (30.27) by

$$\int_{-\infty}^{\infty}\frac{d\omega}{2\pi}\operatorname{Tr}\left[G(i\omega)\frac{\partial}{\partial\omega}\Sigma(i\omega)\right]=0, \tag{30.37}$$

where a matrix multiplication is implied between G and Σ. The analysis from (30.31) to (30.33) is replaced by

$$\begin{aligned}\rho_e&=\frac{i}{N}\int_{-\infty}^{\infty}\frac{d\omega}{2\pi}\frac{\partial}{\partial\omega}\operatorname{Tr}\ln\left[G(i\omega)\right]e^{i\omega 0^+}\\&=\frac{1}{N}\sum_{\alpha}\theta(-\varepsilon_\alpha),\end{aligned} \tag{30.38}$$

where ε_α are the eigenvalues of the matrix $-t_{ij} - \mu\delta_{ij} + \Sigma_{ij}(i0^+)$. In general, we do not know the values of $\Sigma_{ij}(i0^+)$, and so this result is not easy to apply. However, it does yield information on the nature of the quasiparticles, which are the eigenstates with small $|\varepsilon_\alpha|$.

30.2.3 Kondo Impurity Model

The resonant-level model of Section 29.1 satisfies a Friedel sum rule, which is reminiscent of the Luttinger relation. This expresses the change in electron density due to the presence of the impurity in terms of the phase shift of the scattering of the conduction electrons from the impurity.

In terms of the Green's functions of Section 29.1, the change in the electron density induced by the impurity is (with a factor of 2 for spin)

$$\begin{aligned}\delta\rho_e &= 2\int_{-\infty}^{\infty}\frac{d\omega}{2\pi}\left[G_{dd}(i\omega) + \sum_{\mathbf{k}}\left(G(\mathbf{k},\mathbf{k},i\omega) - \frac{1}{i\omega - \varepsilon_{\mathbf{k}}}\right)\right]e^{i\omega 0^+} \\ &= 2\int_{-\infty}^{\infty}\frac{d\omega}{2\pi}i\frac{d}{d\omega}\ln\left[G_{dd}(i\omega)\right]e^{i\omega 0^+} \\ &= -2i\int_{-\infty}^{0}\frac{dz}{2\pi}\frac{\partial}{\partial z}\ln\left[\frac{G_{dd}(z+i0^+)}{G_{dd}(z+i0^-)}\right] \\ &= \frac{-i}{\pi}\ln\left[\frac{G_{dd}(i0^+)}{G_{dd}(i0^-)}\right] \\ &= 1 + \frac{2}{\pi}\tan^{-1}\left[\frac{\mathrm{Re}\,G_{dd}^{-1}(i0^+)}{\mathrm{Im}\,G_{dd}^{-1}(i0^+)}\right].\end{aligned} \tag{30.39}$$

Note the similarity of the manipulations to those in Section 30.2.1. The expression (30.39) is exact, and does not rely on the "flat density of states" approximations used to obtain (29.4). In terms of (29.4) we have

$$\delta\rho_e = 1 - \frac{2}{\pi}\tan^{-1}\left[\frac{\varepsilon_d}{\Gamma}\right]. \tag{30.40}$$

The right-hand side of (30.40) is $1/\pi$ times the scattering phase shift of the conduction electrons at the Fermi level [109]. As expected, the density varies from $\delta\rho_e = 2$, when the d level is far below the Fermi level, to $\delta\rho_e = 0$, when the d level is far above the Fermi level.

We can now extend this result to the interacting Anderson impurity model in (29.10). Adding the interaction U_d to the resonant-level model changes the d-fermion Green's function to that in (29.11). Then, we can see that the analysis in (30.39) acquires an additional contribution

$$\int_{-\infty}^{\infty}\frac{d\omega}{2\pi}G_{dd}(i\omega)\frac{d}{d\omega}\Sigma_{dd}(i\omega) = 0, \tag{30.41}$$

which vanishes because of the existence of the Luttinger–Ward functional, as in (30.27). The self-energy Σ_{dd} in (29.11) does change the electron density from (30.40) to

$$\delta\rho_e = 1 - \frac{2}{\pi}\tan^{-1}\left[\frac{\varepsilon_d^\star}{\Gamma}\right], \tag{30.42}$$

where

$$\varepsilon_d^\star = \varepsilon_d + \mathrm{Re}\left[\Sigma_{dd}(0)\right], \tag{30.43}$$

which is similar to the renormalization in (2.28). Also, as in (2.38) and (30.35), we have $\mathrm{Im}\left[\Sigma_{dd}(0)\right] = 0$ at $T = 0$.

Turning to the Kondo impurity model in (29.13), establishing the Luttinger relation requires the framework of the $1/M$ expansion in Section 29.4. We observe that the large-M saddle point in (29.33) has the same structure as the resonant-level model in (29.1), provided we are at a saddle point with $\overline{P} \neq 0$. As far as the manipulations for the Luttinger relation are concerned, the self-energy corrections to the f fermions in the $1/M$ expansion about such a saddle point play the same role as the self-energy corrections to the d electrons in the Anderson impurity model in (29.10). In other words, once we have $\overline{P} \neq 0$, the f spinons play the same role as electrons. We therefore obtain the same Friedel sum rule as in (30.42).

30.2.4 Kondo Lattice Model

As in the discussion of the Kondo impurity model in Section 30.2.3, it is easier to first establish the Luttinger relation in the context of the Anderson lattice model (30.1) by generalizing the arguments of Section 30.2.1 to the multi-band case. The Anderson lattice Hamiltonian leads to a 2×2 matrix Green's function, and at all orders in perturbation theory in U_d we have

$$\boldsymbol{G}^{-1}(\boldsymbol{k}, i\omega_n) = \begin{pmatrix} i\omega_n - \varepsilon_{\boldsymbol{k}} & w \\ w & i\omega_n - \varepsilon_{\boldsymbol{k}}^d \end{pmatrix} - \boldsymbol{\Sigma}(\boldsymbol{k}, i\omega_n), \tag{30.44}$$

where the self-energy $\boldsymbol{\Sigma}$ is also a 2×2 matrix; this relation generalizes (2.26). We can relate the total density to the size of the Fermi surface by generalizing the identity in (30.26) to the multiband case:

$$\mathrm{Tr}\,\boldsymbol{G}(\boldsymbol{k}, i\omega) = i\frac{\partial}{\partial\omega}\ln\left[\det\boldsymbol{G}(\boldsymbol{k}, i\omega)\right] - i\mathrm{Tr}\left[\boldsymbol{G}(\boldsymbol{k}, i\omega)\frac{\partial}{\partial\omega}\boldsymbol{\Sigma}(\boldsymbol{k}, i\omega)\right]. \tag{30.45}$$

Then, an analysis analogous to that in Section 30.2.1 and 30.2.2 leads to a constraint of $1 + \rho_c$ on the total size of one or more Fermi surfaces, as in (30.13):

$$1 + \rho_c = \frac{2}{V}\sum_{\boldsymbol{k}}\left[\theta(-E_{\boldsymbol{k}}^+) + \theta(-E_{\boldsymbol{k}}^-)\right], \quad T = 0, \tag{30.46}$$

where the quasiparticle dispersions $E_{\boldsymbol{k}}^\pm$ are given by the roots $\omega = E_{\boldsymbol{k}}^\pm$ of the equation

$$\det\boldsymbol{G}(\boldsymbol{k}, \omega) = 0. \tag{30.47}$$

This analysis of the Anderson lattice model appears to leave no room for a metallic state in which the Fermi surface size differs from that implied by a density of $1+\rho_c$. But we will see in Chapter 31 that other metallic states are possible with smaller Fermi surfaces. The above arguments only imply that such states cannot appear in a perturbation theory in U_d, but do not rule out their non-perturbative appearance across a quantum phase transition.

Also, as in the discussion of the Kondo impurity model in Section 30.2.3, we can apply the Luttinger arguments to the Kondo lattice model (30.2) in the context of the $1/M$ expansion. As before, this relies on the large-M saddle point in (30.3) with $\overline{P} \neq 0$, which has the same structure as the non-interacting part of the Anderson lattice model in (30.1). At $M = \infty$, we follow the same route as in the Kondo impurity problem in Section 30.2.3. From the expressions in (30.7)–(30.9) it is easy to explicitly verify that

$$G_{cc}(\boldsymbol{k}, i\omega_n) + G_{ff}(\boldsymbol{k}, i\omega_n) = i\frac{\partial}{\partial\omega} \ln\left[(i\omega_n - \varepsilon_{\boldsymbol{k}})(i\omega_n - \overline{\lambda}) - |\overline{P}|^2\right]. \tag{30.48}$$

Now, proceeding as in the subsections above, we obtain the Luttinger constraint in (30.13). Upon including the fermion self-energies from the $1/M$ expansion, the analysis is a close parallel of the argument above for the Anderson lattice model, and we obtain the generalization of (30.13) to all orders in $1/M$.

As discussed in much detail in Chapter 31, the large-M analysis of the Kondo lattice model also points the way to novel metallic phases that were not apparent in the perturbative expansion of the Anderson lattice model. These phases have non-Luttinger volume Fermi surfaces, and are associated with large-M saddle points with $\overline{P} = 0$.

31 The Fractionalized Fermi Liquid

A possible fate of the Kondo lattice model is that the lattice of spins form a spin-liquid state, while the mobile electrons form a small Fermi surface on their own. This is the fractionalized Fermi liquid, in which neutral spinon excitations coexist with Fermi-liquid-like electronic quasiparticles. Topological arguments are presented for the stability of the modified Luttinger relation in such a metal to arbitrary interactions between the spins and the mobile electrons. A paramagnon fractionalization theory shows that a fractionalized Fermi liquid can also exist in a single-band Hubbard model, and this is proposed as a theory of the pseudogap metal of the cuprates.

Our study of the Anderson lattice model, and the closely related Kondo lattice model in Chapter 30 produced the heavy Fermi liquid (HFL) ground state: a Fermi liquid with a "large" Fermi surface whose enclosed volume counts the total density, $1+\rho_c$, of both the local moments and the conduction electrons, and the quasiparticles on the Fermi surface have a large effective mass. This state has a close connection to the Anderson or Kondo impurity model of Chapter 29, in which the local moment was always screened by the conduction electrons as $T \to 0$, corresponding to the renormalization-group flow of the Kondo coupling $J_K \to \infty$. In the lattice case, the screening of the local moment implies that the spin moment effectively becomes mobile, and then "dissolves" into the Fermi surface.

In this chapter I discuss another possible ground state of the Anderson or Kondo lattice model: this is the fractionalized Fermi liquid (FL*), which has a "small" Fermi surface whose enclosed volume counts only the density ρ_c of the conduction electrons. Clearly, such a state does not obey the Luttinger relation discussed in Section 30.2, and so cannot be contained by adiabatic continuity from the free-fermion state, as in Fermi liquid theory. In other words, the HFL and FL* states must be separated by a quantum phase transition without a symmetry-breaking order parameter [259, 262]. We note a recent observation [169] of a FL* state, along with its phase transition to an HFL state.

The existence of the FL* state can be inferred from an extension of the Kondo lattice model of Chapter 30, the Kondo–Heisenberg model with Hamiltonian

$$H_{KH} = H_{KL} + J \sum_{\langle ij \rangle} \boldsymbol{S}_i \cdot \boldsymbol{S}_j , \tag{31.1}$$

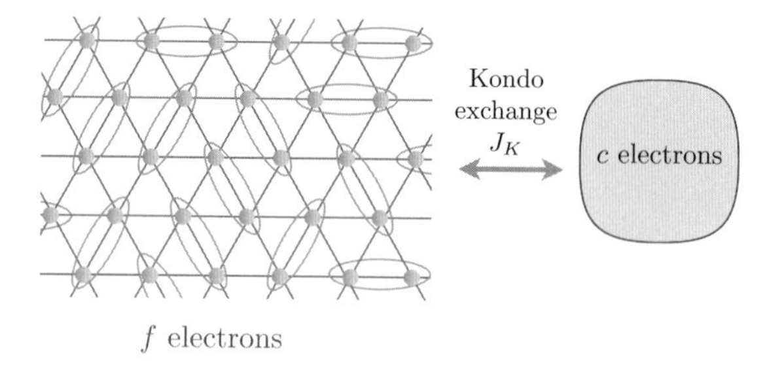

Figure 31.1 A schematic of the FL* state on the Kondo lattice: a spin liquid with fractionalization and emergent gauge fields is coupled by the Kondo coupling J_K to a density ρ_c of c conduction. The small Fermi surface, of size ρ_c, is stable to turning on a non-zero J_K, and this yields a state that does not obey the Luttinger relation of Section 30.2.

where H_{KL} was specified in (30.2). This Hamiltonian now has an antiferromagnetic exchange interaction J between nearest-neighbor sites, which is also generated from a Schrieffer–Wolff transformation of the Anderson lattice model in (30.1). The HFL to FL* phase transition can be generated by increasing the ratio J/J_K. We have obtained the HFL state in the limiting case $J = 0$, but now consider the opposite limiting case $J_K = 0$, but $J \neq 0$. Then the $\boldsymbol{S}_i$ decouple into a spin system, of the type extensively studied in Parts II and IV. Let us assume that the lattice of $\boldsymbol{S}_i$, and the J exchange couplings between them (we allow extensions in which these couplings are non-nearest-neighbor) are such that they form a spin-liquid ground state with fractionalized spinon excitations and emergent gauge fields, as sketched in Fig. 31.1. Meanwhile, the decoupled conduction electrons will form a Fermi liquid on their own with a Fermi volume of ρ_c [11]. Let us now turn on a small J_K. One of the key points of this chapter is that the state of Fig. 31.1 is stable to turning on a non-zero J_K, and the "topological order" of the spin liquid ensures that the Fermi surface volume of the coupled system remains pinned at the decoupled small Fermi surface size, yielding the FL* state. This is in stark contrast to the Kondo impurity model of Chapter 29, where even an infinitesimal antiferromagnetic coupling flows under the renormalization group to strong coupling. For the case of a gapped spin liquid, the stability of perturbation theory in J_K is clear. We consider gapless spin liquids here in the context of the $1/M$ expansion, and show that they can also be stable to a non-zero J_K.

We emphasize here the role of symmetries in allowing the existence of the FL* phase in the Anderson lattice model. The original Anderson lattice model in (30.1) has only a single $U(1)$ global symmetry, that associated with the conservation of the total number of c_α and d_α electrons, which must equal $1+\rho_c$ per unit cell. Consequently, in a perturbation theory in U_d, the Luttinger relation leads to only a single constraint: the total volume enclosed by the Fermi surface must be equivalent to $1+\rho_c$ states, as in the HFL phase. The existence of the FL* phase in the Anderson lattice model becomes evident after we perform a canonical transformation to the Kondo lattice model, which has a much larger emergent symmetry: the total number of c_α electrons is constrained to be ρ_c per unit cell, and the number of f_α spinons (in a fermionic spinon description of the spin liquid) is constrained by (29.15) to be unity at *each* site.

Sections 31.2 and 31.3 describe the role played by these symmetries in obtaining the modified Luttinger* relation in the FL* phase.

While the existence of the FL* state seems evident for the Kondo lattice model by the reasoning in Fig. 31.1, it is not immediately clear whether a FL* state can appear in a single-band model, as in the Hubbard model. Section 31.4 presents arguments for the existence of FL* in single-band models, and describes a theory of "paramagnon fractionalization," which leads to such a metallic phase. The FL* phase of the Hubbard model provides an appealing description of the pseudogap metal phase of the cuprate superconductors.

31.1 The FL* State in the Kondo Lattice

This section applies the large-M method of Section 30.1 to the Kondo–Heisenberg model in (31.1).

A key feature of our discussions of the Kondo impurity model and the Kondo lattice model has been the singular nature of the limit $J_K \to 0$. Even for very small J_K, we have found that the $\overline{P} = 0$ state is unstable to the turning on of an exponentially small $\overline{P}$. In a renormalization-group language, this is the statement that an infinitesimal J_K is a marginally relevant perturbation to the $J_K = 0$ fixed point. The $\overline{P} \neq 0$ state was then found to be a renormalized Fermi liquid, with well-defined quasiparticles obeying the Friedel sum rule for the Kondo impurity model; for the Kondo lattice model, the qusiparticles possess a Luttinger volume Fermi surface, which counts the spins as electrons.

We now consider the case with $J \neq 0$, and find a situation in which the $J_K = 0$ fixed point is stable, and we obtain a novel stable state with $\overline{P} = 0$, and no broken symmetry.

We can proceed with a $1/M$ expansion for H_{KH} by combining the treatment above for the Kondo model H_{KL} with that for the $U(1)$ spin liquid in Chapter 22. This implies, that, in addition to the decoupling field $P_i(\tau)$ used to obtain (30.3), we have the analogs of the decoupling fields $Q_{ij}(\tau)$ of (22.6) between the d sites. This leads here to the following new terms in the Lagrangian

$$\mathcal{L}_Q = \sum_{\langle ij \rangle} \left[\frac{M|Q_{ij}|^2}{J} - Q_{ij} f_{j\alpha}^\dagger f_{i\alpha} - Q_{ij}^* f_{i\alpha}^\dagger f_{j\alpha} \right], \tag{31.2}$$

where Q_{ij} is the link Hubbard–Stratonovich field between the d sites. A crucial role is now played by the emergent $U(1)$ gauge symmetry, which combines the Kondo gauge symmetry of (29.32) with that of the spin-liquid gauge symmetry in (22.10):

$$\begin{aligned} f_{i\alpha} &\to f_{i\alpha} e^{i\phi_i(\tau)}, \\ P_i &\to P_i e^{i\phi_i(\tau)}, \\ \lambda_i &\to \lambda_i - \partial_\tau \phi_i(\tau), \\ Q_{ij} &\to Q_{ij} e^{-i(\phi_i(\tau) - \phi_j(\tau))}. \end{aligned} \tag{31.3}$$

This transformation leaves the full Lagrangian invariant for a gauge transformation $\phi_i(\tau)$, which can have an arbitrary dependence on τ and lattice site i. As in Section 22.2, note that the phase of Q_{ij} transforms just like the vector potential of a $U(1)$ gauge field. So if we write

$$Q_{ij} = |Q_{ij}| \exp(i a_{ij}), \tag{31.4}$$

then

$$a_{ij} \to a_{ij} + \phi_j(\tau) - \phi_i(\tau). \tag{31.5}$$

In the continuum limit, the transformation of a_{ij} is precisely the analog of the vector potential of the Maxwell theory, while that of λ is that of the scalar potential. So a_{ij} and λ_i together realize an emergent, lattice $U(1)$ gauge field. For the a_{ij} to an independent propagating degree of freedom, we do need to expand about a saddle point in which the saddle-point values of $|Q_{ij}|$ are non-zero: we will assume that is the case in the remaining discussion.

Given the identification of a_{ij} and λ_i with a $U(1)$ gauge field, the other fields in (31.3) are easily seen to be *matter* fields, which are charged under the emergent $U(1)$ gauge symmetry. The $f_{i\alpha}$ are fermionic spinons that carry a unit gauge charge, and the P_i are bosons that also carry a unit gauge charge.

All the saddle points we have considered so far had

$$P_i = \overline{P} \neq 0 \quad , \quad \text{HFL}. \tag{31.6}$$

With our new-found identification of P as a gauge-charge boson, we see that any phase of matter satisfying (31.6) is a *Higgs phase* of the emergent $U(1)$ gauge field, with $\overline{P}$ the Higgs condensate. The fluctuations of the gauge fields are quenched by this Higgs condensate, and they become overdamped modes in the particle–hole continuum. This is why we did not have to seriously consider the $U(1)$ gauge field in our analyses so far.

This section considers the possibility of a new saddle point in which

$$P_i = \overline{P} = 0 \quad , \quad \text{FL*}. \tag{31.7}$$

Now, in the language of gauge theories, the $U(1)$ gauge symmetry is unbroken, and there is no Higgs condensate. It is essential to account for the fluctuations of the gauge field in a proper description of such a phase.

Let us examine the possibilities for the saddle point in the context of the mean-field Hamiltonian. Now, the Hamiltonian is generalized to

$$\begin{aligned} \overline{H}_{KH} = {} & \frac{MV|\overline{P}|^2}{J_K} + \sum_k \left[-\overline{P} f_{k\alpha}^\dagger c_{k\alpha} - \overline{P}^* c_{k\alpha}^\dagger f_{k\alpha} + \varepsilon_k c_{k\alpha}^\dagger c_{k\alpha} \right] \\ & + \frac{M\mathfrak{z}V|\overline{Q}|^2}{2J} - \overline{\lambda}\frac{NV}{2} + \sum_k \varepsilon_k^f f_{k\alpha}^\dagger f_{k\alpha}, \end{aligned} \tag{31.8}$$

where $\mathfrak{z}$ is the number of nearest neighbors of each lattice site, we have assumed a spatially uniform saddle-point value of $|Q_{ij}|$ equal to $\overline{Q}$, and the dispersion of the f fermions is given by

$$\varepsilon_k^f = \overline{\lambda} - \overline{Q} \sum_{a=1}^{3} e^{i\boldsymbol{k}\cdot\boldsymbol{e}_a}, \tag{31.9}$$

with $\boldsymbol{e}_a$ vectors connecting nearest-neighbor sites. The saddle-point equations (30.4) and (30.5) are supplemented by an additional equation for $\overline{Q}$

$$\overline{Q} = \frac{2J}{M3V} \sum_{\boldsymbol{k}} \sum_{a=1}^{3} e^{i\boldsymbol{k}\cdot\boldsymbol{e}_a} \left\langle f_{\boldsymbol{k}\alpha}^{\dagger} f_{\boldsymbol{k}\alpha} \right\rangle. \tag{31.10}$$

The equations (30.4), (30.5), and (31.10) can now be solved for the values of $\overline{\lambda}$, $\overline{Q}$, and $\overline{P}$. In the light of (31.6) and (31.7), we pay particular attention to the equation of $\overline{P}$. This can be written in the form (30.12), with the quasiparticle dispersions now given by

$$2E_k^{\pm} = \varepsilon_k + \varepsilon_k^f \pm \left[(\varepsilon_k - \varepsilon_k^f)^2 + 4|\overline{P}|^2 \right]^{1/2}. \tag{31.11}$$

As we are interested in the possibility of a solution like (31.7), let us take the $\overline{P} \to 0$ limit of (30.12), which can be written as

$$\frac{\overline{P}}{J} = \frac{\overline{P}}{V} \sum_{\boldsymbol{k}} \frac{f(\varepsilon_k) - f(\varepsilon_k^f)}{\varepsilon_k^f - \varepsilon_k}. \tag{31.12}$$

Now, the crucial point is that the summation on $\boldsymbol{k}$ in (31.12) is *finite* for a generic dispersion ε_k^f like that in (31.9). In the case $\overline{Q} = 0$ that we considered in Section 30.1, the f band is dispersionless, and then the summation on $\boldsymbol{k}$ in (31.12) is logarithmically divergent (see (30.12)). With $\overline{Q} \neq 0$, and the summation finite, the only solution possible for (31.12) for small J_K is $\overline{P} = 0$. Specifically, an FL* phase is obtained for $J_K < J_{Kc}$ with

$$\frac{1}{J_{Kc}} = \frac{1}{V} \sum_{\boldsymbol{k}} \frac{f(\varepsilon_k) - f(\varepsilon_k^f)}{\varepsilon_k^f - \varepsilon_k}. \tag{31.13}$$

The reader should notice a similarity to the "Stoner criterion" for spin density waves in (9.51), with an inverse interaction strength on the left-hand side, and a fermionic susceptibility on the right-hand side.

We conclude with a brief statement of the physical properties of the FL* phase. With $\overline{P} = 0$, the conduction electrons are decoupled from the Kondo spins at mean-field level. From (30.7), the conduction-electron Green's function is

$$G_{cc}(\boldsymbol{k}, i\omega_n) = \frac{1}{i\omega_n - \varepsilon_k}. \tag{31.14}$$

In comparison to (30.20), there is no shift of $1/(2d(0))$ in the Fermi energy to accommodate the f electrons, and so we have a *small* Fermi surface, as shown in Fig. 30.4a and Fig. 31.2. There is also no corresponding renormalization of the mass of the conduction electrons.

The f spinons are decoupled from the conduction electrons, and form an independent $U(1)$ spin liquid at mean-field level; several realizations of such spin liquids were studied in Parts II and IV, and a possible state with a spinon Fermi surface is illustrated in Fig. 31.2. Beyond the meanfield, the spinons interact with the emergent $U(1)$ gauge

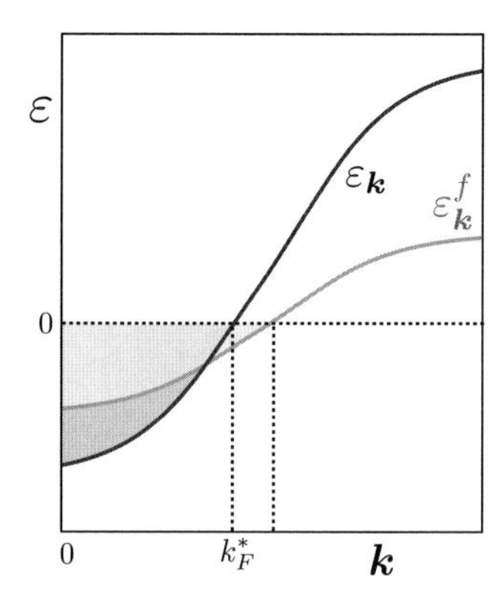

Figure 31.2 Schematic band structures in the FL* phase. There are electron-like quasiparticles in *only* the conduction band up to the wavevector k_F^*. In Fig. 30.4a we showed a perfectly flat f band, but now this acquires the dispersion in (31.9). The f band is occupied by fermionic spinons so that it is half filled (and obeys (30.5)), and decoupled from the conduction band.

field, and this has interesting consequences for the spin spectrum of the spin liquid, as also discussed in Parts II and IV. There are also fluctuations of the hybridization boson P_i about $\overline{P} = 0$, and this leads to some modification of the spinon spectrum from the presence of the conduction electrons. However, this coupling does not lead to a disappearance of the fractionalized spinon excitations, which are stabilized by their charges under the emergent $U(1)$ gauge field.

31.2 Emergent Gauge Fields and Generalized Luttinger Relations

We saw in Chapter 30 that the large-M theory of the HFL phase yielded a "large" Fermi surface corresponding to a density of $1+\rho_c$ electrons. Similarly, in Section 31.1, the large-M theory of the FL* phase yielded a "small" Fermi surface of conduction electrons alone, corresponding to a density of ρ_c. In this section we will show that the results hold to all orders in $1/M$, and also describe the connection to the discussion of the Luttinger relation in Section 30.2.

As we have already noted in Section 30.2, the analysis of the Anderson lattice model appears to leave no room for the small Fermi surface of the FL* phase. However, it must be kept in mind that the above analysis is perturbative in U_d, even though it holds to all orders in U_d.

To understand the FL* phase, we have to turn to the Kondo lattice model (31.1), and understand its $U(1)$ symmetries more carefully. The Kondo lattice model has a global $U(1)$ symmetry of electron-number conservation, which counts *only* the conduction electrons. However, it has an additional $U(1)_{gauge}$ symmetry specified by (31.3), associated with a fixed electron number on each d site. So the total symmetry of the Kondo

Table 31.1 Symmetry charges

Symmetry	f	c	P
$U(1)$	0	1	−1
$U(1)_{gauge}$	1	0	1
$U(1)_{diag}$	1	1	0

lattice model is $U(1) \times U(1)_{gauge}$. The fate of this enlarged symmetry is distinct in the two phases:

(i) HFL phase: the $U(1) \times U(1)_{gauge}$ symmetry is broken to a diagonal $U(1)_{diag}$ symmetry by the condensation of the Higgs boson P. Recall that $P \sim c^{\dagger}_{\alpha} f_{\alpha}$, and so P carries charges of both $U(1)$ and $U(1)_{gauge}$, associated with charges of $c^{\dagger}$ and f. However, the condensation of P leaves $U(1)_{diag}$ unbroken, under which c and f have the same charge:

$$\mathrm{U}_{diag} : c_{\alpha} \to c_{\alpha} e^{i\phi_d}, \quad f_{\alpha} \to f_{\alpha} e^{i\phi_d}. \tag{31.15}$$

The nature of these symmetries is summarized in Table 31.1. The arguments of Section 30.2 now imply that we can only deduce a Luttinger relation for the unbroken $U(1)_{diag}$ symmetry, which counts the number of both c and f fermions. In this manner, we obtain the large Fermi surface of the HFL phase, as already obtained in the Anderson lattice model.

(ii) FL* phase: the $U(1) \times U(1)_{gauge}$ symmetry remains unbroken. The arguments of Section 30.2 now imply that there should be two separate Luttinger relations associated with these two symmetries. Only the c fermions carry the global $U(1)$ charge, and so the usual Luttinger arguments imply a small Fermi surface, as discussed further below. The Luttinger relations also apply to $U(1)_{gauge}$, symmetry and lead to constraints on the spinon excitation structure of the spin liquid. As an example, in the π flux discussed in Section 22.1, the Luttinger relation leads to the presence of massless Dirac fermions. Moreover, there are also Luttinger-like constraints on FL* states with a $\mathbb{Z}_2$ spin liquid; the relations discussed in Section 15.4.2 are intimately connected to the Luttinger relation, as discussed in Section 31.3.2.

Finally, a few further comments on the stability of the small Fermi surface, and the Luttinger (i.e., Luttinger*) relation that applies in the FL* phase. We can compute the conduction-electron Green's function in a $1/M$ expansion of the gauge theory about the $\overline{Q} = 0$ saddle point, and the results can be written as

$$G_{cc}(\boldsymbol{k}, i\omega_n) = \frac{1}{i\omega_n - \varepsilon_{\boldsymbol{k}} - \Sigma_{cc}(\boldsymbol{k}, \omega_n)}. \tag{31.16}$$

As long as we are expanding about the $\overline{Q} = 0$ saddle point, it is not difficult to see that the self-energy in (31.14) is obtained from a Luttinger–Ward functional, and so satisfies

$$\sum_{k} \int_{-\infty}^{\infty} \frac{d\omega}{2\pi} G_{cc}(\boldsymbol{k}, i\omega) \frac{\partial}{\partial \omega} \Sigma_{cc}(\boldsymbol{k}, i\omega) = 0. \tag{31.17}$$

Then, proceeding in the usual route, we obtain the Luttinger* relation

$$\rho_c = \frac{1}{V} \sum_{k} \theta(-\varepsilon_{\boldsymbol{k}}^*), \tag{31.18}$$

with $\varepsilon_{\boldsymbol{k}}^* = \varepsilon_{\boldsymbol{k}} + \Sigma_{cc}(\boldsymbol{k}, 0)$; contrast this to the usual Luttinger relation in (30.46) for the large Fermi surface.

It is also interesting to compare the conduction-electron Green's function in (31.14) with that in (30.7). We can consider the f-electron contribution in (30.7), equal to $|\overline{P}|^2/(i\omega_n - \overline{\lambda})$, as a conduction-electron "self-energy": this self-energy diverges at zero frequency for $\overline{\lambda} = 0$, leading to zeros of the Green's function, which have been the focus of some attention in the literature [7, 30, 61, 64, 70, 147, 243, 274, 321]. We see from our treatment that such zeros are resolved [251] in two possible ways:

(i) In the HFL phase with $\overline{P} \neq 0$, we have $\overline{\lambda}$ non-zero, and given by $\overline{\lambda} = 2d(0)|\overline{P}|^2$ in (30.17). It is this non-zero $\overline{\lambda}$ that leads to the shift in the apparent Fermi surface $\varepsilon_{\boldsymbol{k}} = 0$ to the actual large Fermi surface $\varepsilon_{\boldsymbol{k}} = |\overline{P}|^2/\overline{\lambda}$, as in (30.20).

(ii) In the FL* phase with $\overline{P} = 0$, this apparently divergent contribution to the self-energy is absent, and the actual self-energy, which is Σ_{cc} in (31.16), can be obtained from a Luttinger–Ward functional. Now there is a stable small Fermi surface at $\varepsilon_{\boldsymbol{k}} = 0$ (with $1/M$ fluctuation corrections, at $\varepsilon_{\boldsymbol{k}}^* = 0$) obeying the Luttinger* relation in (31.18).

31.3 Torus Flux Insertion and Generalized Luttinger Relations

Our discussions of the Luttinger and the Luttinger* relations have so far been perturbative, although they expand about different starting points, some of which have non-trivial correlations built in. We employed the U_d expansion of the Anderson lattice model, the J_K expansion of the Kondo lattice model, and the $1/M$ expansions about two different saddle points of the Kondo lattice model. Here, we sketch a non-perturbative approach to deriving these relations. First, we derive the Luttinger relation of the Fermi liquid as obtained by Oshikawa [193], and then turn in Section 31.3.2 to the Luttinger* relation of the FL* phase.

31.3.1 Fermi Liquid

Consider an arbitrary quantum system, of bosons or fermions, defined on (say) a square lattice of unit lattice spacing, and placed on a torus. The size of the lattice is $L_x \times L_y$, and we impose periodic boundary conditions. Assume the system has a global $U(1)$ symmetry, and all the local operators carry integer $U(1)$ charges. We

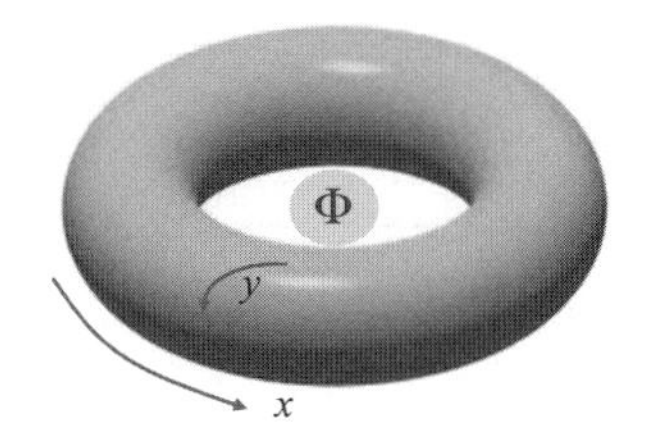

Figure 31.3 Torus geometry with a flux quantum inserted.

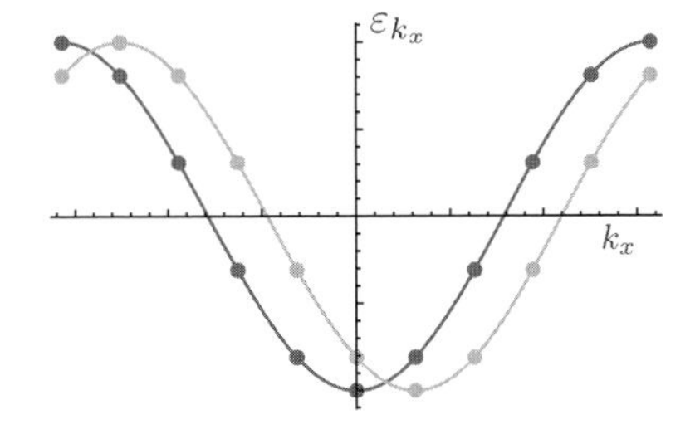

Figure 31.4 Dispersion of a free particle with momentum k_x in a system of size $L_x = 10$. The two plots indicate the allowed values of ε_{k_x} before and after 2π flux insertion. Note that the allowed values of ε_{k_x} coincide, after the shift (31.20).

pick an eigenstate of the Hamiltonian (usually the ground state) $|G\rangle$. Because of the translational symmetry, this state will obey

$$\hat{T}_x |G\rangle = e^{iP_x} |G\rangle, \tag{31.19}$$

where $\hat{T}_x$ is the lattice translational operator by one lattice spacing along the x direction, and P_x is the momentum of the state $|G\rangle$. Note that P_x is only defined modulo 2π. The state $|G\rangle$ will also have a definite total $U(1)$ charge, which we denote by the integer N.

Now we gauge the global $U(1)$ symmetry, and insert one flux quantum (with flux 2π) through one of the cycles of the torus (see Fig. 31.3). Let us consider the consequences for a single particle with crystal momentum $-\pi \leq k_x \leq \pi$ along the x direction with dispersion ε_{k_x}. In the presence of a flux Φ, this dispersion will change to $\varepsilon_{k'_x}$ where (see Fig. 31.4)

$$k'_x = k_x - \frac{\Phi}{L_x}. \tag{31.20}$$

We note that k_x is quantized in integer multiples of $2\pi/L_x$, and so an insertion of flux $\Phi = 2\pi$ yields a system that is gauge equivalent to $\Phi = 0$. Applying the same argument to a non-interacting many-body system, we deduce its crystal momentum P'_x will differ from P_x by ΔP_x with

$$\Delta P_x = \frac{2\pi}{L_x} N \,(\mathrm{mod}\, 2\pi). \tag{31.21}$$

Now, we turn on the interactions between the particles: these cannot change the total momentum, which is conserved (modulo 2π) both by the interactions and the flux insertion; so (31.21) applies also in the presence of interactions.

This argument was somewhat cavalier, so let us derive (31.21) a bit more carefully for a general many-body system. The initial and final Hamiltonians of the flux insertion process are related by a gauge transformation

$$\mathcal{U}_g H_f \mathcal{U}_g^{-1} = H_i \quad , \quad \mathcal{U}_g = \exp\left(i\frac{2\pi}{L_x} \sum_i x_i \hat{n}_i \right) , \tag{31.22}$$

where $\hat{n}_i$ is the integer number operator of the $U(1)$ symmetry. while the wavefunction evolves from $|G\rangle$ to $\mathcal{U}_T |G\rangle$, where $\mathcal{U}_T$ is the time-evolution operator. We want to work in a fixed gauge in which the initial and final Hamiltonians are the same; in this gauge, the final state is $|G'\rangle = \mathcal{U}_G \mathcal{U}_T |G\rangle$. Then we can establish (31.21) using the definitions

$$\hat{T}_x |G\rangle = e^{-iP_x} |G\rangle \quad , \quad \hat{T}_x |G'\rangle = e^{-iP'_x} |G'\rangle , \tag{31.23}$$

and the easily established properties

$$\hat{T}_x \mathcal{U}_T = \mathcal{U}_T \hat{T}_x \quad , \quad \hat{T}_x \mathcal{U}_g = \exp\left(-i2\pi \frac{N}{L_x} \right) \mathcal{U}_g \hat{T}_x . \tag{31.24}$$

So far, we have been quite general, and not specified anything about the many-body system, apart from its translational invariance and global $U(1)$ symmetry. In the subsequent discussion, we make further assumptions about the nature of the ground state and low-lying excitations, and compute ΔP by other methods. Equating such a result to (31.21) will then lead to important constraints on the allowed structure of the many-body ground state.

First, we assume the ground state is a Fermi liquid. So its only low-lying excitations are fermionic quasiparticles around the Fermi surface. For our subsequent discussion, it is important to also include the electron spin index, $\alpha = \uparrow, \downarrow$, and so we will have a Fermi liquid with two global $U(1)$ symmetries, associated respectively with the conservation of electron number and the z component of the total spin S_z. Consequently, there are two Luttinger theorems, one for each global $U(1)$ symmetry. The action for the fermionic quasiparticles $c_{\boldsymbol{k}\alpha}$, with dispersion $\varepsilon(\boldsymbol{k})$, is

$$\mathcal{S}_{FL} = \int d\tau \int \frac{d^2k}{4\pi^2} \sum_{\alpha=\pm 1} c^\dagger_{\boldsymbol{k}\alpha} \left(\frac{\partial}{\partial \tau} - \frac{i}{2} \alpha A^s_\tau - i A^e_\tau + \varepsilon(\boldsymbol{k} - \alpha \boldsymbol{A}^s/2 - \boldsymbol{A}^e) \right) c_{\boldsymbol{k}\alpha} , \tag{31.25}$$

where τ is imaginary time and the gauge coupling $\alpha = \pm 1$. The Fermi surface is defined by $\varepsilon(\boldsymbol{k}) = 0$, and $\mathcal{S}_{FL}$ only applies for $\boldsymbol{k}$ near the Fermi surface, although we have (for notational convenience) written it in terms of an integral over all $\boldsymbol{k}$. We have also coupled the quasiparticles to two probe gauge fields $A^e_\mu = (A^e_\tau, \boldsymbol{A}^e)$ and $A^s_\mu = (A^s_\tau, \boldsymbol{A}^s)$, which couple to the two conserved $U(1)$ currents associated, respectively, with the conservation of electron number and S_z.

We place the Fermi liquid on a torus, and insert a 2π flux of a gauge field that couples only to the spin-up electrons. So we choose $A^s_\mu = 2A^e_\mu \equiv A_\mu$. Then, the general momentum balance in (31.21) requires that

$$\Delta P_x = \frac{2\pi}{L_x} N_\uparrow \,(\text{mod}\, 2\pi) = \frac{2\pi}{L_x} \frac{N}{2} \,(\text{mod}\, 2\pi), \tag{31.26}$$

where we assume equal numbers of spin-up and -down electrons $N_\uparrow = N_\downarrow = N/2$.

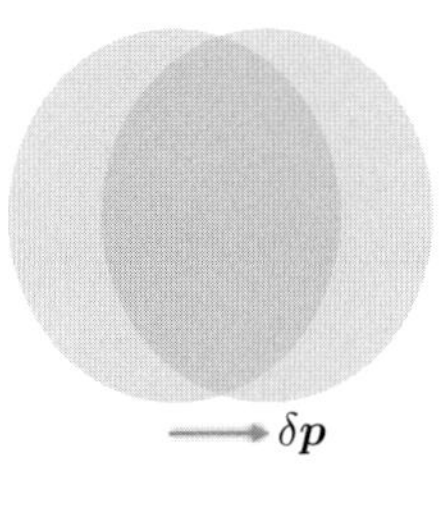

Figure 31.5 Response of a Fermi liquid to flux insertion. The shaded circles represent states occupied by the quasiparticles inside the Fermi surface, before and after the flux insertion. Each quasiparticle near the Fermi surface acquires a momentum shift $\delta \boldsymbol{p} = (\delta p_x, 0)$. The total change in momentum is equal to the difference in the total momenta between the occupied regions within the two Fermi surfaces. This equality assumes quasiparticles exist at all momenta, but this is permissible because the net contribution arises only from the regions near the Fermi surface, where the quasiparticles do exist.

Now, we compute the momentum balance *assuming* that the only low-energy excitations are quasiparticles near the Fermi surface described by $\mathcal{S}_{FL}$, and these react like free particles to a sufficiently slow flux insertion. So each quasiparticle picks up a momentum

$$\delta \boldsymbol{p} \equiv \left(\frac{2\pi}{L_x}, 0 \right) \tag{31.27}$$

(see Fig. 31.5), and then we can write (with δn_p the quasiparticle density excited by the flux insertion)

$$\Delta P_x = \sum_p \delta n_p p_x .$$

Now $\delta n_p = \pm 1$ on a shell of thickness $\delta \boldsymbol{p} \cdot d\boldsymbol{S}_p$ on the Fermi surface (where $d\boldsymbol{S}_p$ is an area element on the Fermi surface). So we can write the above as a surface integral

$$\begin{aligned} \Delta P_x &= \oint_{FS} p_x \left(\frac{L_x L_y}{4\pi^2} \right) \delta \boldsymbol{p} \cdot d\boldsymbol{S}_p \\ &= (\delta \boldsymbol{p} \cdot \hat{x}) \int_{FV} \left(\frac{L_x L_y}{4\pi^2} \right) dV \end{aligned}$$

by the divergence theorem. So

$$\Delta P_x = \frac{2\pi}{L_x} \left(L_x L_y \frac{V_{FS}}{4\pi^2} \right), \tag{31.28}$$

where V_{FS} is the momentum-space area enclosed by the Fermi surface; the factor within the brackets on the right-hand side equals the number of momentum-space points inside the Fermi surface. Note that the entire contribution to the right-hand side of (31.28) comes from the vicinity of the Fermi surface, where the quasiparticles are well-defined; we have merely used a mathematical identity to convert the result to equal the volume, and we are not assuming the existence of quasiparticles far from the Fermi surface.

Now we equate (31.26) and (31.28), along with a corresponding argument along the y direction, and obtain

$$N_\uparrow - L_x L_y \frac{V_{FS}}{4\pi^2} = L_x m_x \quad , \quad N_\uparrow - L_x L_y \frac{V_{FS}}{4\pi^2} = L_y m_y \tag{31.29}$$

for some integers m_x, m_y. By choosing L_x, L_y mutually prime integers we can now show [193, 195]

$$\frac{N_\uparrow}{L_x L_y} = \frac{V_{FS}}{4\pi^2} + m \tag{31.30}$$

for some integer m. This is the Luttinger relation, obtained earlier in (30.33) and (30.46) for $N_\uparrow = N_\downarrow = N/2$.

31.3.2 $\mathbb{Z}_2$ Spin Liquid

We turn now to a non-perturbative discussion of the Luttinger* relation in the FL* phase. We apply the same momentum balance argument by placing the Anderson or Kondo lattice system on the torus in Fig. 31.3. From Fig. 31.1, the small Fermi surface obeying the Luttinger* relation (31.18) contributes a momentum related to the density ρ_c of the conduction electrons only. To satisfy the general relation (31.21), we now need to establish that the spin liquid produces a contribution to the momentum balance equation that is equivalent to a density of one electron per site. We will now establish this for the case of a gapped $\mathbb{Z}_2$ spin liquid, which then constitutes a non-perturbative argument for the Luttinger* theorem [32, 68, 195, 262].

Is convenient to formulate the theory of the $\mathbb{Z}_2$ spin liquid using the $U(1) \times U(1)$ Chern–Simons gauge theory discussed in Chapter 17. We show in Section 17.1.1 that the $\mathbb{Z}_2$ spin liquid is a theory with two gauge fields, a^1_μ and a^2_μ, whose torus line operators in (17.14) obeyed the anti-commutation relations in (17.15), also found for the $\mathbb{Z}_2$ gauge theory in (16.16); we reproduce these here for clarity:

$$\begin{gathered} W_i = \exp\left(i \int_{C_i} a^1_\mu dx_\mu\right), \quad V_i = \exp\left(i \int_{C_i} a^2_\mu dx_\mu\right), \\ W_x V_y = -V_y W_x, \quad W_y V_x = -V_x W_y, \end{gathered} \tag{31.31}$$

where $C_{x,y}$ are contours that encircle the contours of the torus. The bosonic spinons carry unit charge under a^1_μ, while the visons carry unit charge under a^2_μ, as specified by (17.32). We account for the fact that the spinons carry spin $S_z = \pm 1/2$ by the coupling to the external field A^s_μ via the charge vector in (17.36). These considerations yield the Chern–Simons theory

$$\mathcal{S}_{CS} = \int d^2x d\tau \left[\frac{i}{\pi} \varepsilon_{\mu\nu\lambda} a^2_\mu \partial_\nu a^1_\lambda + \frac{i}{2\pi} \varepsilon_{\mu\nu\lambda} A^s_\mu \partial_\nu a^2_\lambda \right]. \tag{31.32}$$

We also need to supplement such a topological field theory by information on the action of translational symmetry. The needed information is obtained in (16.18), and is closely connected to the fact that the visons can accumulate a phase factor upon

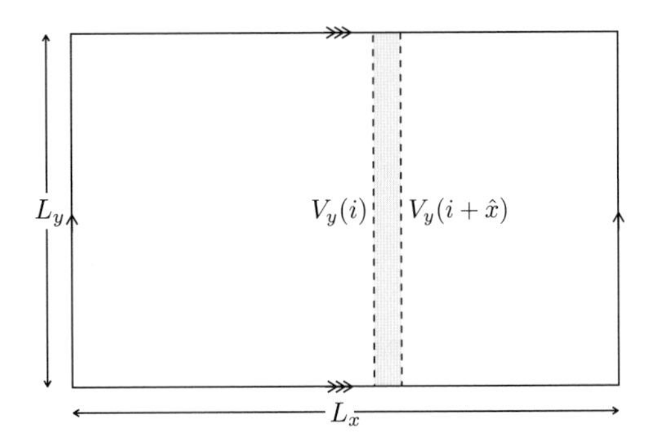

Figure 31.6 Square lattice on a torus. The torus line operator V_y (defined in (17.14)) is translated by one lattice spacing in the $\hat{x}$ direction. Compare with Fig. 16.6.

encircling any site of the lattice, as illustrated in Figs. 16.9 and 15.6, and discussed in Section 15.4.2. We reproduce the required result here, and also illustrate it in Fig. 31.6:

$$T_x V_y = e^{2\pi i S L_y} V_y T_x, \tag{31.33}$$

and there is a second relation with $x \leftrightarrow y$. Here, S is the on-site spin of the underlying spin liquid.

We have now recalled all the information needed to apply the momentum balance argument to the $\mathbb{Z}_2$ spin liquid. The general results in (31.21) and (31.26), describing flux insertion through the cycle of torus, apply to any lattice quantum system with a global $U(1)$ symmetry, and so should also apply to the $\mathbb{Z}_2$ spin liquid. We now show, using (31.33), that (31.21) and (31.26) are indeed satisfied.

As in Section 31.3.1, we insert a flux, Φ, which couples only to the spin-up electrons, which requires choosing $A^s_\mu = 2A^e_\mu \equiv A_\mu$. We work in real time, and thread a flux along the x cycle of the torus. So we have

$$A_x = \frac{\Phi(t)}{L_x}, \tag{31.34}$$

where $\Phi(t)$ is a function that increases slowly from 0 to 2π. In (31.32), the A_x gauge field couples only to a^2_y, and we parameterize

$$a^2_y = \frac{\theta_y}{L_y}. \tag{31.35}$$

Then, from (31.32), the time-evolution operator of the flux-threading operation can be written as

$$\hat{U} = \exp\left(\frac{i}{2\pi}\int dt\, \theta_y \frac{d\Phi}{dt}\right) = e^{i\theta_y} \equiv V_y. \tag{31.36}$$

So the time-evolution operator is simply the torus line operator V_y when acting upon the nearly degenerate topological states of the $\mathbb{Z}_2$ spin liquid on the torus. If the state of the system before the flux threading was $|G\rangle$, the state after the flux threading will be $V_y|G\rangle$. This is illustrated in Fig. 31.7.

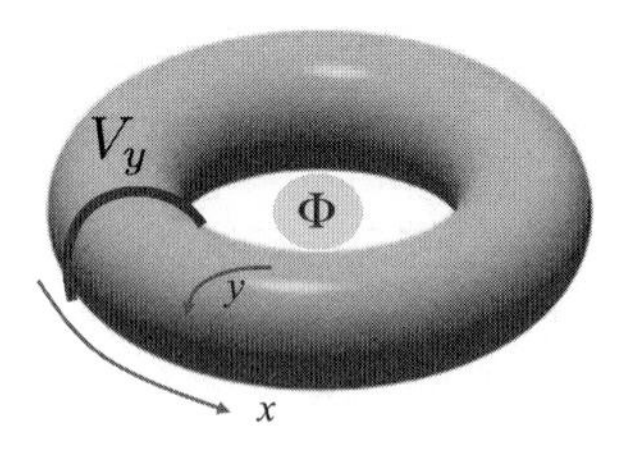

Figure 31.7 For a $\mathbb{Z}_2$ spin liquid, the insertion of a 2π flux in A_x^s is equivalent to the operator V_y acting on the nearly degenerate topological states.

Now we can easily determine the difference in momenta of the states $|G\rangle$ and $V_y|G\rangle$. From (31.33) we obtain

$$\Delta P_x = 2\pi S L_y \,(\mathrm{mod}\, 2\pi) = \frac{2\pi}{L_x}\,(S L_x L_y)\,(\mathrm{mod}\, 2\pi). \tag{31.37}$$

In the second form above, we see that (31.37) is consistent with (31.26) for $N_\uparrow = S L_x L_y$. This is indeed the correct total number of spin-up electrons in a spin-S antiferromagnet. We also note that these results above are closely connected to the relations in (16.19).

Given the abstractness of the above discussion using the Chern–Simons theory, it is useful to present another argument [262] for the relationship illustrated in Fig. 31.7 using the explicit resonating-valence-bond (RVB) wavefunctions described in Sections 13.1 and 13.2. This uses a computation by Bonesteel [33], and also illustrates the close connection to the Lieb–Schultz–Mattis theorem in one dimension [159]. The unitary operator performing the flux insertion of A_x^s from (31.22) is

$$\mathcal{U}_s = \exp\left(i \frac{2\pi}{L_x} \sum_i x_i \hat{S}_{zi} \right), \tag{31.38}$$

where $\hat{S}_{zi}$ is the z component of the spin operator on site i. When acting on any of the dimer components of the RVB wavefunction in (13.1), we have (note, here δ is used as a label for a dimer covering, instead of i in (13.1)) [33],

$$\mathcal{U}_s |D_\delta\rangle = \prod_{d \,\in\, \mathrm{dimers}} (-1)^{\gamma_d} \left[\cos(\Theta_d/2) + 2i \sin(\Theta_d/2) \hat{S}_{zd}\right] |D_\delta\rangle\,, \tag{31.39}$$

where the product is over all dimers in the covering D_δ, $\Theta_d = 2\pi \ell_d / L_x$ with ℓ_d the x component of the length of dimer d, and $\hat{S}_{zd}$ is the z component of the spin on the rightmost site of dimer d. The crucial factor in (31.39) is $(-1)^{\gamma_d}$, which represents the action of V_y on the RVB wavefunction, in a manner similar to Fig. 13.2: $\gamma_d = 1$ for dimers that "cut" a vertical line between the sites with $x = 1$ and $x = L_x$ (this is a vertical line analogous to the horizontal lines in Fig. 13.2), and $\gamma_d = 0$ otherwise. Upon arguing that the factor in square brackets in (31.39) becomes unity in the limit $L_x \to \infty$, when acting on the nearly degenerate topological states on the torus, we obtain the required mapping of Fig. 31.7.

Finally, we note that these arguments on flux insertions are closely connected to the action of translations on the topological ground states on the torus, as discussed near (16.19).

31.4 The FL* State in the Single-Band Hubbard Model

The general phenomenology of the transition from the HFL state to the FL* state in the Kondo lattice model is of a transition from a metal with $1+p$ electronic charge carriers to another metal with p charge carriers. This phenomenology connects to many observations on the heavy-fermion compounds, whose microscopic electronic structure is well described by various forms of the Kondo lattice Hamiltonian. However, remarkably, such a phenomenology is also a good match to observations on the hole-doped cuprates. At first sight, this is quite surprising, because it is not reasonable to describe the cuprates by a Kondo-lattice-type model, that is, there is no natural identification of electrons that can localize in the analog of an f band, and *all* electrons reside in a single band crossing the Fermi level, as is clearly observed in photoemission experiments. This section presents a simple approach to constructing an FL* state in a single-band model, while treating all electrons in this band democratically.

We return to the electron Hubbard model of Chapter 9, and consider the case of the square lattice with electronic density $1-p$, as is relevant to the cuprates doped with holes of density p. At large U, the analog of the HFL state of the Kondo lattice is now the "vanilla metal" state presented in (9.29) [10]. This starts with a wavefunction of free electrons in the single band, and projects out all doubly occupied sites, which are not present as $U \to \infty$. This projection can be treated in a gauge-theoretic framework quite similar to that followed for spin liquids, as we see below in Section 31.4.2. In the simplest approach, we introduce an emergent $U(1)$ gauge field, along with particles carrying charges of the emergent gauge field; these are the spinons f_α of Section 30.1, and also bosonic, spinless "holons" b, which carry the electronic charge. The vanilla metal state is obtained when the b holons condense, and this higgses out the $U(1)$ gauge field. Consequently, the gauge fluctuations are relatively innocuous, and so is the projection operation in (9.29). So we can conclude that the vanilla state has a "large" Fermi surface for all values of p, corresponding to an enclosed volume of $1+p$ holes or $1-p$ electrons. This is not compatible with observations at small p, which clearly show the disappearance of the large Fermi surface state at small p.

This discussion highlights the need for a theory of the FL* state in the Hubbard model at small p. There have been numerous discussions of such states in the literature, by various groups [38, 75, 176, 221, 286, 305, 321], including several by the author and collaborators [128, 185, 210, 211, 212, 227, 232, 240, 251]. I begin our discussion by presenting a simple physical picture of Ref. [211], along the lines of the discussion of gapped spin liquids and RVB states in Chapter 13. We start with the RVB state illustrated in Fig. 13.1b, and remove a density p of electrons to obtain the state shown in Fig. 31.8a. The resonance between the valence bonds can now allow processes in

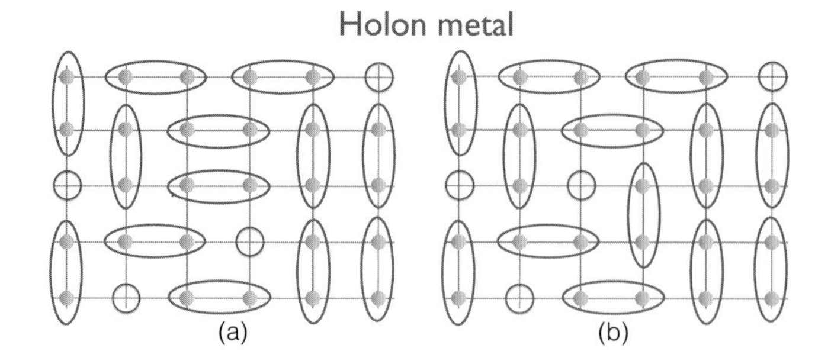

Figure 31.8 (a) Doped spin liquid obtained by removing a density p of electrons from the RVB state in Fig. 13.1b. (b) Resonance between the valence bonds leads to the motion of the vacancy in the center of the figure. The mobile vacancy is a "holon," carrying unit charge but no spin. If the holons have fermionic statistics, such a mobile holon state can realize a holon metal. Only nearest-neighbor valence bonds are shown for simplicity.

which the vacant sites can move, as shown in Fig. 31.8b. As this process now transfers physical charge, the resulting state can be expected to be an electrical conductor. A subtle computation is required to determine the quantum statistics obeyed by the mobile vacancies, but depending upon the parameter regimes, it can be either bosonic or fermionic [139, 215]. Assuming fermionic statistics, we have the possibility that the vacancies will form a Fermi surface, realizing a metallic state. Note that the vacancies do not transport spin, and such spinless charge carriers are often referred to as "holons"; the metallic state we have postulated is a holon metal. The low-energy quasiparticles near the Fermi surface of the holon metal are also holons, carrying unit electrical charge but no spin. Consequently, such quasiparticles are not directly observable in photoemission experiments, which necessarily eject bare electrons with both charge and spin. As low-energy electronic quasiparticles are observed in photoemission studies, the holon metal is not favored as a candidate for the pseudogap state of the cuprates.

To obtain a spinful quasiparticle, we clearly have to attach an electronic spin to each holon. And, as shown in Fig. 31.9, it is not difficult to imagine conditions under which this might be favorable. (i) We break density $p/2$ valence bonds into their constituent spins (Fig. 31.9a); this costs some exchange energy for each valence bond broken. (ii) We move the constituent spins ("spinons") into the neighborhood of the holons. (iii) The holons and spinons form a bound state (Fig. 31.9b), which has both charge $+e$ and spin $S = 1/2$, the same quantum numbers as (the absence of) an electron; this bound-state formation gains energy that can offset the energy cost of (i). We now have a modified RVB state [211], like that in (13.1), but with $|D_i\rangle$ consisting of pairing of sites of the square lattice with two categories of "valence bonds": the elliptical and rectangular dimers in Fig. 31.9b. The first class (elliptical) are the same as the electron singlet pairs found in the Pauling–Anderson RVB state. The second class (rectangular) consists of a single electron resonating between the two sites at the ends of the bond. From their constituents, it is clear that relative to the insulating RVB state, the eilliptical dimers are spinless, charge-neutral bosons, while the rectangular dimers are

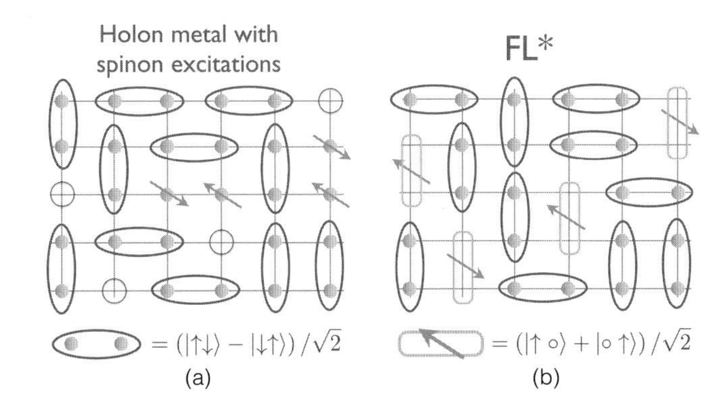

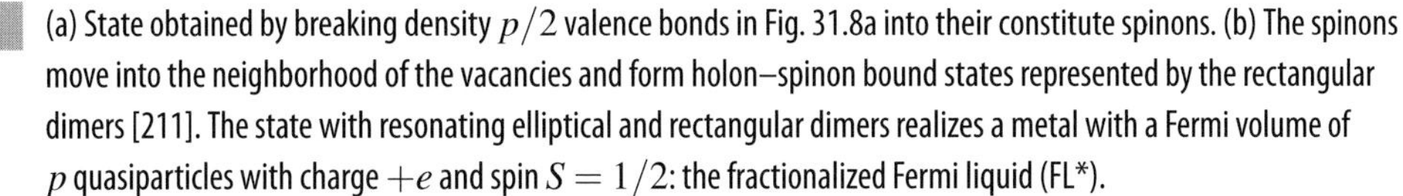
Figure 31.9 (a) State obtained by breaking density $p/2$ valence bonds in Fig. 31.8a into their constitute spinons. (b) The spinons move into the neighborhood of the vacancies and form holon–spinon bound states represented by the rectangular dimers [211]. The state with resonating elliptical and rectangular dimers realizes a metal with a Fermi volume of p quasiparticles with charge $+e$ and spin $S = 1/2$: the fractionalized Fermi liquid (FL*).

spin $S = 1/2$, charge $+e$ fermions. Evidence that the states associated with the elliptical and rectangular dimers dominate the wavefunction of the lightly doped cuprates appears in cluster dynamical mean-field studies [80, 272]. Both classes of dimers are mobile, and the situation is somewhat analogous to ^{4}He–^{3}He mixture. Like the ^{3}He atoms, the rectangular fermions can form a Fermi surface, and an extension of the Luttinger argument to the present situation shows that the Fermi volume is exactly p [210, 259, 262]. However, unlike the ^{4}He–^{3}He mixture, superfluidity is not immediate, because of the close-packing constraint on the elliptical + rectangular dimers; onset of superfluidity will require pairing of the rectangular dimers. So the state obtained [211] by the resonating motion of the dimers in Fig. 31.9b is precisely the FL* metal: it has a Fermi volume of p, with well-defined electron-like quasiparticles near the Fermi surface.

31.4.1 Paramagnon Fractionalization Wavefunction for the FL* State

We can view the description of Fig. 31.9b [211] as a trial wavefunction for the FL* metal. However, it is restricted to a particular class of spin liquids in which the valence bonds and the spinon–holon bound state are short-ranged – the actual situation in the cuprates is almost surely far from this limit. Also, it is not clear from such a construction how we may develop a more complete theory for a transition to the Fermi liquid state, as described by the vanilla wavefunction in (9.29).

I now present a trial wavefunction, based upon the idea of introducing "ancilla" or "hidden" qubits [173, 191, 324, 325], that achieves these objectives. This approach relies upon an important lesson obtained from the theory for the Kondo lattice: we should not fractionalize the mobile electron. For the FL* phase of the Kondo lattice, we only fractionalize the immobile spins in the f band. For the single-band Hubbard

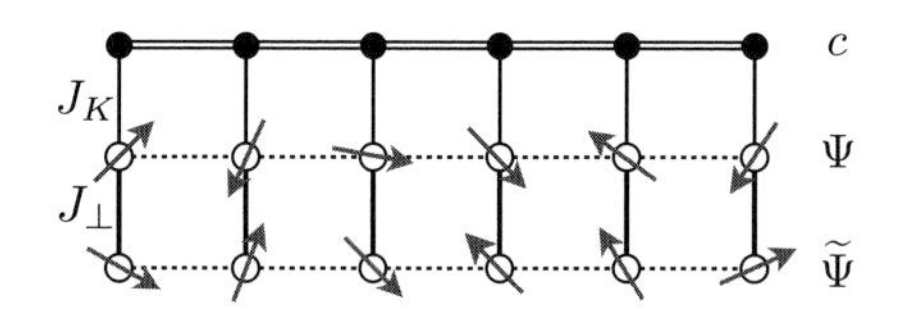

Figure 31.10 The paramagnon rotors in Fig. 9.6 are realized by a pair of ancilla (hidden) qubits (spin-1/2 spins) represented by Schwinger fermions Ψ and $\widetilde{\Psi}$. The antiferromagnetic exchange couplings J_K and $J_\perp$, are indicated and the dashed lines represent exchange interactions within the Ψ and $\widetilde{\Psi}$ layers.

model, a very common approach, which we noted above, is to use the point of view of a doped spin liquid: we begin with a spin liquid at half filling with fermionic spinons f_α, and represent the doping by bosonic holons b. The fractionalization is driven by the exchange interaction J, but there is a stronger counter-effect to reconstructing the electron by the large hopping t, which acts as an attractive potential between the f_α and b. In other words, if we fractionalize the electron with

$$c_\alpha = f_\alpha b^\dagger \text{ or } f b_\alpha^\dagger, \tag{31.40}$$

as in Fig. 21.4, we will have to include fluctuation corrections that bind each holon to a spinon, leading to a small Fermi surface of electrons. Assuming the appearance of such bound states, Ref. [240] presents a theory of interacting electrons and bosonic spinons which yields a pseudogap metal and a quantum phase transition to a Fermi liquid with a large Fermi surface, but I do not describe this theory here.

Instead, we let us turn to the representation of the Hubbard model as a theory of free electrons coupled to a lattice of paramagnon rotors, described near Fig. 9.6. The main new idea is that we should *fractionalize the paramagnon*, which is charge neutral, so that mobile charges are not fractionalized. First, we restrict attention to only the $\ell = 0, 1$ angular momentum states of each rotor in (9.56). We can represent these singlet and triplet states by a pair of $S = 1/2$ spins coupled with an antiferromagnetic exchange coupling $J_\perp$; these are the required ancilla qubits, as illustrated in Fig. 31.10 in which the top physical layer of electrons c, of density $1 - p$, is coupled to two layers of ancilla qubits, replacing the paramagnon rotors in Fig. 9.6. We use a Schwinger fermion representation of ancilla or hidden qubits, as in (29.15) for the Kondo model. So we introduce fermions Ψ with the constraint

$$\sum_a \Psi_{i;a}^\dagger \Psi_{i;a} = 1, \tag{31.41}$$

satisfied on each lattice site i ($a = \pm$ is a pseudospin index) to represent the first layer of ancilla qubits; similarly, we introduce fermions $\widetilde{\Psi}$ to represent the second layer of ancilla qubits. It is important that we add *two* layers of ancilla qubits, because only then are the added layers allowed to form a trivial insulator. (Some earlier descriptions of the FL* phase [185, 212] were obtained by adding a single ancilla band near half filling: this gives a suitable description of the electron spectral function in the FL* phase, but these approaches are difficult to extend into the FL phase.)

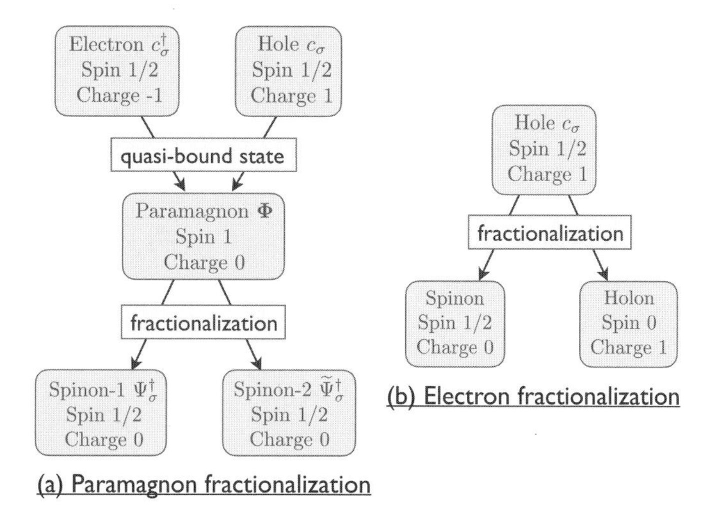

Figure 31.11 (a) The fractionalization of Section 31.4.1. The paramagnon is a quasi-bound state with a finite lifetime because it can decay to a particle–hole pair. (b) The fractionalization of (31.40) and Fig. 21.4. This is not preferred because the holon of Fig. 31.8 is not stable for $t \gg J$, and has not been observed. Note that a mobile charge carrier is not fractionalized in (a), but is fractionalized in (b).

We note, in passing, that we have now complemented each electron c_i with two ancilla fermions Ψ_i and $\widetilde{\Psi}_i$. This $1 \Rightarrow 3$ fermion replacement is similar to that in the successful parton theory of fractional quantum Hall states, as discussed in Section 21.3 and Chapter 19. However, note that here the original electron operator is not fractionalized; rather, we have fractionalized the paramagnon into two ancilla fermions. This fractionalization of collective spin excitations is closer in spirit to that in the Kondo lattice. Fig 31.11 presents a pictorial summary of the paramagnon fractionalization approach, and compares it to the conventional electron fractionalization approach of (31.40).

In the large Fermi surface FL phase, we assume that the non-random and antiferromagnetic coupling $J_\perp$ dominates, and so the ancilla spins are locked into rung singlets, and can be safely ignored in the low-energy theory; then, the c electrons form a conventional Fermi liquid phase, and we obtain a Fermi surface corresponding to electron density $1 - p$, or hole density $1 + p$, as shown in Fig. 31.12. The trial wavefunction for this state is essentially the vanilla wavefunction in (9.29). We also show a triplet paramagnon excitation, which is coupled to the large Fermi surface, just as in the conventional paramagnon theory.

To obtain the small Fermi surface FL* phase, we consider an alternative fate of the coupling of the mobile c electrons to the spins in the ancilla layers. Notice that the Kondo coupling J_K of the spins in the Ψ layer to the c electrons is antiferromagnetic, while the effective Kondo coupling of the spins in the $\widetilde{\Psi}$ layer to the c electrons (mediated by the intermediate Ψ layer) is ferromagnetic. From Fig. 29.5, we know that antiferromagnetic Kondo coupling flows to strong coupling, while the ferromagnetic Kondo coupling is irrelevant. Let us assume that after some renormlization-group flow

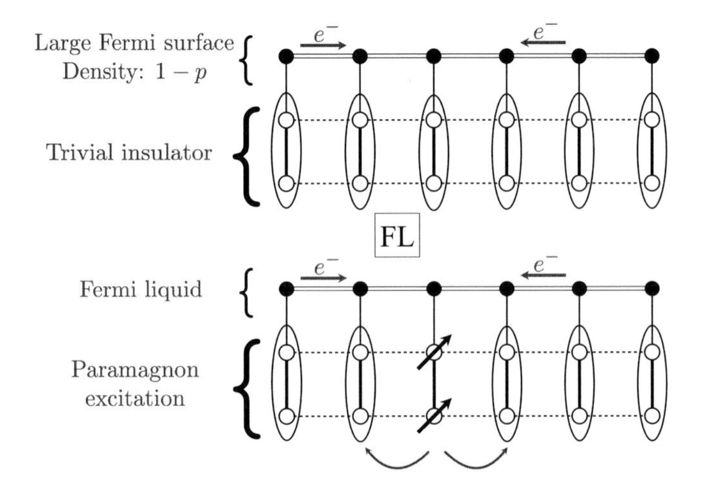

Figure 31.12 In the large Fermi surface FL phase, the ancilla spins lock into rung singlets, while the c electrons are largely decoupled from the ancilla spins, and form a conventional Fermi liquid of electron density $1 - p$ (or hole density $1 + p$). The bottom panel shows a propagating spin–triplet excitation which is the paramagnon.

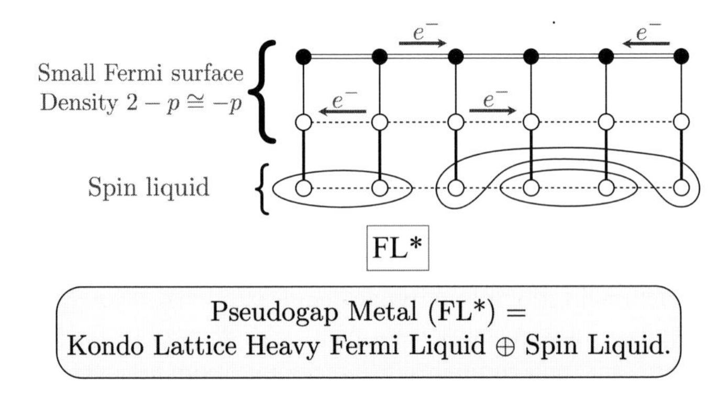

Figure 31.13 In the small Fermi surface FL* phase, the Ψ ancilla spins are Kondo screened by the c electrons to form a Fermi surface with density $2 - p$ electrons, similar to that in the HFL phase of the Kondo lattice. This is equivalent to a small hole-like Fermi surface of size p, consistent with observations in the cuprates at low doping. The $\widetilde{\Psi}$ spins are largely decoupled from the top two layers in the FL* phase, and form a gapless spin liquid with fractionalization, whose presence is required by the generalized Luttinger relation.

to strong coupling, J_K becomes even stronger than $J_\perp$. Then the Kondo effect will cause the Ψ spins to "dissolve" into the Fermi sea of the mobile electrons as shown in Fig. 31.13. By analogy with the HFL phase of the Kondo lattice model in Chapter 30, we conclude that the Fermi surface will correspond to an electron density of $1+(1-p) = 2-p$: this is a small Fermi surface of holes of density p. The second layer of $\widetilde{\Psi}$ spins are now effectively decoupled from the other layers, and we assume their mutual interactions cause them to form a spin liquid. In principle, the $\widetilde{\Psi}$ spin liquid could be any of the spin liquids studied in Parts II and IV, but let us assume for specificity that the spin liquid is one of those studied in Chapter 22. Then we can write down a trial wavefunction for the FL* phase in this theory of paramagnon fractionalization in the single-band Hubbard model [324], which can be viewed as a replacement for the the vanilla wavefunction in (9.29):

$$
\begin{aligned}
|\Phi_{FL*}\rangle = & \left[\text{Projection onto rung singlets of } \Psi, \widetilde{\Psi}\right] \\
& \bowtie \left|\text{Slater determinant of } (c, \Psi)\right\rangle \\
& \otimes \left|\text{Slater determinant of } \widetilde{\Psi}\right\rangle .
\end{aligned} \tag{31.42}
$$

Note that, after the projection onto rung singlets on the right-hand side of (31.42), $|\Phi_{FL*}\rangle$ is a wavefunction dependent only upon the physical c degrees of freedom in the single-band model under consideration. Such a theory yields a good description of photoemission observations in the pseudogap metal [173].

There is an interesting inversion here that is worth noting: at the mean-field level, the *small* Fermi surface FL* phase of the single-band paramagnon fractionalization model maps on to the *large* Fermi surface phase of the two-band Kondo lattice model, where it is the HFL phase of that model. At small doping p, we can refine this to the statement that the FL* phase of the single-band model maps onto a lightly doped Kondo insulator in a Kondo–Heisenberg lattice model. This correspondence, however, does not hold beyond the mean field: in the small Fermi surface FL* phase of the single-band model there are fractionalized spinon excitations arising from the $\widetilde{\Psi}$ ancilla spins, which are required by the generalized Luttinger relation. There are no fractionalized excitations in the large Fermi surface HFL phase of the two-band Kondo lattice model.

This discussion of the pseudogap metal as a FL* phase, can be summarized by the slogan displayed in Fig. 31.13: “Pseudogap Metal = Kondo Lattice Heavy Fermi Liquid + Spin Liquid”.

31.4.2 Gauge Symmetries

Apart from yielding the trial wavefunction in (31.42) for the FL* state, the paramagnon fractionalization approach also allows us to write down a gauge theory for the transition to the Fermi liquid state at large p. As in Parts II and IV, the structure of the gauge theory is dictated by the local constraints associated with the parton construction, followed by an analyses of possible Higgs fields whose condensates can break the gauge symmetry. We show here that the underlying gauge symmetry of the paramagnon fractionalization approach is [325] $(SU(2)_S \times SU(2)_1 \times SU(2)_2)/\mathbb{Z}_2$ (the subscripts are just labels, and do not refer to a Chern–Simons level). However, depending upon the question being addressed, only certain sectors of this rather large gauge group are needed. In the FL* phase, the $(SU(2)_S \times SU(2)_1)/\mathbb{Z}_2$ gauge symmetry is fully broken, and we need only keep track of the $SU(2)_2$ gauge symmetry of the spin liquid on the second ancilla layer. For the FL*-FL transition, the critical theory reduces to a $(SU(2)_S \times SU(2)_1)/\mathbb{Z}_2$ gauge theory in the presence of a decoupled $SU(2)_2$ spin liquid [325].

Let $\boldsymbol{S}_{i;1}$, $\boldsymbol{S}_{i;2}$ be the spin operators acting on the qubits in the two ancilla layers, where i is a lattice site (see Fig. 31.10). We apply the $SU(2)$ gauge-invariant parton representation of Section 22.4 to both layers. So we represent these spin operators with ancilla fermions $F_{i;\sigma}$, $\widetilde{F}_{i;\sigma}$ via

$$
\boldsymbol{S}_{i;1} = \frac{1}{2} F_{i;\sigma}^{\dagger} \boldsymbol{\sigma}_{\sigma\sigma'} F_{i;\sigma'} \quad , \quad \boldsymbol{S}_{i;2} = \frac{1}{2} \widetilde{F}_{i;\sigma}^{\dagger} \boldsymbol{\sigma}_{\sigma\sigma'} \widetilde{F}_{i;\sigma'} , \tag{31.43}
$$

where $\boldsymbol{\sigma}$ are the Pauli matrices. We also define the Nambu pseudospin operators

$$\boldsymbol{T}_{i;1} = \frac{1}{2}\left(F^\dagger_{i;\downarrow}F^\dagger_{i;\uparrow} + F_{i;\uparrow}F_{i;\downarrow}, i\left(F^\dagger_{i;\downarrow}F^\dagger_{i;\uparrow} - F_{i;\uparrow}F_{i;\downarrow}\right), F^\dagger_{i;\uparrow}F_{i;\uparrow} + F^\dagger_{i;\downarrow}F_{i;\downarrow} - 1\right),$$
$$\boldsymbol{T}_{i;2} = \frac{1}{2}\left(\widetilde{F}^\dagger_{i;\downarrow}\widetilde{F}^\dagger_{i;\uparrow} + \widetilde{F}_{i;\uparrow}\widetilde{F}_{i;\downarrow}, i\left(\widetilde{F}^\dagger_{i;\downarrow}\widetilde{F}^\dagger_{i;\uparrow} - \widetilde{F}_{i;\uparrow}\widetilde{F}_{i;\downarrow}\right), \widetilde{F}^\dagger_{i;\uparrow}\widetilde{F}_{i;\uparrow} + \widetilde{F}^\dagger_{i;\downarrow}\widetilde{F}_{i;\downarrow} - 1\right). \quad (31.44)$$

For a more transparent presentation of the symmetries, it is useful to write the fermions as 2×2 matrices (as in (22.19))

$$\boldsymbol{F}_i = \begin{pmatrix} F_{i;\uparrow} & -F^\dagger_{i;\downarrow} \\ F_{i;\downarrow} & F^\dagger_{i;\uparrow} \end{pmatrix}. \quad (31.45)$$

This matrix obeys the relation

$$\boldsymbol{F}^\dagger_i = \sigma^y \boldsymbol{F}^T_i \sigma^y. \quad (31.46)$$

We use a similar representation for $\widetilde{\boldsymbol{F}}$. Now, as in (22.21), we can write the spin and Nambu pseudospin operators as

$$\boldsymbol{S}_{i;1} = \frac{1}{4}\mathrm{Tr}(\boldsymbol{F}^\dagger_i \boldsymbol{\sigma} \boldsymbol{F}_i) \quad , \quad \boldsymbol{T}_{i;1} = \frac{1}{4}\mathrm{Tr}(\boldsymbol{F}^\dagger_i \boldsymbol{F}_i \boldsymbol{\sigma}), \quad (31.47)$$

and similarly for $\boldsymbol{S}_{i;2}$ and $\boldsymbol{T}_{i;2}$ with $\widetilde{\boldsymbol{F}}$. The unit $\boldsymbol{F}$ and $\widetilde{\boldsymbol{F}}$ fermion occupancy constraint can now be stated as the vanishing of the pseudospins on each site of both ancilla layers, as in (22.22)

$$\boldsymbol{T}_{i;1} = 0 \quad , \quad \boldsymbol{T}_{i;2} = 0, \quad (31.48)$$

which implies that the gauge symmetry is $SU(2)$ [153] in the both ancilla layers. Specifically, with the constraint (31.48), note that (31.47) are invariant under (we drop the site index i, as it is common to all fields)

$$\begin{aligned} SU(2)_1 : \quad & \boldsymbol{F} \to \boldsymbol{F}U_1 \quad , \quad \widetilde{\boldsymbol{F}} \to \widetilde{\boldsymbol{F}}; \\ SU(2)_2 : \quad & \boldsymbol{F} \to \boldsymbol{F} \quad , \quad \widetilde{\boldsymbol{F}} \to \widetilde{\boldsymbol{F}}U_2, \end{aligned} \quad (31.49)$$

where $U_{1,2}$ are $SU(2)$ matrices, similar to (22.23).

However, we are not done with the gauge symmetries, as we also need to introduce a gauge symmetry associated with the rung-singlet projection in (31.42). We do this by transforming to a rotating reference frame in spin space [228]. We introduce the fermions $\boldsymbol{\Psi}_i$, $\widetilde{\boldsymbol{\Psi}}_i$ of Fig. 31.10 by the transformation

$$\boldsymbol{F}_i = L_i \boldsymbol{\Psi}_i \quad , \quad \widetilde{\boldsymbol{F}}_i = \widetilde{L}_i \widetilde{\boldsymbol{\Psi}}_i, \quad (31.50)$$

where L, and $\widetilde{L}$ are 2×2 $SU(2)$ matrices, and the $\boldsymbol{\Psi}$ fermions have a decomposition similar to (22.19)

$$\boldsymbol{\Psi}_i = \begin{pmatrix} \Psi_{i;+} & -\Psi^\dagger_{i;-} \\ \Psi_{i;-} & \Psi^\dagger_{i;+} \end{pmatrix}. \quad (31.51)$$

We use indices $a = +,-$ for $\Psi_{i;a}$ rather than $\uparrow,\downarrow$ in (31.51) because the indices do not represent physical spin in the rotated reference frame. Again, an analogous representation for $\widetilde{\Psi}_{ia}$ is used. The transformation (31.50) implies a rotation of the spin operators, but leaves the Nambu pseudospin invariant (and correspondingly for $S^{\alpha}_{i;2}$ and $T^{\alpha}_{i;2}$):

$$S^{\alpha}_{i;1} = \mathcal{L}^{\alpha\beta}_{i} \frac{1}{4} \mathrm{Tr}(\boldsymbol{\Psi}^{\dagger}_{i} \tau^{\beta} \boldsymbol{\Psi}_{i}), \tag{31.52}$$

$$T^{\beta}_{i;1} = \frac{1}{4} \mathrm{Tr}(\boldsymbol{\Psi}^{\dagger}_{i} \boldsymbol{\Psi}_{i} \tau^{\beta}), \tag{31.53}$$

where $\alpha, \beta = x, y, z$ and τ^{β} are Pauli matrices; we are using τ^{β} rather than σ^{β} here to signify that these matrices act on the rotated $a = +,-$ indices. As the pseudospin is invariant, the constraints (31.48) now imply the single occupancy of the $\Psi, \widetilde{\Psi}$ fermions that we noted above in (31.41). The $\mathcal{L}_i$ is a 3×3 $SO(3)$ rotation matrix corresponding to the 2×2 $SU(2)$ rotations:

$$\mathcal{L}^{\alpha\beta}_{i} = \frac{1}{2} \mathrm{Tr}\left(L^{\dagger}_{i} \sigma^{\alpha} L_{i} \tau^{\beta}\right). \tag{31.54}$$

Note that the actions of $SU(2)_1$ and $SU(2)_2$ on $\boldsymbol{F}$, $\widetilde{\boldsymbol{F}}$ in (31.49) translate into corresponding rotations of $\boldsymbol{\Psi}$, $\widetilde{\boldsymbol{\Psi}}$:

$$\begin{aligned} SU(2)_1: &\quad \boldsymbol{\Psi} \to \boldsymbol{\Psi} U_1, \quad \widetilde{\boldsymbol{\Psi}} \to \widetilde{\boldsymbol{\Psi}}, \quad\;\;\; L \to L, \quad \widetilde{L} \to \widetilde{L}; \\ SU(2)_2: &\quad \boldsymbol{\Psi} \to \boldsymbol{\Psi}, \quad\;\;\;\; \widetilde{\boldsymbol{\Psi}} \to \widetilde{\boldsymbol{\Psi}} U_2, \quad L \to L, \quad \widetilde{L} \to \widetilde{L}. \end{aligned} \tag{31.55}$$

We are now ready to discuss the gauge symmetries associated with the transformation to the rotating reference frame in spin space in (31.50). We do *not* wish to impose the analog of the constraints (31.48) in the spin sector, because we don't want vanishing spin on each site of both layers. Rather, we want to couple the layers into spin singlets for each i, corresponding to the $J_{\perp} \to \infty$ limit in Fig. 31.10. This is achieved by the constraints

$$\boldsymbol{S}_{i;1} + \boldsymbol{S}_{i;2} \approx 0. \tag{31.56}$$

The approximate equality is achieved by proceeding as usual with a gauge theory with an exact constraint, and then including a finite bare gauge coupling, that is, including a Maxwell term in the action with a non-zero coefficient. We don't want an exact constraint at infinite coupling because then the ancilla layers would just form rung singlets. We do want to allow for some virtual fluctuations into the triplet sector at each i; otherwise, the ancilla layers would completely decouple from the physical layer at the outset. In contrast, (31.48) is imposed at an infinite bare gauge coupling [153]. In practice, the value of the bare gauge coupling makes little difference, because we are dealing ultimately with the effective low-energy gauge theory.

The mechanism for imposing (31.56) is now straightforward. We transform to a common rotating frame in both layers by identifying

$$\widetilde{L}_i = L_i \quad , \quad \widetilde{\mathcal{L}}_i = \mathcal{L}_i\,, \tag{31.57}$$

so that only states with zero total angular momentum in the two ancilla layers are selected. So the transformation (31.50) introduces only a *single* $SU(2)_S$ gauge symmetry, related to that in Refs. [227, 228, 317], and the analog of (31.55) is

$$SU(2)_S: \quad \boldsymbol{\Psi} \to U_S \boldsymbol{\Psi} \quad , \quad \widetilde{\boldsymbol{\Psi}} \to U_S \widetilde{\boldsymbol{\Psi}} \quad , \quad L \to L U_S^\dagger , \tag{31.58}$$

where U_S is an $SU(2)$ matrix.

We assume $\langle L_i \rangle = 0$ in the whole phase diagram as we are interested in projecting the ancilla layers into rung spin singlets. After that, the crucial transformations for the subsequent discussion are those of the fermions $\boldsymbol{\Psi}$, which we collect here:

$$\begin{aligned} SU(2)_1: &\quad \boldsymbol{\Psi} \to \boldsymbol{\Psi} U_1 \quad , \quad \widetilde{\boldsymbol{\Psi}} \to \widetilde{\boldsymbol{\Psi}}; \\ SU(2)_2: &\quad \boldsymbol{\Psi} \to \boldsymbol{\Psi} \quad , \quad \widetilde{\boldsymbol{\Psi}} \to \widetilde{\boldsymbol{\Psi}} U_2; \\ SU(2)_S: &\quad \boldsymbol{\Psi} \to U_S \boldsymbol{\Psi} \quad , \quad \widetilde{\boldsymbol{\Psi}} \to U_S \widetilde{\boldsymbol{\Psi}}. \end{aligned} \tag{31.59}$$

We need only keep track of (31.59) for the following: the structure of all our effective actions is mainly dictated by the requirements of the gauge symmetries acting on the fermions in (31.59), and on the Higgs fields that will appear in the different cases. The $\mathbb{Z}_2$ divisor in the overall $(SU(2)_1 \times SU(2)_2 \times SU(2)_S)/\mathbb{Z}_2$ gauge symmetry arises from the fact that centers of the two $SU(2)$ tranformations in (31.50) are the same.

We can now write down an effective Hamiltonian for the FL* phase of the single-band Hubbard model in the paramagnon fractionalization theory, analogous to (31.8) for the Kondo lattice. As the second ancilla layer is approximately decoupled into a spin liquid (see Fig. 31.13), we focus on the effective Hamiltonian of the c and Ψ layers. Then the situation is even closer to that in the Kondo lattice, with a close correspondence to the c and f "layers" in (31.8). The main difference is the larger gauge symmetry: we have to consider a $(SU(2)_1 \times SU(2)_S)/\mathbb{Z}_2$ gauge symmetry for the top two layers of Fig. 31.13, while the Kondo lattice only had the U(1) gauge symmetry in (31.3). To understand the fate of this gauge symmetry, it is convenient to transform the 2×2 matrix notation in (31.51) into a four-vector notation:

$$\boldsymbol{\Psi}_i = \left(\Psi_{i;+}, \Psi_{i;-}, -\Psi_{i;-}^\dagger, \Psi_{i;+}^\dagger \right), \tag{31.60}$$

and similarly for the electron operator $\boldsymbol{c}$. Then, as for the Kondo lattice, the J_K interaction can be decoupled by a 4×4 matrix field $\boldsymbol{P}$ (generalizing (29.31)) into terms of the form

$$\sum_i \boldsymbol{\Psi}^\dagger \boldsymbol{P} \boldsymbol{c} + \text{H.c.}. \tag{31.61}$$

The Higgs field $\boldsymbol{P}$ transforms non-trivially under the $(SU(2)_1 \times SU(2)_S)/\mathbb{Z}_2$ gauge symmetry, and also under the global $SU(2)$ spin rotation and $U(1)$ charge symmetries, as can be deduced from requiring the invariance of (31.61). We assume a simple Higgs condensate of the form

$$\boldsymbol{P} = \overline{P} \operatorname{diag}(1, 1, -1, -1), \tag{31.62}$$

and then it is easy to see from (31.61) that the effective Hamiltonian for c and Ψ is *identical* in structure to that for the HFL phase of the Kondo lattice in (31.8) with the Ψ fermions replacing the f fermions. Moreover, the symmetry transformations of $\boldsymbol{P}$ show that the condensate (31.62) breaks the $(SU(2)_1 \times SU(2)_S)/\mathbb{Z}_2$ gauge symmetry completely, and preserves the $SU(2)$ spin rotation and $U(1)$ charge symmetries. The remaining analysis for the mean-field electronic structure of the c and Ψ is therefore very similar to that for the HFL phase in Section 31.1, and was incorporated into the results in Fig. 31.13. The electronic spectrum obtained from (31.62) compares well with photoemission observations in the pseudogap metal [173].

It must be kept in mind that the $\widetilde{\Psi}$ ancilla layer is still active in the FL* phase of the single-band model, although it is decoupled from the other layers in mean-field theory. The $SU(2)_2$ gauge symmetry acts on the $\widetilde{\Psi}$ layer, and its fate determines the nature of the spin liquid in this FL* phase of the single-band Hubbard model. The $SU(2)_2$ gauge symmetry could be unbroken, or it could be broken down to $U(1)$ or $\mathbb{Z}_2$ to obtain any one of the spin liquids in Parts II and IV.

The transition from FL* to FL is described [325] by a theory of the disappearance of the condensate of the Higgs field $\boldsymbol{P}$. This Higgs field carries fundamental gauge charges of $(SU(2)_1 \times SU(2)_S)/\mathbb{Z}_2$, and so the corresponding gauge fields must also be retained, along with the fermions in the ancilla layers.

32 Sachdev–Ye–Kitaev Models

The Sachdev–Ye–Kitaev model describes a metallic phase of matter without quasiparticle excitations. Its solution in the limit of a large number of sites is presented, enabled by random and all-to-all interactions between fermions. The low-energy spectral functions have a conformal structure, and the leading corrections to the conformal solution are described by a time-reparameterization mode. Connections to the physics of charged black holes is briefly noted.

This chapter turns to the further analysis of a phase of quantum matter *without* quasiparticle excitations. We briefly discuss such a phase in Section 11.2.2 at the quantum critical point of a relativistic $O(N)$ scalar field theory in 2+1 dimensions; further discussion can be found in the QPT book [234]. Our interest here is to describe a *solvable* theory of quantum matter without quasiparticle excitations. For this, there is essentially only one class of examples, those based on the Sachdev–Ye–Kitaev (SYK) model, to which we will turn in Section 32.2 by describing the structure of its large-N saddle-point theory. The SYK model also realizes a *compressible* phase of matter: its density can be continuously tuned at $T = 0$ by a chemical potential, much like that of a Fermi liquid, and so qualifies as a metal without quasiparticle excitations (unlike the theory of Section 11.2.2).

At first glance, the theory of one-dimensional quantum gases in Chapter 12, with spectral functions like those in (12.55), might seem like an example of a compressible phase of matter without quasiparticle excitations. However, that is not the case; the one-dimensional quantum gas has no electron-like quasiparticles, but it does have other quasiparticles – these are the free-particle phase or density fluctuation modes of the Luttinger liquid, which describe the entire spectrum of low-energy states.

We will begin in Section 32.1 by discussing a simple theory with quasiparticles: fermions occupying the eigenstates of a random matrix. The contrast to its properties will help highlight the novel features of the large-N solution of the SYK model in Section 32.2, which overlaps with a review article by Chowdhury et al. [46]. Much is known about the structure of the SYK model at finite N; we only note some important results in Section 32.3, and refer the reader to a review article by Chowdhury et al.'s article for further details. These authors also review the connections between the SYK model and the random t–J model – the latter model captures many aspects of the cuprate phase diagram at intermediate temperatures.

32.1 Random Matrix Model: Free Fermions

In the study of charge transport in mesoscopic structures, much experimental effort has focused on electrons moving through "quantum dots." We can idealize a quantum dot as a "billiard," a cavity with irregular walls. The electrons scatter off the walls, before eventually escaping through the leads. If we treat the electron motion classically, we can follow a chaotic trajectory of particles bouncing off the walls of the billiard. Much mathematical effort has been expended on the semi-classical quantization of such non-interacting particles: the "quantum billiard" problem. It was initially a conjecture [31], and now proven [187], that many statistical properties of this quantum billiard can be described by a model in which the electrons hop on a random matrix. It is this random-matrix problem that is described in this section.

Many properties of the random-matrix model are similar to a model of a disordered metal in which the electrons occupy plane-wave eigenstates, which scatter off randomly placed impurities with a short-range potential. However, unlike the random-impurity case, there is no regime in which the eigenstates of a random matrix can be localized. As every site is coupled to every other site, there is no sense of space or distance along which the eigenstate can decay exponentially.

32.1.1 Green's Function

We consider electrons c_i (assumed spinless, for simplicity) hopping between sites labeled $i = 1, \ldots, N$, with a hopping matrix element $t_{ij}/\sqrt{N}$:

$$H_2 = \frac{1}{(N)^{1/2}} \sum_{i,j=1}^{N} t_{ij} c_i^\dagger c_j - \mu \sum_i c_i^\dagger c_i, \tag{32.1a}$$

$$c_i c_j + c_j c_i = 0 \quad , \quad c_i c_j^\dagger + c_j^\dagger c_i = \delta_{ij}, \tag{32.1b}$$

$$\frac{1}{N} \sum_i c_i^\dagger c_i = \mathcal{Q}. \tag{32.1c}$$

The t_{ij} are chosen to be *independent* random complex numbers with $t_{ij} = t_{ji}^*$, $\overline{t_{ij}} = 0$ and $\overline{|t_{ij}|^2} = t^2$. The $1/\sqrt{N}$ scaling of the hopping has been chosen so that the bandwidth of the single-electron eigenstates will be of order unity in the $N \to \infty$ limit, and therefore (as there are N eigenstates) the spacing between the successive eigenvalues with be of order $1/N$. We have also included a chemical potential so that average density of electrons on each site is $\mathcal{Q}$. The subscript ("2") in the Hamiltonian H_2 denotes that it only includes two electron operators.

For a given set of t_{ij}, there is no alternative to numerically diagonalizing the $N \times N$ matrix t_{ij} to solve this problem. However, in the limit of large N, it turns out that certain quantities self-average. In other words, certain observables take the same value on every site, and that value is realized with probability 1 in the $N \to \infty$ limit. We will only be interested in such observables here.

One of these self-averaging observables is the single-particle Green's function, which we define as usual by

$$G_{ij}(\tau) = -T_\tau \left\langle c_i(\tau) c_j^\dagger(0) \right\rangle. \tag{32.2}$$

In the limit of large N, we have the self-averaging result

$$G_{ij}(\tau) \quad \Rightarrow \quad G(\tau)\delta_{ij}. \tag{32.3}$$

The simplest way to see this is to evaluate averages of G_{ij} order by order in a perturbation theory in t_{ij}. To zeroth-order, the Green's function is simply

$$G_{ij}^0(i\omega_n) = \frac{\delta_{ij}}{i\omega_n + \mu}, \tag{32.4}$$

where $i\omega_n$ are imaginary (Matsubara) frequencies. The Feynman graph expansion consists of a single-particle line, with an infinite set of possible products of G_{ij}^0 and t_{ij}. We now average each graph over the distribution of t_{ij}, and take the limit $N \to \infty$. Then only a simple set of graphs survive, and the average Green's function is a solution of the following set of equations

$$G(i\omega_n) = \frac{1}{i\omega_n + \mu - \Sigma(i\omega_n)} \quad , \quad \Sigma(\tau) = t^2 G(\tau), \tag{32.5a}$$

$$G(\tau = 0^-) = \mathcal{Q}. \tag{32.5b}$$

The solution of (32.5a) reduces to solving a quadratic equation for $G(i\omega)$, and so we obtain, for a complex frequency z,

$$G(z) = \frac{1}{2t^2}\left[z + \mu \pm \sqrt{(z+\mu)^2 - 4t^2}\right]. \tag{32.6}$$

The sign in front of the square root is chosen so that $G(z)$ has the correct analytic properties as $z \to \infty$:

- $G(|z| \to \infty) = 1/z$,
- $\mathrm{Im}\, G(\omega + i0^+) < 0$ for real ω,
- $\mathrm{Im}\, G(\omega + i0^-) > 0$ for real ω.

All of these constraints can be obtained from the spectral representation of the Green's function. We can also define the density of single-particle states as

$$\rho(\omega) = -\frac{1}{\pi}\mathrm{Im}\, G(\omega - \mu + i0^+) = \frac{1}{2\pi t^2}\sqrt{4t^2 - \omega^2} \tag{32.7}$$

for $\omega \in [-2t, 2t]$, and $\rho(\omega) = 0$ otherwise. This is the famous semicircle density of states for the random matrix.

The chemical potential is fixed by requiring that (32.5b) is satisfied, which can be written as

$$\int_{-2t}^{2t} d\omega\, \rho(\omega) f(\omega - \mu) = \mathcal{Q}, \tag{32.8}$$

where $f(\varepsilon) = 1/(e^{\varepsilon/T}+1)$ is the Fermi function. Performing a Sommerfeld expansion of the left-hand side for $T \ll t$, we obtain

$$\int_{-2t}^{\mu} d\omega \rho(\omega) + \frac{\pi^2 T^2}{6}\rho'(\mu) = \mathcal{Q}. \tag{32.9}$$

where $\rho'(\omega) = d\rho/d\omega$, This equation must be satisfied at all T, and depending upon the particular ensemble, it requires variation of μ or $\mathcal{Q}$ with T. In particular, if we keep $\mathcal{Q}$ fixed and vary T, then

$$\mu(T) = \mu_0 - \frac{\rho'(\mu_0)}{\rho(\mu_0)}\frac{\pi^2 T^2}{6}, \tag{32.10}$$

where $\mu_0 = \mu(T=0)$.

32.1.2 Many-Body Density of States

A quantity that will play an important role in our subsequent discussion of the SYK model is the many-body density of states, $\mathcal{N}(E)$. Unlike the single-particle density of states $\rho(\omega)$, this is not an intensive quantity, but is typically exponentially large in N, because there is an exponentially large number of ways of making states within a small window of an energy $E \sim N$. In the grand canonical ensemble, we can relate the grand potential $\Omega(T)$ to $\mathcal{N}(E)$ via an expression for the grand partition function

$$Z = \exp\left(-\frac{\Omega(T)}{T}\right) = \int_{-\infty}^{\infty} dE\, \mathcal{N}(E) e^{-E/T}. \tag{32.11}$$

Note that we have absorbed a contribution $-\mu N \mathcal{Q}$ into the definition of the energy E, just as is frequently done in Fermi liquid theory. So we can obtain $\mathcal{N}(E)$ by an inverse Laplace transform of $\Omega(T)$.

First, let us evaluate $\Omega(T)$. By the standard Sommerfeld expansion for free fermions, we have

$$\begin{aligned}\Omega(T) &= -T\int_{-2t}^{2t} d\omega \rho(\omega) \ln\left(1+e^{-(\omega-\mu)/T}\right)\\ &= \int_{-2t}^{\mu} d\omega (\omega-\mu)\rho(\omega) - \frac{\pi^2 T^2}{6}\rho(\mu)\\ &\equiv E_0 - \frac{\pi^2 T^2}{6}\rho(\mu).\end{aligned} \tag{32.12}$$

This results implies an entropy that vanishes linearly as $T \to 0$:

$$S = \gamma T, \tag{32.13}$$

with, as in a Fermi liquid in (2.12),

$$\gamma = \frac{\pi^2}{3}\rho(\mu). \tag{32.14}$$

We now have to insert (32.12) into (32.11) and determine $\mathcal{N}(E)$. Rather than perform the inverse Laplace transform, we make a guess of the form of $\mathcal{N}(E)$. First, it is not

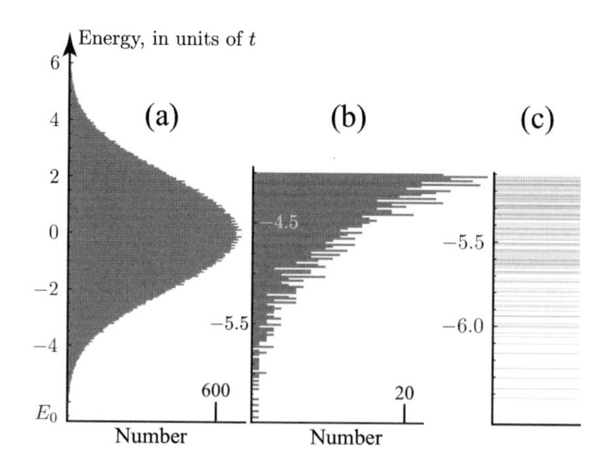

Figure 32.1 65 536 Many-body eigenvalues of a $N = 32$ Majorana matrix model with random fermion hopping terms. $\mathcal{N}(E)$ is plotted in (a) and (b) in 200 and 100 bins, (b) and (c) zoom into the bottom of the band. Individual energy levels are shown in (c), and these are expected to have spacing $1/(N\rho(\mu))$ at the bottom of the band as $N \to \infty$. Figure by G. Tarnopolsky.

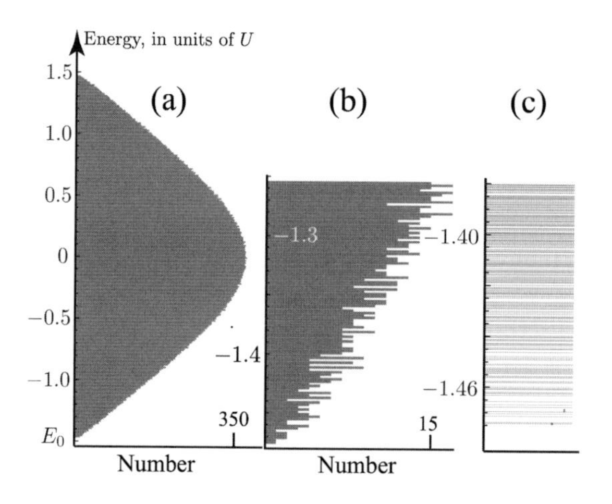

Figure 32.2 65 536 Many-body eigenvalues of a $N = 32$ Majorana SYK Hamiltonian with random $q = 4$ fermion terms. $\mathcal{N}(E)$ is plotted in (a) and (b) in 200 and 100 bins, (b) and (c) zoom into the bottom of the band. Individual energy leves are shown in (c), and these are expected to have spacing e^{-NS} at the bottom of the band as $N \to \infty$. Compare to Fig. 32.1 for the random-matrix model, which has a much sparser spacing $\sim 1/N$ at the bottom of the band. Figure by G. Tarnopolsky.

difficult to see that $\mathcal{N}(E < E_0) = 0$. Next, we expect $\mathcal{N}(E)$ to be exponentially large in N when $E - E_0 \sim N$. So we make a guess:

$$\mathcal{N}(E) \sim \exp\left(aN[(E - E_0)/N]^b\right) \quad , \quad E > E_0 \tag{32.15}$$

for some constants a and b. Then, we insert (32.15) into (32.11), and perform the integral over E by the steepest descent method in the large-N limit. Matching the result to the left-hand side of (32.11), we obtain the main result of this section:

$$\mathcal{N}(E) \sim \exp(S(E)), \qquad (32.16)$$
$$S(E) = \begin{cases} \sqrt{2N\gamma(E-E_0)} & , \quad E > E_0 \\ 0 & , \quad E < E_0 \end{cases},$$

where $S(E)$ is the entropy as a function of the grand energy. Consideration of the derivation shows that this result is valid for

$$1 \ll \rho(\mu)(E-E_0) \ll N, \qquad (32.17)$$

in the limit of large N. Note that the entropy vanishes as $E \searrow E_0$ in (32.16). Numerical results for $\mathcal{N}(E)$ are shown for a closely related random Majorana fermion model in Fig. 32.1. When $E - E_0 \sim N$, the entropy $S(E)$ is extensive, the energy-level spacing is exponentially small, $\sim e^{-aN}$, with $a > 0$, and $\mathcal{N}(E) \sim e^{aN}$ is exponentially large. However, when $E - E_0 \sim 1/N$, we expect the many-particle eigenstates to be a few single-particle excitations with energies $\sim 1/(N\rho(\mu))$, and so $\mathcal{N}(E) \sim N$. This rapid dropoff in $\mathcal{N}(E)$ near the bottom of the band is clearly evident in Fig. 32.1a from the "tails" in the density of states. A more complete analysis of the finite-N corrections is needed to understand the behavior of the $\mathcal{N}(E)$ at low energy, along the lines of recent analyses [157, 158].

We also show in Fig. 32.2 the corresponding results for the Majorana SYK model. These results are discussed further in Section 32.3, but for now the reader should note the striking absence of the tails in $\mathcal{N}(E)$ in Fig. 32.2a in comparison to Fig. 32.1a.

There is an interesting interpretation of (32.16), which gives us some insight into the structure of the random matrix eigenenergies, and also highlights a key characteristic of many-body systems *with* quasiparticle excitations. It is known that the eigenvalues of a random matrix undergo level repulsion and their spacings obey Wigner–Dyson statistics [175]. For a zeroth-order picture, let us assume that the random-matrix eigenvalues are rigidly equally spaced, with energy-level spacing (near the chemical potential) of $1/(N\rho(\mu))$. Now we ask for the number of ways to create a many body excitation with energy $E - E_0$. This many-body excitation energy is the sum of particle–hole excitations, each of which has an energy equal to an integer times the level spacing $1/(N\rho(\mu))$:

$$N\rho(\mu)(E-E_0) = n_1 + n_2 + n_3 + n_4 + \cdots, \qquad (32.18)$$

where the n_i are the excitation numbers of the particle–hole excitations (this mapping is the essence of bosonization in one dimension). So, we estimate that the number of such excitations is equal to the number of partitions of the integer $N\rho(\mu)(E-E_0)$. Now we use the Hardy–Ramanujan result that the number of partitions of an integer n is $p(n) \sim \exp(\pi\sqrt{2n/3})$ at large n. This immediately yields (32.16). Note that the special case with exactly equally spaced quasiparticle levels (which is the case in Section 12.1) has many-body levels with a spacing $\sim 1/N$ but an exponentially large degeneracy; in contrast, the generic random-matrix case has no degeneracy but an exponentially small many-body-level spacing.

This argument highlights a key feature of the many-body spectrum: it is just the sum of single-particle excitation energies. We expect that if we add weak interactions to the random-matrix model, we will obtain quasiparticle excitations in a Fermi liquid state whose energies add to give many-particle excitations; therefore, we expect the general form of (32.16) to continue to hold even with interactions. However, we see at the end of Section 32.3 that such a decomposition to quasiparticle excitations does not hold for the SYK model.

We can also estimate the lifetime of the quasiparticles by a perturbative computation based on Fermi's golden rule. The computation in (2.36) now applies essentially exactly, as there is no momentum dependence and we can perform the integrals over the energies using the density of states in (32.7); we obtain $1/\tau \sim U^2T^2/t^3$ at low T for interactions with root-mean-square strength U (see Problem 32.1). As this is parametrically smaller than a quasiparticle excitation energy $\sim T$, they remain well-defined excitations.

32.2 Large-N Theory of the SYK Model

As in the random-matrix model, we consider electrons (assumed spinless for simplicity) that occupy sites labeled $i = 1,2,\ldots,N$. However, instead of a random one-particle hopping t_{ij}, we now have only a random two-particle interaction $U_{ij;k\ell}$:

$$H_4 = \frac{1}{(2N)^{3/2}} \sum_{ijk\ell=1}^{N} U_{ij;k\ell}\, c_i^\dagger c_j^\dagger c_k c_\ell - \mu \sum_i c_i^\dagger c_i,$$
$$c_i c_j + c_j c_i = 0 \quad , \quad c_i c_j^\dagger + c_j^\dagger c_i = \delta_{ij},$$
$$\mathcal{Q} = \frac{1}{N} \sum_i c_i^\dagger c_i, \tag{32.19a}$$

where the subscript "4" emphasizes that the coupling depends randomly on four indices. We choose the couplings $U_{ij;k\ell}$ to be *independent* random variables with zero mean $\overline{U_{ij;k\ell}} = 0$, while satisfying $U_{ij;k\ell} = -U_{ji;k\ell} = -U_{ij;\ell k} = U^*_{k\ell;ij}$. All the random variables have the same variance $\overline{|U_{ij;k\ell}|^2} = U^2$.

A model similar to H_4 appears in nuclear physics, where it is called the two-body random ensemble [37], and studied numerically. The existence and structure of the large-N limit was understood [91, 92, 196, 242] in the context of a closely related model. More recently, a Majorana version was introduced [136], and the large-N limit of H_4 was obtained [235].

The useful self-averaging properties of the random-matrix model as $N \to \infty$ also apply to the SYK model (32.19a). Indeed, the self-averaging properties are *much* stronger, as the average takes place over the many-body Hilbert space of size $e^{\alpha N}$, rather than the single-particle Hilbert space of size N. Proceeding just as in the random-matrix model, we perform a Feynman graph expansion in $U_{ij;k\ell}$, and then average graph by graph. In the large-N limit, only the so-called "melon graphs" survive (Fig. 32.3),

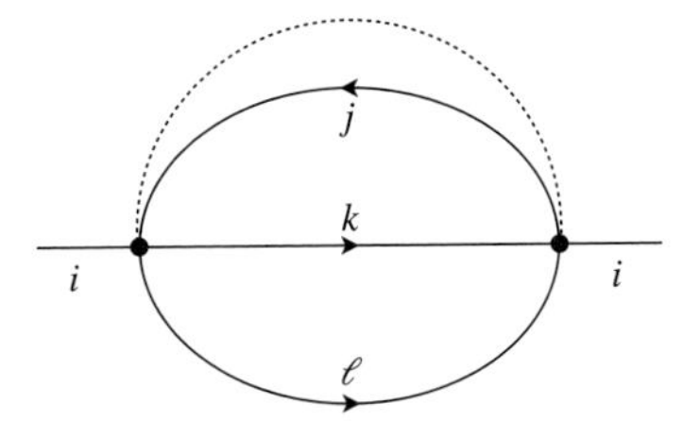

Figure 32.3 The "melon graph" for the electron self-energy $\Sigma(\tau)$ in (32.20b). Solid lines denote fully dressed-electron Green's functions. The dashed line represents the disorder averaging associated with the interaction vertices (denoted as solid circles), $\overline{|U_{ij;k\ell}|^2}$.

and the determination of the on-site Green's function reduces to the solution of the following equations:

$$G(i\omega_n) = \frac{1}{i\omega_n + \mu - \Sigma(i\omega_n)}, \tag{32.20a}$$

$$\Sigma(\tau) = -U^2 G^2(\tau) G(-\tau), \tag{32.20b}$$

$$G(\tau = 0^-) = \mathcal{Q}. \tag{32.20c}$$

Unlike the random-matrix equations, these equations cannot be solved analytically, and a full solution can only be obtained numerically. However, it is possible to make significant analytic progress at frequencies and temperatures $\ll U$, as described in the following subsections.

Before embarking on a general low-energy solution of (32.20a)–(32.20c), let us note a remarkable feature that can be deduced on general grounds [242]: any non-trivial solution (i.e., with $\mathcal{Q} \neq 0, 1$) must be gapless. Let us suppose otherwise, and assume there is a gapped solution with $\operatorname{Im} G(\omega) = 0$ for $|\omega| < E_G$. Then, by an examination of the spectral decomposition of the equation for the self-energy in (32.20b), we can establish that $\operatorname{Im}\Sigma(\omega) = 0$ for $|\omega| < 3E_G$. Inserting this back into Dyson's equation (32.20a), we obtain the contradictory result that $\operatorname{Im} G(\omega) = 0$ for $|\omega| < 3E_G$. So the only possible value is $E_G = 0$.

We also mention briefly a model that combines the interaction term in (32.19a) with the hopping term in (32.1a), namely

$$H_{24} = \frac{1}{(N)^{1/2}} \sum_{i,j=1}^{N} t_{ij} c_i^\dagger c_j + \frac{1}{(2N)^{3/2}} \sum_{ijk\ell=1}^{N} U_{ij;k\ell}\, c_i^\dagger c_j^\dagger c_k c_\ell - \mu \sum_i c_i^\dagger c_i. \tag{32.21}$$

The large-N equations for this model merely combine the self-energy contributions in (32.20b) and (32.5a) so that

$$\Sigma(\tau) = t^2 G(\tau) - U^2 G^2(\tau) G(-\tau), \tag{32.22}$$

while (32.20a) and (32.20c) remain the same. The solution of (32.20a), (32.20c), and (32.22) are presented later in Section 33.3.3, where the same equations appear in a different context. For now, we simply state that the t^2 term is more important than the

U^2 term at low frequencies, and so the low-energy solution reduces to a renormalized free-fermion solution similar to that in Section 32.1.

We now return to a consideration of H_4 in (32.19a) for the remainder of this chapter.

32.2.1 Low-Energy Solution at $T = 0$

Knowing that the solution must be gapless, let us assume that we have a power-law singularity at zero frequency. So we assume [242]

$$G(z) = C\frac{e^{-i(\pi\Delta+\theta)}}{z^{1-2\Delta}} \quad , \quad \mathrm{Im}(z) > 0, \ |z| \ll U \,. \tag{32.23}$$

We have a prefactor $C > 0$, a power-law singularity determined by the exponent $\Delta > 0$, and a spectral asymmetry angle θ, which yields distinct density of states for particle and hole excitations. We now have to insert the ansatz (32.23) into (32.20a) and (32.20b) and find the values of C, Δ, and θ for which there is a self-consistent solution. Of course, the solution also has to satisfy the constraint arising from the spectral representation $\mathrm{Im}\, G(\omega + i0^+) < 0$; for (32.23) this translates to

$$-\pi\Delta < \theta < \pi\Delta \,. \tag{32.24}$$

We now wish to obtain the Green's function as a function of imaginary time τ. For this purpose, we write the the spectral representation using the density of states $\rho(\Omega) = -(1/\pi)\mathrm{Im}\, G(\omega + i0^+) > 0$, so that

$$G(z) = \int_{-\infty}^{\infty} d\Omega \frac{\rho(\Omega)}{z - \Omega} \,. \tag{32.25}$$

We can take a Fourier transform and obtain

$$G(\tau) = \begin{cases} -\displaystyle\int_0^{\infty} d\Omega\, \rho(\Omega) e^{-\Omega\tau} & , \quad \text{for } \tau > 0 \\ \displaystyle\int_0^{\infty} d\Omega\, \rho(-\Omega) e^{\Omega\tau} & , \quad \text{for } \tau < 0 \end{cases} \,. \tag{32.26}$$

Using (32.26) we obtain in τ space

$$G(\tau) = \begin{cases} -\dfrac{C\Gamma(2\Delta)\sin(\pi\Delta+\theta)}{\pi|\tau|^{2\Delta}} & , \quad \text{for } \tau \gg 1/U \\ \dfrac{C\Gamma(2\Delta)\sin(\pi\Delta-\theta)}{\pi|\tau|^{2\Delta}} & , \quad \text{for } \tau \ll -1/U \end{cases} \,. \tag{32.27}$$

This expression makes it clear that θ determines the particle–hole asymmetry, associated with the fermion propagation forward and backward in time. For our later purpose, it is also useful to parametrize the asymmetry in terms of a real number $-\infty < \mathcal{E} < \infty$ so that

$$G(\tau) \sim \begin{cases} -\dfrac{e^{\pi\mathcal{E}}}{|\tau|^{2\Delta}} & , \quad \text{for } \tau \gg 1/U \\ \dfrac{e^{-\pi\mathcal{E}}}{|\tau|^{2\Delta}} & , \quad \text{for } \tau \ll -1/U \end{cases} \,, \tag{32.28}$$

and then we have

$$e^{2\pi\mathcal{E}} = \frac{\sin(\pi\Delta+\theta)}{\sin(\pi\Delta-\theta)}, \tag{32.29}$$

and $\mathcal{E}=\theta=0$ is the particle–hole symmetric case.

We also use the spectral representation for the self-energy

$$\Sigma(z) = \int_{-\infty}^{\infty} d\Omega \frac{\rho(\Omega)}{z-\Omega}. \tag{32.30}$$

Using (32.20b) and (32.27) to obtain $\Sigma(\tau)$, and performing the inverse Laplace transform as for $G(\tau)$, we obtain

$$\rho(\Omega) = \begin{cases} \Upsilon(\Delta)\,[\sin(\pi\Delta+\theta)]^2\,[\sin(\pi\Delta-\theta)]\,|\Omega|^{6\Delta-1} \\ \qquad \text{for } \Omega>0 \\ \\ \Upsilon(\Delta)\,[\sin(\pi\Delta+\theta)]\,[\sin(\pi\Delta-\theta)]^2\,|\Omega|^{6\Delta-1} \\ \qquad \text{for } \Omega<0 \end{cases}, \tag{32.31}$$

where

$$\Upsilon(\Delta) = \frac{\pi^2 U^2}{\Gamma(6\Delta)}\left[\frac{C\Gamma(2\Delta)}{\pi}\right]^3. \tag{32.32}$$

Finally, we have to insert the $\Sigma(i\omega_n)$ obtained from (32.30) and (32.31) back into (32.20a). To understand the structure of the solution, let us first assume that $0 < 6\Delta - 1 < 1$; we will find soon that this is indeed the case, and no other solution is possible. Then, as $|\omega_n| \to 0$, the frequency dependence in $\Sigma(i\omega_n)$ is much larger than that from the $i\omega_n$ term in (32.20a). Also, we have $1-2\Delta>0$, and so $G(z)$ in (32.23) diverges as $|z| \to 0$. So we find that a solution of (32.20a) is only possible under two conditions:

$$\begin{aligned} &\mu - \Sigma(0) = 0, \\ &1-2\Delta = 6\Delta - 1 \quad \Rightarrow \quad \Delta = \frac{1}{4}. \end{aligned} \tag{32.33}$$

Matching the divergence in the coefficient of $G(z)$ as $z \to 0$, we also obtain the value of C:

$$C = \left(\frac{\pi}{U^2\cos(2\theta)}\right)^{1/4}. \tag{32.34}$$

The value of the asymmetry angle θ remains undetermined by the solution (32.20a) and (32.20b). As we see in Section 32.2.2, the value of θ is fixed by a generalized Luttinger's theorem, which relates it to the value of the fermion density $\mathcal{Q}$. But without further computation we can conclude that, at the particle–hole symmetric point with $\mathcal{Q}=1/2$, we have $\mathcal{E}=\theta=0$.

The main result of this section is therefore summarized in (32.28). The fermion has the "dimension" $\Delta = 1/4$ and its two-point correlator decays as $1/\sqrt{\tau}$; there is an unknown particle–hole asymmetry determined by $\mathcal{E}$. This should be contrasted with the corresponding features of the random-matrix model with a Fermi liquid ground state: the two-point fermion correlator decays as $1/\tau$, and the leading decay is particle–hole symmetric.

32.2.2 Luttinger Relation

Section 30.2.2 presented a discussion of the Luttinger relation for a disordered Fermi liquid. We argued in Section 30.2.2 that a Luttinger relation can be obtained for any quantum system with an unbroken $U(1)$ symmetry that is compressible, that is, for which the $U(1)$ charge density can be varied continuously by varying parameters in the Hamiltonian. The SYK model is another example of such a system, and we obtain its Luttinger relation below; it turns out to be different from that for a disordered Fermi liquid because of the absence of quasiparticles, and the resulting singular nature of the low-frequency Green's function.

The Luttinger relation for the SYK model [91] relates the angle θ characterizing the particle–hole asymmetry at long times in (32.23), to the fermion density $\mathcal{Q}$, which is an equal-time fermion correlator. As in the conventional Luttinger analysis in Section 30.2, we start by manipulating the expression for $\mathcal{Q}$ into two terms

$$
\begin{aligned}
\mathcal{Q} - 1 &= \int_{-\infty}^{\infty} \frac{d\omega}{2\pi} G(i\omega) e^{-i\omega 0^+} = I_1 + I_2, \\
I_1 &= i \int_{-\infty}^{\infty} \frac{d\omega}{2\pi} \frac{d}{d\omega} \ln\left[G(i\omega)\right] e^{-i\omega 0^+}, \\
I_2 &= -i \int_{-\infty}^{\infty} \frac{d\omega}{2\pi} G(i\omega) \frac{d}{d\omega} \Sigma(i\omega) e^{-i\omega 0^+}.
\end{aligned} \tag{32.35}
$$

In all the cases considered so far in Section 30.2 and 31.2, I_2 vanishes because of the existence of the Luttinger–Ward functional [2], while I_1 is easily evaluated because it is a total derivative, and this yields the Luttinger relation. The situation is more complicated for the SYK model because of the singular nature of $G(\omega)$ as $|\omega| \to 0$. Indeed, both I_1 and I_2 are logarithmically divergent at small $|\omega|$, although, naturally, their sum is well defined. Nevertheless, the separation of $\mathcal{Q}$ into I_1 and I_2 is useful because it allows us to use the special proprties of the Luttinger–Ward functional to account for the unknown high-frequency behavior of the Green's function. We define $I_{1,2}$ by a regularization procedure, and it is then important that the same regularization be used for both I_1 and I_2. We employ the symmetric principle value, with

$$
\int_{-\infty}^{\infty} d\omega \Rightarrow \lim_{\eta \to 0} \left[\int_{-\infty}^{-\eta} d\omega + \int_{\eta}^{\infty} d\omega \right]. \tag{32.36}
$$

Now we evaluate I_1 using the usual procedure of Section 30.2.1: we distort the contour of integration to the real frequency axis and obtain

$$
\begin{aligned}
I_1 &= i \lim_{\eta \to 0} \int_0^{\infty} \frac{d\omega}{2\pi} \frac{d}{d\omega} \ln\left[\frac{G(\omega + i\eta)}{G(\omega - i\eta)}\right] \\
&= -\frac{1}{\pi} \lim_{\eta \to 0} \left[\arg G(\infty + i\eta) - \arg G(i\eta)\right].
\end{aligned} \tag{32.37}
$$

In a Fermi liquid, I_1 now evaluates to unity outside the Fermi surface, and vanishes inside the Fermi surface. In the present case, using (32.23), we obtain

$$I_1 = -\frac{1}{2} - \frac{\theta}{\pi}. \tag{32.38}$$

In the evaluation of I_2 we must substitute the expression (32.20b) for Σ into I_2, because then we ensure cancellations at high frequencies arising from the existence of the Luttinger–Ward functional:

$$\Phi_{LW}[G] = -\frac{U^2}{4} \int d\tau\, G^2(\tau) G^2(-\tau). \tag{32.39}$$

Using $\Sigma = \delta\Phi_{LW}/\delta G$, and ignoring the singularity at $\omega = 0$, we obtain, as in Fermi liquid theory, $I_2 = -i\int_{-\infty}^{\infty} d\omega (d/d\omega)\Phi_{LW} = 0$. So the entire contribution to I_2 arises from the regularization of singularity near $\omega = 0$. We can therefore evaluate I_2 by using (32.20b) for Σ, the regularization in (32.36), and the low-frequency spectral density in (32.31), and ignore the high-frequency contribution to I_2. After a somewhat involved evaluation of such an integral [91, 97], we obtain

$$I_2 = -\frac{\sin(2\theta)}{4}. \tag{32.40}$$

Combining (32.35,32.38,32.40), we obtain our generalized Luttinger theorem [62, 91, 97],

$$\mathcal{Q} = \frac{1}{2} - \frac{\theta}{\pi} - \frac{\sin(2\theta)}{4}. \tag{32.41}$$

This expression evaluates to the limiting values $\mathcal{Q} = 1, 0$ for the limiting values of $\theta = -\pi/4, \pi/4$ in (32.24), and decreases monotonically in between; $\mathcal{Q}$ is also a monotonically decreasing function between these limits of $-\infty < \mathcal{E} < \infty$, via (32.29).

All our results have so far been obtained by an analytic analysis of the low-energy behavior. A numerical analysis is needed to ensure that such low-energy solutions have high-energy continuations that also obey (32.20a) and (32.20b). Such analyses show that complete solutions exist only for a range of values around $\mathcal{Q} = 1/2$ [16]; for values of $\mathcal{Q}$ close to 0,1, there is phase separation into the trivial $\mathcal{Q} = 0, 1$ state, and densities closer to half filling. However, this conclusion is only for the specific microscopic Hamiltonian in (32.19a); other Hamiltonians, with additional q-fermion terms, with $q > 4$, could have solutions with the same low-energy behavior described so far for a wider range of $\mathcal{Q}$, because these higher q terms are irrelevant at low energy.

32.2.3 Non-zero Temperatures

It turns out to be possible to extend the solutions for $T = 0$ Green's functions obtained so far to non-zero $T \ll U$. This is done most cleanly using a subtle argument employing conformal invariance. However, here we take a pedestrian approach, look for a solution directly from the defining equations (32.20a) and (32.20b), and show that we can *guess* a solution that works.

We limit the considerations of this section to the particle–hole symmetric case with $\mathcal{Q} = 1/2$ and $\theta = 0$. We use the similarity to multichannel Kondo problems [197] to generalize the τ dependence of the Green's function in (32.27) to [196]

$$G(\tau) = -\frac{B}{U^{1/2}}\operatorname{sgn}(\tau)\left|\frac{\pi T}{\sin(\pi T\tau)}\right|^{1/2}, \quad T, |\tau|^{-1} \ll U, \tag{32.42}$$

where B is a dimensionless constant whose value can be deduced from (32.27), to which (32.42) reduces for $1/U \ll |\tau| \ll 1/T$.

Then, the self-energy is

$$\Sigma(\tau) = -U^{1/2}B^3\operatorname{sgn}(\tau)\left|\frac{\pi T}{\sin(\pi T\tau)}\right|^{3/2}, \quad T, |\tau|^{-1} \ll U.$$

Taking Fourier transforms, we have, as a function of the Matsubara frequency ω_n,

$$G(i\omega_n) = \left[\frac{-iB}{U^{1/2}}\right]\frac{T^{-1/2}\Gamma\left(\frac{1}{4}+\frac{\omega_n}{2\pi T}\right)}{\Gamma\left(\frac{3}{4}+\frac{\omega_n}{2\pi T}\right)}, \tag{32.43a}$$

$$\Sigma_{sing}(i\omega_n) = \left[-i4\pi U^{1/2}B^3\right]\frac{T^{1/2}\Gamma\left(\frac{3}{4}+\frac{\omega_n}{2\pi T}\right)}{\Gamma\left(\frac{1}{4}+\frac{\omega_n}{2\pi T}\right)}, \tag{32.43b}$$

where we have dropped a less-singular term in $\Sigma(i\omega_n)$. Now, the singular part of Dyson's equation is

$$G(i\omega_n)\Sigma_{sing}(i\omega_n) = -1. \tag{32.44}$$

Remarkably, the Γ functions in (32.43a) and (32.43b) appear with just the right arguments, so that they can indeed obey (32.44) for all ω_n.

A deeper understanding of the origin of (32.42), and its generalization to the particle–hole asymmetric case, can be obtained by analyzing the low-energy limit of the original saddle-point equations (32.20a) and (32.20b). These equations are characterized by a remarkably large set of emergent symmetries, which are described in Section 32.3.1. The final result for the Green's function in imaginary time away from the particle–hole symmetric point is

$$G(\tau) = -C\frac{e^{-2\pi\mathcal{E}T\tau}}{\sqrt{1+e^{-4\pi\mathcal{E}}}}\left(\frac{T}{\sin(\pi T\tau)}\right)^{1/2} \tag{32.45}$$

for $0 < \tau < 1/T$. This can be extended to all real τ using the antisymmetry of the fermion Green's function. Performing a Fourier transform, and analytically continuing to real frequencies leads to the Green's function [196, 235]

$$G(\omega + i0^+) = \frac{-iCe^{-i\theta}}{(2\pi T)^{1/2}}\frac{\Gamma\left(\frac{1}{4}+i\mathcal{E}-\frac{i\omega}{2\pi T}\right)}{\Gamma\left(\frac{3}{4}+i\mathcal{E}-\frac{i\omega}{2\pi T}\right)}. \tag{32.46}$$

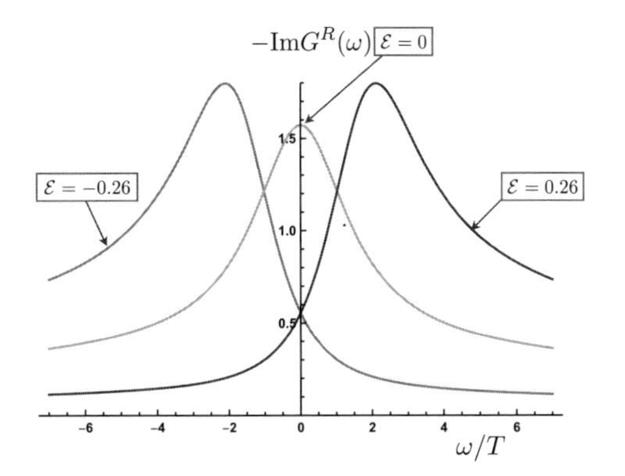

Figure 32.4 Plot of the electron spectral density in the SYK model, obtained from the imaginary part of (32.46). The $\mathcal{E}=0$ curve is the particle–hole symmetric case with $\mathcal{Q}=1/2$, while $\mathcal{E}$ positive (negative) corresponds to $\mathcal{Q}<1/2$ ($\mathcal{Q}>1/2$).

We show a plot of the imaginary part of the Green's function in Fig. 32.4.

For later comparison with other models, let us note that these results imply that the singular part of the electron self-energy obeys the scaling form

$$\Sigma(\omega,T)=U^{1-\alpha}T^{\alpha}\Phi\left(\frac{\hbar\omega}{k_BT}\right), \tag{32.47}$$

with $\alpha=1/2$ and Φ is a universal scaling function with a known dependence on the particle–hole asymmetry parameter $\mathcal{E}$. The universal dependence of the self-energy on the "Planckian ratio," $\hbar\omega/(k_BT)$, implies the absence of electronic quasiparticles; the characteristic lifetime of the excitations $\sim\hbar/(k_BT)$ is of the same order as their energy $\sim\hbar\omega$, and so quasiparticles are not well defined. Contrast this with the behavior of the random-matrix model in Section 32.1.2, where the self-energy was negligible at low T.

32.2.4 Computation of the $T\to 0$ Entropy

We have now presented detailed information on the nature of the Green's function of the SYK model at low T. We will proceed next to use this information to compute some key features of the low-T thermodynamics.

First, we establish some properties of the behavior of the chemical potential μ as $T\to 0$ at fixed $\mathcal{Q}$. Recall that for the random matrix model, and more generally for any Fermi liquid, there was a $\sim T^2$ correction to the chemical potential, which depended upon the derivative of the density of single-particle states. For the SYK model, the leading correction is much stronger: the correction is $\sim T$, which is universally related to parameters in the Green's function [91].

A simple way to determine the linear T dependence of μ is to examine the particle–hole asymmetry of the Green's function at $T>0$. From (32.28) and (32.45), this is given by the ratio

$$\lim_{T \to 0} \frac{G(\tau)}{G(1/T - \tau)} = e^{2\pi\mathcal{E}}, \tag{32.48}$$

where the limit is taken at a fixed $\tau \gg 1/U$. We now use a crude picture of the low-energy theory and imagine that all the low-energy degrees of freedom are essentially at zero energy, compared to U. So we compare (32.48) with the corresponding ratio for a zero-energy fermion whose chemical potential has been shifted by $\delta\mu$

$$G_0(0 < \tau < 1/T) = -\frac{e^{\delta\mu\tau}}{1 + e^{\delta\mu/T}} \quad , \quad \frac{G_0(\tau)}{G_0(1/T - \tau)} = e^{-\delta\mu(1/T - 2\tau)}. \tag{32.49}$$

From this comparison, we conclude that there is a linear-in-T dependence of the chemical potential that keeps the particle–hole asymmetry fixed as $T \to 0$:

$$\mu - \mu_0 = \delta\mu = -2\pi\mathcal{E}T + \text{terms vanishing as } T^p \text{ with } p > 1, \tag{32.50}$$

with μ_0 a non-universal constant. Note that the density of the zero-energy fermion $= 1/(e^{-\delta\mu/T} + 1)$ remains fixed as $T \to 0$, and so (32.50) applies at fixed $\mathcal{Q}$.

A more formal analysis [91, 197, 235], leading to the same result for the T dependence of μ, relates the long-time conformal Green's function (valid for $\tau \gg 1/U$) to its short-time behavior. In particular, at $|\omega_n| \gg U$, we have

$$G(i\omega_n) = \frac{1}{i\omega_n} - \frac{\mu}{(i\omega_n)^2} + \cdots, \tag{32.51}$$

which implies, for the spectral density of the Green's function $\rho(\Omega)$,

$$\mu = -\int_{-\infty}^{\infty} d\Omega \, \Omega \rho(\Omega), \tag{32.52}$$

and this makes it evident that μ depends only upon the particle–hole asymmetric part of the spectral density. Next, using the spectral relations, we can relate the Ω integrals to the derivative of the imaginary-time correlator:

$$\mu = -\partial_\tau G(\tau = 0^+) - \partial_\tau G(\tau = (1/T)^-). \tag{32.53}$$

We pull out an explicitly particle–hole asymmetric part of $G(\tau)$ by defining

$$G(\tau) \equiv e^{-2\pi\mathcal{E}T\tau} G_c(\tau) \quad , \quad 0 < \tau < \frac{1}{T}, \tag{32.54}$$

where G_c is given by a particle–hole symmetric conformal form at low T and low ω. Then we obtain

$$\begin{aligned}\mu &= 2\pi\mathcal{E}T \left[G(\tau = 0^+) + G(\tau = (1/T)^-)\right] \\ &\quad + \text{terms dependent on } G_c \\ &= -2\pi\mathcal{E}T + \text{terms dependent on } G_c.\end{aligned}$$

It can be shown that all the terms dependent upon G_c have a T dependence that is weaker than linear in T provided $\mathcal{Q}$ is held fixed. Hence, we obtain (32.50).

Now we can deduce the T dependence of the entropy by the Maxwell relation

$$\left(\frac{\partial\mu}{\partial T}\right)_{\mathcal{Q}} = -\frac{1}{N}\left(\frac{\partial S}{\partial \mathcal{Q}}\right)_T; \tag{32.55}$$

the $1/N$ is needed because we define S to be the total extensive entropy, and so we must use the total number $N\mathcal{Q}$ in the Maxwell relation. Applying this to (32.50) we obtain

$$\frac{1}{N}\left(\frac{\partial S}{\partial \mathcal{Q}}\right)_T = 2\pi\mathcal{E} \neq 0 \;\text{ as } T \to 0. \tag{32.56}$$

In Section 32.2.2, we obtained an "extended" Luttinger relationship between the density $\mathcal{Q}$ and the particle–hole asymmetry parameter $\mathcal{E}$. Assuming that $S = 0$ at $\mathcal{Q} = 0$, we can now integrate (32.56) to obtain for the entropy S [91]:

$$S(T \to 0) = N\mathcal{S} \quad , \quad \mathcal{S} = 2\pi \int_0^{\mathcal{Q}} d\mathcal{Q}\mathcal{E}(\mathcal{Q}), \tag{32.57}$$

where the function $\mathcal{E}(\mathcal{Q})$ is determined by eliminating θ between (32.29) and (32.41).

The remarkable feature of this result is that the entropy S is extensive, that is, proportional to N, as $T \to 0$. Specifically, we have

$$\lim_{T\to 0}\lim_{N\to\infty} \frac{S}{N} \neq 0. \tag{32.58}$$

The order of limits is crucial here; the above order of limits defines the zero temperature entropy density, in which the thermodynamic limit is taken before the zero-temperature limit. If we had taken the other order of limits, we would obtain the ground-state entropy density, which does indeed vanish.

32.2.5 Corrections to Scaling

Our analysis of the large $N = \infty$ theory has so far focused on the leading scaling behavior at $T \ll U$. Given the gapless nature of the theory, we expect that all corrections to this leading behavior will scale with powers of T/U. In this subsection we determine the possible powers of T of these subleading terms, and the ratios of the coefficients of the first subleading term.

To understand the structure of the possible corrections, we postulate that the low-energy corrections can be computed from an effective action of the following form:

$$I = I_* + \sum_h g_h \int_0^{\beta} d\tau\, O_h(\tau), \tag{32.59}$$

where O_h are a set of scaling operators with scaling dimension h. One of our tasks for this subsection is to determine the possible values of h, and we will accomplish this shortly. The term I_* is the leading critical theory that leads to the results described so far; in particular to the Green's function in (32.23) and (32.46), and the entropy in (32.57). We normalize the perturbing operators by the two-point correlator

$$\langle O_h(\tau) O_h(0) \rangle = \frac{1}{|\tau|^{2h}}; \tag{32.60}$$

then, the coefficient g_h is fully specified. In general, the g_h are a set of non-universal numbers of order U^{1-h}, whose precise values depend upon the details of the underlying theory, for example, on possible higher-order fermion interaction terms we can add to the SYK Hamiltonian.

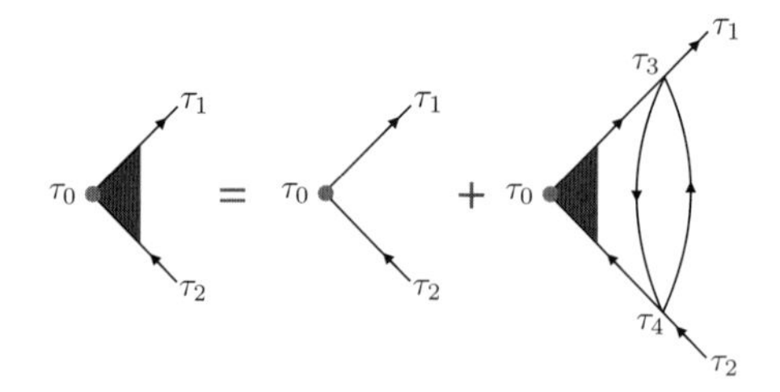

Figure 32.5 Large-N equation satisfied by the three-point correlator in (32.64). The filled circle represents the operator O_h.

Given (32.59), we can use the scaling dimension of O_h to estimate the form of the corrections to the grand potential $\Omega(T)$:

$$\Omega(T) = E_0 - N\mathcal{S}T + \sum_h \Omega_h T^h, \tag{32.61}$$

where E_0 is the ground-state energy, $\mathcal{S}$ is the entropy in (32.24), and the set of coefficients Ω_h are determined by

$$\langle O_h \rangle_{T*} = \frac{\Omega_h}{g_h} T^h, \tag{32.62}$$

where the expectation value is evaluated at a temperature T in I_*. Similarly, we can write the corrections to the Green's function in (32.27) from the O_h perturbations:

$$G(\tau) = G_*(\tau)\left(1 + \sum_h \frac{\alpha_h}{|\tau|^{h-1}}\right), \quad G_*(\tau) = -B\frac{\text{sgn}(\tau)}{\sqrt{U|\tau|}}, \tag{32.63}$$

where we now use G_* to denote the leading-order result in (32.42). Here, and below, we limit ourselves to the particle–hole symmetric case with $\theta = 0$, $\mu = 0$, $\mathcal{E} = 0$, and refer to Ref. [279] for the general case. Like g_h, the coefficients Ω_h and α_h are non-universal numbers of order U^{1-h}. However, we expect the ratio of the coefficients of the corrections in the observables in (32.61) and (32.63), α_h/Ω_h, to be universal, and we describe its determination below.

Our remaining tasks are to determine the allowed values of h, and then determine the ratio α_h/Ω_h.

Our evaluation of h follows Refs. [96, 142, 143], and we only consider the "antisymmetric" operators O_h, which are represented at short times by $O_{h_n} = c_i^\dagger \partial_\tau^{2n+1} c_i$ with $n = 0, 1, 2, \ldots$. The needed information is contained in the three-point functions

$$v_h(\tau_1, \tau_2, \tau_0) = \langle c(\tau_1) c^\dagger(\tau_2) O_h(\tau_0) \rangle. \tag{32.64}$$

In the large-N limit, this three-point function obeys the integral equation shown in Fig. 32.5. In the long-time scaling limit, we can drop the bare first term on the right-hand side, and then Fig. 32.5 reduces to the eigenvalue equation [96]

$$k(h)v(\tau_1,\tau_2,\tau_0) = \int d\tau_3 d\tau_4 K(\tau_1,\tau_2;\tau_3,\tau_4) v_h(\tau_3,\tau_4,\tau_0), \tag{32.65}$$

where the kernel K is

$$K(\tau_1,\tau_2;\tau_3,\tau_4) = -3U^2 G_*(\tau_{13}) G_*(\tau_{24}) G_*(\tau_{34})^2, \tag{32.66}$$

with $\tau_{ij} \equiv \tau_i - \tau_j$, and we have introduced an eigenvalue $k(h)$ by hand, which must obey

$$k(h) = 1. \tag{32.67}$$

The solution of (32.65) is aided by an assumption of conformal symmetry, which implies that the three-point functions obey the functional form [96]

$$v(\tau_1,\tau_2,\tau_0) = \frac{c_h B \,\mathrm{sgn}(\tau_{12})}{U^{1/2}|\tau_{12}|^{1/2-h}|\tau_{10}|^h|\tau_{20}|^h}, \tag{32.68}$$

where we have introduced a set of dimensionless "structure constants" c_h, which are described further below. Inserting (32.66) and (32.68) into (32.65), and evaluating the integrals over τ_3 and τ_4, it can be verified that (32.68) is indeed a solution of (32.65). Our interest is mainly to determine $k(h)$, and then we don't actually need the full form of v in (32.68). We can use the limit $\tau_0 \to \infty$, where we can assume $v \sim \mathrm{sgn}(\tau_{12})/|\tau_{12}|^{1/2-h}$; then, evaluation of (32.65) yields the eigenvalue

$$k(h) = -\frac{3\tan(\pi h/2 - \pi/4)}{2h-1}. \tag{32.69}$$

The solution of (32.67) and (32.69) finally yields the needed values of h. There are an infinite number of solutions, and the lowest values are $h = 2, 3.77354, \ldots, 5.567946, \ldots, 7.63197, \ldots, \ldots$. Only the lowest value $h = 2$ is an integer, and all higher values are irrational numbers. We have a particular interest in the $h = 2$ operator in the remaining discussion in this chapter.

To determine the prefactors of the correction to scaling, we need more information on the values of the structure constants c_h. The name alludes to its appearance in the operator product expansion

$$\frac{1}{N}\sum_i c_i(\tau_1)c_i^\dagger(\tau_2) = \frac{B\,\mathrm{sgn}(\tau_{12})}{\sqrt{U|\tau_{12}|}} + \sum_h \frac{c_h B\,\mathrm{sgn}(\tau_{12})}{U^{1/2}|\tau_{12}|^{1/2-h}} O_h(\tau_2) + \cdots \tag{32.70}$$

as $\tau_{12} \to 0$. It is easy to show that (32.70) is consistent with (32.68). The structure constants were also computed using the conformal structure of the theory [96, 170], and they are given by

$$c_h^2 = \frac{1}{3U^{1/2}B} \cdot \frac{(h-1/2)}{\pi\tan(\pi h/2)} \frac{\Gamma(h)^2}{\Gamma(2h)} \cdot \frac{1}{k'(h)}. \tag{32.71}$$

An important, and initially surprising, feature of (32.71) is that c_h^2 has a pole at the lowest scaling dimension $h = 2$. We will see that this pole cancels out in the universal ratio of the $N = \infty$ theory, α_2/Ω_2 that we are trying to compute. However, the presence of this pole is an indication that the $1/N$ corrections from the $h = 2$ operator are singular, and need to be resummed – we refer the reader to the review by Chowdhury et al.

[46] for further discussion. There is no pole in (32.71) for the higher scaling dimensions h, and their fluctuation corrections are not as important.

We can now relate both α_h and Ω_h to c_h and g_h for $h = 2$, and hence obtain the needed universal ratio. To compute α_h, we directly compute $G(\tau)$ by expanding to first order in g_h in (32.59) and then, using (32.68), we obtain from the correction to scaling of the Green's function

$$-G_*(\tau_{12})\frac{\alpha_h}{|\tau_{12}|^{h-1}} = \frac{g_h c_h B}{U^{1/2}}\int_{-\infty}^{+\infty} d\tau_3 \frac{\text{sgn}(\tau_{12})}{|\tau_{12}|^{1/2-h}|\tau_{13}|^h|\tau_{23}|^h}. \tag{32.72}$$

Evaluating the integral by analytically continuing from the values $1/2 < h < 1$ for which it converges, we obtain

$$\alpha_h = -g_h c_h \frac{\pi \tan(\pi h/2)\sec(\pi h)\Gamma(1-h)}{\Gamma(2-2h)\Gamma(h)}. \tag{32.73}$$

To obtain Ω_h, we need $\langle O_h \rangle_{T*}$, which we can constrain by taking the expectation value of the operator product expansion in (32.70) using (32.62)

$$G(\tau) = -B\frac{\text{sgn}(\tau)}{\sqrt{U|\tau|}}\left[1 + \sum_h c_h \frac{\Omega_h}{g_h}|T\tau|^h + \cdots\right]. \tag{32.74}$$

On the other hand, we know from (32.42) that

$$G(\tau) = -B\frac{\text{sgn}(\tau)}{\sqrt{U|\tau|}}\left[1 + \frac{\pi^2}{12}|T\tau|^2 + \cdots\right]. \tag{32.75}$$

Combining (32.74) and (32.75), we obtain, for $h = 2$,

$$\Omega_2 = \frac{g_2}{c_2}\frac{\pi^2}{12}. \tag{32.76}$$

Finally, from (32.73) and (32.76) we have the required amplitude ratio

$$\frac{\Omega_2}{\alpha_2} = \frac{\pi^2}{4}Bk'(2). \tag{32.77}$$

Note that the pole at $h = 2$ in (32.71) has cancelled against the zero in (32.73). This implies that Ω_h and α_h observables characterizing the corrections to scaling at $N = \infty$ remain finite as $h \to 2$, while the strength of the normalized O_2 operator diverges as $g_h \sim |h-2|^{-1/2}$.

To conclude this section, we reiterate the important implication of these results for the low-temperature entropy and the fermion Green's function. Taking the T derivative of (32.61), we obtain a sharper version of (32.57)

$$S(T \to 0) = N(\mathcal{S} + \gamma T). \tag{32.78}$$

So entropy vanishes linearly with temperature, as does the corresponding contribution to the specific heat. The linear-in-T coefficient of the specific heat γ (compare to (2.12) in a Fermi liquid, and (32.14) in the random-matrix model) is now given by (32.77):

$$\gamma = -\frac{\pi^2}{2}Bk'(2)\alpha_2, \tag{32.79}$$

where $k'(2) = -(2/3+\pi)$ from (32.69). The value of γ depends upon the coefficient of the leading correction to the Green's function in (32.63):

$$G(\tau) = -\frac{B\,\text{sgn}(\tau)}{\sqrt{|\tau|}}\left(1+\frac{\alpha_2}{|\tau|}+\cdots\right). \tag{32.80}$$

But the value of $B\alpha_2$ itself is not universal, and depends upon the precise microscopic model under consideration.

32.3 G–Σ Effective Action

In this section, we obtain the large-N theory described so far as the saddle point of an effective action. This is an essential step towards understanding finite-N corrections to the above results. It turns out that there are interesting emergent symmetries at low energy in the effective action, and this enables an exact resummation of the most singular fluctuation corrections. We describe the emergent symmetries here, but refer the reader to the review by Chowdhury et al. [46] for further details of the fluctuation corrections.

We begin with a path-integral representation of the underlying SYK Hamiltonian (32.19a). To treat the random couplings, we need to perform a quenched average using the replica method. Further discussion of the replica approach is deferred to Chapter 33. The SYK model is strongly self-averaging, and so we can work directly with the averaged theory, ignoring replicas for simplicity. So, after averaging over the $U_{ijk\ell}$, the path integral becomes

$$\overline{\mathcal{Z}} = \int \mathcal{D}c_i(\tau)\exp\left[-\sum_i\int_0^\beta d\tau\, c_i^\dagger\left(\frac{\partial}{\partial\tau}-\mu\right)c_i + \frac{U^2}{4N^3}\int_0^\beta d\tau d\tau'\left|\sum_i c_i^\dagger(\tau)c_i(\tau')\right|^4\right], \tag{32.81}$$

where $\beta = 1/T$. We now introduce the following "trivial" identity in the path integral,

$$\begin{aligned}1 = &\int \mathcal{D}G(\tau_1,\tau_2)\mathcal{D}\Sigma(\tau_1,\tau_2)\\ &\times\exp\left[-N\int_0^\beta d\tau_1 d\tau_2\Sigma(\tau_1,\tau_2)\left(G(\tau_2,\tau_1)+\frac{1}{N}\sum_i c_i(\tau_2)c_i^\dagger(\tau_1)\right)\right]\end{aligned} \tag{32.82}$$

and interchange the orders of integration. Then, the partition function can be written as a "G–Σ" theory, a path integral with an action $I[G,\Sigma]$ for the Green's function and the self-energy analogous to a Luttinger–Ward functional [91, 137, 170]:

$$\begin{aligned}\overline{\mathcal{Z}} &= \int \mathcal{D}G(\tau_1,\tau_2)\mathcal{D}\Sigma(\tau_1,\tau_2)\exp(-NI[G,\Sigma]),\\ I[G,\Sigma] &= -\ln\det\left[(\partial_{\tau_1}-\mu)\delta(\tau_1-\tau_2)+\Sigma(\tau_1,\tau_2)\right]\end{aligned}$$

$$- \mathrm{Tr}\,(\Sigma \cdot G) - \frac{U^2}{4} \mathrm{Tr}\left(G^2 \cdot G^2\right). \tag{32.83}$$

We have integrated over the fermions to obtain the $\ln \det$ term. This is an exact representation of the averaged partition function. Notice that it involves G and Σ as bilocal fields that depend upon two times, and we have introduced a compact notation for such fields:

$$\mathrm{Tr}\,(f \cdot g) \equiv \int d\tau_1 d\tau_2\, f(\tau_2, \tau_1) g(\tau_1, \tau_2). \tag{32.84}$$

Evaluating the variational derivatives of $I[G, \Sigma]$ with respect to G and Σ, we can now verify that we obtain the large-N saddle-point equations in (32.20a) and (32.20b).

32.3.1 Emergent Time-Reparameterization and Gauge Symmetries

It is possible to make progress in evaluating the path integral in (32.83) by exploiting its remarkable emergent symmetries at low energies, which are described here. These emergent symmetries also clarify the origin of the $T > 0$ solution of the saddle-point equations in (32.42) and (32.45).

We can describe the emergent symmetries by directly analyzing the action (32.83), but it is a bit simpler to discuss them in terms of the saddle-point equations. So we return to the original equations (32.20a) and (32.20b), and simplify them in the low-energy limit. As we saw above (32.33), at frequencies $\ll U$, the $i\omega + \mu$ can be dropped, because $\mu - \Sigma(0) = 0$ and the $i\omega_n$ term is smaller than the singular frequency dependence in $\Sigma(i\omega_n)$. After Fourier transforming to the time domain, we can rewrite the original saddle-point equations as

$$\int_0^\beta d\tau_2\, \Sigma_{sing}(\tau_1, \tau_2) G(\tau_2, \tau_3) = -\delta(\tau_1 - \tau_3), \tag{32.85a}$$

$$\Sigma_{sing}(\tau_1, \tau_2) = -U^2 G^2(\tau_1, \tau_2) G(\tau_2, \tau_1), \tag{32.85b}$$

where Σ_{sing} is the singular part of Σ. Also the saddle-point Green's functions and self-energies are functions only of time differences, such as $\tau_1 - \tau_2$. Nevertheless, we have written them as a function of two independent times, because that is the form they appear in the action (32.83). Such a bilocal in time formulation turns out to be essential for an understanding of the emergent symmetries.

It is now not difficult to verify that (32.85a) and (32.85b) are invariant under the following transformation:

$$\tau = f(\sigma), \tag{32.86a}$$

$$G(\tau_1, \tau_2) = \left[f'(\sigma_1) f'(\sigma_2)\right]^{-1/4} \frac{g(\sigma_1)}{g(\sigma_2)} \widetilde{G}(\sigma_1, \sigma_2), \tag{32.86b}$$

$$\Sigma(\tau_1, \tau_2) = \left[f'(\sigma_1) f'(\sigma_2)\right]^{-3/4} \frac{g(\sigma_1)}{g(\sigma_2)} \widetilde{\Sigma}(\sigma_1, \sigma_2), \tag{32.86c}$$

where $f(\sigma)$ and $g(\sigma)$ are arbitrary functions. Here, $f(\sigma)$ is a time reparametrization, and $g(\sigma)$ is a $U(1)$ gauge transformation in imaginary time. These are emergent symmetries because the form of the equations obeyed by $\widetilde{G}(\sigma_1,\sigma_2)$ and $\widetilde{\Sigma}(\sigma_1,\sigma_2)$ is the same as (32.85a) and (32.85b) obeyed by $G(\tau_1,\tau_2)$ and $\Sigma(\tau_1,\tau_2)$.

We obtain the non-zero temperature solution by choosing the time reparametrization in (32.86a) as the conformal map

$$\tau = \frac{1}{\pi T}\tan(\pi T \sigma), \tag{32.87}$$

where σ is the periodic imaginary-time coordinate with period $1/T$. Applying this map to (32.27), we obtain

$$G(\pm\sigma) = \mp C g(\pm\sigma)\sin(\pi/4+\theta)\left(\frac{T}{\sin(\pi T\sigma)}\right)^{1/2}, \tag{32.88}$$

for $0 < \pm\sigma < 1/T$. The function $g(\sigma)$ is so far undetermined apart from a normalization choice $g(0)=1$. We can now determine $g(\sigma)$ by imposing the Kubo–Martin–Schwinger condition

$$G(\sigma+1/T) = -G(\sigma), \tag{32.89}$$

which implies

$$g(\sigma) = \tan(\pi/4+\theta)g(\sigma+1/T). \tag{32.90}$$

The solution is clearly

$$g(\sigma) = e^{-2\pi\mathcal{E}T\sigma}, \tag{32.91}$$

where the new parameter $\mathcal{E}$ and the angle θ are related as in (32.29). This yields the final expression for $G(\sigma)$ in (32.45).

32.3.2 Symmetries of the Saddle Point

We have shown that (32.85a) and (32.85b) have a very large set of symmetries when expressed in terms of bilocal correlators of two times. However, the actual solution in (32.45) of the saddle-point equations is a function only of time differences. Now we ask a somewhat different question: what subgroup of the symmetries apply to the thermal solution in (32.45). In other words, how are the emergent low-energy time-reparametrization and gauge symmetries broken by the low-T thermal state?

First, let us consider the simplest case with particle–hole symmetry at $T=0$, when we can schematically represent the large-N solutions in Section 32.2.1 as

$$G_c(\tau_1-\tau_2) \sim (\tau_1-\tau_2)^{-1/2},$$
$$\Sigma_c(\tau_1-\tau_2) \sim (\tau_1-\tau_2)^{-3/2}.$$

The saddle point will be invariant under a reparameterization $f(\tau)$ when choosing $G(\tau_1,\tau_2) = G_c(\tau_1-\tau_2)$ leads to a transformed $\widetilde{G}(\sigma_1,\sigma_2) = G_c(\sigma_1-\sigma_2)$ (and similarly for Σ). It turns out this is true only for the $SL(2,R)$ transformations under which

$$f(\tau) = \frac{a\tau + b}{c\tau + d} \quad , \quad ad - bc = 1. \tag{32.92}$$

So the (approximate) reparametrization symmetry is spontaneously broken down to $SL(2,R)$ by the saddle point.

Now let us consider the most general case with $T > 0$ and no particle–hole symmetry. We write (32.86b) as

$$\begin{aligned} G(\tau_1, \tau_2) &= [f'(\tau_1) f'(\tau_2)]^{1/4} \\ &\quad \times G_c(f(\tau_1) - f(\tau_2)) e^{i\phi(\tau_1) - i\phi(\tau_2)}, \end{aligned} \tag{32.93}$$

where $G_c(\tau)$ is the conformal saddle-point solution given in (32.45). Here, we have parameterized $g(\tau) = e^{-i\phi(\tau)}$ in terms of a phase field ϕ.

It can now be shown that the $G(\tau_1, \tau_2)$ obtained from (32.93) equals $G_c(\tau_1 - \tau_2)$ only if the transformations $f(\tau)$ and $\phi(\tau)$ satisfy

$$\begin{aligned} \frac{\tan(\pi T f(\tau))}{\pi T} &= \frac{a \dfrac{\tan(\pi T \tau)}{\pi T} + b}{c \dfrac{\tan(\pi T \tau)}{\pi T} + d} \quad , \quad ad - bc = 1, \\ -i\phi(\tau) &= -i\phi_0 + 2\pi \mathcal{E} T (\tau - f(\tau)). \end{aligned} \tag{32.94}$$

The transformation of $f(\tau)$ looks rather mysterious, but we can simplify it as follows: we define

$$z = e^{2\pi i T \tau} \quad , \quad z_f = e^{2\pi i T f(\tau)} \tag{32.95}$$

and then the transformation in (32.94) is between unimodular complex numbers representing the thermal circle

$$z_f = \frac{w_1 z + w_2}{w_2^* z + w_1^*} \quad , \quad |w_1|^2 - |w_2|^2 = 1, \tag{32.96}$$

where $w_{1,2}$ are complex numbers. In this form, we have a $SU(1,1)$ transformation, a group that is isomorphic to $SL(2,R)$.

Finally, we note that the symmetries in (32.96) are also the isometries of two-dimensional anti-de Sitter space AdS_2, and this is an important ingredient of the connection between the SYK model and two-dimensional quantum gravity, as reviewed by Chatterjee et al. [46]. The metric of AdS_2 is

$$ds^2 = \frac{d\tau^2 + d\zeta^2}{\zeta^2} \tag{32.97}$$

in terms of coordinates τ and ζ. This metric is invariant under isometries that are $SL(2,R)$ transformations, as in (32.92). It is easy to verify that the coordinate change

$$\tau' + i\zeta' = \frac{a(\tau + i\zeta) + b}{c(\tau + i\zeta) + d}, \quad ad - bc = 1, \tag{32.98}$$

with a, b, c, d real, leaves the metric (32.97) invariant.

32.3.3 Finite-N Corrections

The emergent symmetries of the action (32.83) described in Sections 32.3.1 and 32.3.2 have led to remarkable progress in evaluating the corrections to the large-N SYK theory described so far. An important consequence of the time-reparameterization and gauge symmetries in (32.86c) is that the singular terms in the path integral in (32.83) are identical to the singular terms in the path integral of Einstein–Maxwell gravity and electromagnetism about a Reissner–Nördstrom black hole (the consequences of this correspondence are summarized below in Fig. 32.6). At low T, the most singular contribution of this path integral is dominated by the "dangerously irrelevant" contributions of the $h = 2$ operator of (32.69), which is just a linearized generator of time-reparameterization symmetry. I refer the reader to Chowdhury et al. [46] for further details and original references, and just present the main results here.

The leading finite-N correction to the free energy F at fixed $\mathcal{Q}$ is

$$-\frac{F}{T} = \ln \overline{\mathcal{Z}} = -\frac{E_0}{T} + N\mathcal{S} + \frac{N\gamma T}{2} - \frac{3}{2}\ln\left(\frac{U}{T}\right), \tag{32.99}$$

where E_0 is the non-universal ground-state energy which is order N in the large-N limit. We thus have a non-trivial finite-N correction to the entropy in (32.78) of the SYK model: the $-(3/2)\ln(1/T)$ correction to the logarithm of the partition function. It is also useful to compare (32.99) to our earlier large-N result for $-T\ln Z$ in the random-matrix model in (32.12). That had a leading $N\gamma T/2$ term, but there was no T-independent term proportional to N, as the random-matrix model does not have an extensive entropy in the zero-temperature limit. Comparing the last two terms in (32.99), we see that the $1/N$ expansion breaks down when $T \sim U\exp(-N\gamma T/3)$. This is an exponentially small T of order the many-body-level spacing (as we see below), which we do not expect a thermodynamic description to apply.

The $-(3/2)\ln(1/T)$ correction to (32.99) has important consequences for the many-body density of states at fixed $\mathcal{Q}$, $\mathcal{N}_{\mathcal{Q}}(E)$. We define this by

$$\overline{\mathcal{Z}}(T) = \int_{E_0}^{\infty} dE\, \mathcal{N}_{\mathcal{Q}}(E) e^{-E/T}, \tag{32.100}$$

where $\mathcal{N}_{\mathcal{Q}}(E) = 0$ for $E < E_0$. It turns out to be possible to determine $\mathcal{N}_{\mathcal{Q}}(E)$ by performing the inverse Laplace transform exactly using the value in (32.99). This yields [137, 275]

$$\mathcal{N}_{\mathcal{Q}}(E) \propto e^{N\mathcal{S}} \sinh\left(\sqrt{2N\gamma(E - E_0)}\right). \tag{32.101}$$

It is easier to insert the result (32.101) into (32.100), perform the E integral, and verify that we obtain (32.99).

The result (32.101) is accurate for $E \ll NU$, and even down to $E \sim U/N$. Near the lower bound it predicts a many-body density of states $\sim e^{N\mathcal{S}}$, in sharp contrast to the random-matrix model of Section 32.1, which did not have an exponentially large density of states at such low energies. We show numerical plots of the many-body density of states [54, 89, 93] for a closely related Majorana fermion model in Fig. 32.2. Notice the much larger density of states, and much smaller-level spacing near the bottom of the band, in comparison to the free-fermion random-matrix model in Fig. 32.1 of the same size. This is also evident from a comparision of the Schwarzian result in (32.101), with the free-fermion result in (32.16): the most important difference is the presence of the prefactor of $e^{N\mathcal{S}}$ in (32.101).

We now recall our discussion at the end of Section 32.1.2 where we argued that the low-lying many-body eigenstates at excitation energies of order $1/N$ could be interpreted as the sums of quasiparticle energies. In the SYK model we have order $\sim e^{N\mathcal{S}}$ energy levels even within energy $\sim 1/N$ above the many-body ground states. It is impossible to construct so many many-body eigenstates from order $\sim N$ quasiparticle states. This is therefore strong evidence that there is no quasiparticle decomposition of the many-body eigenstates of the SYK model. Note that the presence of an extensive entropy as $T \to 0$ (the non-zero value of $\mathcal{S}$) is a *sufficient*, but not a *necessary*, condition for the absence of quasiparticles; the models we study in Chapter 34 do not have quasiparticles, but do not have an extensive entropy as $T \to 0$.

32.3.4 From the SYK Model to Charged Black Holes

Figure 32.6 relates the properties of the SYK model described in the present chapter to those of a charged black hole in (3+1)-dimensional Minkowski space, a connection first pointed out in Ref. [233]. Details are reviewed elsewhere [46].

The top line of row A displays the Bekenstein–Hawking entropy of such a black hole at low T, expressed in terms of A_0, the area of its horizon at $T = 0$; also, c is the velocity of light, and G is Newton's gravitational constant. The second line of row A displays the common leading logarithmic correction to this large-N result at low T, obtained from the path integral over the time-reparameterization mode. The same effective action is obtained for the time-reparameterization mode, starting either from the SYK path integral in (32.83), or from the path integral over the spacetime metric and electromagnetic field with the Einstein–Maxwell action [236]. Row A also

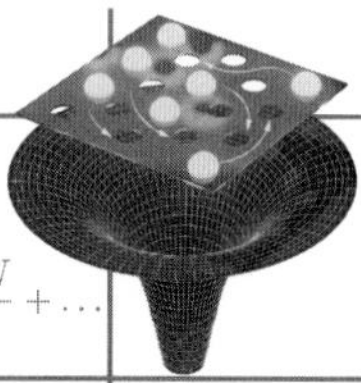

	SYK model	Charged black holes
A	$\frac{S(T)}{k_B} = N(\mathcal{S} + \gamma k_B T) - \frac{3}{2}\ln\left(\frac{U}{k_B T}\right) - \frac{\ln N}{2} + \ldots$	$\frac{S(T)}{k_B} = \frac{1}{\hbar G}\left(\frac{A_0 c^3}{4} + \frac{\sqrt{\pi} A_0^{3/2} c^2}{2}\frac{k_B T}{\hbar}\right) - \frac{3}{2}\ln\left(\frac{(\hbar c^5/G)^{1/2}}{k_B T}\right) - \frac{559}{180}\ln\left(\frac{A_0 c^3}{\hbar G}\right) + \ldots$
B	$G(\tau) \sim e^{-2\pi\mathcal{E}T\tau}\left(\frac{T}{\sin(\pi T\tau)}\right)^{2\Delta}$	$G(\tau) \sim e^{-2\pi\mathcal{E}T\tau}\left(\frac{T}{\sin(\pi T\tau)}\right)^{2\Delta}$
C	$\frac{1}{k_B}\frac{d\mathcal{S}}{d\mathcal{Q}} = 2\pi\mathcal{E}$	$\frac{1}{k_B}\frac{d\mathcal{S}}{d\mathcal{Q}} = 2\pi\mathcal{E}$
D	$\mathcal{N}_{\mathcal{Q}}(E) \sim \frac{1}{N}\exp(N\mathcal{S})\sinh\left(\sqrt{2N\gamma E}\right)$	$\mathcal{N}_{\mathcal{Q}}(E) \sim \left(\frac{A_0 c^3}{\hbar G}\right)^{-347/90}\exp\left(\frac{A_0 c^3}{4\hbar G}\right)\sinh\left(\left[\sqrt{\pi} A_0^{3/2}\frac{c^3}{\hbar G}\frac{E}{\hbar c}\right]^{1/2}\right)$

Figure 32.6 Correspondence between the SYK model and a charged black hole in asymptotically (3+1)-dimensional Minkowski space described by the Einstein–Maxwell action of gravity and electromagnetism.

contains logarithmic corrections which are independent of T: these differ between the SYK model and black holes - those for the SYK model were computed in Ref. [97], and those for black holes were computed in Ref. [116].

Row B of Fig. 32.6 shows that the Green's function of a probe fermion in a charged black hole background [74] is exactly the same as that of a fermion in the SYK model [233, 235]. In the black hole case, the spectral asymmetry parameter has the additional interpretation as a dimensionless measure of the electric field on the black hole horizon.

In row C of Fig. 32.6, the entropy $\mathcal{S}$ of the black hole is defined to be the $T = 0$ term in the Bekenstein–Hawking entropy in the first line of row A, and the total charge of the black hole is $\mathcal{Q}$. With these definitions, the black hole relation in row C is obtained from the Einstein–Maxwell action, to be compared with the relation (32.56) obtained from the structure of the large-N SYK saddle point [235].

Given the common behavior of the entropies of the SYK model and the charged black hole in row A of Fig. 32.6, the common behavior of the many-body density of states in row D follows from the definition in (32.100). The black hole result in row D [236] is a rare formula that combines Planck's constant $\hbar$ with Newton's gravitational constant G: the exponential term was obtained by Hawking, and the sinh and the prefactor follows from developments ensuing from the solution of the SYK model. All terms depend only upon the $T = 0$ area of the black hole horizon A_0 and fundamental constants of nature.

Problem

32.1 Compute the imaginary part of the self-energy of the model (32.21) to second order in U. This involves evaluating the graph in Fig. 32.3 using the zeroth-order Green's function in (32.6).

33 Random Quantum Spin Liquids and Spin Glasses

The classical, infinite-range Ising spin glass is described, and extended to the quantum rotor spin glass. The quantum model exhibits a gapped quantum paramagnet ground state, along with a spin-glass state similar to that of the classical model. The model with quantum spins with a Berry phase also has a similar spin-glass phase, but the paramagnet is a gapless spin liquid, which realizes a Sachdev–Ye–Kitaev state of spinons.

The Sachdev–Ye–Kitaev (SYK) model discussed in Chapter 32 provides a valuable example of a compressible metallic state without quasiparticle excitations. However, its Hamiltonian misses an important characteristic of correlated materials that display such phases: there is no strong local repulsion between the fermions, as there is in the electron Hubbard model of Chapter 9. Consequently, there is no analog of the Mott insulator.

The original model proposed by Sachdev and Ye [242] does focus on such local correlations. They considered a random version of the Heisenberg spin model, which was the focus of our attention in Parts II and IV. We recall the Hamiltonian in (9.14) and (15.1) in the form

$$H_J = \frac{1}{\sqrt{N}} \sum_{1 \leq i < j \leq N} J_{ij} \boldsymbol{S}_i \cdot \boldsymbol{S}_j, \tag{33.1}$$

describing $S = 1/2$ $SU(2)$ spins on sites i with Heisenberg exchange couplings J_{ij}. In this chapter, we turn our attention to (33.1) for the case where J_{ij} are independent random numbers with zero mean and variance J, and act between *any* pair of sites i, j.

Unlike the SYK model, the Hamiltonian (33.1) is not solvable in the limit $N \to \infty$. However, as pointed out in Ref. [242], it is solvable when we generalize it to $SU(M)$ spins, and take the $M \to \infty$ limit after the $N \to \infty$ limit. Then, as described in Section 33.3, we obtain a solution that is formally identical to that of the SYK model in Chapter 32. As we note in Section 21.2, the SYK non-Fermi liquid is now a metallic state of partons, which are fermionic spinons, without quasiparticle excitations. Thus, the $SU(M \to \infty)$ model realizes compressible state of spins (sometimes called a "Bose metal"), and the spin density is continuously variable by an applied Zeeman field; this justifies the study of (33.1) in Part V on correlated metals.

An important question for the application to realistic materials is the extent to which the $SU(M \to \infty)$ solution applies to the $SU(2)$ case. As discussed in

Section 33.3, the SYK non-Fermi liquid applies down to an energy or temperature scale

$$T \sim J \exp(-\sqrt{\pi M}) \tag{33.2}$$

below which the $SU(M)$ generalization of (33.1) is unstable to the appearance of spin glass order. Note that this energy scale is exponentially small for large M, in which case the non-Fermi liquid behavior of spinons is visible over a wide intermediate energy scale. As we see in Section 33.3.4, there is clear numerical evidence for the non-Fermi liquid behavior of spinons even for the $M = 2$ case, when the exponential factor in (33.2) is ≈ 0.08, which is reasonably small.

With the appearance of spin-glass order in (33.1), it is useful to begin our discussion by first reviewing spin-glass theory in simpler classical and quantum models that do not include the Berry phase terms in (18.23) and (A.38): this we do in Sections 33.1 and 33.2.

Finally, we note that it is possible to extend the analysis of the random Heisenberg exchange model in (33.1) to the doped case of a random t–J model [48, 280]: I refer the reader to the review by Chowdhury et al. [46] for further discussion of this.

33.1 Classical Ising Spin Glass

We begin our discussion of spin glasses by considering the celebrated Sherrington–Kirkpatrick model of Ising spins with all-to-all and random couplings. This is defined by

$$\begin{aligned} H_{SK} &= \frac{1}{2\sqrt{N}} \sum_{i,j=1}^{N} J_{ij} \sigma_i \sigma_j, \\ \mathcal{Z}(J_{ij}) &= \sum_{\sigma_i = \pm 1} e^{-H_{SK}/T}, \\ \overline{J_{ij}} = 0, \quad \overline{J_{ij}^2} = J^2, &\quad \text{different } J_{ij} \text{ uncorrelated.} \end{aligned} \tag{33.3}$$

We have emphasized here that the partition function $\mathcal{Z}$ depends upon the values of the specific J_{ij} couplings.

At high T, the Ising spins σ_i are in a random thermal state, with the thermal expectation value $\langle \sigma_i \rangle = 0$ for each J_{ij} realization. As we lower the temperature, as in a non-random Ising model, we expect a phase transition below a critical $T = T_{sg}$ to magnetic order in the σ_i, and now we will have $\langle \sigma_i \rangle \neq 0$. However, the values of $\langle \sigma_i \rangle$ for different i depend sensitively on the particular J_{ij} chosen. In particular, if we average over the J_{ij}, the random order will average to zero, and so

$$\overline{\langle \sigma_i \rangle} \equiv \int \mathcal{D} J_{ij} P(J_{ij}) \langle \sigma_i \rangle = 0, \tag{33.4}$$

where $P(J_{ij})$ is the probability distribution of the J_{ij}.

The Edwards–Anderson order parameter evades this problem by averaging the square of the magnetic order

$$q_{EA} \equiv \overline{\langle\sigma_i\rangle^2} \equiv \int \mathcal{D}J_{ij}P(J_{ij})\langle\sigma_i\rangle^2, \tag{33.5}$$

and we have $q_{EA} \neq 0$ for $T < T_{sg}$, the spin-glass transition temperature. More explicitly, let us write out the expression for q_{EA} as

$$q_{EA} = \int \mathcal{D}J_{ij}P(J_{ij})\frac{1}{[\mathcal{Z}(J_{ij})]^2}\left(\sum_{\sigma_i=\pm1}\sigma_i e^{-H_{SK}/T}\right)^2. \tag{33.6}$$

The difficulty in evaluating (33.6) lies in taking the average over the dependence of J_{ij} in the denominator $[\mathcal{Z}(J_{ij})]^2$. The replica method gets around this difficulty by formally moving the denominator to the numerator by writing

$$q_{EA} = \lim_{n\to0}\int \mathcal{D}J_{ij}P(J_{ij})[\mathcal{Z}(J_{ij})]^{n-2}\left(\sum_{\sigma_i=\pm1}\sigma_i e^{-H_{SK}/T}\right)^2. \tag{33.7}$$

Now, it is relatively easier to evaluate the expression in (33.7) for integer $n \geq 2$. The replica method evaluates (33.7) for all integer $n \geq 2$, and then analytically continues the result to $n = 0$. We can clean up the notation in (33.7) by introducing replicas of the Ising spins, σ_i^a, with $a = 1,\ldots,n$ and $i = 1,\ldots,N$ and defining the replicated partition function and its correlator

$$\overline{\mathcal{Z}^n} = \int \mathcal{D}J_{ij}P(J_{ij})\sum_{\sigma_i^a=\pm1}\exp\left(-\frac{1}{2T\sqrt{N}}\sum_{a=1}^{n}\sum_{i,j=1}^{N}J_{ij}\sigma_i^a\sigma_j^a\right),$$

$$q_{bc} = \int \mathcal{D}J_{ij}P(J_{ij})\sum_{\sigma_i^a=\pm1}\sigma_i^b\sigma_i^c\exp\left(-\frac{1}{2T\sqrt{N}}\sum_{a=1}^{n}\sum_{i,j=1}^{N}J_{ij}\sigma_i^a\sigma_j^a\right), \tag{33.8}$$

where b and c are any pair of replicas chosen from $a = 1,\ldots,n$. Then, (33.7) is equivalent to

$$q_{EA} = \lim_{n\to0}\frac{1}{n(n-1)}\sum_{a\neq b}q_{ab}. \tag{33.9}$$

The advantage of expressions like (33.8) is that the average over the J_{ij} can be readily evaluated. We have

$$\overline{\mathcal{Z}^n} = \sum_{\sigma_i^a=\pm1}\exp\left(\frac{J^2}{4T^2N}\sum_{a,b=1}^{n}\left[\sum_{i=1}^{N}\sigma_i^a\sigma_i^b\right]^2\right). \tag{33.10}$$

And now with the aid of a Hubbard–Stratonovich transformation, we can reduce the partition function to that of a single $n \times n$ matrix q_{ab}, with the number of sites N appearing only as a parameter

$$\overline{\mathcal{Z}^n} = \int \mathcal{D}q_{ab}\exp\left(-\frac{NJ^2}{2T^2}q_{ab}^2\right)[\mathcal{Z}_s(q_{ab})]^N. \tag{33.11}$$

Here, $\mathcal{Z}_s(q_{ab})$ is a *single-site* partition function of n Ising spins σ^a coupled to q_{ab}:

$$\mathcal{Z}_s(q_{ab}) = \sum_{\sigma^a=\pm 1} \exp\left(\frac{J^2}{T^2} q_{ab}\sigma^a\sigma^b\right). \tag{33.12}$$

We can also see that the expectation value $\langle q_{bc}\rangle$ under the partition function in (33.11) co-incides with that in (33.8), validating our choice of the q_{ab} as the Hubbard–Stratonovich field.

The large-N limit of (33.11) is now easily taken, and we obtain a free energy per site for n replicas as a function of q_{ab}:

$$F(q_{ab}) = -T\ln\overline{\mathcal{Z}^n} = \frac{J^2}{2T} q_{ab}^2 - T\ln\left[\mathcal{Z}_s(q_{ab})\right]. \tag{33.13}$$

Our remaining task is to solve the saddle-point equations

$$\frac{\partial F}{\partial q_{ab}} = 0 \quad \Rightarrow \quad q_{ab} = \left\langle \sigma^a\sigma^b \right\rangle_{\mathcal{Z}_s} \tag{33.14}$$

for the optimal self-consistent value of the $n\times n$ matrix q_{ab}, and analytically continue the result to $n=0$.

The task defined by (33.13) and (33.14) was solved by Parisi, and requires the introduction of replica symmetry breaking to obtain a solution in the spin-glass phase $T < T_{sg}$. We do not need to enter into the technical complexity of replica symmetry breaking here because the quantum replica symmetry breaking is very similar to that in the classical problem, and only indirectly influences the quantum excitation spectrum we are interested in. We refer the reader to the book by Fischer and Hertz [81] for further details on the solution of (33.13) and (33.14). Section 33.2 describes a closely related quantum spin-glass model, which we solve also in the classical limit; this provides a simpler realization of spin-glass saddle-point equations similar to (33.13) and (33.14), with the solution in (33.41), and this is sufficient for our purposes.

33.2 Quantum Rotor Spin Glass

The simplest way to extend the Sherrington–Kirkpatrick Ising model in (33.3) to a quantum model is to add a transverse field on each Ising spin:

$$H = \frac{1}{2\sqrt{N}} \sum_{i,j=1}^{N} J_{ij}\sigma_i^z\sigma_j^z - g\sum_i \sigma_i^x. \tag{33.15}$$

Here, we have replaced the Ising spin $\sigma_i = \pm 1$ with the Pauli matrix σ_i^z, and added a field g acting along the x direction in spin space. The two terms in (33.15) do not commute with each other, and so we obtain a model of a quantum spin glass.

By considering the behavior at large and small g, we are led to propose the phase diagram in Fig. 33.1. At $g=0$, (33.15) reduces to the classical Ising spin glass studied in Section 33.1, and so we have a high-temperature paramagnetic phase, and a low-temperature spin-glass phase. As we see below, the spin-glass order is stable to a small

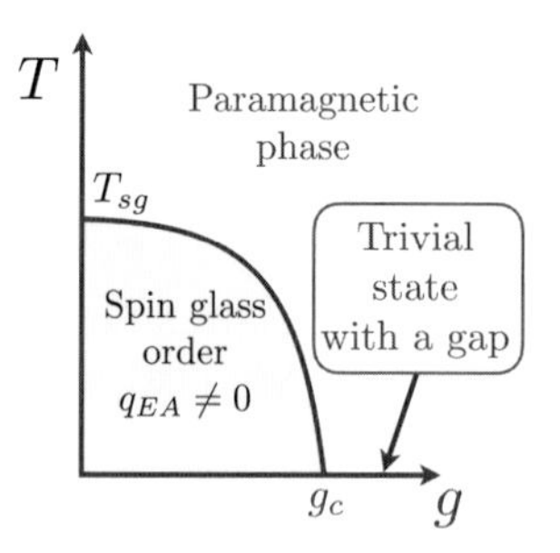

Figure 33.1 Phase diagram of the quantum Ising spin-glass Hamiltonian (33.15) or the quantum rotor model (33.18) as a function of temperature T and transverse field g.

non-zero g. At $T = 0$ and large g, we can also see that (33.15) has a non-degenerate ground state

$$|\Rightarrow\rangle = \prod_i |\rightarrow\rangle_i, \tag{33.16}$$

where $|\rightarrow\rangle_i$ is the eigenstate of σ_i^x with eigenvalue $+1$. It is also not difficult to see that there is a gap to all excitations for sufficiently large g. This trivial "quantum paramagnet" is similar to that in Section 16.4.1. So, there is a $T = 0$ quantum phase transition at a critical value $g = g_c$ between the spin glass and the paramagnet, which is described below.

We could now proceed with a replica analysis of the Ising model in (33.15), but will turn our attention to a closely related model that has the same basic structure of the phases, and the phase diagram of Fig. 33.1 continues to apply. This is the quantum spin glass of M-component rotors, $n_\mu(i)$ on N sites i, $\mu = 1,\ldots,M$. We take rotors to obey the fixed length constraint

$$\sum_{\mu=1}^{M} [n_\mu(i)]^2 = M \tag{33.17}$$

on each site i. The Lagrangian of these rotors is

$$\mathcal{L} = \frac{1}{2g}\sum_i \left(\frac{\partial n_\mu(i)}{\partial \tau}\right)^2 + \frac{1}{\sqrt{N}}\sum_{i<j} J_{ij} n_\mu(i) n_\mu(j), \tag{33.18}$$

where the J_{ij} are random couplings with a zero mean and a root-mean-square value J. We can now verify that the large-g ground state of this model is also trivial, and is a product of zero angular momentum states of each rotor. At small g and low T, we expect a spin glass state where the $O(M)$ rotational symmetry is broken in a fixed realization of J_{ij}. The resulting phase diagram is essentially identical to Fig. 33.1, and we will shortly show this in a full solution in the large M limit.

As we see below, the theory of the quantum rotor model is expressed in terms of the quantum generalization of the spin-glass order parameter in (33.14), in which the components of the replica matrix are also functions of imaginary time τ, τ'

$$Q_{ab,\mu\nu}(\tau,\tau') = \langle n_{a\mu}(\tau) n_{b\nu}(\tau')\rangle . \tag{33.19}$$

Here, $n_{a\mu}$ is the replicated version of the quantum rotor coordinate n_μ, and Q also has $O(M)$ indices. Time-translation invariance of the quantum problem imposes some important constraints on the structure of Q. The quantum analog of the replica identity in (33.5) and (33.9) is now

$$\overline{\langle n_\mu(\tau)\rangle \langle n_\nu(\tau')\rangle} = \lim_{n\to 0} \frac{1}{n(n-1)} \sum_{a\neq b} Q_{ab,\mu\nu}(\tau,\tau')\,. \tag{33.20}$$

The left-hand side of (33.20) is independent of τ and τ' because the one-point expectation values are independent of time for each realization of the J_{ij}; so $Q_{ab,\mu\nu}$ is independent of τ and τ' for $a \neq b$, and we can identify these time-independent values with the spin-glass order parameter. We also expect the spin-glass state to have a statistical $O(M)$ symmetry, and so we identify

$$Q_{ab,\mu\nu}(\tau,\tau') = q_{ab}\delta_{\mu\nu}\,, \quad a \neq b\,, \tag{33.21}$$

where q_{ab} plays exactly the same role as in the classical Ising model in Section 33.1. Unlike the classical spin glass, the diagonal components of Q also play an important role, as they contain information on the excitation spectrum of the model. The analog of (33.20) for the diagonal components is

$$\overline{\langle n_\mu(\tau) n_\nu(\tau')\rangle} = \lim_{n\to 0} \frac{1}{n} \sum_{a} Q_{aa,\mu\nu}(\tau,\tau')\,. \tag{33.22}$$

Now, the quantum dynamics in any fixed realization of J_{ij} implies that the left-hand side of (33.22) is a function only of $\tau - \tau'$, and so we generalize (33.21) to

$$Q_{ab,\mu\nu}(\tau,\tau') = \delta_{\mu\nu}\, Q(\tau-\tau')\,, \quad a = b\,. \tag{33.23}$$

We see from (33.22) that $Q(\tau)$ is an autocorrelation function of the rotor coordinate.

The task of the theory is to determine the values of q_{ab} and $Q(\tau)$ as a function of g and T for the Lagrangian in (33.18). At $T = 0$, there is an important consistency requirement on these values, which are satisfied by our results below. Given the rotor autocorrelation, we can also identify the Edwards–Anderson order parameter by its long-time limit, as that is another way to characterize the broken symmetry. So we have from (33.5) and (33.23)

$$q_{EA} = \lim_{|\tau|\to\infty} Q(\tau)\,, \quad T = 0\,. \tag{33.24}$$

Thus, there are two ways to obtain the Edwards–Anderson order parameter for a quantum spin glass at $T = 0$, from (33.9) and from (33.24), and these must equal each other.

33.2.1 Effective Action

Let us now proceed with a replica analysis of the Lagrangian in (33.18), and obtain a combined effective action for q_{ab} and $Q(\tau)$. We introduce replicas $a = 1, \ldots, n$, and average over J_{ij} to obtain the replicated partition function

$$\overline{\mathcal{Z}^n} = \int \mathcal{D}n_{a\mu}(i,\tau)\mathcal{D}\lambda_a(i,\tau)\exp\left[-S_n - S_J\right],$$
$$S_n = \frac{1}{2g}\sum_i \int d\tau \left[\left(\frac{\partial n_{a\mu}(i)}{\partial \tau}\right)^2 + i\lambda_a(i)\left([n_{a\mu}(i)]^2 - M\right)\right],$$
$$S_J = -\frac{J^2}{4N}\int d\tau d\tau' \left[\sum_i n_{a\mu}(i,\tau)n_{b\nu}(i,\tau')\right]^2. \quad (33.25)$$

The constraint (33.17) has been imposed by a Lagrange multiplier λ. We can now decouple S_J with a Hubbard–Stratonovich field $Q_{ab,\mu\nu}(\tau,\tau')$, just as in (33.10), and take the large-N limit. Then, the problem reduces to finding saddle points of the action

$$\frac{\mathcal{S}[Q]}{N} = \frac{J^2}{4}\int d\tau d\tau' [Q_{ab,\mu\nu}(\tau,\tau')]^2 - \ln \mathcal{Z}_n[Q], \quad (33.26)$$

where $\mathcal{Z}_n[Q]$ is the single-site partition function:

$$\mathcal{Z}_n[Q] = \int \mathcal{D}n_{a\mu}(\tau)\mathcal{D}\lambda_a(\tau)\exp\left[-S_n\right],$$
$$S_n = \frac{1}{2g}\int d\tau \left[\left(\frac{\partial n_{a\mu}}{\partial \tau}\right)^2 + i\lambda_a\left(n_{a\mu}^2 - M\right)\right]$$
$$- \frac{J^2}{2}\int d\tau d\tau' Q_{ab,\mu\nu}(\tau,\tau')n_{a\mu}(\tau)n_{b\nu}(\tau'). \quad (33.27)$$

There is a close similarity to the structure of the classical Ising model in (33.10), and now we have to consider a single-site quantum problem that generalizes the classical problem in (33.12). There is no remaining path integral over Q because we have taken the large-N limit, and we simply have to find the saddle points of $\mathcal{S}[Q]$ in (33.26). Let us assume that the saddle point does not break spin rotation symmetry; this is true in both the spin-glass, and spin-liquid phases. So we employ the ansatz

$$Q_{ab,\mu\nu}(\tau,\tau') = \delta_{\mu\nu}\, Q_{ab}(\tau-\tau'), \quad (33.28)$$

where $Q_{ab}(\tau)$ is a real function. The saddle-point equation for Q is

$$Q_{ab}(\tau-\tau') = \frac{1}{M}\sum_\mu \left\langle n_{a\mu}(\tau)n_{b\mu}(\tau')\right\rangle_{\mathcal{Z}_n[Q]}. \quad (33.29)$$

The different μ components are now decoupled, and we can also perform the path integral over $n_{a\mu}$ in (33.27) and obtain a compact expression for $\mathcal{Z}_n[Q]$:

$$\mathcal{Z}_n[Q] = \int \mathcal{D}\lambda_a(\tau)\exp\left[-S_\lambda[Q]\right],$$
$$S_\lambda[Q] = \frac{M}{2}\ln\det\left[-\delta''(\tau-\tau')\delta_{ab} + i\lambda_a(\tau)\delta(\tau-\tau')\delta_{ab} - gJ^2 Q_{ab}(\tau,\tau')\right]$$
$$- \frac{iM}{2g}\int d\tau \lambda_a(\tau). \quad (33.30)$$

From (33.26), the large-N saddle-point equations determining $Q_{ab}(\tau)$ are

$$J^2 Q_{ab}(\tau) = \frac{\delta \ln \mathcal{Z}_n[Q]}{\delta Q_{ab}(\tau)}. \quad (33.31)$$

It now remains to perform the path integral over $\lambda_a(\tau)$ in (33.30) and solve (33.31).

So far, our analysis has been valid for all M. However, the rest of Section 33.2 works in the large-M limit in order to simplify solutions of (33.30) and (33.31). In this limit, the path integral over $\lambda_a(\tau)$ can be evaluated also via a saddle point, and we can set $i\lambda_a = \overline{\lambda}$. Then, after a Fourier transform to Matsubara frequences, the equations for $Q_{ab}(i\omega_n)$ and $\overline{\lambda}$ reduce to the following algebraic equations:

$$Q_{ab}(i\omega_n) = g\left[(\omega_n^2 + \overline{\lambda})\delta_{ab} - gJ^2 Q_{ab}(i\omega_n)\right]^{-1}, \tag{33.32}$$

$$T\sum_{\omega_n} Q_{aa}(i\omega_n) = 1. \tag{33.33}$$

We solve (33.32) and (33.33) in the paramagnetic and spin-glass regions of Fig. 33.1 in the following subsections.

33.2.2 Paramagnetic Phase

In the paramagnetic phase, we can assume that Q_{ab} is a replica diagonal matrix. Then, we can solve (33.32) in closed form

$$Q_{ab}(i\omega_n) = \frac{2g\delta_{ab}}{\omega_n^2 + \overline{\lambda} + \left[\left(\omega_n^2 + \overline{\lambda}\right)^2 - 4g^2 J^2\right]^{1/2}}. \tag{33.34}$$

After analytic continuation to real ω, this implies a non-zero spectral weight for $\overline{\lambda} - 2Jg < \omega^2 < \overline{\lambda} + 2Jg$. We require $\overline{\lambda} > 2Jg$ for a consistent paramagnetic solution.

The value $\overline{\lambda}$ is determined by solving (33.33). This has to be done numerically, in general, and leads to a restricted region of the g, T plane where we have a solution with $\overline{\lambda} > 2Jg$. This defines the region of stability of the paramagnetic phase, which is schematically sketched in Fig. 33.1. At $T = 0$, the paramagnetic phase is present for $g > g_c$, and we can determine g_c by setting $\overline{\lambda} = 2Jg_c$ in (33.33) at $T = 0$. This yields

$$g_c = \frac{9\pi^2 J}{16}. \tag{33.35}$$

It is also useful to compute the imaginary part of the dynamic spin susceptibility in the paramagnetic phase

$$\chi''(\omega) = \mathrm{Im}\, Q_{aa}(\omega + i0^+). \tag{33.36}$$

A plot of $\chi''(\omega)$ in the $T = 0$ paramagnetic phase and at the critical point $g = g_c$ in Fig. 33.2. Note that there is a excitation energy gap $= (\overline{\lambda} - 2gJ)^{1/2}$ that vanishes continuously as g approaches g_c from above. At the critical point, we have a linear frequency dependence

$$\chi''(\omega) = \frac{4}{3\pi J^2}\omega + \cdots, \quad g = g_c, T = 0 \tag{33.37}$$

at small ω.

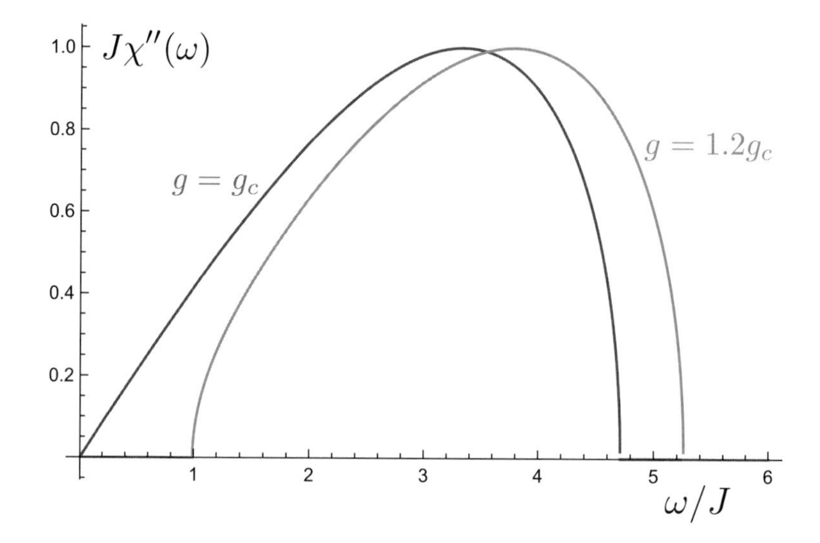

Figure 33.2 Dynamic spin susceptibility of the quantum rotor model at $T = 0$ in the paramagnetic phase ($g > g_c$) and at the critical point ($g = g_c$).

33.2.3 Spin-Glass Phase

To obtain a sensible solution of (33.32) and (33.33) in the spin-glass phase, we have to include the replica off-diagonal components of Q_{ab}. Fortunately, it turns out that the stable solution is replica, which means that all the off-diagonal components of Q_{ab} are equal to each other. This is a special feature of the large-M limit: we do have to allow for replica symmetry breaking in the solution of (33.30) and (33.31) at finite values of M, as has been discussed at length in Ref. [220]. There have also been studies of quantum rotor models with multi-spin interactions, which display replica symmetry breaking already in the large-M limit [12, 55].

Assuming replica symmetry, we can now make a suitable ansatz for $Q_{ab}(i\omega_n)$ to solve (33.32) and (33.33). We set

$$Q_{ab}(i\omega_n) = \beta q_{EA} \delta_{\omega_n,0}, \quad a \neq b, \tag{33.38}$$

where $\beta = 1/T$ and q_{EA} is the Edwards–Anderson order parameter. For the replica diagonal components, it is convenient to define

$$Q_{aa}(i\omega_n) = \beta q_{EA} \delta_{\omega_n,0} + Q_r(i\omega_n). \tag{33.39}$$

At the moment, there is no prescribed frequency dependence for Q_r, and so this ansatz, which includes the βq_{EA} offset at zero frequency, can be made without loss of generality. However, as we see below, this ansatz is convenient because it leads to solutions in which Q_r is a smooth function of frequency, consistent with the discussion around (33.24).

Now we insert the ansatz (33.38) and (33.39) into (33.32) and (33.33). We invert the matrix in (33.32) for general integers n, and obtain the equations

$$
\begin{aligned}
\widetilde{\lambda} &\equiv \overline{\lambda} - gJ^2 Q_r(0), \\
\beta q_{EA} + Q_r(0) &= g \frac{-gJ^2(n-1)\beta q_{EA} + \widetilde{\lambda}}{-gJ^2 n \widetilde{\lambda} \beta q_{EA} + \widetilde{\lambda}^2}, \\
\beta q_{EA} &= g \frac{gJ^2 \beta q_{EA}}{-gJ^2 n \widetilde{\lambda} \beta q_{EA} + \widetilde{\lambda}^2}, \\
Q_r(i\omega_n) &= \frac{g}{\omega_n^2 + \overline{\lambda} - gJ^2 Q_r(i\omega_n)}, \qquad \omega_n \neq 0, \\
q_{EA} + T \sum_{\omega_n} Q_r(i\omega_n) &= 1.
\end{aligned} \tag{33.40}
$$

These equations are complicated for general n, but we can analytically continue to $n \to 0$ by setting $n = 0$ in (33.40). Then the equations simplify considerably, and have a remarkably simple solution for all g and T

$$
\begin{aligned}
\overline{\lambda} &= 2Jg, \quad \widetilde{\lambda} = Jg, \\
Q_r(i\omega_n) &= \frac{2g}{\omega_n^2 + 2Jg + |\omega_n| \sqrt{\omega_n^2 + 4gJ}}, \\
q_{EA} &= 1 - T \sum_{\omega_n} Q_r(i\omega_n).
\end{aligned} \tag{33.41}
$$

We require $q_{EA} > 0$, and the solution of the equation $q_{EA} = 0$ determines the boundary of the spin-glass phase in Fig. 33.1.

The $|\omega_n|$ in $Q_r(i\omega_n)$ indicates a gapless spectrum in the spin-glass phase. Indeed, evaluating the dynamic spin susceptibility, we obtain

$$
\chi_r''(\omega) = \frac{1}{J\sqrt{gJ}} \omega + \cdots, \qquad g < g_c, T = 0 \tag{33.42}
$$

at small ω in the spin-glass phase, similar to the linear ω behavior in (33.37) at the critical point. The full dynamic spin susceptibility $\chi''(\omega)$ of the spin glass also contains a delta function at zero frequency associated with the first term in (33.39). For the dynamic structure factor, this term implies

$$
S(\omega) = 2\pi q_{EA} \delta(\omega) + \cdots, \tag{33.43}
$$

and then using the fluctuation–dissipation theorem in (11.12) we have

$$
\chi''(\omega) = \pi \beta \omega q_{EA} \delta(\omega) + \chi_r''(\omega). \tag{33.44}
$$

33.3 Random Heisenberg Magnet

We are now finally ready to study the Heisenberg spin model in (33.1). Initially, the analysis closely parallels that for the quantum rotor model in Section 33.2. The main difference is that the different components of $\boldsymbol{S}_i$ do not commute with each other, and so we have to use the coherent-state path integral for spins discussed in Appendix A.2 instead of the configuration path integral in (33.27). In other words, the Berry phase terms in (18.23) and (A.38) are absent in the rotor model. Closely related to this difference is that fact that each site with an $S = 1/2$ spin $\boldsymbol{S}_i$ has a doubly degenerate state, whereas the rotor model has a non-degenerate state with angular momentum $\ell = 0$ as its ground state. Consequently, for the rotor model, we could take the large-g limit with a trivial ground state, and then reduce the value of g to obtain the spin glass. On the other hand, for the Heisenberg spin model (33.1) there is no regime with a trivial ground state, and we will see that the structure of the paramagnet state is much richer and more subtle, and closely connected to the SYK model.

Proceeding with replica analysis of the averaged partition function following the same steps as in Section 33.2.1, the large-N physics of (33.1) reduces evaluating the following path integral for n quantum $S = 1/2$ spins $\boldsymbol{S}_a$, which is the analog of (33.27)

$$\begin{aligned} \mathcal{Z}_J[Q] &= \int \mathcal{D}\boldsymbol{S}_a(\tau)\delta(\boldsymbol{S}_a^2 - 1)e^{-\mathcal{S}_B - \mathcal{S}_J}, \\ \mathcal{S}_B &= \frac{i}{2}\int_0^1 du \int d\tau\, \boldsymbol{S}_a \cdot \left(\frac{\partial \boldsymbol{S}_a}{\partial \tau} \times \frac{\partial \boldsymbol{S}_a}{\partial u}\right), \\ \mathcal{S}_J &= -\frac{J^2}{2}\int d\tau d\tau' Q_{ab}(\tau - \tau')\boldsymbol{S}_a(\tau)\cdot\boldsymbol{S}_b(\tau'). \end{aligned} \tag{33.45}$$

This is a coherent-state path integral, and $\mathcal{S}_B$ is the geometric Berry phase of (A.38) and (18.23), closely connected to the spin commutation relations: this Berry phase was absent in the quantum rotor model of Section 33.2, and is responsible here for the key feature of the absence of a trivial ground state noted in the previous paragraph. The spin has a temporal self-interaction with itself, represented by the function $Q_{ab}(\tau)$. The value of $Q_{ab}(\tau)$ is to be determined self-consistently by computing the correlator,

$$Q_{ab}(\tau - \tau') \equiv \frac{1}{3}\left\langle \boldsymbol{S}_a(\tau)\cdot\boldsymbol{S}_b(\tau')\right\rangle_{\mathcal{Z}_J}, \tag{33.46}$$

which is the analog of (33.29).

The Berry phase term in (33.45) makes the direct evaluation of the spin path integral prohibitively difficult. So we resort to the same method succesfully employed in Parts II and IV for non-random spin systems: we represent the spin in terms of fermionic partons $f_a^\alpha(i)$ as in Chapter 22 and elsewhere, with the fermionic partons having acquired an additional replica label.

33.3.1 G–Σ–Q theory

We will study the $SU(M)$ generalization of the $SU(2)$ model (33.1), following Ref. [47],

$$H = \frac{1}{\sqrt{NM}}\sum_{i<j=1}^{N}\sum_{\alpha,\beta=1}^{M} J_{ij} S_\beta^\alpha(i) S_\alpha^\beta(j). \tag{33.47}$$

Here, $\mathcal{S}^{\alpha}_{\beta}(i) = [\mathcal{S}^{\beta}_{\alpha}(i)]^{\dagger}$ are generators of $SU(M)$ on each site i, with $\alpha, \beta = 1, \ldots, M$. Each site contains states corresponding to the *antisymmetric* product of $M/2$ (integer) fundamentals, and these are realized by fermionic spinons with

$$\mathcal{S}^{\alpha}_{\beta}(i) = f^{\dagger}_{\beta}(i) f^{\alpha}(i) - \frac{1}{2}\delta^{\alpha}_{\beta}, \quad \sum_{\alpha} f^{\dagger}_{\alpha}(i) f^{\alpha}(i) = \frac{M}{2}, \tag{33.48}$$

with fermions $f^{\alpha}(i)$ on each site i. The Hamiltonian in (33.47) reduces to the $S = 1/2$ case of the $SU(2)$ Hamiltonian in (33.1) for $M = 2$ (apart from an overall factor of $1/\sqrt{2}$).

We introduce replicas $a = 1, \ldots, n$, and average over J_{ij} to obtain the averaged, replicated partition function as in (33.25)

$$\begin{aligned}
\overline{\mathcal{Z}^n} &= \int \mathcal{D} f^{\alpha}_{a}(i,\tau) \mathcal{D}\lambda_a(i,\tau) \exp\left[-S_B - S_J\right], \\
S_B &= \sum_i \int d\tau \left[f^{\dagger}_{a\alpha}(i) \partial_\tau f^{\alpha}_{a}(i) + i\lambda_a(i) \left(f^{\dagger}_{a\alpha}(i) f^{\alpha}_{a}(i) - \frac{M}{2} \right) \right], \\
S_J &= -\frac{J^2}{4NM} \int d\tau d\tau' \left[\sum_i \mathcal{S}^{\alpha}_{a\beta}(i,\tau) \mathcal{S}^{\gamma}_{b\delta}(i,\tau') \right] \left[\sum_j \mathcal{S}^{\beta}_{a\alpha}(j,\tau) \mathcal{S}^{\delta}_{b\gamma}(j,\tau') \right].
\end{aligned} \tag{33.49}$$

We can now decouple $\mathcal{S}_J$ with a Hubbard–Stratonovich field $Q^{\alpha\gamma}_{ab,\beta\delta}(\tau,\tau')$ and take the large-N limit. Then the problem reduces to finding saddle points of the single-site action analogous to (33.26):

$$\frac{\mathcal{S}[Q]}{N} = \frac{J^2}{4M} \int d\tau d\tau' |Q^{\alpha\gamma}_{ab,\beta\delta}(\tau,\tau')|^2 - \ln \mathcal{Z}_f[Q], \tag{33.50}$$

where $\mathcal{Z}_f[Q]$ is the single-site partition function analogous to (33.27):

$$\begin{aligned}
\mathcal{Z}_f[Q] &= \int \mathcal{D} f^{\alpha}_{a}(\tau) \mathcal{D}\lambda_a(\tau) \exp\left[-S_B - S_f\right], \\
S_B &= \int d\tau \left[f^{\dagger}_{a\alpha} \partial_\tau f^{\alpha}_{a} + i\lambda_a \left(f^{\dagger}_{a\alpha} f^{\alpha}_{a} - \frac{M}{2} \right) \right], \\
S_f &= -\frac{J^2}{2M} \int d\tau d\tau' Q^{\alpha\gamma}_{ab,\beta\delta}(\tau,\tau') \left[f^{\dagger}_{a\alpha}(\tau) f^{\beta}_{a}(\tau) - \frac{\delta^{\beta}_{\alpha}}{2} \right] \left[f^{\dagger}_{b\gamma}(\tau') f^{\delta}_{b}(\tau') - \frac{\delta^{\delta}_{\gamma}}{2} \right].
\end{aligned} \tag{33.51}$$

Note that now there is no remaining path integral over Q. We simply have to find the saddle points of the action $\mathcal{S}[Q]$ in (33.50).

Let us assume that the saddle point does not break spin-rotation symmetry; this is true in both the spin-glass, and quantum spin-liquid phases. So we make the ansatz analogous to (33.28) [242]:

$$Q^{\alpha\gamma}_{ab,\beta\delta}(\tau,\tau') = \delta^{\alpha}_{\delta} \delta^{\gamma}_{\beta} Q_{ab}(\tau - \tau'). \tag{33.52}$$

where $Q_{ab}(\tau)$ is a real function. Then (33.50) is replaced by

$$\frac{\mathcal{S}[Q]}{N} = \frac{J^2 M}{4} \int d\tau d\tau' [Q_{ab}(\tau - \tau')]^2 - \ln \mathcal{Z}_f[Q], \tag{33.53}$$

while S_f in (33.51) is replaced by

$$S_f = -\frac{J^2}{2M}\int d\tau d\tau' Q_{ab}(\tau-\tau')\left[f^{\dagger}_{a\alpha}(\tau)f^{\beta}_{a}(\tau)f^{\dagger}_{b\beta}(\tau')f^{\alpha}_{b}(\tau') - \frac{M}{4}\right]. \tag{33.54}$$

At this point, the analysis diverges from that for the quantum rotor model. While the path integral over the matter field $n_{a\mu}$ in $\mathcal{Z}_n[Q]$ in (33.27) is Gaussian and can be formally performed, that over the matter field f^{α}_{a} in $\mathcal{Z}_f[Q]$ in (33.51) is not. Instead, we proceed in close analogy with the method employed for the SYK model in Section 32.3, and express $\mathcal{Z}_f[Q]$ as a G–Σ theory. We define the spinon Green's function

$$G_{ab}(\tau,\tau') = -\frac{1}{M}\sum_{\alpha} f^{\alpha}_{a}(\tau)f^{\dagger}_{b\alpha}(\tau'). \tag{33.55}$$

Then, we can write

$$\mathcal{Z}_f[Q] = \exp\left(-\frac{k^2J^2}{2}\int d\tau d\tau' \sum_{a,b} Q_{ab}(\tau-\tau')\right) \times \int \mathcal{D}G_{ab}(\tau,\tau')\mathcal{D}\Sigma_{ab}(\tau,\tau')\mathcal{D}\lambda_a(\tau)\exp\left[-MI[Q]\right], \tag{33.56}$$

where the action $I[Q]$ is

$$\begin{aligned} I[Q] &= -\ln\det\left[-\delta'(\tau-\tau')\delta_{ab} - i\lambda_a(\tau)\delta(\tau-\tau')\delta_{ab} - \Sigma_{ab}(\tau,\tau')\right] - ik\int d\tau \lambda_a(\tau) \\ &+ \int d\tau d\tau'\left[-\Sigma_{ab}(\tau,\tau')G_{ba}(\tau',\tau) + \frac{J^2}{2}Q_{ab}(\tau-\tau')G_{ab}(\tau,\tau')G_{ba}(\tau',\tau)\right]. \end{aligned} \tag{33.57}$$

We note that (33.56) and (33.57) constitute an exact formulation of the theory for all M. The action in (33.57) is the advertized G–Σ–Q action of the random Heisenberg magnet, similar to the G–Σ action for the SYK model in (32.83). In Section 33.3.2 we discuss the large-M limit of (33.57), whence it will become identical to the SYK action.

Our remaining task is to evaluate the path integral over $G_{ab}(\tau,\tau')$, $\Sigma_{ab}(\tau,\tau')$, and $\lambda_a(\tau)$ in (33.56), and then determine the saddle-point solutions for $Q_{ab}(\tau)$ in (33.53). The saddle-point equations for Q from (33.53), (33.54), and (33.57) are

$$\begin{aligned} Q_{ab}(\tau-\tau') &= \frac{1}{M^2}\left\langle f^{\dagger}_{a\alpha}(\tau)f^{\beta}_{a}(\tau)f^{\dagger}_{b\beta}(\tau')f^{\alpha}_{b}(\tau')\right\rangle_{\mathcal{Z}_f[Q]} - \frac{1}{4M} \\ &= -\left\langle G_{ab}(\tau,\tau')G_{ba}(\tau',\tau)\right\rangle_{\mathcal{Z}_f[Q]} - \frac{1}{4M}, \end{aligned} \tag{33.58}$$

but we will find it more convenient to obtain them directly from the functional form of $\mathcal{S}[Q]$ in (33.53).

33.3.2 SYK Spin Liquid

We now discuss the evaluation of the path integral in (33.56) order by order in $1/M$. In such an evaluation we will find a gapless paramagnetic state, which can be viewed as a SYK state of fermionic spinons; we noted this connection between the SYK model and the random Heisenberg magnet in Fig. 21.2. This gapless, fractionalized, spin-liquid state is in stark contrast with the trivial, gapped, paramagnet of the rotor model in Section 33.2.2. The random Heisenberg magnet also has a spin-glass state, but obtaining this state requires considerations that are non-perturbative in M, and are deferred to Section 33.3.3.

Assuming first a general $Q_{ab}(\tau)$, the large-M limit of the path integral in (33.56) leads to the following saddle-point equations for the fermion Green's function and self-energy

$$\begin{aligned}\Sigma_{ab}(\tau) &= J^2 Q_{ab}(\tau) G_{ab}(\tau),\\ G_{ab}(i\omega) &= \left[i\omega\delta_{ab} - \Sigma_{ab}(i\omega)\right]^{-1},\end{aligned} \tag{33.59}$$

where $\lambda_a = 0$ at the saddle point because of particle–hole symmetry. However, using the analog of (33.20), we can conclude that there cannot be any off-diagonal components of the fermion Green's function at the saddle point, because it is not possible for fermions to condense (but off-diagonal fermion Green's functions do need to be included in the theory of fluctuations [47]). So we write

$$G_{ab}(\tau, \tau') = G_Q(\tau - \tau')\delta_{ab} \quad , \quad M = \infty, \tag{33.60}$$

and similarly for Σ_{ab}. From the large-N saddle-point equation for Q_{ab} in (33.58), we see that Q_{ab} must also be replica diagonal,

$$Q_{ab}(\tau) = Q(\tau)\delta_{ab} \quad , \quad M = \infty, \tag{33.61}$$

and so there is no spin-glass order at $M = \infty$ [242]. The large-M saddle-point equations (33.59) therefore reduce to

$$\begin{aligned}\Sigma_Q(\tau) &= J^2 Q(\tau) G_Q(\tau),\\ G_Q(i\omega) &= \left[i\omega - \Sigma_Q(i\omega)\right]^{-1}.\end{aligned} \tag{33.62}$$

These equations hold for general $Q(\tau)$, and we have emphasized this by the subscript Q on G and Σ. Upon including the large-N saddle-point equation for Q in (33.58), we obtain

$$Q(\tau) = -G_Q(\tau)G_Q(-\tau), \quad M = \infty. \tag{33.63}$$

The combination of (33.62) and (33.63) yields precisely the large-N equations of the fermion of the complex SYK model [242] in (32.20a) and (32.20b) at $\mu = 0$.

For completeness, we also present the large-M expressions for the path integral in (33.57):

$$\begin{aligned}-\frac{\ln \mathcal{Z}_f[Q]}{Mn} &= \frac{I[Q]}{n} + \frac{J^2}{8Mn}\int d\tau d\tau' \sum_{a,b} Q_{ab}(\tau - \tau'),\\ \frac{I[Q]}{n} &= -\ln\det\left[-\delta'(\tau - \tau') - \Sigma_Q(\tau - \tau')\right]\\ &\quad + \int d\tau d\tau' \left[-\Sigma_Q(\tau - \tau')G_Q(\tau' - \tau) + \frac{J^2}{2} Q(\tau - \tau')G_Q(\tau - \tau')G_Q(\tau' - \tau)\right].\end{aligned} \tag{33.64}$$

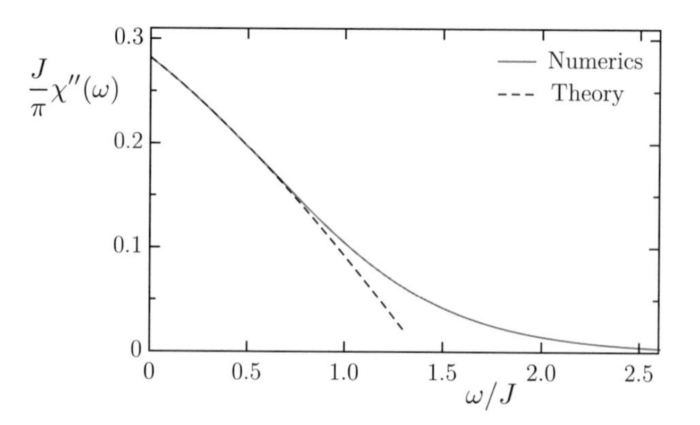

Figure 33.3 Numerical results for the dynamic spin susceptibility of the SYK spin liquid from Ref. [279]. The solution of (33.62) and (33.63) is compared with an expansion that extends (33.66). Compare to the dynamic spin susceptibility of the quantum rotor model in Fig. 33.2. Reprinted with permission from APS.

Employing (33.63) we see that (33.64) is identical to the G–Σ action for the SYK model in (32.83) at $\mu = 0$.

Given the identity of (33.62) and (33.63) to the SYK equations, we can now read off results from Chapter 32 to solve them. From (32.80), we obtain the long-time behavior of the spin autocorrelation in this spin liquid

$$Q(\tau) = \frac{B^2}{|\tau|}\left(1 + \frac{2\alpha_2}{|\tau|} + \cdots\right), \quad T = 0. \tag{33.65}$$

After a Fourier transform, we obtain the dynamic spin susceptibility

$$\chi''(\omega) = \pi B^2 \text{sgn}(\omega)\left(1 - 2\alpha_2|\omega| + \cdots\right), \quad T = 0. \tag{33.66}$$

We compare results from the full numerical solution of (33.62) and (33.63) with an analytic expression that extends (33.66) to higher orders [279] in Fig. 33.3. Note the large density of states at low energy, with a discontinuity at $\omega = 0$; this is to be compared with the gapped density of states in the paramagnet for the rotor model in Fig. 33.2.

We can also extend (33.66) to $T > 0$, using methods developed for the SYK model, and find "Planckian" dissipation, with damping on the scale $\sim k_B T/\hbar$, and independent of J [279]:

$$\chi''(\omega) = \pi B^2 \tanh\left(\frac{\hbar\omega}{2k_B T}\right)\left[1 - 2\alpha_2\,\omega \tanh\left(\frac{\hbar\omega}{2k_B T}\right) + \cdots\right]. \tag{33.67}$$

Our results for the spin liquid have so far been obtained at $M = \infty$. However, it is possible to extend some of them to all orders in $1/M$ by using connections to the Bose Kondo model studied in Section 29.5, and to the analysis of fluctuations of the SYK model. Specifically, we now argue that both exponents associated with powers of $|\tau|$ in (33.65) acquire no corrections in the expansion in $1/M$.

To see that there are no corrections to the leading $1/|\tau|$ power in (33.65), it is useful to refer back to the formulation of the problem in (33.45), and note its close connection to the analysis of the Bose Kondo problem of Section 29.5. Observe that if we integrate out the ϕ_a field in (29.48), we get exactly the action in (33.45) with $Q(\tau)$ having the power-law form in (29.52) (we are assuming that the theory is replica diagonal). Then, the self-consistency condition in (33.46) can only be satisfied if the power law in (29.53) equals the power law in (29.52). Using the all-orders result in (29.54), we obtain $2-\varepsilon=\varepsilon$, or $\varepsilon=\alpha=1$ for all M. This is indeed the power of the leading term in (33.65). There has been recent evidence that an $SU(2)$ Bose Kondo fixed point is not present for larger values of ε [23, 57, 189, 300], and this is consistent with the appearance of spin-glass order in the $SU(2)$ model at the energy scale in (33.2).

For the subleading term in (33.65), we refer back to its connection to the $h=2$ operator in (32.63). This $h=2$ operator is closely related to the time-reparameterization symmetry noted in Section 32.3, and reviewed by Chowdhury et al. [46] – its connection to this symmetry implies that the value $h=2$ is protected.

33.3.3 Spin Glass

We have so far analyzed the theory (33.57) for the random Heisenberg spin glass in a $1/M$ expansion about the $M=\infty$ saddle point and found a gapless paramagnetic phase that can be interpreted as a SYK quantum liquid of fermionic spinons. Here we show, following Ref. [91], that there is a non-perturbative instability to spin-glass order at a low-energy scale of order (33.2). This instability will quench the entropy of the SYK spin liquid, so that the entropy is not extensive as $T\to 0$. However, it has been argued [47] that the entropy is replaced by an extensive "complexity" of the spin-glass state.

First, let us place a bound on the magnitude of the spin-glass order in the $SU(M)$ Hamiltonian defined by (33.47) and (33.48). The state with maximum order has the spins frozen in a state in which the fermions occupy the states with, say, $\alpha=1,\ldots,M/2$, while the other values of α are empty. Evaluating (33.58) on such a state, and using the definition (33.24), we obtain

$$q_{EA}\leq\frac{1}{4M}. \tag{33.68}$$

Note that (33.68) vanishes as $M\to\infty$, and q_{EA} is at most $\mathcal{O}(1/M)$ in the large-M limit.

As a first step to understanding the instability to spin-glass order, let use expand the action in (33.53) to quadratic order in the off-diagonal components of $Q_{ab}(\tau)=q_{ab}$ with $a\neq b$. A direct computation from (33.53) shows that

$$\frac{S[Q]}{NM}=\frac{\beta^2J^2}{4}\left(\sum_{a\neq b}q_{ab}^2\right)\left[1-\frac{J^2}{M}\chi_{loc}^2\right]+\cdots, \tag{33.69}$$

where χ_{loc} is the local spin susceptibility. In the SYK spin-liquid state, we obtain from the results in (33.65) and Section 32.2.1

$$\chi_{loc}=\int_0^\beta Q(\tau)d\tau=\frac{1}{J\sqrt{\pi}}\ln(\beta J), \tag{33.70}$$

which diverges logarithmically as $T\to 0$, and so the term in square brackets in (33.69) turns negative. This indicates an instability to spin-glass order at the scale in (33.2).

Determination of the value of q_{ab} requires consideration of the higher-order terms in the expansion in (33.69), and this has been discussed in Ref. [47]. Here, we assume that such considerations have determined the appropriate values of q_{ab} and q_{EA}, and ask for the feedback of the spin-glass order on the fermion and spin excitation spectrum. The leading singular effect at low frequency arises from the long-time limit of $Q(\tau)$ in (33.24), and so we argue that we can replace the equation (33.63) of the spin liquid-phase by

$$Q(\tau) = -G_Q(\tau)G_Q(-\tau) + q_{EA}, \tag{33.71}$$

with no change to the equations in (33.62).

The equations (33.62) and (33.71) are precisely those that appeared in (32.22) in the context of the SYK model with additional random hopping terms in (32.21), after setting $U = J$ and $t = J\sqrt{q_{EA}}$. We now present a numerical solution of these equations [47] showing that they exhibit a crossover from SYK non-Fermi liquid behavior to Fermi liquid behavior at a coherence energy scale [196, 269]

$$\omega_* = \frac{t^2}{J} = Jq_{EA}. \tag{33.72}$$

This crossover determines the structure of the low-frequency spectrum in the spin-glass phase when $q_{EA} \ll 1$. For the spinon spectral density, the crossover is described by the crossover function Φ_ρ with

$$\rho(\omega) = -\frac{1}{\pi}\mathrm{Im}\, G_Q(\omega) = \frac{1}{\pi\sqrt{J\omega_*}}\Phi_\rho(\omega/\omega_*), \tag{33.73}$$

where ω_* is given by (33.72). The result for $\rho(\omega)$ is presented in Fig. 33.4, comparing with the low-frequency scaling scaling in (33.73). In the context of the model in (32.21), the scaling function Φ_ρ crosses over from the Fermi liquid behavior of Section 32.1 at low frequencies with $\Phi_\rho(0) = 1$, to the non-Fermi liquid behavior of the SYK model at higher frequencies with $\Phi_\rho(\overline{\omega} \gg 1) \sim 1/\sqrt{\overline{\omega}}$. In the present context of the random Heisenberg magnet, the crossover is from the excitations characteristic of the spin-glass state for $\omega < \omega_*$, to the fractionalized excitations of the spin liquid for $\omega > \omega_*$.

Similarly, for the spin spectral density we have

$$\chi''(\omega) = \mathrm{Im}\, Q(\omega) = \pi\beta\omega q_{EA}\,\delta(\omega) + \frac{1}{J}\Phi_\chi(\omega/\omega_*), \tag{33.74}$$

and the scaling function Φ_χ is shown in Fig. 33.5. Note that the spin-glass condensate contributes the zero-frequency delta-function contribution in (33.74), as in (33.44) for the quantum rotor model. At $\omega > \omega_*$, the spin spectral density is given by the fractionalized spin liquid behavior shown earlier in Fig. 33.3. At lower $\omega < \omega_*$ we have behavior characteristic of the spin-glass phase with

$$\chi''(\omega) = \frac{\omega}{\omega_* \pi J} + \cdots \quad , \quad 0 < |\omega| < \omega_*, \; T = 0. \tag{33.75}$$

Note that this behavior is identical to that appearing in (33.42) for the spin-glass regime of the quantum rotor model, suggesting a universality of a linear in ω spectrum in infinite-range quantum spin glasses. In the present situation, this linear-in-ω spectrum arises from the “Fermi liquid” regime of the spinon spectral density in Fig. 33.4.

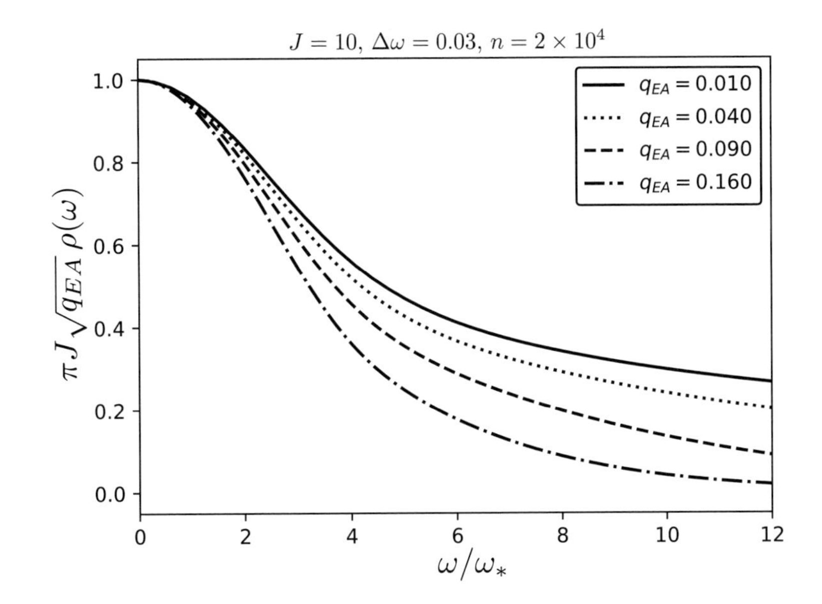

Figure 33.4 Numerical results for the spinon spectral density [47] obtained by the solution of (33.62) and (33.71). The results scale as in (33.73) for small q_{EA}. The solutions were obtained with n frequency points. Adapted by Maine Christos, and with permission from APS.

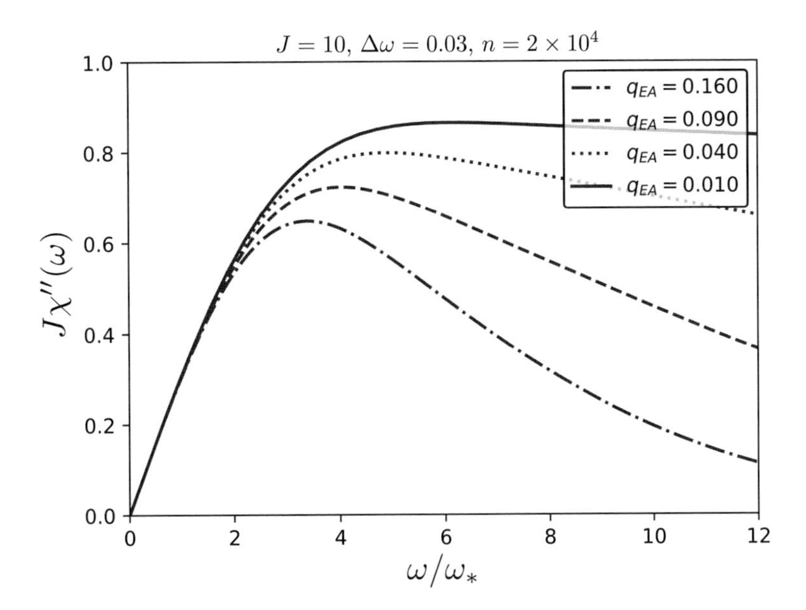

Figure 33.5 Numerical results for the spin spectral density [47] obtained by the solution of (33.62) and (33.71). The results scale as in (33.74) for small q_{EA}. Adapted by Maine Christos, and with permission from APS.

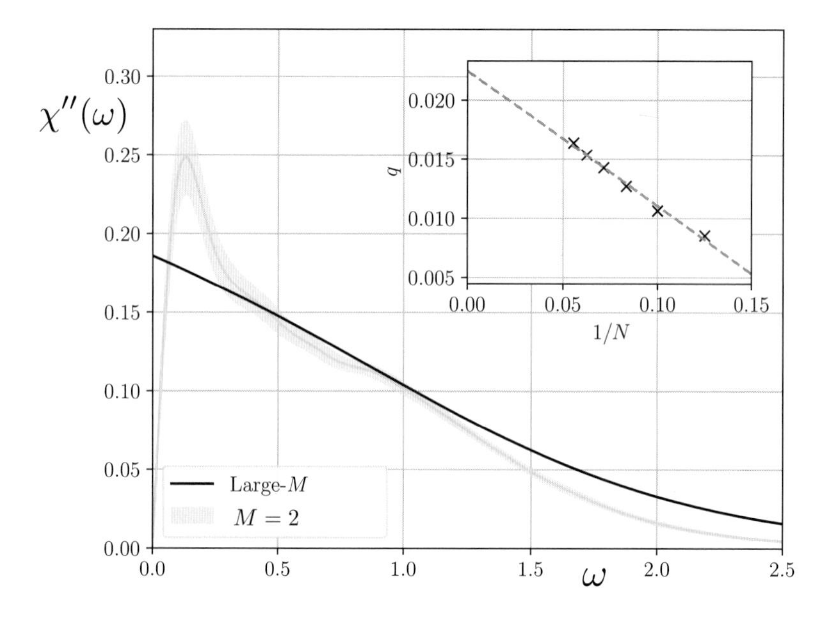

Figure 33.6 Spin spectral density obtained from exact diagonalization [263] of the random Heisenberg magnet in (33.1). Reprinted with permission from APS.

33.3.4 Numerical Results

We now compare the above analyses of the spin-liquid and spin-glass states with exact diagonalization results on the Hamiltonian in (33.1). Figure 33.6 shows results obtained by Shackleton et al. [263] for the spin spectral density. These were obtained by averages over samples with up to $N = 18$ $S = 1/2$ spins.

The results show a peak at low frequency, which increases in height and sharpens in frequency as N becomes larger; this is interpreted as a signal of spin-glass order, and estimates of the spin-glass order parameter are shown in the inset.

At larger ω, the results show a spectrum that compares well with a rescaled plot (while preserving total spectral density) of the large-M spin-liquid numerical result in Fig. 33.3. Recall the characteristic behavior of the large-M theory in (33.66), and our arguments that the exponents of both terms in (33.66) will not be renormalized by higher orders in $1/M$. There is clear evidence of the behavior of (33.66) in Fig. 33.6, implying the presence of a fractionalized SYK spin liquid at intermediate frequencies, above spin-glass order at low frequency.

34 Fermi Surfaces without Quasiparticles

The general theory of a two-dimensional Fermi surface of quasiparticles coupled to a gapless scalar is presented. A systematic large-N expansion is possible when the fermion–scalar Yukawa coupling is random in flavor space. Such a theory is shown to exhibit a Fermi surface that is sharp in momentum space, but broad in frequency because of the absence of coherent quasiparticle excitations. A model with the an additional spatial randomness in the Yukawa coupling has linear-temperature resistivity at the lowest temperatures.

The Sachdev–Ye–Kitaev (SYK) model of Chapter 32 has provided significant insights into the structure of metallic phases without quasiparticle excitations. However, such a theory has no spatial structure, and so no Fermi-surface-like feature similar to that observed in the strange-metal phase of the cuprates. This chapter draws upon the insights gained in Chapter 32, and describes more realistic models of metals without quasiparticle excitations with spatial structure. In the presence of full translational symmetry, such models do have sharp Fermi surfaces in momentum space at $T = 0$. The absence of quasiparticles only makes them diffuse in energy space, but the location of the Fermi surface is well defined in momentum space, it is still given by (2.39). We also consider the influence of spatial disorder on the sharp Fermi surface: this makes the Fermi surface diffuse also in momentum space, and is essential for a theory of the transport properties.

One of our main results is the form of the Green's function in (34.23) for the Fermi surface without quasiparticles in two spatial dimensions in the absence of spatial disorder. We note that this Green's function is very different from that in (12.55) for the one-dimensional Tomonaga–Luttinger liquid. This is evidence that it is not valid to think of the higher-dimensional Fermi surface as a collection of independent one-dimensional quantum systems along each direction orthogonal to the Fermi surface. A more appropriate description is in terms of overlapping patches at points on the Fermi surface, as shown in Section 34.1.2. The structure of the Green's function in (34.23) is much closer to that of the SYK model, with a purely frequency-dependent *local* self-energy in the large-N limit of Section 34.1.1; we only have to add a smooth momentum-dependent bare energy to a purely local SYK-like self-energy.

I present the discussion in the context of a simple model for the onset of Ising ferromagnetism in a two-dimensional metal which is introduced Section 34.1. However, the results are far more general, and apply to a wide class of models in which the Fermi surface is coupled to a gapless bosonic mode in two spatial dimensions. This includes: (i) the onset of Ising-nematic order in a Fermi liquid, (ii) the $U(1)$ spin liquid with a spinon Fermi surface that we briefly noted below (22.7), in which the Fermi surface excitations are coupled to a $U(1)$ gauge field, and (iii) the Halperin–Lee–Read state of a half-filled Landau level, which was noted in Sections 24.4 and 27.4. The extension to these cases is discussed in Section 34.3.

Our main tool for analyzing these problems is a recently introduced large-N approach, which is directly inspired by the SYK model. This method is described in Section 34.1.1, and leads to the analog of a G–Σ theory with a large-N saddle point. Section 34.1.2 then describes how an exact low-energy solution of the saddle-point equations can be obtained for the case without spatial disorder; this solution involves a sharp Fermi surface without quasiparticle excitations.

The other sections detail further properties of metals without quasiparticles. Section 34.2 shows that the volume enclosed by the Fermi surface obeys the usual Luttinger relation, despite the absence of quasiparticles. Section 34.4 considers pairing instabilities of the sharp Fermi surface, using methods closely related to those presented in Section 32.2.5 for the SYK model. Section 34.5 contains a brief discussion of electrical transport, where the Fermi surface cannot be treated within the patch approximation: it presents the argument of Ref. [200] that spatial randomness is required to obtain the linear temperature resistivity.

34.1 Onset of Ising Ferromagnetism

As our simplest example of a Fermi surface without quasiparticles, we consider the onset of ferromagnetic order in a two-dimensional metal. We assume that spin–orbit couplings render the spin correlations anisotropic in spin space, so that we can focus on only the z (say) component of the ferromagnetic order. Let us use the framework of the paramagnon theory employed in Section 9.4 to describe the onset of spin density wave order at a wavevector $\boldsymbol{K} = (\pi, \pi)$, as in (9.61). In its original formulation [27, 63], the paramagnon theory was introduced as a theory of ferromagnetic spin fluctuations in liquid ^{3}He, and in such a theory we should take $\boldsymbol{K} = (0, 0)$. This requires that the underlying band structure and density of the electrons is such that the Lindhard susceptibility in (9.49) has a maximum at zero wavevector. We account for the anisotropy in spin space by including only the field $\phi \equiv \Phi_z$ in our low-energy theory. Recent quantum Monte Carlo studies [318, 319] have examined an Ising model in a transverse field coupled to Fermi surfaces of electrons, and observed the onset of Ising magnetic order at a continuous quantum phase transition; the theory presented here is expected to describe such a transition.

The field theory for such a transition is obtained by the same route as that followed in Section 9.4. We combine the free-fermion theory in (2.2) with the scalar field theory for ϕ in (10.2) to obtain the Lagrangian

$$\mathcal{L} = \sum_{k\alpha} c_{k\alpha}^\dagger \left[\frac{\partial}{\partial \tau} + \varepsilon(\boldsymbol{k}) \right] c_{k\alpha} + \int d^2r \left\{ \frac{1}{2} \left[(\nabla\phi)^2 + (\partial_\tau \phi)^2 + s\phi^2 \right] + \frac{u}{4!}\phi^4 \right\}$$
$$- \int d^2r \, g\, \phi \, c_\alpha^\dagger \sigma_{\alpha\beta}^z c_\beta . \quad (34.1)$$

We have allowed for an arbitrary dispersion of the electrons $c_{k\alpha}$ in momentum space, with a Fermi surface at $\varepsilon_k = 0$. However, we only include long-wavelength fluctuations in ϕ and so have performed a gradient expansion in its Lagrangian. The electrons are coupled to ϕ via the Yukawa coupling g, with σ^z the Pauli matrix. A crucial property of this Yukawa coupling is that it acts at zero momentum, unlike the non-zero-momentum shift in (9.61). Other cases with a zero-momentum order parameter lead to essentially the same results, as described below.

There has been a great deal of work [155] on theory (34.1), based essentially on a renormalized expansion in powers of g, supplemented by a large number of fermion flavors. This work has led to numerous insights on the properties of (34.1), but not to a formulation in terms of a saddle-point theory that can be used to systematically classify the nature of higher-order corrections.

34.1.1 Large-N Theory

Following the example of the SYK model, it was argued [5, 71, 72] that problems of fermions coupled to a critical boson could also be addressed by examining ensembles of theories with different Yukawa couplings. It is also possible to choose the ensemble so that the couplings are spatially independent, and this maintains full translational symmetry in each member of the ensemble. If most members of the ensemble flow to the same universal low-energy theory, then we can access the low-energy behavior by studying the average over the ensemble. We also obtain the added benefit of a G–Σ action with a large-N prefactor, which allows for a systematic treatment of the theory.

Let us consider the following generalization of the theory (34.1):

$$\mathcal{L} = \sum_{\alpha=1}^{N} \sum_{k} c_{k,\alpha}^\dagger \left[\frac{\partial}{\partial \tau} + \varepsilon(\boldsymbol{k}) \right] c_{k,\alpha} + \int d^2r \sum_{\gamma=1}^{M} \left\{ \frac{1}{2} \left[(\nabla\phi_\gamma)^2 + (\partial_\tau \phi_\gamma)^2 + s\phi_\gamma^2 \right] \right\}$$
$$- \int d^2r \sum_{\gamma=1}^{M} \sum_{\alpha,\beta=1}^{N} \frac{g_{\alpha\beta\gamma}}{N} \phi_\gamma c_\alpha^\dagger c_\beta . \quad (34.2)$$

Here, the fermion has N components, the boson has M components, and we take the large-N limit with

$$\lambda = \frac{M}{N} \quad (34.3)$$

fixed. The Yukawa coupling is taken to be a random function of the flavor indices with

$$\overline{g_{\alpha\beta\gamma}} = 0, \quad g_{\alpha\beta\gamma}^* = g_{\beta\alpha\gamma}, \quad \overline{|g_{\alpha\beta\gamma}|^2} = g^2 . \quad (34.4)$$

We have dropped the quartic self-coupling u of the the scalar field for simplicity; it is unimportant for the leading critical behavior, but is needed for certain subleading

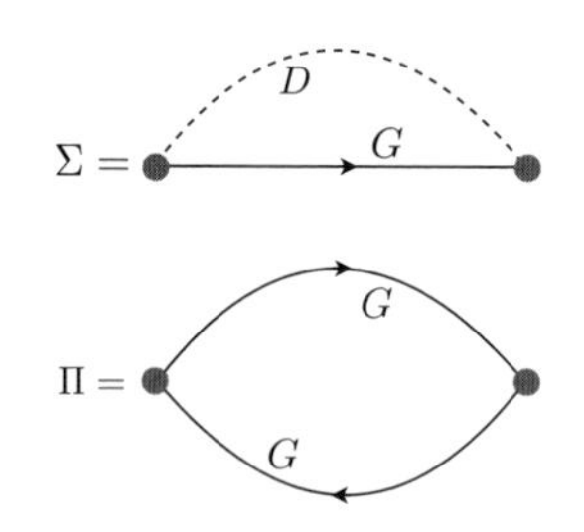

Figure 34.1 Saddle-point equations for the fermion self-energy Σ and boson self-energy Π, expressed in terms of the renormalized fermion Green's function G and boson Green's function D. The filled circle is the Yukawa coupling $g_{\alpha\beta\gamma}$.

effects at non-zero temperature [71]. The original theory in (34.1) has a $\phi \to -\phi$ symmetry, which is only statistically present in (34.2); we can maintain this symmetry in each member of the ensemble by dividing the indices into groups of two, but we avoid this complexity because it does not modify the large-N results. We consider an ensemble of complex couplings because it simplifies the analysis, but real couplings lead to essentially the same results.

We can now proceed with the large-N analysis following the script of the SYK model. As in Section 32.2, the large-N saddle-point equations are most easily obtained by a diagrammatic perturbation theory in g, in which we average each graph order by order. In the large-N limit, only the graphs shown in Fig. 34.1 survive, and yield the following saddle-point equations

$$\begin{aligned}
\Sigma(\boldsymbol{r},\tau) &= g^2\lambda D(\boldsymbol{r},\tau)G(\boldsymbol{r},\tau), \\
\Pi(\boldsymbol{r},\tau) &= -g^2 G(-\boldsymbol{r},-\tau)G(\boldsymbol{r},\tau), \\
G(\boldsymbol{k},i\omega_n) &= \frac{1}{i\omega_n - \varepsilon(\boldsymbol{k}) - \Sigma(\boldsymbol{k},i\omega_n)}, \\
D(\boldsymbol{q},i\Omega_m) &= \frac{1}{\Omega_m^2 + q^2 + s - \Pi(\boldsymbol{q},i\Omega_m)}.
\end{aligned} \tag{34.5}$$

Here, G is the Green's function for the fermion c, and Σ its self-energy; and D is the Green's function for the boson f, and Π is its self-energy.

The equations (34.5) are the analog of the SYK equations in (32.20a)–(32.20c), but the Green's functions now involve both spatial and temporal arguments. Remarkably, as we see in Section 34.1.2, an exact solution of the low-energy scaling behavior is possible for (34.5), just as it was for the SYK model.

For completeness, we also write down the path integral of the averaged theory using bilocal Green's functions, the analog of (32.83) for the SYK model. We introduce the spacetime coordinate $X \equiv (\tau,x,y)$, and all Green's functions and self-energies in the path integral are functions of two spacetime coordinates X_1 and X_2. Then we have

$$\begin{aligned}
\overline{\mathcal{Z}} = \int & \mathcal{D}G(X_1,X_2)\mathcal{D}\Sigma(X_1,X_2)\mathcal{D}D(X_1,X_2) \\
& \times \mathcal{D}\Pi(X_1,X_2)\exp\left[-NI(G,\Sigma,D,\Pi)\right].
\end{aligned} \tag{34.6}$$

The G–Σ–D–Π action is now

$$
\begin{aligned}
I(G,\Sigma,D,\Pi) &= \frac{g^2\lambda}{2}\mathrm{Tr}\left(G\cdot[GD]\right) - \mathrm{Tr}(G\cdot\Sigma) + \frac{\lambda}{2}\mathrm{Tr}(D\cdot\Pi) \\
&\quad - \ln\det\left[(\partial_{\tau_1} + \varepsilon(-i\nabla_1))\,\delta(X_1 - X_2) + \Sigma(X_1,X_2)\right] \\
&\quad + \frac{\lambda}{2}\ln\det\left[\left(-\partial_{\tau_1}^2 - \nabla_1^2 + s\right)\delta(X_1 - X_2) - \Pi(X_1,X_2)\right],
\end{aligned}
\tag{34.7}
$$

where we have introduced notation analogous to (32.84):

$$
\mathrm{Tr}\left(f\cdot g\right) \equiv \int dX_1 dX_2\, f(X_2,X_1)g(X_1,X_2)\,. \tag{34.8}
$$

Note the crucial prefactor of N before I in the path integral. It can be verified that the saddle-point equations of (34.7) reduce to (34.5).

34.1.2 Patch Solution

This subsection presents an exact solution of the saddle-point equations (34.5) in the low–energy scaling limit. We can obtain this solution for an arbitrary $\varepsilon(\boldsymbol{k})$, and for a general shape of the Fermi surface. The key to the solution is the observation that the singular behavior at any point on the Fermi surface is determined only by a small momentum-space patch around it, as well as that of the anti-podal point. We do need to include the curvature of the Fermi surface though, and it is not sufficient to think of the Fermi surface as a set of one-dimensional chiral fermions at each point on the Fermi surface. Although this patch approach correctly captures the behavior of the Green's functions, it runs into difficulties in computations of transport properties [98], as noted in Section 34.5.

We begin by evaluating Π in (34.5) using the bare-fermion Green's function. This yields the Lindhard susceptibility in (9.49) and (9.50)

$$
\begin{aligned}
\Pi(\boldsymbol{q}, i\Omega_m) &= -g^2 T\sum_{\omega_n}\int \frac{d^2k}{4\pi^2}\frac{1}{(i(\omega_n+\Omega_m) - \varepsilon(\boldsymbol{k}+\boldsymbol{q}))(i\omega_n - \varepsilon(\boldsymbol{k}))} \\
&= g^2\int \frac{d^2k}{4\pi^2}\frac{f(\varepsilon(\boldsymbol{k}+\boldsymbol{q})) - f(\varepsilon(\boldsymbol{k}))}{i\Omega_m + \varepsilon(\boldsymbol{k}) - \varepsilon(\boldsymbol{k}+\boldsymbol{q})},
\end{aligned}
\tag{34.9}
$$

where $f(\varepsilon)$ is the Fermi function. We are interested in the behavior of Π for small $\boldsymbol{q}$ and Ω_m at low T. On the real frequency axis, the real part of Π is not universal , and depends in a complicated manner on the entire fermion dispersion. However, the behavior of the imaginary part of Π is much simpler and universal. We have

$$
\begin{aligned}
\mathrm{Im}\,\Pi(\boldsymbol{q},\Omega) &= -\pi g^2\int\frac{d^2k}{4\pi^2}\left[f(\varepsilon(\boldsymbol{k}+\boldsymbol{q})) - f(\varepsilon(\boldsymbol{k}))\right]\delta\left(\Omega + \varepsilon(\boldsymbol{k}) - \varepsilon(\boldsymbol{k}+\boldsymbol{q})\right) \\
&= \pi g^2\Omega\int\frac{d^2k}{4\pi^2}\,\delta\left(\varepsilon(\boldsymbol{k})\right)\delta\left(\Omega + \varepsilon(\boldsymbol{k}) - \varepsilon(\boldsymbol{k}+\boldsymbol{q})\right) \quad \text{as } T\to 0.
\end{aligned}
\tag{34.10}
$$

The last expression contains an integral over two-dimensional momentum space of $\boldsymbol{k}$, along with two delta functions containing arguments that are functions of $\boldsymbol{k}$.

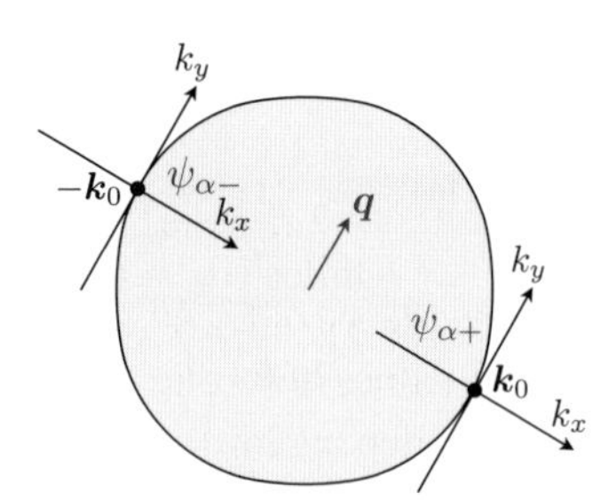

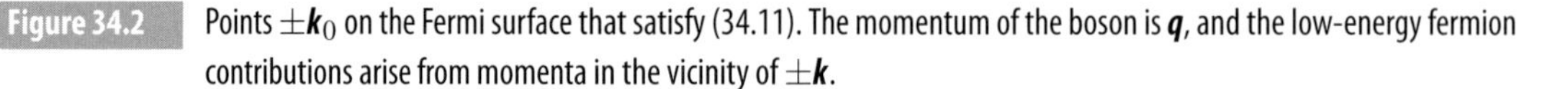
Figure 34.2 Points $\pm\boldsymbol{k}_0$ on the Fermi surface that satisfy (34.11). The momentum of the boson is $\boldsymbol{q}$, and the low-energy fermion contributions arise from momenta in the vicinity of $\pm\boldsymbol{k}$.

Generically, both delta functions are satisfied only at isolated points in momentum space. For $|\boldsymbol{q}|, |\Omega| \to 0$, the isolated points are solutions of

$$\varepsilon(\boldsymbol{k}) = 0 \quad \text{and} \quad \boldsymbol{q} \cdot \nabla_{\boldsymbol{k}} \varepsilon(\boldsymbol{k}) = 0 . \tag{34.11}$$

The solution of (34.11) is illustrated in Fig. 34.2; for a simply connected, convex Fermi surface, each direction of $\boldsymbol{q}$ is identified with the two anti-podal points $\pm\boldsymbol{k}_0$ on the Fermi surface, where $\boldsymbol{q}$ is parallel to the tangent to the Fermi surface. Note that the value of $\boldsymbol{k}_0$ is fully determined by $\boldsymbol{q}$, but we leave this dependence implicit.

As illustrated in Fig. 34.2, we choose our momentum-space axes so that $\boldsymbol{q} = (0, q_y)$. In the vicinity of $\boldsymbol{k}_0$ we write the fermion dispersion near the Fermi surface patch at $\boldsymbol{k}_0$ as

$$\boldsymbol{k} = \boldsymbol{k}_0 + (k_x, k_y), \quad \varepsilon(\boldsymbol{k}) = v_F k_x + \frac{\kappa}{2} k_y^2 , \tag{34.12}$$

whereas near $-\boldsymbol{k}_0$ we have

$$\boldsymbol{k} = -\boldsymbol{k}_0 + (k_x, k_y), \quad \varepsilon(\boldsymbol{k}) = -v_F k_x + \frac{\kappa}{2} k_y^2 . \tag{34.13}$$

Here, v_F is the Fermi velocity, and κ is the curvature of the Fermi surface. The values of v_F and κ depend upon $\boldsymbol{k}_0$, which in turn depends upon $\boldsymbol{q}$, and they will vary as $\boldsymbol{k}_0$ moves around the Fermi surface, but we have not explicitly indicated that; our results remain valid even in the presence of such variation. We can now insert (34.12) into (34.10) and obtain the Landau damping result

$$\begin{aligned} \operatorname{Im}\Pi(\boldsymbol{q}, \Omega) &= 2\pi g^2 \Omega \int \frac{d^2k}{4\pi^2} \, \delta\left(v_F k_x + \kappa k_y^2/2\right) \delta\left(\kappa k_y q_y + q_y^2/2 - \Omega\right) \\ &= \frac{g^2 \Omega}{2\pi v_F \kappa |q_y|} , \end{aligned} \tag{34.14}$$

where the leading factor of two is from the sum over the anti-podal points. Note that the curvature κ appears in the denominator, and so it is not valid to take the $\kappa \to 0$ limit, and no description in terms of purely linearly dispersing excitations around the Fermi surface is possible.

Let us now turn to an evaluation of Π in (34.5) using the fully renormalized Green's function. Remarkably, as we now show, the result in (34.14) remains largely unchanged.

We anticipate that a full solution of (34.5) leads to a fermion Green's function of the following form:

$$\Sigma(\boldsymbol{k}, i\omega_n) = \Sigma_0(\boldsymbol{k}) + \Sigma(i\omega_n). \tag{34.15}$$

The momentum dependence of $\Sigma_0(\boldsymbol{k})$ is non-singular, and we assume it can be absorbed by redefinition of the values of v_F and κ; it is therefore not included in the computations below. The frequency-dependent part $\Sigma(i\omega_n)$ can be singular (as we see below) but it has no dependence on k_x and k_y; however, it will depend upon the choice of $\boldsymbol{k}_0$, via the implicit $\boldsymbol{k}_0$ dependence of v_F and κ. We now insert $\Sigma(i\omega_n)$ into the first expression in (34.9) and use the dispersion (34.12) to obtain

$$\begin{aligned}\Pi(\boldsymbol{q}, i\Omega_m) = &-2g^2 T \sum_{\omega_n} \int \frac{d^2k}{4\pi^2} \frac{1}{(i\omega_n - v_F k_x - \kappa q_y^2/2 - \Sigma(i\omega_n))} \\ &\times \frac{1}{(i(\omega_n + \Omega_m) - v_F k_x - \kappa(k_y + q_y)^2/2 - \Sigma(i\omega_n + i\Omega_m))}.\end{aligned} \tag{34.16}$$

At this point in (34.9) we evaluated the summation over the frequency ω_n, but we are unable to do that here because of the unknown frequency dependence in $\Sigma(i\omega_n)$. So I have instead decided to focus only on the contribution of the patches near $\pm\boldsymbol{k}_0$, and linearized the fermion dispersion accordingly. In this situation the dependence of the integrand on k_x and k_y is simple. Performing the integral over k_x in (34.16), we obtain

$$\begin{aligned}\Pi(\boldsymbol{q}, i\Omega_m) = &\frac{-ig^2 T}{v_F} \sum_{\omega_n} \int \frac{dk_y}{(2\pi)} [\operatorname{sgn}(\omega_n + \Omega_m) - \operatorname{sgn}(\omega_n)] \\ &\times \frac{1}{i\Omega_m - \kappa q_y^2/2 - \kappa q_y k_y + \Sigma(i\omega_n) - \Sigma(i\omega_n + i\Omega_m)}.\end{aligned} \tag{34.17}$$

We have assumed here that $\operatorname{sgn}(\omega_n - \Sigma(i\omega_n)/i) = \operatorname{sgn}(\omega_n)$, and this always turns out to be the case from the positivity requirements of the fermion spectral weight. The next step is the evaluation of the q_y integral in (34.17). The real part of this integral is logarithmically divergent at large q_y, but then we are no longer in a regime where it is valid to keep the linearized dispersion. We assume that the divergent pieces only yield a non-singular contribution, and keep the singular imaginary part of the integral. In this manner, we obtain from (34.17)

$$\begin{aligned}\Pi(\boldsymbol{q}, i\Omega_m) &= \frac{g^2 T}{2\kappa v_F |q_y|} \sum_{\omega_n} \operatorname{sgn}(\Omega_m) [\operatorname{sgn}(\omega_n + \Omega_m) - \operatorname{sgn}(\omega_n)] \\ &= -\frac{g^2 |\Omega_m|}{2\pi\kappa v_F |q_y|}.\end{aligned} \tag{34.18}$$

This agrees precisely with (34.14), and all dependence on Σ has dropped out, as we claimed.

The final step in the exact solution of (34.5) is the evaluation of $\Sigma(i\omega_n)$ at the point $\boldsymbol{k}_0$ on the Fermi surface. As we noted earlier, the parameters v_F and κ are smooth functions of the value of $\boldsymbol{k}_0$, and this is the only momentum dependence in the singular part of the fermion self-energy. A careful evaluation first proceeds by the real frequency method used for Π in (34.10), and we can follow that method for the imaginary part of the $\Sigma(\omega)$ on the real frequency axis. Such an evaluation shows that the result is dominated by the fermions in the vicinity of $\boldsymbol{k}_0$, and with boson momentum $q_y \gg q_x$, which is nearly tangent to the Fermi surface. However, we proceed directly to the second method used for Π below (34.16), in which we integrate over momenta before integrating over frequency; this has the advantage of allowing us to include $\Sigma(i\omega_n)$ in the fermion propagator. From the first equation in (34.5), using the linearized dispersion and result above, we have

$$\Sigma(\boldsymbol{k}, i\omega_n) = g^2\lambda \int \frac{d^2q}{(2\pi)^2} T \sum_{\Omega_m} \frac{1}{q_y^2 + s + \dfrac{g^2|\Omega_m|}{2\pi v_F \kappa |q_y|}} \times \frac{1}{i(\Omega_m + i\omega_n) - v_F(k_x + q_x) - \kappa(k_y + q_y)^2/2 - \Sigma(i\Omega_m + i\omega_n)}, \quad (34.19)$$

where we have dropped q_x in the boson propagator. We can now perform the integral over q_x, and observe that the expression is indeed independent of $\boldsymbol{k}$, and the $\Sigma(i\Omega_m + i\omega_n)$ in the denominator. So we have our closed-form expression for the fermion self-energy:

$$\Sigma(i\omega_n) = -i\frac{g^2\lambda}{2v_F} \int \frac{dq_y}{2\pi} T \sum_{\Omega_m} \frac{\operatorname{sgn}(\omega_n + \Omega_m)}{q_y^2 + s + \dfrac{g^2|\Omega_m|}{2\pi v_F \kappa |q_y|}}. \quad (34.20)$$

We are interested in the singular behavior of this fermion self-energy at the critical point $s = 0$. At $T > 0$, we have to account for thermal effects arising from the boson self-interaction u in (34.1), which make the renormalized s temperature dependent. We do not discuss these subtle issues [5, 59, 71, 298] here, and limit ourselves below to $T = 0$.

For $s > 0$ and $T = 0$, evaluation of the integrals over q_y and Ω in (34.20) shows that $\operatorname{Im}\Sigma(i\omega) \sim -(\omega/s)^2 \ln(1/|\omega|)$, which is the expected behavior for a two-dimensional Fermi liquid (see QPT book). At the critical point $s = 0$, and at $T = 0$, we perform the q_y integral, and then the frequency integral to obtain

$$\begin{aligned} \Sigma(i\omega) &= -i\frac{g^2\lambda}{3v_F\sqrt{3}} \left(\frac{2\pi v_F \kappa}{g^2}\right)^{1/3} \int \frac{d\Omega}{2\pi} \frac{\operatorname{sgn}(\omega + \Omega)}{|\Omega|^{1/3}} \\ &= -iB \operatorname{sgn}(\omega)|\omega|^{2/3} \quad s = 0, T = 0, \end{aligned} \quad (34.21)$$

with

$$B = \frac{g^2\lambda}{2\pi v_F\sqrt{3}} \left(\frac{2\pi v_F \kappa}{g^2}\right)^{1/3}. \quad (34.22)$$

It is instructive to examine the frequency and momentum dependence of the $T = 0$ fermion Green's function across the Fermi surface. In the scaling limit, we can write the real frequency axis Green's function near the Fermi surface as

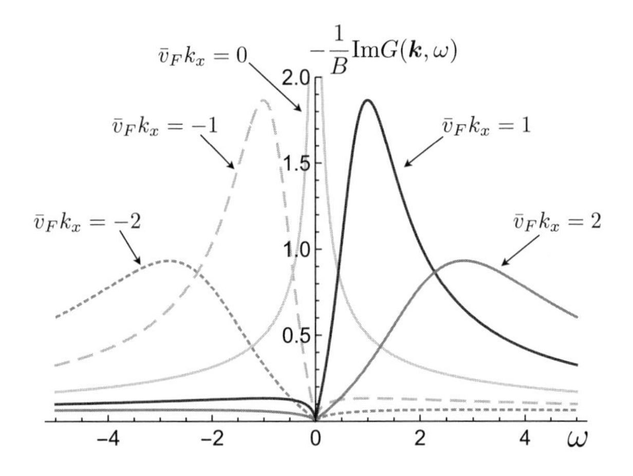

Figure 34.3 Plot of fermion spectral density from (34.23) at wavevectors $\boldsymbol{k} = \boldsymbol{k}_0 + (k_x, 0)$ across the Fermi surface without quasiparticles. Here, $\bar{v}_F = v_F/B$.

$$G(\boldsymbol{k},\omega) = \frac{1}{-v_F k_x - \kappa k_y^2/2 + iBe^{-i\pi \mathrm{sgn}(\omega)/3}|\omega|^{2/3}}. \tag{34.23}$$

As in the SYK model, we can drop the bare ω term in G^{-1} because it is subleading with respect to the frequency-dependent self-energy. Note also the distinction in the singularity structure from (12.55) for the one-dimensional Tomonaga–Luttinger liquid – the singularity here is entirely in the frequency dependence of the self-energy, as in the SYK model. A plot of $-\mathrm{Im}G$ is shown in Fig. 34.3. On the Fermi surface $k_x = 0$, $k_y = 0$ we have $\mathrm{Im}G \sim -1/|\omega|^{2/3}$, which is similar to the $\mathrm{Im}G \sim -1/|\omega|^{1/2}$ behavior of the SYK model. Unlike the Fermi liquid, there is no delta function in ω on the Fermi surface, indicating the absence of quasiparticles. Away from the Fermi surface, $\mathrm{Im}G$ actually vanishes on the Fermi surface (see Fig. 34.3), and there is a broad spectral feature that disperses as $\omega = [(2v_F/(\sqrt{3}B))k_x]^{2/3}$. Note that the position of the Fermi surface is still given by the vanishing of the inverse Green's function at zero frequency, as in (2.39).

We can compute the momentum distribution function of the electrons from (34.23), and it leads to a result similar in form to that of a Tomonaga–Luttinger liquid in (12.57):

$$n(\boldsymbol{k}) \sim -\mathrm{sgn}(v_F k_x + \kappa k_y^2/2)|v_F k_x + \kappa k_y^2/2|^{1/2}, \tag{34.24}$$

with a power-law singularity on the Fermi surface. But recall that the frequency-dependent form of (34.23) is quite different from (12.55) for the one-dimensional electron gas.

At non-zero T, the SYK model displays simple ω/T scaling in its spectral function. There are "quantum" contributions that do indeed scale as ω/T for the critical Fermi surface, but there are also additional corrections that arise from classical thermal fluctuations of ϕ, which are important. So the $T > 0$ situation is rather complex [5, 59, 71, 298], as was noted above.

34.1.3 Patch Field Theory

Having obtained the analytic solution of the large-N saddle-point equations in (34.5) by the asymptotic low-energy analysis above, it is natural to ask if the asymptotic analysis can be performed directly on the theory (34.2) so that we can understand the solution in terms of a more conventional scaling analysis of a quantum field theory. The analysis of the saddle-point equations makes it clear that all the singular effects arise from the vicinity of the points $\pm\boldsymbol{k}_0$ on the Fermi surface for the case of a boson fluctuation in the direction $\boldsymbol{q}$, as shown in Fig. 34.2. So we introduce fermion fields $\psi_{\alpha\pm}$ in the vicinity of these points, and expand their dispersion in gradients according to (34.12) and (34.13). This yields the action [71, 154, 177]

$$
\begin{aligned}
\mathcal{S} &= \int dxdyd\tau\,\mathcal{L}, \\
\mathcal{L} &= \sum_{\alpha=1}^{N}\left\{\psi_{\alpha+}^{\dagger}\left[\partial_\tau - i\partial_x - (\kappa/2)\partial_y^2\right]\psi_{\alpha+} + \psi_{\alpha-}^{\dagger}\left[\partial_\tau + i\partial_x - (\kappa/2)\partial_y^2\right]\psi_{\alpha-}\right\} \\
&\quad + \frac{1}{2}\sum_{\gamma=1}^{M}\left(\partial_y\phi_\gamma\right)^2 + \sum_{\gamma=1}^{M}\sum_{\alpha,\beta=1}^{N}\frac{g_{\alpha\beta\gamma}}{N}\phi_\gamma\left[\psi_{\alpha+}^{\dagger}\psi_{\beta+} + \psi_{\alpha-}^{\dagger}\psi_{\beta-}\right].
\end{aligned}
\tag{34.25}
$$

We have dropped the x and τ gradient terms of ϕ in (34.2), anticipating that they are irrelevant in the scaling analysis we now present.

We analyze the behavior of (34.25) under the rescaling transformation

$$
x \to x/b, \quad y \to y/b^{1/2}, \quad \tau \to \tau/b^z, \tag{34.26}
$$

where the rescaling of x and y leaves the fermion dispersion invariant, but we leave the dynamic critical exponent undetermined for now. Then the $(\partial_y\phi_\gamma)^2$ term is invariant if we choose

$$
\phi \to \phi\, b^{(1+2z)/4}. \tag{34.27}
$$

Similarly, the spatial gradient terms of ψ are invariant if we choose

$$
\psi \to \psi b^{(1+2z)/4}. \tag{34.28}
$$

At this point, it is conventional to fix z by demanding the invariance of temporal gradient terms. However, we saw in our analysis that the bare ω term in G^{-1} was irrelevant, and we dropped it in (34.23), and so this is not the appropriate way to proceed. Instead, we examine the scaling of the Yukawa coupling in (34.25), which is

$$
g \to g b^{(3-2z)/4}. \tag{34.29}
$$

At a critical fixed point, we expect g to be invariant, and this yields the value

$$
z = \frac{3}{2}. \tag{34.30}
$$

This is precisely the value we would have obtained by comparing the k_x and ω terms in (34.23).

The lesson is that we have to study the theory (34.25) at *fixed* g, and it is not permissible to expand in powers of g. We can regard this fixed-g requirement as the analog of a non-linear sigma model constraint in more conventional quantum field theories.

The quantum field theory (34.25) can be used to compute corrections beyond the large-N saddle-point theory presented in Section 34.1.2 (although not for the conductivity [98], as noted earlier). This has not yet been computed within the large-N method of Section 34.1.1, but in an uncontrolled method that examines certain three-loop graphs [177]; this leads to a small fermion anomalous dimension, and hence a breakdown of the purely local form of the singular electron self-energy. It is interesting to note that finite-N corrections discussed in Section 32.3.3 also lead to a breakdown of the local scaling of the SYK model, although from a different mechanism involving the time-reparameterization mode (there is no time-reparameterization soft model for the Fermi surface being discussed here [71]).

Related scaling analyses can also be used in higher dimensions, and in particular for $d = 3$. A key feature in $d = 2$ is that both the fermion Green's function in (34.23), and the Landau-damped boson Green's function implied by (34.18) are characterized by the same dynamic critical exponent $z = 3/2$. A perturbative computation of the corresponding Green's functions in general d shows that the boson Green's function still has $z_b = 3/2$, while the fermion Green's function has $z_f = 3/d$ (see QPT book). For $d > 2$ we have $z_f < z_b$, and so at any given small wavevector, fermionic excitations are higher in energy than bosonic excitations; this implies that the fermions can be safely integrated out, and a perturbative analysis of the effective bosonic theory is valid.

34.2 Luttinger Relation

The strong damping and breakdown of quasiparticles implied by (34.21) and (34.22) nevertheless does not remove the sharp Fermi surface. There is no singular momentum dependence in these expressions, and the frequency dependence still obeys (2.38). Consequently, there is still a Fermi surface specified by (2.39).

We now show that this Fermi surface obeys the same Luttinger relation as that of a Fermi liquid. The argument proceeds just as in Section 30.2.1. The evaluation of (30.31) proceeds as before, as the self-energy has all the needed properties. We only need to examine more carefully the fate of the Luttinger–Ward term in (30.27): in the SYK model, the corresponding term I_2 in (34.35) did not vanish. Here, the Green's function is momentum dependent, and the expression for I_2 has an additional momentum integral

$$I_2 = -i \int_{-\infty}^{\infty} \int \frac{d^2k}{4\pi^2} \frac{d\omega}{2\pi} G(\boldsymbol{k}, i\omega) \frac{d}{d\omega} \Sigma(i\omega) e^{-i\omega 0^+}. \tag{34.31}$$

As the self-energies of the SYK model and the critical Fermi surface both obey (32.47) with $\alpha < 1$, it is possible that there is an anomalous contribution at $\omega = 0$ that leads to a non-vanishing I_2. However, that is not the case here because the singularity of the Green's function is much weaker as a result of its momentum dependence; now the low-energy Green's function is

$$G^{-1}(\boldsymbol{k}, i\omega) = -v_F k_x - \frac{\kappa}{2} k_y^2 - \Sigma(i\omega), \tag{34.32}$$

and this diverges at $\omega = 0$ only on the Fermi surface $v_F k_x + \kappa k_y^2/2 = 0$. Indeed, with this form, the local density of states is a constant at the Fermi level. Consequently, there is no anomaly at $T = 0$, and $I_2 = 0$ from the Luttinger–Ward functional analysis. Incidentally, we note that the Luttinger–Ward functional in the large-N limit is just the first term in the action I in (34.7), similar to the SYK model.

To complete this discussion, I add a few remarks on the structure of the Luttinger–Ward functional, and its connection to global $U(1)$ symmetries [53, 209]. Consider the general case where there are multiple Green's functions (of bosons or fermions) $G_\alpha(k_\alpha, \omega_\alpha)$. Let the α-th particle have a charge q_α under a global $U(1)$ symmetry. Then, for each such $U(1)$ symmetry, the Luttinger–Ward functional will obey the identity

$$\Phi_{LW}\left[G_\alpha(\boldsymbol{k}_\alpha, \omega_\alpha)\right] = \Phi_{LW}\left[G_\alpha(\boldsymbol{k}_\alpha, \omega_\alpha + q_\alpha \Omega)\right]. \tag{34.33}$$

Here, we are regarding Φ_{LW} as a functional of two distinct sets of functions $f_{1,2\alpha}(\omega_\alpha)$, with $f_{1\alpha}(\omega_\alpha) \equiv G_\alpha(k_\alpha, \omega_\alpha + q_\alpha \Omega)$ and $f_{2\alpha}(\omega) \equiv G_\alpha(k_\alpha, \omega_\alpha)$, and Φ_{LW} evaluates to the same value for these two sets of functions. Expanding (34.33) to first order in Ω, and integrating by parts, we establish the corresponding $I_2 = 0$.

34.3 Fermi Surface Coupled to a Gauge Field

As was noted in the beginning of this chapter, the problem of a Fermi surface coupled to a gauge field in 2+1 dimensions leads to properties very similar to those of the Ising ferromagnet described by (34.1). This becomes clear when we reduce the field theory to a two-patch theory along the lines of Section 34.1.2, as I now describe. These results are applicable to the $U(1)$ spin liquid with a spinon Fermi surface noted below (22.7), and the Halperin–Lee–Read state in the half-filled Landau level, noted in Sections 24.4 and 27.4.

Following the procedure in Section 22.2, we can describe the problem of a $U(1)$ gauge field coupled to a Fermi surface by the following general Lagrangian (replacing (34.1)):

$$\mathcal{L} = \sum_{\boldsymbol{k},\alpha} c_{\boldsymbol{k},\alpha}^\dagger \left[\frac{\partial}{\partial \tau} + \varepsilon(\boldsymbol{k} - g\boldsymbol{a})\right] c_{\boldsymbol{k},\alpha} + \int d^2 r \left\{\frac{K_\tau}{2}(\partial_\tau \boldsymbol{a})^2 + \frac{K}{2}(\boldsymbol{\nabla} \times \boldsymbol{a})^2\right\}. \tag{34.34}$$

We focus only on the spatial components of the gauge field, as the temporal components are screened by the background charge density. We also work in the Landau gauge:

$$\boldsymbol{\nabla} \cdot \boldsymbol{a} = 0. \tag{34.35}$$

We can now perform an analysis of the gauge field polarization from the fermion loop diagram using an analysis closely related to that in Section 34.1.2. As in Fig. 34.2, we find that a gauge field fluctuation at wavevector $\boldsymbol{q}$ is damped only by fermion excitations at the anti-podal points $\pm \boldsymbol{k}_0$. Using the condition (34.35) at this point, we can write the gauge field in terms of its single transverse component

$$\boldsymbol{a} = (\phi, 0) . \tag{34.36}$$

Now we take the long-wavelength limit, following the mapping from (34.2) to (34.25). The theory (34.34) yields the Lagrangian density

$$\begin{aligned}\mathcal{L} = \sum_{\alpha} \Big\{ \psi_{\alpha+}^{\dagger} \left[\partial_\tau - i\partial_x - (\kappa/2)\partial_y^2 \right] \psi_{\alpha+} + \psi_{\alpha-}^{\dagger} \left[\partial_\tau + i\partial_x - (\kappa/2)\partial_y^2 \right] \psi_{\alpha-} \Big\} \\ + \frac{K}{2} (\partial_y \phi)^2 + g \sum_{\alpha} \phi \left[\psi_{\alpha+}^{\dagger} \psi_{\alpha+} - \psi_{\alpha-}^{\dagger} \psi_{\alpha-} \right] .\end{aligned} \tag{34.37}$$

The key difference between (34.37) and (34.25) is in the relative sign of the two terms in the Yukawa coupling. This sign makes no difference to the analyses in Section 34.1.1, and so all previous results apply also to (34.34). However, we see that this sign does make a crucial difference in the considerations of fermion pairing in Section 34.4.

34.4 Pairing Correlations

We now study possible pairing instabilities of the non-Fermi liquid states, analogous to the Bardeen–Cooper–Schrieffer pairing instability of Fermi liquids in Chapter 4. As we are dealing with critical states without quasiparticle excitations, we now consider the pairing correlations by a method analogous to that used to study composite operators of the SYK model in Section 32.2.5. We examine a large-N equation analogous to Fig. 32.5, and compute the scaling dimension of the Cooper-pairing operator. If the value of the scaling dimension is real, this gives us information on the correlation functions of the pairing operator in the non-Fermi liquid state. However, we find that under suitable conditions the scaling dimension is complex. Following Refs. [130, 141], we interpret the complex scaling dimension as in indication of an instability to a paired state.

To begin with, we can ignore the absence of quasiparticles, and consider pairing by exchange of ϕ between ψ_+ and ψ_-, along the lines of the paramagnon exchange in Section 9.4.3. Such considerations show that the interaction is attractive (repulsive) between parallel (anti-parallel) spin particles for the Ising ferromagnetic case, and repulsive for arbitrary spin particles for the gauge field case.

To go beyond such leading-order results, and self-consistently include the absence of quasiparticles, it is important to work with a systematic large-N limit. So, we generalize the patch theories in (34.25) and (34.37) to a theory with N flavors of fermions, M_1 flavors of bosons that mediate an attractive interaction (in the pairing channel) between anti-podal points on the Fermi surface, and M_2 flavors of bosons that mediate a repulsive interaction. By rescaling the bosons, we normalize the mean-square Yukawa coupling for both classes of bosons with the same value of g; the value of g will drop out in the scaling equations considered in this section. Having obtained the same Yukawa coupling, we do have to consider the coefficient of the $(\partial_y \phi)^2$ term in

(34.25) more carefully. We take this coefficient to equal K_1 and K_2 for the two bosons, and we see below that the ratio K_1/K_2 influences the critical exponents. For the gauge field case, the values of $K_{1,2}$ are equal to the corresponding diamagnetic susceptibility of the system [324], and this depends upon the lattice scale properties. So we have the theory

$$\mathcal{L} = \sum_{s=\pm 1}\sum_{\alpha=1}^{N} \psi^\dagger_{\alpha s}\left[\partial_\tau - is\partial_x - \partial_y^2\right]\psi_{\alpha s} + \sum_{a=1,2}\frac{K_a}{2}\sum_{\gamma=1}^{M_a}\left(\partial_y\phi_{\gamma a}\right)^2$$
$$+ \sum_{s=\pm 1}\sum_{a=1}^{2} s^{3-a}\sum_{\gamma=1}^{M_a}\sum_{\alpha,\beta=1}^{N}\frac{g^a_{\alpha\beta\gamma}}{N}\psi^\dagger_{\alpha s}\psi_{\beta s}\phi_{\gamma a}\,. \tag{34.38}$$

Here, $s = \pm 1$ is the index of the two anti-podal patches (see Fig. 34.2), and $a = 1,2$ represents the attractive and repulsive bosons, respectively. Also, it will be necessary to take the random couplings $g^a_{\alpha\beta\gamma}$ to now be real independent variables.

Let us now recompute the boson and fermion self-energies of Section 34.1.2 for the theory (34.38). The self-energy of the boson $\phi_{\gamma a}$ is still equal to (34.18), while the self-energy of the fermion in (34.21) becomes

$$\Sigma(i\omega) = -i\frac{g^2}{2\pi\sqrt{3}}\left(\frac{M_1K_1^{-2/3} + M_2K_2^{-2/3}}{N}\right)\left(\frac{2\pi v_F\kappa}{g^2}\right)^{1/3}\operatorname{sgn}(\omega)|\omega|^{2/3}. \tag{34.39}$$

We now consider the scaling dimension of the composite operator $\psi^\dagger_{\alpha+}\psi^\dagger_{\alpha-}$ along the lines of Section 32.2.5. The large-N limit leads to an integral equation for the pairing vertex, analogous to that in Fig. 32.5 and (32.65), shown in Fig. 34.4. We first consider the internal loop, and evaluate the integral over momenta along the lines of the analysis in Section 34.1.2, while assuming momentum independence of the vertex – we see below that this assumption is consistent:

$$\int\frac{dk_xdk_y}{4\pi^2}\frac{1}{(i\omega - v_F(k_x+p_x) - \kappa(k_y+p_y)^2/2 - \Sigma(i\omega))}$$
$$\times\frac{1}{(-i\omega - v_F(k_x+p_x) - \kappa(k_y+p_y)^2/2 - \Sigma(-i\omega))(K_ak_y^2 + \Pi(k_y, i\omega - i\Omega))}\,. \tag{34.40}$$

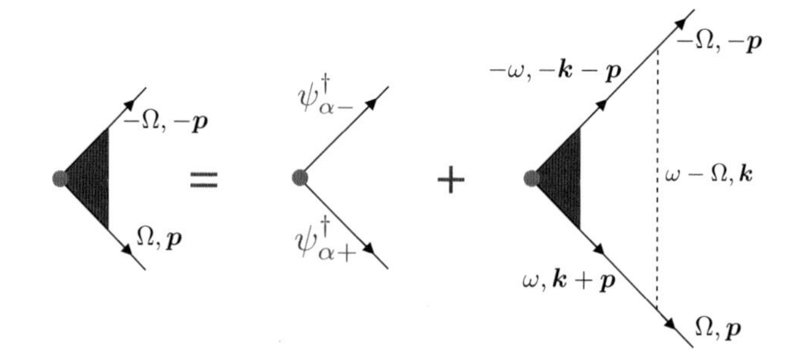

Figure 34.4 Large-N equation for the scaling dimension of the composite operator $\psi^\dagger_{\alpha+}\psi^\dagger_{\alpha-}$, leading to (34.42) after integration over the momentum in the loop of the diagram on the right. The filled triangle is the pairing vertex Δ.

Notice that the k_x term appears with the same sign in the two fermion propagators, while the frequencies have opposite signs, corresponding to the pairing between antipodal patches. This structure is crucial to the non-vanishing result of the k_x integral in (34.40), which yields

$$\int \frac{dk_y}{4\pi v_F} \frac{1}{|\omega + i\Sigma(i\omega)|(K_a k_y^2 + \Pi(k_y, i\omega - i\Omega))}. \tag{34.41}$$

This result is independent of the external momentum $\boldsymbol{p}$ – this implies we can consistently take the pairing vertex to be independent of momentum. The pairing vertex depends only upon frequency, just like the fermion self-energy, and the situation is now essentially identical to that for the SYK model in Section 32.2.5, with the composite operator also having only local correlations. We can perform the k_y integral in (34.41), and then Fig. 34.4 yields the following integral equation for the pairing vertex in frequency space alone

$$E\Delta(i\Omega) = -\sum_a \frac{M_a \zeta_a g^2}{3N\sqrt{3}} \int \frac{d\omega}{2\pi} \frac{\Delta(i\omega)}{|\omega + i\Sigma(i\omega)|} \frac{(4\pi)^{1/3}}{(gK_a)^{2/3}|\omega - \Omega|^{1/3}}. \tag{34.42}$$

Here, $a = 1,2$ sums over the attractive and repulsive bosons and $\zeta_a = 2a - 3 = -1\,(+1)$ for the attractive (repulsive) interactions. Solutions of this equation with eigenvalue $E = 1$ will determine the scaling of $\Delta(i\Omega)$, as in (32.67). At low energies and $T = 0$, where we drop the bare ω term in the right-hand side of (34.42), because it is irrelevant in the infrared, we obtain a universal equation *independent* of the coupling g:

$$E\Delta(i\Omega) = \frac{\mathcal{K}}{3} \int \frac{d\omega}{2\pi} \frac{2\pi\Delta(i\omega)}{|\omega|^{2/3}|\omega - \Omega|^{1/3}}, \tag{34.43}$$

where the dimensionless constant

$$\mathcal{K} \equiv \frac{M_1 K_2^{2/3} - M_2 K_1^{2/3}}{M_1 K_2^{2/3} + M_2 K_1^{2/3}} \tag{34.44}$$

determines the balance between the attractive and repulsive interactions. The Ising ferromagnet case has $M_1 = 1$, $M_2 = 0$, $\mathcal{K} = 1$, while the gauge field case has $M_1 = 0$, $M_2 = 1$, $\mathcal{K} = -1$. The equation (34.43) has the same form as that for the $\gamma = 1/3$ case of the γ model of quantum-critical pairing studied by Chubukov and collaborators [1, 49, 184, 315].

As in Section 32.2.5, we assume the eigenvector has the form

$$\Delta(i\Omega) = \frac{1}{|\Omega|^{\alpha}}. \tag{34.45}$$

We assume $0 < \text{Re}\,[\alpha] < 1/3$ to ensure a convergent integral in (34.43), and then we have

$$E(\alpha) = \mathcal{K} \frac{\pi^2 \left(3\cot\left(\frac{\pi\alpha}{2}\right) + \sqrt{3}\right)\sec\left(\pi\left(\alpha + \frac{1}{6}\right)\right)}{9\Gamma\left(\frac{1}{3}\right)\Gamma(1-\alpha)\Gamma\left(\alpha + \frac{2}{3}\right)}, \tag{34.46}$$

a result analogous to (32.69). The solution of $E(\alpha) = 1$ is shown in Fig. 34.5. For $\mathcal{K} = 1$, setting $E(\alpha) = 1$ indicates a complex scaling dimension $\alpha = 1/6 \pm i \times 0.53734\ldots$, which

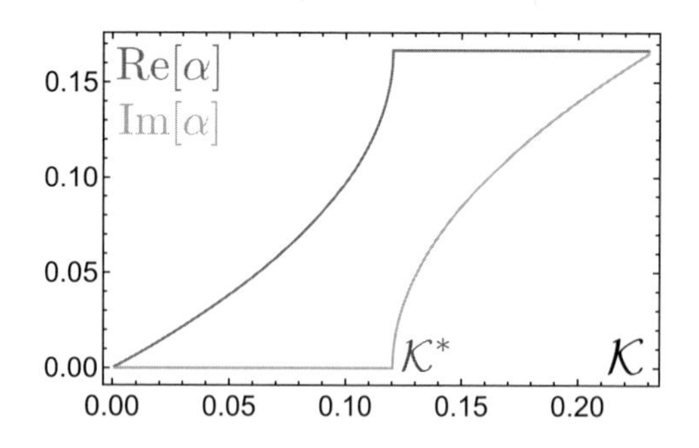

Figure 34.5 Plot [71] of $\text{Re}[\alpha]$ and $\text{Im}[\alpha]$, for the solutions which have $\text{Re}[\alpha] \leq 1/6$ and $\text{Im}[\alpha] > 0$, as a function of $\mathcal{K}$. The critical Fermi surface is unstable to pairing for $\mathcal{K}^* < \mathcal{K} < 1$, where $\text{Re}[\alpha] = 1/6$ and $\text{Im}[\alpha] \neq 0$. For $0 < \mathcal{K} < \mathcal{K}^*$, the pairing operator has a non-trivial scaling dimension determined by $\text{Re}[\alpha]$. Reprinted with permission from APS.

implies that a pairing instability exists and the ground state is superconducting. As the value of $\mathcal{K}$ is reduced, the magnitude of the imaginary part of α also reduces, going to zero at $\mathcal{K} = \mathcal{K}^* = 0.12038\ldots$, at which point $\alpha = 1/6$ exactly. For $0 < \mathcal{K} < \mathcal{K}^*$, $E = 1$ has two solutions with purely real α: α_1, with $0 < \alpha_1 < 1/6$ and $\alpha_2 = 1/3 - \alpha_1$, indicating the absence of a superconducting instability arising purely out of the relevant operators in the low-energy critical theory, when the repulsive interaction is strong enough. Arguments have been made [71] that α_1 is the correct choice for the scaling dimension. For $\mathcal{K} < 0$, there is no solution for $E = 1$; therefore, there is no superconducting instability, and the scaling dimension of the pairing vertex equals its bare value.

To summarize, the above results imply a pairing instability for the Ising ferromagnet with purely attractive interactions ($\mathcal{K} = 1$), but no pairing instability for the gauge field case with purely repulsive interactions ($\mathcal{K} = -1$). If we have a combination of repulsive and attractive interactions, $\mathcal{K}$ interpolates between 1 and -1, and there is no pairing instability for $\mathcal{K} < \mathcal{K}^* = 0.12038$. The critical Fermi surface state is stable for all $\mathcal{K} < \mathcal{K}^*$, and has a non-trivial dimension for pairing fluctuations in the regime $0 < \mathcal{K} < \mathcal{K}^*$ shown in Fig. 34.5.

I conclude by noting that it is possible to also consider the scaling of other composite operators from the product of two fermion fields, as in Section 32.2.5 for the SYK model. It turns out there is no non-trivial scaling of operators made by the product of $\psi_+^\dagger$ and ψ_+ because the analog of the k_x integral in (34.40) vanishes. However, there is interesting behavior in $\psi_+^\dagger \psi_-$, which is an operator with wavevector $2k_F$: this has been studied in Ref. [71].

34.5 Transport

One of the primary motivations for the intense study of critical Fermi surfaces has been the hope that it can lead to a theory of the iconic linear-T resistivity of strange metals (see Section 1.5). However, the theory presented so far cannot describe such observations. The key point is that the important singular processes in such a theory can all be expressed in terms of a continuum field theory, such as that in (34.25), which conserves total momentum. In the absence of particle–hole symmetry, any state with a non-zero momentum has a non-zero current and vice versa; consequently if we set up a state with a non-zero current, the non-zero total momentum of such a state prevents the current from decaying to zero. In other words, the resistivity will vanish [66, 102, 103, 104, 174].

This effect can be viewed as one analogous to phonon drag [202, 203]. However, because of the weak electron–phonon coupling, phonon drag is significant only in the cleanest samples [110]. On the other hand, the coupling in the critical Fermi surface is so strong that the individual fermions and bosons lose their identity and there are no quasiparticle excitations. Thus, we cannot separately consider the momentum carried by the fermions and bosons.

A computation of the conductivity in the large-N limit described above requires a summation of ladder diagrams, which is described in Ref. [98]. Such an analysis leads only to a delta function in the conductivity at $T = 0$:

$$\mathrm{Re}\,\sigma(\omega) = D_1\,\delta(\omega)\,, \tag{34.47}$$

and the anomalous self-energy of the electron in (34.21) does not show up. A subtle but important point is that this analysis of the conductivity cannot be carried out in the patch field theory of Section 34.1.3, and it is necessary to account for scattering between adjacent patches of the Fermi surface [98]. To remove the delta function in (34.47), we need a mechanism to relax the momentum. Studies have examined the influence of Umklapp scattering [152, 297], but here we focus on the promising results [98, 200] obtained by including spatial disorder.

The most important source of spatial disorder in the theory of disordered Fermi liquids is potential scattering, and so it is natural to include that here in the present theory. A form amenable to the large-N limit described here is the random potential action

$$\begin{aligned} \mathcal{S}_v &= \frac{1}{\sqrt{N}} \sum_{\alpha,\beta=1}^{N} \int d^2r d\tau\, v_{\alpha\beta}(\boldsymbol{r}) \psi_\alpha^\dagger(\boldsymbol{r},\tau)\psi_\beta(\boldsymbol{r},\tau), \\ \overline{v_{\alpha\beta}(\boldsymbol{r})} &= 0\,, \quad \overline{v_{\alpha\beta}^*(\boldsymbol{r}) v_{\gamma\delta}(\boldsymbol{r}')} = v^2\,\delta(\boldsymbol{r}-\boldsymbol{r}')\delta_{\alpha\gamma}\delta_{\beta\delta}. \end{aligned} \tag{34.48}$$

The solution of the corresponding large-N saddle-point equations shows [200] that the boson polarizibility in (34.18) is replaced by

$$\Pi(\boldsymbol{q}, i\Omega_n) \sim -\frac{g^2}{v^2}|\Omega_n|, \tag{34.49}$$

which leads to $z = 2$ behavior in the boson propagator, with $[D(\boldsymbol{q}, i\Omega_n)]^{-1} \sim q^2 + \gamma|\Omega_n|$. The corresponding fermion self-energy is modified from (34.21): there is a familiar elastic impurity scattering contribution Σ_v also present in a disordered Fermi liquid, along with an inelastic term Σ_g [98] with the "marginal Fermi liquid" form [285]:

$$\Sigma_v(i\omega_n) \sim -iv^2 \text{sgn}(\omega_n), \quad \Sigma_g(i\omega_n) \sim -\frac{g^2}{v^2}\omega_n \ln(1/|\omega_n|). \tag{34.50}$$

Despite the promising singularity in Σ_g, (34.50) does not translate [98] into interesting behavior in the transport; the scattering is mostly forward, and the resistivity is Fermi-liquid-like, with $\rho(T) = \rho(0) + AT^2$.

Much more interesting and appealing behavior results when we add spatial randomness in the Yukawa coupling. Such randomness is generated by the potential randomness $v_{\alpha\beta}(x)$ considered above, but it has to be included at the outset in the large-N limit. More explicitly, we recall that the Yukawa coupling invariably arises from a Hubbard–Stratonovich decoupling of a four-fermion interaction; we can decouple such an interaction via a ϕ^2 term that is spatially uniform, and then all the spatial disorder is transferred to the Yukawa term.

So we *add* to the spatially independent Yukawa couplings $g_{\alpha\beta\gamma}$ in (34.2) a second coupling $g'_{\alpha\beta\gamma}(\boldsymbol{r})$, which has both spatial and flavor randomness with action

$$\mathcal{S}_{g'} = \frac{1}{N}\int d^2r d\tau\, g'_{\alpha\beta\gamma}(\boldsymbol{r})\psi^\dagger_\alpha(\boldsymbol{r},\tau)\psi_\beta(\boldsymbol{r},\tau)\phi_\gamma(\boldsymbol{r},\tau), \tag{34.51}$$
$$\overline{g'_{\alpha\beta\gamma}(\boldsymbol{r})} = 0, \quad \overline{g'^*_{\alpha\beta\gamma}(\boldsymbol{r})g'_{\delta\rho\sigma}(\boldsymbol{r}')} = g'^2\,\delta(\boldsymbol{r}-\boldsymbol{r}')\delta_{\alpha\delta}\delta_{\beta\rho}\delta_{\gamma\sigma}.$$

Then we obtain additional contributions to the boson and fermion self-energies [200]:

$$\Pi_{g'}(\boldsymbol{q}, i\Omega_n) \sim -g'^2|\Omega_n|, \quad \Sigma_{g'}(i\omega_n) \sim -ig'^2\omega_n \ln(1/|\omega_n|). \tag{34.52}$$

Now the marginal Fermi liquid self-energy does contribute significantly to transport [200], with a linear-T resistivity $\sim g'^2 T$, while the residual resistivity is determined primarily by v. It is notable that it is the disorder in the interactions, v, which determines the slope of the linear-T resistivity, while it is the potential scattering disorder that determines the residual resistivity. Other attractive features of this theory are that it has an anomalous optical conductivity $\sigma(\omega)$ with $\text{Re}[1/\sigma(\omega)] \sim \omega$ and a $T\ln(1/T)$ specific heat [200].

Appendix A Coherent-State Path Integral

To avoid inessential indices, I present the derivation of the coherent-state path integral by focusing on a single site, and drop the site index. We will first derive the result in a general notation, to allow subsequent application to quantum spin systems. So we consider a general Hamiltonian $H(\hat{\boldsymbol{S}})$, dependent upon operators $\hat{\boldsymbol{S}}$ that need not commute with each other. So, for the boson Hubbard model, $\hat{\boldsymbol{S}}$ is a two-dimensional vector of operators $\hat{b}$ and $\hat{b}^\dagger$, which obey (8.1). When we apply the results to quantum spin systems, $\hat{\boldsymbol{S}}$ represents the usual spin operators $\hat{S}_{x,y,z}$.

The first step is to introduce the coherent states. These are an infinite set of states $|\mathbf{N}\rangle$, labeled by the a continuous vector $\boldsymbol{N}$ (in two or three dimensions for the two cases above). They are normalized to unity,

$$\langle \boldsymbol{N}|\boldsymbol{N}\rangle = 1, \tag{A.1}$$

but are not orthogonal $\langle \boldsymbol{N}|\boldsymbol{N}'\rangle \neq 0$ for $\boldsymbol{N} \neq \boldsymbol{N}'$. They do, however, satisfy a completeness relation

$$\mathcal{C}_N \int d\boldsymbol{N}\, |\boldsymbol{N}\rangle\langle \boldsymbol{N}| = 1, \tag{A.2}$$

where $\mathcal{C}_N$ is a normalization constant. Because of their non-orthogonality, these states are called "over-complete." Finally, they are chosen with a useful property: the diagonal expectation values of the operators $\hat{\boldsymbol{S}}$ are very simple:

$$\langle \boldsymbol{N}|\hat{\boldsymbol{S}}|\boldsymbol{N}\rangle = \boldsymbol{N}. \tag{A.3}$$

This property implies that the vector $\boldsymbol{N}$ is a classical approximation to the operators $\hat{\boldsymbol{S}}$. The relations (A.1), (A.2), and (A.3) define the coherent states, and are all we need here to set up the coherent-state path integral.

We also need the diagonal matrix elements of the Hamiltonian in the coherent-state basis. Usually, it is possible to arrange the operators such that

$$\langle \boldsymbol{N}|H(\hat{\boldsymbol{S}})|\boldsymbol{N}\rangle = H(\boldsymbol{N}); \tag{A.4}$$

that means, $H(\boldsymbol{N})$ has the same functional dependence upon $\boldsymbol{N}$ as the original Hamiltonian has on $\boldsymbol{S}$. For the boson Hubbard model, this corresponds, as we see, to normal-ordering the creation and annihilation operators. In any case, the right-hand side could have a distinct functional dependence on $\boldsymbol{N}$, but we just refer to the diagonal matrix element as above.

We proceed to the derivation of the coherent-state path integral for the partition function

$$\mathcal{Z} = \mathrm{Tr}\exp(-H(\hat{\boldsymbol{S}})/T). \tag{A.5}$$

We break up the exponential into a large number of exponentials of infinitesimal time evolution operators

$$\mathcal{Z} = \lim_{M\to\infty}\prod_{i=1}^{M}\exp(-\Delta\tau_i H(\hat{\boldsymbol{S}})), \tag{A.6}$$

where $\Delta\tau_i = 1/MT$, and a set of coherent states are inserted between each exponential by using the identity (A.2); we label the state inserted at a "time" τ by $|\boldsymbol{N}(\tau)\rangle$. We can then evaluate the expectation value of each exponential by use of the identity (A.3)

$$\begin{aligned}
&\langle \boldsymbol{N}(\tau)|\exp(-\Delta\tau H(\hat{\boldsymbol{S}}))|\boldsymbol{N}(\tau-\Delta\tau)\rangle \\
&\quad\approx \langle \boldsymbol{N}(\tau)|(1-\Delta\tau H(\hat{\boldsymbol{S}}))|\boldsymbol{N}(\tau-\Delta\tau)\rangle \\
&\quad\approx 1-\Delta\tau\langle \boldsymbol{N}(\tau)|\frac{d}{d\tau}|\boldsymbol{N}(\tau)\rangle - \Delta\tau H(\boldsymbol{N}) \\
&\quad\approx \exp\left(-\Delta\tau\langle \boldsymbol{N}(\tau)|\frac{d}{d\tau}|\boldsymbol{N}(\tau)\rangle - \Delta\tau H(\boldsymbol{N})\right).
\end{aligned} \tag{A.7}$$

In each step we have retained expressions correct to order $\Delta\tau$. Because the coherent states at time τ and $\tau+\Delta\tau$ can in principle have completely different orientations, a priori, it is not clear that expanding these states in derivatives of time is a valid procedure. This is a subtlety that afflicts all coherent-state path integrals and has been discussed more carefully by Negele and Orland [190]. The conclusion of their analysis is that except for the single "tadpole" diagram where a point-splitting of time becomes necessary, this expansion in derivatives of time always leads to correct results. In any case, the resulting coherent-state path integral is a formal expression that cannot be directly evaluated, and in case of any doubt one should always return to the original discrete-time product in (A.6).

Keeping in mind the above caution, we insert (A.7) into (A.6), take the limit of small $\Delta\tau$, and obtain the following functional integral for $\mathcal{Z}$:

$$\mathcal{Z} = \int_{\boldsymbol{N}(0)=\boldsymbol{N}(1/T)} \mathcal{D}\boldsymbol{N}(\tau)\exp\left\{-\mathcal{S}_B - \int_0^{1/T} d\tau H(\boldsymbol{N}(\tau))\right\}, \tag{A.8}$$

where

$$\mathcal{S}_B = \int_0^{1/T} d\tau \langle \boldsymbol{N}(\tau)|\frac{d}{d\tau}|\boldsymbol{N}(\tau)\rangle \tag{A.9}$$

and $H(S\boldsymbol{N})$ is obtained by replacing every occurrence of $\hat{\boldsymbol{S}}$ in the Hamiltonian by $S\boldsymbol{N}$. The promised Berry phase term is $\mathcal{S}_B$, and it represents the overlap between the coherent states at two infinitesimally separated times. It can be shown straightforwardly from the normalization condition, $\langle \boldsymbol{N}|\boldsymbol{N}\rangle = 1$, that $\mathcal{S}_B$ is purely imaginary.

A.1 Boson Coherent States

We now apply the general formalism above to the boson Hubbard model. As before, we drop the site index i.

For the state label, we replace the two-dimensional vector $\boldsymbol{N}$ by a complex number ψ, and so the coherent states are $|\psi\rangle$, with one state for every complex number. A state with the properties (A.1), (A.2), and (A.3) turns out to be

$$|\psi\rangle = e^{-|\psi|^2/2} \exp\left(\psi \hat{b}^\dagger\right) |0\rangle, \tag{A.10}$$

where $|0\rangle$ is the boson vacuum state (one of the states in (8.3)). This state is normalized as required by (A.1), and we can now obtain its diagonal matrix element

$$\begin{aligned}\langle\psi|\hat{b}|\psi\rangle &= e^{-|\psi|^2} \frac{\partial}{\partial \psi^*} \langle 0|\, e^{\psi^* \hat{b}} e^{\psi \hat{b}^\dagger} |0\rangle \\ &= e^{-|\psi|^2} \frac{\partial}{\partial \psi^*} e^{|\psi|^2} = \psi,\end{aligned} \tag{A.11}$$

which satisfies the requirement (A.3). For the complete relation, we evaluate

$$\begin{aligned}\int d\psi d\psi^* |\psi\rangle\langle\psi| &= \sum_{n=0}^{\infty} \frac{|n\rangle\langle n|}{n!} \int d\psi d\psi^* |\psi|^{2n} e^{-|\psi|^2} \\ &= \pi \sum_{n=0}^{\infty} |n\rangle\langle n|,\end{aligned} \tag{A.12}$$

where $|n\rangle$ are the number states in (8.3), $d\psi d\psi^* \equiv d\mathrm{Re}(\psi) d\mathrm{Im}(\psi)$, and we have picked only the diagonal terms in the double sum over number states because the off-diagonal terms vanish after the angular ψ integration. This result identifies $C_N = 1/\pi$. So we have satisfied the properties (A.1), (A.2), and (A.3) required of all coherent states.

For the path integral, we need the Berry phase term in (A.9). This is a path integral over trajectories in the complex plane, $\psi(\tau)$, and we have

$$\langle\psi(\tau)|\frac{d}{d\tau}|\psi(\tau)\rangle = e^{-|\psi(\tau)|^2} \langle 0| e^{\psi^*(\tau)\hat{b}} |\frac{d}{d\tau}| e^{\psi(\tau)\hat{b}^\dagger} |0\rangle = \psi^* \frac{d\psi}{d\tau}. \tag{A.13}$$

We are now ready to combine (A.13) and (A.8) to obtain the coherent-state path integral of the boson Hubbard model in (8.18).

A.2 Spin Coherent States

We deal in this section with a single spin and will therefore drop the site index. Now $\hat{\boldsymbol{S}}$ is the vector of spin operators of representation S. We construct the states $|\boldsymbol{N}\rangle$ explicitly below, where we choose $\boldsymbol{N}$ to be a unit vector with $\boldsymbol{N}^2 = 1$. Thus, the coherent states are labeled by points on the unit sphere. With this definition, (A.4) is modified here to

$$\langle \boldsymbol{N}|\hat{\boldsymbol{S}}|\boldsymbol{N}\rangle = S\boldsymbol{N}. \tag{A.14}$$

The completeness relation (A.2) takes the form

$$\int \frac{dN}{2\pi} |N\rangle\langle N| = 1 = \sum_{m=-S}^{S} |S,m\rangle\langle S,m|, \tag{A.15}$$

where the integral of $\boldsymbol{N}$ is over the unit sphere. The state $|\boldsymbol{N}\rangle$ is almost like a classical spin of length S pointing in the $\boldsymbol{N}$ direction; indeed, the spin coherent states are the minimum uncertainty states localized as much in the $\boldsymbol{N}$ direction as the principles of quantum mechanics will allow, and, in the large-S limit, $|\boldsymbol{N}\rangle$ reduces to a classical spin in the $\boldsymbol{N}$ direction.

Let us now explicitly construct the spin coherent states. For $\boldsymbol{N} = (0,0,1)$, the state $|\boldsymbol{N}\rangle$ is easy to determine; we have

$$|\boldsymbol{N} = (0,0,1)\rangle = |S, m = S\rangle \equiv |\Psi_0\rangle. \tag{A.16}$$

We have labeled this particular coherent state as a reference state $|\Psi_0\rangle$ as it will be needed frequently in the following. It should be clear that for other values of $\boldsymbol{N}$ we can obtain $|\boldsymbol{N}\rangle$ simply by acting on $|\Psi_0\rangle$ by an operator that performs an $SU(2)$ rotation from the direction $(0,0,1)$ to the direction $\boldsymbol{N}$. In this manner we obtain the following explicit representation for the coherent state $|\boldsymbol{N}\rangle$:

$$|\boldsymbol{N}\rangle = \exp(z\hat{S}_+ - z^*\hat{S}_-)|\Psi_0\rangle, \tag{A.17}$$

where the complex number z is related to the vector $\boldsymbol{N}$. This relationship is simplest in spherical coordinates; if we parameterize $\boldsymbol{N}$ as

$$\boldsymbol{N} = (\sin\theta\cos\phi, \sin\theta\sin\phi, \cos\theta), \tag{A.18}$$

then

$$z = -\frac{\theta}{2}\exp(-i\phi). \tag{A.19}$$

I leave it as an exercise for the reader to verify that (A.17) satisfies (A.1), (A.2), and (A.3); this verification is aided by the knowledge that the value of the expression $\exp(-i\boldsymbol{a}\cdot\hat{\boldsymbol{S}})\hat{\boldsymbol{S}}\exp(i\boldsymbol{a}\cdot\hat{\boldsymbol{S}})$, where $\boldsymbol{a}$ is some vector, is determined solely by the spin commutation relations

$$[\hat{S}_\alpha, \hat{S}_\beta] = i\varepsilon_{\alpha\beta\gamma}S_\gamma, \tag{A.20}$$

and can therefore be worked out by temporarily assuming that the $\hat{\boldsymbol{S}}$ are 1/2 times the Pauli matrices; the result, when expressed in terms of $\hat{\boldsymbol{S}}$, is valid for arbitrary S.

It is useful for our subsequent formulation to rewrite the above results in a somewhat different manner, making the $SU(2)$ symmetry more manifest. The 2×2 matrix of operators $\hat{\mathcal{S}}$ is defined by

$$\hat{\mathcal{S}} = \begin{pmatrix} \hat{S}_z & \hat{S}_x - i\hat{S}_y \\ \hat{S}_x + i\hat{S}_y & -\hat{S}_z \end{pmatrix}. \tag{A.21}$$

Then (A.16) can be rewritten as

$$\langle \boldsymbol{N}|\hat{\mathcal{S}}_{\alpha\beta}|\boldsymbol{N}\rangle = SW_{\alpha\beta}, \tag{A.22}$$

where the matrix W is

$$W = \begin{pmatrix} N_z & N_x - iN_y \\ N_x + iN_y & -N_z \end{pmatrix} \equiv \boldsymbol{N} \cdot \vec{\sigma}, \tag{A.23}$$

and $\vec{\sigma}$ are the Pauli matrices. So instead of labeling the coherent states with the unit vector $\boldsymbol{N}$, we could equally well use the traceless Hermitean matrix W. Furthermore, there is a simple relationship between W and the complex number z. In particular, if we use the spin-1/2 version of the operator in (A.17):

$$U = \exp\left[\begin{pmatrix} 0 & z \\ -z^* & 0 \end{pmatrix}\right] \tag{A.24}$$

(U is thus a 2×2 matrix), we find

$$W = U\sigma_z U^\dagger. \tag{A.25}$$

Now let us apply these results to the path-integral representation in (A.8) and (A.9). Clearly, the τ dependence of $\boldsymbol{N}(\tau)$ implies a τ-dependent $z(\tau)$ through (A.19). From (A.17) we have therefore

$$\frac{d}{d\tau}|\boldsymbol{N}(\tau)\rangle = \frac{d}{d\tau}\exp(z(\tau), \hat{S}_+ - z^*(\tau)\hat{S}_-)|\Psi_0\rangle. \tag{A.26}$$

Taking this derivative is, however, not so simple. Notice that if an operator $\hat{O}$ does not commute with its derivative $d\hat{O}/d\tau$ then

$$\frac{d}{d\tau}\exp(\hat{O}) \neq \frac{d\hat{O}}{d\tau}\exp(\hat{O}). \tag{A.27}$$

The correct form of this result is in fact

$$\frac{d}{d\tau}\exp(\hat{O}) = \int_0^1 du \exp(\hat{O}(1-u))\frac{d\hat{O}}{d\tau}\exp(\hat{O}u), \tag{A.28}$$

where u is just a dummy integration variable. This result can be checked by expanding both sides in powers of $\hat{O}$ and verifying that they agree, term by term. More constructively, a "hand-waving" derivation can be given as follows:

$$\begin{aligned}
\frac{d}{d\tau}\exp(\hat{O}) &= \frac{d}{d\tau}\exp\left(\hat{O}\int_0^1 du\right) \\
&= \lim_{M\to\infty}\frac{d}{d\tau}\exp\left(\sum_{i=1}^M \hat{O}\Delta u_i\right) \quad \text{with } \Delta u_i = 1/M \\
&\approx \lim_{M\to\infty}\frac{d}{d\tau}\prod_{i=1}^M \exp(\hat{O}\Delta u_i) \\
&\approx \lim_{M\to\infty}\sum_{j=1}^M\prod_{i=1}^j \exp(\hat{O}\Delta u_i)\frac{d\hat{O}}{d\tau}\Delta u_j \prod_{i=j+1}^M \exp(\hat{O}\Delta u_i).
\end{aligned} \tag{A.29}$$

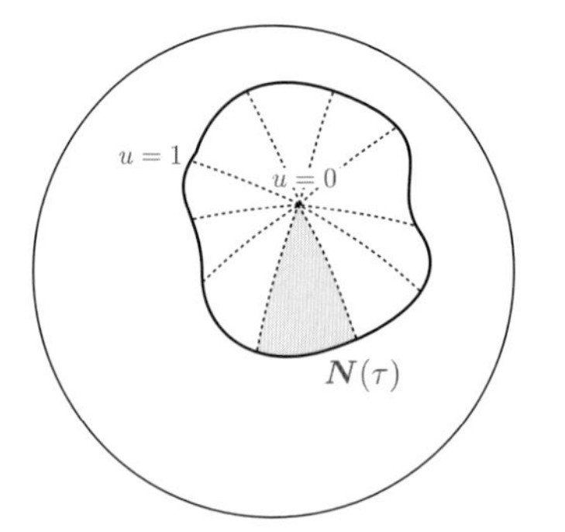

Figure A.1 Strings (dashed lines) connecting the spin orientation $\boldsymbol{N}(\tau)$ at $u=1$ to the north pole.

Finally, taking the limit $M \to \infty$, we obtain the required result (A.28). Now, using (A.26) and (A.28), we find

$$\begin{aligned} \mathcal{S}_B &= \int_0^{1/T} d\tau \langle \boldsymbol{N}(\tau)|\frac{d}{d\tau}|\boldsymbol{N}(\tau)\rangle \\ &= \int_0^{1/T} d\tau \int_0^1 du \langle \boldsymbol{N}(\tau,u)| \left(\frac{\partial z}{\partial \tau}\hat{S}_+ - \frac{\partial z^*}{\partial \tau}\hat{S}_- \right) |\boldsymbol{N}(\tau,u)\rangle, \end{aligned} \tag{A.30}$$

where $\boldsymbol{N}(\tau,u)$ is defined by

$$|\boldsymbol{N}(\tau,u)\rangle = \exp(u(z(\tau)\hat{S}_+ - z^*(\tau)\hat{S}_-))|\Psi_0\rangle. \tag{A.31}$$

From this definition, three important properties of $\boldsymbol{N}(\tau,u)$ should be apparent:

$$\begin{aligned} &\boldsymbol{N}(\tau,u=1) \equiv \boldsymbol{N}(\tau), \\ &\boldsymbol{N}(\tau,u=0) = (0,0,1), \\ &\text{and } \boldsymbol{N}(\tau,u) \text{ moves with } u \text{ along the great circle} \\ &\text{between } \boldsymbol{N}(\tau,u=0) \text{ and } \boldsymbol{N}(\tau,u=1). \end{aligned} \tag{A.32}$$

We can visualize the dependence on u by imagining a *string* connecting the physical value of $\boldsymbol{N}(\tau) = \boldsymbol{N}(\tau,u=1)$ to the north pole, along which u decreases to 0; see Fig. A.1. Associated with each $\boldsymbol{N}(\tau,u)$ we can also define a u-dependent $W(\tau,u)$ as in (A.23); the analog of (A.32) is $W(\tau,u=1) \equiv W(\tau)$ and $W(\tau,u=1) = \sigma_z$. A simple explicit expression for $W(\tau,u)$ is also possible: We simply generalize (A.24) to

$$U(\tau,u) = \exp\left[u\begin{pmatrix} 0 & z \\ -z^* & 0 \end{pmatrix}\right]; \tag{A.33}$$

then, the relationship (A.25) gives us $W(\tau,u)$. Now we can use the expression (A.22) to rewrite (A.30) as

$$\mathcal{S}_B = S\int_0^{1/T} d\tau \int_0^1 du \left[\frac{\partial z}{\partial \tau}W_{21}(\tau,u) - \frac{\partial z^*}{\partial \tau}W_{12}(\tau,u)\right]. \tag{A.34}$$

As everything is a periodic function of τ, we may freely integrate this expression by parts and obtain

$$\mathcal{S}_B = -S\int_0^{1/T} d\tau \int_0^1 du \mathrm{Tr}\left[\begin{pmatrix} 0 & z(\tau) \\ -z^*(\tau) & 0 \end{pmatrix} \partial_\tau W(\tau,u)\right], \tag{A.35}$$

where the trace is over the 2×2 matrix indices. The definitions (A.25) and (A.33) can be used to easily establish the identity

$$\begin{pmatrix} 0 & z(\tau) \\ -z^*(\tau) & 0 \end{pmatrix} = -\frac{1}{2}W(\tau,u)\frac{\partial W(\tau,u)}{\partial u}, \tag{A.36}$$

which when inserted into (A.35) yields the expression for $\mathcal{S}_B$ in one of its final forms:

$$\mathcal{S}_B = \int_0^{1/T} d\tau \int_0^1 du \left[\frac{S}{2}\mathrm{Tr}\left(W(\tau,u)\frac{\partial W(\tau,u)}{\partial u}\frac{\partial W(\tau,u)}{\partial \tau}\right)\right]. \tag{A.37}$$

An expression for $\mathcal{S}_B$ solely in terms of $\boldsymbol{N}(\tau,u)$ can be obtained by substituting in (A.23); this yields the final expression for $\mathcal{S}_B$, which when inserted in (A.8) gives us the coherent-state path integral for a spin:

$$\mathcal{S}_B = iS\int_0^{1/T} d\tau \int_0^1 du\, \boldsymbol{N}\cdot\left(\frac{\partial \boldsymbol{N}}{\partial u}\times\frac{\partial \boldsymbol{N}}{\partial \tau}\right). \tag{A.38}$$

This expression has a simple geometric interpretation. The function $\boldsymbol{N}(\tau,u)$ is a map from the rectangle $0\le\tau\le 1/T$, $0\le u\le 1$ to the unit sphere. As $\boldsymbol{N}$ moves from $\boldsymbol{N}(\tau)$ to $\boldsymbol{N}(\tau+\Delta\tau)$ it drags along the string connecting it to the north pole represented by the u dependence of $\boldsymbol{N}(\tau,u)$ (recall (A.32)). It is easy to see that the contribution to $\mathcal{S}_B$ of this evolution is simply iS times the oriented area swept out by the string (see Fig. A.1). The value of this area clearly depends upon the fact that the $u=0$ end of the string was pinned at the north pole. This was a "gauge" choice and, by choosing the phases of the coherent states differently, we could have pinned the point $u=0$ anywhere on the sphere. However, when we consider the complete integral over τ in (A.38), the boundary condition $\boldsymbol{N}(1/T)=\boldsymbol{N}(0)$ (required by the trace in (A.5)) shows that $\boldsymbol{N}(\tau)$ sweeps out a closed loop on the unit sphere. Then, the total τ integral in (A.38) is the area contained within this loop and is independent of the choice of the location of the $u=0$ point. Actually, this last statement is not completely correct: The "inside" of a closed loop is not well defined and the location of the $u=0$ point makes the oriented area uncertain modulo 4π (which is the total area of the unit sphere). Thus, the net contribution of $e^{\mathcal{S}_B}$ is uncertain up to a factor of $e^{i4\pi S}$. For consistency, we can now demand that this arbitrary factor always equal unity, which, of course, leads to the familiar requirement that $2S$ is an integer.

Appendix B Grassman Path Integral

by **Rhine Samajdar**

This appendix describes the construction of coherent states for fermions. The operators a_i annihilate fermions with label i, and

$$a_i a_j^\dagger + a_j^\dagger a_i = \delta_{ij}\,, \quad a_i a_j + a_j a_i = 0\,. \tag{B.1}$$

We proceed as for bosons in Appendix A.1, and introduce, introduce states $|\eta\rangle$ so that

$$a_i|\eta\rangle = \eta_i|\eta\rangle. \tag{B.2}$$

However, the anti-commutation relations in (B.1) imply that we must have

$$\eta_i\eta_j = -\eta_j\eta_i. \tag{B.3}$$

Hence, the objects η_i cannot be ordinary numbers as was the case for bosons; this leads us to generalize the concept of numbers (or fields) to Grassmann variables.

The mathematical structure underlying the definition (B.3) is an *algebra*, which is simply a vector space endowed with a multiplication rule $\mathcal{A}\times\mathcal{A}\to\mathcal{A}$. We can construct an algebra by beginning with a set of elements $\eta_i \in \mathcal{A}$, $i = 1,\ldots,N$ – which are the generators of the algebra – and imposing two properties:

(i) First, the vector space $\mathcal{A}$ is closed under addition and scalar multiplication i.e., the elements η_i can be added and multiplied by complex numbers and the result of any such operation also belongs to the algebra:

$$\alpha_0 + \alpha_i\eta_i + \alpha_j\eta_j \in \mathcal{A} \quad \forall\, \alpha_0, \alpha_i, \alpha_j \in \mathbb{C}. \tag{B.4}$$

(ii) The product map $\mathcal{A}\times\mathcal{A}\to\mathcal{A}$ defined as $(\eta_i,\eta_j)\to\eta_i\eta_j$ is associative and, as a consequence of (B.3), anti-commutative. Note that products of an odd number of generators anti-commute while all other combinations commute.

Then, the set $\mathcal{A}$ of all linear combinations

$$\alpha_0 + \sum_{n=1}^{\infty}\sum_{i_1,i_2,\ldots,i_n=1}^{N} \alpha_{i_1,i_2,\cdots,i_n}\eta_{i_1}\eta_{i_2}\ldots\eta_{i_n}, \quad \alpha_0, \alpha_{i_1,i_2,\ldots,i_n} \in \mathbb{C} \tag{B.5}$$

spans a finite-dimensional associative algebra of dimension 2^N, which is called the Grassmann algebra. The generators $\{\eta_i\}$ and their products are members of the algebra and are called Grassmann numbers.

Grassmann Calculus

While (B.5) provides a formal definition of Grassmann numbers, before proceeding further, we need to introduce the functional operations, or calculus, of these variables. As we outline below, it turns out that Grassmann calculus is actually much simpler than the calculus of ordinary numbers.

- Functions of Grassmann numbers are defined via their Taylor expansions:

$$f(\eta_1,\eta_2,\ldots,\eta_k)=\sum_{n=0}^{\infty}\sum_{i_1,i_2,\ldots,i_n=1}^{k}\frac{1}{n!}\frac{\partial^n f}{\partial\eta_1\,\partial\eta_2\cdots\partial\eta_k}\bigg|_{\eta=0}\eta_{i_n}\cdots\eta_{i_2}\eta_{i_1}, \tag{B.6}$$

  where $\{\eta_1,\eta_2,\ldots,\eta_k\}$ is some subset of the Grassmann generators ($k\leq N$). A particularly convenient feature of this expansion is that the series necessarily terminates after a finite number of terms. For instance, when $N=1$, $f(\eta)=f(0)+f'(0)\,\eta$ since $\eta^2=0$.
- Differentiation is defined as $\partial_{\eta_i}\eta_j=\delta_{ij}$. By virtue of (B.3), the differential operator itself has to be anti-commutative, that is, $\partial_{\eta_i}\eta_j\eta_i=-\eta_j$ for $i\neq j$.
- Integration is defined by the rules

$$\int \mathrm{d}\eta_i=0,\quad \int \mathrm{d}\eta_i\,\eta_i=1. \tag{B.7}$$

  As an example, consider the integral

$$\int \mathrm{d}\eta\, f(\eta)=\int \mathrm{d}\eta\,\left[f(0)+f'(0)\eta\right]=f'(0)=\partial_\eta f(\eta). \tag{B.8}$$

  In other words, the actions of Grassmann differentiation and integration are one and the same.

Construction of Coherent States

In order to construct fermionic coherent states, we need one final ingredient, namely, the enlargement of the algebra to allow for the (commutative) multiplication of Grassmann numbers and fermion operators such that $[\eta_i,a_j]=0$. Then, we define

$$|\eta\rangle=\exp\left[-\sum_i\eta_i a_i^\dagger\right]|0\rangle, \tag{B.9}$$

which is easily shown to be a coherent state. The corresponding ket is defined as

$$\langle\eta|=\langle 0|\exp\left[-\sum_i a_i\bar{\eta}_i\right]=\langle 0|\exp\left[\sum_i\bar{\eta}_i a_i\right]. \tag{B.10}$$

However, η_i and $\bar{\eta}_i$ are independent numbers unlike in the bosonic case, for which ψ_i and $\bar{\psi}_i$ were related by complex conjugation.

The various properties of both the bosonic and fermionic coherent states are summarized in Table B.1.

Table B.1 Coherent states for bosons ($z=1$, $\psi_i \in \mathbb{C}$) and fermions ($z=-1$, $\psi_i \in \mathcal{A}$).

Definition	$\|\psi\rangle = \exp\left[z\sum_i \psi_i a_i^\dagger\right]\|0\rangle.$
Action of a_i	$a_i\|\psi\rangle = \psi_i\|\psi\rangle, \quad \langle\psi\|a_i = \partial_{\bar\psi_i}\langle\psi\|$
Action of $a_i^\dagger$	$a_i^\dagger\|\psi\rangle = z\partial_{\psi_i}\|\psi\rangle, \ \langle\psi\|a_i^\dagger = \langle\psi\|\bar\psi_i$
Overlap	$\langle\psi'\|\psi\rangle = \exp\left[\sum_i \bar\psi_i'\psi_i\right]$
Completeness	$\int\prod_i \frac{d\bar\psi_i d\psi_i}{\pi^{(1+z)/2}} e^{-\sum_i\bar\psi_i\psi_i}\|\psi\rangle\langle\psi\| = \mathbb{1}$

Grassmann Gaussian Integrals

Gaussian integrals play a recurring role in field theory. With regard to Grassmann numbers, the fundamental Gaussian integral is

$$\int d\bar\eta\, d\eta\, e^{-\bar\eta\alpha\eta} = \alpha \quad \forall\, \alpha \in \mathbb{C}, \tag{B.11}$$

which can be regarded as the extension of the famous identity

$$\int_{-\infty}^{\infty} dx e^{-ax^2/2} = \frac{2\pi}{a}; \quad \text{Re}\, a > 0,$$

to Grassmann variables. The generalization of (B.11) to multidimensional integrals is also straightforward. Suppose ψ and $\bar\psi$ are N-dimensional vectors of Grassmann variables. Then,

$$\int d(\bar\psi,\psi)\, e^{-\bar\psi^T A\psi} = \det A, \quad \int d(\bar\psi,\psi)\, e^{-\bar\psi^T A\psi + \bar\lambda^T\cdot\psi + \bar\psi^T\cdot\lambda} = \det A\, e^{\bar\lambda^T A^{-1}\lambda}, \tag{B.12}$$

where $d(\bar\psi,\psi) \equiv \prod_i d\bar\psi_i d\psi_i$. These equations are useful for deriving several important identities. For instance, defining $\langle\cdots\rangle \equiv (\det A)^{-1} d(\bar\psi,\psi)\, e^{-\bar\psi^T A\psi}(\cdots)$, one can obtain the correlators

$$\langle\psi_j\bar\psi_i\rangle = A_{ji}^{-1} \quad \text{and} \quad \langle\psi_{j_1}\psi_{j_2}\cdots\psi_{j_n}\bar\psi_{i_n}\cdots\bar\psi_{i_2}\bar\psi_{i_1}\rangle = \sum_P (\text{sgn}P) A^{-1}_{j_1 i_{P_1}}\cdots A^{-1}_{j_n i_{P_n}}, \tag{B.13}$$

where the sign of the permutation P keeps track of the sign changes resulting from the anti-commuting interchange of Grassmann variables.

B.1 Application: Partition Function

Having established the powerful machinery of coherent states and Grassmann calculus, we turn to a demonstration of its utility. To this end, we now derive the field integral for the partition function. The quantum partition function is given by

$$Z = \text{Tr}\, e^{-\beta(\hat{\mathcal{H}}-\mu\hat{\mathcal{N}})} = \sum_n \langle n|e^{-\beta(\hat{\mathcal{H}}-\mu\hat{\mathcal{N}})}|n\rangle, \tag{B.14}$$

where $\beta = 1/T$ is the inverse temperature, and the sum runs over a complete set of Fock-space states $|n\rangle$. Next, we insert the resolution of the identity (from the last line of Table B.1) as

$$Z = \int d(\bar{\psi}, \psi)\, e^{-\sum_i \bar{\psi}_i \psi_i} \sum_n \langle n|\psi\rangle\langle\psi|e^{-\beta(\hat{\mathcal{H}}-\mu\hat{\mathcal{N}})}|n\rangle. \tag{B.15}$$

As with the usual path-integral construction, we now want to use the property that $\sum_n |n\rangle\langle n| = \mathbb{1}$ to eliminate the summation over n. So far, our construction has been completely general and applies equally well to bosons and fermions. However, recall that, for fermionic coherent states, $\langle n|\psi\rangle\langle\psi|n\rangle = \langle -\psi|n\rangle\langle n|\psi\rangle$. So,

$$\begin{aligned} Z &= \int d(\bar{\psi}, \psi)\, e^{-\sum_i \bar{\psi}_i \psi_i} \sum_n \langle z\psi|e^{-\beta(\hat{\mathcal{H}}-\mu\hat{\mathcal{N}})}|n\rangle\langle n|\psi\rangle \\ &= \int d(\bar{\psi}, \psi)\, e^{-\sum_i \bar{\psi}_i \psi_i} \langle z\psi|e^{-\beta(\hat{\mathcal{H}}-\mu\hat{\mathcal{N}})}|\psi\rangle, \end{aligned} \tag{B.16}$$

with the statistics encoded, as in Table B.1 previously, by taking $z = 1$ ($z = -1$) for bosons (fermions).

Let us now consider a generic Hamiltonian

$$\hat{\mathcal{H}}(a, a^\dagger) = \sum_{i,j} h_{ij} a_i^\dagger a_j + \sum_{i,j,k,l} V_{ijkl}\, a_i^\dagger a_j^\dagger a_k a_l, \tag{B.17}$$

where we have written the interaction term in its normal-ordered form. Now, the standard path-integral construction proceeds by dividing the "time" interval β into M steps of width $\delta = \beta/M$ and inserting the resolution of the identity between each step. Adopting the shorthand $\psi^n = \{\psi_i^n\}$, we find

$$\begin{aligned} Z &= \int_{\substack{\bar{\psi}^0 = z\bar{\psi}^M \\ \psi^0 = z\psi^M}} \prod_{n=0}^{M} d(\bar{\psi}^n, \psi^n) \\ &\times \exp\left(-\delta \sum_{n=0}^{M-1} \left[\delta^{-1}(\bar{\psi}^n - \bar{\psi}^{n+1})\cdot\psi^n + \mathcal{H}(\bar{\psi}^{n+1}, \psi^n) - \mu\mathcal{N}(\bar{\psi}^{n+1}, \psi^n)\right]\right), \end{aligned} \tag{B.18}$$

with

$$\mathcal{H}(\bar{\psi}, \psi') \equiv \frac{\langle\psi|\hat{\mathcal{H}}(a, a^\dagger)|\psi'\rangle}{\langle\psi|\psi'\rangle} = \sum_{i,j} h_{ij} \bar{\psi}_i \psi'_j + \sum_{i,j,k,l} V_{ijkl}\, \bar{\psi}_i \bar{\psi}_j \psi'_k \psi'_l, \tag{B.19}$$

and similarly for $\mathcal{N}(\bar{\psi}, \psi')$. Finally, taking the limit $M \to \infty$, we obtain the continuum version of the path integral:

$$Z = \int \mathcal{D}(\bar{\psi}, \psi)\, e^{-S[\bar{\psi}, \psi]}; \quad S[\bar{\psi}, \psi] = \int_0^\beta d\tau\, [\bar{\psi}\partial_\tau\psi + \mathcal{H}(\bar{\psi}, \psi) - \mu\mathcal{N}(\bar{\psi}, \psi)], \tag{B.20}$$

where $\mathcal{D}(\bar{\psi}, \psi) = \lim_{M\to\infty} \prod_{n=1}^{M} d(\bar{\psi}^n, \psi^n)$ and the fields satisfy the periodic or antiperiodic boundary conditions $\bar{\psi}^0 = z\bar{\psi}^M$, $\psi^0 = z\psi^M$. Therefore, on expanding out the terms in (B.20), we see that the action associated with the most general pair Hamiltonian is

$$\mathcal{S} = \int_0^\beta \mathrm{d}\tau \left[\sum_{i,j} \bar{\psi}_i(\tau) \left[(\partial_\tau - \mu)\delta_{ij} + h_{ij} \right] \psi_j(\tau) + \sum_{i,j,k,l} V_{ijkl}\, \bar{\psi}_i(\tau) \bar{\psi}_j(\tau) \psi_k(\tau) \psi_l(\tau) \right] \tag{B.21}$$

or equivalently, after Fourier transforming to Matsubara frequencies, by

$$\mathcal{S} = \sum_{i,j,n} \bar{\psi}_{in} \left[(-i\omega_n - \mu)\delta_{ij} + h_{ij} \right] \psi_{jn} + \frac{1}{\beta} \sum_{i,j,k,l} V_{ijkl}\, \bar{\psi}_{in_1} \bar{\psi}_{jn_2} \psi_{kn_3} \psi_{ln_4} \delta_{n_1+n_2, n_3+n_4}. \tag{B.22}$$

This path integral provides an alternate approach to the analysis in Section 6.1, and was directly employed in Section 9.4.

Appendix C From Spin Berry Phases to Background Gauge Charges

This appendix derives the Berry phase term in (26.3), (26.4), and (26.37) for the $U(1)$ gauge theory of square-lattice antiferromagnets, starting from the spin Berry phase in (18.23) or (A.38).

C.1 Single Spin

We begin by examining a single spin in a magnetic field $\boldsymbol{h}$ with the Hamiltonian $-\boldsymbol{h}\cdot\boldsymbol{S}$. From (18.23) and Section A.2, the partition function at a temperature T is

$$\operatorname{Tr}\exp\left(\boldsymbol{h}\cdot\boldsymbol{S}/T\right) = \int \mathcal{D}\boldsymbol{N}(\tau)\exp\left(i2S\mathcal{A}[\boldsymbol{N}(\tau)] + S\int_0^{1/T} d\tau \boldsymbol{h}\cdot\boldsymbol{N}(\tau)\right). \tag{C.1}$$

Here, S is the angular momentum of the spin $\boldsymbol{S}$ and $\boldsymbol{N}(\tau)$ is a unit 3-vector with $\boldsymbol{N}(0) = \boldsymbol{N}(1/T)$. So the above path integral is over all closed curves on the surface of a sphere. The first term in the action of the path integral is the Berry phase: $\mathcal{A}[\boldsymbol{N}(\tau)]$ is *half* the oriented area enclosed by the curve $\boldsymbol{N}(\tau)$ (the reason for choosing the half normalization here will become clear momentarily). Note that this area is only defined modulo 4π, the surface area of a unit sphere.

The half-area $\mathcal{A}[\boldsymbol{N}(\tau)]$ is a global object defined by the whole curve $\boldsymbol{N}(\tau)$, and we would like to break it up into local contributions. We proceed as illustrated in Fig A.1: discretize imaginary time, choose a fixed arbitrary point on the sphere (say the north pole), and thus write the area as the sum of a large number of spherical triangles. Now each triangle is associated with a local portion of the curve $\boldsymbol{N}(\tau)$.

We now need an expression for $\mathcal{A}(\boldsymbol{N}_1, \boldsymbol{N}_2, \boldsymbol{N}_3)$, defined as half the area of the spherical triangle with vertices $\boldsymbol{N}_1$, $\boldsymbol{N}_2$, $\boldsymbol{N}_3$. Complicated expressions for this appear in treatises on spherical trigonometry, but a far simpler expression is obtained after transforming to spinor variables [25]. Let us write

$$\boldsymbol{N}_j \equiv z_{j\alpha}^* \boldsymbol{\sigma}_{\alpha\beta} z_{j\beta}, \tag{C.2}$$

where $\alpha, \beta = \uparrow, \downarrow$, $\boldsymbol{\sigma}_{\alpha\beta}$ are the Pauli matrices, and $z_{j\uparrow}$, $z_{j\downarrow}$ are complex numbers obeying $|z_{j\uparrow}|^2 + |z_{j\downarrow}|^2 = 1$. Note that knowledge of $\boldsymbol{N}_j$ only defines $z_{j\alpha}$ up to a $U(1)$ gauge transformation under which

$$z_{j\alpha} \to z_{j\alpha} e^{i\phi_j}. \tag{C.3}$$

Then, associated with each pair of vertices $\boldsymbol{N}_i, \boldsymbol{N}_j$ we define

$$\mathcal{A}_{ij} \equiv \arg\left[z_{i\alpha}^* z_{j\alpha}\right]. \tag{C.4}$$

Under the gauge transformation (C.3) we have

$$\mathcal{A}_{ij} \rightarrow \mathcal{A}_{ij} - \phi_i + \phi_j, \tag{C.5}$$

that is, $\mathcal{A}_{ij}$ behaves like a $U(1)$ gauge field. Note also that $\mathcal{A}_{ij}$ is only defined modulo 2π, and that $\mathcal{A}_{ji} = -\mathcal{A}_{ij}$. We also mention the following identity, which follows from (C.2) and (C.4):

$$z_{i\alpha}^* z_{j\alpha} = \left(\frac{1 + \boldsymbol{N}_i \cdot \boldsymbol{N}_j}{2}\right)^{1/2} e^{i\mathcal{A}_{ij}}. \tag{C.6}$$

The classical result for the half-area of the spherical triangle can be written in the simple form in terms of the present $U(1)$ gauge variables:

$$\mathcal{A}(\boldsymbol{N}_1, \boldsymbol{N}_2, \boldsymbol{N}_3) = \mathcal{A}_{12} + \mathcal{A}_{23} + \mathcal{A}_{31}. \tag{C.7}$$

We chose $\mathcal{A}$ as a half-area earlier mainly because then the expressions (C.4) and (C.7) come out without numerical factors. It is satisfying to observe that this total area is invariant under (C.5), and that the half-area is ambiguous modulo 2π.

Using (C.7), we can now write down the needed expression for $\mathcal{A}[\boldsymbol{N}(\tau)]$. We assume that imaginary time is discretized into times τ_j separated by intervals $\Delta\tau$. Also, we denote by $j+\tau$ the site at time $\tau_j + \Delta\tau$, and define $\mathcal{A}_{j,j+\tau} \equiv \mathcal{A}_{j\tau}$. Then, the Berry phase is proportional to the sum of the areas of all the spherical triangles in Fig. A.1, and so

$$\mathcal{A}[\boldsymbol{N}(\tau)] = \sum_j \mathcal{A}_{j\tau}. \tag{C.8}$$

Note that this expression is a gauge-invariant function of the $U(1)$ gauge field $\mathcal{A}_{j\tau}$, and is analogous to the quantity sometimes called the Polyakov loop.

C.2 Square-Lattice Antiferromagnet

We apply the single-spin formulation to a lattice antiferromagnet with the following steps:

(i) Discretize spacetime into a cubic lattice of points j.

(ii) On each spacetime point j, we represent the quantum spin operator $\boldsymbol{S}_j$ by

$$\boldsymbol{S}_j = \eta_j S \boldsymbol{N}_j, \tag{C.9}$$

where $\boldsymbol{N}_j$ is a unit vector and $\eta_j = \pm 1$ is the sublattice staggering factor. We have chosen to include η_j because of the expected local antiferromagnetic correlations of the spins. So in a quantum-fluctuating Néel state, we can reasonably expect $\boldsymbol{N}_j$ to be a slowly varying function of j.

(iii) Associated with each $\boldsymbol{N}_j$, define a spinor $z_{j\alpha}$ by (C.2).
(iv) With each link of the cubic lattice, we use (C.4) to associate with it $\mathcal{A}_{j\mu} \equiv \mathcal{A}_{j,j+\mu}$. Here $\mu = x, y, \tau$ extends over the three spacetime directions.

With these preliminaries in hand, we can motivate the following effective action for fluctuations of the square-lattice antiferromagnet:

$$\widetilde{\mathcal{Z}} = \prod_{j\alpha} \int dz_{j\alpha} \prod_j \delta\left(|z_{j\alpha}|^2 - 1\right) \exp\left(\frac{1}{\widetilde{g}} \sum_{\langle ij \rangle} \boldsymbol{N}_i \cdot \boldsymbol{N}_j + i2S \sum_j \eta_j \mathcal{A}_{j\tau}\right). \quad \text{(C.10)}$$

Here, the summation over $\langle ij \rangle$ extends over nearest neighbors on the cubic lattice. The integrals are over the $z_{j\alpha}$, and the $\boldsymbol{N}_j$ and $\mathcal{A}_{j\tau}$ are *dependent* variables defined via (C.2) and (C.4). Note that both terms in the action are invariant under the gauge transformation (C.3); consequently, we could equally well have rewritten $\widetilde{\mathcal{Z}}$ as an integral over the $\boldsymbol{N}_j$, but it turns out to be more convenient to use the $z_{j\alpha}$ and to integrate over the redundant gauge degree of freedom. The first term in the action contains the energy of the Hamiltonian H_s, and acts to prefer nearest neighbors $\boldsymbol{N}_j$ that are parallel to each other – this "ferromagnetic" coupling between the $\boldsymbol{N}_j$ in spacetime ensures, via (C.9), that the local quantum spin configurations are as in the Néel state. The second term in the action is simply the Berry phase required in the coherent-state path integral, as obtained from (C.1) and (C.8); the additional factor of η_j compensates for that in (C.9). The dimensionless coupling $\widetilde{g}$ controls the strength of the local antiferromagnetic correlations: it is like a "temperature" for the ferromagnet in spacetime. So for small $\widetilde{g}$ we expect $\widetilde{\mathcal{Z}}$ to be in the Néel phase, while for large $\widetilde{g}$ we can expect a "quantum-disordered" state, as in Fig. 15.1.

While it is possible to proceed with $\widetilde{\mathcal{Z}}$, it is convenient to work with a very closely related alternative model. The proposed theory for the quantum-fluctuating antiferromagnet in its final form is [229, 239]

$$\mathcal{Z} = \prod_{j\mu} \int_0^{2\pi} \frac{da_{j\mu}}{2\pi} \prod_{j\alpha} \int dz_{j\alpha} \prod_j \delta\left(|z_{j\alpha}|^2 - 1\right)$$
$$\exp\left(\frac{1}{g} \sum_{j\mu} \left(z^*_{j\alpha} e^{-ia_{j\mu}} z_{j+\mu,\alpha} + \text{c.c.}\right) + i2S \sum_j \eta_j a_{j\tau}\right). \quad \text{(C.11)}$$

Just as in the analysis for the XY model in (14.21), we have introduced a new field $a_{j\mu}$, on each link of the cubic lattice, which is integrated over. Like $\mathcal{A}_{i\mu}$, this is also a $U(1)$ gauge field because all terms in the action above are invariant under the analog of (C.5):

$$a_{j\mu} \to a_{j\mu} - \phi_j + \phi_{j+\mu}. \quad \text{(C.12)}$$

The very close relationship between $\mathcal{Z}$ and $\widetilde{\mathcal{Z}}$ may be seen [239] by explicitly integrating over the $a_{j\mu}$ in (C.11), similar to (14.22). This integral can be done exactly because the integrand factorizes into terms on each link that depend only on a single $a_{j\mu}$. After inserting (C.6) into (C.11), the integral over the $j\mu$ link is

$$\int_0^{2\pi} \frac{da_{j\mu}}{2\pi} \exp\left(\frac{(2(1+\boldsymbol{N}_j\cdot\boldsymbol{N}_{j+\mu}))^{1/2}}{g}\cos(\mathcal{A}_{j\mu}-a_{j\mu})+i2S\eta_j\delta_{\mu\tau}a_{j\mu}\right)$$
$$= I_{2S\delta_{\mu\tau}}\left[\frac{(2(1+\boldsymbol{N}_j\cdot\boldsymbol{N}_{j+\mu}))^{1/2}}{g}\right]\exp\left(i2S\eta_j\delta_{\mu\tau}\mathcal{A}_{j\mu}\right), \tag{C.13}$$

where the result involves either the modified Bessel function I_0 (for $\mu = x, y$) or I_{2S} (for $\mu = \tau$). We can use the identity (C.13) to perform the integral over $a_{j\mu}$ on each link of (C.11), and so obtain a partition function, denoted $\mathcal{Z}'$, as an integral over the $z_{j\alpha}$ only. This partition function $\mathcal{Z}'$ has essentially the same structure as $\widetilde{\mathcal{Z}}$ in (C.10). The Berry phase term in $\mathcal{Z}'$ is identical to that in $\widetilde{\mathcal{Z}}$. The integrand of $\mathcal{Z}'$ also contains a real action expressed solely as a sum over functions of $\boldsymbol{N}_i\cdot\boldsymbol{N}_j$ on nearest-neighbor links: in $\widetilde{\mathcal{Z}}$ this function is simply $\boldsymbol{N}_i\cdot\boldsymbol{N}_j/\widetilde{g}$, but the corresponding function obtained from (C.11) is more complicated (it involves the logarithm of a Bessel function), and has distinct forms on spatial and temporal links. We do not expect this detailed form of the real action function to be of particular importance for universal properties; the initial simple nearest-neighbor ferromagnetic coupling between the $\boldsymbol{N}_j$ in (C.10) was chosen arbitrarily anyway. So we may safely work with the theory $\mathcal{Z}$ in (C.11) henceforth.

One of the important advantages of (C.11) is that we no longer have to keep track of the complicated non-linear constraints associated with (C.2) and (C.4); this was one of the undesirable features of (C.10). In $\mathcal{Z}$, we simply have free integration over the independent variables $z_{j\alpha}$ and $a_{j\mu}$. The theory $\mathcal{Z}$ in (C.11) has some resemblance to the so-called CP^{N-1} model from the particle physics literature [25, 58, 311]: our indices α, β take only two possible values, but the general model is obtained when $\alpha, \beta = 1, \ldots, N$. The case of general N describes $SU(N)$ and $USp(N)$ antiferromagnets on the square lattice [217, 218]. Note also that it is essential for our purposes that the theory is invariant under $a_{j\mu} \to a_{j\mu} + 2\pi$, and so the $U(1)$ gauge theory is *compact*. Finally, our model contains a Berry phase term that is not present in any of the particle physics analyses.

The properties of $\mathcal{Z}$ are quite evident in the limit of small g. Here, the partition function is strongly dominated by configurations in which the real part of the action is a minimum. In a suitable gauge, these are the configurations in which $z_{j\alpha} =$ constant, and, by (C.2), we also have $\boldsymbol{N}_j$ a constant. This obviously corresponds to the Néel phase. A Gaussian fluctuation analysis about such a constant saddle point is easily performed, and we obtain the expected spectrum of a doublet of gapless spin waves obtained in Section 9.2.2.

For large g, we expect a quantum-disordered phase with $\langle \boldsymbol{S}_j\rangle = \langle \boldsymbol{N}_j\rangle = 0$. Here, we can integrate out the z_α from (C.11) in a $1/g$ expansion, and so obtain (26.4).

Appendix D Emergent $\mathbb{Z}_2$ Gauge Theories

This appendix summarizes the various $\mathbb{Z}_2$ gauge theories employed to describe spin-S antiferromangets in two spatial dimensions.

In Chapter 14, we considered XY models in which the XY order parameter represents the antiferromagnetic order of an easy-plane antiferromagnet. We fractionalized the XY order parameter in (14.14)

$$\Psi_i \equiv H_i e^{2i\varphi_i}, \tag{D.1}$$

and condensed the $U(1)$ gauge charge Higgs field H. This reduced the gauge symmetry to $\mathbb{Z}_2$, and allowed us to replace the $U(1)$ gauge field $a_{i\mu}$ by a $\mathbb{Z}_2$ gauge field in (14.31):

$$a_{i\mu} = 0, \pi \quad , \quad e^{ia_{i\mu}} \equiv Z_{i,i+\mu} = \pm 1 . \tag{D.2}$$

Then we obtained a $\mathbb{Z}_2$ gauge theory for the parton field $e^{i\varphi_i}$, which carries a $\mathbb{Z}_2$ gauge charge in (14.32); we write this in quantum form on the sites of the square lattice using (16.7)

$$\mathcal{H} = \mathcal{H}_\phi + \mathcal{H}_{\mathbb{Z}_2}, \tag{D.3}$$

$$\mathcal{H}_\phi = U\sum_i L_i^2 - J_2 \sum_{\langle ij \rangle} Z_{ij}\cos(\varphi_i - \varphi_j), \tag{D.4}$$

$$\mathcal{H}_{\mathbb{Z}_2} = -K\sum_{\square}\prod_{\ell\in\square} Z_\ell - g\sum_\ell X_\ell . \tag{D.5}$$

We have introduced here an integer-valued "angular-momentum" variable L_i conjugate to φ_i with the commutation relation

$$[L_i, \varphi_j] = -i\delta_{ij} . \tag{D.6}$$

The global $U(1)$ charge of the underlying XY model is $\mathcal{Q} = (1/2)\sum_i L_i$. The theory (D.3) also has a conserved $\mathbb{Z}_2$ gauge charge on each site, which generalizes (16.9) to

$$G_i = (-1)^{L_i}\prod_{\ell\in +} X_\ell , \tag{D.7}$$

and it was argued that an easy-plane spin-S antiferromagnet, or interacting bosons of density S, are described by the theory with

$$G_i = (-1)^{2S}. \tag{D.8}$$

The matter in (D.4) is at zero density, hence this theory falls into the category of (16.13).

In Chapter 15, we fractionalized the spin operator of an antiferromagnet into bosonic partons in (15.6)

$$\boldsymbol{S}_i = \frac{1}{2} s^{\dagger}_{i\alpha} \boldsymbol{\sigma}^{\alpha}_{\beta} s^{\beta}_i . \tag{D.9}$$

Then we condensed the boson-pair field $\mathcal{Q}_{ij}$ in (15.22)

$$\bar{\mathcal{Q}}_{ij} = \langle \varepsilon_{\alpha\beta} s^{\alpha}_i s^{\beta}_j \rangle , \tag{D.10}$$

and this broke the $U(1)$ gauge symmetry to $\mathbb{Z}_2$, allowing us to replace $\mathcal{Q}_{ij}$ by a $\mathbb{Z}_2$ gauge field in (16.2):

$$\mathcal{Q}_{ij} \Rightarrow Z_{ij} . \tag{D.11}$$

We then obtained the Hamiltonian of the $\mathbb{Z}_2$ gauge theory coupled to spinons in (16.8)

$$\mathcal{H}^{vs}_{\mathbb{Z}_2} = \mathcal{H}_{\mathbb{Z}_2} + \mathcal{H}_s \tag{D.12}$$

$$\mathcal{H}_s = \sum_{\langle ij \rangle} \left(-\tilde{J}_{ij} Z_{ij} \varepsilon_{\alpha\beta} s^{\alpha}_i s^{\beta}_j + H.c. \right) + \sum_i \bar{\lambda}_i (s^{\dagger}_{i\alpha} s^{\alpha}_i - n_s) , \tag{D.13}$$

where $\mathcal{H}_{\mathbb{Z}_2}$ is as in (D.5). The conserved $\mathbb{Z}_2$ gauge charges that commute with $\mathcal{H}^{vs}_{\mathbb{Z}_2}$ are in (16.11):

$$G^{vs}_i = \exp\left(i\pi s^{\dagger}_{i\alpha} s^{\alpha}_i \right) \prod_{\ell \in +} X_{\ell} . \tag{D.14}$$

In the full theory $\mathcal{H}^{vs}_{\mathbb{Z}_2}$ with the active s^{α}_i degrees of freedom we have to impose the constraint $G^{vs}_i = 1$ in (16.12). However, if we work in an effective pure $\mathbb{Z}_2$ gauge theory $\mathcal{H}_{\mathbb{Z}_2}$ in which we integrate out the gapped spinon degrees of freedom, the boson constraint (15.4) implies the $\mathbb{Z}_2$ gauge constraint in (16.13)

$$G_i \equiv \prod_{\ell \in +} X_{\ell} = (-1)^{2S} . \tag{D.15}$$

Alternatively, for antiferromagnets with global $SO(3)$ spin-rotation symmetry, we can keep the spinons in the "relativistic" form in (15.49) "without matter Berry phases," which is the $O(4)$ global symmetry generalization of the $O(2)$ global symmetry in (D.4). To write (15.49) in Hamiltonian form, it is convenient to decompose z_{α} in terms of a real field m_a ($a = 1, \ldots, 4$):

$$z_{\uparrow} = m_1 + i m_2 \quad , \quad z_{\downarrow} = m_3 + i m_4 \quad , \quad \sum_{a=1}^{4} m_a^2 = 1 . \tag{D.16}$$

Then, the Hamiltonian is $\mathcal{H} = \mathcal{H}_m + \mathcal{H}_{\mathbb{Z}_2}$, with $\mathcal{H}_m$ generalizing $\mathcal{H}_{\phi}$ in (D.4) to

$$\mathcal{H}_m = U \sum_i \sum_{a \neq b} L^2_{i,ab} - J_2 \sum_{\langle ij \rangle} \sum_a Z_{ij} m_{ia} m_{ja} , \tag{D.17}$$

where

$$L_{ab} = m_a \frac{\partial}{\partial m_b} - m_b \frac{\partial}{\partial m_a} . \tag{D.18}$$

The conserved $\mathbb{Z}_2$ gauge charge generalizing (D.7) is

$$G_i = (-1)^{\ell_i} \prod_{\ell \in +} X_\ell , \tag{D.19}$$

where ℓ_i is the hyperspherical harmonic angular-momentum quantum number [307] of the state on site i. The gauge constraint remains as in (D.8). Ultimately, the constraints in (D.8) and (D.15) are the consequences of the spin Berry phases in Section 18.2 on the $\mathbb{Z}_2$ liquid and its vicinity, as discussed in Section 26.2.

References

[1] A. Abanov and A.V. Chubukov, Interplay between superconductivity and non-Fermi liquid at a quantum critical point in a metal. I. The γ model and its phase diagram at $T = 0$: The case $0 < \gamma < 1$, *Phys. Rev. B* **102** (2020) 024524 [arXiv:2004.13220].

[2] A. Abrikosov, L. Gorkov, and I. Dzyaloshinskii, *Methods of Quantum Field Theory in Statistical Physics*, Prentice-Hall (1963).

[3] I. Affleck, T. Kennedy, E.H. Lieb and H. Tasaki, Rigorous results on valence-bond ground states in antiferromagnets, *Phys. Rev. Lett.* **59** (1987) 799.

[4] I. Affleck and J.B. Marston, Large-n limit of the Heisenberg–Hubbard model: Implications for high-T_c superconductors, *Phys. Rev. B* **37** (1988) 3774.

[5] E.E. Aldape, T. Cookmeyer, A.A. Patel, and E. Altman, Solvable theory of a strange metal at the breakdown of a heavy Fermi liquid, *Phys. Rev. B* **105** (2022) 235111 [arXiv:2012.00763].

[6] J. Alicea, New directions in the pursuit of Majorana fermions in solid state systems, *Rep. Prog. Phys.* **75** (2012) 076501 [arXiv:1202.1293].

[7] B.L. Altshuler, A.V. Chubukov, A. Dashevskii, A.M. Finkel'stein, and D.K. Morr, Luttinger theorem for a spin-density-wave state, *Europhys. Lett.* **41** (1998) 401 [cond-mat/9703120].

[8] P.W. Anderson, Resonating valence bonds: A new kind of insulator?, *Mater. Res. Bull.* **8** (1973) 153.

[9] P.W. Anderson, The resonating valence bond state in La_2CuO_4 and superconductivity, *Science* **235** (1987) 1196.

[10] P.W. Anderson, P.A. Lee, M. Randeria, T.M. Rice, N. Trivedi, and F.C. Zhang, The physics behind high-temperature superconducting cuprates: The "plain vanilla" version of RVB, *J Phys. Condens. Matter* **16** (2004) R755 [cond-mat/0311467].

[11] N. Andrei and P. Coleman, Cooper instability in the presence of a spin liquid, *Phys. Rev. Lett.* **62** (1989) 595.

[12] T. Anous and F.M. Haehl, The quantum p-spin glass model: A user manual for holographers, *J. Stat. Mech.* **2111** (2021) 113101 [arXiv:2106.03838].

[13] D.P. Arovas and A. Auerbach, Functional integral theories of low-dimensional quantum Heisenberg models, *Phys. Rev. B* **38** (1988) 316.

[14] D.P. Arovas, E. Berg, S.A. Kivelson, and S. Raghu, The Hubbard model, *Annu. Rev. Condens. Matter Phys.* **13** (2022) 239 [arXiv:2103.12097].

[15] J.K. Asbóth, L. Oroszlány, and A. Pályi, A short course on topological insulators: Band-structure topology and edge states in one and two dimensions, *Lecture Notes in Physics* (2015) [`arXiv:1509.02295`].

[16] T. Azeyanagi, F. Ferrari, and F.I. Schaposnik Massolo, Phase diagram of planar matrix quantum mechanics, tensor, and Sachdev–Ye–Kitaev models, *Phys. Rev. Lett.* **120** (2018) 061602 [`arXiv:1707.03431`].

[17] F.A. Bais, P. van Driel, and M. de Wild Propitius, Quantum symmetries in discrete gauge theories, *Phys. Lett. B* **280** (1992) 63.

[18] L. Balents, L. Bartosch, A. Burkov, S. Sachdev, and K. Sengupta, Putting competing orders in their place near the Mott transition, *Phys. Rev. B* **71** (2005) 144508 [`cond-mat/0408329`].

[19] L. Balents, M.P. Fisher, and S.M. Girvin, Fractionalization in an easy-axis Kagome antiferromagnet, *Phys. Rev. B* **65** (2002) 224412 [`cond-mat/0110005`].

[20] M. Barkeshli, E. Berg, and S. Kivelson, Coherent transmutation of electrons into fractionalized anyons, *Science* **346** (2014) 722 [`arXiv:1402.6321`].

[21] M. Barkeshli and J. McGreevy, Continuous transition between fractional quantum Hall and superfluid states, *Phys. Rev. B* **89** (2014) 235116 [`arXiv:1201.4393`].

[22] G. Baskaran and P.W. Anderson, Gauge theory of high-temperature superconductors and strongly correlated Fermi systems, *Phys. Rev. B* **37** (1988) 580.

[23] M. Beccaria, S. Giombi, and A.A. Tseytlin, *Wilson loop in general representation and RG flow in 1d defect QFT*, `arXiv:2202.00028`.

[24] J.G. Bednorz and K.A. Müller, Possible high T_c superconductivity in the Ba–La–Cu–O system, *Z. Phys. B Condens. Matter* **64** (1986) 189.

[25] B. Berg and M. Lüscher, Definition and statistical distributions of a topological number in the lattice $O(3)$ σ-model, *Nucl. Phys. B* **190** (1981) 412.

[26] C. Bergemann, S.R. Julian, A.P. MacKenzie, S. Nishizaki and Y. Maeno, Detailed topography of the Fermi surface of Sr_2RuO_4, *Phys. Rev. Lett.* **84** (2000) 2662 [`cond-mat/9909027`].

[27] N.F. Berk and J.R. Schrieffer, Effect of ferromagnetic spin correlations on superconductivity, *Phys. Rev. Lett.* **17** (1966) 433.

[28] B. Bernevig and T. Hughes, *Topological Insulators and Topological Superconductors*, Princeton University Press (2013).

[29] H. Bernien, S. Schwartz, A. Keesling, H. Levine, A. Omran, H. Pichler, et al., Probing many-body dynamics on a 51-atom quantum simulator, *Nature* **551** (2017) 579 [`arXiv:1707.04344`].

[30] C. Berthod, T. Giamarchi, S. Biermann, and A. Georges, Breakup of the Fermi surface near the Mott transition in low-dimensional systems, *Phys. Rev. Lett.* **97** (2006) 136401 [`cond-mat/0602304`].

[31] O. Bohigas, M.J. Giannoni and C. Schmit, Characterization of chaotic quantum spectra and universality of level fluctuation laws, *Phys. Rev. Lett.* **52** (1984) 1.

[32] P. Bonderson, M. Cheng, K. Patel, and E. Plamadeala, *Topological enrichment of Luttinger's theorem*, *arXiv e-prints* (2016) [`arXiv:1601.07902`].

[33] N.E. Bonesteel, Valence bonds and the Lieb–Schultz–Mattis theorem, *Phys. Rev. B* **40** (1989) 8954.

[34] V. Borokhov, A. Kapustin, and X.-k. Wu, Topological disorder operators in three-dimensional conformal field theory, *JHEP* **11** (2002) 049 [hep-th/0206054].

[35] J. Bricmont and J. Fröhlich, Statistical mechanical methods in particle structure analysis of lattice field theories, *Commun. Math. Phys.* **98** (1985) 553.

[36] W.F. Brinkman and T.M. Rice, Application of Gutzwiller's variational method to the metal–insulator transition, *Phys. Rev. B* **2** (1970) 4302.

[37] T.A. Brody, J. Flores, J.B. French, P.A. Mello, A. Pandey, and S.S.M. Wong, Random-matrix physics: Spectrum and strength fluctuations, *Rev. Mod. Phys.* **53** (1981) 385.

[38] J. Brunkert and M. Punk, Slave-boson description of pseudogap metals in t–J models, *Phys. Rev. Res.* **2** (2020) 043019 [arXiv:2002.04041].

[39] H. Bruus and K. Flensberg, *Many-Body Quantum Theory in Condensed Matter Physics: An Introduction*, Oxford University Press (2004).

[40] M. Büttiker, Absence of backscattering in the quantum Hall effect in multiprobe conductors, *Phys. Rev. B* **38** (1988) 9375.

[41] C.G. Callan and J.A. Harvey, Anomalies and fermion zero modes on strings and domain walls, *Nucl. Phys. B.* **250** (1985) 427.

[42] M. Campanino, D. Ioffe and Y.v. Velenik, Ornstein–Zernike theory for finite range Ising models above T_c, *Probability Theory and Related Fields* **125** (2003) 305.

[43] L. Capriotti, F. Becca, A. Parola, and S. Sorella, Resonating valence bond wave functions for strongly frustrated spin systems, *Phys. Rev. Lett.* **87** (2001) 097201 [cond-mat/0107204].

[44] W. Chen, M.P.A. Fisher, and Y.-S. Wu, Mott transition in an anyon gas, *Phys. Rev. B* **48** (1993) 13749.

[45] X. Chen, Z.-C. Gu, Z.-X. Liu, and X.-G. Wen, Symmetry protected topological orders and the group cohomology of their symmetry group, *Phys. Rev. B* **87** (2013) 155114 [arXiv:1106.4772].

[46] D. Chowdhury, A. Georges, O. Parcollet and S. Sachdev, Sachdev–Ye–Kitaev models and beyond: Window into non-Fermi liquids, *Rev. Mod. Phys.* **94** (2022) 035004 [*arXiv:2109.05037*].

[47] M. Christos, F.M. Haehl, and S. Sachdev, Spin liquid to spin glass crossover in the random quantum Heisenberg magnet, *Phys. Rev. B* **105** (2022) 085120 [arXiv:2110.00007].

[48] M. Christos, D.G. Joshi, S. Sachdev, and M. Tikhanovskaya, *Critical metallic phase in the overdoped random t–J model*, arXiv:2203.16548.

[49] A.V. Chubukov and A. Abanov, Pairing by a dynamical interaction in a metal, *Sov. J. Exp. Thoer. Phys.* **132** (2021) 606 [arXiv:2012.11777].

[50] A.V. Chubukov, S. Sachdev, and T. Senthil, Quantum phase transitions in frustrated quantum antiferromagnets, *Nucl. Phys. B* **426** (1994) 601 [cond-mat/9402006].

[51] A.V. Chubukov, T. Senthil, and S. Sachdev, Universal magnetic properties of frustrated quantum antiferromagnets in two dimensions, *Phys. Rev. Lett.* **72** (1994) 2089 [cond-mat/9311045].

[52] P. Coleman, *Introduction to Many-Body Physics*. Cambridge: Cambridge University Press (2015). https://dx.doi.org/10.1017/CBO9781139020916.

[53] P. Coleman, I. Paul, and J. Rech, Sum rules and Ward identities in the Kondo lattice, *Phys. Rev. B* **72** (2005) 094430 [cond-mat/0503001].

[54] J.S. Cotler, G. Gur-Ari, M. Hanada, J. Polchinski, P. Saad, S.H. Shenker, et al., Black holes and random matrices, *JHEP* **05** (2017) 118 [arXiv:1611.04650].

[55] L.F. Cugliandolo, D.R. Grempel, and C.A. da Silva Santos, From second to first order transitions in a disordered quantum magnet, *Phys. Rev. Lett.* **85** (2000) 2589 [cond-mat/0003268].

[56] Y. Cui, L. Liu, H. Lin, K.-H. Wu, W. Hong, X. Liu, et al., *Deconfined quantum criticality and emergent symmetry in* $SrCu_2(BO_3)_2$, arXiv:2204.08133.

[57] G. Cuomo, Z. Komargodski, M. Mezei, and A. Raviv-Moshe, Spin impurities, Wilson lines and semiclassics, JHEP **06** (2022) 112 [arXiv:2202.00040].

[58] A. D'Adda, M. Lüscher, and P. Di Vecchia, A $1/n$ expandable series of non-linear σ-models with instantons, *Nucl. Phys. B* **146** (1978) 63.

[59] J.A. Damia, M. Solís, and G. Torroba, How non-Fermi liquids cure their infrared divergences, *Phys. Rev. B* **102** (2020) 045147 [arXiv:2004.05181].

[60] C. Dasgupta and B.I. Halperin, Phase transition in a lattice model of superconductivity, *Phys. Rev. Lett.* **47** (1981) 1556.

[61] K.B. Dave, P.W. Phillips, and C.L. Kane, Absence of Luttinger's theorem due to zeros in the single-particle Green function, *Phys. Rev. Lett.* **110** (2013) 090403 [arXiv:1207.4201].

[62] R.A. Davison, W. Fu, A. Georges, Y. Gu, K. Jensen and S. Sachdev, Thermoelectric transport in disordered metals without quasiparticles: The Sachdev–Ye–Kitaev models and holography, *Phys. Rev. B* **95** (2017) 155131 [arXiv:1612.00849].

[63] S. Doniach and S. Engelsberg, Low-temperature properties of nearly ferromagnetic Fermi liquids, *Phys. Rev. Lett.* **17** (1966) 750.

[64] I. Dzyaloshinskii, Some consequences of the Luttinger theorem: The Luttinger surfaces in non-Fermi liquids and Mott insulators, *Phys. Rev. B* **68** (2003) 085113.

[65] S. Ebadi, T.T. Wang, H. Levine, A. Keesling, G. Semeghini, A. Omran et al., Quantum phases of matter on a 256-atom programmable quantum simulator, *Nature* **595** (2021) 227 [arXiv:2012.12281].

[66] A. Eberlein, I. Mandal, and S. Sachdev, Hyperscaling violation at the Ising-nematic quantum critical point in two dimensional metals, *Phys. Rev. B* **94** (2016) 045133 [arXiv:1605.00657].

[67] S. Elitzur, G. Moore, A. Schwimmer, and N. Seiberg, Remarks on the canonical quantization of the Chern–Simons–Witten theory, *Nucl. Phys. B* **326** (1989) 108.

[68] D.V. Else and T. Senthil, Strange metals as ersatz Fermi liquids, *Phys. Rev. Lett.* **127** (2021) 086601 [arXiv:2010.10523].

[69] A.M. Essin and M. Hermele, Classifying fractionalization: Symmetry classification of gapped Z_2 spin liquids in two dimensions, *Phys. Rev. B* **87** (2013) 104406 [arXiv:1212.0593].

[70] F.H. Essler and A.M. Tsvelik, Weakly coupled one-dimensional Mott insulators, *Phys. Rev. B* **65** (2002) 115117 [cond-mat/0108382].

[71] I. Esterlis, H. Guo, A.A. Patel, and S. Sachdev, Large N theory of critical Fermi surfaces, *Phys. Rev. B* **103** (2021) 235129 [arXiv:2103.08615].

[72] I. Esterlis and J. Schmalian, Cooper pairing of incoherent electrons: An electron–phonon version of the Sachdev–Ye–Kitaev model, *Phys. Rev. B* **100** (2019) 115132 [arXiv:1906.04747].

[73] Y. Fang, G. Grissonnanche, A. Legros, S. Verret, F. Laliberte, C. Collignon, et al., Fermi surface transformation at the pseudogap critical point of a cuprate superconductor, *Nat. Phys.* (2022) [arXiv:2004.01725].

[74] T. Faulkner, H. Liu, J. McGreevy, and D. Vegh, Emergent quantum criticality, Fermi surfaces, and AdS(2), *Phys. Rev.* **D83** (2011) 125002 [arXiv:0907.2694].

[75] J. Feldmeier, S. Huber, and M. Punk, Exact solution of a two-species quantum dimer model for pseudogap metals, *Phys. Rev. Lett.* **120** (2018) 187001 [arXiv:1712.01854].

[76] P. Fendley, K. Sengupta and S. Sachdev, Competing density-wave orders in a one-dimensional hard-boson model, *Phys. Rev. B* **69** (2004) 075106 [cond-mat/0309438].

[77] Z. Feng, Z. Li, X. Meng, W. Yi, Y. Wei, J. Zhang, et al., Gapped spin-1/2 spinon excitations in a new kagome quantum spin liquid compound $Cu_3Zn(OH)_6FBr$, *Chinese Phys. Lett.* **34** (2017) 077502 [arXiv:1702.01658].

[78] F. Ferrari and F. Becca, Spectral signatures of fractionalization in the frustrated Heisenberg model on the square lattice, *Phys. Rev. B* **98** (2018) 100405 [arXiv:1805.09287].

[79] F. Ferrari and F. Becca, Gapless spin liquid and valence-bond solid in the J_1–J_2 Heisenberg model on the square lattice: Insights from singlet and triplet excitations, *Phys. Rev. B* **102** (2020) 014417 [arXiv:2005.12941].

[80] M. Ferrero, P.S. Cornaglia, L. de Leo, O. Parcollet, G. Kotliar, and A. Georges, Pseudogap opening and formation of Fermi arcs as an orbital-selective Mott transition in momentum space, *Phys. Rev. B* **80** (2009) 064501 [arXiv:0903.2480].

[81] K. Fischer and J. Hertz, *Spin Glasses*, Cambridge University Press (1993).

[82] D.S. Fisher and J.D. Weeks, Shape of crystals at low temperatures: Absence of quantum roughening, *Phys. Rev. Lett.* **50** (1983) 1077.

[83] M.P.A. Fisher, P.B. Weichman, G. Grinstein, and D.S. Fisher, Boson localization and the superfluid–insulator transition, *Phys. Rev. B* **40** (1989) 546.

[84] E. Fradkin, D.A. Huse, R. Moessner, V. Oganesyan, and S.L. Sondhi, Bipartite Rokhsar Kivelson points and Cantor deconfinement, *Phys. Rev. B* **69** (2004) 224415 [cond-mat/0311353].

[85] E. Fradkin, Roughening transition in quantum interfaces, *Phys. Rev. B* **28** (1983) 5338.

[86] E. Fradkin and S.A. Kivelson, Short range resonating valence bond theories and superconductivity, *Mod. Phys. Lett. B* **04** (1990) 225.

[87] E. Fradkin and S.H. Shenker, Phase diagrams of lattice gauge theories with Higgs fields, *Phys. Rev. D* **19** (1979) 3682.

[88] M. Freedman, C. Nayak, K. Shtengel, K. Walker, and Z. Wang, A class of P, T-invariant topological phases of interacting electrons, *Ann. Phys.* **310** (2004) 428 [cond-mat/0307511].

[89] W. Fu and S. Sachdev, Numerical study of fermion and boson models with infinite-range random interactions, *Phys. Rev. B* **94** (2016) 035135 [arXiv:1603.05246].

[90] S. Gazit, M. Randeria, and A. Vishwanath, Emergent Dirac fermions and broken symmetries in confined and deconfined phases of $\mathbb{Z}_2$ gauge theories, *Nat. Phys.* **13** (2017) 484 [arXiv:1607.03892].

[91] A. Georges, O. Parcollet, and S. Sachdev, Quantum fluctuations of a nearly critical Heisenberg spin glass, *Phys. Rev. B* **63** (2001) 134406 [cond-mat/0009388].

[92] A. Georges, O. Parcollet, and S. Sachdev, Mean field theory of a quantum Heisenberg spin glass, *Phys. Rev. Lett.* **85** (2000) 840 [cond-mat/9909239].

[93] H. Gharibyan, M. Hanada, S.H. Shenker, and M. Tezuka, Onset of random matrix behavior in scrambling systems, *JHEP* **07** (2018) 124 [arXiv:1803.08050].

[94] P. Giraldo-Gallo, J.A. Galvis, Z. Stegen, K.A. Modic, F.F. Balakirev, J.B. Betts, et al., Scale-invariant magnetoresistance in a cuprate superconductor, *Science* **361** (2018) 479 [arXiv:1705.05806].

[95] M. Greiner, O. Mandel, T. Esslinger, T.W. Hänsch, and I. Bloch, Quantum phase transition from a superfluid to a Mott insulator in a gas of ultracold atoms, *Nature* **415** (2002) 39.

[96] D.J. Gross and V. Rosenhaus, A generalization of Sachdev–Ye–Kitaev, *JHEP* **02** (2017) 093 [arXiv:1610.01569].

[97] Y. Gu, A. Kitaev, S. Sachdev, and G. Tarnopolsky, Notes on the complex Sachdev–Ye–Kitaev model, *JHEP* **02** (2020) 157 [arXiv:1910.14099].

[98] H. Guo, A.A. Patel, I. Esterlis, and S. Sachdev, Large-N theory of critical Fermi surfaces. II. Conductivity, *Phys. Rev. B* **106** (2022) 115151 [arXiv:2207.08841].

[99] F.D.M. Haldane, $O(3)$ Nonlinear σ model and the topological distinction between integer- and half-integer-spin antiferromagnets in two dimensions, *Phys. Rev. Lett.* **61** (1988) 1029.

[100] B.I. Halperin and P.C. Hohenberg, Hydrodynamic theory of spin waves, *Phys. Rev.* **188** (1969) 898.

[101] T.H. Hansson, V. Oganesyan, and S.L. Sondhi, Superconductors are topologically ordered, *Ann. Phys.* **313** (2004) 497 [cond-mat/0404327].

[102] S.A. Hartnoll, P.K. Kovtun, M. Muller, and S. Sachdev, Theory of the Nernst effect near quantum phase transitions in condensed matter, and in dyonic black holes, *Phys. Rev. B* **76** (2007) 144502 [arXiv:0706.3215].

[103] S.A. Hartnoll, A. Lucas and S. Sachdev, *Holographic Quantum Matter*, MIT Press (2016) [arXiv:1612.07324].

[104] S.A. Hartnoll, R. Mahajan, M. Punk, and S. Sachdev, Transport near the Ising-nematic quantum critical point of metals in two dimensions, *Phys. Rev.* **B89** (2014) 155130 [arXiv:1401.7012].

[105] M.B. Hastings, Lieb–Schultz–Mattis in higher dimensions, *Phys. Rev. B* **69** (2004) 104431 [cond-mat/0305505].

[106] Y.-C. He, S. Bhattacharjee, F. Pollmann, and R. Moessner, Kagome chiral spin liquid as a gauged $U(1)$ symmetry protected topological phase, *Phys. Rev. Lett.* **115** (2015) 267209 [arXiv:1509.03070].

[107] Y.-C. He, J. Rong, and N. Su, Non-Wilson–Fisher kinks of $O(N)$ numerical bootstrap: From the deconfined phase transition to a putative new family of CFTs, *SciPost Phys.* **10** (2021) 115 [arXiv:2005.04250].

[108] M. Hermele, T. Senthil, M.P.A. Fisher, P.A. Lee, N. Nagaosa, and X.-G. Wen, Stability of $U(1)$ spin liquids in two dimensions, *Phys. Rev. B* **70** (2004) 214437 [cond-mat/0404751].

[109] A.C. Hewson, *The Kondo Problem to Heavy Fermions*, Cambridge University Press (1997).

[110] C.W. Hicks, A.S. Gibbs, A.P. Mackenzie, H. Takatsu, Y. Maeno, and E.A. Yelland, Quantum oscillations and high carrier mobility in the delafossite $PdCoO_2$, *Phys. Rev. Lett.* **109** (2012) 116401 [arXiv:1207.5402].

[111] P.C. Hohenberg and B.I. Halperin, Theory of dynamic critical phenomena, *Rev. Mod. Phys.* **49** (1977) 435.

[112] W.-J. Hu, F. Becca, A. Parola, and S. Sorella, Direct evidence for a gapless $\mathbb{Z}_2$ spin liquid by frustrating Néel antiferromagnetism, *Phys. Rev. B* **88** (2013) 060402 [arXiv:1304.2630].

[113] Y. Huh, L. Fritz, and S. Sachdev, Quantum criticality of the kagome antiferromagnet with Dzyaloshinskii-Moriya interactions, *Phys. Rev. B* **81** (2010) 144432 [arXiv:1003.0891].

[114] Y. Huh, M. Punk, and S. Sachdev, Vison states and confinement transitions of $\mathbb{Z}_2$ spin liquids on the kagome lattice, *Phys. Rev. B* **84** (2011) 094419 [arXiv:1106.3330].

[115] L.-Y. Hung and Y. Wan, K matrix construction of symmetry-enriched phases of matter, *Phys. Rev. B* **87** (2013) 195103 [arXiv:1302.2951].

[116] L.V. Iliesiu, S. Murthy and G.J. Turiaci, *Revisiting the Logarithmic Corrections to the Black Hole Entropy*, arXiv:2209.13608.

[117] D. Ivanov, Non-abelian statistics of half-quantum vortices in p-wave superconductors, *Phys. Rev. Lett.* **86** (2001) 268 [cond-mat/0005069].

[118] J.K. Jain, Incompressible quantum Hall states, *Phys. Rev. B* **40** (1989) 8079.

[119] R.A. Jalabert and S. Sachdev, Spontaneous alignment of frustrated bonds in an anisotropic, three-dimensional Ising model, *Phys. Rev. B* **44** (1991) 686.

[120] C.-M. Jian, A. Thomson, A. Rasmussen, Z. Bi, and C. Xu, Deconfined quantum critical point on the triangular lattice, *Phys. Rev. B* **97** (2018) 195115 [arXiv:1710.04668].

[121] J.V. José, L.P. Kadanoff, S. Kirkpatrick, and D.R. Nelson, Renormalization, vortices, and symmetry-breaking perturbations in the two-dimensional planar model, *Phys. Rev. B* **16** (1977) 1217.

[122] D.G. Joshi, C. Li, G. Tarnopolsky, A. Georges, and S. Sachdev, Deconfined critical point in a doped random quantum Heisenberg magnet, *Phys. Rev. X* **10** (2020) 021033 [arXiv:1912.08822].

[123] V. Kalmeyer and R.B. Laughlin, Equivalence of the resonating-valence-bond and fractional quantum Hall states, *Phys. Rev. Lett.* **59** (1987) 2095.

[124] C.L. Kane and M.P.A. Fisher, Impurity scattering and transport of fractional quantum Hall edge states, *Phys. Rev. B* **51** (1995) 13449 [cond-mat/9409028].

[125] C.L. Kane and E.J. Mele, $\mathbb{Z}_2$ Topological order and the quantum spin Hall effect, *Phys. Rev. Lett.* **95** (2005) 146802 [cond-mat/0506581].

[126] C.L. Kane and E.J. Mele, Quantum spin Hall effect in graphene, *Phys. Rev. Lett.* **95** (2005) 226801 [cond-mat/0411737].

[127] A. Karch and D. Tong, Particle–vortex duality from 3d bosonization, *Phys. Rev. X* **6** (2016) 031043 [arXiv:1606.01893].

[128] R.K. Kaul, Y.B. Kim, S. Sachdev, and T. Senthil, Algebraic charge liquids, *Nat. Phys.* **4** (2008) 28 [arXiv:0706.2187].

[129] A. Keesling, A. Omran, H. Levine, H. Bernien, H. Pichler, S. Choi, et al., Quantum Kibble–Zurek mechanism and critical dynamics on a programmable Rydberg simulator, *Nature* **568** (2019) 207 [arXiv:1809.05540].

[130] J. Kim, I.R. Klebanov, G. Tarnopolsky, and W. Zhao, Symmetry breaking in coupled SYK or tensor models, *Phys. Rev. X* **9** (2019) 021043 [arXiv:1902.02287].

[131] A.D. King, J. Carrasquilla, J. Raymond, I. Ozfidan, E. Andriyash, A. Berkley, et al., Observation of topological phenomena in a programmable lattice of 1,800 qubits, *Nature* **560** (2018) 456 [arXiv:1803.02047].

[132] A. Kitaev and J. Preskill, Topological entanglement entropy, *Phys. Rev. Lett.* **96** (2006) 110404 [hep-th/0510092].

[133] A.Y. Kitaev, Fault-tolerant quantum computation by anyons, *Ann. Phys.* **303** (2003) 2 [quant-ph/9707021].

[134] A. Kitaev, Anyons in an exactly solved model and beyond, *Annals Phys.* **321** (2006) 2 [cond-mat/0506438].

[135] A. Kitaev, Periodic table for topological insulators and superconductors, *AIP Conf. Proc.* **1134** (2009) 22 [arXiv:0901.2686].

[136] A. Kitaev, A simple model of quantum holography, talk given at KITP program: Entanglement in Strongly-Correlated Quantum Matter, *University of California, Santa Barbara* (2015).

[137] A. Kitaev and S.J. Suh, The soft mode in the Sachdev–Ye-Kitaev model and its gravity dual, *JHEP* **05** (2018) 183 [arXiv:1711.08467].

[138] S. Kivelson, Statistics of holons in the quantum hard-core dimer gas, *Phys. Rev. B* **39** (1989) 259.

[139] S.A. Kivelson, D.S. Rokhsar, and J.P. Sethna, Topology of the resonating valence-bond state: Solitons and high-T_c superconductivity, *Phys. Rev. B* **35** (1987) 8865.

[140] S.A. Kivelson, D.S. Rokhsar, and J.P. Sethna, $2e$ or not $2e$: Flux quantization in the resonating valence bond state, *Europhysics Letters* **6** (1988) 353.

[141] I.R. Klebanov, A. Milekhin, G. Tarnopolsky, and W. Zhao, Spontaneous breaking of $U(1)$ symmetry in coupled complex SYK models, *JHEP* **11** (2020) 162 [arXiv:2006.07317].

[142] I.R. Klebanov, F. Popov, and G. Tarnopolsky, TASI lectures on large N tensor models, *PoS* **TASI2017** (2018) 004 [arXiv:1808.09434].

[143] I.R. Klebanov and G. Tarnopolsky, Uncolored random tensors, melon diagrams, and the Sachdev–Ye–Kitaev models, *Phys. Rev. D* **95** (2017) 046004 [arXiv:1611.08915].

[144] K.v. Klitzing, G. Dorda, and M. Pepper, New method for high-accuracy determination of the fine-structure constant based on quantized Hall resistance, *Phys. Rev. Lett.* **45** (1980) 494.

[145] J. Koepsell, D. Bourgund, P. Sompet, S. Hirthe, A. Bohrdt, Y. Wang, et al., Microscopic evolution of doped Mott insulators from polaronic metal to Fermi liquid, *Science* **374** (2021) 82 [arXiv:2009.04440].

[146] J.B. Kogut, An introduction to lattice gauge theory and spin systems, *Rev. Mod. Phys.* **51** (1979) 659.

[147] R.M. Konik, T.M. Rice, and A.M. Tsvelik, Doped spin liquid: Luttinger sum rule and low temperature order, *Phys. Rev. Lett.* **96** (2006) 086407 [cond-mat/0511268].

[148] S.-P. Kou, M. Levin, and X.-G. Wen, Mutual Chern–Simons theory for $\mathbb{Z}_2$ topological order, *Phys. Rev. B* **78** (2008) 155134 [arXiv:0803.2300].

[149] P.E. Lammert, D.S. Rokhsar, and J. Toner, Topology and nematic ordering. I. A gauge theory, *Phys. Rev. E* **52** (1995) 1778 [cond-mat/9501101].

[150] P.E. Lammert, D.S. Rokhsar, and J. Toner, Topology and nematic ordering, *Phys. Rev. Lett.* **70** (1993) 1650.

[151] R.B. Laughlin, Anomalous quantum Hall effect: An incompressible quantum fluid with fractionally charged excitations, *Phys. Rev. Lett.* **50** (1983) 1395.

[152] P.A. Lee, Low-temperature T-linear resistivity due to Umklapp scattering from a critical mode, *Phys. Rev. B* **104** (2021) 035140 [arXiv:2012.09339].

[153] P.A. Lee, N. Nagaosa, and X.-G. Wen, Doping a Mott insulator: Physics of high-temperature superconductivity, *Rev. Mod. Phys.* **78** (2006) 17 [cond-mat/0410445].

[154] S.-S. Lee, Low-energy effective theory of Fermi surface coupled with $U(1)$ gauge field in 2+1 dimensions, *Phys. Rev. B* **80** (2009) 165102 [arXiv:0905.4532].

[155] S.-S. Lee, Recent developments in non-Fermi liquid theory, *Annu. Rev. Condens. Matter Phys.* **9** (2018) 227 [arXiv:1703.08172].

[156] M. Levin and X.-G. Wen, Detecting topological order in a ground state wave function, *Phys. Rev. Lett.* **96** (2006) 110405 [cond-mat/0510613].

[157] Y. Liao and V. Galitski, Emergence of many-body quantum chaos via spontaneous breaking of unitarity, *Phys. Rev. B* **105** (2022) L140202 [arXiv:2104.05721].

[158] Y. Liao, A. Vikram, and V. Galitski, Many-body level statistics of single-particle quantum chaos, *Phys. Rev. Lett.* **125** (2020) 250601 [arXiv:2005.08991].

[159] E. Lieb, T. Schultz, and D. Mattis, Two soluble models of an antiferromagnetic chain, *Ann. Phys.* **16** (1961) 407.

[160] C.-J. Lin, V. Calvera, and T.H. Hsieh, Quantum many-body scar states in two-dimensional Rydberg atom arrays, *Phys. Rev. B* **101** (2020) 220304 [arXiv:2003.04516].

[161] W.-Y. Liu, S.-S. Gong, Y.-B. Li, D. Poilblanc, W.-Q. Chen, and Z.-C. Gu, Gapless quantum spin liquid and global phase diagram of the spin-1/2 J_1-J_2 square antiferromagnetic Heisenberg model, *Sci. Bull.* **67** (2022) 1034 [arXiv:2009.01821].

[162] Y.-M. Lu and A. Vishwanath, Classification and properties of symmetry enriched topological phases: A Chern–Simons approach with applications to $\mathbb{Z}_2$ spin liquids, *Phys. Rev. B* **93** (2016) 155121 [arXiv:1302.2634].

[163] Y.-M. Lu, G.Y. Cho, and A. Vishwanath, Unification of bosonic and fermionic theories of spin liquids on the kagome lattice, *Phys. Rev. B* **96** (2017) 205150 [arXiv:1403.0575].

[164] Y.-M. Lu and A. Vishwanath, Theory and classification of interacting 'integer' topological phases in two dimensions: A Chern–Simons approach, *Phys. Rev. B* **86** (2012) 125119 [arXiv:1205.3156].

[165] Luther, A. and Emery, V. J., Backward Scattering in the One-Dimensional Electron Gas, *Phys. Rev. Lett.* **33** (1974) 10, *American Physical Society* 589–592, https://link.aps.org/doi/10.1103/PhysRevLett.33.589.

[166] J. Ma, Y. Kamiya, T. Hong, et al., Static and dynamical properties of the spin-1/2 equilateral triangular-lattice antiferromagnet $Ba_3CoSb_2O_9$, *Phys. Rev. Lett.* **116** (2016) 087201 [arXiv:1507.05702].

[167] R. Ma and C. Wang, Theory of deconfined pseudocriticality, *Phys. Rev. B* **102** (2020) 020407 [arXiv:1912.12315].

[168] A.H. MacDonald, S.M. Girvin, and D. Yoshioka, t/U Expansion for the Hubbard model, *Phys. Rev. B* **37** (1988) 9753.

[169] N. Maksimovic, D.H. Eilbott, T. Cookmeyer, F. Wan, J. Rusz, V. Nagarajan, et al., Evidence for a delocalization quantum phase transition without symmetry breaking in $CeCoIn_5$, *Science* **375** (2022) 76.

[170] J. Maldacena and D. Stanford, Remarks on the Sachdev–Ye–Kitaev model, *Phys. Rev. D* **94** (2016) 106002 [arXiv:1604.07818].

[171] J.M. Maldacena, G.W. Moore and N. Seiberg, D-brane charges in five-brane backgrounds, *JHEP* **10** (2001) 005 [hep-th/0108152].

[172] M. Mariño, Chern–Simons theory and topological strings, *Rev. Mod. Phys.* **77** (2005) 675.

[173] E. Mascot, A. Nikolaenko, M. Tikhanovskaya, Y.-H. Zhang, D.K. Morr, and S. Sachdev, Electronic spectra with paramagnon fractionalization in the single band Hubbard model, *Phys. Rev. B* **105** (2022) 075146 [arXiv:2111.13703].

[174] D.L. Maslov, V.I. Yudson, and A.V. Chubukov, Resistivity of a non-Galilean-invariant Fermi liquid near Pomeranchuk quantum criticality, *Phys. Rev. Lett.* **106** (2011) 106403 [arXiv:1012.0069].

[175] M.L. Mehta, *Random Matrices*, Elsevier (2004).

[176] J.-W. Mei, S. Kawasaki, G.-Q. Zheng, Z.-Y. Weng, and X.-G. Wen, Luttinger-volume violating Fermi liquid in the pseudogap phase of the cuprate superconductors, *Phys. Rev. B* **85** (2012) 134519 [arXiv:1109.0406].

[177] M.A. Metlitski and S. Sachdev, Quantum phase transitions of metals in two spatial dimensions. I. Ising-nematic order, *Phys. Rev. B* **82** (2010) 075127 [arXiv:1001.1153].

[178] B. Miksch, A. Pustogow, M.J. Rahim, A.A. Bardin, K. Kanoda, J.A. Schlueter, et al., Gapped magnetic ground state in quantum spin liquid candidate κ-(BEDT-TTF)$_2$Cu$_2$(CN)$_3$, *Science* **372** (2021) 276 [arXiv:2010.16155].

[179] R. Moessner and S.L. Sondhi, Ising models of quantum frustration, *Phys. Rev. B* **63** (2001) 224401 [cond-mat/0011250].

[180] R. Moessner and S.L. Sondhi, Resonating valence bond phase in the Triangular Lattice Quantum Dimer Model, *Phys. Rev. Lett.* **86** (2001) 1881 [cond-mat/0007378].

[181] R. Moessner, S.L. Sondhi, and P. Chandra, Two-dimensional periodic frustrated Ising models in a transverse field, *Phys. Rev. Lett.* **84** (2000) 4457 [cond-mat/9910499].

[182] R. Moessner, S.L. Sondhi, and P. Chandra, Phase diagram of the hexagonal lattice quantum dimer model, *Phys. Rev. B* **64** (2001) 144416 [cond-mat/0106288].

[183] R. Moessner, S.L. Sondhi, and E. Fradkin, Short-ranged resonating valence bond physics, quantum dimer models, and Ising gauge theories, *Phys. Rev. B* **65** (2002) 024504 [cond-mat/0103396].

[184] E.-G. Moon and A. Chubukov, Quantum-critical pairing with varying exponents, *J. Low Temp. Phys.* **161** (2010) 263 [arXiv:1005.0356].

[185] E.G. Moon and S. Sachdev, Underdoped cuprates as fractionalized Fermi liquids: Transition to superconductivity, *Phys. Rev. B* **83** (2011) 224508 [arXiv:1010.4567].

[186] O.I. Motrunich and T. Senthil, Exotic order in simple models of bosonic systems, *Phys. Rev. Lett.* **89** (2002) 277004 [cond-mat/0205170].

[187] S. Müller, S. Heusler, A. Altland, P. Braun and F. Haake, Periodic-orbit theory of universal level correlations in quantum chaos, *New. J. Phys.* **11** (2009) 103025 [arXiv:0906.1960].

[188] A. Nahum, Note on Wess–Zumino–Witten models and quasiuniversality in 2+1 dimensions, *Phys. Rev. B* **102** (2020) 201116 [arXiv:1912.13468].

[189] A. Nahum, *Fixed point annihilation for a spin in a fluctuating field*, arXiv:2202.08431.

[190] J. Negele and H. Orland, *Quantum Many-Particle Systems*, Addison-Wesley (1988).

[191] A. Nikolaenko, M. Tikhanovskaya, S. Sachdev, and Y.-H. Zhang, Small to large Fermi surface transition in a single band model, using randomly coupled ancillas, *Phys. Rev. B* **103** (2021) 235138 [arXiv:2103.05009].

[192] Y. Nomura and M. Imada, Dirac-type nodal spin liquid revealed by refined quantum many-body solver using neural-network wave function, correlation ratio, and level spectroscopy, *Phys. Rev. X* **11** (2021) 031034 [arXiv:2005.14142].

[193] M. Oshikawa, Topological approach to Luttinger's theorem and the Fermi surface of a Kondo lattice, *Phys. Rev. Lett.* **84** (2000) 3370 [cond-mat/0002392].

[194] M. Oshikawa, Commensurability, excitation gap, and topology in quantum many-particle systems on a periodic lattice, *Phys. Rev. Lett.* **84** (2000) 1535 [cond-mat/9911137].

[195] A. Paramekanti and A. Vishwanath, Extending Luttinger's theorem to $\mathbb{Z}_2$ fractionalized phases of matter, *Phys. Rev. B* **70** (2004) 245118 [cond-mat/0406619].

[196] O. Parcollet and A. Georges, Non-Fermi-liquid regime of a doped Mott insulator, *Phys. Rev. B* **59** (1999) 5341 [cond-mat/9806119].

[197] O. Parcollet, A. Georges, G. Kotliar, and A. Sengupta, Overscreened multichannel $SU(N)$ Kondo model: Large-N solution and conformal field theory, *Phys. Rev. B* **58** (1998) 3794 [cond-mat/9711192].

[198] K. Park and S. Sachdev, Bond and Néel order and fractionalization in ground states of easy-plane antiferromagnets in two dimensions, *Phys. Rev. B* **65** (2002) 220405 [cond-mat/0112003].

[199] A.A. Patel, D. Chowdhury, A. Allais, and S. Sachdev, Confinement transition to density wave order in metallic doped spin liquids, *Phys. Rev. B* **93** (2016) 165139 [arXiv:1602.05954].

[200] A.A. Patel, H. Guo, I. Esterlis, and S. Sachdev, Universal, low temperature, T-linear resistivity in two-dimensional quantum-critical metals from spatially random interactions, arXiv:2203.04990.

[201] L. Pauling, A resonating-valence-bond theory of metals and intermetallic compounds, *Proc. R. Soc. London, Ser. A* **196** (1949) 343.

[202] R. Peierls, Zur Theorie der elektrischen und thermischen Leitfähigkeit von Metallen, *Annal. Phys.* **396** (1930) 121.

[203] R. Peierls, Zur Frage des elektrischen Widerstandsgesetzes für tiefe Temperaturen, *Annal. Phys.* **404** (1932) 154.

[204] N.M.R. Peres, F. Guinea, and A.H. Castro Neto, Electronic properties of disordered two-dimensional carbon, *Phys. Rev. B* **73** (2006) 125411 [cond-mat/0512091].

[205] M. Platé, J.D.F. Mottershead, I.S. Elfimov, D.C. Peets, R. Liang, D.A. Bonn, et al., Fermi surface and quasiparticle excitations of overdoped $Tl_2Ba_2CuO_{6+\delta}$, *Phys. Rev. Lett.* **95** (2005) 077001 [cond-mat/0503117].

[206] D. Podolsky and S. Sachdev, Spectral functions of the Higgs mode near two-dimensional quantum critical points, *Phys. Rev. B* **86** (2012) 054508 [arXiv:1205.2700].

[207] A.M. Polyakov, Quark confinement and topology of gauge theories, *Nucl. Phys. B* **120** (1977) 429.

[208] M. Potthoff, Non-perturbative construction of the Luttinger–Ward functional, *Condens. Mat. Phys* **9** (2006) 557 [cond-mat/0406671].

[209] S. Powell, S. Sachdev, and H.P. Büchler, Depletion of the Bose–Einstein condensate in Bose–Fermi mixtures, *Phys. Rev. B* **72** (2005) 024534 [cond-mat/0502299].

[210] M. Punk and S. Sachdev, Fermi surface reconstruction in hole-doped t–J models without long-range antiferromagnetic order, *Phys. Rev. B* **85** (2012) 195123 [arXiv:1202.4023].

[211] M. Punk, A. Allais and S. Sachdev, A quantum dimer model for the pseudogap metal, *Proc. Nat. Acad. Sci.* **112** (2015) 9552 [arXiv:1501.00978].

[212] Y. Qi and S. Sachdev, Effective theory of Fermi pockets in fluctuating antiferromagnets, *Phys. Rev. B* **81** (2010) 115129 [arXiv:0912.0943].

[213] S. Raghu, S.A. Kivelson, and D.J. Scalapino, Superconductivity in the repulsive Hubbard model: An asymptotically exact weak-coupling solution, *Phys. Rev. B* **81** (2010) 224505 [arXiv:1002.0591].

[214] Y. Ran and X.-G. Wen, Continuous quantum phase transitions beyond Landau's paradigm in a large-N spin model, arXiv:cond-mat/0609620.

[215] N. Read and B. Chakraborty, Statistics of the excitations of the resonating-valence-bond state, *Phys. Rev. B* **40** (1989) 7133.

[216] N. Read and S. Sachdev, Some features of the phase diagram of the square lattice SU(N) antiferromagnet, *Nucl. Phys. B* **316** (1989) 609.

[217] N. Read and S. Sachdev, Valence-bond and spin-Peierls ground states of low-dimensional quantum antiferromagnets, *Phys. Rev. Lett.* **62** (1989) 1694.

[218] N. Read and S. Sachdev, Spin-Peierls, valence-bond solid, and Néel ground states of low-dimensional quantum antiferromagnets, *Phys. Rev. B* **42** (1990) 4568.

[219] N. Read and S. Sachdev, Large N expansion for frustrated quantum antiferromagnets, *Phys. Rev. Lett.* **66** (1991) 1773.

[220] N. Read, S. Sachdev and J. Ye, Landau theory of quantum spin glasses of rotors and Ising spins, *Phys. Rev. B* **52** (1995) 384 [cond-mat/9412032].

[221] N.J. Robinson, P.D. Johnson, T.M. Rice, and A.M. Tsvelik, Anomalies in the pseudogap phase of the cuprates: Competing ground states and the role of Umklapp scattering, *Rep. Prog. Phys.* **82** (2019) 126501 [arXiv:1906.09005].

[222] D.S. Rokhsar and S.A. Kivelson, Superconductivity and the quantum hard-core dimer gas, *Phys. Rev. Lett.* **61** (1988) 2376.

[223] K. Rommelse and M. den Nijs, Preroughening transitions in surfaces, *Phys. Rev. Lett.* **59** (1987) 2578.

[224] K. Roychowdhury, S. Bhattacharjee, and F. Pollmann, Z_2 topological liquid of hard-core bosons on a kagome lattice at 1/3 filling, *Phys. Rev. B* **92** (2015) 075141 [arXiv:1505.05998].

[225] C. Rüegg, B. Normand, M. Matsumoto, A. Furrer, D.F. McMorrow, K.W. Krämer, et al., Quantum magnets under pressure: Controlling elementary excitations in $TlCuCl_3$, *Phys. Rev. Lett.* **100** (2008) 205701 [arXiv:0803.3720].

[226] S. Ryu, A.P. Schnyder, A. Furusaki, and A.W.W. Ludwig, Topological insulators and superconductors: Tenfold way and dimensional hierarchy, *New. J. Phys.* **12** (2010) 065010 [arXiv:0912.2157].

[227] S. Sachdev, Topological order, emergent gauge fields, and Fermi surface reconstruction, *Rep. Prog. Phys.* **82** (2019) 014001 [arXiv:1801.01125].

[228] S. Sachdev, M.A. Metlitski, Y. Qi, and C. Xu, Fluctuating spin density waves in metals, *Phys. Rev. B* **80** (2009) 155129 [arXiv:0907.3732].

[229] S. Sachdev and K. Park, Ground states of quantum antiferromagnets in two dimensions, *Ann. Phys.* **298** (2002) 58 [cond-mat/0108214].

[230] S. Sachdev and M. Vojta, Translational symmetry breaking in two-dimensional antiferromagnets and superconductors, *J. Phys. Soc. Jpn* ***69**, Supp. B, 1* (1999) [cond-mat/9910231].

[231] S. Sachdev, Kagome and triangular-lattice Heisenberg antiferromagnets: Ordering from quantum fluctuations and quantum-disordered ground states with unconfined bosonic spinons, *Phys. Rev. B* **45** (1992) 12377.

[232] S. Sachdev, Quantum phases of the Shraiman–Siggia model, *Phys. Rev. B* **49** (1994) 6770 [cond-mat/9311037].

[233] S. Sachdev, Holographic metals and the fractionalized Fermi liquid, *Phys. Rev. Lett.* **105** (2010) 151602 [arXiv:1006.3794].

[234] S. Sachdev, *Quantum Phase Transitions*, Cambridge University Press, Cambridge (2011).

[235] S. Sachdev, Bekenstein–Hawking entropy and strange metals, *Phys. Rev. X* **5** (2015) 041025 [arXiv:1506.05111].

[236] S. Sachdev, Universal low temperature theory of charged black holes with AdS_2 horizons, *J. Math. Phys.* **60** (2019) 052303 [arXiv:1902.04078].

[237] S. Sachdev and R.N. Bhatt, Bond-operator representation of quantum spins: Mean-field theory of frustrated quantum Heisenberg antiferromagnets, *Phys. Rev. B* **41** (1990) 9323.

[238] S. Sachdev, C. Buragohain, and M. Vojta, Quantum impurity in a nearly critical two-dimensional antiferromagnet, *Science* **286** (1999) 2479 [cond-mat/0004156].

[239] S. Sachdev and R. Jalabert, Effective lattice models for two-dimensional antiferromagnets, *Modern Physics Letters B* **04** (1990) 1043.

[240] S. Sachdev, H.D. Scammell, M.S. Scheurer, and G. Tarnopolsky, Gauge theory for the cuprates near optimal doping, *Phys. Rev. B* **99** (2019) 054516 [arXiv:1811.04930].

[241] S. Sachdev, K. Sengupta, and S.M. Girvin, Mott insulators in strong electric fields, *Phys. Rev. B* **66** (2002) 075128 [cond-mat/0205169].

[242] S. Sachdev and J. Ye, Gapless spin-fluid ground state in a random quantum Heisenberg magnet, *Phys. Rev. Lett.* **70** (1993) 3339 [cond-mat/9212030].

[243] S. Sakai, Y. Motome, and M. Imada, Evolution of electronic structure of doped Mott insulators: Reconstruction of poles and zeros of Green's function, *Phys. Rev. Lett.* **102** (2009) 056404 [arXiv:0809.0950].

[244] R. Samajdar, W.W. Ho, H. Pichler, M.D. Lukin, and S. Sachdev, Complex density wave orders and quantum phase transitions in a model of square-lattice Rydberg atom arrays, *Phys. Rev. Lett.* **124** (2020) 103601 [arXiv:1910.09548].

[245] R. Samajdar, W.W. Ho, H. Pichler, M.D. Lukin, and S. Sachdev, *Quantum phases of Rydberg atoms on a kagome lattice*, *Proc. Nat. Acad. Sci.* **118** (2021) e2015785118 [arXiv:2011.12295].

[246] R. Samajdar, D.G. Joshi, Y. Teng, and S. Sachdev, *Emergent $\mathbb{Z}_2$ gauge theories and topological excitations in Rydberg atom arrays*, arXiv:2204.00632.

[247] A.W. Sandvik, Evidence for deconfined quantum criticality in a two-dimensional Heisenberg model with four-spin interactions, *Phys. Rev. Lett.* **98** (2007) 227202 [cond-mat/0611343].

[248] M. Sato and Y. Ando, Topological superconductors: A review, *Rep. Prog. Phys.* **80** (2017) 076501 [arXiv:1608.03395].

[249] D.J. Scalapino, E. Loh, and J.E. Hirsch, d-Wave pairing near a spin-density-wave instability, *Phys. Rev. B* **34** (1986) 8190.

[250] A.O. Scheie, E.A. Ghioldi, J. Xing, J.A.M. Paddison, N.E. Sherman, M. Dupont, et al., *Witnessing quantum criticality and entanglement in the triangular antiferromagnet $KYbSe_2$*, arXiv:2109.11527.

[251] M.S. Scheurer, S. Chatterjee, W. Wu, M. Ferrero, A. Georges, and S. Sachdev, Topological order in the pseudogap metal, *Proc. Nat. Acad. Sci.* **115** (2018) E3665 [arXiv:1711.09925].

[252] M. Schuler, S. Whitsitt, L.-P. Henry, S. Sachdev and A.M. Läuchli, Universal signatures of quantum critical points from finite-size torus spectra: A window into the operator content of higher-dimensional conformal field theories, *Phys. Rev. Lett.* **117** (2016) 210401 [arXiv:1603.03042].

[253] N. Seiberg, T. Senthil, C. Wang, and E. Witten, A duality web in 2+1 dimensions and condensed matter physics, *Annals Phys.* **374** (2016) 395 [arXiv:1606.01989].

[254] G. Semeghini, H. Levine, A. Keesling, S. Ebadi, T.T. Wang, D. Bluvstein, et al., Probing topological spin liquids on a programmable quantum simulator, *Science* **374** (2021) 1242 [arXiv:2104.04119].

[255] A.M. Sengupta, Spin in a fluctuating field: The Bose(+Fermi) Kondo models, *Phys. Rev. B* **61** (2000) 4041 [cond-mat/9707316].

[256] T. Senthil, L. Balents, S. Sachdev, A. Vishwanath, and M.P.A. Fisher, Quantum criticality beyond the Landau–Ginzburg–Wilson paradigm, *Phys. Rev. B* **70** (2004) 144407 [cond-mat/0312617].

[257] T. Senthil and M.P.A. Fisher, $\mathbb{Z}_2$ gauge theory of electron fractionalization in strongly correlated systems, *Phys. Rev. B* **62** (2000) 7850 [cond-mat/9910224].

[258] T. Senthil and O. Motrunich, Microscopic models for fractionalized phases in strongly correlated systems, *Phys. Rev. B* **66** (2002) 205104 [cond-mat/0201320].

[259] T. Senthil, S. Sachdev, and M. Vojta, Fractionalized Fermi liquids, *Phys. Rev. Lett.* **90** (2003) 216403 [cond-mat/0209144].

[260] T. Senthil, D.T. Son, C. Wang, and C. Xu, Duality between $(2+1)d$ quantum critical points, *Physics Reports* **827** (2019) 1 [arXiv:1810.05174].

[261] T. Senthil, A. Vishwanath, L. Balents, S. Sachdev, and M.P.A. Fisher, Deconfined quantum critical points, *Science* **303** (2004) 1490 [cond-mat/0311326].

[262] T. Senthil, M. Vojta, and S. Sachdev, Weak magnetism and non-Fermi liquids near heavy-fermion critical points, *Phys. Rev. B* **69** (2004) 035111 [cond-mat/0305193].

[263] H. Shackleton, A. Wietek, A. Georges and S. Sachdev, Quantum phase transition at nonzero doping in a random t–J model, *Phys. Rev. Lett.* **126** (2021) 136602 [arXiv:2012.06589].

[264] K.M. Shen, F. Ronning, D.H. Lu, F. Baumberger, N.J.C. Ingle, W.S. Lee, et al., Nodal quasiparticles and antinodal charge ordering in $Ca_{2-x}Na_xCuO_2Cl_2$, *Science* **307** (2005) 901.

[265] A. Sommerfeld, Zur elektronentheorie der metalle auf grund der fermischen statistik, *Z. Phys.* **47** (1928) 1.

[266] D.T. Son, Is the composite fermion a Dirac particle?, *Phys. Rev. X* **5** (2015) 031027 [arXiv:1502.03446].

[267] D.T. Son, The Dirac composite fermion of the fractional quantum Hall effect, *PTEP* **2016** (2016) 12C103 [arXiv:1608.05111].

[268] X.-Y. Song, Y.-C. He, A. Vishwanath, and C. Wang, From spinon band topology to the symmetry quantum numbers of monopoles in Dirac spin liquids, *Phys. Rev. X* **10** (2020) 011033 [arXiv:1811.11182].

[269] X.-Y. Song, C.-M. Jian, and L. Balents, Strongly correlated metal built from Sachdev–Ye–Kitaev models, *Phys. Rev. Lett.* **119** (2017) 216601 [arXiv:1705.00117].

[270] X.-Y. Song, C. Wang, A. Vishwanath and Y.-C. He, Unifying description of competing orders in two dimensional quantum magnets, *Nat. Commun.* **10** (2019) 4254 [arXiv:1811.11186].

[271] Z.-D. Song and B.A. Bernevig, MATBG as topological heavy fermion: I. Exact mapping and correlated insulators, arXiv:2111.05865.

[272] G. Sordi, K. Haule, and A.-M.S. Tremblay, Mott physics and first-order transition between two metals in the normal-state phase diagram of the two-dimensional Hubbard model, *Phys. Rev. B* **84** (2011) 075161 [arXiv:1102.0463].

[273] G. Sreejith, S. Powell, and A. Nahum, Emergent $SO(5)$ symmetry at the columnar ordering transition in the classical cubic dimer model, *Phys. Rev. Lett.* **122** (2019) 080601 [arXiv:1803.11218].

[274] T.D. Stanescu and G. Kotliar, Fermi arcs and hidden zeros of the Green function in the pseudogap state, *Phys. Rev. B* **74** (2006) 125110 [cond-mat/0508302].

[275] D. Stanford and E. Witten, Fermionic localization of the Schwarzian theory, *JHEP* **10** (2017) 008 [arXiv:1703.04612].

[276] H.T.C. Stoof, Breaking up a superfluid, *Nature* **415** (2002) 25.

[277] D.J. Thouless, Fluxoid quantization in the resonating-valence-bond model, *Phys. Rev. B* **36** (1987) 7187.

[278] D.J. Thouless, M. Kohmoto, M.P. Nightingale, and M. den Nijs, Quantized Hall conductance in a two-dimensional periodic potential, *Phys. Rev. Lett.* **49** (1982) 405.

[279] M. Tikhanovskaya, H. Guo, S. Sachdev, and G. Tarnopolsky, Excitation spectra of quantum matter without quasiparticles I: Sachdev–Ye–Kitaev models, *Phys. Rev. B* **103** (2021) 075141 [arXiv:2010.09742].

[280] M. Tikhanovskaya, H. Guo, S. Sachdev, and G. Tarnopolsky, Excitation spectra of quantum matter without quasiparticles II: random t–J models, *Phys. Rev. B* **103** (2021) 075142 [arXiv:2012.14449].

[281] J. Toner, P.E. Lammert, and D.S. Rokhsar, Topology and nematic ordering. II. Observable critical behavior, *Phys. Rev. E* **52** (1995) 1801 [cond-mat/9501100].

[282] D.C. Tsui, H.L. Stormer, and A.C. Gossard, Two-dimensional magnetotransport in the extreme quantum limit, *Phys. Rev. Lett.* **48** (1982) 1559.

[283] C.J. Turner, A.A. Michailidis, D.A. Abanin, M. Serbyn, and Z. Papić, Quantum scarred eigenstates in a Rydberg atom chain: Entanglement, breakdown of thermalization, and stability to perturbations, *Phys. Rev. B* **98** (2018) 155134 [arXiv:1806.10933].

[284] D. Vanderbilt, *Berry Phases in Electronic Structure Theory: Electric Polarization, Orbital Magnetization and Topological Insulators*, Cambridge University Press (2018).

[285] C.M. Varma, P.B. Littlewood, S. Schmitt-Rink, E. Abrahams, and A.E. Ruckenstein, Phenomenology of the normal state of Cu–O high-temperature superconductors, *Phys. Rev. Lett.* **63** (1989) 1996.

[286] B. Verheijden, Y. Zhao, and M. Punk, Solvable lattice models for metals with Z_2 topological order, *SciPost Physics* **7** (2019) 074 [arXiv:1908.00103].

[287] R. Verresen, private communication.

[288] R. Verresen, M.D. Lukin, and A. Vishwanath, Prediction of toric code topological order from Rydberg blockade, *Phys. Rev. X* **11** (2021) 031005 [arXiv:2011.12310].

[289] J. Villain, Theory of one- and two-dimensional magnets with an easy magnetization plane. II. The planar, classical, two-dimensional magnet, *J. de Phys. (Paris)* **36** (1975) 581.

[290] A. Vishwanath, L. Balents, and T. Senthil, Quantum criticality and deconfinement in phase transitions between valence bond solids, *Phys. Rev. B* **69** (2004) 224416 [cond-mat/0311085].

[291] J. Voit, Charge–spin separation and the spectral properties of Luttinger liquids, *J Phys. Condens. Mat.* **5** (1993) 8305 [cond-mat/9310048].

[292] J. Voit, One-dimensional Fermi liquids, *Rep. Prog. Phys.* **58** (1995) 977 [cond-mat/9510014].

[293] M. Vojta, C. Buragohain, and S. Sachdev, Quantum impurity dynamics in two-dimensional antiferromagnets and superconductors, *Phys. Rev. B* **61** (2000) 15152 [cond-mat/9912020].

[294] C. Wang, A. Nahum, M.A. Metlitski, C. Xu, and T. Senthil, Deconfined quantum critical points: Symmetries and dualities, *Phys. Rev. X* **7** (2017) 031051 [`arXiv:1703.02426`].

[295] F. Wang, S.A. Kivelson, and D.-H. Lee, Nematicity and quantum paramagnetism in FeSe, Nat. Phys. **11** (2015) 959 [`arXiv:1501.00844`].

[296] L. Wang and A.W. Sandvik, Critical level crossings and gapless spin liquid in the square-lattice spin-1 /2 J_1–J_2 heisenberg antiferromagnet, *Phys. Rev. Lett.* **121** (2018) 107202 [`arXiv:1702.08197`].

[297] X. Wang and E. Berg, Scattering mechanisms and electrical transport near an Ising nematic quantum critical point, *Phys. Rev. B* **99** (2019) 235136 [`arXiv:1902.04590`].

[298] Y. Wang, A. Abanov, B.L. Altshuler, E.A. Yuzbashyan, and A.V. Chubukov, Superconductivity near a quantum-critical point: The special role of the first Matsubara frequency, *Phys. Rev. Lett.* **117** (2016) 157001 [`arXiv:1606.01252`].

[299] Z. Wang, M.P. Zaletel, R.S.K. Mong, and F.F. Assaad, Phases of the (2+1) dimensional $SO(5)$ nonlinear sigma model with topological term, *Phys. Rev. Lett.* **126** (2021) 045701 [`arXiv:2003.08368`].

[300] M. Weber and M. Vojta, $SU(2)$-symmetric spin-boson model: Quantum criticality, fixed-point annihilation, and duality, `arXiv:2203.02518`.

[301] F.J. Wegner, Duality in generalized Ising models and phase transitions without local order parameters, *J. Math. Phys.* **12** (1971) 2259.

[302] X.G. Wen, Chiral Luttinger liquid and the edge excitations in the fractional quantum Hall states, *Phys. Rev. B* **41** (1990) 12838.

[303] X.G. Wen, Mean-field theory of spin-liquid states with finite energy gap and topological orders, *Phys. Rev. B* **44** (1991) 2664.

[304] X.-G. Wen, Quantum orders in an exact soluble model, *Phys. Rev. Lett.* **90** (2003) 016803 [`quant-ph/0205004`].

[305] X.-G. Wen and P.A. Lee, Theory of underdoped cuprates, *Phys. Rev. Lett.* **76** (1996) 503 [`cond-mat/9506065`].

[306] X.-G. Wen, Quantum orders and symmetric spin liquids, *Phys. Rev. B* **65** (2002) 165113 [`cond-mat/0107071`].

[307] Z.-Y. Wen and J. Avery, Some properties of hyperspherical harmonics, *Journal of Mathematical Physics* **26** (1985) 396.

[308] S. Whitsitt and S. Sachdev, Transition from the Z_2 spin liquid to antiferromagnetic order: Spectrum on the torus, *Phys. Rev. B* **94** (2016) 085134 [`arXiv:1603.05652`].

[309] J. Wildeboer, A. Seidel, and R.G. Melko, Entanglement entropy and topological order in resonating valence-bond quantum spin liquids, *Phys. Rev. B* **95** (2017) 100402 [`arXiv:1510.07682`].

[310] K.G. Wilson and M.E. Fisher, Critical exponents in 3.99 dimensions, *Phys. Rev. Lett.* **28** (1972) 240.

[311] E. Witten, Instantons, the quark model, and the $1/N$ expansion, *Nucl. Phys. B* **149** (1979) 285.

[312] E. Witten, Quantum field theory and the Jones polynomial, *Commun. Math. Phys.* **121** (1989) 351.

[313] E. Witten, Fermion path integrals and topological phases, *Rev. Mod. Phys.* **88** (2016) 035001 [arXiv:1508.04715].

[314] C. Wu, B.A. Bernevig, and S.-C. Zhang, Helical liquid and the edge of quantum spin Hall systems, *Phys. Rev. Lett.* **96** (2006) 106401 [cond-mat/0508273].

[315] Y.-M. Wu, A. Abanov, Y. Wang, and A.V. Chubukov, Interplay between superconductivity and non-Fermi liquid at a quantum critical point in a metal. II. The γ model at a finite T for $0 < \gamma < 1$, *Phys. Rev. B* **102** (2020) 024525 [arXiv:2006.02968].

[316] C. Xu and J.E. Moore, Stability of the quantum spin Hall effect: Effects of interactions, disorder, and $\mathbb{Z}_2$ topology, *Phys. Rev. B* **73** (2006) 045322 [cond-mat/0508291].

[317] C. Xu and S. Sachdev, Majorana liquids: The complete fractionalization of the electron, *Phys. Rev. Lett.* **105** (2010) 057201 [arXiv:1004.5431].

[318] X.Y. Xu, A. Klein, K. Sun, A.V. Chubukov, and Z.Y. Meng, Identification of non-Fermi liquid fermionic self-energy from quantum Monte Carlo data, *npj Quantum Mater.* **5** (2020) 65 [arXiv:2003.11573].

[319] X.Y. Xu, K. Sun, Y. Schattner, E. Berg, and Z.Y. Meng, Non-Fermi liquid at (2+1)D ferromagnetic quantum critical point, *Phys. Rev. X* **7** (2017) 031058 [arXiv:1612.06075].

[320] Z. Yan, R. Samajdar, Y.-C. Wang, S. Sachdev, and Z.Y. Meng, Triangular lattice quantum dimer model with variable dimer density, *Nature Commun.* **13** (2022) 5799 [arXiv:2202.11100].

[321] K.-Y. Yang, T.M. Rice, and F.-C. Zhang, Phenomenological theory of the pseudogap state, *Phys. Rev. B* **73** (2006) 174501 [cond-mat/0602164].

[322] M. Yue, Z. Wang, B. Mukherjee, and Z. Cai, Order by disorder in frustration-free systems: Quantum Monte Carlo study of a two-dimensional PXP model, *Phys. Rev. B* **103** (2021) L201113 [arXiv:2103.15287].

[323] M.E. Zayed, C. Rüegg, J. Larrea J., et al., 4-Spin plaquette singlet state in the Shastry–Sutherland compound $SrCu_2(BO_3)_2$, *Nat. Phys.* **13** (2017) 962 [arXiv:1603.02039].

[324] Y.-H. Zhang and S. Sachdev, Deconfined criticality and ghost Fermi surfaces at the onset of antiferromagnetism in a metal, *Phys. Rev. B* **102** (2020) 155124 [arXiv:2006.01140].

[325] Y.-H. Zhang and S. Sachdev, From the pseudogap metal to the Fermi liquid using ancilla qubits, *Phys. Rev. Res.* **2** (2020) 023172 [arXiv:2001.09159].

[326] Zhang, S. C. and Hansson, T. H. and Kivelson, S., Effective-Field-Theory Model for the Fractional Quantum Hall Effect, *Phys. Rev. Lett.* **62** (1989) 1 *American Physical Society* 82–85, https://link.aps.org/doi/10.1103/PhysRevLett.62.82

[327] W. Zheng and S. Sachdev, Sine-Gordon theory of the non-Néel phase of two-dimensional quantum antiferromagnets, *Phys. Rev. B* **40** (1989) 2704.

Index